The Minerals, Metals & Materials Series

Dmytro Orlov · Vineet Joshi
Kiran N. Solanki · Neale R. Neelameggham
Editors

# Magnesium Technology 2018

TMS

Springer

*Editors*
Dmytro Orlov
Lund University
Lund
Sweden

Kiran N. Solanki
Arizona State University
Tempe, AZ
USA

Vineet Joshi
Pacific Northwest National Laboratory
Richland, WA
USA

Neale R. Neelameggham
IND LLC
South Jordan, UT
USA

ISSN 2367-1181 ISSN 2367-1696 (electronic)
The Minerals, Metals & Materials Series
ISBN 978-3-319-72331-0 ISBN 978-3-319-72332-7 (eBook)
https://doi.org/10.1007/978-3-319-72332-7

Library of Congress Control Number: 2017960920



Printed on acid-free paper

This Springer imprint is published by Springer Nature
The registered company is Springer International Publishing AG
The registered company address is: Gewerbestrasse 11, 6330 Cham, Switzerland

# Preface

By now, magnesium alloys have become an integral part of our daily life improving its quality. From spectacular entrance as fireworks in festive moments, this material has gained ground as a strong contender to steels, Al and Ti alloys, polymers and composites in lightweight mobility, biomedical, and energy sectors. Its unique properties address many demanding challenges of modern society including the reduction of carbon footprint and building of circular economies. Such properties include low density, making Mg the lightest structural metal, excellent capacity for vibration damping and electromagnetic shielding, perfect biocompatibility along with bio-resorbability, and many others. At the same time, real-life applications of magnesium demand overcoming significant challenges in end-product fabrication technologies and tailoring performance characteristics. Such demands are relevant to rather limited primary supply sources, alloying capacity and technological plasticity, often excessive chemical reactivity, creep properties, and closed-loop recycling. Therefore, many prominent researchers in academia, industrial engineers, and economists, as well as state policy makers are involved in the very dynamic area of magnesium technology development globally.

This proceedings volume, which is the 19th annual volume of papers on all aspects of basic research, simulation, and magnesium technology development, reflects the worldwide efforts and latest advancements in all aspects of magnesium research. In line with long-standing tradition, the proceedings consist of papers from all oral and selected poster presenters in the Magnesium Technology Symposium as well as from Magnesium Alloy Development: An LMD Symposium in Honor of Karl Kainer, both held during the 147th TMS Annual Meeting & Exhibition in Phoenix, Arizona, USA, March 11–15, 2018. All papers in this compilation were peer-reviewed by experts in the fields of magnesium and lightweight metals development, characterization, and simulation.

The lectures contributed by researchers and engineers from 17 countries were nominally split for presentation during the meeting into 11 sessions including a Keynote and a unique Poster-pitch sessions. These were arranged in accordance with the material lifecycle including environmental challenges and sustainability. Starting from primary Mg extraction, alloy development, casting, and solidification, the session topics further covered thermo-mechanical processing; deformation mechanisms and mechanical behavior; applications in biomedical, energy, and mobility sectors; and finally degradation and recycling. Magnesium Alloy Development: An LMD Symposium in Honor of Karl Kainer had three sessions additionally.

The volume also includes papers and expert commentaries from renowned leaders in various areas of magnesium technology development presented as keynote lectures. The symposium was opened by Prof. Dr. Karl Ulrich Kainer, Director of Magnesium Innovation Centre (MagIC) at Helmholtz-Zentrum Geesthacht who presented his vision on “Magnesium Alloys: Challenges and Achievements in Controlling Performance, and Future Application Perspectives.” This was followed by other distinguished keynote speakers including Professor and Head of the Laboratory for Multiscale Mechanics Modeling at École Polytechnique Fédérale de Lausanne (EPFL), Prof. William A. Curtin who presented a talk on “Solute/Stacking Fault Energies in Mg and Implications for Ductility.” Prof. Nick Birbilis,

Woodside Innovation Chair, Professor and Head of the Department of Materials Science and Engineering at Monash University presented an overview on "Recent Developments in Magnesium Alloy Corrosion Research," and Prof. Dr. Mikhail Zheludkevich, Head of Department of Corrosion and Surface Technology at Magnesium Innovation Centre (MagIC), Helmholtz-Zentrum Geesthacht, gave a talk "Towards Active Corrosion Protection of Mg Alloys Using Corrosion Inhibition Approaches," closing the Keynote session. Invited speakers from Japan opened following sessions having more specific focuses with lectures "Study on Metal Smelting Process under Microwave Irradiation" by Prof. Satoshi Fujii (Tokyo Institute of Technology) and "Material Design for Enhancing Toughness of Mg Alloy and Application for Biodegradable Devices" by Prof. Toshiji Mukai (Kobe University).

Magnesium Alloy Development: An LMD Symposium in Honor of Karl Kainer was organized by Norbert Hort (Magnesium Innovation Centre MagIC within the Helmholtz-Zentrum Geesthacht) and Alan Luo (Ohio State University). There were 30 presentations including 4 keynote lectures: "Solutions for Next Generation Automotive Lightweight Concepts Based on Material Selection and Functional Integration" by Horst E. Friedrich, Elmar Beeh, and Carmen S Roider; "Recent Developments in the Application of the Interdependence Model of Grain Formation and Refinement" by David St John; "Magnesium Pistons in Engines: Fiction or Fact?" by Norbert Hort, Hajo Dieringa, and Karl Ulrich Kainer; and "Degradable Magnesium Implants—Assessment of the Current Situation" by Regine Willumeit-Roemer, Nezha Ahmad Agha, Berengere Luthringer.

Last but not least, the 2017–2018 Magnesium Committee expresses deep appreciation to all authors for contributing to the success of the symposia, our panel of distinguished keynote speakers for sharing their valuable thoughts on most recent achievements in the area and the future of magnesium technology, the reviewers for their hard work in reviewing the manuscripts, the session chairs, judges, TMS staff members, and other volunteers for their impeccable support.

Dmytro Orlov, Chair
Vineet Joshi, Vice Chair
Kiran N. Solanki, Past Chair
Neale R. Neelameggham, Advisor

# Contents

# About the Editors

## Lead Editor

**Dmytro Orlov** is Professor and Head of the Division of Materials Engineering at the Faculty of Engineering (LTH) in Lund University, Lund, Sweden. Among other professional activities, at present he has a joint appointment as a Senior Scientist at the University of Nova Gorica in Slovenia and serves as a Chair of a Magnesium Committee of The Minerals, Metals & Materials Society (TMS), USA.

He obtained his graduate degrees at Donetsk National Technical University in Ukraine. During his Ph.D. studies, he joined a research institute within National Academy of Sciences, Ukraine, where he spent ten years, and then almost ten years on postdoctoral and senior research positions in world-renowned laboratories at Osaka, Kyoto, and Ritsumeikan Universities in Japan, Monash University in Australia, and University of Nova Gorica in Slovenia. In the latter university, he also received habilitation. To date, his track record includes 20+ research projects, 7 patents, 70+ research papers and book contributions, and approximately as many lectures at international meetings among which 20+ were invited.

His background is in the engineering of thermo-mechanical processing technologies for metallic materials fabrication with a core expertise in the design of deformation processing-based techniques. To date, his contributions in the field include the development of Twist Extrusion technique for imparting large plastic deformations to materials without changing the net shape of a workpiece, analysis of microstructure, and texture evolution in metallic materials under large deformations with strain reversals as well as dependence of crystallographic orientation of surface on degradation phenomena in magnesium. The primary scope of his laboratory within LTH is the engineering of novel hybrid, composite, and mono-materials with hierarchical structures architectured from atomic through to macro-scales. His present research interests and ongoing research projects are focused on the design of Mg alloys for biomedical and lightweight mobility applications, multiscale-architectured structures with topological control of their heterogeneity, and the development of relevant in-situ characterization techniques at large-scale facilities.

## Magnesium Technology 2018 Editors

**Vineet Joshi** is a Materials Scientist in the Energy and Environment Directorate at the Pacific Northwest National Laboratory located in Richland, Washington, USA. Mr. Joshi currently leads several projects at the laboratory related to development of lightweight structural materials and processing of metallic nuclear fuels. He holds an M.S. degree in Materials Science and Engineering from Alfred University in New York, which he earned in 2010. He is currently pursuing his Ph.D. in Materials Science and Engineering from Washington State University.

He has made numerous original and important contributions through his research related to lightweight structural materials and nuclear materials processing and has published over 40 articles and delivered as many talks at international meetings or conferences. His primary forte is thermo-mechanical processing of materials and development of structures that addresses unique performance requirements. Over the years, he has developed a process to make high-strength titanium alloy, the results of which were published in *Nature Communications*. He and his team have been developing Shear Assisted Processing and Extrusion technique to form magnesium and other lightweight structures. Earlier, he was involved in the development of a new reactive air brazing technique, which enabled joining of mixed ionic and electronic conducting ceramics to metals in air.

He currently serves as the Vice Chair of the Magnesium Committee of The Minerals, Metals & Materials Society (TMS) and was the *JOM* representative for the Magnesium Committee. He is the recipient of several awards at the laboratory. He has been the member of TMS since his undergraduate days in India.

**Kiran N. Solanki** is an Associate Professor of Mechanical Engineering in the SEMTE at Arizona State University. Prior to coming to ASU, he was an Associate Director for the Center for Advanced Vehicular Systems at Mississippi State University. Dr. Solanki received his Ph.D. from Mississippi State University in December 2008. His research interest is at the interface of solid mechanics and material science, with a focus on characterizing and developing microstructure-based structure property relationships across multiple length and time scales. To date, he has co-authored >60 journal articles, four book chapters, and >35 conference proceedings with faculty and students at ASU and MSU. In addition, his article published *in Engineering Fracture Mechanics* was recognized as one of the most highly cited papers from 2002 to 2005. For his efforts to promote the education of engineering students in the area of fatigue technology, he was awarded the SAE Henry O. Fuch Award by the SAE Fatigue Design & Evaluation Committee. In 2011, he received the TMS Light Metals Magnesium Best Fundamental

Research Paper Award for his work on predicting deformation and failure behavior in magnesium alloys using a multiscale modeling approach. He received the 2013 TMS Light Metals Division Young Leaders Professional Development Award; the 2013 Air Force Office of Scientific Research Young Investigator Research Award; the 2013 ASME "Orr Award" for Early Career Excellence in Fatigue, Fracture, and Creep; and the 2016 Science Award from ECI/ONR.

**Neale R. Neelameggham** is "The Guru" at IND LLC, involved in international consulting in the field of metals and associated chemicals (boron, magnesium, titanium, and lithium and rare earth elements), thiometallurgy, energy technologies, soil biochemical reactor design, etc. He was a visiting expert at Beihang University of Aeronautics and Astronautics, Beijing, China. He was a plenary speaker at the Light Metal Symposium in South Africa—on low carbon dioxide emission processes for magnesium. He has more than 38 years of expertise in magnesium production and was involved in process development of its start-up company NL Magnesium through to the present US Magnesium LLC, UT until 2011. Neelameggham and Brian Davis authored the ICE-JNME award-winning (2016) paper "21st Century Global Anthropogenic Warming Convective Model," which notes that constrained air mass warming is independent of the energy conversion source—fossil or renewable energy. He is presently developing Agricoal™ and agricoalture to improve arid soils. He holds 16 patents and patent applications and has published several technical papers. He has served in the Magnesium Committee of the Light Metals Division (LMD) of TMS since its inception in 2000, chaired it in 2005, and in 2007, he was made a permanent co-organizer for the Magnesium Symposium. He has been a member of the Reactive Metals Committee, Recycling Committee, Titanium Committee, and Program Committee Representative of LMD and LMD council. He was the inaugural chair, when in 2008, LMD and EPD (Extraction & Processing Division) created the Energy Committee, and has been a co-editor of the energy technology symposium proceedings through the present. He received the LMD Distinguished Service Award in 2010. While he was the chair of Hydrometallurgy and Electrometallurgy Committee, he initiated the rare metal technology symposium in 2014. He is co-editor for the 2018 proceedings for the symposia on Magnesium Technology, Energy Technology, Rare Metal Technology, and Solar Cell Silicon. He is the lead organizer and editor of the 2018 symposium Stored Renewable Energy in Coal.

## Magnesium Alloy Development: An LMD Symposium in Honor of Karl Kainer Editors

**Norbert Hort** is the head of the Magnesium Processing Department at the Magnesium Innovation Centre (MagIC) within the Helmholtz-Zentrum Geesthacht Zentrum für Material- und Küstenforschung, Geesthacht, Germany (formerly the GKSS Research Centre). Concurrently he is lecturer at the Leuphana University, Lüneburg, Germany. He studied Materials Sciences at the Clausthal University of Technology (CUT), Germany, where he was involved in magnesium research since the early 1990s. His diploma thesis (1994) dealt with gas atomized, hot extruded magnesium alloys. During 1994–1999 he worked as a researcher at the Institute of Materials Sciences (CUT) and he joined the HZG in 2000. He got his Ph.D. degree in Materials Sciences in 2002 from the Clausthal University of Technology. In the MagIC he is responsible for the development of new creep resistant magnesium alloys, grain refinement and the castability of magnesium alloys (viscosity and density of melts, fluidity, and mold filling). This also covers in situ observations of solidification behavior using synchrotron diffraction and tomography. These activities are accompanied by the development of new magnesium alloys for biodegradable implants in a close collaboration with biologists and clinicians. He is co-author of more than 170 peer-reviewed journal articles and more than 200 contributions to conference proceedings. In recent years he was involved in organization of Magnesium Technology sessions at TMS annual meetings (2012–2014). Additionally, he was a member of the organizing committees of the conference series Magnesium Alloys and Their Applications (2009, 2012, 2015), and the conference Light Metal Technologies 2011. Since 2009 he has been the chairman of the technical committee "Magnesium" of the German Society of Materials (DGM).

**Alan Luo** is Professor of Materials Science and Engineering and Professor of Integrated Systems Engineering (Manufacturing) at The Ohio State University (OSU) in Columbus, Ohio, USA. He is also Director of OSU Light Metals and Manufacturing Research Laboratory (LMMRL) and a steering board member of OSU Center for Simulation Innovation and Modeling (SIMCenter). He is an elected Fellow of ASM (American Society of Metals) International and SAE (Society for Automotive Engineers) International. He is presently chair of the TMS Light Metals Division and past chair of the SAE Materials Engineering Division. Prior to joining OSU in July 2013, he was a GM Technical Fellow at General Motors Global Research and Development Center (Warren, Michigan, USA) with 20 years of industrial experience. He has 18 patents and more than 220 technical publications in advanced materials, manufacturing, and applications. He won two John M. Campbell Awards for his fundamental research and three

Charles L. McCuen Awards for research applications at GM. He received the TMS (The Minerals, Metals & Materials Society) Brimacombe Medalist Award and SAE Forest R. McFarland Award in 2013, USCAR (United States Council for Automotive Research) Special Recognition Award in 2009, and ASM Materials Science Research Silver Medal in 2008. His research is also recognized by several Best Paper awards from TMS, SAE, AFS (American Foundry Society), and CALPHAD (CALculation of PHAse Diagrams) journal.

# Session Chairs

## Magnesium Technology

**Keynote Session**
Dmytro Orlov, Lund University
Vineet Joshi, Pacific Northwest National Laboratory—PNNL

**Corrosion and Surface Protection**
Mikhail Zheludkevich, Helmholtz-Zentrum Geesthacht (HZG)
Nick Birbilis, Monash University

**Alloy Design**
Sean Agnew, University of Virginia
Suveen Mathaudhu, University of California—Riverside

**Primary Production and Casting**
Neale R. Neelameggham, IND LLC
Elsa Olivetti, MIT

**Deformation Mechanisms**
Vineet Joshi, Pacific Northwest National Laboratory—PNNL
Kristián Máthis, Charles University

**Thermo-Mechanical Processing**
Kiran Solanki, Arizona State University
Vineet Joshi, Pacific Northwest National Laboratory—PNNL

## Magnesium Alloy Development: An LMD Symposium in Honor of Karl Kainer

**Cast Alloys**
Norbert Hort, MagIC (Magnesium Innovation Centre in the Helmholtz-Zentrum Geesthacht)
Alan Luo, The Ohio State University

**Wrought Alloys**
J. Li, University of Leoben
Mert Celikin, McGill University

**Degradation and Microstructure**
Dietmar Letzig, MagIC (Magnesium Innovation Centre in the Helmholtz-Zentrum Geesthacht)
Mark Easton, RMIT University

# Reviewer Pool

## Magnesium Technology 2018

Neale R. Neelameggham, IND LLC
Chamini Mendis, Brunel University London
Raymond Decker, Thixomat Inc
Michele Manuel, University of Florida
Ilaksh Adlakha, Indian Institute of Technology—Madras
Chaitanya Kale, Arizona State University
Tracy Berman, University of Michigan
Ishak Karakaya, Middle East Tech University, Turkey
Tian Yang, Kunming University of Science and Technology in China
Vineet Joshi, Pacific Northwest National Laboratory
Sean Agnew, University of Virginia
Jishnu Bhattacharya, University of Virginia
Victoria Miller, North Carolina State University
Eric Nyberg, Brunel University London
Kiran Solanki, Arizona State University
Alok Singh, National Institute for Materials Science
Dmytro Orlov, Lund University
Pulkit Garg, Arizona State University
Brian Jordon, The University of Alabama
Scott Turnage, Arizona State University

## Magnesium Alloy Development: An LMD Symposium in Honor of Karl Kainer

Nobert Hort, MagIC (Magnesium Innovation Centre in the Helmholtz-Zentrum Geesthacht)
Alan Luo, The Ohio State University
Mark Easton, RMIT University
Serge Gavras, MagIC (Magnesium Innovation Centre in the Helmholtz-Zentrum Geesthacht)

# Part I

# Magnesium Technology 2018

# Mg Alloys: Challenges and Achievements in Controlling Performance, and Future Application Perspectives

Hajo Dieringa, Norbert Hort, Dietmar Letzig, Jan Bohlen, Daniel Höche, Carsten Blawert, Mikhail Zheludkevich, and Karl Ulrich Kainer

**Abstract**

In recent years, Mg alloys have made inroads into applications for transport industries. The favorable property profile of Mg promotes increased usage. Despite magnesium alloys being used for years, there is still a lack of knowledge about the potential of Mg alloys. New or optimized alloys and processes are creating new ideas for substituting traditional materials. High-pressure die-casting (HPDC) is the predominant technology, while other casting and wrought processes are of secondary importance. Developments in the last decade have led to an improvement of the property profile and effectiveness of magnesium wrought alloys. Additive manufacturing has opened new opportunities for tailoring of the property profile and functionality. In addition, Mg as material for battery anodes adds a new field of application in the energy sector. This presentation will provide an overview of the status of modern process and alloy development, and discuss the challenges to extending the use of magnesium alloys in various applications.

**Keywords**

Wrought and cast alloys • Biomaterials • Creep
Corrosion • Battery • Coating • Nanocomposite

## Introduction

Over the last decade the areas of application for magnesium alloys have expanded considerably, because the economic pressures to build vehicles and aircraft more easily have risen sharply. This applies to casting alloys for which the trend towards higher application temperatures is still a driver. The reduction of flammability by adding special alloying elements is also a relevant topic. The banning of SF6 by EU regulations has prompted the development of an alternative protective gas. In the case of wrought alloys the trend is towards magnesium sheets, but their production must become cheaper in order to enter the market. Furthermore, fundamental topics such as optimization of the deformation behavior during rolling and the deep-drawing capability of the sheet metal require further investigation. Some applications for magnesium sheets in the automotive industry have already been successfully introduced. Corrosion resistance and the question of coating are important considerations, especially for sheet metal. However, this also applies to all magnesium applications. New topics such as magnesium-based nanocomposites broaden the possible fields of application. Degradable implants for surgical uses are important considerations in the development of new alloys under completely different conditions from those of the automotive industry. Magnesium-based batteries will also gain importance in the future because of their low weight and very good energy density.

## Wrought Alloys

For enhanced, lightweight performance in industrial applications the use of wrought magnesium is an obvious choice. The challenge in hand is the initial low formability of magnesium and its alloys compared to other material classes, especially steels. The prospect focuses on the property development of magnesium sheets and extrusions as semi-finished products for the production of formed parts and profiles by massive forming processes [1].

Conventional wrought magnesium alloy sheets, such as of the alloy AZ31, exhibit comparably poor formability. The formation of strong textures with a typical strong alignment of basal planes parallel to the sheet plane, a basal texture, has been identified as the main obstacle to improved formability [2, 3]. In a simple approach such textures limit the ability for

H. Dieringa · N. Hort · D. Letzig · J. Bohlen · D. Höche · C. Blawert · M. Zheludkevich · K. U. Kainer (✉)
MagIC—Magnesium Innovation Centre, Helmholtz-Zentrum Geesthacht, Max-Planck-Straße 1, 21502 Geesthacht, Germany
e-mail: karl.kainer@hzg.de


D. Orlov et al. (eds.), *Magnesium Technology 2018*, The Minerals, Metals & Materials Series, https://doi.org/10.1007/978-3-319-72332-7_1

strain accommodation by basal slip and therefore limit the ability for work hardening of the sheets as well [4]. Therefore, changing the distinct textures by way of texture weakening or a qualitative texture change by tilting the basal planes out of the forming direction is expected to improve forming properties for further processing. This would mean higher ductility even at room temperature as well as improved formability could be achieved [3].

One major finding in the last two decades is that microstructure and texture development during sheet rolling is distinctly influenced by using alloy modifications with specific alloying elements, i.e. the addition of RE elements. Multifold approaches have been used to explain the corresponding texture changes [5]. This includes discussions on variations of activity in deformation mechanisms, mainly changes to the contributions of slip modes and their influence on orientation changes, but also different types of twin activities and other inhomogeneous deformation mechanisms such as shear bands [4, 6–8]. Furthermore, changes to recrystallization behavior have been addressed [9, 10]. Growth restrictions of grains due to boundary pinning effects can be associated with the presence of solute RE or precipitates with RE and these allow changes to the growth of grains with different orientations. Such sheets, also in a combination of RE and Zn, are more favorable for enhanced forming operations. An example of this improvement is illustrated in Fig. 1, where the temperature dependence of two magnesium sheet alloys, AZ31 and ZE10, is shown for various temperatures. At all temperatures studied, an increase in the formability is visible. Especially the stretch formability, visible on the right hand side of the forming limit diagram, is increased in ZE10 compared to AZ31. Similar textures and therefore improved formability can also be achieved with Ca [11, 12] instead of RE. Additional alloying elements such as Zn or Mn can have a further effect on the texture formation, as illustrated by the examples in Fig. 2.

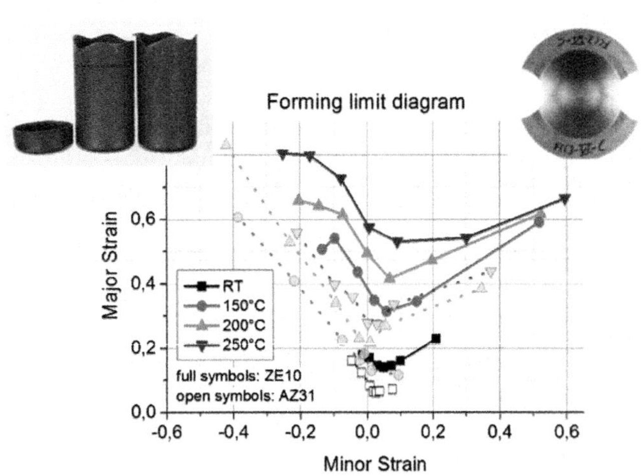


**Fig. 1** Temperature dependence of formability of two magnesium alloy sheets (in accordance with [3])

## TRC-Twin Roll Casting

In the case of sheet materials, it is recognized that the feedstock for a warm-rolling process needs to have the form of thin bands, such as those produced by twin roll casting, if thin magnesium sheets are to become competitive industrial products [13]. For this reason, HZG installed the twin roll caster shown in Fig. 3.

The twin roll casting line consists of a furnace (Striko-Westofen) and a twin roll caster (Novelis, Jumbo 3CM). The twin roll caster is designed to produce thin strips of magnesium with a maximum width of 650 mm and a thickness in the range 4–12 mm and a maximum rolling speed of 6 m/min. The furnace line is configured to allow manual loading of raw materials in ingots and the possibility to vary the alloy composition, i.e. cast strips can be made from different alloys. The furnace line includes an ingot pre-heater to remove humidity from the ingots and accelerate fusion in the melting furnace. After pre-heating, the feedstock is transferred to the melting furnace. A second transfer device conveys the molten metal to the cleaning furnace. From the cleaning furnace the melt is transferred to the head-box which is connected to the tip. The liquid metal flows by gravity through the tip into the gap of the rolls of the twin roll caster. The metal exiting the tip solidifies on the rolls into a strip that is further deformed by the rolls. Important features of the strip, such as microstructure and texture, are influenced by the position of the solidification front. The position of the solidification front is controlled mainly by the melt temperature, the height of the rolling gap and the strip speed. Its position influences the degree of deformation in the strip. If the solidification front is located at the exit of the tip, the degree of deformation increases, because the strip is fully solid before entering the rolls. If the solidification front moves to the kissing point of the rolls, the strip is not completely solid and the degree of solid-state deformation is reduced.

The simultaneous development of new wrought alloys such as aluminum-free Mg-alloys containing rare-earth elements (RE) and the magnesium twin roll casting (TRC) process has permitted the production of high strength magnesium sheet with good formability at moderate temperatures. This new Mg-Zn-RE alloy was developed and tested: Zn was used to confer improved strength properties, while the RE-element additions enable the development of a weak non-basal texture of the rolled sheets, which is favorable for good formability [14]. By using the TRC process, the alloy was cast into strips of 5.6 mm thickness

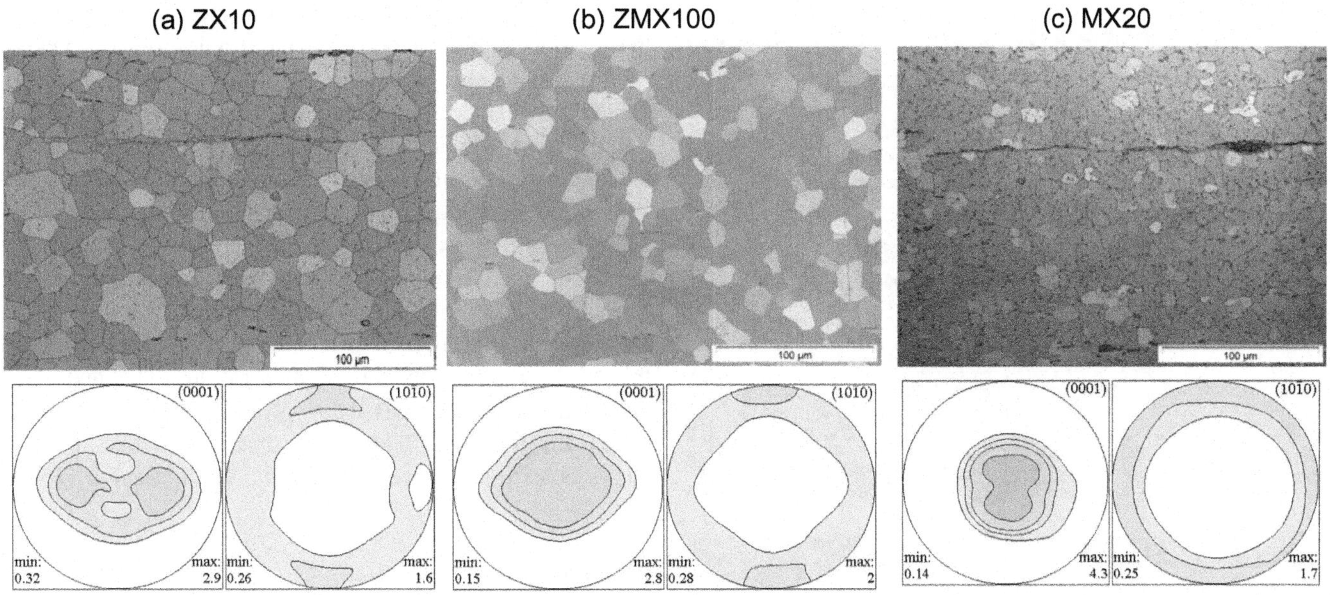


**Fig. 2** Microstructure and corresponding texture of magnesium alloy sheets containing Ca [11]

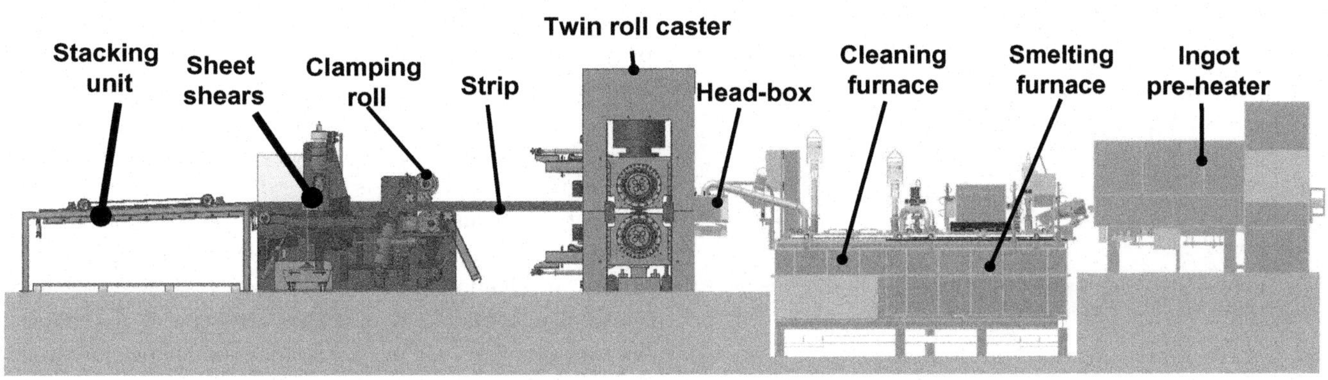


**Fig. 3** Twin roll casting equipment at HZG

and 300 mm width. Prior to rolling the strips were homogenized for 16 h at 400 °C. The rolling procedure was carried out at a sheet temperature of 400 °C and initially included four passes with a degree of deformation $\varphi$ of 0.1 each. In the second stage, three additional passes with $\varphi = 0.2$ were applied, resulting in a final gauge of 2 mm. Between each pass the sheets were intermediately annealed at the rolling temperature for 20 min to maintain the rolling temperature. For further processing the sheets were heat treated for 1 h at 400 °C and then formed into the final shape by a deep drawing procedure.

The twin roll cast strip shows a homogeneous microstructure with slightly squeezed equiaxed grains. An average grain size of 20 μm was determined by means of the line intersection method. Clear centerline segregation was revealed, as well as numerous precipitates located especially at the grain boundaries. Figure 4 compares the microstructures of AZ31 and the Mg-Zn-RE alloy alloys. The AZ31 strip exhibits coarse and a partially columnar microstructure. Conversely, the fine-grained microstructure of the Mg-Zn-RE alloy strip is favorable for good rollability. The homogenization treatment did not affect grain sizes, but allows for resolving the precipitates at the grain boundaries. After subsequent rolling the Mg-Zn-RE alloy sheets exhibit a typical deformed microstructure with squeezed grains, twins and intermetallic phases that mostly consist of small particles aligned with the rolling direction (RD). The homogenizing heat treatment for 30 min at 400 °C leads to a recrystallized microstructure with an average grain size of approximately 10 μm.

Throughout the process, the texture of the strips and sheets develops from a broad and weak texture, with components in transverse direction (TD) of the strips with maximum intensity of (002) pole figure of 2.8 m.r.d., to a typical RE-texture of similar intensity. The basal pole figure exhibits two poles towards the TD and a pronounced TD

(a)

(b)

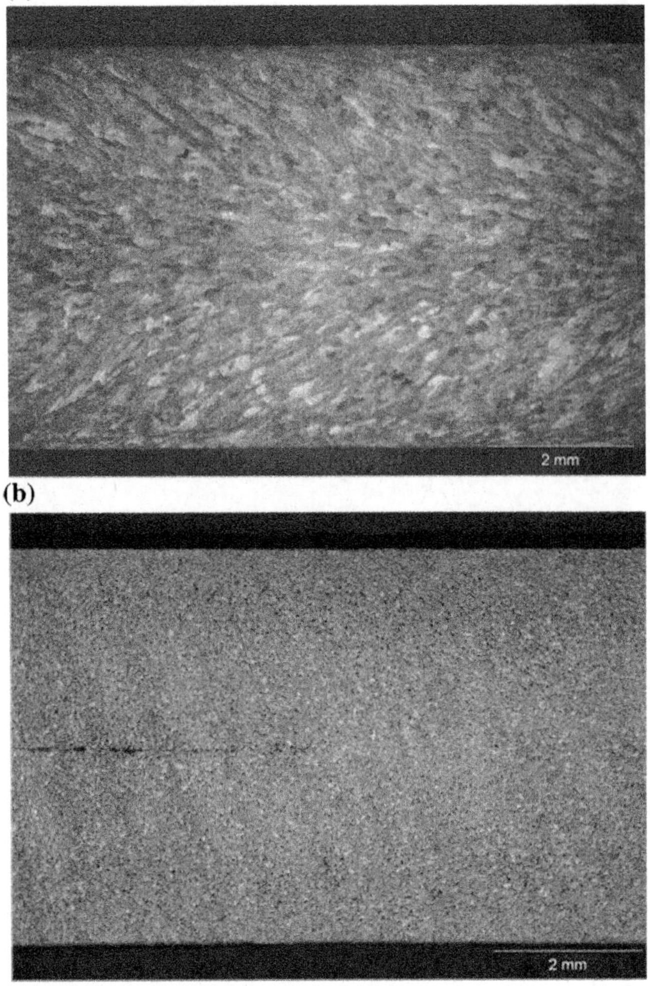


**Fig. 4** **a** Coarse microstructure of the twin roll cast AZ31 strip with columnar structures near the surfaces and an equiaxed area in the center of the strip. **b** Fine grained microstructure of the twin roll cast Mg-Zn-RE alloy strip with local centerline segregation. Both strips are shown in the as-cast condition

component. The comparative mechanical properties of AZ31 and the Mg-Zn-RE alloy sheets are listed in Table 1. First of all it should be noted that the Mg-Zn-RE alloy exhibits, both in RD and TD, significantly higher elongations compared to those of AZ31 alloy. Moreover, it is also worth noting that the yield stress of the Mg-Zn-RE alloy sheets is higher in RD than in TD, which can be attributed to the typical RE-texture, with a spread of the basal poles towards TD [4].

**Table 1** Mechanical properties of twin roll cast sheets of AZ31 and Mg-Zn-RE

| Alloy | AZ31 | | Mg-Zn-RE | |
|---|---|---|---|---|
| Direction[a] | RD | TD | RD | TD |
| TYS [MPa] | 130 | 176 | 174 | 122 |
| UTS [MPa] | 276 | 266 | 238 | 221 |
| Elongation [%] | 12 | 11 | 26 | 33 |

[a] *RD* rolling direction, *TD* transverse direction

The global aim of one actual project at the HZG is the development of Mg sheet materials containing calcium and free of rare earth elements with advantageous properties via the twin roll casting process, which are currently only possible with alloys containing rare earth elements.

The results of this project has shown that Ca additions in the TRC strip leads to significant advantages in terms of formability and strength of the sheets. These new alloys form a very fine grained and homogeneous microstructure free of local deformed areas after rolling. The texture of these Zn-based (RE-free) sheets is similar to conventionally rolled ZE10 sheets. Thanks to these properties the new alloys boast a high formability and a good strength [15].

## Extrusion

Feasibility of the extrusion of magnesium alloys to profiles has been shown in many application-driven works. This well-established forming technology allows the production of long shaped profiles with uniform cross-sections. Achievements with respect to R&D on magnesium alloys focuses on improvement of mechanical properties and therefore develops in a similar way to that described above for magnesium alloy sheets. The extrusion process is used to identify the behavior of new alloy compositions by their microstructure and texture development as well as their related mechanical properties. In this context extrusion alloys that contain RE elements are certainly of interest, because they have different behaviors compared to conventional alloys [8, 10]. Ductility can be increased and the yield asymmetry in tension and compression be lowered significantly, due to the texture weakening in those alloys. An example is shown in Fig. 5.

Extrusion is also considered a possibility for improving the microstructures of extrudates by process-related grain refining, resulting in an increase of the strength and ductility of extrudates compared to their cast counterparts. This process is conducted as a single-step deformation with a resulting microstructure based on dynamic thermo-mechanical processes. Processing limits are based on the capacity of the respective press, with a tendency of the alloy to show brittle behavior on the one hand and incipient melting (hot cracking) on the other [16].

## Creep Resistant Alloys

For HPDC, magnesium alloys are used that contain 4–10 wt% of aluminum and other elements. These alloys show good room temperature properties, corrosion resistance, castability

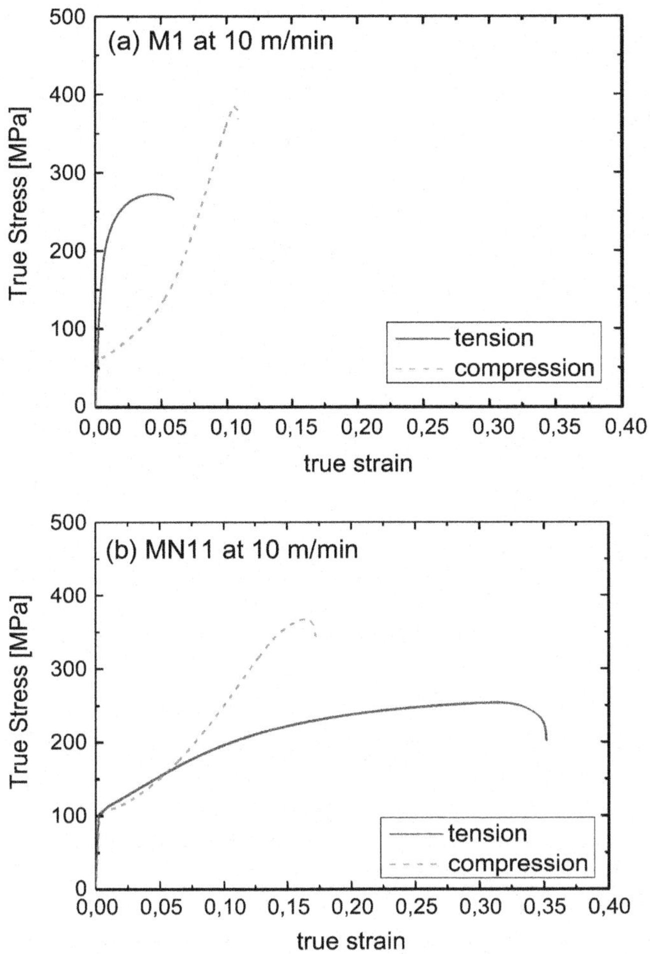


**Fig. 5** Mechanical properties of extruded bars after indirect extrusion at 10 m/min; **a** alloy M1, **b** experimental alloy MN11 containing Nd [17]

and machinability, but when exposed to the higher temperatures of an engine or gearbox housing, they suffer from creep susceptibility. Only HPDC is cost efficient for mass production and therefore alloys are needed that withstand stresses at higher temperatures, but at the same time can be die cast. During solidification a Mg–Al β-phase $Mg_{17}Al_{12}$ forms, which is responsible for the moderate creep resistance and its weakening at higher temperatures [18]. For high-temperature applications special alloys are thereby developed that are Al-free, to avoid the β-phase. However, alloys that contain aluminum have been developed with additional alloying elements, which prefer to precipitate with aluminum during solidification; reducing the amount of β-phase and additional precipitates form that strengthen the alloys.

Al-free alloys were developed for sand casting and permanent mold casting. One of these is AM-SC1, a heat treatable, aluminum-free magnesium alloy for application in automotive engine blocks [19]. As described in the patent [20] it contains in wt% 1.4–1.9 Nd, 0.8–1.2 rare earth elements other than Nd, 0.4–0.7 Zn, 0.3–1.0 Zr, 0–0.3 Mn, and 0–0.1 oxidation-inhibiting elements. In creep tests the alloy has shown excellent creep resistance, which can be attributed to a combination of thermally stable precipitates at the grain boundary, with triple point and fine dispersed precipitates at the grain boundary dislocations. Diffusion controlled creep seems to be the rate-controlling deformation mechanism.

Gibson et al. developed an improved HPDC alloy AM-HP2 for power train applications [21, 22]. It does not contain aluminum, but 3.5% rare earth elements (Ce, La, and Nd) and 0.5% zinc [23]. It was found that AM-HP2 has a higher yield strength than the Al alloy A380, but its ductility is relatively low. This effect is true for several creep resistant magnesium alloys, e.g. MRI230D. In creep tests at 150 and 175 °C, AM-HP2 also proved its creep strength is comparable to that of A380.

Gavras et al. found [24] that additions of neodymium, yttrium or gadolinium to Mg–La alloys result in a different creep response up to an addition of 0.2 wt%. Ternary alloys containing Gd and Y had an order of magnitude lower minimum creep rate compared to alloys containing Nd. After microstructure investigation it was found that the most important influence on creep resistance was achieved by the ability of the alloy to develop fine dispersed precipitates.

Barium as an alloying element for magnesium alloys has rarely been used. In combination with aluminum and calcium it promises good solid solution strengthening, because its atomic radius is 1.36 times larger than that of magnesium. Binary phases of barium with aluminum and magnesium are known, which promise to act as precipitates for Orowan strengthening. The alloy DieMag422 containing 4 wt% Al, 2 wt% Ba and 2 wt% Ca was cast in a direct chill process. Compressive creep tests were performed at temperatures between 150 and 240 °C at constant stresses between 60 and 120 MPa [25]. For comparison, creep resistant magnesium alloys AE42 (4 wt% Al and 2 wt% rare earths) and MRI230D (7 wt% Al, 0.3 wt% Mn, 2.1 wt% Ca, 0.5 wt% Sn and 0.3 wt% Sr) were chosen. Figure 6 shows the creep curves of tests performed at 200 °C and 60 MPa. The minimum creep rate of DieMag422 is nearly one order of magnitude lower than the one of MRI230D.

Three different HPDC alloys containing Ba with a nominal constant aluminum/barium/calcium weight ratio of 2:1:1 were die cast in a 250-tonne Toshiba cold chamber with HPDC and were subsequently tested in terms of creep resistance at 200 °C (Fig. 7) and mechanical properties at RT and 150 °C (Table 2) [26]. The alloys showed good castability, excellent creep resistance and high temperature strength.

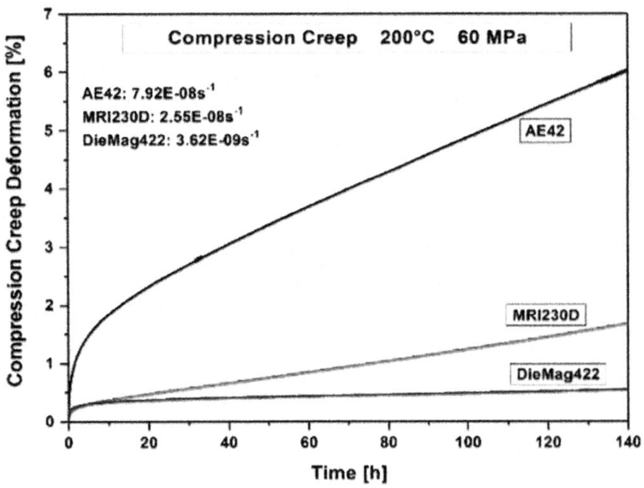


**Fig. 6** Compression creep curves of AE42, MRI230D, and DieMag422 from tests at 200 °C and 60 MPa

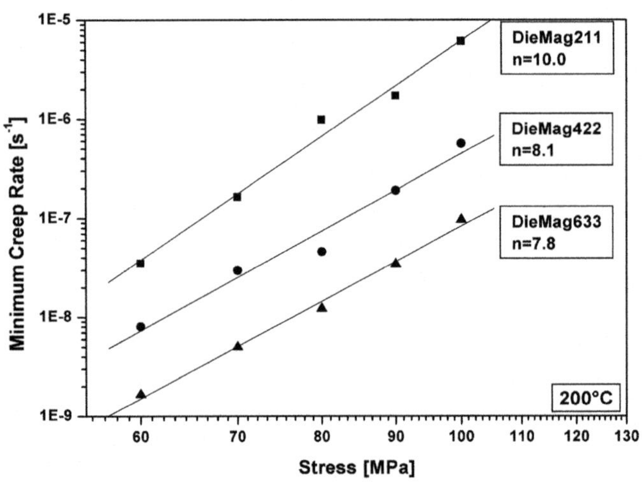


**Fig. 7** Plots of creep rate as a function of applied stress with resulting stress exponents n

## Mg-Based Nanocomposites

The main reason for an increase in research into and consequently publications about light alloy based metal matrix nanocomposites (MMNCs) is the significantly reduced price of nanoparticles. Figure 8 shows the increased number of publications in Web of Science on "magnesium" and "nanocomposites" over the last two decades. It is an advantage that a smaller amount is needed to improve mechanical properties compared to micron-sized particles. The reason for this is the much higher interparticle spacing between large particles. Only a very small distance between particles is able to achieve Orowan strengthening and this is best attained with randomly distributed particles of sizes

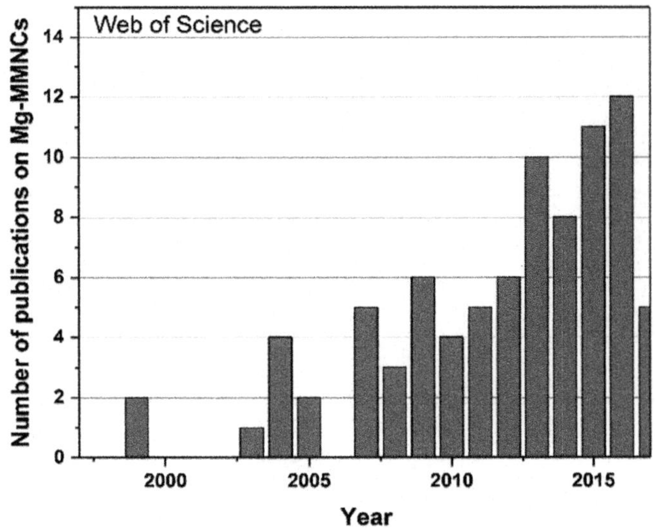


**Fig. 8** Number of publications in WoS on "magnesium" and "nanocomposite"

between 50 and 200 nm. An increase in strength due to Orowan strengthening is therefore inversely proportional to the mean free path between particles.

Orowan strengthening is assumed to be responsible for the increase in yield strength in metal matrix nanocomposites (MMNCs). The reason for this strength increase is the resistance of hard second phase particles to the movement of dislocations through a matrix alloy. Zhang and Chen proposed a description for the Orowan contribution $\Delta\sigma_{OR}$ to strengthening [27] given by the following equation Eq. 1:

$$\Delta\sigma_{OR} = \frac{0.13bG_m}{\lambda}\ln\frac{d_p}{2b} \tag{1}$$

$$\text{where}\quad \lambda = d_p\left[\left(\frac{1}{2V_p}\right)^{1/3} - 1\right] \tag{2}$$

$V_p$ is the volume fraction of nanoparticles, $G_m$ is the shear modulus of the matrix alloy, $d_p$ is the average diameter of the nanoparticles, and $b$ the Burgers vector of the matrix.

Grain refinement or Hall-Petch strengthening positively influences the mechanical properties of MMNCs. Figure 9 shows micrographs of an AM60 and an AM60+ 1wt% AlN nanoparticle MMNC [28]. Grain size is significantly refined from 1277 to 85 µm and the tensile yield strength is improved from 44.9 to 91.2 MPa by reinforcing the AM60 with AlN nanoparticles. Nanoparticles were added under ultrasound-assisted stirring. Cavitation is helpful to de-agglomerate particle clusters and acoustic streaming facilitates a convection of the melt, transporting the released particles.

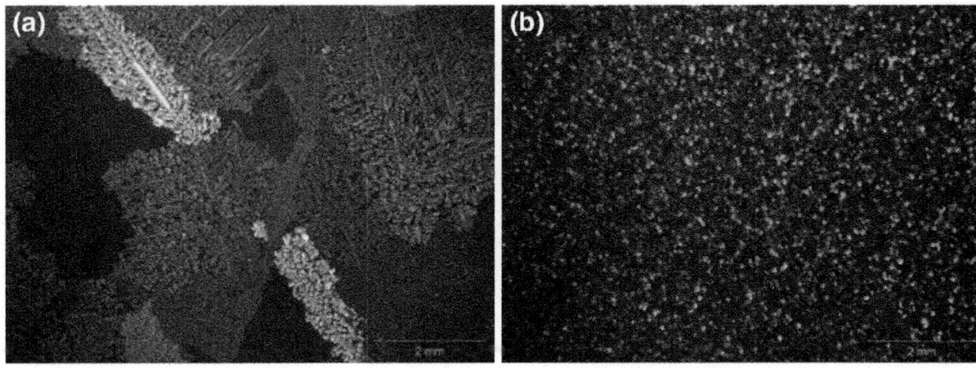


**Fig. 9** Microstructure of **a** AM60, and **b** AM60 + AlN [28]

**Table 2** Mechanical properties of HPDC alloys containing barium [26]

| Alloy | Hardness [HV5] | RT | | 150 °C | |
|---|---|---|---|---|---|
| | | YS [MPa] | UTS [MPa] | YS [MPa] | UTS [MPa] |
| DieMag211 | 54.1 ± 2.4 | 140.6 | 165.9 | 111.5 | 144.0 |
| DieMag422 | 73.3 ± 4.3 | 172.6 | 196.9 | 142.6 | 182.1 |
| DieMag633 | 85.3 ± 9.7 | 202.6 | 229.6 | 160.0 | 196.1 |

## Corrosion and Corrosion Protection

Magnesium is the most electronegative metal used for structural applications. This fact explains the high thermodynamic driving force of Mg metal to corrode when exposed to a corrosive environment. However, when the metal is relatively free of impurities and is used in less aggressive environments it can stay reasonably passive. The main concern for the corrosion resistance of Mg alloys is related to heavy metal impurities, which can be introduced during the primary alloy production or by subsequent forming steps. The latter can be addressed by the development of proper cleaning treatments to remove surface contaminations. Bulk contamination within the alloys is more difficult to deal with and the best way is to prevent them by producing alloys with contamination levels below the well-known tolerance limits. Properly produced Mg alloys normally show reasonable to good corrosion resistance. Alloying elements with lower electrochemical potential differences and/or good passive film formation abilities are preferred (e.g. Al, RE elements) due to internal micro-galvanic corrosion issues between intermetallic phases and the Mg matrix.

There are only a few coating systems for magnesium that have reached commercial and industrial interest in recent decades. In common with many other metallic coating systems, the main strategy is to develop a high barrier layer against corrosive environments while maintaining good adhesion of the coating to the metal substrate. To promote adhesion with paint systems the most relevant are conversion and anodizing treatments. These treatments normally consist of several steps that includes cleaning, etching (activation), conversion or anodizing prior to the final paint application. The latter is normally a multi-layer solution as shown in Fig. 10. In the case of multi-material structures, such as a car body, special care is needed to protect magnesium from galvanic corrosion (coating thickness, number of layers, sealing of all the joint areas and constructive measures). Ideally the Mg parts are electrically decoupled from other metallic components, because during use the barrier protection layers are often disrupted, opening a path for strong galvanic corrosion attack. However, in most of the cases the current joining technologies do not offer full insulation.

The surface treatment of Mg alloys sometimes requires special attention during the different treatment steps. This will be addressed in the following section:

- Cleaning and etching

Various commercial or experimental solutions are available and these do the job. Degreasing is normally done in alkaline solutions. Problems can arise due to severe contamination from production lubricants (especially during wrought processing), but also from different alloy qualities regarding the impurity levels. Therefore sufficient material has to be removed and that can only be guaranteed when acidic etching solutions are applied.

- Conversion coatings

Nowadays, several Cr(VI)-free alternatives are available:
e.g. Bonderite products (Henkel)
Magpass (AHC),
Gardobond and Oxsilan products (Chemetall)

They are used to provide temporary corrosion protection or as pretreatment for paint applications. The chromate-free alternatives often require a better and more uniform cleaning and etching, because they are less fault-tolerant than the old chromate treatments. In combination with powder and KTL paints, the new conversion coatings offer good corrosion resistance.

- Anodizing

By contrast to conversion treatments, the surface is modified by applying an electrical current, converting the surface into ceramic like films with and without the help of spark discharges as in the case of Plasma Electrolytic Oxidation (PEO).

Several companies do offer commercial PEO treatments (Keronite, AHC, Tagnite, Henkel, Chemetall (AMTS)) but there is not much transparency regarding the processing parameters used and treatment baths.

As open porosity is related to the PEO process the coatings do not offer long-term corrosion protection, especially in cases of galvanic corrosion. Notwithstanding, open porosity gives them excellent adhesion for coating systems and migration under the paint is reduced by a large degree compared to conversion coatings.

- Paint systems

State-of-the-art are conversion coatings for promoting adhesion with a multi-layer paint solution. Magnesium components in automotive applications require protection by paint systems if they are to pass through conventional phosphate baths together with aluminum or steel components. The result is a multi-layer system as shown in Fig. 10. Up to the powder coating, the layer system is specific for Mg and the rest is the same for the other metals. This protection is comparable to other materials used in automotive industry, but galvanic corrosion issues still require special care, mainly in the design, additional sealing and/or coating selection.

Considering the variety of potential applications of Mg-based materials, passive corrosion protection approaches are not always sufficient, especially when barrier coatings suffer from premature failure as a result of strong mechanical or environmental impacts. The introduction of corrosion inhibitors into the coating system is a promising solution in this case, offering increased fault tolerance and adding an active corrosion protection for defects [29]. Recently a new class of corrosion inhibitors for Mg alloys was discovered. Their main inhibition mechanism is based on complexation of cations originated from the nobler impurities present in the alloy and reducing in turn the negative effect of these impurities [30]. Thus further development of strategies for proper introduction of new corrosion inhibitors into protective coating schemes becomes an indispensable duty.

## Mg Biomaterials

Magnesium is an essential element for the human body and a well-balanced system controls its intake and excretion. One apparent disadvantage of Mg and its alloys, poor corrosion behavior, is actually a tremendous advantage when it comes to degradable implant materials. This advantage was discovered more than a century ago, when an American surgeon, E. C. Huse, used Mg wires to block bleeding vessels in 1878 [32]. In the following decades, further research took place, mainly in Europe, and the first implants were

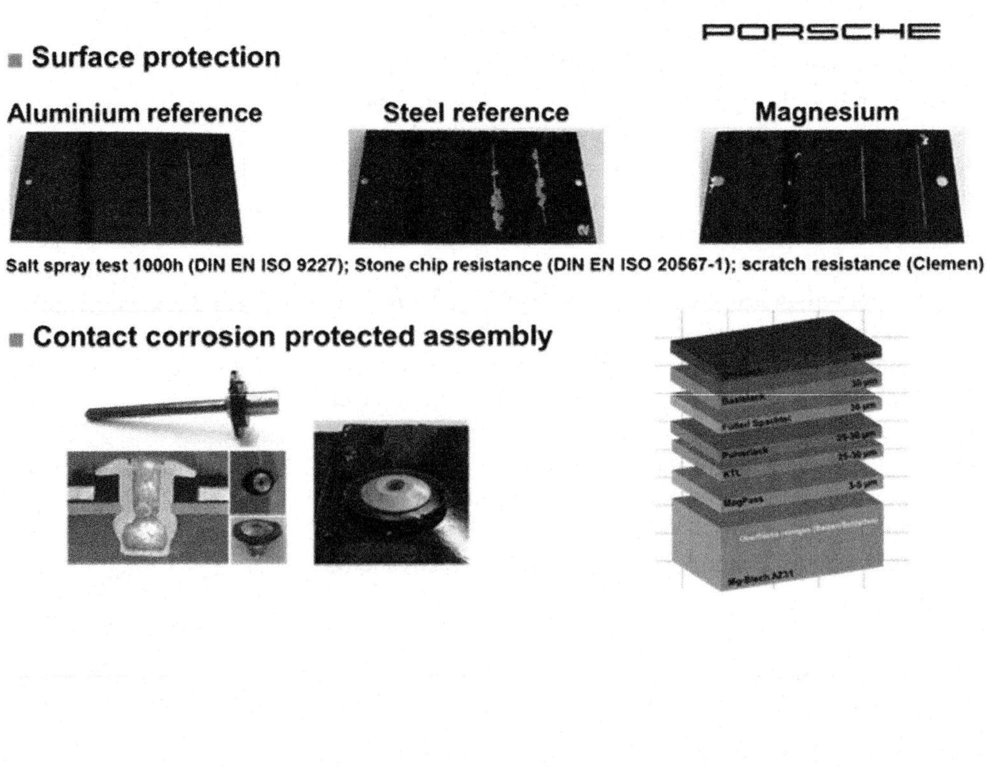


**Fig. 10** Typical paint system, comparison of performance and joining to other metals for AZ31 sheet (results from project "M3—Mobil mit Magnesium") [31]

developed and even tested in humans. F. Witte reported the beginnings and further development until 2010 in his review paper [33]. However, Mg reacts with aqueous body fluids and every gram of Mg basically generates around a liter of hydrogen gas. Fast degradation and gas formation was also observed in the early days of Mg implants and still can be a problem. In the early decades of the 20th century stainless steels and Co–Cr alloys were also developed and tested as implant materials and were favored [34]. It took almost 70 years before Mg and its alloys again gained interest as biodegradable implant materials.

Properties result from the microstructure. Microstructure is a result of the combination of alloying elements and processing. Processing also includes the ultimate steps with regard to surface conditioning. This means that in the first proper alloying, the elements have to be selected. To produce an alloy, melting and casting are the first steps and the alloying elements already have impact on it. In the steps that follow, wrought processes like extrusion, rolling, forging etc. are applied, often in combination with heat treatments. Again, the alloying elements used have an impact on these processing steps and the processing has an influence over the microstructure.

Today almost any alloy composition is under investigation. Quite often, researchers claim their alloy could be used as degradable implant material simply because it is based on magnesium. Mechanical tests are performed, as well as simple corrosion tests, and in conclusion most magnesium based materials are declared suitable. However, this simplistic approach is not appropriate. Prior to this, all the alloying elements in combination with the matrix and also intermetallic phases should be tested for their suitability. With respect to biological performance, only a few elements remain viable as alloying elements (Fig. 11). However, even Al is often among the alloying elements for degradable implants, with disregard to the fact that it may be responsible for Alzheimer's disease, and other elements are also under discussion. It remains unclear what elements in what amount should be used to achieve the required properties and biocompatibility.

Quite often it is mentioned that degradation of Mg and its alloys is too fast. Comparing the requirements of a stent and a bone implant for a child makes it clear that no general statement can be applied. A stent with a strut thickness of 150–200 µm has to resist corrosion for minimum of six months. A bone implant for children has a thickness ranging around 1 mm and can disappear completely after a few weeks. Clearly when the same material is used for both applications we can assume it will fail for one of the applications. A similar comparison can be made for strength and ductility. A stent is first crimped on the catheter (compressive loading), applied to the appropriate place in the vessel and then inflated by the catheter (tensile loading). A bone implant may be bent to fit the bone. However, the mechanical requirements for the stent material are much more demanding than those for the bone implant. It is obviously necessary to develop different alloys for these different applications. When a fast degrading material must serve in an environment where a slow degradation is in fact necessary, perhaps degradable coatings are a solution.

For the testing of implant materials, the ISO 10993 set entails a series of standards (parts 1–20) for evaluating the biocompatibility of medical devices. Unfortunately, this series of standards is not designed for degradable materials and in particular for magnesium and its alloys. Some tests lead to false results and based on them it could be concluded that Mg is not a suitable implant material. Therefore, testing methods need to be adapted to fulfill the requirements of a degradable metallic system. Even when tremendous progress can be observed over the years, proper testing of magnesium alloys as degradable implant materials is still a challenge.

Today there are two CE certified applications on the market. Syntellix achieved the CE mark in 2013 for its Hallux Valgus screw (Magnezix®) and derived a family of products from this [35]. Biotronik also achieved a CE certificate for their Magmaris® stent in 2016 [36]. Interestingly,

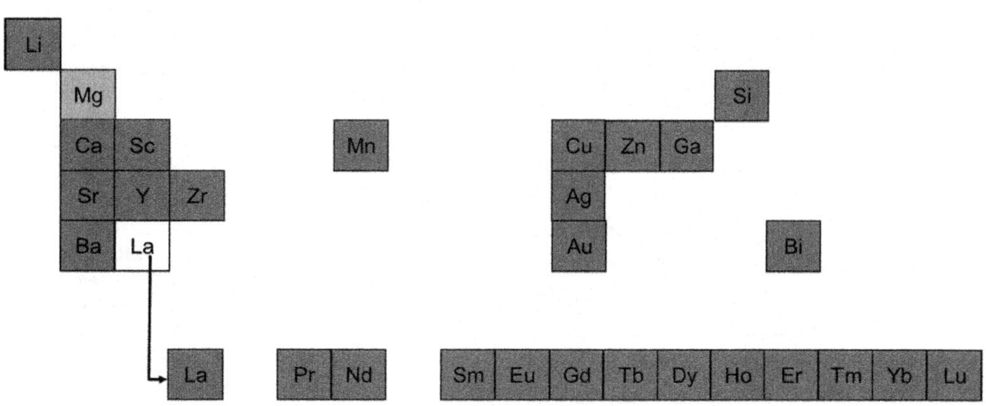


**Fig. 11** Alloying elements for Mg based implants

both implants are based on Mg-Y-RE (WE) alloys, but serve in two different applications.

## Mg Batteries

Nowadays the development of long-lasting, green energy storage devices is a topic with high social impact. Innovative technologies and the electrification of daily life demand reliable and efficient power sources that are ecologically harmless, yet cost-effective. In this context batteries based on Mg might offer a promising future alternative to conventional Li systems [37–41]. They have greater power storage capacity, a few times more than the best performing Li-ion battery. For example, the theoretical volumetric capacity of magnesium metal anodes is 3833 mAh $cm^{-3}$, about four times greater than typical graphite anode in a Li-ion battery. Unfortunately, the commercialization of Mg-based energy storage devices is very limited due to self-corrosion issues of anode materials for primary (aqueous) cells or lacking electrolytes in cases of rechargeable batteries. Additional research progress on stable working cathode materials is required as well, since existing systems are not capable or not adapted to Mg.

Nevertheless, there are some success stories available. Back in 1943 water-activated silver chloride/Mg battery was commercially accessible [42] however, it felt out of favor due to its costs compared with nickel-metal hybrid and lithium batteries. Such aqueous primary Mg batteries have several advantages for marine applications and are in use, for example, in rescue systems or for underwater vehicles [43, 44]. To become a more applicable battery alternative, a stable battery voltage and discharge behavior has to be achieved. A controlled degradation of the Mg anode surface and release of $Mg^{2+}$ ions are required. One approach is alloying. Very recently Chinese scientists started undertaking a more dedicated search for different magnesium alloys as enhanced anodes [45, 46]. Another concept is based on the control of self-corrosion of anodes via electrolyte additives [47]. All these strategies need to be further developed and performance limits are still yet to be defined.

Secondary Mg batteries are of greater interest. Rechargebale Mg cells might become a real competitor to Li systems. Recently research progress for suitable electrolytes was achieved. Two novel systems were developed. The first relies on the Mg–S system, which is stable until 3.9 V and is based on modified complexants [48]. In the second the electrolyte is based on weakly coordinated Mg salts. It is free of chlorides and even stable in air [49]. However, an applicable full cell does not exist yet. In particular cathode development is required. For example, intercalation based cathodes [50], for Mg–Li hybrids need to be further explored [51], and well-functioning cathodes for the Mg–sulfur system established [52]. The very attractive Mg-air system is also a focus of research [53]. In this case the interaction of advanced Mg-based anode materials with oxygen in a non-aqueous electrolyte needs to be understood.

Not only is the economical aspect, but also the ecological aspect an issue for all systems. For processing of battery and electrodes alike, consideration of the up-scaling aspects needs to be developed and improved [54]. Other aspects that must be considered include critical raw materials [55, 56] the possibility of toxicity [57, 58] and the entire eco-balance [59, 60] including life-cycle-analysis [61]. Generally speaking, there is still much research and development for all battery components required. Anode development and the related material functionalization and design, adaption of e.g. ionic liquids or electrolyte additives with advanced cell design might enable this technology. Industrial transfer might become possible in niche applications, which have a tailored power-consumption requirement profile, such as marine technologies.

## Conclusions

Although there are many new developments and research projects in the field of magnesium-based materials, there is a lack of implementation in the market. The automotive industry is only partially prepared to break new ground, because well-trodden paths are always easier to follow. This can only be changed by intensive work, e.g. in the field of magnesium sheets. Potential medical applications of magnesium alloys as biodegradable implants are very challenging. Much of the work of physicians and materials scientists still remains to be done in this respect. Overall, it can be said that the potential of magnesium materials is still not fully exploited and there is probably a need for even greater cost pressure in order to bring more applications to market.

## References


1. H. Friedrich and S. Schumann: Research for a “new age of magnesium” in the automotive industry; J. Mater. Process. Tech. 117 (2001) 276–281.
2. C.E. Dreyer, W.V. Chiu, R.H. Wagoner, S.R. Agnew: Formability of a more randomly textured magnesium alloy sheet: Application of an improved warm sheet formability test; J. Mat. Proc. Techn. 210 (2010) 37–47.
3. L. Stutz, J. Bohlen, G. Kurz, D. Letzig, K.U. Kainer: Influence of the Processing of Magnesium Alloys AZ31 and ZE10 on the Sheet Formability at Elevated Temperature; Key Eng. Mater. 473 (2011) 335–342.
4. J. Bohlen, M.R. Nürnberg, J.W. Senn, D. Letzig, S.R. Agnew: The texture and anisotropy of magnesium-zinc-rare earth alloy sheets; Acta Materialia 55 (2007) 2101–2112.
5. L.W.F. Mackenzie, M.O. Pekguleryuz: The recrystallization and texture of magnesium–zinc–cerium alloys; Scripta Materialia 59 (2008) 665–668.

6. A. Styczynski, C. Hartig, J. Bohlen, D. Letzig: Cold rolling textures in AZ31 wrought magnesium alloy; Scripta Materialia 50 (2004) 943–947.
7. S.R. Agnew, M.H. Yoo, C.N. Tomé: Application of texture simulation to understanding mechanical behavior of Mg and solid solution alloys containing Li or Y; Acta Materialia 49 (2001) 4277–4289.
8. N. Stanford, M.R. Barnett: The origin of "rare earth" texture development in extruded Mg-based alloys and its effect on tensile ductility; Mater. Sci. Eng. A 496 (2008) 399–408.
9. T. Al-Samman, X. Li: Sheet texture modification in magnesium-based alloys by selective rare earth alloying; Mater. Sci. Eng. A528 (2011) 3809–3820.
10. N. Stanford: The effect of rare earth elements on the behaviour of magnesium-based alloys: Part 2—recrystallisation and texture development; Mater. Sci. Eng. A565 (2013) 469–475.
11. J. Bohlen, S. Yi, J. Victoria Hernandez, G. Kurz, D. Letzig, K.U. Kainer: Proceedings of Mg 2015, Jeju 2015.
12. S. Yi, D. Letzig, K. U. Kainer, O. D. Kwon, J. H. Park, J. J. Kim: Proceedings of Mg 2015, Jeju Island 2015.
13. F. Basson, D. Letzig: Aluminium twin roll casting transfers benefits to magnesium; Aluminium International Today, November/December (2010) 19–21.
14. K. Hantzsche, J. Bohlen, J. Wendt, K.U. Kainer, S.B. Yi, D. Letzig: Effect of rare earth additions on microstructure and texture development of magnesium alloy sheets; View Point Paper, Scripta Materialia 63 (2010) 725–730.
15. D. Letzig, J. Bohlen, G. Kurz, J. Victoria-Hernandez, R. Hoppe, S. Yi: Development of Magnesium Sheets; Magnesium Technology, Proceeding of the TMS 2018 (2018) in press.
16. A.G. Beer: Enhancing the extrudability of wrought magnesium alloys; in: C.J. Bettles, M.R. Barnett (Ed.) "Advances in wrought magnesium alloys", Woodhead Publishing, Cambridge, UK (2012), 304.
17. J. Bohlen, J. Swiostek, D. Letzig, K.U. Kainer: Magnesium Technology, Proceedings of the TMS 2009 (2009) 225–230.
18. M. Regev, A. Rosen, M. Bamberger: Qualitative model for creep of AZ91D magnesium alloy; Met. Mat. Trans. 32A (2001) 1335–1345.
19. C.J. Bettles, M.A. Gibson, S.M. Zhu: Microstructure and mechanical behaviour of an elevated temperature Mg-rare earth based alloy; Mat. Sci. Eng. A505 (2009) 6–12.
20. C.J. Bettles, C.T. Forwood: Creep resistant magnesium alloy; US patent 20050002821.
21. M.A. Gibson, C.J. Bettles, M.T. Murray, G.L. Dunlop: AM-HP2: A new magnesium high pressure diecasting alloy for automotive powertrain applications; Magnesium Technology 2006 A.A. Luo, N.R. Neelameggham, R.S. Beals (Eds.) TMS (The Minerals, Metals & Materials Society) (2006) 327–331.
22. M.A. Gibson, M. Easton, V. Tyagi, M. Murray, G. Dunlop: Further improvements in HPDC Mg alloys for power train applications; Magnesium Technology 2008, M.O. Pekguleryuz, N.R. Neelameggham, R.S. Beals, E.A. Nyberg (Eds.) TMS, The Minerals, Metals and Materials Society [2008] 227–232.
23. S. Zhu, M.A. Easton, T.B. Abbott, J.-F. Nie, M.S. Dargusch, N. Hort, M.A. Gibson: Evaluation of Magnesium Die-Casting Alloys for Elevated Temperature Applications: Microstructure, Tensile Properties, and Creep Resistance; Met. Mat. Trans. A 46 (2015) 3543–3554.
24. S. Gavras, S.M. Zhu, F.F. Nie, M.A. Gibson, M.A. Easton: On the microstructural factors affecting creep resistance of die-cast Mg–La-rare earth (Nd, Y or Gd) alloys; Mat. Sci. Eng. A675 (2016) 65–75.
25. H. Dieringa, Y. Huang, P. Wittke, M. Klein, F. Walther, M. Dikovits, C. Poletti: Compression-creep response of magnesium alloy DieMag422 containing barium compared with the commercial creep-resistant alloys AE42 and MRI230D; Mat. Sci. Eng. A 585 (2013) 430–438.
26. H. Dieringa, D. Zander, M.A. Gibson: Creep behaviour under compressive stresses of calcium and barium containing mg-al-based die casting alloys; Mat. Sci. Forum 765 (2013) 69–73.
27. Z. Zhang, D.L. Chen: Consideration of Orowan strengthening effect in particulate-reinforced metal matrix nanocomposites: A model for predicting their yield strength; Scr. Mat. 54 (2006) 1321–1326.
28. H. Dieringa, L. Katsarou, R. Buzolin, G. Szakács, M. Horstmann, M. Wolff, C. Mendis, S. Vorozhtsov, D. StJohn: Ultrasound assisted casting of an AM60 based metal matrix nanocomposite, its properties and recyclability; submitted to Metals, accepted.
29. D.K. Ivanou, K.A. Yasakau, S. Kallip, A.D. Lisenkov, M. Starykevich, S.V. Lamaka, M.G.S. Ferreira, M.L. Zheludkevich: Active corrosion protection coating for a ZE41 magnesium alloy created by combining PEO and sol–gel techniques; RSC Advances 6 (2016) 12553–12560.
30. S.V. Lamaka, D. Höche, R.P. Petrauskas, C. Blawert, M.L. Zheludkevich: A new concept for corrosion inhibition of magnesium: Suppression of iron re-deposition; Electrochemistry Communications 62 (2016) 5–8.
31. D. Letzig: New Developments in the Field of Wrought Magnesium Alloys, International Workshop on "Processing-Microstructure-Property Relationships & Deformation Mechanisms of Magnesium Alloys", Madrid, 21–24 Mai 2013.
32. E. C. Huse: A new ligature?; Chicago Medical Journal and Examiner 37 (1878) 171–172.
33. F. Witte: The history of biodegradable magnesium implants: A review; Acta Biomaterialia 6 (2010) 1680–1692.
34. Q. Chen, G. A. Thouas: Metallic implant biomaterials; Materials Science and Engineering R 87 (2015) 1–57.
35. http://www.syntellix.de/en/products/product-overview/all.html.
36. http://www.magmaris.com/en.
37. Y. Orikasa et al.: High energy density rechargeable magnesium battery using earth-abundant and non-toxic elements; Scientific Reports 4 (2014) 5622.
38. H. D. Yoo et al.: Mg rechargeable batteries: an on-going challenge; Energy & Environmental Science 6 (2013) 2265–2279.
39. M. Armand, J. M. Tarascon: Building better batteries; Nature 451 (2008) 652–657.
40. G. E. Blomgren: Electrochemistry: Making a potential difference; Nature 407 (2000) 681–682.
41. D. Aurbach et al.: Prototype systems for rechargeable magnesium batteries; Nature 407 (2000) 724–727.
42. I. Blake: Silver Chloride-Magnesium Reserve Battery; Journal of the Electrochemical Society 99 8 (1952) 202C–203C.
43. F. Cheng and J. Chen: Metal–air batteries: from oxygen reduction electrochemistry to cathode catalysts; Chemical Society Reviews 41 6 (2012) 2172–2192.
44. Ø. Hasvold, T. Lian, E. Haakaas, N. Størkersen, O. Perelman and S. Cordier: CLIPPER: a long-range, autonomous underwater vehicle using magnesium fuel and oxygen from the sea; Journal of Power Sources 136 (2004) 232–239.
45. H. Xiong, K. Yu, X. Yin, Y. Dai, Y. Yan, H. Zhu: Effects of microstructure on the electrochemical discharge behavior of Mg-6 wt% Al-1 wt% Sn alloy as anode for Mg-air primary battery; Journal of Alloys and Compounds 708 (2017) 652–661.
46. J. Li, K. Wan, Q. Jiang, H. Sun, Y. Li, B. Hou, L. Zhu, M. Liu: Corrosion and Discharge Behaviors of Mg-Al-Zn and Mg-Al-Zn-In Alloys as Anode Materials; Metals 6 3 (2016) 65.
47. D. Höche, S. V. Lamaka, M. L. Zheludkevich: European Patent Office, HZG, Ed. (2016), EP16187152.0.
48. Zh. Zhao-Karger et al.: Magnesium sulfur battery with improved non-nucleophilic electrolyte; Adv. Energy Mater. 5 3 (2015) 1401155.

49. Zh. Zhao-Karger et al.: New class of non-corrosive, highly efficient electrolytes for rechargeable magnesium batteries; J. Materials Chem. A5 (2017) 10815–10820.
50. C. Bonatto Minella et al.: Interlayer-Expanded VOCl as Electrode Material for Magnesium-Based Batteries; ChemElectroChem 4 (2017) 738–745.
51. C. Bonatto Minella et al.: Li-Mg Hybrid Battery with Vanadium Oxychloride as Electrode Material; submitted (2017).
52. B. P. Vinayan et al.: Performance study of magnesium–sulfur battery using a graphene based sulfur composite cathode electrode and a non-nucleophilic Mg electrolyte; Nanoscale 8 (2016) 3296–3306.
53. C.-S. Li, Y. Sun, F. Gebert, S.-L. Chou: Adv. Energy Mater. (2017) 1700869.
54. B. Simon et al.: Proposal of a framework for scale-up life cycle inventory: A case of nanofibers for lithium iron phosphate cathode applications; Integrated Environmental Assessment and Management 12 (2016) 465–477.
55. M. Weil et al.: Recycling of traction batteries as a challenge and chance for future lithium availability. In: Pistoia, G. (Eds.): Lithium-Ion Batteries: Advances and applications; Elsevier (2014), 509–528.
56. B. Simon et al.: Potential metal requirement of active materials in lithium-ion battery cells of electric vehicles and its impact on reserves: Focus on Europe; Resources, Conservation and Recycling 104 (2015) 300–310.
57. F. Gschwind et al.: Chloride Ion Battery Review: Theoretical Calculations, State of the Art, Safety, Toxicity, and an Outlook towards Future Developments; Inorganic Chemistry 21 (2017) 2784–2799.
58. F. Gschwind et al.: Fluoride ion batteries: Theoretical performance, safety, toxicity, and a combinatorial screening of new electrodes; Fluorine Chemistry 182 (2016) 76–90.
59. J. Peters et al.: The environmental impact of Li-Ion batteries and the role of key parameters—A review; Renewable and Sustainable Energy Reviews 67 (2017) 491–506.
60. J. Peters et al.: Life cycle assessment of sodium-ion batteries; Energy Environ. Sci. 9 (2016) 1744–1751.
61. M. Baumann et al.: CO2 footprint and life cycle costs of electrochemical energy storage for stationary grid applications. Energy Technology 5 (2016) 1071–1083.

# Solute/Stacking Fault Energies in Mg and Implications for Ductility

Binglun Yin, Zhaoxuan Wu, and W. A. Curtin

**Abstract**

Mg is the lightest structural metal but pure Mg has low ductility due to strong plastic anisotropy and to a transition of <c+a> pyramidal dislocations to a sessile basal-oriented structure [1]. Alloying generally improves ductility, but the mechanisms of the enhancement are not yet known. Mg-3 wt% RE (RE = Y, Tb, Dy, Ho, Er) show high ductility [2], as compared to most commercial Mg–Al–Zn alloys at similar grain size. To investigate possible proposed mechanisms of ductility in alloys, and differences between Al, Zn, and Y solutes, first-principles density functional theory (DFT) calculations are used to compute all relevant stacking fault (SF) energies as a function of solute type (Y, Al, Zn) and concentration in the dilute limit.

In DFT calculations, we compute the solute-SF interaction energy $E_{int}(d_i)$ versus solute-SF distance $d_i$. Accurate energies requires the use of large supercells. For the pyramidal II plane, a single solute may induce migration of the SF. Constraints, and corrections for the constraints, are thus needed. In the random alloy, every atom site has a probability $c$ (in at.) to be occupied by a solute atom. The value of the SF energy of the alloy, at small $c$, is then $\gamma^A = \gamma^{Mg} + \frac{c}{A_0}\sum_i E_{int}(d_i)$.

The stacking fault energies for basal and pyramidal faults versus concentration are shown in Fig. 1 for the solutes Y, Al and Zn. From these results, we can draw some conclusions regarding ductility in Mg alloys. First, the proposed role of the $I_1$ basal SF in ductility enhancement in Mg–Y [2] is not supported. The effects of Y can be achieved at twice the concentration using Al. However, neither Mg–Al or Mg–Zn alloys show significantly enhanced ductility. Second, using the $I_1$ and pyr. II SFs for Y, elasticity calculations show that Y does not appear to significantly alter the energetics of the detrimental pyramidal-to-basal orientation transformation. Third, the only unique property of Y, compared to Al and Zn, is the much larger reduction of the pyramidal I SF energy. This suggests new mechanisms for enhanced ductility that will be discussed and supported by further results on other solutes.

**Keywords**

Alloys • Stacking faults • Ductility • Dislocations

B. Yin · Z. Wu · W. A. Curtin (✉)
École Polytechnique Fédérale de Lausanne, EPFL STI IGM, Institute of Mechanical Engineering, Station 9, 1015 Lausanne, Switzerland
e-mail: william.curtin@epfl.ch

Z. Wu
Institute of High Performance Computing, 1 Fusionopolis Way #16-16, Connexis, Singapore 138632, Singapore


D. Orlov et al. (eds.), *Magnesium Technology 2018*, The Minerals, Metals & Materials Series, https://doi.org/10.1007/978-3-319-72332-7_2

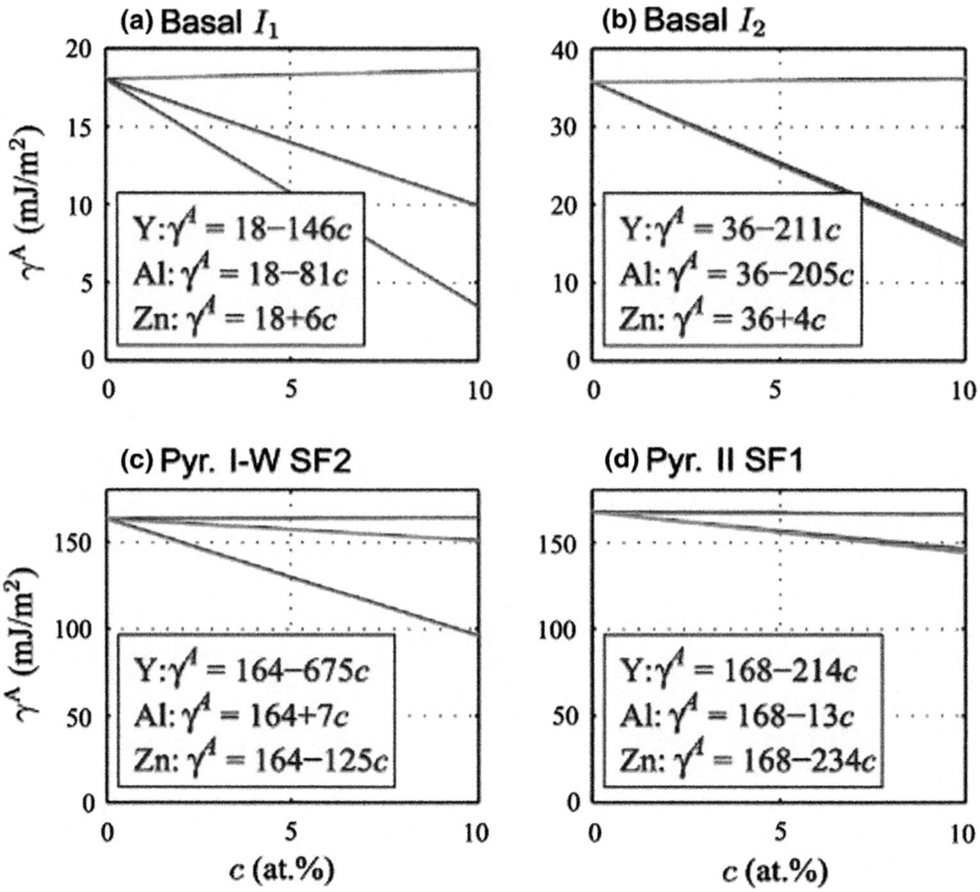


**Fig. 1** Predicted SF energies versus concentration of three solutes [3]

## References


1. Z. Wu, W.A. Curtin, Nature, 526 (2015) 62–67.
2. S. Sandlobes, et al., Acta Materialia, 59 (2011) 429–439; Acta Materialia 70 (2014) 92–104.
3. B. Yin, Z. Wu, and W. A. Curtin, Acta Materialia, 136 (2017) 249–261.

# Recent Developments in Magnesium Alloy Corrosion Research

Nick Birbilis, R. L. Liu, Y. Yan, and O. Gharbi

**Abstract**

When exposed to atmospheric or aqueous conditions, magnesium (Mg) undergoes corrosion, and so do Mg alloys. In recent years, there has been a transition from accepting the corrosion of magnesium alloys, to proactive research aimed at (1) clarifying magnesium corrosion mechanisms to devise methods for restricting magnesium corrosion, and (2) demonstration of methodologies to produce more corrosion resistant magnesium alloys.

Herein, a number of examples will be shown, selected on the basis that they represent a design directed approach. Two key systems will be highlighted, including the Mg-*X* systems, where *X* represents group 14 and group 15 elements. It was recently demonstrated that arsenic (As) has the ability to stifle the cathodic reaction kinetics (via cathodic poisoning mechanism) upon Mg. Given the toxicity of As, alternatives with similar chemical properties are explored, with particular focus given to germanium (Ge) additions. The addition of Ge was shown to decrease corrosion rates significantly, by a range of short and long-term tests, which also resulting in a concomitant lower rate of hydrogen evolution.

In contrast, the Mg–Li alloy system is also explored as a corrosion resistant Mg-alloy. A critical concentration of Li additions to Mg (>10.5 wt% Li) have been demonstrated to allow the dynamic development of a corrosion resistant surface film consisting of lithium carbonate. Such a lithium carbonate film is stable in aqueous electrolytes, and insoluble, providing a basis for significant reduction in the dissolution (and corrosion) of Mg.

The above alloying additions as functional additions for promoting appreciable corrosion resistance to Mg-alloys is elaborated in the context of future Mg alloys.

N. Birbilis (✉) · R. L. Liu · Y. Yan · O. Gharbi
Department of Materials Science and Engineering, Monash University, Clayton, VIC, Australia
e-mail: nick.birbilis@monash.edu


D. Orlov et al. (eds.), *Magnesium Technology 2018*, The Minerals, Metals & Materials Series, https://doi.org/10.1007/978-3-319-72332-7_3

# Towards Active Corrosion Protection of Mg Alloys Using Corrosion Inhibition Approaches

M. L. Zheludkevich, S. V. Lamaka, Y. Chen, D. Hoeche, C. Blawert, and K. U. Kainer


**Abstract**

The high susceptibility of Mg alloys to corrosive degradation is calling for new efficient corrosion protection solutions. In the present paper we discuss the approach based on the introduction of new generation of corrosion inhibitors into the composite protective coatings aiming at an additional active corrosion protection.



**Keywords**

Magnesium • Corrosion coating • Inhibitor


One of the main obstacles for active uptake of Mg based materials by different industries is related to their relatively high corrosion susceptibility in aggressive environments. Therefore adequate corrosion protection strategies have to be applied in order to reduce the corrosion impact of the Mg-based structures or multi-material assemblies where Mg is one of the partners. The traditional approach towards corrosion protection of Mg alloy surfaces relies on the creation of superior barrier coatings and the extensive application of sealants to joint areas. Nevertheless the passive protection can be partially lost during the service-life of the structure because of exploitation induced defects such as pinholes, scratches and cracks which appear in the coatings. The disruption of the barrier can lead to fast propagation of the corrosion processes in such local areas. The alternative approach is to introduce the corrosion inhibitors into the protection scheme. The corrosion inhibitors locally leached from the damaged coating can actively suppress the corrosion rate in the defects increasing the fault-tolerance of the protective coatings. However the choice of efficient environmentally-friendly inhibitors for Mg was extremely limited.

The detailed knowledge on the corrosion mechanism is required to design the corrosion inhibitors which can be used as a part of protection scheme. However there is still an active discussion in the corrosion community concerning the main mechanistic details of Mg corrosion. Recently it was demonstrated that noble impurities present in the Mg-based materials can play a detrimental role for corrosion resistance. One of the important mechanisms is related to the enhanced cathodic activity at the corrosion front as a result of increased cathodically active area caused by re-deposition of thin iron film on Mg surface [1]. The Fe cathode film is plated be reduction of $Fe^{2+}/Fe^{3+}$ cations generated during corrosive dissolution of detached Fe impurity particles present in the alloy. This effect drastically accelerates corrosion of impurity containing Mg.

In the present work a new approach based on the complexation of Fe ions [2] and consequently preventing re-deposition is employed to select compounds which can offer high corrosion inhibition efficiency. The newly discovered corrosion inhibitors are integrated within protective coatings in order to obtain an active corrosion protection effect for Mg alloys.

The composite two layer system was used to apply a protective coating on a AZ91 magnesium alloy. The first layer is created by short PEO (plasma electrolytic oxidation) treatment. It leads to formation of relatively thin (about 4 μm) porous layer composed mainly Mg oxide/hydroxide species. Such a layer can play a double function improving the stability of the interface and offering a good reservoir to load the corrosion inhibitors. The developed porosity can ensure a sufficient capacity for the inhibitor loading and a good anchoring to the polymer top coating. Hybrid sol-gel polymer layer was applied here in order to seal the inhibitor loaded PEO pores and provide an efficient additional barrier against corrosive species. The sol-gel formulation was composed by GPTMS and TPOT dissolved in ethanol and controllably hydrolyzed as described elsewhere [3].

The corrosion protection performance of the obtained coatings was evaluated using combination of integral and


M. L. Zheludkevich (✉) · S. V. Lamaka · Y. Chen · D. Hoeche · C. Blawert · K. U. Kainer
MagIC—Magnesium Innovation Centre, Helmholtz-Zentrum Geesthacht, Max-Planck Str. 1, 21502 Geesthacht, Germany
e-mail: mikhail.zheludkevich@hzg.de





D. Orlov et al. (eds.), *Magnesium Technology 2018*, The Minerals, Metals & Materials Series, https://doi.org/10.1007/978-3-319-72332-7_4

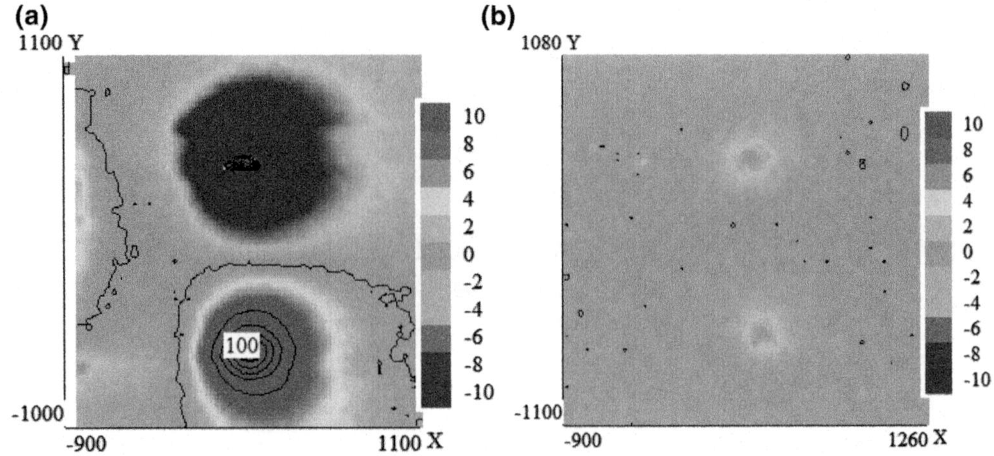


**Fig. 1** SVET maps on AZ91 coated with composite protective coatings immersed in 0.05 M NaCl for 36 h. **a** blank coating; **b** coating with 4-aminosalycilic acid

localized electrochemical techniques. The evolution of impedance spectra of the coated AZ31 immersed in 0.5% NaCl shows that in the case of inhibitor-loaded coatings the impedance decrease in whole frequency range is significantly slower than that for the inhibitor-free reference system. The active corrosion protection was evaluated by SVET (scanning vibrating electrode technique). This method allows measuring the local anodic and cathodic currents at corroding interface with a high spatial resolution. Figure 1 presents the SVET maps measured over the surface of coated Mg alloys immersed in 0.05 M NaCl solution for 36 h. Two artificial defects were introduced in each coating and development of local corrosion currents in these defects was monitored. After 36 h of corrosion test the corrosion current density in both defects on blank coating shows high values. While only a minor activity is detected in the case of inhibitor-containing system.

The obtained electrochemical results clearly demonstrated a superior performance of the coating loaded with 4-aminosalycilic acid as a corrosion inhibitor. The addition of inhibitor to the coating system leads to a significant increase of its fault tolerance. Well defined active corrosion protection in the defects is demonstrated. The inhibitor-loaded protective coating on Mg alloy can be considered as a promising solution for different structural applications.

## References

1. D. Höche, C. Blawert, S.V. Lamaka, N. Scharnagl, C. Mendis, M.L. Zheludkevich, Phys. Chem. Chem. Phys., 18 (2016) 1279–1291.
2. S.V. Lamaka, D. Höche, R.P. Petrauskas, C. Blawert, M.L. Zheludkevich, Electrochemistry Communications, 62 (2016) 5–8.
3. D.K. Ivanou, M. Starykevich, A.D. Lisenkov, M.L. Zheludkevich, H.B. Xue, S.V. Lamaka, M.G.S. Ferreira, Plasma anodized ZE41 magnesium alloy sealed with hybrid epoxy-silane coating, Corrosion Science, 73 (2013) 300.

# Ni-P-MWNTs Composite Coatings on Magnesium Alloys AZ31 Part 1: MWNTs Content in Coating

Dong Guo, Haiwang Wu, Sheng Wang, Yongjuan Dai, Shiqing Sun, Sen Qin, and Kai Fu

**Abstract**

In the present paper, Ni-P-MWNTs composite coating was prepared on AZ31 magnesium alloy by electroless plating. Surface modification of carbon nanotubes by cationic surfactant DTAB were studied using IR spectrum. The morphology and the coating composition on the electrode surface were analyzed (SEM and X-ray diffraction). The results show that the prepared Ni-P-MWNTs coating is uniform and compact. The relationship among the concentration of surfactant DTAB, the concentration of MWNTs in bath and coating were analyzed.

**Keywords**

Magnesium alloys • Electroless plating • Ni-P-MWNTs coating • The concentration of MWNTs

Magnesium alloys are among the best lightweight structural materials with a relatively high strength to weight ratio and excellent technological properties [1, 2]. The requirement to reduce the weight of car components has triggered renewed interest in magnesium alloys. In addition, the use of magnesium alloys in other fields, such as aerospace, electronic and telecommunication components has also increased steadily in recent years [3, 4]. However, magnesium is intrinsically highly reactive and its alloys usually have relatively poor corrosion resistance, which is actually one of the main obstacles to the application of magnesium alloys [5, 6]. In order to meet the requirement for practical applications, suitable surface treatment techniques should be applied to magnesium alloys to improve their corrosion resistance [7–10]. Among various surface treatments, electroless nickel plating for magnesium alloys was investigated extensively because of high corrosion resistance, good wear resistance, high hardness, acceptable lubricating characteristics [11]. In order to improve the properties of Ni-P coating on magnesium alloy surface, composite coatings were formed by introducing secondary particles into Ni-P electroless plating bath.

Carbon nanotubes (MWNTs) are a new type of carbon material discovered in the early 90s of last century. Its high length/diameter ratio, strength, elastic modulus, flexibility, stiffness, large energy absorbing capacity, unique conductivity and chemical stability along with other excellent properties have led to the use of the carbon nanotubes as a novel fiber for a variety of composite materials [12]. Alishahi M [13] prepared Ni-P-MWNTs composite coating on copper metal substrate. It was found that addition of MWNTs in both before and after heat treatment improves the hardness of the coating. Wu YC [14] obtained Ni-Cu-P/carbon nanotubes quaternary composite coatings on low carbon steel by electroless plating. With increasing CNTs concentration, both the CNTs content in the composite coatings and the hardness of composite coatings increased at first and then decreased. He YD [15] developed a mechanical attrition (MA)-assisted electroless plating technique to deposit Ni-P-multiwalled carbon nanotubes composite coatings on carbon steel. It showed that the MA-assisted coating possessed 59 wt% CNTs, which was much higher than that of 22 wt% in conventional coating. Arai S [16] fabricated Ni-P alloy/multiwalled carbon nanotube composite films on acrylonitrile butadiene styrene resin by electroless plating.

In this study, we described our successful attempt to achieve modified MWNTs and employ chemical deposition to prepare Ni-P-MWNTs composite coatings on AZ31 magnesium alloys. Using this method, we achieved a uniform distribution of MWNTs in the Ni-P matrix and good adhesion to the matrix. Moreover, we paid more attention to

D. Guo (✉) · H. Wu · S. Wang · Y. Dai · S. Sun · S. Qin · K. Fu
School of Materials Science and Engineering, Hebei University of Science and Technology, Shijiazhuang, People's Republic of China
e-mail: guodongbill@hotmail.com

D. Guo · H. Wu · S. Wang · Y. Dai · S. Sun · S. Qin · K. Fu
Hebei Education Department, Hebei Key Laboratory of Material Near-Net Forming Technology, Shijiazhuang, People's Republic of China


D. Orlov et al. (eds.), *Magnesium Technology 2018*, The Minerals, Metals & Materials Series, https://doi.org/10.1007/978-3-319-72332-7_5

investigate MWNTs concentration in the bath and MWNTs content in the composite coatings were investigated.

## Experimental

The substrate material used was as extruded AZ31 magnesium alloy. The chemical composition of the alloy is given in Table 1. The procedure of electroless plating on magnesium alloys was: polishing sample (1500 grade SiC paper) → degreasing → alkaline etching → preparing nickel metal film → electroless plating Ni-P-MWNTs composite coating (water rinse each step). They were polished successively to a 1200 and 2400 grit finish before pretreatment. The procedures of pretreatment were processed by the procedure given by Guo D et al. [17]. Composition of the electroless plating bath and its operating conditions were given in Table 2.

Multi-walled carbon nanotubes (MWNTs) were purchased form Shenzhen Nanotech Port Co., Ltd. Its size is 1–2 μm in length, 20–40 nm in outer diameter. The crude material were purified by immersing in a 3:1 mixture of concentrated $H_2SO_4$ and $HNO_3$ and refluxed for 0.5 h at boiling, subsequently suspended and refluxed in 5 mol $L^{-1}$ HCl solution for 2 h at boil. After washing with de-ionized water and drying at 80 °C, the MWNTs samples were characterized by field emission scanning electron microscope (Hitachi S-4800) and Infrared spectroscopy (300E Jasco spectrometer and KBr disk).

After pretreatment, samples were placed in the electroless plating bath. The composition and operation parameters of electroless plating bath were shown in Table 2. After accomplishment of electroless plating, samples were removed from plating bath, washed by distilled water and dried.

Scanning electron microscopy (Shimadzu SSX-550) with energy dispersive spectroscopy (EDS) was used to assess deposit morphologies and composition distribution. The crystal structures of the electroless composite coatings were analyzed by XRD (Cu $K_{a1}$).

## Results and Discussion

Figure 1a showed a SEM image of the crude MWNTs and impurities. After purification by acid treatment, the metal particles, the oxide particles, and the non-nanotubes carbon materials have been almost eliminated completely (as shown in Fig. 1b). Most of MWNTs twine with each other, which has a negative effect on dispersing of MWNTs in the bath. In order to improve the dispersity of MWNTs in electroless plating bath, the MWNTs were mechanically milled. The SEM image of carbon nanotubes after being mechanical milled for 8 h is shown in Fig. 1c. It was evident from this figure that carbon nanotubes become shorter and the twine is alleviated. This feature contributed to the subsequent dispersion of MWNTs in the plating bath.

Figure 2 showed the infrared spectra of the acid treated MWNTs. It can be seen that there are many groups on the surface of the acid-treated MWNTs with mixture acid, oxygen-hydrogen bonds at about 1320 and 3670 $cm^{-1}$, carbonyl groups at about 1500 $cm^{-1}$ and C = C bonds at about 1390 $cm^{-1}$, respectively. These groups were of great benefit to absorbing sodium dodecyl sulfate (SDS). The addition of surfactant significantly enhanced the suspension stability of MWNTs with electrostatic repulsion. These phenomena led us to conclude that steric stabilization plays an essential role for dispersion of MWNTs. The surfactants conquered the van der Waal's interaction through steric hindrance and helped to separate MWNTs from each other.

**Table 1** Chemical compositions of AZ31 magnesium alloy (mass, %)

| Mg | Al | Na | Zn | Si |
|---|---|---|---|---|
| 95.729 | 2.073 | 1.437 | 0.582 | 0.179 |

**Table 2** Compositions of bath and operating conditions of nickel deposition

| Bath constituents and parameters | Condition |
|---|---|
| Nickel sulfate, $NiSO_4{\cdot}6H_2O$ | 18 g $L^{-1}$ |
| Sodium hypophosphite, $NaH_2PO_2{\cdot}H_2O$ | 22 g $L^{-1}$ |
| Hydrofluoric acid, HF(40%) | 10 ml $L^{-1}$ |
| Ammonium bifluoride, $NH_4HF_2$ | 20 g $L^{-1}$ |
| Citric acid, $C_6H_8O_7{\cdot}H_2O$ | 5 g $L^{-1}$ |
| Thiourea | 1 mg $L^{-1}$ |
| Ammonia solution (25%) | To adjust pH = 6.5 |
| Temperature and time | 80 ± 2 °C, 60 min |
| Modified MWNTs | 0.5–2 g $L^{-1}$ |

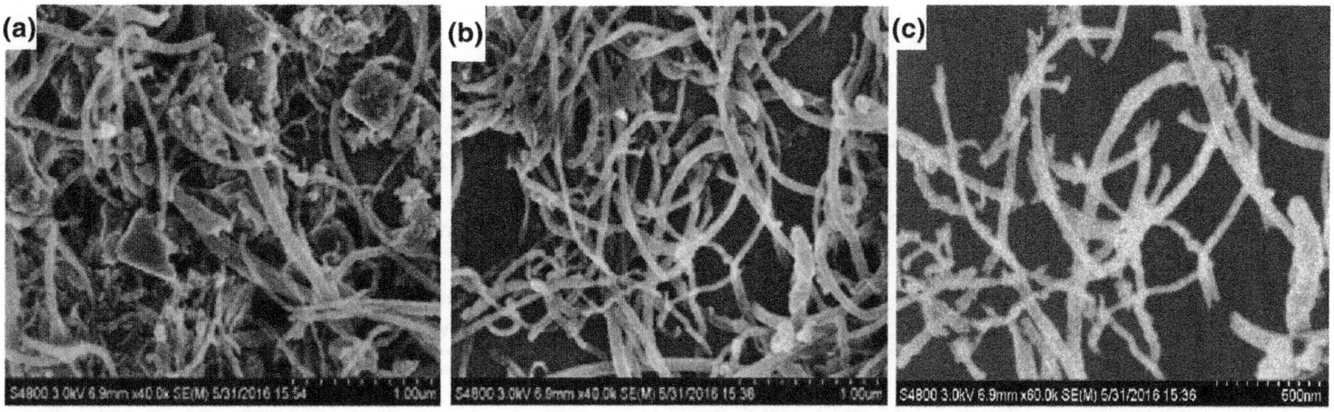


**Fig. 1** SEM images of: **a** crude MWNTs; **b** purified MWNTs; **c** ball milled MWNTs

As shown in Fig. 3, Ni-P-MWNTs composite coating on AZ31 magnesium alloys showed a dark black colour. It was semibright, smooth, uniform and compact. Plating rate can reach 12 μm/h. Composite coating adhesion was checked by thermal shock test and file test. Samples were heat treated at 250 ± 10 °C for 1 h in the furnace, and then immediately plunged into cold water. The surface of the samples remained intact without peeling and flaking. A non-important part of the surface of the samples was filed off at an angle of 45°, and then the interface of substrate and coating examined but no peeling off was revealed. The results showed good adhesion of the composite coating.

Figure 4 showed the XRD patterns of the composite coatings before and after annealing treatment for 3 h. It was found that their crystalline structures change from amorphous state to crystalline state, and $Ni_3P$ phase forms in the Ni matrix. In this work, the peak of carbon nanotube was still indexed in Fig. 4. It was confirmed that the MWNTs were embedded in the matrix.

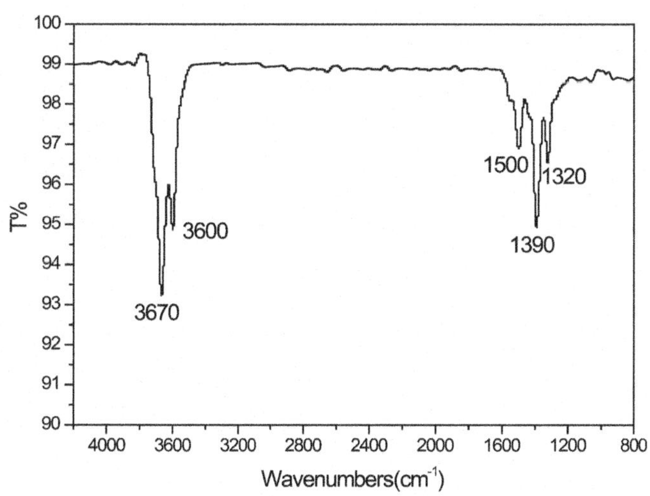


**Fig. 2** Spectrum of acid treated MWNTs by FTIR

**Fig. 3** SEM surface morphology of Ni-P-MWNTs composite coatings

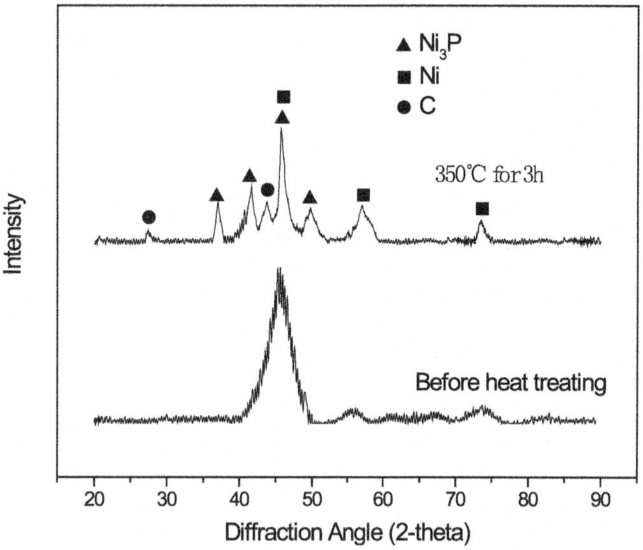


**Fig. 4** XRD patterns of Ni-P-MWNTs composite coatings before and after annealing

As shown in Fig. 5, it was relationship among MWNTs content in composite coating, MWNTs concentration and DTAB/MWNTs rate in bath. When the MWNTs content in bath was at a low level, the content of MWNTs in composite coating increased with the increase of MWNTs content in bath. When the content of MWNTs in bath reached 1.2 g/L, the content of MWNTs in coating reached the highest value. Then the MWNTs content in coating decreased with the increase of MWNTs content in bath. When the DTAB/MWNTs ratio was at a high level, the MWNTs content in coating increased with the decrease of DTAB/MWNTs. When DTAB/MWNTs ratio reached 1, MWNTs content in coating reached the highest value. Then the MWNTs content in coating decreased with the increase of DTAB/MWNTs. When the MWNTs content of in bath was 1.2 g/L and DTAB/MWNTs ratio was 1, the content of MWNTs in coating reached the maximum value(2.5 wt%). This is because when the content of MWNTs in bath was low or DTAB/MWNTs ratio was high, the MWNTs in bath showed good dispersibility. With the increase of MWNTs content, the probability that the MWNT particles reached sample surface and embedded in coating increased. But on the other hand, excessive DTAB adsorbed on sample surface hindered the diffusion of $Ni^{2+}$ and $H_2PO_2^-$, and hindered the normal deposition of Ni-P coating.

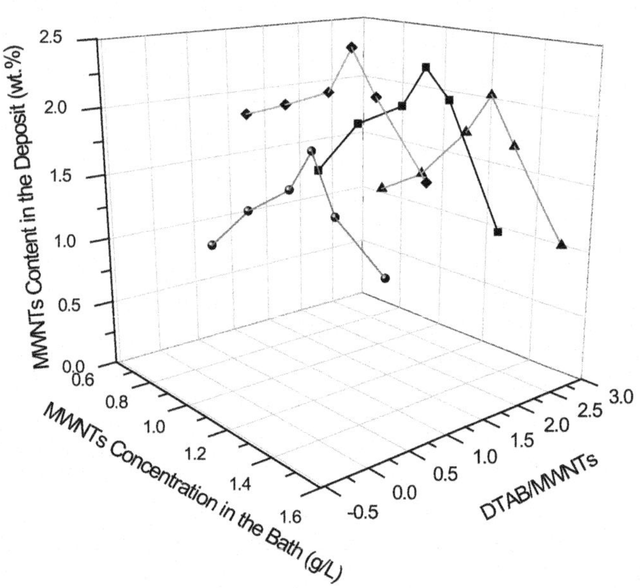


**Fig. 5** Relationship among MWNTs content in deposit, MWNTs concentration and DTAB/MWNTs rate in bath

## Conclusion

(1) After surface modification, the MWNTs solution with good dispersion was obtained. Ni-P-MWNTs composite coating on AZ31 magnesium alloy was prepared by electroless plating. It showed a dark black color, semibright, smooth, uniform and compact. Plating rate can reach 12 μm/h. The test results showed good adhesion of the composite coating.

(2) When the MWNTs content in bath was at a low level, the content of MWNTs in composite coating increased with the increase of MWNTs content in bath. But with continued increase of the content of MWNTs in bath, the content of MWNTs in coating decreased.

(3) When the DTAB/MWNTs ratio was at a high level, the MWNTs content in coating increased with the decrease of DTAB/MWNTs. But with continued decrease of DTAB/MWNTs, the MWNTs content in coating decreased with the increase of DTAB/MWNTs.

**Acknowledgements** This work was supported by the National Natural Science Foundation of China [grant number 51774108], the Natural Science Foundation of Hebei Province [grant number E2014208159] and the Research Foundation of Education Bureau of Hebei [grant number ZD20131003].

## References

1. Pan FS, Yang MB, Chen XH (2016) A Review on Casting Magnesium Alloys: Modification of Commercial Alloys and Development of New Alloys, J MATER SCI TECHNOL, 32 (12): 1211–1221
2. Pan HC, Ren YP, Fu H, Zhao H, Wang LQ, Meng XY, Qin GW (2016) Recent developments in rare-earth free wrought magnesium alloys having high strength: A review, J ALLOY COMPD, 663(5): 321–331
3. Song JF, Pan FS, Jiang B, Atrens A, Zhang MX, Lu Y (2016) A review on hot tearing of magnesium alloys, J Magnes Alloy, 4(3): 151–172
4. Song B, Guo N, Liu TT, Yang QS (2014) Improvement of formability and mechanical properties of magnesium alloys via pre-twinning: A review, MATER DESIGN, 62: 352–360
5. Esmaily M, Svensson JE, Fajardo S, Birbilis N, Frankel GS, Virtanen S, Arrabal R, Thomas S, Johansson LG (2017) Fundamentals and advances in magnesium alloy corrosion, PROG MATER SCI, 89: 92–193
6. Agarwal S, Curtin J, Duffy B, Jaiswal S (2016) Biodegradable magnesium alloys for orthopaedic applications: A review on corrosion, biocompatibility and surface modifications, Mater Sci Eng C, 68: 948–963

7. Zuleta AA, Correa E, Castaño JG, Echeverría F, Baron-Wiecheć A, Skeldon P, Thompson GE (2017) Study of the formation of alkaline electroless Ni-P coating on magnesium and AZ31B magnesium alloy, SURF COAT TECH 321: 309–320
8. Zhou LF, Liu ZY, Wu W, Li XG, Jiang B (2017) Stress corrosion cracking behavior of ZK60 magnesium alloy under different conditions, INT J HYDROGEN ENERG, 42(41): 26162–26174
9. PHUONG NV, GUPTA M, MOON S (2017) Enhanced corrosion performance of magnesium phosphate conversion coating on AZ31 magnesium alloy, T NONFERR METAL SOC, 27(5): 1087–1095
10. Cao FY, Song GL, Atrens A (2016) Corrosion and passivation of magnesium alloys, Corros Sci, 111: 835–845
11. Zeng LY, Yang SW, Zhang W, Guo YH, Yan CW (2010) Preparation and characterization of a double-layer coating on magnesium alloy AZ91D, ELECTROCHIM ACTA, 55(9): 3376–3383
12. Liew KM, Kai MF, Zhang LW (2016) Carbon nanotube reinforced cementitious composites: An overview, COMPOS PART A-APPL S, 91(1): 301–323
13. Alishahi M, Monirvaghefi SM, Saatchi A, Hosseini SM (2012) The effect of carbon nanotubes on the corrosion and tribological behavior of electroless Ni-P-CNT composite coating, APPL SURF SCI, 2012, 258(7): 2439–2446
14. Wu YC, Ren R, Wang FT, Yuan ZS, Wang TG, Hu XY (2008) Preparation and characterization of Ni-Cu-P/CNTs quaternary electroless composite coating, MATER RES BULL, 2008, 43(12): 3425–3432
15. Zhao GH, Ren C, He YD (2012) Ni-P-multiwalled carbon nanotubes composite coatings prepared by mechanical attrition (MA)-assisted electroless plating, SURF COAT TECH, 206(11–12): 2774–2779
16. Arai S, Sato T, Endo M (2011) Fabrication of various electroless Ni-P alloy/multiwalled carbon nanotube composite films on an acrylonitrile butadiene styrene resin, SURF COAT TECH, 205 (10): 3175–3181
17. Guo D, Wang B, Mao L, Fan ZG (2011) Novel process of electroless Ni-P plating with replacement pretreatment on magnesium alloys, T I MET FINISH, 89(2): 109–112

# Ni-P-MWNTs Composite Coatings on Magnesium Alloys AZ31 Part 2: Tribological Behavior and MWNTs Content in Coating

Dong Guo, Sheng Wang, Yongjuan Dai, Shiqing Sun, Sen Qin, and Kai Fu

**Abstract**

In this study, Ni-P-MWNTs composite coating was successfully deposited on the surface of AZ31 magnesium alloys by electroless plating. The electrochemical properties of the composite coatings were studied by electrochemical workstation system. The corrosion behavior of the composite coatings was evaluated by polarization curves in 3.5 wt% NaCl aqueous solution at room temperature. Its corrosion resistance was improved significantly than AZ31 magnesium alloys. The wear behavior of the coatings was investigated using friction and wear test method. The results indicated that the incorporation of carbon nanotubes in the coating improved both tribological behavior and corrosion resistance. Comparing with Ni-P coating, Ni-P-MWNTs composite coatings showed not only higher wear resistance but also lower friction coefficient. These improvements have been attributed to superior mechanical properties, unique topological structure and high chemical stability of nanotubes.

**Keywords**

Magnesium alloys • Electroless plating • Ni-P-MWNTs coating • Tribological behavior

Magnesium alloy exhibits an attractive properties for automotive industry but its low corrosion resistance has limited its use [1–3]. Electroless plating is an important technique to improve the corrosion resistance of magnesium alloys [4]. Coatings containing solid particles such as SiC, $Al_2O_3$, WC and diamond, etc. have been developed for better wear resistance or dispersion hardening [5, 6]. Composite coating has better wear resistance, corrosion resistance and high temperature resistance than single coating [7]. Recently, carbon nanotubes are increasingly attracting scientific and technological interest by virtue of their unique chemical and physical properties after discovered by Iijima [8–10]. Goel [11] investigated Co-P-CNT coating on the magnetic properties of grain oriented electrical steel. The results of wear test reveal that Co-P-CNT coating reduced the surface roughness and enhanced the magnetic properties. Zeng [12] electrodeposited chromium-multiwalled carbon nanotubes composite coatings. The introduction of MWNTs obviously improved the hardness, toughness and tribological performance of Cr coatings. Meng [13] prepared Ni-P-MWNTs composite coatings by electroless plating on 45# steel. The results of wear test reveal that the Ni-P-MWNTs composite coatings posses much better friction reduction and anti-wear performances when compared with Ni-P coating. The excellent tribological performances of the composite coatings can be attributed to the introduction of MWNTs, which play both roles of reinforcements and solid lubricant during the wear process. Allahkara [14] found that corrosion resistance increasing with incorporation of CNTs of Ni-P coatings. Chen [15] compared Ni-P-Carbon nanotube (CNT) composite coatings as well as Ni-P-SiC and Ni-P-graphite coatings were prepared by electroless plating. The Ni-P-CNT composite coatings exhibited not only high wear resistance but also low friction coefficient compared with the Ni-P-SiC and Ni-P-graphite composite coatings.

In the present work, the corrosion resistance and dry friction behavior of Ni-P-MWNTs composite coatings on AZ31 magnesium alloys were investigated.

## Experimental

Preparation of Ni-P-MWNTs composite coatings on AZ31 magnesium alloys refer to "Ni-P-MWNTs Composite Coatings on Magnesium Alloys AZ31 part 1: MWNTs

D. Guo (✉) · S. Wang · Y. Dai · S. Sun · S. Qin · K. Fu
School of Materials Science and Engineering, Hebei University of Science and Technology, Shijiazhuang, People's Republic of China
e-mail: guodongbill@hotmail.com

D. Guo · S. Wang · Y. Dai · S. Sun · S. Qin · K. Fu
Hebei Key Laboratory of Material Near-Net Forming Technology, Shijiazhuang, People's Republic of China


D. Orlov et al. (eds.), *Magnesium Technology 2018*, The Minerals, Metals & Materials Series, https://doi.org/10.1007/978-3-319-72332-7_6

content in coating". The thickness of composite coating is about 10–15 μm. For studying the corrosion resistance of coated samples, polarisation curves of Ni-P-MWNTs composition coating were obtained using the electrochemical measurement system Princeton PARSTAT 2273 at room temperature. Linear sweep voltammetry experiments were carried out in a corrosive environment of 3.5 wt% NaCl aqueous solution (pH 6.9) using a classic three-electrode cell with a platinum plate (Pt) as a counter electrode and a calomel reference electrode. The working electrode and the samples were cleaned in acetone agitated ultrasonically, and rinsed in deionised water before the electrochemical test. The coated samples were masked with epoxy resin (EP 651) so that only 1 $cm^2$ area was exposed to the electrolyte.

Wear resistance and dynamic friction coefficient of Ni-P-MWNTs composite coatings and Ni-P coatings on AZ31 magnesium alloys were examined by vertical universal friction and wear tester. Results were recorded as the weight loss versus the sliding distance and friction coefficient variations. It was adopted ball-chip friction pairs. Test under dry friction condition, ball speed was 75 r $min^{-1}$, test load was 15 N. The friction torque was obtained by measuring the output power of the DC motor. Then the coefficient of friction was obtained. The coefficient of friction in test process was automatically recorded by the device. The experimental temperature was room temperature, and samples were washed by deionized water, ultrasonic cleaned by ethanol and vacuum dried before and after experiment. The friction coefficient was calculated as follows:

$$\mu = \frac{F}{N} = \frac{\frac{T}{3r\cos\alpha}}{\frac{P}{3r\sin\alpha}} = \frac{T}{P\cdot r} tg\alpha = \frac{T}{P\times 6.35} tg54.77 = 0.233\frac{T}{P}$$

*T* was the friction torque (*N mm*), *P* was the loading force (*N*), $\mu$ was the coefficient of friction (*mm*).

## Results and Discussion

### Electrochemical Corrosion Resistance of Ni-P-MWNTs Composite Coating

After the surface potential of the sample had tended to be stable, corrosion potential and potentiodynamic polarization curve of the sample were measured in 3.5% NaCl solution. As shown in Fig. 1, the corrosion potential of AZ31 magnesium alloy matrix was about −1.4 V. In the stage of anodic polarization, the corrosion current increased rapidly. The sample was dissolved, and hydrogen on the electrode and sample surface were generated. By comparison, the corrosion potential of Ni-P-MWNTs composite coatings on AZ31 magnesium alloy was about −0.5 V. Its corrosion resistance was improved significantly than the former. After potentiodynamic scanning into the anode area, it created passivate layer on Ni-P-MWNTs composite coatings, and the passivation interval was about 300 mV. Finally, pitting corrosion formed on the composite coating and corrosion current increased. Therefore, corrosion resistance of Ni-P-MWNTs composite coating was very good for AZ31 magnesium alloys.

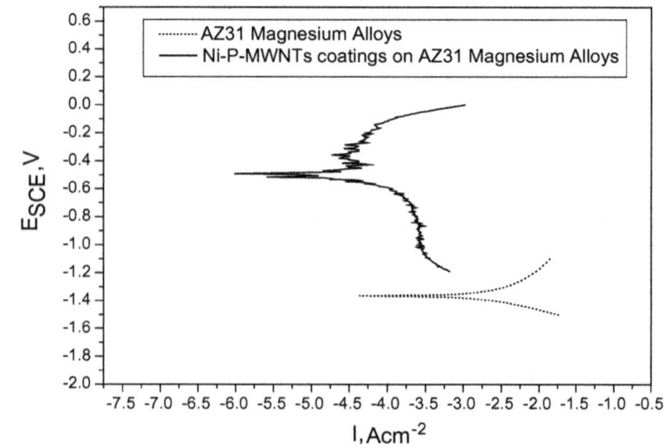


**Fig. 1** Potentiodynamic polarization curves for bare alloy and Ni-P-MWNTs composite coating on AZ31 magnesium alloys

### Friction Properties of Ni-P-MWNTs Composite Coating

Figure 2 showed relationship between friction coefficient and friction time of Ni-P coating and Ni-P-MWNTs composite coating with different MWNTs content in coating on AZ31 magnesium alloys. With the increase of friction time, the friction coefficient increased continuously. Compared with

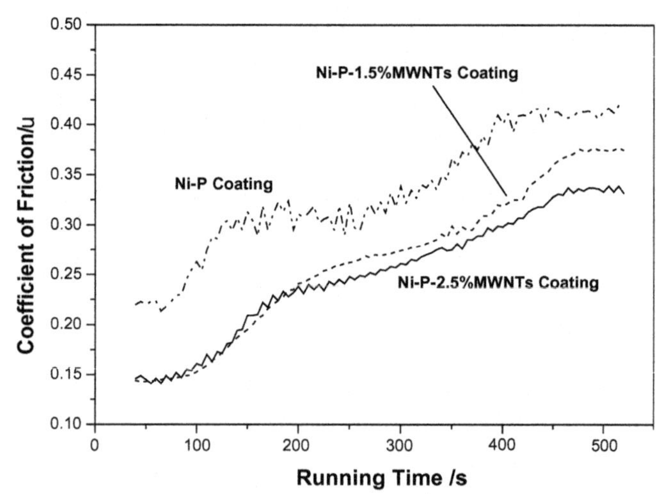


**Fig. 2** Relationship between friction coefficient and friction time of Ni-P-MWNTs composite coatings

**Fig. 3** Wear scar surface morphologies: **a** Ni-P, **b** Ni-P-1.5% MWNTs, **c** Ni-P-2.5% MWNTs

Ni-P-MWNTs composite coatings, the friction coefficient of Ni-P coating fluctuated more greatly. Its average friction coefficient was 0.34, and the friction coefficients of Ni-P-1.5% MWNTs composite coatings and Ni-P-2.5%MWNTs composite coatings were lower respectively, 0.27 and 0.25.

Wear scar surface morphologies of Ni-P coating and Ni-P-MWNTs composite coating were shown in Fig. 3 under the same conditions. The exposed matrix in the wear scars of Ni-P coating was the largest and deepest among the three. With the addition of MWNTs, the wear scars of Ni-P-MWNTs composite coating became narrower and shallower. Compared with the wear scars of Ni-P-1.5% MWNTs composite coatings, the wear scars of Ni-P-2.5% MWNTs composite coatings was smaller, no matter in depth or width. The above results showed that for composite coating, the addition of MWNTs could bring not only lower friction coefficient, but also better wear resistance.

Because of lower loading capacity of Ni-P coating, stress concentration and temperature rising caused by friction led to brittle fracture of the coating and peeled off in the form of abrasive dust after undergone a certain friction distance. Grinding and adhesion of abrasive dust increased friction coefficient. Wear of Ni-P coating was mainly dominated by adhesive wear and plough mechanism. Ploughing wear of abrasive dust aggravated the wear effect of the coating. Peeling and ploughing wear were main causes of friction coefficient fluctuating and high wear rate of Ni-P coating.

For Ni-P-MWNTs composite coatings, the addition of MWNTs particles led to the decline of hardness of the coating, but its friction coefficient remained at a low level and stable. This means that almost no larger particles generated during friction process. The molecular structure of MWNTs determined its high crystallization and weak van der Waals force between MWNTs molecules. This helped to transfer shear friction effect to friction surface. Uniform distribution in coatings of MWNTs in a network connected state, which played a role of dispersion strengthening. According to polymer friction theory [16], when soft polymers ground against hard materials, polymers usually transferred to surface of hard materials. A polymer film should be formed on surface of hard materials. Once the film is formed, subsequent interactions should be carried out between the polymer and the film so as to achieve lubrication and wear reduction effect.

## Conclusion

(1) Corrosion behavior of Ni-P-MWNTs composite coatings on AZ31 magnesium alloys was evaluated. Its corrosion resistance was improved significantly than AZ31 magnesium alloys.
(2) Tribological properties of Ni-P coating and Ni-P-MWNTs composite coating on AZ31 magnesium alloy were investigated. Its was found that MWNTs solid lubricant particles incorporated into Ni-P-MWNTs composite coating could significantly reduce the friction coefficient. The fluctuation of Ni-P coating friction coefficient was more than Ni-P-MWNTs composite coating.
(3) The improvement of wear resistance and friction properties of Ni-P-MWNTs composite coatings stemmed from high strength, high toughness and self-lubricating properties of MWNTs, as well as their fiber morphology and uniform distribution in coating.

**Acknowledgements** This work was supported by the National Natural Science Foundation of China [grant number 51774108], the Natural Science Foundation of Hebei Province [grant number E2014208159] and the Research Foundation of Education Bureau of Hebei [grant number ZD20131003].

## References

1. Pan HC, Ren YP, Fu H, Zhao H, Wang LQ, Meng XY, Qin GW (2016) Recent developments in rare-earth free wrought magnesium alloys having high strength: A review, J ALLOY COMPD, 663(5): 321–331
2. Hu RG, Zhang S, Bu JF, Lin CJ, Song GL (2012) Recent progress in corrosion protection of magnesium alloys by organic coatings, PROG ORG COAT, 73(2–3): 129–141

3. Zhong C, Liu F, Wu YT, Le JJ, Liu L, He MF, Zhu JC, Hu WB (2012) Protective diffusion coatings on magnesium alloys: A review of recent developments, J ALLOY COMPD, 520: 11–21
4. Rajabalizadeh Z, Seifzadeh D (2016) Strontium phosphate conversion coating as an economical and environmentally-friendly pretreatment for electroless plating on AM60B magnesium alloy, SURF COAT TECH, 304: 450–458
5. Yang B, Lu WZ, Feng W, Yang X, Zuo DW (2017) Adsorption and deposition of micro diamond particles in preparing diamond magnetic abrasives by electroless composite plating, DIAM RELAT MATER, 73: 137–142
6. NGUYEN VH, NGO TAT, PHAM HH, NGUYEN NP (2013) Nickel composite plating with fly ash as inert particle, T NONFERR METAL SOC, 23(8): 2348–2353
7. Guo L, Xiao LR, Zhao XJ, Song YF, Cai ZY, Wang HJ, Liu CB (2017) Preparation of WC/Co composite powders by electroless plating, CERAM INT, 43(5): 4076–4082
8. Iijima S (1991) Carbon nanotubes, Nature, 354: 56–58
9. Yamakawa A, Suzuki S, Oku T, Enomoto K, Kitamura S (2017) Nanostructure and physical properties of cellulose nanofiber-carbon nanotube composite films, CARBOHYD POLYM, 171: 129–135
10. Kingston C, Zepp R, Andrady A, Boverhof D, Wohlleben W (2014) Release characteristics of selected carbon nanotube polymer composites, Carbon, 68: 33–57
11. Goel V, Anderson P, Hall J, Robinson F, Bohm S (2016) Electroless Co-P-Carbon Nanotube composite coating to enhance magnetic properties of grain-oriented electrical steel, J MAGN MAGN MATER, 407: 42–45
12. Liu B, Zeng ZX, Lin YM (2009) Mechanical properties of hard Cr-MWNT composite coatings. SURF COAT TECH, 203(23): 3610–3613
13. MENG ZQ, Li XB, Xiong YJ, Zhan J (2012) Preparation and tribological performances of Ni-P-multi-walled carbon nanotubes composite coatings, T NONFERR METAL SOC, 22(11): 2719–2725
14. Zarebidaki A, Allahkara SR (2011) Effect of surfactant on the fabrication and characterization of Ni-P-CNT composite coatings, J ALLOY COMPD, 509(5): 1836–1840
15. Chen WX, Tu JP, Gan HY, Xu ZD, Wang QG, Lee JY, Liu ZL, Zhang XB (2002) Electroless preparation and tribological properties of Ni-P-Carbon nanotube composite coatings under lubricated condition. SURF COAT TECH, 160 68–73
16. Hutchings IM (1992) Tribology: Friction and wear of Engineer Materials, Butterworth-Heinemann, London

# Adding Dimensions to the Immersion Testing of Magnesium Corrosion

Lars Wadsö and Dmytro Orlov

**Abstract**

With the versatility of structural performance in magnesium alloys, Achilles hill remains to be their susceptibility to corrosion. The Mg community agrees that traditional methods are insufficient for revealing the root cause of difficulties in controlling Mg degradation rate. Therefore, developing new methods allowing simultaneous assessment of several characteristics is of great importance now. We designed an advanced cell for immersion testing allowing simultaneous assessment of two complementary characteristics of Mg corrosion in aqueous environments: isothermal calorimetry and pressure. Isothermal calorimetry monitors in situ heat production rate during chemical reactions, which can be recalculated to corrosion rate if the enthalpy of a process is known. Pressure monitoring allows alternative quantification of corrosion rate through hydrogen production. The proof-of-concept testing presented here reveals details of a corrosion process depending on electrolyte.

**Keywords**

Magnesium • Corrosion • Isothermal calorimetry
Pressure measurements

L. Wadsö
Division of Building Materials, LTH, Lund University,
P.O. Box 118 22100 Lund, Sweden
e-mail: lars.wadso@byggtek.lth.se

D. Orlov (✉)
Division of Materials Engineering, LTH, Lund University,
P.O. Box 118 22100 Lund, Sweden
e-mail: dmytro.orlov@material.lth.se

D. Orlov
Materials Research Laboratory, University of Nova Gorica,
Vipavska 11c, 5270 Ajdovscina, Slovenia


D. Orlov et al. (eds.), *Magnesium Technology 2018*, The Minerals, Metals & Materials Series, https://doi.org/10.1007/978-3-319-72332-7_7

## Introduction

The benefits of magnesium (Mg) use are well known to the experts working with this material as well as to the general public [1]. At the same time, it has some weaknesses, also rather well known to these communities. A major weakness considered now as the main challenge on the way to a wide Mg adoption in many structural applications is its corrosion performance. Although the existence of such a challenge is well known, it still cannot be adequately addressed. One of the reasons for that is the relative uniqueness of Mg corrosion process in many aspects including difficulties in forming protective passive film on the surface from corrosion products and so-called 'negative difference effect' (or anodic hydrogen evolution). Recent achievements in the studies of Mg corrosion and advances in respective instrumentation are comprehensively reviewed by Esmaily et al. in [2]. Among other aspects, this review suggests that new instrumentation capable of 'multimodal' correlative assessment of Mg corrosion is necessary now. This paper addresses exactly that.

In the present paper, we introduce isothermal calorimetry as an interesting and potent techniques for the study of magnesium corrosion. In our laboratory-scale experimental set up, we have combined this method with pressure measurements, a gravimetric technique commonly used in magnesium corrosion studies [2]. Together, these techniques provide at least two independent sources of in situ process monitoring allowing the evaluation of Mg corrosion rate in real time with excellent temporal resolution.

## Materials

Commercial purity magnesium received as a hot-extruded bar 25 mm in diameter and having a mean grain size of 30 μm was used in this study. Specimens for the investigation were cut with a diamond wire saw from the centre of the bar to the dimension of 10 × 10 × 2.5 mm$^3$ giving the total surface area of approximately 300 $mm^2$. Afterwards, they were ground on all surfaces with a SiC paper to a final finish of P1200, washed in ethanol and dried with hot air immediately before testing. After this, the actual surface area of each specimen was measured with sufficient accuracy. All the measurements were then normalised by the actual surface area of specimens tested. Four identical specimens were measured in parallel at 20 °C.

The solution of 1% NaCl (sodium chloride of pro analysis quality) in deionized water ($H_2O$ type 1) having a concentration of $Cl^-$ ions similar to that in simulated body fluid (SBF) was used as an electrolyte for primary screening. Furthermore, a SBF was prepared according to a recipe in [3], see Table 1. The pH-value of the SBF solution was further adjusted to a required level by adding HCl immediately before testing.

## Methods

### Isothermal calorimetry

The main experimental technique used in this study was isothermal calorimetry, in which the heat production rate (thermal power) from a process is measured. This technique has been used for a long time in such fields as cement chemistry [4], pharmaceutical science [5] and microbiology [6]. However, it also has uses in almost every other field of science and technology, as it is a generic tool to study the rate of processes from the heat they produce. The measured thermal power $P$ (W) equals the rate of a process $dn/dt$ (mol $s^{-1}$) times the process enthalpy $\Delta H$ (J $mol^{-1}$):

$$P = \frac{dn}{dt}\Delta H. \qquad (1)$$

The time integral of thermal power is heat $Q$ (J), proportional to the amount of reacted material:

$$Q = (n_0 - n_t)\Delta H. \qquad (2)$$

Here, $n_0$ and $n_t$ (mol) are the amounts of material at time zero and at time $t$, respectively. In deriving the above two equations, we assumed the activity of a single process, but a similar reasoning can be applied also for more complex situations.

Isothermal calorimetry measures $P$ as a function of time and integration of $P$ gives $Q$. A measurement can thus give information on both the rate of a process and its extent. It is thus a powerful method to investigate process kinetics, for example of a corrosion process. However, there are no published studies of corrosion by isothermal calorimetry except an application note from a calorimeter manufacturer [7], to the best of our knowledge.

In the present study, a TAM Air (Thermometric, Järfälla, Sweden; now TA Instruments) was used. This is an instrument with eight heat conduction calorimeters for 20 mL vials [4]. The set-up for each calorimeter is shown schematically in Fig. 1. In the present measurements, plastic (polyethylene) vials were used. On the sample side, each of these contained 20 mL of an aqueous solution and the magnesium specimen. On the reference side, 20 mL water was placed in a similar plastic vial. The calorimeters were calibrated electrically and baselines were taken with 20 mL water in the sample vials. The magnesium specimens were hanging on nylon wires, so that they could be in contact with the solution on all sides.

### Pressure measurements

It is well known that the degradation of magnesium in aqueous solutions is associated with the release of hydrogen gas. Therefore, it is common by now to use the amount of evolved hydrogen gas as the measure of the amount of corroded magnesium because of 1:1 molar relationship between reacted magnesium and produced hydrogen:

$$Mg(s) + H_2O(l) \rightarrow MgO(aq/s) + H_2(aq/g). \qquad (3)$$

This reaction can be followed by the production of magnesium hydroxide:

$$MgO(aq/s) + H_2O(l) \rightarrow Mg(OH)_2(aq/s), \qquad (4)$$

but this reaction does not produce gas. In the above reactions, we have specified the states of some of the substances

**Table 1** The composition of simulated body fluid used in this work [3]

| Compound | Volume of concentrated salt solution in 1 l SBF [ml] |
|---|---|
| KCl | 5 |
| NaCl | 50 |
| $NaHCO_3$ | 50 |
| $MgSO_4 \cdot 7H_2O$ | 5 |
| $CaCl_2 \cdot 2H_2O$ | 25 |
| TRIS | 50 |
| $KH_2PO_4$ | 5 |

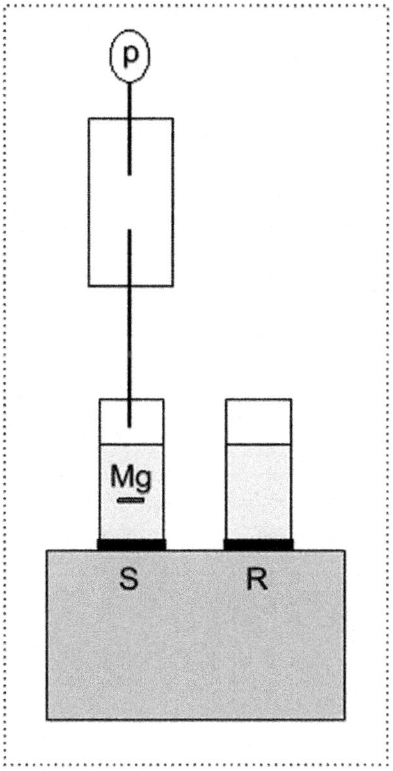


**Fig. 1** The measurement system. The calorimetric vial contains 20 mL of the solution and the sample (Mg). The sample vial (S) is in contact with a sample heat flow sensor (black) that registers the heat flow from the corrosion process. A second vial with water (R) is a reference with 20 mL water. The output signal is the sample voltage minus the reference voltage; in this way noise is decreased. The headspace of the sample vial is connected to a larger volume with a pressure sensor

as either "aq" (dissolved in water) or as "s" (solid) or "g" (gas). This is treated in the Discussion section.

In the present study, the pressure was measured with sensors (Motorola MPX5100) connected to the head space of the calorimetric vials. As the volume of the head space is small—and gas evolution would give high pressure changes in small volumes—an increased volume was created by inserting a 100 mL glass vial in the system, as is shown in Fig. 1. The pressure sensors were calibrated by changing the volume of the system with the help of a syringe, and calculating the change in output from the sensor electronics per change in pressure. The calibration coefficients have units of mV/Pa.

The amount of gas produced was calculated from the ideal gas law:

$$pV = nRT, \tag{5}$$

where $p$ (Pa) is pressure, $V$ (m$^3$) is volume, $n$ (mol) is the amount of gas, $R$ (8.314 J mol$^{-1}$ K$^{-1}$) is the gas constant, and $T$ (K) is temperature. As the thermal power is proportional to the rate of reaction, and thus also to the rate of gas production, the following equation was used:

$$\frac{dn}{dt} = \frac{dp}{dt} \cdot \frac{V}{RT}. \tag{6}$$

In the present case, the volume was 118 mL, and the temperature was 293 K.

## Results

In Fig. 2, we show the results of 30 h measurement on four samples of pure Mg in 1% NaCl solution at 20 °C. It is seen that they all show similar results, both for the thermal power and for the pressure change rate. The calculated enthalpy of the process is also rather constant after 5 h. In Fig. 3, the same type of results is shown for pure magnesium in SBF at 20 °C (only for three samples as one vial leaked). All the diagrams in Figs. 2 and 3 have minute signal variations having the appearance of noise. Nevertheless, these signal variations actually reveal intrinsic features of the corrosion process, as demonstrated below. Therefore, they were intentionally not filtered out but presented in the figures.

It is seen that both the thermal power and the pressure change rates are much lower for SBF than for the NaCl solution. Such a difference can be explained by significantly lower corrosive property of the SBF. The thermal power diagram and the corrosion rate diagram look similar, and this results in rather constant enthalpies. However, the 'noise' type signal seen in these two curves have different origin. In the thermal power curve, Fig. 2a, what looks like noise is actually variations in the intensity of the corrosion process (as elaborated in Fig. 4 and the Discussion section below), while the noise in the corrosion rate, Fig. 2c comes from the limited resolution and stability of the pressure sensors, and from the derivation of the pressure signal. The situation in Fig. 3 is similar, but the lower corrosion rate makes the measurements more uncertain. In the evaluations for Figs. 2 and 3, we used a data point interval of 500 s and a 10 point (5000 s) moving average filter for the corrosion rate, while the thermal power was not filtered. The noise seen in the enthalpy should be seen as a random noise since the remaining noise in the corrosion rate is random.

## Discussion

The present results show that isothermal calorimetry and pressure measurements are an interesting combination for the study of magnesium corrosion. Both techniques give continuous results viz. in situ monitoring of the process, although the calorimetry has the advantage of being a technique giving the direct measurement of rate. To calculate the enthalpy, it is needed to take the time derivative of the

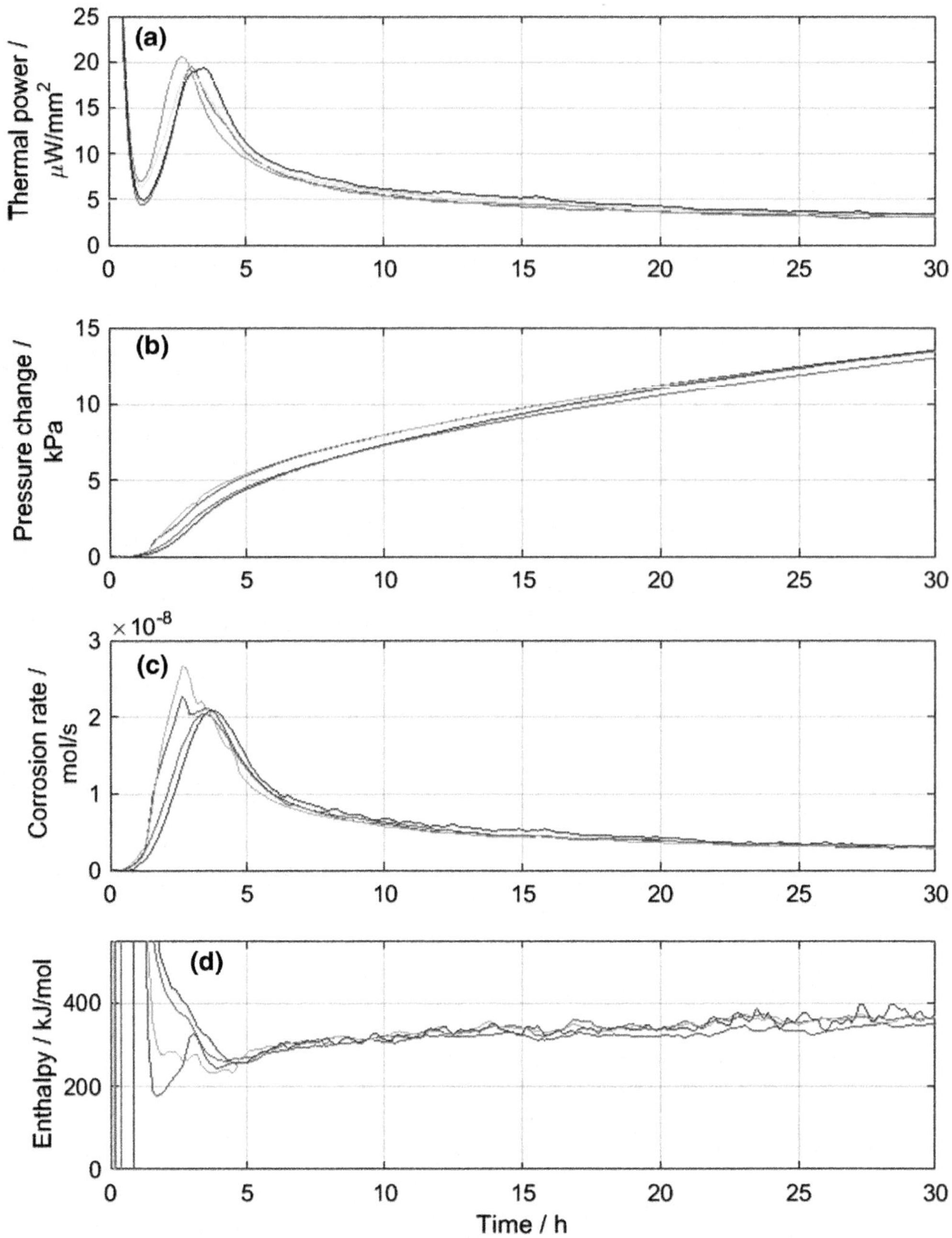


**Fig. 2** The results of measurements on four samples of pure magnesium in a 1% NaCl solution: **a** The calorimetric results. **b** The measured pressure. **c** The corrosion rate calculated from the pressure change rate assuming a 1:1 relation (in moles) between consumed Mg and produced $H_2$. **d** The enthalpy calculated from the data in plots **a** and **c**

pressure. The noise in the enthalpy curves, Figs. 2d and 3d, mainly comes from this derivation.

The calorimetric equipment used in this study was appropriate for detecting the entire range of thermal power release and sufficient temporal resolution. However, the measurements of pressure can be significantly improved by optimising the range of a sensor for each measurement by changing the second volume of the pressure measurement system (Fig. 1).

It can also be seen that the enthalpy calculated from the present measurement with SBF has significantly higher noise compared to the measurement with 1% NaCl. This is natural as SBF is much less corrosive than 1% NaCl solution and gives lower signals. At the same time, it should be noted that although the enthalpy calculated for the SBF results is noisy, it is still in the correct range (see below).

In Fig. 4, a part of the calorimetric result from Fig. 2a is magnified and presented together with the results from a baseline measurement with water (no heat production). Although the signal in Fig. 4a looks 'noisy', the baseline measurements in Fig. 4b indicate that these signal variations are two orders of magnitude higher than actual noise in the calorimeter system (note the differences in the thermal power scales between the diagrams in Figs. 4a and 4b). It means

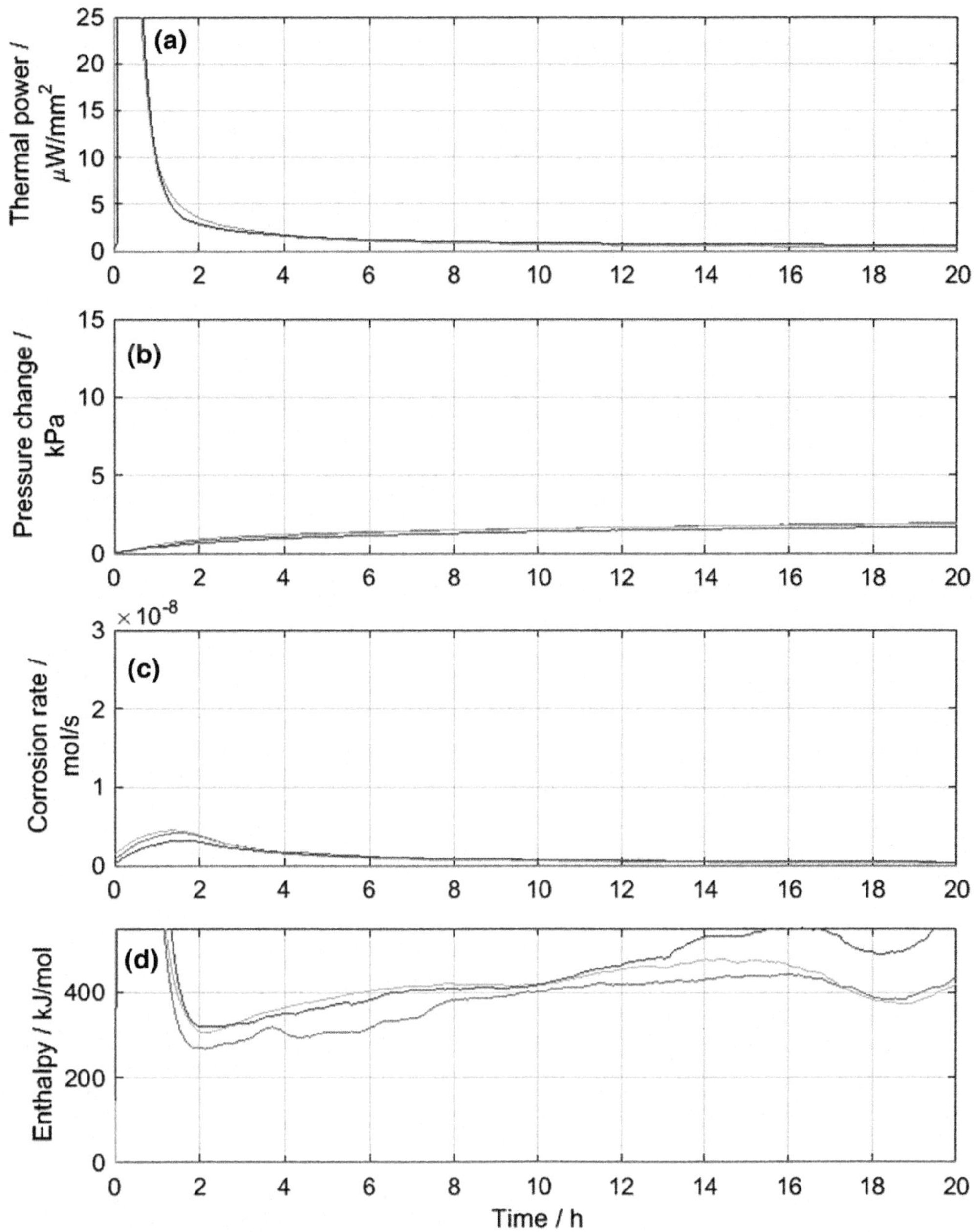


**Fig. 3** The results from measurements on three samples of pure magnesium in SBF: **a** The calorimetric results. **b** The measured pressure. **c** The corrosion rate calculated from the pressure change rate assuming a 1:1 relation (in moles) between consumed Mg and produced $H_2$. **d** The enthalpy calculated from the data in plots **a** and **c**

that the deviations from a smooth line in Figs. 2a and 3a actually indicate relatively minor events in the process of corrosion. A speculation can be made here that these variations indicate the events associated with e.g. (i) a cracking MgO and $Mg(OH)_2$ films and their delamination from the base Mg metal surface, and/or (ii) the formation and following release of $H_2$ bubbles from Mg surface.

During a measurement on magnesium corrosion, two main processes take place: Mg oxidation and the following Hydroxylation, as described by Eqs. (3) and (4), respectively. In these equations, we indicated that many of the compounds can have more than one state, e.g. solid and dissolved MgO. We will not carry out the detailed calculations of the different possibilities here since this is out of scope in present report, but it is worth noting that the reaction Mg $\rightarrow$ $Mg(OH)_2$ produces approximately 350 kJ/mol, which is at the same order of magnitude as our experimental results. The enthalpies of dissolution are comparatively small and can be neglected in an approximate evaluation.

Hydrogen gas evolution is a robust method to quantify the amount of corroded magnesium as hydrogen has a low solubility in water. Although it is rather common in Mg corrosion studies to use $H_2$ collection, the use of pressure sensors for evaluating hydrogen release is novel. However, $H_2$ is a small molecule, and care must be taken to prevent its

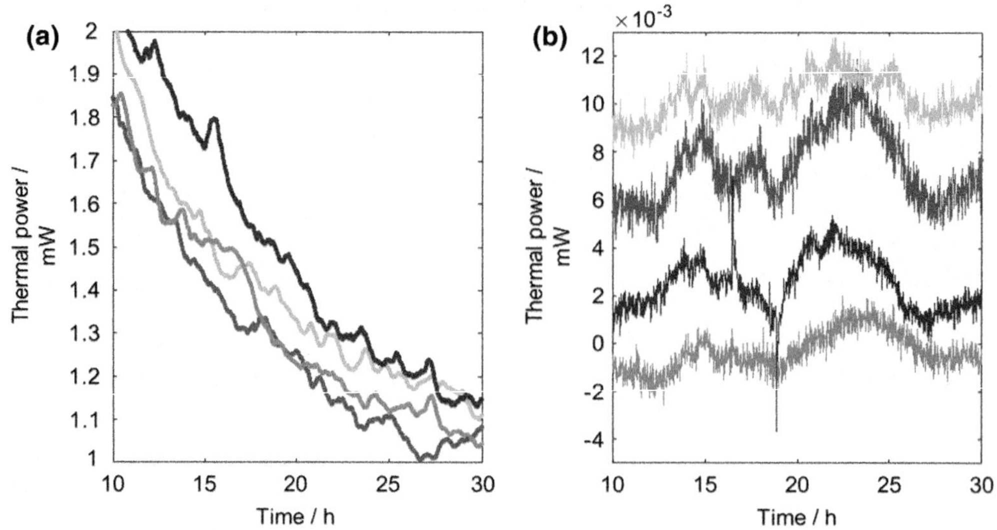


**Fig. 4** Comparison between a corrosion measurement (**a**) and a baseline measurement (**b**) with the same calorimeters. Note the differences in the thermal power scales in **a** and **b**. It is seen that the noise in the calorimeter itself is approximately 1 μW, while the deviations of signal from a smooth curve seen in the case of corrosion are approximately 100 μW

leakage through, e.g., polymeric materials. The tightness of the present system was tested with air, and found to be adequate. However, hydrogen is a much smaller molecule than air and it is thus possible that some leakage can take place during measurements. We will continue developing this instrumentation by refining the measurement of pressure and the data analysis tools.

## Conclusions

The combination of isothermal calorimetry and continuous pressure measurements has been introduced as a method to study the kinetics of magnesium corrosion. The high temporal resolution of the thermal power allows for the detection of even minute variations in the corrosion process, while the pressure measurement makes it possible to relate the thermal power to the corrosion rate, i.e., calculate the process enthalpy. This should make such a combination of methods attractive in the field of magnesium corrosion.

The capabilities of the system are illustrated by the study of the corrosion processes of pure Mg in 1% NaCl solution in water and a simulated body fluid. The pressure measurement needs to be improved to match the sensitivity and temporal resolution of the calorimeter.

**Acknowledgements** We thank Samuel Biteau and Alexandre Lansel from the University of Nantes for performing the experiments during their internship at Lund University in 2017. DO also gratefully acknowledges Slovenian Research Agency (ARRS) for the financial support through Research Project J2-7157.

## References

1. Tresa M. Pollock, *Science* 2010, vol. 328, pp. 986–987.
2. M. Esmaily, J. E. Svensson, S. Fajardo, N. Birbilis, G. S. Frankel, S. Virtanen, R. Arrabal, S. Thomas and L. G. Johansson, *Progress in Materials Science* 2017, vol. 89, pp. 92–193.
3. Lenka Müller and Frank A. Müller, *Acta Biomaterialia* 2006, vol. 2, pp. 181–189.
4. L Wadsö, *Cement Int.* 2005, pp. 94–101.
5. G Buckton, *Thermochim. Acta* 1995, vol. 248, pp. 117–129.
6. L Gustafsson, *Thermochim. Acta* 1991, vol. 193, pp. 145–171.
7. L E Paulsson, L G Svensson and C K Forsgren, (Thermometric AB, Järfälla, Sweden).

# Effect of Fluoride Ion on the Microstructure and Properties of Permanganate Conversion Coating on AZ91D Magnesium Alloy

Shih-An Yang and Chao-Sung Lin

**Abstract**

Because of the toxicity of chromate, it is necessary to develop alternatives to chromate conversion coatings. Permanganate conversion coating treatment is one of the methods that are potentially alternative to the chromate treatment on magnesium alloys. This study investigated the effect of fluoride ion in the permanganate solution on the microstructure and corrosion resistance of the conversion coating on AZ91D alloy. Experimental results showed that the presence of fluoride ion impeded the conversion coating reaction and accordingly retarded the formation of the conversion coating, which resulted in a thinner and more uniform coating. Therefore, the permanganate conversion coating formed in the fluoride-containing solution was nearly crack-free and displayed higher corrosion resistance than the coating formed in the solution without fluoride ion.

**Keywords**

Magnesium alloy • Permanganate conversion coating
Corrosion resistance • EIS

## Introduction

Magnesium is the lightest metallic structural material, with a density of 1.74 g/cm$^3$, which is only 66% of aluminum, 38% of titanium, and 25% of iron. Magnesium alloys also have high specific strength, stiffness, electromagnetic shielding property, and good heat dissipation [1, 2]. Therefore, magnesium alloys have found many applications in automotive, aerospace, and electronics industries [3, 4]. Moreover, magnesium alloys are easily recycled and considered as an environmental–friendly material. However, magnesium alloys are prone to corrosion in most application environments due to the high chemical activity of magnesium.

Chromate conversion coating is known for its excellent corrosion resistance and unique self-healing properties [5, 6]. However, chromate conversion treatments are now being restricted because hexavalent chromium is extremely harmful to human health and the environment. Permanganate/phosphate conversion coating treatment is one of the potential treatments alternative to the chromate conversion treatment [7]. Jian et al. [8] studied the permanganate conversion coatings on AZ31 magnesium alloys in $KMnO_4$ solution with potassium dihydrogen phosphate ($KH_2PO_4$) and manganese(II) nitrate ($Mn(NO_3)_2$) additives and found that the coating formed via the Guyard reaction $\left(2MnO_4^- + 3Mn^{2+} + 2H_2O \rightarrow 5MnO_2 + 4H^+\right)$ and the phosphate ion can impede the autocatalysis of the Guyard reaction, which drastically reduced the presence of cracks on the conversion coating; thereby, the corrosion resistance of the conversion coating was markedly enhanced.

It has been generally recognized that the incorporated $MnO_2$ in permanganate/phosphate conversion coatings on magnesium alloys is essential to improve the corrosion resistance of the coating [8]. In addition, the Guyard reaction is the very feasible route for the precipitation of $MnO_2$. Phosphate ions are added in the permanganate solution to enhance the stability of the solution [9]. Adamson [10] reported that the presence of fluoride retarded the precipitation of $MnO_2$ from the Guyard reaction by the formation of stable manganic complex. Chiu et al. [11] showed that the fluoride conversion coating treatment on pure magnesium effectively increased the corrosion resistance due to the dense $MgF_2$ conversion coating. The presence of fluoride ions in the permanganate conversion solution can thus enhance the stability of the solution, and, on the other hand, facilitate the precipitation of $MgF_2$ in the conversion coating.

S.-A. Yang (✉) · C.-S. Lin
Department of Materials Science and Engineering, National Taiwan University, Taipei, Taiwan
e-mail: R05527030@ntu.edu.tw

C.-S. Lin
e-mail: csclin@ntu.edu.tw


D. Orlov et al. (eds.), *Magnesium Technology 2018*, The Minerals, Metals & Materials Series, https://doi.org/10.1007/978-3-319-72332-7_8

In the present study, the effect of fluoride ions in permanganate/manganese (II) solutions on the microstructure and properties of permanganate conversion coatings on AZ91D magnesium alloys was detailed.

## Experimental

### Conversion Treatment

A die-cast AZ91D plate with a thickness of approximately 1 mm was employed for this study. Coupons of 40 mm × 50 mm were cut from the as-received AZ91D and ground using sandpaper up to 4000 grit, followed by polishing using 1 μm alumina solution. The coupons were thoroughly cleaned using de-ionized water and dried with compressed air.

The conversion coating solution was composed of 0.1 M potassium permanganate ($KMnO_4$) and 0.025 M manganese (II) nitrate ($Mn(NO_3)_2$) with and without the addition of 0.02 M potassium fluoride (KF). The presence of 0.02 M KF slightly increased the pH from 1.51 to 1.77. The as-polished coupon was immersed in the conversion solution at 25 °C for 60 s. After removal from the solution, the coupon was immediately immersed in de-ionized water to stop the conversion coating reaction. Again, the coupon was dried by compressed air after being thoroughly cleaned. Finally, the coupon was left at the ambient atmosphere for 24 h before characterization.

### Titration

The remaining permanganate concentration of the conversion solution was measured using titration with oxalic and sulfuric acid solution. The titration was conducted at an interval of 24 h up to 192 h. The effect of phosphate ions on the stability of the permanganate/manganese (II) solution was also studied for comparison.

### Microstructure Characterization

The surface morphology of the conversion coating was characterized using scanning electron microscopy (JEOL JSM6510) to study the uniformity and surface morphology of the conversion coating. The composition of the coating was also measured using the X-ray energy dispersive spectroscopy equipped in the SEM. To prepare the cross-sectional specimen, two slices of the conversion-coated AZ91D were glued face to face using the G1 glue and cured at 150 °C for 30 min. The glued specimen was cold mounted in an epoxy. After curing, the specimen was mechanically ground and polished using the same procedure for the as-received AZ91D. Then, the specimen was characterized using a field-emission SEM (Nova NanoSEM 450).

### Electrochemical Measurement

The corrosion resistance of the AZ91D with and without the permanganate conversion coating was studied using AC impedance spectroscopy (EIS) in 0.05 M sodium chloride (NaCl) and 0.1 M sodium sulfate ($Na_2SO_4$) solution. The AZ91D with an exposure area of around 1.8 $cm^2$ was employed as the working electrode, and a saturated calomel electrode (SCE) and platinum plate were used as the reference and counter electrodes, respectively. When the open circuit potential (OCP) of the tested specimen had become stable, the EIS response was measured at a sinusoidal amplitude of 10 mV imposed on the OCP in the frequency ranging from $10^5$ to $10^{-2}$ Hz using an EG&G 263A Potentiostat together with a frequency response detector (FRD100).

## Results and Discussion

### Titration of the Permanganate Conversion Solution

Figure 1 shows the remaining permanganate concentration of conversion solutions as a function of time. For the solution without $F^-$ and $H_2PO_4^-$, the $MnO_4^-$ concentration of the as-prepared solution had declined to approximately 0.096 M right after titration. Then, the $MnO_4^-$ concentration declined linearly with time and only 50% $MnO_4^-$ left after 192 h. This is consistent with the autocatalytic reaction characteristic of $MnO_4^-$ and $Mn^{2+}$ to form $MnO_2$ precipitates, which is well known as the Guyard reaction [12]. In contrast, the $MnO_4^-$ concentration of the solution with the addition of $H_2PO_4^-$ hardly changed up to 192 h, in good consistence with the results that phosphate ions effectively inhibit the Guyard reaction via their absorption on $MnO_2$ colloids [13, 14]. The $F^-$ ions also contributed to the stability of the solution, but its effectiveness was inferior to that of $H_2PO_4^-$. Previous studies have shown that the $F^-$ retards the Guyard reaction when the intermediate products of the Guyard reaction form stable complexes with $F^-$ [10]. However, in the presence of $F^-$, the $MnO_4^-$ concentration underwent a slight decrease only after 48 h. Because the permanganate conversion coating was conducted in fresh solution, the effect of $F^-$ studied can be regarded as the presence and absence of 0.02 M $F^-$ in the solution with a slight difference in $MnO_4^-$ concentration, i.e., less than 0.01 M, as shown by the 24 h data in Fig. 1.

## Surface Morphology of the Permanganate Conversion Coating

Figure 2 shows the surface morphology of the permanganate conversion coating formed in the solution without the addition of $F^-$. The coating was dotted with uniform white-contrast particles under low magnifications (Fig. 2a). Close-up observations revealed that scratches resulting from mechanical grinding and polishing disappeared after conversion coating, indicating the conversion reaction prevails to significant extents after 60 s of immersion in the permanganate/manganese (II) solution free of $F^-$. Moreover, cracks were frequently observed on the relatively-large particles, whose average size was around 3 μm. EDS analyses of the area with and without particles, which are marked as 1 and 2, respectively, in Fig. 2b, found that the relatively-large particles were mainly composed of Mn and O (Table 1), suggesting the precipitation of $MnO_2$. As a result, the Guyard reaction prevails in the solution free of $F^-$, which is launched mainly via a rise in interfacial pH accompanying the oxidation and dissolution of magnesium substrate.

Figure 3 shows the surface morphology of the permanganate conversion coating formed in the solution with the addition of 0.02 M $F^-$. The coating was smoother than that formed in the solution in the absence of $F^-$, especially free of white-contrast particles and cracks. Because scratches were still discernible, the coating formed with $F^-$ was thinner than that formed without $F^-$. Moreover, F was detected in the coating formed with $F^-$. The precipitation of $MgF_2$ was thus evident because the solubility product of $MgF_2$ is as low as $5.16 \times 10^{-11}$ [15].

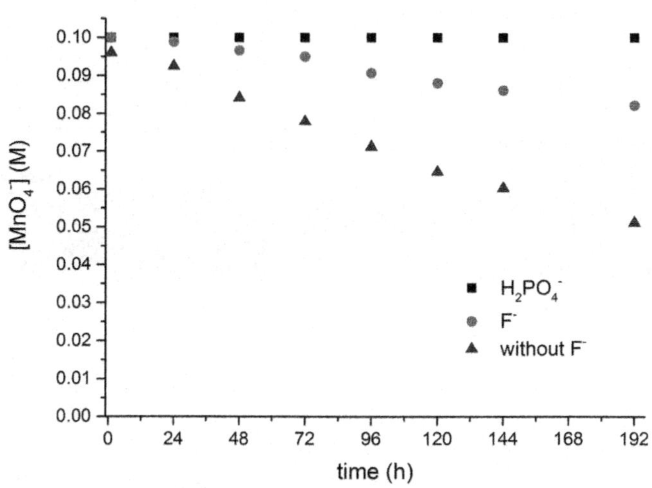


**Fig. 1** The remaining permanganate concentration of conversion solutions as a function of time

## Cross-section Characterization of the Permanganate Conversion Coating

Figure 4 shows the cross-sectional SEM micrograph of the permanganate conversion coating on AZ91D formed in the solution without the addition of $F^-$. The structure composed of α and β phases was readily distinguished on the cross section, specifically the β phases showed up with a white contrast and with network structure encompassing the α phase. For the coating formed without $F^-$ (Fig. 4a), uniform coating with a thickness of approximately 250 nm was observed on the α phases, meanwhile large particles were dotted on top of the β phases. Cracks were present in the large particle. Conversely, the coating formed with $F^-$ was continuous and uniform in thickness (~190 nm) across the α/β interface (Fig. 4b). The results of cross-sectional characterization agree well with those of surface morphology observation. The presence of $F^-$ in the solution slows down the conversion coating reaction and ameliorates the uniformity of the coating on the α and β phases.

## Formation Mechanism of the Permanganate Conversion Coating

To further elucidate how the dual-phase structure of the AZ91D influences the evolution of the permanganate conversion coating, the surface morphology of the coating formed during early stage of immersion in the solution without $F^-$ was studied, as shown in Fig. 5. The presence of $MnO_2$ particles was noticed after 30 s of immersion (Fig. 5a). Back scattered electron image further showed the $MnO_2$ particles were distributed mainly on the β phases, which generally displays a brighter contrast (Fig. 5b). Moreover, the EDS analyses taken from the surface of the coating shows that more Mn was detected in the coating on the β phases (Table 2). This indicates that the $MnO_2$ is present mainly on the outer part of the coating.

The formation mechanism of the permanganate conversion coating on the AZ91D is proposed as the follows. Upon immersion in acidic permanganate/manganese (II) solution, the oxidation and dissolution of the AZ91D substrate proceeds mainly at the α phase, which is chemically more active than the β phase [16], as shown by Eqs. (1) and (2). The reduction reactions then take place via Eqs. (3)–(5), including the hydrogen discharge and the reduction of $MnO_4^-$ to $Mn^{2+}$ or $MnO_2$. These reduction reactions prevail at the β phases, as evident from the result that the relatively large $MnO_2$ particles mainly reside on the β phase. Meanwhile, hydrogen discharge results in a pH rise at the AZ91D/solution interface, which, in turn, triggers the Guyard reaction and the precipitation of $Mg(OH)_2$. As a

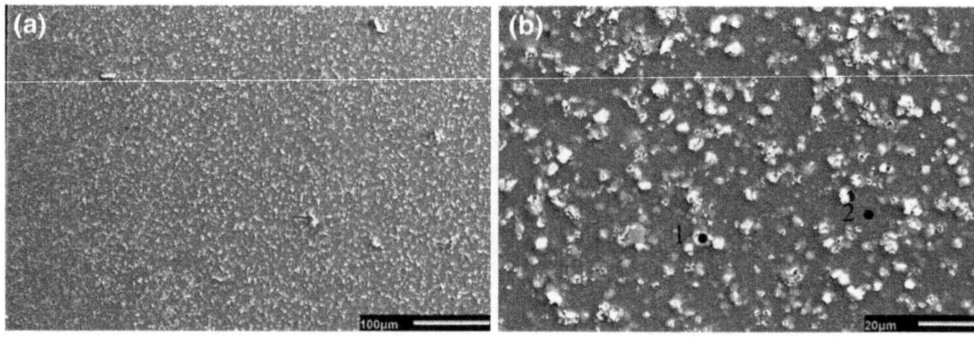


**Fig. 2** Surface morphology of the permanganate conversion coating formed in the solution without the addition of $F^-$ under **a** low magnification, **b** magnified view

**Table 1** The EDS results of the spot marked by 1, 2, 3, and 4 in Figs. 1 and 2

| Spot | O | Mg | Al | Mn | F |
|---|---|---|---|---|---|
| 1 | 35.28 | 32.17 | 6.71 | 25.83 | – |
| 2 | 17.32 | 68.42 | 3.10 | 11.16 | – |
| 3 | 12.94 | 58.33 | 23.04 | 4.36 | 1.33 |
| 4 | 4.57 | 89.69 | 3.62 | 0.86 | 1.25 |

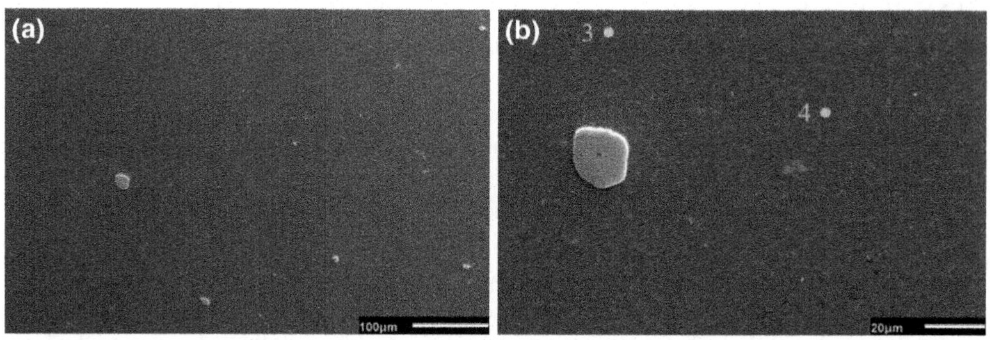


**Fig. 3** Surface morphology of the permanganate conversion coating formed in the solution with the addition of $F^-$ under **a** low magnification, **b** magnified view

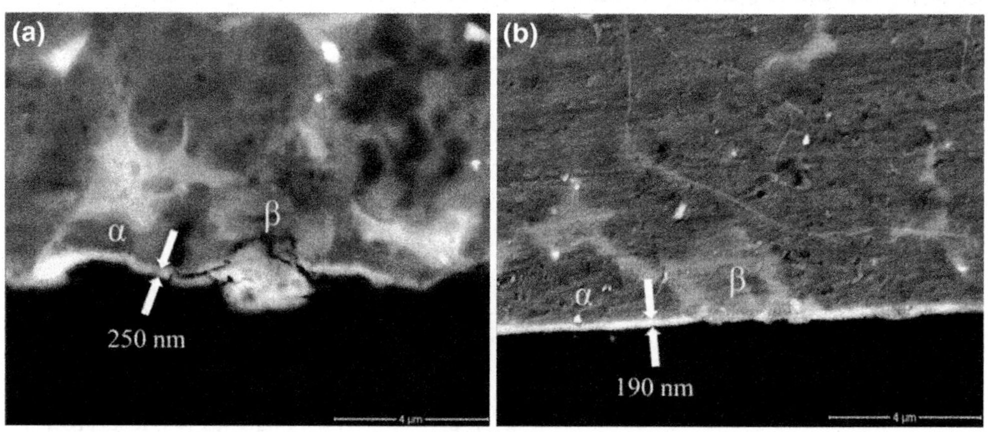


**Fig. 4** The cross-sectional SEM micrograph of the permanganate conversion coating on AZ91D formed in the solution **a** without and **b** with the addition of $F^-$

result, the coating on the β phase contains more $MnO_2$ particles than that on the α phases.

$$Mg \rightarrow Mg^{2+} + 2e^- \quad (1)$$

$$Al \rightarrow Al^{3+} + 3e^- \quad (2)$$

$$2H^+ + 2e^- \rightarrow H_2 \quad (3)$$

$$MnO_4^- + 8H^+ + 5e^- \rightarrow Mn^{2+} + 4H_2O \quad (4)$$

$$MnO_4^- + 4H^+ + 3e^- \rightarrow MnO_2 + 2H_2O \quad (5)$$

The dissolved $Mg^{2+}$, in one hand, results in the precipitation of $Mg(OH)_2$. On the other hand, in the solution containing $F^-$, $Mg^{2+}$ can react with $F^-$ to form $MgF_2$ [17]. It is likely that the precipitation of $MgF_2$ occurs mainly at the α

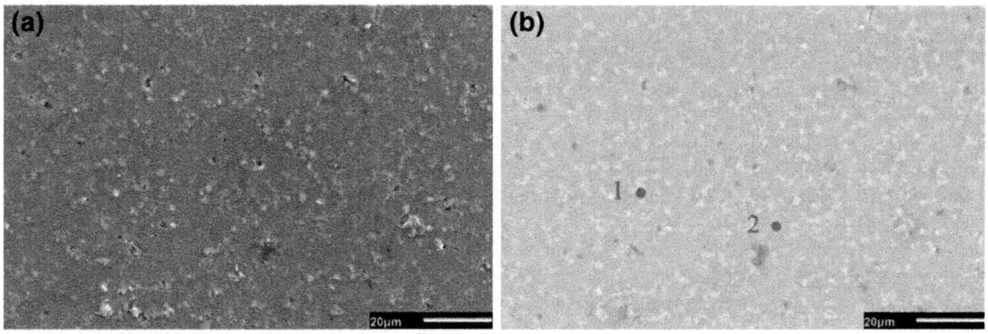


**Fig. 5** Surface morphology of the AZ91D treated in the conversion solution without the addition of $F^-$ for 30 s under **a** secondary electron image mode and **b** back scattered electron image mode

**Table 2** The EDS results of the spot marked by 1, and 2 in Fig. 5

| Spot | O | Mg | Al | Mn | F |
|---|---|---|---|---|---|
| 1 | 32.46 | 45.71 | 11.16 | 10.07 | – |
| 2 | 11.41 | 81.10 | 4.25 | 3.24 | – |

phases, i.e., anodic sites compared to the β phase. The protective $MgF_2$ reduces the chemical activity difference between the α and β phases. Consequently, the galvanic corrosion appearing between the α and β phases is retarded to some extent, which results in a thin, continuous permanganate conversion coating on both the α and β phases.

## Corrosion Resistance of the Permanganate Conversion Coating

Figure 6 shows the Nyquist plots of the bare AZ91D and the permanganate conversion coating on AZ91D formed with and without $F^-$. The Nyquist plot of the bare AZ91D displayed two well-defined capacitive loops appearing at high and medium frequency, respectively, characteristic of the corrosion behavior of magnesium alloys covered with corrosion products [18]. The Nyquist plot of the permanganate conversion coating on AZ91D formed without $F^-$ had a diameter larger than that of the bare AZ91D, suggesting the permanganate conversion coating improves the corrosion resistance of the AZ91D. A further increase in the diameter was noted for the permanganate conversion coating on AZ91D formed with $F^-$. This further increase in the total impedance is apparently related the presence of continuous, crack-free conversion coating.

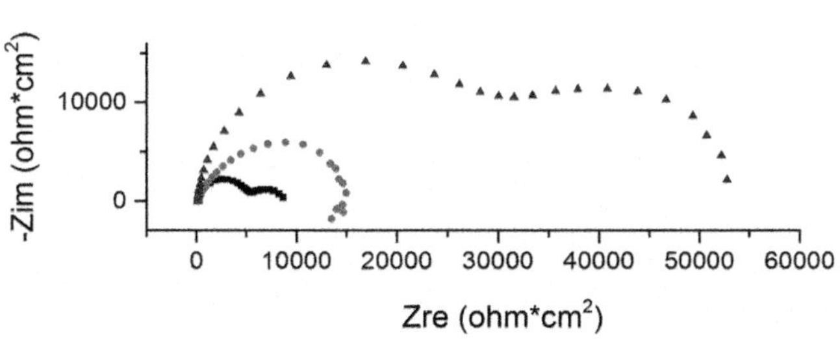


**Fig. 6** The Nyquist plots of the bare AZ91D plate (black square) and the permanganate-coated AZ91D plates treated in the solution without (red circle) and with (blue triangle) $F^-$

## Conclusions

This study investigated the effect of $F^-$ in the permanganate/manganese (II) solution on the microstructure and corrosion resistance of the conversion coating on the AZ91D. Based on the titration results, the addition of $F^-$ to the permanganate/manganese (II) solution enhanced the stability of the solution. Moreover, the SEM characterization showed that the permanganate conversion coating formed in the presence of $F^-$ was more continuous and free of cracks compared to the coating formed without $F^-$. As a result, the resulting coating had larger total impedance than the counterpart formed in the absence of $F^-$. It is likely that $MgF_2$ precipitates on top of the α phase and alleviates the galvanic corrosion between the α and β phases. A uniform, continuous permanganate conversion coating can thus form on the AZ91D.

## References

1. M. M. Avedesian and H. Baker, *ASM specialty handbook: magnesium and magnesium alloys*. ASM international, 1999.
2. E. Ghali, W. Dietzel, and K.-U. Kainer, "General and localized corrosion of magnesium alloys: a critical review," *Journal of Materials Engineering and Performance,* vol. 13, no. 1, pp. 7–23, 2004.

3. E. Aghion, B. Bronfin, and D. Eliezer, "The role of the magnesium industry in protecting the environment," *Journal of materials processing technology,* vol. 117, no. 3, pp. 381–385, 2001.
4. A. A. Aal, "Protective coating for magnesium alloy," *Journal of materials science,* vol. 43, no. 8, pp. 2947–2954, 2008.
5. L. Xia and R. L. McCreery, "Chemistry of a Chromate Conversion Coating on Aluminum Alloy AA2024-T3 Probed by Vibrational Spectroscopy," *Journal of the Electrochemical Society,* vol. 145, no. 9, pp. 3083–3089, 1998.
6. X. Zhang, C. Van den Bos, W. Sloof, A. Hovestad, H. Terryn, and J. De Wit, "Comparison of the morphology and corrosion performance of Cr (VI)-and Cr (III)-based conversion coatings on zinc," *Surface and coatings technology,* vol. 199, no. 1, pp. 92–104, 2005.
7. Y. L. Lee, Y. R. Chu, W. C. Li, and C. S. Lin, "Effect of permanganate concentration on the formation and properties of phosphate/permanganate conversion coating on AZ31 magnesium alloy," *Corrosion Science,* vol. 70, pp. 74–81, 2013.
8. S.-Y. Jian, Y.-R. Chu, and C.-S. Lin, "Permanganate conversion coating on AZ31 magnesium alloys with enhanced corrosion resistance," *Corrosion Science,* vol. 93, pp. 301–309, 2015.
9. M. Orban and I. R. Epstein, "Systematic design of chemical oscillators. 59. Minimal permanganate oscillator: the Guyard reaction in a CSTR," *Journal of the American Chemical Society,* vol. 111, no. 22, pp. 8543–8544, 1989.
10. A. W. Adamson, "The Kinetics of the Manganous–Permanganate Reaction," *The Journal of Physical Chemistry,* vol. 55, no. 2, pp. 293–303, 1951.
11. K. Chiu, M. Wong, F. Cheng, and H. Man, "Characterization and corrosion studies of fluoride conversion coating on degradable Mg implants," *Surface and Coatings Technology,* vol. 202, no. 3, pp. 590–598, 2007.
12. F. Tompkins, "The kinetics of the reaction between manganous and permanganate ions," *Transactions of the Faraday Society,* vol. 38, pp. 131–139, 1942.
13. M. Mrákavová and Ł. Treindl, "Kinetics of guyard's reaction with regard to permanganate chemical oscillators," *Collection of Czechoslovak chemical communications,* vol. 55, no. 12, pp. 2898–2903, 1990.
14. M. J. Polissar, "The kinetics of the reaction between permanganate and manganous ions," *The Journal of Physical Chemistry,* vol. 39, no. 8, pp. 1057–1066, 1935.
15. J. Booster, A. Van Sandwijk, and M. Reuter, "Conversion of magnesium fluoride to magnesium hydroxide," *Minerals engineering,* vol. 16, no. 3, pp. 273–281, 2003.
16. W. Zhou, D. Shan, E.-H. Han, and W. Ke, "Structure and formation mechanism of phosphate conversion coating on die-cast AZ91D magnesium alloy," *Corrosion Science,* vol. 50, no. 2, pp. 329–337, 2008.
17. Y. Huang, Y. Lee, and C. Lin, "Acid pickling pretreatment and stannate conversion coating treatment of AZ91D magnesium alloy," *Journal of The Electrochemical Society,* vol. 158, no. 9, pp. C310–C317, 2011.
18. G. Song, A. L. Bowles, and D. H. StJohn, "Corrosion resistance of aged die cast magnesium alloy AZ91D," *Materials Science and Engineering: A,* vol. 366, no. 1, pp. 74–86, 2004.

# Corrosion Characteristics of Two Rare Earth Containing Magnesium Alloys

M. AbdelGawad, B. Mansoor, and A. U. Chaudhry

**Abstract**

Magnesium (Mg) based alloys have received prevalent attention, especially in the biomedical, aerospace and automotive industries due to their low density, moderate strength and natural ability to degrade. However, widespread use of Mg-based alloys as degradable biomedical implants still remains a significant technological challenge because of their rapid corrosion kinetics that leads to premature loss of mechanical integrity. Addition of certain alloying elements such as Zinc and Rare Earth Elements (REs) improve the mechanical and corrosion response of Mg alloys. In this paper, the mechanical and corrosion characteristics of two commercially available RE containing Mg alloys, ZE41 and EZ33 are studied. Results obtained using hydrogen evolution, weight-loss measurements, and electrochemistry (PD and EIS) in comparison with pure Mg data are presented here. These initial findings indicate that although, ZE41 and EZ33 have the same primary alloying elements, due to the addition of REs, differences in composition, nature of precipitates and passivating layer play a key role in positively impacting the corrosion behavior and hence, the corrosion rate.

**Keywords**

Magnesium • Corrosion • Rare earths • ZE41
EZ33 • Electrochemical

M. AbdelGawad · B. Mansoor (✉) · A. U. Chaudhry
Mechanical Engineering Program, Texas A&M University at Qatar, Doha, Qatar
e-mail: marwa.abdelgawad@qatar.tamu.edu

M. AbdelGawad · B. Mansoor
Mechanical Engineering Department, Texas A&M University, College Station, TX, USA

B. Mansoor
Materials Science and Engineering Program, Texas A&M University, College Station, TX, USA

## Introduction

Magnesium (Mg) and its alloys are light-weight metals that are characterized by their high strength-to-weight ratios, high specific mechanical strength and fracture toughness. They make competitive alternatives to traditional structural materials especially in aerospace and automotive industries. In orthopedic applications, the need to address the two major challenges associated with permanent internal implants i.e. stress shielding and removal surgeries, triggers the use of temporary implants made from biocompatible and biodegradable or bioabsorbable materials with the stiffness close to that of bones (3–20 GPa). As a result, the attention towards Mg alloys as biodegradable implants has increased over the last few years. Mg is present in the human body and its low density (1.74 g/cm$^3$) is close to the density of the human cortical bone (1.75 g/cm$^3$). However, the unfortunate complication is that it can corrode too quickly in the physiological pH (7.4–7.6) and high chloride environment of a physiological system, which in turn leads to rapid loss of mechanical integrity earlier than desired to enable fracture healing. Commercial Mg alloys designed primarily for structural applications have been considered but they exhibit rapid degradation kinetics and cytotoxic characteristics. Magnesium alloys with biocompatible alloying elements are currently under development. Therefore, to aid such alloy design efforts it is essential to understand the effect of certain key alloying additions such as Rare Earths (REs) on the corrosion characteristics of Mg alloys in chloride environments.

The corrosion response of pure Mg has been studied extensively in several papers [1–4]. Pure Mg has a negative free-corrosion potential and forms a Mg hydroxide $Mg(OH)_2$ film that can provide some protection against corrosion [2]. A phenomenon, known as the negative difference effect (NDE), has been identified to uniquely impact its corrosion characteristics. In typical metals, as the applied potential increases, anodic reaction rates increase and cathodic


D. Orlov et al. (eds.), *Magnesium Technology 2018*, The Minerals, Metals & Materials Series, https://doi.org/10.1007/978-3-319-72332-7_9

reaction rates decrease. The corrosion reaction of Mg results in hydrogen evolution during the cathodic partial reaction. However, it was found that for Mg, increase in applied potential leads to an increase in the cathodic hydrogen evolution rate rather than decrease [1–3]. Mg alloys also experience the NDE during corrosion [1, 4]. The susceptibility of the alloy to corrosion depends on the quality of the surface film being formed. To improve the stability of the surface film, several different strategies are employed such as surface treatment, mechanical processing, and/or introducing new alloying elements.

Several alloying elements have been identified to improve the corrosion resistance of Mg. Specifically, aluminum, zirconium, zinc, and REs have shown to influence the microstructure, texture, mechanical properties, and corrosion characteristics of Mg alloys [2, 5, 6]. REs are elements that are found in the Earth's crust, where they tend to exist in the same ore deposits and exhibit similar chemical properties [5]. Initially, rare earths were found to improve the mechanical properties of Mg alloys and hence were used extensively in automotive, aerospace and biomedical applications. Tekumalla et al. provided an extensive overview of the different Mg-RE systems available in literature and their respective tensile and compressive properties [7]. Siebert-trimmer et al. discussed how the addition of rare earths resulted in improved creep performance of EZ33 (3 wt% RE, 3 wt% Zn) when to compared to AX30 (3 wt% Al, 0.5 wt% Ca) [8]. Chen et al. studied the yield strength and elongation at failure of biodegradable Mg-alloys and showed that Mg-RE alloys usually exhibited higher strength and better elongation when compare to other Mg alloys [9]. Ding et al. reported that Cerium, Lanthanum and Neodymium, enhanced the corrosion resistance when added to Mg alloys due to the formation of secondary phases that contributed to the improvement of their corrosion resistance [5]. Although, there are many detailed studies on the mechanical behavior of Mg-RE alloys, the role of RE in influencing the corrosion kinetics of Mg alloys is rather less reported.

In this initial study, the corrosion response of two commercially available RE containing Mg alloys, ZE41 and EZ33 and pure Mg was investigated in 3.5 wt% NaCl solution. The main goal was to analyze the addition of rare-earths on microstructure and corrosion resistance, as well the formation and characteristics of respective passivating films. The corrosion response was evaluated by conducting immersion testing using weight loss and hydrogen evolution methods. In addition, to complement the immersion study, electrochemical techniques including open circuit potential (OCP), Electrochemical Impedance Spectroscopy (EIS) and Potentiodynamic Polarization (PD) were employed.

## Experimental Methods

### Materials

The alloys used in this work were commercially obtained in the form of 12.7 mm thick, cast plates of ZE41 and EZ33 in T5 temper. The commercially pure Mg was obtained in homogenized condition, as 2 mm thick sheets with 99.93% purity. The nominal compositions of all materials are given in Table 1 [10, 11].

### Microstructure and Mechanical Properties

The microstructure and grain size of pure Mg, ZE41 and EZ33 were examined using optical microscopy. Specimens were taken from the as-received plates and sheets and mounted in an epoxy resin. The metallographic preparation was carried out in the usual manner, which involved mechanical grinding successively up to 1200 grit SiC paper,

**Table 1** Nominal compositions of ZE41A-T5 and EZ33A-T5 [10, 11]

| Element | Composition, wt% | | | |
|---|---|---|---|---|
| | ZE41A-T5 | | EZ33A-T5 | |
| | Minimum | Maximum | Minimum | Maximum |
| Zinc | 3.5 | 5.0 | 2.0 | 3.1 |
| Total rare earths[a] | 0.75 | 1.75 | 2.5 | 4.0 |
| Zirconium | 0.4 | 1.0 | 0.4 | 1.0 |
| Manganese | – | 0.15 | – | – |
| Copper | – | 0.10 | – | 0.1 |
| Nickel | – | 0.01 | – | 0.01 |
| Other elements | – | 0.3 | – | 0.3 |
| Magnesium | Remainder | | Remainder | |

[a]*Note* Main rare earth elements include Cerium and Lanthanum [12, 13]

followed by polishing using 3 and 1 micron diamond suspensions. Samples were then washed with ethanol and dried in air. Etching was done using acetic-picral solution (4.2 g picric acid, 70 ml ethanol, 10 ml acetic acid and 10 ml distilled water).

Compression tests were performed for ZE41 and EZ33 using electromechanical MTS Insight 30 kN machine with a strain rate of $10^{-4}$/s at ambient temperature. Vickers micro-hardness tests were conducted as well with a load of 100 gf for pure Mg and ZE41 and 50 gf for EZ33 with a dwell time of 10 s.

## Weight Loss and Hydrogen Evolution Tests

Weight loss and hydrogen evolution experiments were conducted simultaneously using the fishing line method as described by Shi et al. [14]. Metallographic preparation of the samples was performed using the same procedure as described above for microstructure sample preparation. However, before mounting, the sides of the samples were also grinded to avoid crevice corrosion occurring at the edges which could influence weight loss results. A reference was also prepared out of epoxy with a similar size as the mounted samples but without a test specimen. The weight loss readings from the reference was used to eliminate any influence due to the epoxy mount. A 1.5 mm hole was drilled towards the top of each of the mounted samples and a fishing line was used to suspend the specimens inside a beaker filled with the electrolyte (3.5 wt% NaCl). An inverted funnel fixed with a burette on top of the suspended specimen was used to collect the hydrogen gas. The burette was initially filled with solution, and the amount of hydrogen evolution read and converted into corrosion rate using the following formula [14]:

$$CR(\text{mm/yr}) = 2.279 * \textit{Volume of hydrogen evolved}\left(\text{ml/cm}^2/\text{day}\right) \quad (1)$$

After hydrogen evolution readings were taken, samples were removed and rinsed carefully with distilled water, air dried and weighed. Weight loss was calculated and corrosion rate determined using the formula below [15]:

$$CR(\text{mm/yr}) = \left[\frac{\textit{weight loss}(\text{g})}{\textit{surface area}(\text{mm}^2) * \textit{density}\left(\text{g/mm}^3\right) * \textit{time}(\text{h})}\right] * 8760\ \text{yr}^{-1} \quad (2)$$

Weight loss and hydrogen evolution readings for ZE41 and EZ33 were taken every 24 h over a period of seven days. However, for pure Mg, weight loss readings were taken every 3 h while hydrogen evolution readings were taken every hour.

## Electrochemical Tests

Samples for electrochemical tests were prepared by cutting 1.5 cm by 1.5 cm coupons out of the as-received sheets and plates and grinding all sides up to 600 grit, while grinding one side up to 1200 grit SiC paper. The finely grinded side was then polished using 3 micron diamond suspensions and exposed as the working electrode.

Corrosion tests were performed under naturally aerated conditions using Gamry series G 300™ Potentiostat/Galvanostat/ZRA. The electrochemical behavior of pure Mg, ZE41 and EZ33 were studied in 3.5 wt% NaCl aqueous solution in a closed Gamry Paracell™ with each of the prepared samples as the working electrode, a graphite plate as the counter electrode, and Ag/AgCl as the reference electrode. The working volume of the corrosive solution was kept as 350 ml and exposed areas for working and counter electrode were 1 $cm^2$. Different corrosion studies were performed, starting from open circuit potential (OCP) followed by Electrochemical Impedance Spectroscopy (EIS) and Potentiodynamic Polarization (PD) for different periods of immersion up to 3 h. EIS was conducted by sweeping the frequency from $10^5$ to $10^{-2}$ Hz at 10 mV AC amplitude whereas for PD, electrodes were scanned from −2.5 to +0.5 V with scan rate of 1 mV/s.

# Results and Discussion

## Microstructure Analysis

Microstructures of pure Mg, ZE41 and EZ33 are shown in Fig. 1 and their respective grain sizes summarized in Table 2. Structure of pure Mg is formed predominantly of equiaxed grains with some twinning occurring within the grains. Both ZE41 and EZ33 are considered part of the same alloy group, Mg–Zn–RE–Zr and therefore are expected to

**Fig. 1** Microstructure of pure Mg (top), ZE41A (bottom left) and EZ33A (bottom right)

**Table 2** Summary of microstructural and mechanical properties of ZE41 and EZ33

| Alloy | Yield strength (MPa) | Compressive strength (MPa) | Compressive strain (%) | Hardness (HV) | Grain size (μm) |
|---|---|---|---|---|---|
| Pure Mg | – | – | – | 34.92 | 49.5 |
| ZE41 | 130 | 345 | 24 | 62.99 | 39.5 |
| EZ33 | 90 | 359 | 32 | 53.34 | 40.1 |

have similar microstructures [16]. In T5 condition, both alloys have equiaxed α-Mg matrix structure and a eutectic microconstituent referred to as the T-phase located along and/or adjacent to the boundaries of the α-Mg. The composition of the eutectic microconstituent is $Mg_7Zn_3RE$, which has been widely reported in literature [12, 17, 18]. However, for EZ33, the phase found at the grain boundaries is believed to be an intermetallic phase of $Mg_{12}RE$ and doesn't contain any Zn [16]. This difference could be attributed to the higher content of rare earths found in EZ33; EZ33 has about 3 wt% REs compared to 1 wt% in ZE41 (Table 1). Furthermore, the volume fraction of intermetallic phases found adjacent to the grain boundaries was noticed to be higher in EZ33 than ZE41. Ding et al. concluded that the increased volume fraction of the secondary phases was due to the increase in the concentration of rare earths [18]. In addition, some particles were found within ZE41 grains with a dark region around it. Neil et al. analyzed these particles and identified them as Zr-rich intermetallic particles playing a significant role in the corrosion mechanism of ZE41, as will be shown later in the immersion test results [12, 17, 18]. However further studies need to be done using energy dispersive X-ray spectroscopy (EDS) to confirm the composition of these particles.

## Mechanical Properties

Figure 2 shows the stress-strain curves of ZE41 and EZ33 under compressive loading, and their respective mechanical properties summarized in Table 2. The ultimate compressive strength (UCS) and ductility were both higher for EZ33 compared to ZE41, however, the yield strength was lower. The increase in ductility and UCS are possibly linked to the weakening of basal texture due to the REs. However, the impact of REs on texture of Mg alloys is not the focus of this study. In addition to texture, as reported in literature, rare earths do tend to have a positive effect on the strength and ductility of Mg alloys due to the formation of the hard eutectic phases and secondary intermetallic phases [7].

The Vickers microhardness values for pure Mg was 34.92. The microhardness values for ZE41 and EZ33 were 62.99 and 53.34 respectively, indicating that the addition of REs to pure Mg increased the hardness due to presence of the eutectic and secondary phases, as explained earlier. With an increased volume fraction of REs, EZ33 showed a lower hardness value than ZE41 which was in agreement with its lower compressive yield strength in comparison with ZE41.

## Immersion Test Results

Corrosion rates via weight loss are shown in Fig. 3 below for pure Mg, ZE41 and EZ33. Pure Mg was shown to exhibit a rigorous corrosion behavior with an average corrosion rate of 478 mm/yr. The rapid corrosion of pure Mg was believed to be due to the presence of chloride ions in the electrolyte. The oxide film that initially forms on Mg is usually vulnerable to corrosion when exposed to chloride-containing solutions resulting in rapid initiation of pitting corrosion [2, 4, 6]. In contrast, ZE41 and EZ33 both showed a much slower corrosion behavior with the corrosion rate of ZE41 being about four times faster than that of EZ33. For both alloys, corrosion rates were increasing and then stabilized after 4 days. The same trend was also seen in hydrogen evolution results (Fig. 4). Pitting corrosion was observed in Mg during the first hour of sample immersion and within 36 h, the sample was completely dissolved. Therefore, it was essential to take readings every hour instead of every 24 h as in the cases of ZE41 and EZ33. In summary, the corrosion rate of pure Mg was two orders of magnitude higher than both ZE41 and EZ33, while among the two Mg-RE alloys, ZE41 corrosion rate was around four times higher than EZ33.

Although, ZE41 and EZ33 showed similarities in their microstructure, they exhibited distinct corrosion characteristics. Figure 5 shows the progression of corrosion behavior of ZE41 and EZ44 samples after 24 h of immersion and at the end of the seven days. Overall, the corrosion of ZE41 showed more pitting than that of EZ33. The active corrosion area in EZ33 was relatively more uniform, with localized pitting corrosion around one of the edges. Neil et al. proposed a corrosion mechanism for as-received ZE41 in chloride-containing solution [17, 19]. Corrosion was believed to initiate close to the eutectic T-phase around the grain boundaries. As the alloy continues to be exposed to the electrolyte, corrosion starts to happen around the intermetallic Zr-rich particles and eventually produces large "crates" which can be clearly seen in the stereoscopic images of the corroded samples at the end of the immersion tests (Fig. 6). This behavior was not seen in EZ33 which is an indication that the EZ33 developed a surface film that possibly was able to resist the initiation of corrosion better than ZE41. Electrochemical tests were performed to study the development and progression of the surface films developed on each of the materials.

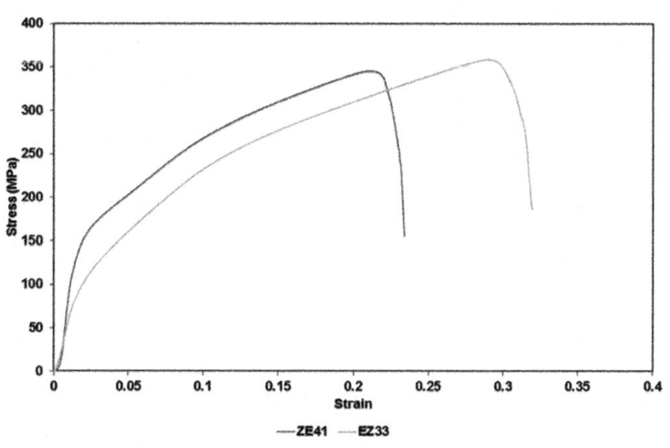


**Fig. 2** Stress-strain curve of ZE41 and EZ33 under compressive loading

## Electrochemical Analysis

Figure 7 below represents the measured open circuit potential (OCP) for pure Mg, as-received ZE41 and EZ33. For pure Mg, the potential increased for the first 30 min which indicated the initial formation of a passive layer, Mg $(OH)_2$. As time progressed, the potential decreased as the protective layer started to disintegrate since it was being attacked by $Cl^-$ ions. ZE41 showed a relatively uniform corrosion rate, which was also seen during the hydrogen evolution and weight loss analysis. It should be noted that during the first 10 min of immersion, ZE41 was at a higher potential than pure Mg, which indicated that a passive layer had already started to form. However, this surface layer was unstable and slowly began to disintegrate (as shown in EIS results in the next section) but it still offered some minimal

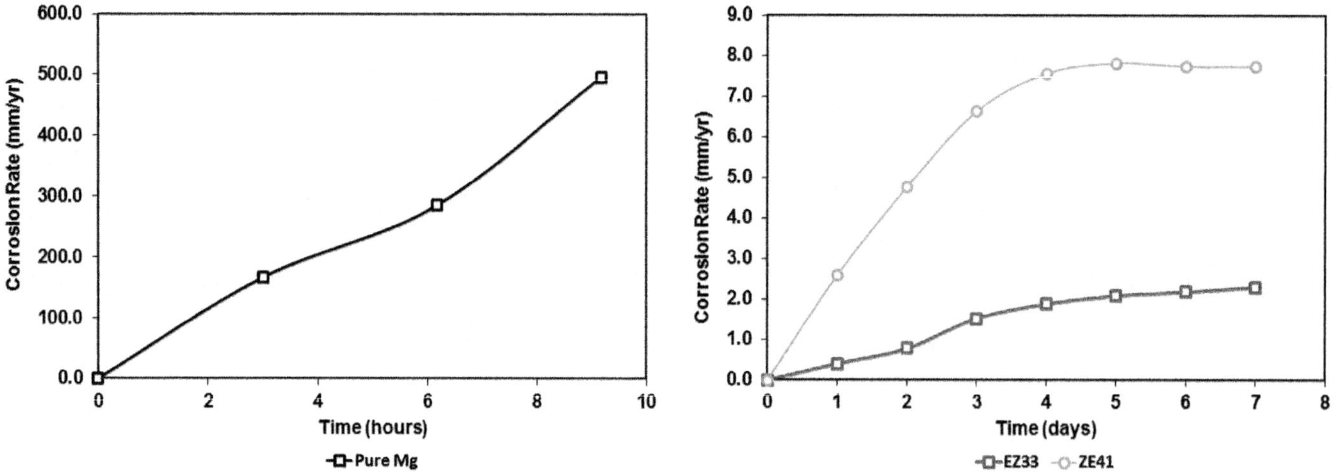


**Fig. 3** Corrosion rates of pure Mg (left), ZE41A and EZ33A (right) determined using the weight loss method

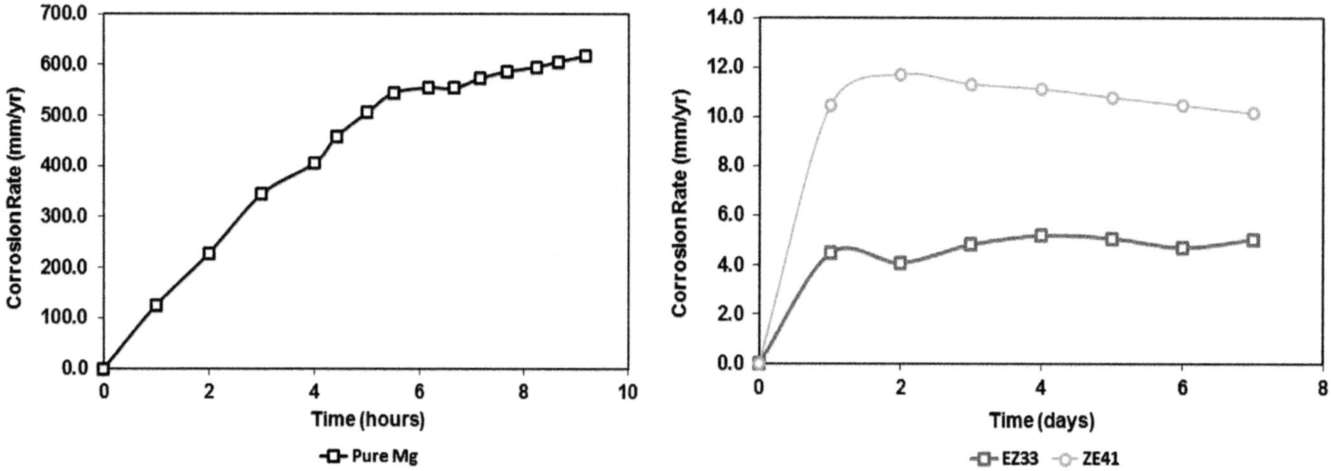


**Fig. 4** Corrosion rates of pure Mg (left), ZE41A and EZ33A (right) determined using the hydrogen evolution method

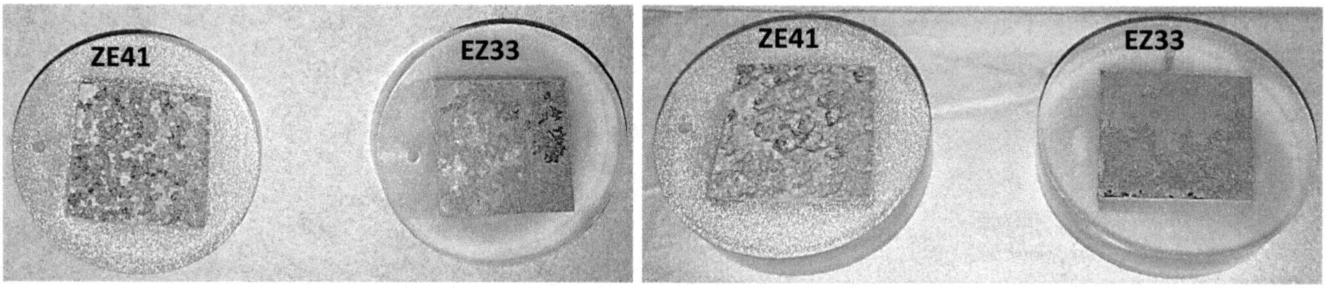


**Fig. 5** Images of ZE41 and EZ33 samples after 24 h (left), and after seven days (right)

protection that resulted in improved corrosion rates when compared to pure Mg.

On the other hand, the OCP of EZ33 was higher than both pure Mg and ZE41 and continued to increase proving that a more stable, protective surface film was being formed and its thickness was increasing. After about 60 min of immersion, the potential still increased but at a much lower rate and was still higher than pure Mg and ZE41. Therefore, although the decrease in rate indicated that the thickness of the film was decreasing, EZ33 still showed a higher OCP and hence lower corrosion rates than the two other materials.

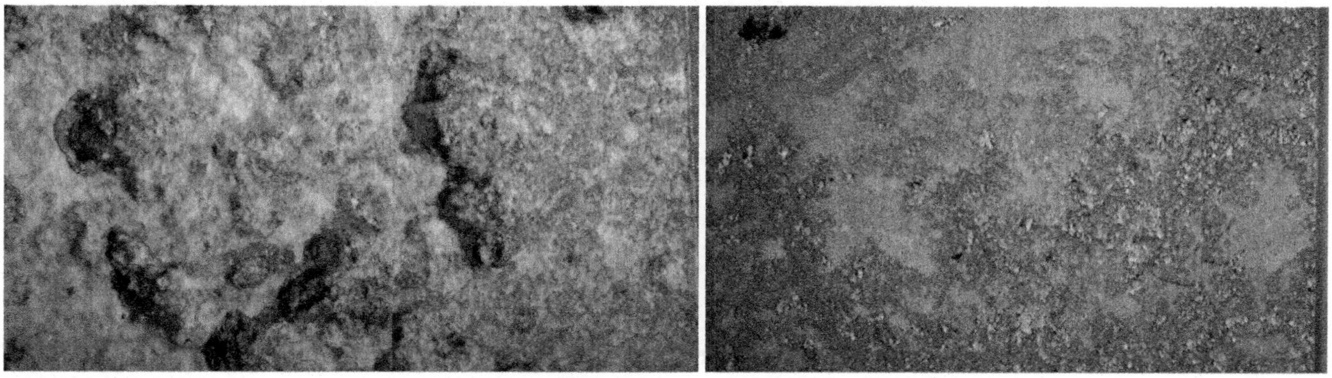

**Fig. 6** Stereoscopic Images of ZE41 (left) and EZ33 (right) samples after seven days of immersion

Overall, the addition and increase in vol.% of rare earths seemed to provoke an increase in OCP. As the concentration of rare earths was increased in EZ33 (EZ33 has about 3 wt% REs compared to 1 wt% in ZE41), the OCP showed more positive values which strongly indicated the formation of a surface film. In addition, secondary phases in Mg alloys tend to be more stable than the α-Mg matrix and in the case of ZE41 and EZ33 have a higher positive potential than the matrix [5]. Therefore, it is expected that EZ33 and ZE41 would have higher OCP potentials than pure Mg, and specifically EZ33 would have the higher OCP due to the higher vol.% of RE component in the solid solution, as well as a higher volume fraction of intermetallic secondary phases, as explained earlier.

Potentiodynamic polarization results for all three materials are shown in Fig. 8 below and are in agreement with the OCP results presented earlier. ZE41 and EZ33, both have increased corrosion potentials compared to pure Mg due to the presence of rare earths, similar to what was shown in Fig. 7. Due to the higher content of rare earths in EZ33, its corrosion potential showed a more positive value than ZE41. In order to further understand the role of rare earths alloying additions, on the corrosion of Mg, Gusieva et al. demonstrated the electrochemical impact of different alloying elements on the polarization curve of Mg [6]. Their study concluded that Cerium and Lanthanum, which are the main rare earths found in ZE41 and EZ33, tend to increase the corrosion potential while enhancing the cathodic kinetics, when added to Mg. This trend is consistent with the present findings from the PD plots shown in in Fig. 8.

The values of the corrosion current were determined using Tafel extrapolation from the PD curves in Fig. 8. Table 3 compares the corrosion rates obtained by weight loss and hydrogen evolution measurements (Figs. 3 and 4) and Tafel extrapolation. The corrosion rates obtained using Tafel extrapolation are quite different than those obtained using hydrogen evolution or weight loss measurements. However, this difference has been experienced and widely

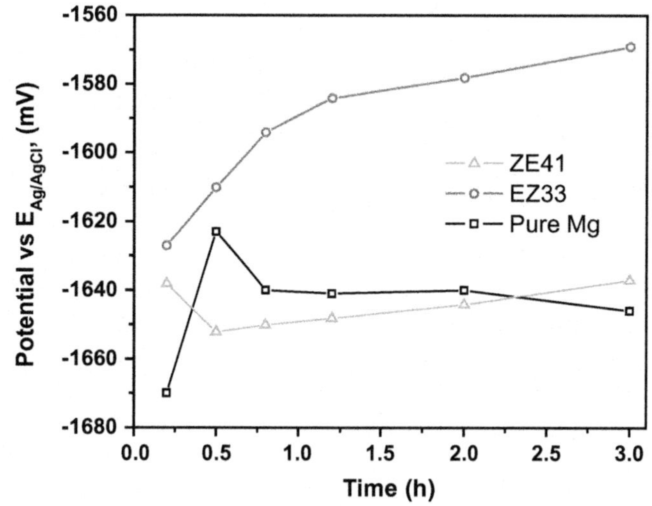


**Fig. 7** Evolution of open circuit potential of pure Mg and alloys during 3 h immersion in 3.5 wt% NaCl

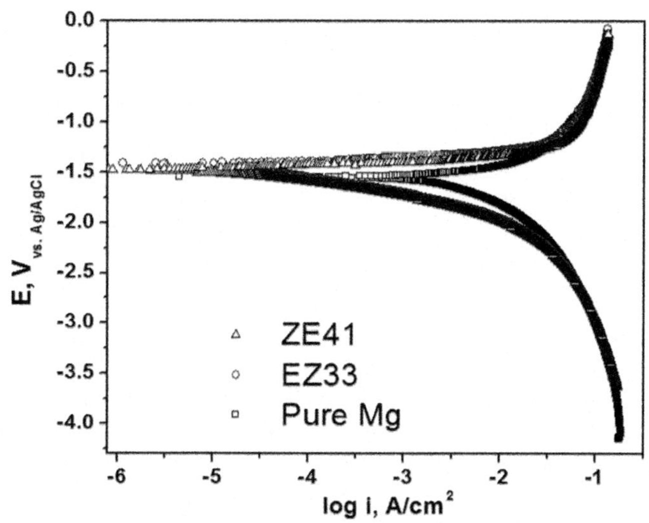


**Fig. 8** Potential versus log (i) profiles of pure Mg and alloys obtained after 3 h of immersion in 3.5 wt% NaCl

**Table 3** Comparison of corrosion rates obtained through different measurement techniques for Pure Mg, ZE41, and EZ33 in 3.5 wt% NaCl solution

| Alloy | Hydrogen evolution rate (ml/cm$^2$) | Corrosion rate (mm/yr) | Average weight loss (g) | Corrosion rate (mm/yr) | Tafel extrapolation (Icorr) (mA/cm$^2$) | Corrosion rate (mm/yr) |
|---|---|---|---|---|---|---|
| Pure Mg | 209.87 | 478.30 | 0.12 | 316.03 | 2.82 | 63.1 |
| ZE41 | 4.77 | 10.87 | 0.061 | 6.41 | 0.128 | 2.9 |
| EZ33 | 2.09 | 4.77 | 0.015 | 1.59 | 0.348 | 7.9 |

discussed in literature especially during the assessment of Mg corrosion [12, 20, 21]. In order for the Tafel extrapolation method to give a correct estimate for the corrosion rate, a single anodic or cathodic reaction has to occur [12]. During the corrosion of Mg and its alloys, the surface film begins to breakdown when the applied potential increases. As the potential reaches the pitting potential, the hydrogen evolution reaction is also increased resulting in the negative difference effect. This does not allow for the accurate measurement of the corrosion currents from the Tafel extrapolation, hence impacting the estimated corrosion rates. Nonetheless, the corrosion rates obtained using Tafel extrapolation were in agreement with those found in the literature [12, 19].

In order to provide further insight on the nature and properties of the surface films being formed and the corrosion characteristics of the three materials, results from Electrochemical Impedance Spectroscopy (EIS) were analyzed using Nyquist plots and relevant model circuits. Figure 9 shows the evolution of the Nyquist spectra of ZE41 and EZ33 as time progresses. Generally, EZ33 revealed higher impedance levels than ZE41 and therefore higher corrosion resistance. This agrees with previously obtained results from weight loss, hydrogen evolution and other electrochemical tests. From the OCP and immersion results, it was concluded that ZE41 exhibits a uniform corrosion with the possibility of a surface film forming during the first 10 min but starts to slowly degenerate as time increases. Impedance levels for ZE41 were at their highest values at 10 min, even higher than EZ33 (Fig. 10) which correlates with the existence of a passive film. However, soon after, it started to decrease continuously until the end of 180 min indicating that the film is less protective and slowly deteriorating. On the other hand, the impedance of EZ33 continued to increase until 50 min, after which it started to decrease until it reached its lowest value at 180 min. The increase in impedance values until 50 min showed the thickening of the protective film being formed at the surface of the alloy which was also noticed in OCP results. The decrease in the impedance values indicated that the film being formed is saturated with aggressive $Cl^-$ ions and becomes partially protective since the impedance values are still higher than ZE41.

Furthermore, the impedance spectra for the three materials were fitted with equivalent circuits. Figure 10 shows the impedance results with their respective fits for the pure Mg, ZE41 and EZ33 after 10 min and after 180 min. The

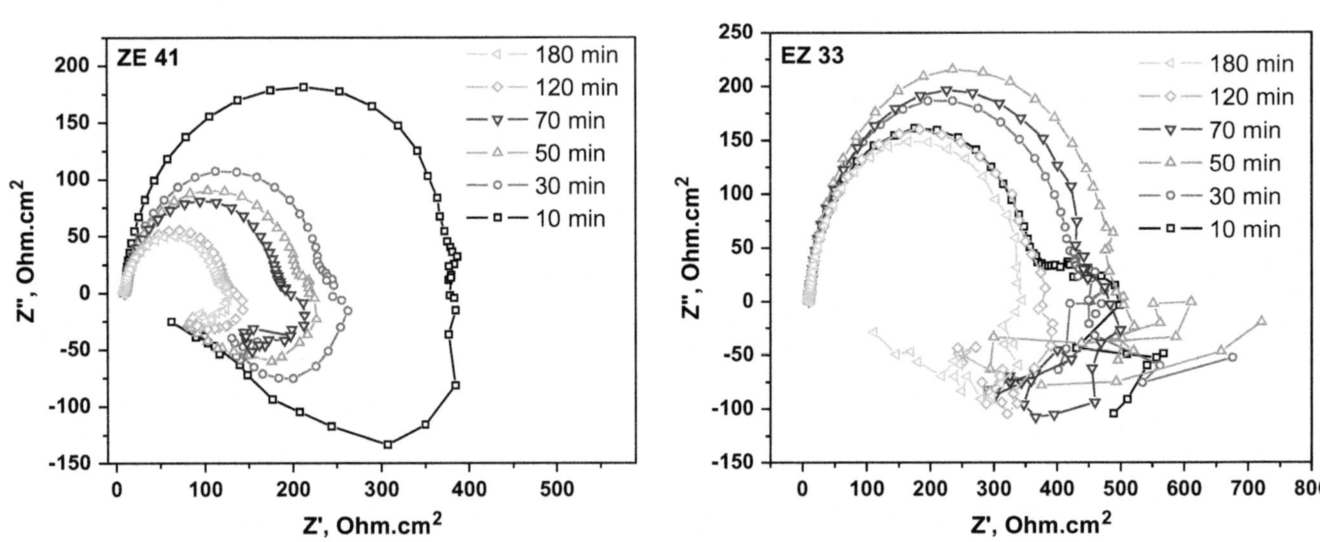


**Fig. 9** Nyquist plots of ZE41 and EZ33 immersed in 3.5 wt% NaCl after 10, 30, 50, 70, 120 and 180 min

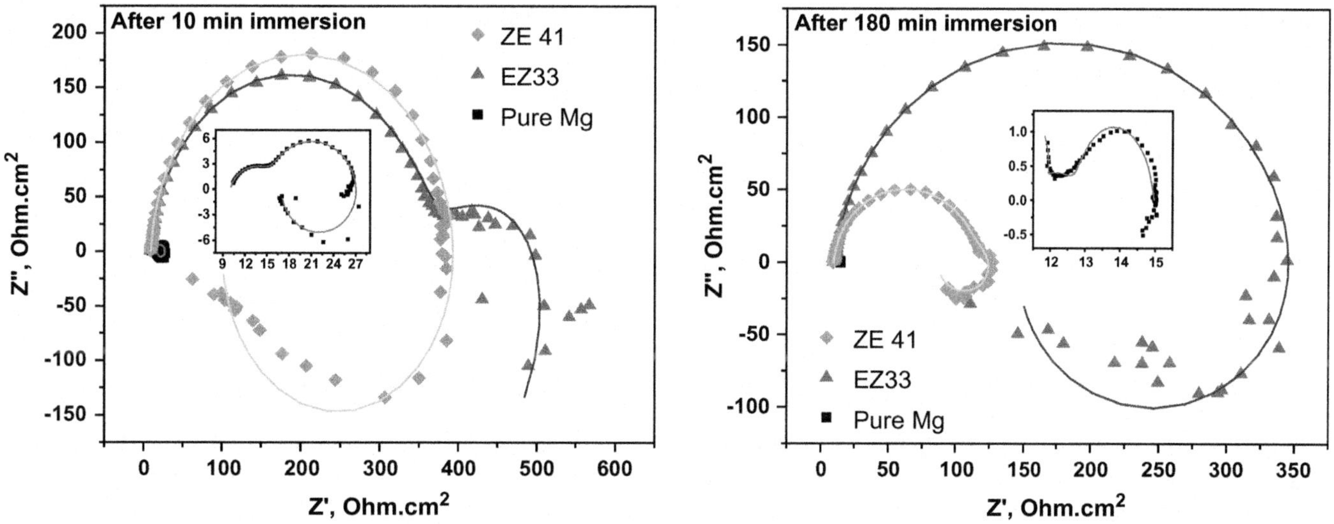


**Fig. 10** Nyquist plots of pure Mg, ZE41 and EZ33 immersed in 3.5 wt% NaCl after 10 min (left) and 180 min (right) along with the respective fits from equivalent circuits

equivalent circuits developed to produce the fitted impedances are shown in Fig. 11 with the numerical results of the fitting procedure presented in Table 4.

The equivalent circuit used for pure Mg (Fig. 11a) was proposed in [21]. CPE1 (constant phase element—Y1 and a1) and C2 represent the capacitance of the oxide/hydroxide layer and the double layer capacitance respectively. R1 and RL represent resistances that describe the environmental changes occurring around the alloy, such the presence of bubbles, L represents the inductance

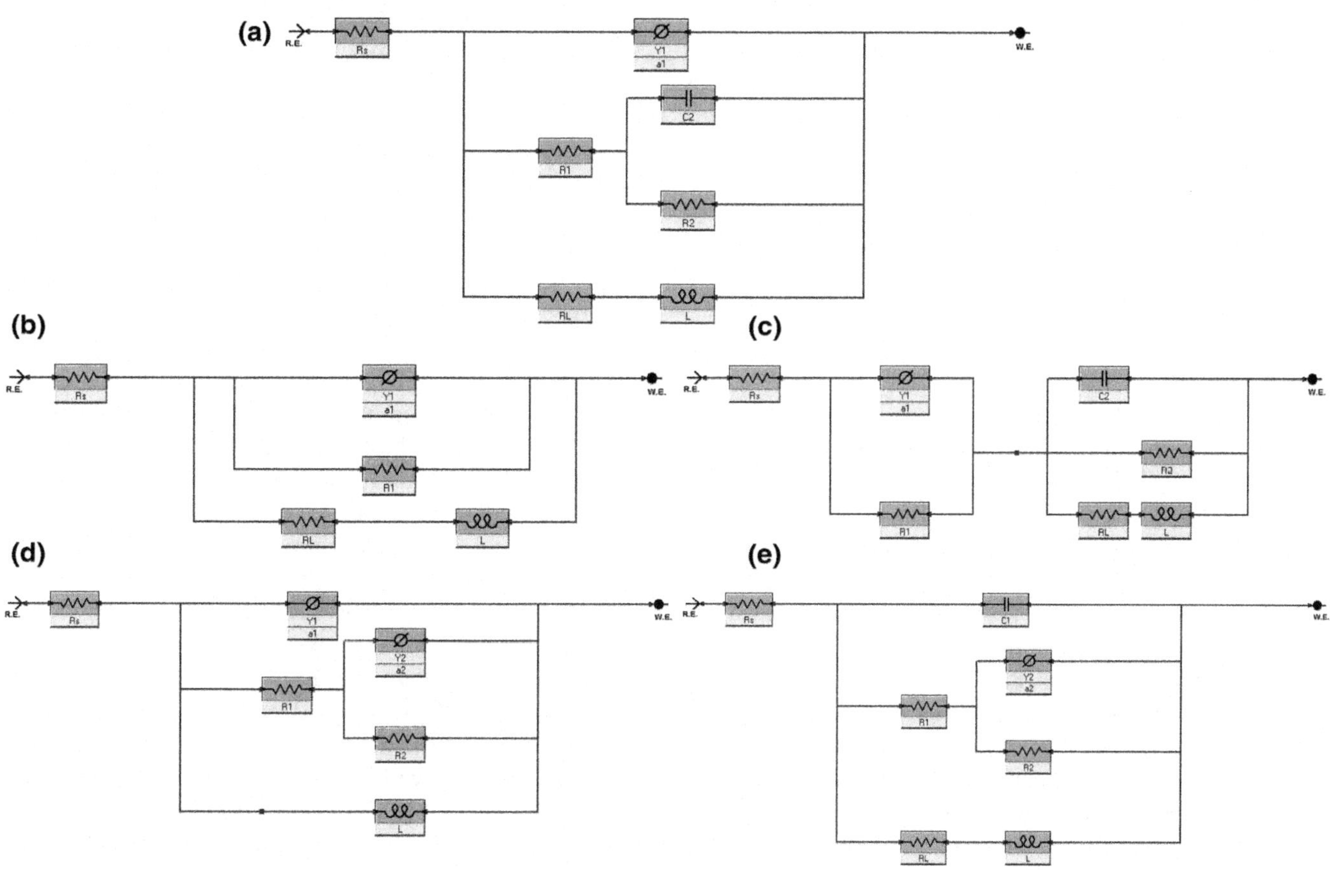


**Fig. 11** Equivalent circuits used for the fitting of EIS data for **a** pure Mg after 10 and 180 min, **b** ZE41 after 10 min, **c** ZE41 after 180 min, **d** EZ33 after 10 min and **e** EZ33 after 180 min of immersion in 3.5 wt% NaCl solution

**Table 4** Fitting results for Electrochemical Impedance Measurements made on Pure Mg, ZE41, and EZ33 in 3.5 wt% NaCl solution at open circuit after 10 and 180 mins, as per the equivalent circuits shown in Fig. 9

| | Pure Mg | | EZ33 | | ZE41 | |
|---|---|---|---|---|---|---|
| Immersion time | 10 min | 180 min | 10 min | 180 min | 10 min | 180 min |
| $R_1$ (Ohm cm$^2$) | 7 | 10.3 | 351 | 200 | 385 | 92.51 |
| $R_2$ (Ohm cm$^2$) | 10.1 | 4.5 | 195 | 125 | – | 31.05 |
| $C_1$ (F/cm$^2$) | $6.39 \times 10^{-05}$ | $4.22 \times 10^{-03}$ | $2.24 \times 10^{-05}$ | $1.42 \times 10^{-04}$ | $1.59 \times 10^{-05}$ | $8.22 \times 10^{-02}$ |
| $C_2$ (F/cm$^2$) | $1.51 \times 10^{-04}$ | $4.46 \times 10^{-04}$ | $2.27 \times 10^{-03}$ | $2.72 \times 10^{-03}$ | – | $9.67 \times 10^{-05}$ |
| $R_L$ (Ohm cm$^2$) | 11.9 | 24 | – | 227 | 120 | 100.3 |
| L (H cm$^2$) | 53 | 4.6 | $2.39 \times 10^{04}$ | $1.39 \times 10^{03}$ | 578 | 745 |

associated with the variation in the anodic regions, and finally, R2 represents the charge transfer resistance related to the hydrogen evolution reaction [21]. The same circuit was found to fit the impedance results for EZ33 after 10 and 180 min of immersion with the exception of RL (Figs. 11d, e). More insight is needed as to why RL exists after 180 min and not after 10 min. For ZE41, after 10 min of immersion, one capacitive loop and one inductive loop was found to be sufficient to provide a good fit with the measured results (Fig. 11b) where R1 represents the charge transfer resistance related to the hydrogen evolution reaction. After 180 min, ZE41 impedance values were best fitted with Voight circuit (Fig. 11c). This specific circuit model usually implies that the passive film is composed of two distinct layers. Currently, there is not enough data to proof this and so further analysis is required. In general, the fitted impedance values and measured values were closely similar as shown in the Nyquist plots in Fig. 10. It should be noted that the capacitive elements were sometimes represented using constant phase elements and they showed a good fit with experimental data. The error percentage between the simulated and measured data was found to be approximately 10%.

## Conclusions

The following are the conclusions of this work:

1. Compression testing revealed that the increase in volume fraction of REs helps in improving the mechanical properties. EZ33 showed higher ultimate compressive strength as compared to ZE41, but the YS of ZE41 was higher and so was the Vickers microhardness.
2. ZE41 and EZ33 both exhibited slow corrosion rates compared to Pure Mg during immersion testing. The presence of $Cl^-$ ions were believed to attack the oxide film being formed on pure Mg, resulting in pitting corrosion and the accelerated corrosion response.
3. ZE41 had a corrosion rate four times faster than EZ33 and displayed more pitting corrosion during immersion testing. Meanwhile, the active corrosion area in EZ33 was relatively more uniform. Stereoscopic images revealed that the corrosion mechanism of ZE41 resulted in large crates due to corrosion occurring around intermetallic Zr-rich particles. This behavior was not seen in EZ33 which indicated that a surface film was formed resisting the initiation of corrosion due to the higher content of rare earths.
4. The presence of rare earths provoked an increase in OCP to more positive values indicating the formation of a surface film. Therefore, ZE41 and EZ33 had higher OCP potentials than pure Mg with EZ33 having the most positive OCP due to the larger volume fraction of intermetallic secondary phases. A similar trend of potential values were also observed in PD results.
5. Tafel extrapolation was used to calculate the corrosion rates from PD plots and were compared with results from immersion testing. The distinct difference found in the results from Tafel extrapolation was related to the negative difference effect that caused the Tafel method to produce inaccurate estimates, as shown in the literature.
6. Nyquist plots were able to show the initiation and progression of the surface films being formed on all three alloys. Overall, EZ33 showed higher impedance levels than ZE41 indicating higher corrosion resistance and a stronger surface film. Initially, a passive film was formed on ZE41 since the impedance levels were at their highest during the first 10 min of immersion. However, the film was not able to sustain its integrity, as the impedance values continued to decrease. On the other hand, the thickness of the surface film on EZ33 was increasing until 50 min of immersion, but the values also started to decrease indicating that the film being formed was becoming partially protective.
7. The fitted parameters derived from the equivalent circuits for the three alloys provided preliminary understanding of the structure of the developed surface films and presented a good fit with experimental data with an error of approximately 10%. However, further investigation is needed on the reasons why certain equivalent circuits produced good fits with the experimental data.

**Acknowledgements** This research was performed with support from the Qatar Foundation under the National Priorities Research Program grant# NPRP 8-856 - 2 - 364. The authors acknowledge this financial support with gratitude.

## References


1. Song G, Atrens A. (2003) Understanding Magnesium Corrosion—A Framework for Improved Alloy Performance. Adv Eng Mater 5:837–858. https://doi.org/10.1002/adem.200310405
2. Song GL, Atrens A (1999) Corrosion Mechanisms of Magnesium Alloys. Adv Eng Mater 1:11–33. https://doi.org/10.1002/(SICI)1527-2648(199909)1:1<11::AID-ADEM11>3.0.CO;2-N
3. Curioni M (2014) The behaviour of magnesium during free corrosion and potentiodynamic polarization investigated by real-time hydrogen measurement and optical imaging. Electrochim Acta 120:284–292. https://doi.org/10.1016/j.electacta.2013.12.109
4. Yang Y, Scenini F, Curioni M (2016) A study on magnesium corrosion by real-time imaging and electrochemical methods: Relationship between local processes and hydrogen evolution. Electrochim Acta 198:174–184. https://doi.org/10.1016/j.electacta.2016.03.043
5. Ding Y, Wen C, Hodgson P, Li Y (2014) Effects of alloying elements on the corrosion behavior and biocompatibility of biodegradable magnesium alloys: a review. J Mater Chem B 2:1912–1933. https://doi.org/10.1039/C3TB21746A
6. Gusieva K, Davies CHJ, Scully JR, Birbilis N (2015) Corrosion of magnesium alloys: the role of alloying. Int Mater Rev 60:169–194. https://doi.org/10.1179/1743280414Y.0000000046
7. Tekumalla S, Seetharaman S, Almajid A, et al (2015) Mechanical Properties of Magnesium-Rare Earth Alloy Systems: A Review. Metals (Basel) 5:1–39. https://doi.org/10.3390/met5010001
8. Siebert-Timmer A, Fletcher M, Bichler L, Sediako D (2013) Creep performance of wrought AX30 and EZ33 magnesium alloys. Can Metall Q 52:430–438. https://doi.org/10.1179/1879139513Y.0000000069
9. Chen Y, Xu Z, Smith C, Sankar J (2014) Recent advances on the development of magnesium alloys for biodegradable implants. Acta Biomater 10:4561–4573. https://doi.org/10.1016/j.actbio.2014.07.005
10. (2012) Magnesium Alloy Castings 4.2Zn - 1.2 Rare Earths - 0.7Zr (ZE41A-T5) Precipitation Heat Treated. https://doi.org/10.4271/AMS4439G
11. (2012) Magnesium Alloy Sand Castings 3.2 Rare Earths - 2.5Zn - 0.70Zr (EZ33A-T5) Precipitation Heat Treated. https://doi.org/10.4271/AMS4442G
12. Zhao MC, Liu M, Song GL, Atrens A (2008) Influence of microstructure on corrosion of As-cast ZE41. Adv Eng Mater 10:104–111. https://doi.org/10.1002/adem.200700246
13. Augusto R, Pinto DA (2008) Electrochemical Behaviour of Magnesium Alloys Study on the influence of rare earths as alloying elements.
14. Shi Z, Atrens A (2011) An innovative specimen configuration for the study of Mg corrosion. Corros Sci 53:226–246. https://doi.org/10.1016/j.corsci.2010.09.016
15. Standard Guide for Determining Synergism Between Wear and Corrosion 1
16. Kocurek R, Adamiec J (2013) Evaluation of structure and properties of welded joint of magnesium alloy EZ33A-T5. 13:65–71
17. Neil WC, Forsyth M, Howlett PC, et al (2009) Corrosion of magnesium alloy ZE41—The role of microstructural features. Corros Sci 51:387–394. https://doi.org/10.1016/j.corsci.2008.11.005
18. Ding R, Chung C, Chiu Y, Lyon P (2010) Effect of ECAP on microstructure and mechanical properties of ZE41 magnesium alloy. Mater Sci Eng A 527:3777–3784. https://doi.org/10.1016/j.msea.2010.02.030
19. Neil WC, Forsyth M, Howlett PC, et al (2011) Corrosion of heat treated magnesium alloy ZE41. Corros Sci 53:3299–3308. https://doi.org/10.1016/j.corsci.2011.06.005
20. Bland LG, King AD, Birbilis N, Scully JR (2015) Assessing the corrosion of commercially pure magnesium and commercial AZ31B by electrochemical impedance, mass-loss, hydrogen collection, and inductively coupled plasma optical emission spectrometry solution analysis. Corrosion 71:128–145. https://doi.org/10.5006/1419
21. Curioni M, Scenini F, Monetta T, Bellucci F (2015) Correlation between electrochemical impedance measurements and corrosion rate of magnesium investigated by real-time hydrogen measurement and optical imaging. Electrochim Acta 166:372–384. https://doi.org/10.1016/j.electacta.2015.03.050

# Surface and Interfacial Energies of $Mg_{17}Al_{12}$–Mg System

Fangxi Wang and Bin Li

**Abstract**

For upscale simulation and modeling of magnesium alloys, data of surface and interfacial energies are critical. In this work, we calculated the surface energies of $Mg_{17}Al_{12}$ β-phase with different surface configurations by using molecular dynamic simulations. Surface terminations were carefully selected to calculate the energy of β-phase. The lowest energy surface for each crystallographic plane was determined by varying the surface termination. The results show that surfaces occupied by higher fraction of magnesium atoms generate lower surface energies. The interfacial energy for $Mg_{17}Al_{12}$ β-phase and Mg matrix was calculated as well based on the Burger's orientation relationship. We found that the lowest energy surface of $Mg_{17}Al_{12}$ does not generate the lowest interfacial energy. The interfacial energy for $Mg_{17}Al_{12}$ β-phase and a $\{10\bar{1}2\}$ twin was also calculated. The interfacial energy increases by $\sim 250$ mJ/m$^2$ due to the change in orientation relationship between $Mg_{17}Al_{12}$ and the matrix after twinning.

**Keywords**

$Mg_{17}Al_{12}$ β-phase • Surface termination • Surface energies • Interfacial energy • Atomistic calculations

## Introduction

Magnesium (Mg) alloys have great potential for applications in automobile, aerospace, and other industries [1] due to their relative low densities and high specific strength [2, 3]. Magnesium–aluminum alloys are the most common commercial Mg alloys and have been widely used as model alloys for research [4]. In these alloys, β-phase ($Mg_{17}Al_{12}$) is the primary equilibrium precipitates in the AZ series Mg alloys. From the literature [5], there are two types of morphology for $Mg_{17}Al_{12}$: continuous precipitation and discontinuous precipitation. The discontinuous precipitation is usually occurred at high grain boundary, and the precipitate growth cellularly to give alternating layers of β-phase and matrix. The continuous precipitation forms large plate β-$Mg_{17}Al_{12}$ at the rest area of the matrix where is no discontinuous precipitation [6], and the β-$Mg_{17}Al_{12}$ prefers the Burgers orientation relationship (OR) with magnesium matrix, i.e., $(0001)_{Mg}||(011)_P$, $[2\bar{1}\bar{1}0]_{Mg}||[1\bar{1}1]_P$ [5–7].

It is well known that $Mg_{17}Al_{12}$ precipitates influence the mechanical behavior of Mg alloys [8]. Robson et al. discussed the effect of precipitates on strengthening considering the precipitate hardening against slip and twinning [9]. They calculated precipitate hardening effect based on Orowan's mechanism. They found that the basal plate precipitates were inefficient to block basal slip, but it hindered twinning growth, because precipitates provided the maximum back-stress and prevented plastic relaxation in the twin. Liao et al. studied the interaction between both prismatic slip and basal slip with a $Mg_{17}Al_{12}$ precipitate in magnesium using molecular dynamics simulations. Their results indicated that both a basal dislocation is able to pass the precipitate without strong interaction, whereas a prismatic dislocation may cut through the precipitate [10, 11]. They also showed that the interface between the precipitate and the matrix was incoherent and the interfacial strength was weak to hinder dislocation slip. Li and Zhang showed that the twinning shear for $\{10\bar{1}2\}\langle 10\bar{1}\bar{1}\rangle$ mode should be zero [12]. Consequently, twin-precipitate interaction should be minimal, which explained why precipitate hardening in magnesium alloy is not as effective as alloys with cubic structures [8]. It was also reported that $Mg_{17}Al_{12}$ phase influences the corrosion behavior [13, 14] because of the free corrosion potential of the β-phase is relative more positive in the electrolyte. Mg/$Mg_{17}Al_{12}$ interfaces can act as a source of crack initiation [15].

F. Wang (✉) · B. Li
Department of Chemical and Materials Engineering, University of Nevada, Reno, NV 89557, USA
e-mail: fangxi.wang@nevada.unr.edu


D. Orlov et al. (eds.), *Magnesium Technology 2018*, The Minerals, Metals & Materials Series, https://doi.org/10.1007/978-3-319-72332-7_10

Despite the extensive research on precipitation in Mg alloys, key information of surface and interfacial energies is missing. For upscale simulation and modeling, such energies between $Mg_{17}Al_{12}$ and Mg matrix are critically important. Han et al. studied morphological evolution of $Mg_{17}Al_{12}$ phase using phase field simulation. Anisotropy of interfacial energy and interface mobility, and elastic strain energy was considered [16–18]. How the energy data was calculated was not shown [19]. Hutchinson calculated the interfacial energies in [5], but only the average interfacial energies for $Mg_{17}Al_{12}$/matrix was considered, but the effect of surface termination of $Mg_{17}Al_{12}$ was not considered. Xiao et al. [20] calculated the surface energies of the precipitate for several planes using atomistic simulation but surface termination was not considered as well.

In this paper, we calculated the surface energies of $Mg_{17}Al_{12}$ β-phase with different surface configurations by using atomistic simulation. Surface terminations were carefully selected to calculate the energy of $Mg_{17}Al_{12}$. The interfacial energy between $Mg_{17}Al_{12}$ and Mg matrix was calculated as well based on the Burger's orientation relationship. Because twinning changes the orientation relationship, the interfacial energy between $Mg_{17}Al_{12}$ and matrix after $\{10\bar{1}2\}$ twinning was also calculated.

## Simulation Method

The EAM [21] interatomic potential for Mg and Al developed by Liu et al. [22] was used in our molecular dynamics simulations. This potential has been used in extensive atomistic simulations of physical properties of Mg and Mg alloys [23–26]. Simulation package XMD was used to perform the calculations, and visualization program Ovito [27] was used for graphic presentations.

The Dimensions of the simulation system are 42(X) × 42(Y) × 11(Z) $nm^3$ (Fig. 1a). The orientations are $X-[100], Y-[010], Z-[001]$. The time step size is 3 fs. The system was relaxed for 10,000 time steps (30 ps) to the local minimum potential energy before the surface energy calculation. The temperature for the system was constant at 10 K. Free surfaces were applied to the system. The total number of atoms of the system was about 928,000. To calculate the surface energy of the bottom xy-surface (bottom (001) plane), a box A with the dimension 20(X) × 20(Y) × 3(Z) $nm^3$ (35,811 Mg atoms and 25,486 Al atoms) was selected in the center of the above relaxed simulation cell, and the average potential energy per atom for each typeof atoms was calculated. Another box B with dimension 20(X) × 20(Y) × 5(Z) $nm^3$ (59,019 Mg atoms and 41,880 Al atoms) which contains part of the bottom -surface was selected, and the average potential energy per atom was calculated. The surface energy $\gamma$ was given by

$$\gamma = \frac{(E'_{Mg} - E_{Mg}) \times N_{Mg} + (E'_{Al} - E_{Al}) \times N_{Al}}{A}$$

Where $E_{Mg}$ and $E_{Al}$ are the average potential energies per atom for Mg and Al from box A; $E'_{Mg}$ and $E'_{Al}$ are the average potential energies per atom for Mg and Al from box B; $N_{Mg}$ and $N_{Al}$ are the number of Mg and Al atoms in box B; $A$ is the area of the precipitate bottom xy-surface in box B. Figure 1b shows the schematic view of the above calculation methods. The surface energy of each termination was calculated. Then, the system was reoriented to the directions $X-[\bar{2}\bar{1}1], Y-[1\bar{1}1], Z-[011]$, and the surface energies were calculated similarly. The system shows periodicity in all three axes. To find the lowest surface energy termination, we removed the outermost atoms of each surface layer by layer until periodicity is reached, and then calculated the surface energies of different terminations using the above methods. The lowest energy surface termination, as well as the highest energy surface termination can then be determined. Figure 2a shows the (010) surface configuration of the precipitate and Fig. 2b shows the reoriented $(\bar{2}\bar{1}1)$ surface configuration.

In the calculation of the interfacial energy of the bottom xy-surface (bottom (011) plane) of the reoriented precipitate and the Mg matrix, the dimensions of the system are 29(X) × 29(Y) × 56(Z) $nm^3$. The system was constructed in the Burger's OR (Fig. 3), and six surfaces of the precipitate were the lowest surface energy termination. The system was relaxed for 20,000 time-steps. Free surfaces were applied to the system. The total number of atoms was about 2,244,481. After relaxation, box A and box B with the dimension 10(X) × 10(Y) × 10(Z) $nm^3$ were selected inside the precipitate and matrix, and the average potential energy per atom for each type of atoms was calculated. Another box C with dimension 20(X) × 20(Y) × 7(Z) which contains the matrix/precipitate interface was selected, and the average potential energy per atom for each type of atoms was calculated as well. The interfacial energy was given by

$$\text{Interfacial enegy} = \frac{(E'_{Mg(m)} - E_{Mg(m)}) \times N_{Mg(im)} + (E'_{Mg(p)} - E_{Mg(p)}) \times N_{Mg(ip)} + (E'_{Al(p)} - E_{Al(p)}) \times N_{Al(ip)}}{A}$$

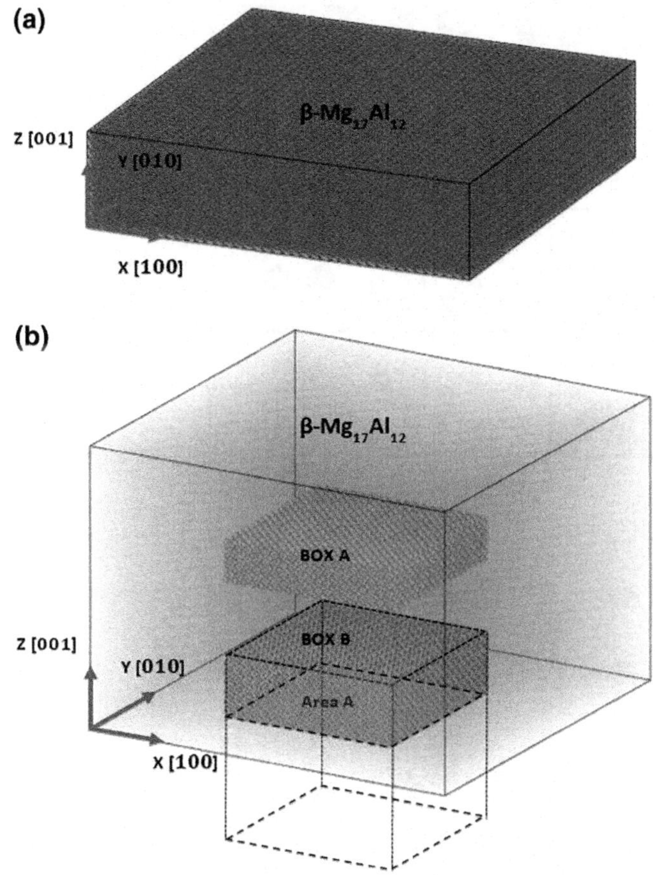


**Fig. 1** a The 3-D view of the simulation system for $Mg_{17}Al_{12}$. **b** The schematic view of the calculation method for surface energies

where $E_{Mg(p)}$ and $E_{Al(p)}$ are the average potential energies per atom for Mg and Al in box A; $E_{Mg(m)}$ is the average potential energy per atom for Mg in box B; $E'_{Mg(p)}$ and $E'_{Al(p)}$ are the average potential energies per atom for Mg and Al in the precipitate in box C; $E'_{Mg(m)}$ is the average potential energy per atom for Mg in the matrix in box C; $N_{Mg(ip)}$ and $N_{Al(ip)}$ are the number of Mg and Al atoms in the precipitate in box C; $N_{Mg(im)}$ is the number of Mg from the matrix in box C;; $A$ is the area of the matrix/precipitate interface in box C (Fig. 4). Then, the matrix was rotated along the $[2\bar{1}\bar{1}0]$ for 90°, and the new interfacial free energies were calculated as well.

## Results and Discussion

The surface termination for the original $Mg_{17}Al_{12}$ β-phase simulation system shows the same periodicity along the three axes. Each repetition contains 12 atomic layers (Fig. 2a) (Mg atoms in red and Al atoms in blue). Therefore, we calculated the surface energies for all the 12 layers as bottom surface. We found that the lowest surface energy termination appears when the 8th atomic layer becomes the surface for this structure (Fig. 5a), and the surface energy equals 742.7 mJ/m$^2$. The highest surface energy termination gives the surface energy about 980 mJ/m$^2$ on the 10th atomic layer (Fig. 5b).

For the reoriented β-phase, it shows periodicity along each axis (Fig. 2b). Along the positive $[1\bar{1}1]$ direction, the structure repeats every 9 atomic layers. However, this case does not show central symmetry as the structure before reorientation. Therefore, the layer structure underneath a specific top or bottom atomic surface will be different, and so will be the surface energies. Figure 6 shows such differences in detail. Considering the 6th atomic layer becomes the bottom $(1\bar{1}1)$ surface (Fig. 6a), the underneath atom layers will be layer 7–9 in one period. This termination shows the lowest surface energy of about 775.6 mJ/m$^2$. For the top $(1\bar{1}1)$ surface, the lowest energy termination is the second layer (Fig. 6c). The surface energies results are summaries in Table 1.

For the (011) surface, there is only one result because the atom layers show axial symmetry. The (011) surface energy shows a value close to Xiao's result [20] which is 716 mJ/m$^2$. Other surface energies reported by Xiao [20] fall within the range of our calculations. Because no details how surface termination was chosen was provided, discrepancies may incur. Ning et al. calculated the surface energy for (001) plane with atomic layer 1 as the surface by using

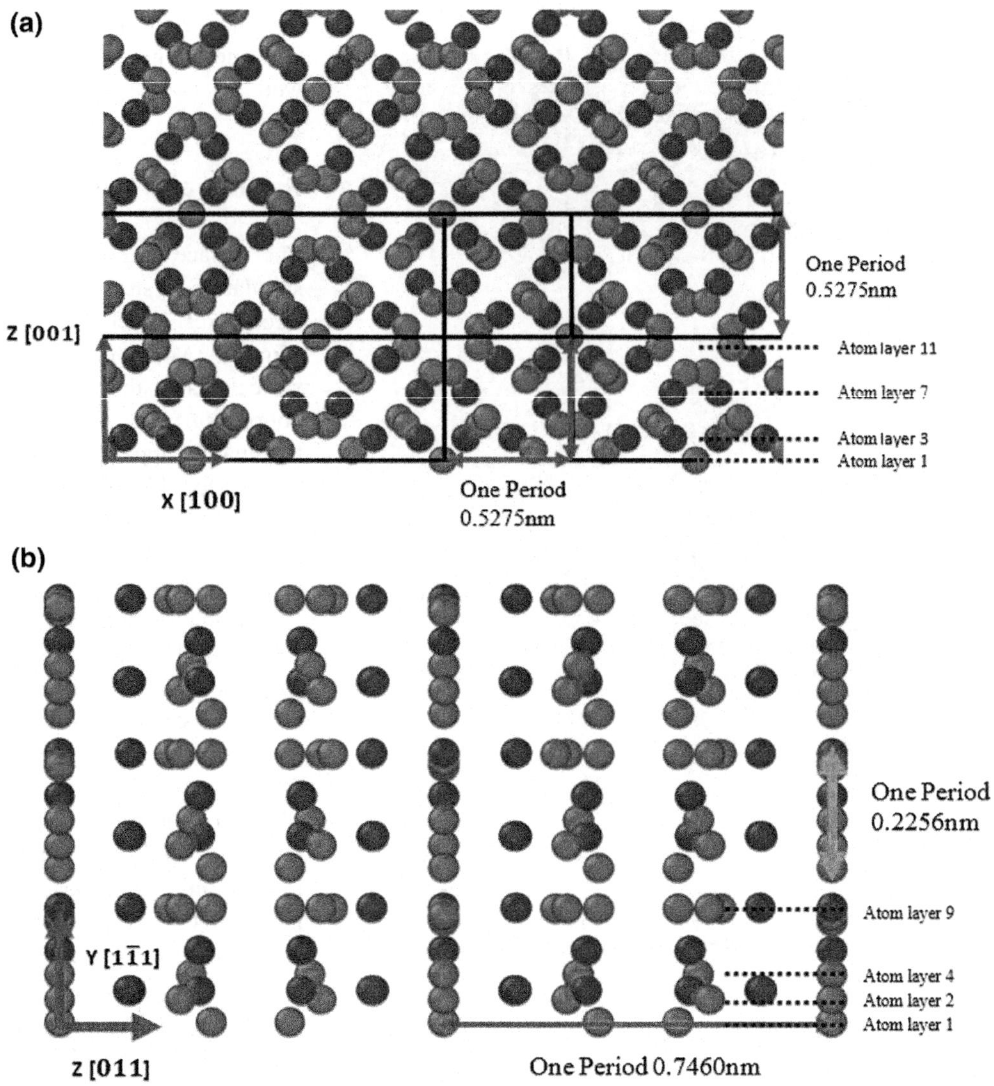


**Fig. 2** **a** The (010) surface configuration of the simulation system for $Mg_{17}Al_{12}$. **b** The reoriented ($\bar{2}\bar{1}1$) surface configuration

density functional theory calculation [28]. The value is 799 mJ/m$^2$, which is slightly higher than the result in our calculation (783 mJ/m$^2$). For BCC and FCC structures, the closed packed plane surface will usually give the lowest surface energies [29, 30]. $Mg_{17}Al_{12}$ has a BCC structure, the (011) surface indeed provide the lowest surface energy of 715.7 mJ/m$^2$. In our result, Fig. 6a shows the atom layers 6–9 which has a high atomic density gives the lowest surface energies for bottom (1$\bar{1}$1) plane, but Fig. 6c indicates the lowest energy surface termination is not provide by the same layers for top (1$\bar{1}$1) plane. Therefore, the atom layers with higher atomic density do not necessarily give the lowest

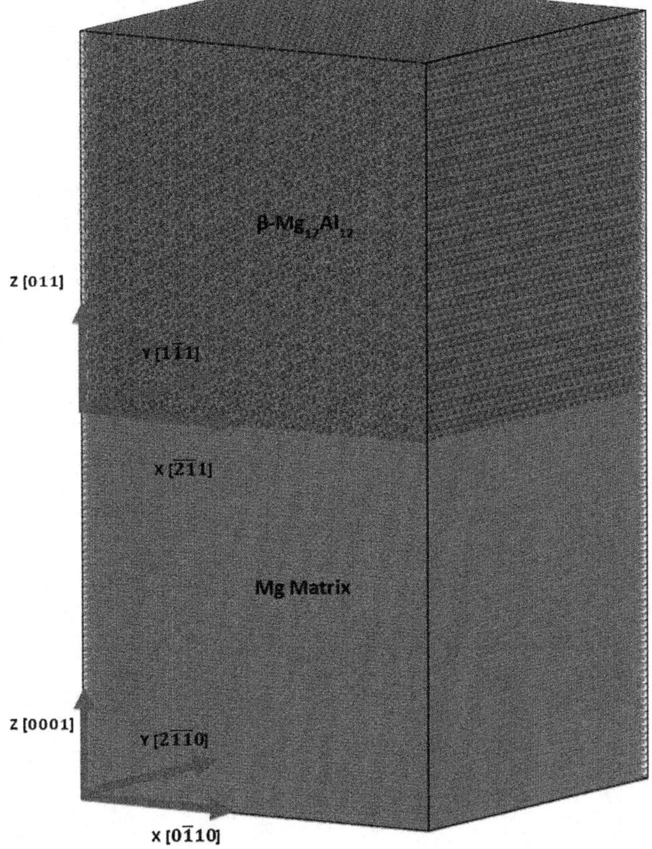


**Fig. 3** Schematic 3-D view of the simulation system for interfacial energies calculation with the Burger's OR

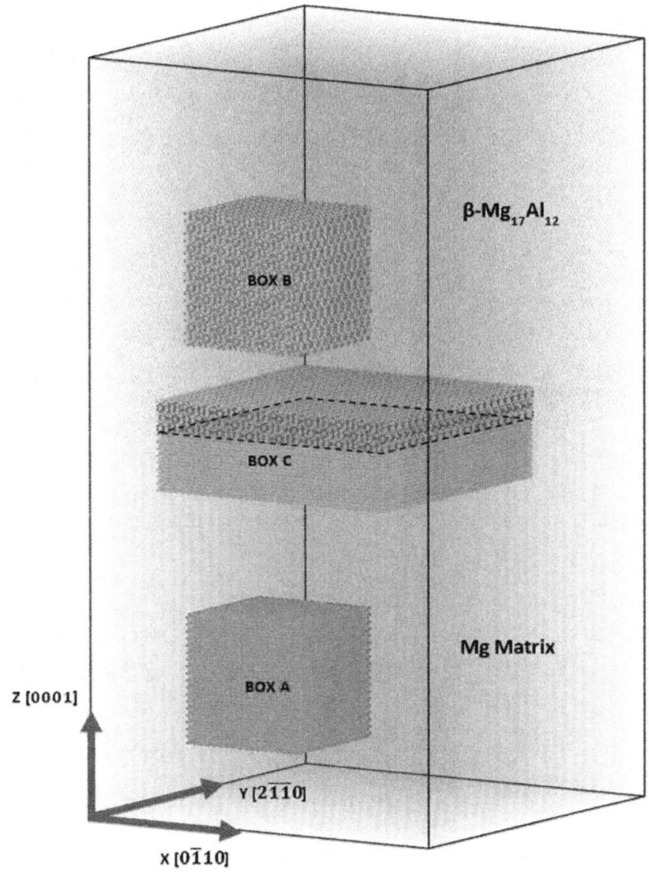


**Fig. 4** The schematic view of the calculation methods for interfacial energies calculation

surface energies. Additionally, Figs. 5b and 6b, d show the highest surface energy terminations and these surfaces have more Al atoms. In contrast, surfaces occupied by a higher fraction of magnesium atoms generate lower surface energies.

Lowest surface energies termination is then used to calculate the interfacial energies based on the Burger's OR. After $\{10\bar{1}2\}$ twinning, the OR changes because the Mg matrix was reoriented by $\sim 90°$ around the zone axis $\langle 2\bar{1}\bar{1}0\rangle$. The interfacial energies between $Mg_{17}Al_{12}$ and a twinned matrix was calculated as well [31]. The results are shown in Table 2.

In this case, the lowest energies surface termination generates a relative low interfacial energy about 250 mJ/m$^2$, which falls in the range of 140–390 mJ/m$^2$ in [17]. In [5], the interfacial energy was 430 mJ/m$^2$, and the author showed the effective interfacial energy of nucleation of $Mg_{17}Al_{12}$ was 114 mJ/m$^2$ by using assessed thermodynamics. Li et al. assumed the interfacial energy of semi-coherent interfaces parallel to the basal planes was 60

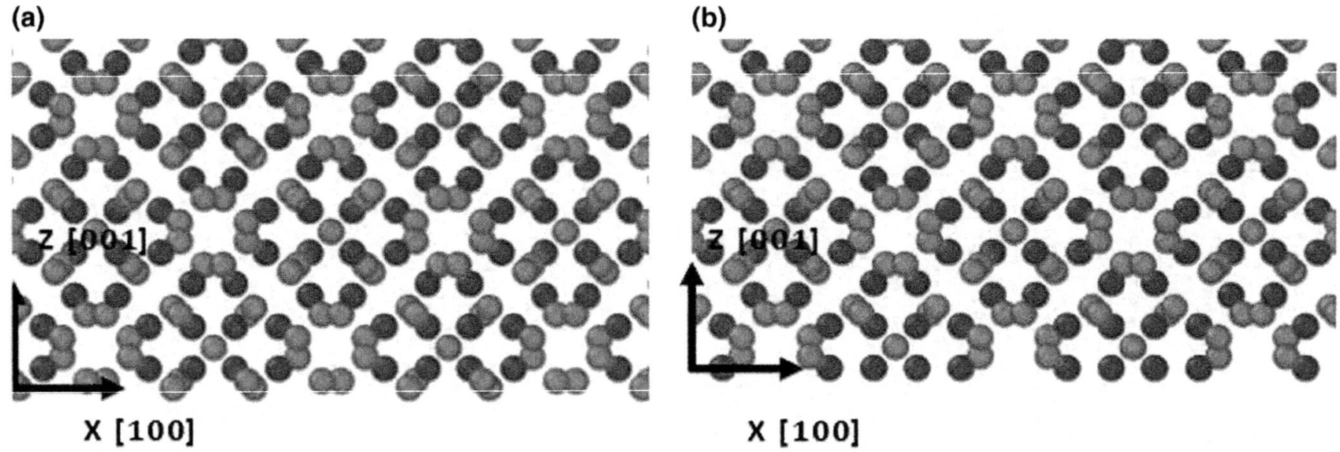


**Fig. 5** The side view of the surface termination for $Mg_{17}Al_{12}$ β-phase without reorienting, (001) plane. **a** Lowest surface energy termination (atom layer 8; bottom (001) plane); **b** highest surface energy termination (atom layer 10; bottom (001) plane)

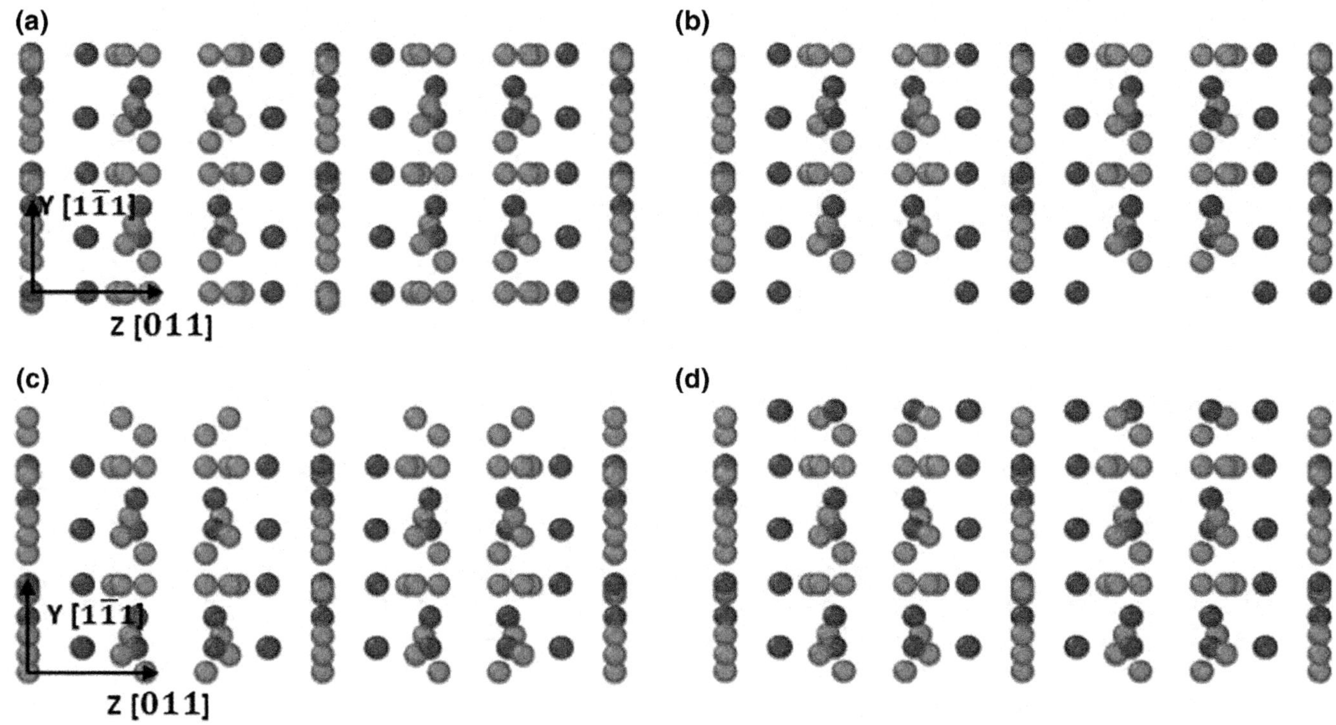


**Fig. 6** The side view of the surface termination for reoriented $Mg_{17}Al_{12}$ β-phase, $(1\bar{1}1)$ plane. **a** Lowest surface energy termination (atom layer 6; bottom $(1\bar{1}1)$ plane) for reoriented β-$Mg_{17}Al_{12}$; **b** highest surface energy termination (atom layer 9; bottom $(1\bar{1}1)$ plane) for reoriented β-$Mg_{17}Al_{12}$; **c** lowest surface energy termination (atom layer 2; top $(1\bar{1}1)$ plane) for reoriented β-$Mg_{17}Al_{12}$; **d** highest surface energy termination (atom layer 3; top $(1\bar{1}1)$ plane) for reoriented β-$Mg_{17}Al_{12}$

**Table 1** Surface energies results for $Mg_{17}Al_{12}$ β-phase

| | Plane | Direction (layer removing) | Lowest (mJ/m$^2$) | Highest (mJ/m$^2$) |
|---|---|---|---|---|
| $Mg_{17}Al_{12}$ without reorient | (001) | [001] | 742.7 | 980.0 |
| $Mg_{17}Al_{12}$ reoriented | $(\bar{2}\bar{1}1)$ | $[\bar{2}\bar{1}1]$ | 789.4 | 1052.5 |
| | $(\bar{2}\bar{1}1)$ | $[21\bar{1}]$ | 762.5 | 952.7 |
| | $(1\bar{1}1)$ | $[1\bar{1}1]$ | 775.6 | 959.6 |
| | $(1\bar{1}1)$ | $[\bar{1}1\bar{1}]$ | 761.3 | 898.2 |
| | (011) | [011] | 715.7 | 1008.6 |

**Table 2** Interfacial energies results for $Mg_{17}Al_{12}$ β-phase/Mg matrix

| | Plane (based on β-phase) | Direction (correspond to the surface in Table 1) | Interfacial energy (mJ/m$^2$) |
|---|---|---|---|
| $Mg_{17}Al_{12}$/Matrix | (011) | [011] | 248–252 |
| $Mg_{17}Al_{12}$/Twin matrix | (011) | [011] | 492–496 |

and 100 mJ/m$^2$ for incoherent interfaces [19]. Our results show that, for the incoherent interfaces with the Burger's OR, the interfacial energy should be higher than the reported values. Additionally, the lowest energy surface of $Mg_{17}Al_{12}$ does not generate the lowest interfacial energy. The interfacial energy the (011) surface plane of $Mg_{17}Al_{12}$ increases by ~250 mJ/m$^2$, which indicates that the prismatic plane of the matrix has higher bond strength with the β-phase (011) surface than that of basal plane.

## Conclusion

In this work, we provided detailed calculations of the surface energies and interfacial energies between $Mg_{17}Al_{12}$ and Mg matrix, using molecular dynamic simulations. Atomic layers were carefully selected to find the lowest surface energy termination for each crystallographic plane of $Mg_{17}Al_{12}$ β-phase. The surface energies vary as the surface structure changes. Generally, surfaces with higher fraction of magnesium atoms generate lower surface energies. The lowest surface energies termination of $Mg_{17}Al_{12}$ does not generate the lowest interfacial energies. The interfacial energy increases by ~250 mJ/m$^2$ for $Mg_{17}Al_{12}$ β-phase (011) plane after a $\{10\bar{1}2\}$ twinning.

**Acknowledgements** Bin Li gratefully thanks support from the U.S. National Science Foundation (CMMI-1635088).

## References


1. Hirsch J, Al-Samman T (2013) Superior light metals by texture engineering: Optimized aluminum and magnesium alloys for automotive applications. Acta Mater 61:818–843. https://doi.org/10.1016/j.actamat.2012.10.044
2. Mordike BL, Ebert T (2001) Magnesium: properties—applications—potential. Mater Sci Eng A 302:37–45
3. Smola B, Stulíková I, von Buch F, Mordike BL (2002) Structural aspects of high performance Mg alloys design. Mater Sci Eng A 324:113–117. https://doi.org/10.1016/S0921-5093(01)01291-6
4. Frank C (2011) Magnesium Alloys-Design, Processing and Properties. Intech: India
5. Hutchinson CR, Nie J-F, Gorsse S (2005) Modeling the precipitation processes and strengthening mechanisms in a Mg-Al-(Zn) AZ91 alloy. Metall Mater Trans A 36:2093–2105
6. Celotto S (2000) TEM study of continuous precipitation in Mg–9 wt%Al–1 wt%Zn alloy. Acta Mater 48:1775–1787. https://doi.org/10.1016/S1359-6454(00)00004-5
7. Zhang M-X, Kelly PM (2003) Crystallography of $Mg_{17}Al_{12}$ precipitates in AZ91D alloy. Scr Mater 48:647–652. https://doi.org/10.1016/S1359-6462(02)00555-9
8. Nie J-F (2012) Precipitation and Hardening in Magnesium Alloys. Metall Mater Trans A 43:3891–3939. https://doi.org/10.1007/s11661-012-1217-2
9. Robson JD, Stanford N, Barnett MR (2013) Effect of Precipitate Shape and Habit on Mechanical Asymmetry in Magnesium Alloys. Metall Mater Trans A 44:2984–2995. https://doi.org/10.1007/s11661-012-1466-0
10. Liao M, Li B, Horstemeyer MF (2013) Interaction between prismatic slip and a $Mg_{17}Al_{12}$ precipitate in magnesium. Comput Mater Sci 79:534–539. https://doi.org/10.1016/j.commatsci.2013.07.016
11. Liao M, Li B, Horstemeyer MF (2014) Interaction Between Basal Slip and a $Mg_{17}Al_{12}$ Precipitate in Magnesium. Metall Mater Trans A 45:3661–3669. https://doi.org/10.1007/s11661-014-2284-3
12. Li B, Zhang XY (2016) Twinning with zero twinning shear. Scr Mater 125:73–79. https://doi.org/10.1016/j.scriptamat.2016.07.004
13. Salman SA, Ichino R, Okido M (2010) A Comparative Electrochemical Study of AZ31 and AZ91 Magnesium Alloy. Int J Corros 2010:1–7. https://doi.org/10.1155/2010/412129
14. Zhao M-C, Liu M, Song G, Atrens A (2008) Influence of the β-phase morphology on the corrosion of the Mg alloy AZ91. Corros Sci 50:1939–1953. https://doi.org/10.1016/j.corsci.2008.04.010
15. Lü YZ, Wang QD, Ding WJ, et al (2000) Fracture behavior of AZ91 magnesium alloy. Mater Lett 44:265–268

16. Han G, Han Z, Alan AL, et al (2013) PHASE field simulation on morphology of continuous precipitate $Mg_{17}Al_{12}$ in Mg–Al alloy. Acta Metall Sin 49:277. https://doi.org/10.3724/SP.J.1037.2012.00531
17. Han G, Han Z, Luo AA, Liu B (2014) Three-Dimensional Phase-Field Simulation and Experimental Validation of β-Mg17Al12 Phase Precipitation in Mg-Al-Based Alloys. Metall Mater Trans A 46:948–962. https://doi.org/10.1007/s11661-014-2674-6
18. Han Z, Han G, Luo AA, Liu B (2015) Large-scale three-dimensional phase-field simulation of multi-variant β-$Mg_{17}Al_{12}$ in Mg–Al-based alloys. Comput Mater Sci 101:248–254. https://doi.org/10.1016/j.commatsci.2015.01.038
19. Li M, Ruijie Z, John A (2010) Modeling Casting and Heat Treatment Effects on Microstructure in Super Vacuum Die Casting (SVDC) AZ91 Magnesium Alloy. In: Magnes. Technol. 2010 Proc. Symp. Held Tms 2010 Annu. Meet. Exhib. Minerals, Metals and Materials Society, pp 623–627
20. Xiao W, Zhang X, Geng WT, Lu G (2013) Atomistic study of plastic deformation in Mg–Al alloys. Mater Sci Eng A 586:245–252. https://doi.org/10.1016/j.msea.2013.07.093
21. Foiles SM, Baskes MI, Daw MS (1986) Embedded-atom-method functions for the fcc metals Cu, Ag, Au, Ni, Pd, Pt, and their alloys. Phys Rev B 33:7983
22. Liu X-Y, Ohotnicky PP, Adams JB, et al (1997) Anisotropic surface segregation in Al–Mg alloys. Surf Sci 373:357–370
23. Curtin WA, Olmsted DL, Hector LG (2006) A predictive mechanism for dynamic strain ageing in aluminium-magnesium alloys. Nat Mater Lond 5:875–80. http://dx.doi.org/10.1038/nmat1765
24. Olmsted DL, Jr LGH, Curtin WA, Clifton RJ (2005) Atomistic simulations of dislocation mobility in Al, Ni and Al/Mg alloys. Model Simul Mater Sci Eng 13:371. https://doi.org/10.1088/0965-0393/13/3/007
25. Jones R, Baer D, Danielson M, Vetrano J (2001) Role of Mg in the stress corrosion cracking of an Al–Mg alloy. Metall Mater Trans A 32:1699–1711
26. Jelinek B, Groh S, Horstemeyer MF, et al (2012) Modified embedded atom method potential for Al, Si, Mg, Cu, and Fe alloys. Phys Rev B 85:245102. https://doi.org/10.1103/PhysRevB.85.245102
27. Stukowski A (2010) Visualization and analysis of atomistic simulation data with OVITO–the Open Visualization Tool. Model Simul Mater Sci Eng 18:015012
28. Ning H, Zhou Z, Zhang Z, et al (2017) Hydrogen dissociation and incorporation on $Mg_{17}Al_{12}$(100) surface: A density functional theory study. Appl Surf Sci 396:851–856. https://doi.org/10.1016/j.apsusc.2016.11.041
29. Zhang J-M, Ma F, Xu K-W (2003) Calculation of the surface energy of bcc metals by using the modified embedded-atom method. Surf Interface Anal 35:662–666. https://doi.org/10.1002/sia.1587
30. Zhang J-M, Ma F, Xu K-W (2004) Calculation of the surface energy of FCC metals with modified embedded-atom method. Appl Surf Sci 229:34–42. https://doi.org/10.1016/j.apsusc.2003.09.050
31. Liu B-Y, Wang J, Li B, et al (2014) Twinning-like lattice reorientation without a crystallographic twinning plane. Nat Commun. https://doi.org/10.1038/ncomms4297

# Effect of Ca on the Microstructure and Mechanical Properties in Mg Alloys

E. I. Andritsos, G. C. G. Skinner, and A. T. Paxton

**Abstract**

Rare Earth (RE)-free Mg alloys suffer from low formability due to strong textures and highly anisotropic deformation modes. In the present study, we examine the effects of Ca addition on microstructure and mechanical properties of Mg–Li–Ca and Mg–Zn–Ca alloys. Based on experimental observations, Ca is reported as the element that should solid-solution strengthen Mg–Li alloys due to its significant size mismatch and weaken the texture in Mg–Zn alloys, similarly to the RE contribution in Mg alloys. Using the density functional theory (DFT) we examine the intrinsic type II stacking faults in the basal and pyramidal I planes. We try different alloy compositions in order to understand the solid-solution effect on the different stacking faults and reduce the high plastic anisotropy in Mg alloys mechanical properties.

**Keywords**

Magnesium alloys • Anisotropy • Stacking fault energy Solid solution strengthening • Calcium • Density functional theory

## Introduction

With the growing interest of the automotive and aerospace sector in lightweight materials that demonstrate high strength and ductility, magnesium and its alloys are the main candidates for replacing existing alloys. Magnesium is a very light structural metal, lighter than steel or aluminium, with better strength-to-weight ratio than any other commonly used structural metal. Mg and Mg alloys have high specific strength and thermal conductivity and excellent damping capacity but they exhibit low ductility, high plastic anisotropy and strong texture. Both high strength and high ductility in the same material is difficult to be achieved [1] as low ductility makes the material more brittle. The low ductility is a major restriction for commercial applications.

It is known that the low tensile ductility and high plastic anisotropy in Mg alloys is associated to Mg's hexagonal closed-packed (hcp) formation. hcp structures have a limited number of slip planes with only one closed-packed, the basal plane. Mg is a highly anisotropic material with properties that could have more than 50 times higher value in the pyramidal plane $\langle c + a \rangle$ than in the basal $\langle a \rangle$ (e.g. critical resolved shear stress [2, 3]). Hence, it is important to lower the anisotropy in order to increase formability. Most commercial Mg alloys can improve their poor formability at high temperatures through grain refinement or solid-solution [4–6].

The majority of commercial Mg alloys contain RE elements. Depending on the preparation, the use of a small amount of RE elements in Mg alloys has a wide range of benefits, such as improved high-temperature strength, grain refinement, creep and corrosion resistance and weaker texture [7–13]. Despite the great benefits of RE elements in Mg alloys, their relatively high cost and the limited sources throughout the world reduces the interest of using them in commercial applications. Nevertheless, the benefits of RE are mainly known empirically but the underlying mechanisms in microscopic level are yet to be understood.

Lately there is an increasing interest for designing RE-free Mg alloys with properties similar to those RE produce. Amongst other elements, calcium is considered a great candidate as a RE substitute. Over the last decades there has been a lot of research on the mechanical properties of low concentration Ca in Mg alloys, especially the commercial die cast alloy AZ91 and the ternary Mg–Zn–Ca. Low concentration Ca in Mg alloys, either with or without RE elements,

E. I. Andritsos (✉) · G. C. G. Skinner · A. T. Paxton
King's College London, Strand, London, WC2R 2LS, UK
e-mail: lefteri.andritsos@kcl.ac.uk

G. C. G. Skinner
e-mail: guy.c.skinner@kcl.ac.uk

A. T. Paxton
e-mail: tony.paxton@kcl.ac.uk


D. Orlov et al. (eds.), *Magnesium Technology 2018*, The Minerals, Metals & Materials Series, https://doi.org/10.1007/978-3-319-72332-7_11

have been found to improve creep resistance and strength, provide enhanced precipitation hardening, refine the microstructure, increase ductility, weaken the basal texture and enhance stretch formability [14–22]. In our research, we examine low concentration Ca RE-free Mg alloys in order to study the plastic anisotropy of three major slip planes of the hexagonal structure and understand the effect of Ca.

To understand the effect of anisotropy in pure Mg and Mg alloys we study the deformation fault (intrinsic) $I_2$. The stacking sequence of the $I_2$ type deformation is $ABABCACA$ and is formed by slip of a B plane by $1/3[10\bar{1}0]$ for the basal plane $\{0001\}$ and by approximately $44/300[\bar{1}\bar{1}23]$ for the pyramidal I $\{10\bar{1}1\}$ plane. The stacking faults are a type of defects which play an important role in the understanding of more complex defects such as dislocations and grain boundaries, therefore accurate prediction is needed. Dislocations often split into partials with the formation of a stacking fault connecting the partials. Therefore, the stacking fault region can modify the properties of a dislocation and help in the better understanding of properties like dislocation mobility. Various mechanical properties such as strength, toughness and fracture are severely affected by the stacking faults due to their direct influence on the slip mode or deformation twinning.

The stacking fault energy (SFE) is defined as the energy difference between the crystal with the fault and the energy of the perfect crystal over the sheared area. The SFE represents the energy dependency of rigidly shearing a crystal along a direction and it highly depends on the temperature and the chemical composition. It is known that in cases where the SFE is high the dissociation of a full dislocation into two partials is energetically unfavourable and the material deforms only by dislocation glide. In cases with low SFE, materials display wider stacking faults and have more difficulties for cross-slip and climb. SFE curves are not always experimentally accessible or their experimental values could largely vary, as in the case for pure Mg where the measured SFEs are $78 \pm 15$ mJ/m$^2$ [23] (60 mJ/m$^2$ from reference therein), 50–280 mJ/m$^2$ from references within Ref. [24]. This large uncertainty in the experimental values of the SFE has led many scientists to use atomistic simulations. In our approach we employ the DFT in order to achieve high accuracy calculations. With the DFT we study the SFE of pure Mg, Mg–Zn–Ca and Mg–Li–Ca in the basal and pyramidal I planes in order to determine the effect of Ca in plastic anisotropy of Mg alloys.

## Computational Details

For the calculation of the SFE and the $\gamma$ surface we use the supercell model and apply homogeneous shear boundary conditions (HSBC). For each discrete shear parallel to the $\gamma$ surface the system undergoes atomic relaxation only along the direction perpendicular to it, while the $c/a$ ratio is kept fixed. In the standard supercell model, two supercells are required; one is the structure of the perfect crystal and the other contains the stacking fault. With the HSBC only the supercell with the structure of the perfect crystal is required. To calculate the gamma surface we choose a vector $\mathbf{T}$ which lies in the plane spanned by the vectors $\mathbf{b}_1$ and $\mathbf{b}_2$, which are parallel to the $\gamma$ surface. The new vector perpendicular to the $\gamma$ surface is $\mathbf{b_3} = \mathbf{b_3} + a\mathbf{T}$, where $a$ is a number between 0 and 1. We apply periodic boundary conditions towards all directions of the system.

We study four Mg–Zn–Ca and four Mg–Li–Ca alloys in various compositions, as described in Table 1. To design the ternary alloys we use the special quasi-random structure (SQS) approach [25]. The supercell size for the simulation of the basal and pyramidal I planes is 128 atoms. The lattice constant we use is $a = 6.041$ au (3.197 Å) and the $c/a = 1.627$. The supercell dimensions for the basal plane are $2\sqrt{3}a \times 4a \times 4c$ and the pyramidal I $2\langle c + a\rangle \times 2a \times \sim 12.2a$. The pyramidal I structure has been constructed following Frank's four-dimensional vectors method [26] in which the $\mathbf{a}_3$ corresponds to $\mathbf{a_3} = \cos(\theta)\left|\mathbf{a}_3^{\mathrm{pI}}\right|/|\mathbf{a}_1|$, where $\left|\mathbf{a}_3^{\mathrm{pI}}\right| = \cos(\theta^{\mathrm{pI}})c/a$. $\theta^{\mathrm{pI}}$ is the characteristic angle for the pyramidal I plane and $\theta$ is the angle the projection of $\mathbf{a}_3$ forms along the $\mathbf{a}_1$. We create multiple structures and position the Ca atom in specific positions, either at the interface (stacking fault) or in the middle of the structure as shown in Fig. 1.

For the calculation of the SFE at 0 K we use the full-potential linear muffintin orbital (FP-LMTO) method as implemented in the Questaal suite. We calculate the exchange and correlation potential using the generalised gradient approximation (GGA) with the PBE functional, using single basis set. For the sampling of the Brillouin zone we use a $10 \times 10 \times 10$ $k$-mesh, using the Methfessel-Paxton sampling integration. We perform non-self consistent (nSC) total energy calculations ($\mathrm{E}_{\mathrm{tot}}^{\mathrm{nSC}}$) with the Harris-Foulkes (HF) functional [27, 28]. In the HF approximation the total energy can be evaluated just by a given input density $\rho^{\mathrm{in}}$, thus there is no need for an iteration cycle of input densities as in the full self consistent (SC) calculations. Therefore the density convergence criterion does not need to be fulfilled as in the standard SC Hohenberg-Kohn (HK) functional [29, 30], leading to computational efficiency.

The HF theorem shows that the $E_{\mathrm{tot}}^{\mathrm{nSC}}$ deviates from the SC ground-state $E_{\mathrm{tot}}^{\mathrm{SC}}$ by an amount of second order in the deviation of $\rho^{\mathrm{in}}$ from the SC ground-state output densities $\rho^{\mathrm{out}}$. The validity of the HF functional has been investigated many times [31–37] and it has been found to produce accurate total energies for bulk and surface systems, with some cases to be even more accurate than the HK energies [38]. To test the efficiency of the two methods we compare the SFEs of the basal plane for a 24-atom Mg supercell using the method described earlier. For high quality calculations we use a very large k-point mesh of 40 × 40 × 12. In Fig. 2 we show the SFE along the basal plane for shear up to 0.4 lattice vector in the $[10\bar{1}0]$ direction, calculated with the two different methods. The unstable ($\gamma_{usf}$) and stable ($\gamma_{sf}$) SFEs are formed at 1/6 and 1/3 in the $[10\bar{1}0]$ direction. The energy predicted at those points is almost the same with both methods ($\gamma^{SC}_{usf} - \gamma^{nSC}_{usf} = 1.90\ \mathrm{mJ/m^2}$, $\gamma^{SC}_{sf} - \gamma^{nSC}_{sf} = 0.89\ \mathrm{mJ/m^2}$). As it is know from literature the nSC energies are lower than the SC energies [32] but the difference in this case is almost negligible, validating the predictive power of the HF functional.

The main benefit of the HF approximation is its efficiency. Although both functionals have similar computational cost per iteration (every diagonalisation of the Hamiltonian), the SC method is at least six times more computationally expensive than the nSC over the whole relaxation calculation. The efficiency of the nSC method has been tested before [37] with same order of magnitude results. Therefore we can conclude that the nSC method produces as accurate results as the SC for the calculation of the SFE and it is highly computationally efficient. We take advantage of the high efficiency of this method to calculate the SFE of the Mg based ternary alloys.

## Results and Discussion

We summarise the results of the SFEs for Mg and the Mg-based ternary alloys in Table 2. We present two different values for the SFEs depending on the position of the Ca atom in the system for both basal and pyramidal I planes. All values refer to the stable SFE of intrinsic type II fault and are in mJ/m$^2$. Where possible values from DFT calculations in literature are included. Alloys A-C and E-G contain low concentration Zn or Li and Ca, while the alloys D and H contain near or above the solubility limit of Zn (2.5 at.% at 340 °C [39]) or Li (around 16 at.% at 273 K [40]) and low concentration Ca (limited solubility or <1 at.% [39, 41]).

As we observe in Table 2, the Mg–Zn–Ca alloys exhibit lower SFEs at the basal plane for increasing Zn composition when the Ca atom is positioned at the interface, while there is no major change when the Ca atom is at the middle of the crystal. This shows that Ca segregation is favoured in the basal plane for Mg–Zn alloys as the Zn composition increases. From the literature we know that low concentration Zn in Mg does not have a major effect on the SFE [46, 49, 50]. On the other hand low concentration Ca in Mg lowers the basal SFE of the system [49, 51], thus we can assume that in the ternary Mg–Zn–Ca system the Ca addition is the main contributor in the SFE. The Ca contribution fades when it is positioned in the middle of the crystal. The $\gamma_{usf}$ has been studied for Mg–Zn, Mg–Ca and Mg–Zn–Ca (the same composition as alloy A), showing a decrease in the basal unstable SFE from 94 to 76 mJ/m$^2$ for the ternary alloy and similar behaviour as in literature for the binary alloys [52].

It has been demonstrated that a wide range of RE elements strongly segregate to grain boundaries due to the large atomic size misfit with magnesium [53]. In the Mg–Zn–Ca alloys we observe a similar mechanism. The replacement of Mg atoms with Zn, which have smaller atomic radius than Mg, introduces empty space in the lattice allowing the significantly larger Ca atom to segregate easier. This leads to lower basal SFE in system.

The pyramidal I plane SFEs for the Mg–Zn–Ca alloys exhibit different behaviour than basal plane SFEs. They increase considerably when the Ca atom is positioned at the interface, regardless of the Zn concentration. On the other hand, for low Zn concentration and Ca position in the middle of the crystal, the SFEs remain similar to this of pure Mg.

**Table 1** The studied Mg–Zn–Ca and Mg–Li–Ca alloys in various compositions

| Name | at.% | wt.% | No. of atoms per 128 atoms |
|---|---|---|---|
| A | Mg–0.78Zn–0.78Ca | Mg–2.06Zn–1.27Ca | Mg–1Zn–1Ca |
| B | Mg–1.56Zn–0.78Ca | Mg–4.07Zn–1.25Ca | Mg–2Zn–1Ca |
| C | Mg–2.34Zn–0.78Ca | Mg–6.04Zn–1.23Ca | Mg–3Zn–1Ca |
| D | Mg–17.19Zn–0.78Ca | Mg–35.69Zn–0.99Ca | Mg–22Zn–1Ca |
| E | Mg–0.78Li–0.78Ca | Mg–0.22Li–1.29Ca | Mg–1Li–1Ca |
| F | Mg–1.56Li–0.78Ca | Mg–0.45Li–1.30Ca | Mg–2Li–1Ca |
| G | Mg–3.13Li–0.78Ca | Mg–0.91Li–1.31Ca | Mg–4Li–1Ca |
| H | Mg–13.28Li–0.78Ca | Mg–4.17Li–1.42Ca | Mg–17Li–1Ca |

In order are the reference name for each alloy, the atomic and weight percentage and the number of atoms in the simulation box

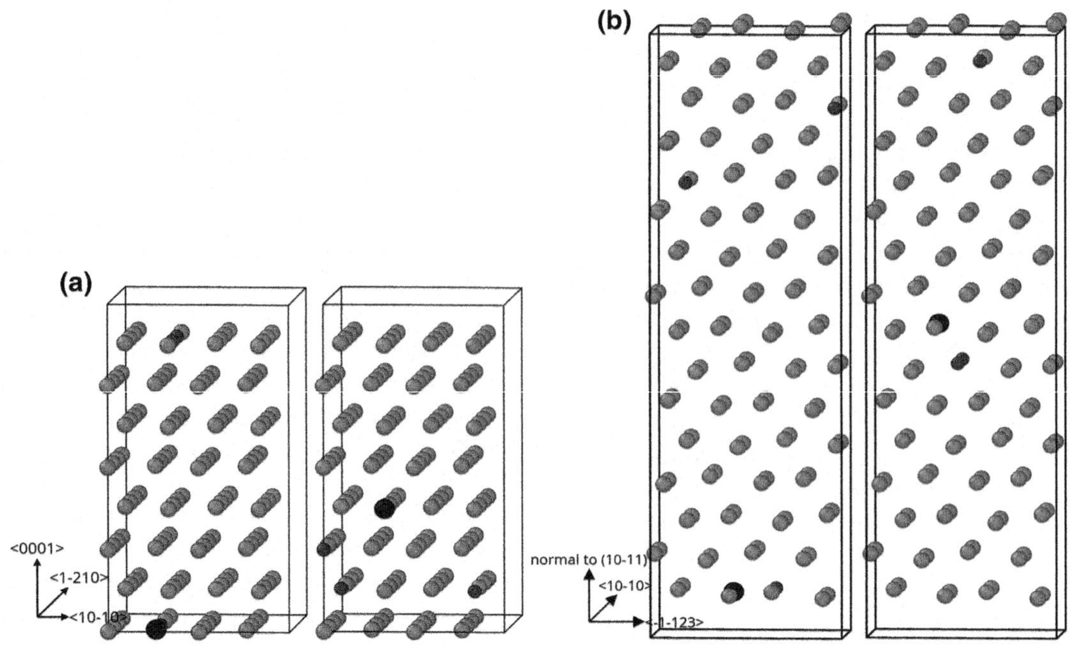


**Fig. 1** Schematic representation of the alloy C (Mg–2.34Zn–0.78Ca at.%) alloy for the basal (**a**) and pyramidal I (**b**) structure. The grey spheres represent the Mg atoms, the purple the Zn atoms and the red the Ca atom. The structures have been created with the SQS method and the Ca atom has been positioned either at the interface (top and bottom of the figures) or the middle of the crystal

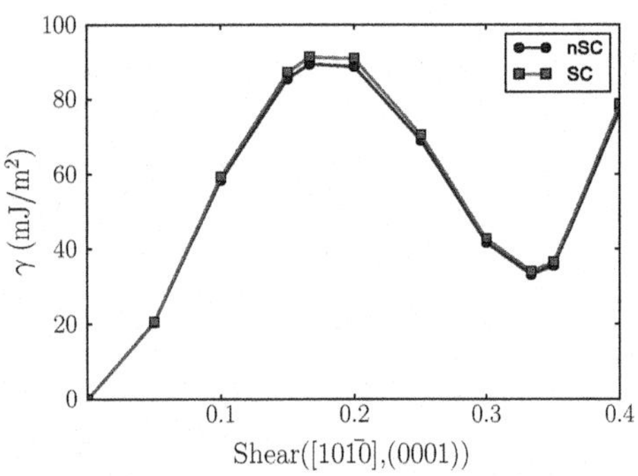


**Fig. 2** The SFE of Mg for the basal plane calculated with the nSC (blue circles) and SC (red squares) methods. The energy difference of the two methods of the $\gamma_{sf}$ is less than 1 mJ/m$^2$, showing the good predictive power of the nSC method

We can observe a great decrease in the SFE only for the D alloy, which has high Zn concentration. This shows that high concentration of Zn in Mg tends to reduce the SFE but the position of Ca has a major effect in the further reduction of it.

For the Mg–Li–Ca alloys we observe an increase in the basal SFE with increasing Li composition in the system. The increase in the basal SFE is even higher when the Ca atom is located at the middle of the crystal. From the literature we know that low concentration Li in Mg results in higher basal SFE [50, 54]. This shows that basal Ca segregation in Mg–Li–Ca alloys is not favourable. According to Robson [53], the predicted maximum possible concentration of RE solute on boundaries at equilibrium occurs for high solubility additions such as the Y-subgroup, since these elements have both a large misfit and relatively high solubility in bulk magnesium. Despite lithium's higher solubility than zinc, its atomic radius is significantly smaller, something that could explain the unfavourable segregation of Ca at the interface. Slip in the pyramidal plane is not highly affected for the alloys E-H. Despite the small increase or decrease when the Ca atom is positioned at the interface or in the middle of the crystal accordingly, the Li concentration has small effect in the SFE.

Despite the effort to minimise the errors in our calculations a number of factors can affect the SFE. The supecell size, the number of *k*-points, the basis set and the convergence criteria are the major factors. An indication of the size of the error in the calculations is the deviation of the $\gamma_{sf}$ of an alloy with the low concentration solid-solutions positioned in the middle of the crystal from the $\gamma_{sf}$ of pure Mg. The Zn or Li affect the total energy of the system but when they are

**Table 2** Stable SFEs ($\gamma_{sf}$) of Mg and various ternary alloys, as described in Table 1, for displacement along the basal and pyramidal I planes

| Alloy | Basal $[10\bar{1}0]$ (0001) | | Pyramidal I $1/3[\bar{1}\bar{1}23](10\bar{1}1)$ | | $\gamma_{sf}^{PyrI}/\gamma_{sf}^{Bas}$ | $\gamma_{sf}^{PyrI}/\gamma_{sf}^{Bas}$ |
|---|---|---|---|---|---|---|
| | Interface | Middle | Interface | Middle | Interface | Middle |
| Mg | 32.1 (33.0[a], Lit.[b]) | | 220.3 (Lit.[c]) | | 6.9 | |
| A | 31.2 | 36.5 | 322.1 | 208.5 | 10.3 | 5.7 |
| B | 26.6 | 36.3 | 318.0 | 205.1 | 12.0 | 5.7 |
| C | 28.0 | 36.0 | 317.3 | 202.3 | 11.3 | 5.6 |
| D | 13.1 | 31.6 | 271.0 | 126.3 | 20.7 | 4.0 |
| E | 30.1 | 36.9 | 248.0 | 203.5 | 8.2 | 5.5 |
| F | 40.7 | 45.2 | 228.2 | 213.9 | 5.6 | 4.7 |
| G | 30.5 | 45.3 | 241.6 | 192.5 | 7.9 | 4.2 |
| H | 50.7 | 64.1 | 261.7 | 189.8 | 5.2 | 3.0 |

The two different values for each plane, interface and middle, refer to the Ca position in the alloy. We have calculated the SFE with the nSC method based on the HF approximation. All the values are in mJ/m$^2$

[a]Calculated with the SC method

[b]Literature: 44 [42], 36 [43], 34.1 [44], 34 [45], 33.8 [46], 37 [47]

[c]Literature: 180 [48]

placed in the middle of the bulk the SFE should not be affected. Although in our simulations we have placed the Ca atom by hand, with the SQS method the Zn or Li atoms are placed at a quasi-random position. These positions are energetically favourable in the system and they could force the Zn or Li atoms either near the interface or in the middle. This could affect the results of our calculations and it might be the case for the anomaly of the basal SFE of alloy F. Comparing the structure of alloys F and G, the alloy F has both Li atoms at the interface while G has only one atom. This anomaly is considered within the standard error in our calculations.

In Table 2 we calculate the ratio of the stacking fault energy of the pyramidal I plane over the basal plane for the two different calculations where the Ca atom is at the interface or in the middle ($\gamma_{sf}^{PyrI}/\gamma_{sf}^{Bas}$). Although this ratio is evaluated for the stable stacking fault, it essentially represents a generic factor of the plastic anisotropy for any mechanical property related to the SFs (e.g. yield stress, dislocation core energy etc.). Increase in the $\gamma_{sf}$ usually suggests an increase in the $\gamma_{usf}$ and vice versa. Despite the fact this increase is not linear and the $\gamma_{sf}^{PyrI}/\gamma_{sf}^{Bas}$ ratio will differ from the $\gamma_{usf}^{PyrI}/\gamma_{usf}^{Bas}$, for the same material it will correctly predict the qualitative change in anisotropy.

This ratio is estimated at 6.9 for pure Mg. For increasing concentration of Zn in alloys A-D we observe an increasing anisotropy when the Ca atom is placed at the stacking fault. When the Ca atom is positioned in the middle of the crystal the anisotropy is at the same level for all low Zn concentration alloys, while for high Zn concentration (alloy D) the anisotropy is significantly lower. Similarly for the alloys E-H, the anisotropy is lower when the Ca atom is positioned in the middle of the crystal and is reduced with increasing concentration of Li. The alloy H achieves the highest reduction in anisotropy, which is less than half of this of pure Mg. It is interesting to mention that for all alloys when the Ca atom is positioned in the middle the anisotropy is lower than in pure Mg.

## Conclusions

We have calculated the SFEs of various Mg–Zn–Ca and Mg–Li–Ca alloys for the basal and pyramidal I plane. We studied the effect of Ca in the slip behaviour and the plastic anisotropy by positioning it either at the interface of the SF or in the middle of the crystal. We observe a significant difference in the SFE depending first on the position of Ca and second on the concentration of Zn or Li in the system. Ca segregation is favoured in the basal plane in the Mg–Zn alloys but not for the pyramidal I plane or in the Mg–Li alloys. Moreover, Ca in the middle of the system decreases notably the SFE for pyramidal I plane for high concentration Zn or Li, while it either does not affect or increases the SFE in the basal plane. The highest reduction in plastic anisotropy is for the two high concentration alloys, D and H, when the Ca is in the middle of the crystal. Glide along the pyramidal I plane is more difficult than in the basal plane since the interlayer distance is smaller. The glide is even more difficult when the Ca is at the interface than in the middle of the pyramidal I plane, showing a possible size effect of the solid-solutions.

Similar composition ternary alloys have been studied experimentally for their mechanical properties. Neither experimentally or computationally is easy to study the whole spectra of compositions unless specific thermodynamics models are used (e.g. CALPHAD), although their predictive power sometimes is limited. Although it is possible, it is either too expensive or time consuming and only a small number of different compositions is usually chosen. A systematic method of studying various alloys needs to take into consideration not only the composition but also the design of the alloys. As we concluded from our results, not only the composition but the design method can highly affect the prediction of the mechanical properties and it should be taken into consideration.

Moreover it is important to examine the mechanical properties along the pyramidal II plane. The calculated SFE of pure Mg for the pyramidal II plane is higher than those of the basal and pyramidal I planes, showing higher anisotropy in the mechanical properties. Recent findings show that energetically stabilizing the easy-glide $\langle c + a \rangle$ dislocations on pyramidal II plane could lower ductility [55]. Using solid-solution to lower the SFE energy on the pyramidal II plane could reduce anisotropy and lead to low ductility Mg alloys.

**Acknowledgements** This work was supported by the EPSRC [EP/L025213/1].

## References


1. K. Lu, L. Lu, and S. Suresh. Strengthening materials by engineering coherent internal boundaries at the nanoscale. *Science*, 324(5925):349–352, 2009.
2. S. Ando et al. {1122} ⟨1123⟩ slip in magnesium single crystal. *Journal of Japan Institute of Light Metals*, 42(12):765–771, 1992.
3. W.B. Hutchinson and M.R. Barnett. Effective values of critical resolved shear stress for slip in polycrystalline magnesium and other hcp metals. *Scripta Materialia*, 63(7):737–740, 2010. Viewpoint set no. 47 Magnesium Alloy Science and Technology.
4. M.M. Avedesian, H. Baker, and A.S.M.I.H. Committee. *ASM Specialty Handbook: Magnesium and Magnesium Alloys*. ASM International, 1999.
5. W.J. Kim, H.G. Jeong, and H.T. Jeong. Achieving high strength and high ductility in magnesium alloys using severe plastic deformation combined with low-temperature aging. *Scripta Materialia*, 61(11):1040–1043, 2009.
6. O. Kulyasova et al. Microstructure and mechanical properties of ultrafinegrained Mg–Zn–Ca alloy. *IOP Conference Series: Materials Science and Engineering*, 63(1):012142, 2014.
7. I. J. Polmear. Magnesium alloys and applications. *Materials Science and Technology*, 10(1):1–16, Jan 1994.
8. L.L. Rokhlin. *Magnesium Alloys Containing Rare Earth Metals: Structure and Properties*. Advances in Metallic Alloys. CRC Press, 2003.
9. Y. Wang and J. C. Huang. Texture analysis in hexagonal materials. *Materials Chemistry and Physics*, 81(1):11–26, JUL 20 2003.
10. S.R. Agnew and J.F. Nie. Preface to the viewpoint set on: The current state of magnesium alloy science and technology. *Scripta Materialia*, 63(7):671–673, 2010. Viewpoint set no. 47 Magnesium Alloy Science and Technology.
11. A. Chapuis and J. H. Driver. Temperature dependency of slip and twinning in plane strain compressed magnesium single crystals. *Acta Materialia*, 59(5):1986–1994, Mar 2011.
12. J. Hirsch and T. Al-Samman. Superior light metals by texture engineering: Optimized aluminum and magnesium alloys for automotive applications. *Acta Materialia*, 61(3):818–843, Feb 2013.
13. D. Griffiths. Explaining texture weakening and improved formability in magnesium rare earth alloys. *Materials Science and Technology*, 31(1):10–24, 2015.
14. W. Qudong et al. Effects of ca addition on the microstructure and mechanical properties of AZ91 magnesium alloy. *Journal of Materials Science*, 36(12):3035–3040, Jun 2001.
15. A.A. Luo. Recent magnesium alloy development for elevated temperature applications. *International Materials Reviews*, 49(1):13–30, 2004.
16. K. Hirai et al. Effects of Ca and Sr addition on mechanical properties of a cast AZ91 magnesium alloy at room and elevated temperature. *Materials Science and Engineering: A*, 403(1):276–280, 2005.
17. L. Han, H. Hu, and D. O. Northwood. Effect of Ca additions on microstructure and microhardness of an as-cast Mg-5.0 wt.% Al alloy. *Materials Letters*, 62(3):381–384, 2008.
18. S.W. Xu et al. High temperature tensile properties of as-cast MgAlCa alloys. *Materials Science and Engineering: A*, 509 (1):105–110, 2009.
19. L. Geng et al. Microstructure and mechanical properties of Mg–4.0Zn–0.5Ca alloy. *Materials Letters*, 63(5):557–559, 2009.
20. J. Jayaraj et al. Enhanced precipitation hardening of MgCa alloy by Al addition. *Scripta Materialia*, 63(8):831–834, 2010.
21. B.P. Zhang et al. Enhanced mechanical properties in fine-grained Mg–1.0Zn–0.5Ca alloys prepared by extrusion at different temperatures. *Scripta Materialia*, 63(10):1024–1027, 2010.
22. Y. Chino et al. Effects of Ca on tensile properties and stretch formability at room temperature in Mg–Zn and Mg–Al alloys. *Materials Transactions*, 52(7):1477–1482, 2011.
23. D.H. Sastry, Y.V.R.K. Prasad, and K.I. Vasu. On the stacking fault energies of some close-packed hexagonal metals. *Scripta Metallurgica*, 3(12):927–929, 1969.
24. Y. Wang et al. First-principles calculations of twin-boundary and stacking-fault energies in magnesium. *Scripta Materialia*, 62(9):646–649, 2010.
25. A Zunger et al. Special quasirandom structures. *Phys. Rev. Lett.*, 65:353–356, Jul 1990.
26. F. C. Frank. On Miller-Bravais indices and four-dimensional vectors. *Acta Crystallographica*, 18(5):862–866, May 1965.
27. J. Harris. Simplified method for calculating the energy of weakly interacting fragments. *Phys. Rev. B*, 31:1770–1779, Feb 1985.
28. W. Matthew C. Foulkes and R. Haydock. Tight-binding models and density-functional theory. *Phys. Rev. B*, 39:12520–12536, Jun 1989.
29. P. Hohenberg and W. Kohn. Inhomogeneous electron gas. *Phys. Rev.*, 136:B864-B871, Nov 1964.
30. W. Kohn and L. J. Sham. Self-consistent equations including exchange and correlation effects. *Phys. Rev.*, 140:A1133–A1138, Nov 1965.
31. A J Read and R J Needs. Tests of the harris energy functional. *Journal of Physics: Condensed Matter*, 1(41):7565, 1989.
32. H. M. Polatoglou and M. Methfessel. Comparison of the harris and the Hohenberg-Kohn-Sham functionals for calculation of structural and vibrational properties of solids. *Phys. Rev. B*, 41:5898–5903, Mar 1990.
33. M. Schilfgaarde et al. Recent advances in non self-consistent total energy calculations in alloys. In *MRS Proceedings*, volume 186, 1990.

34. F. W. Averill and G. S. Painter. Harris functional and related methods for calculating total energies in density-functional theory. *Phys. Rev. B*, 41:10344–10353, May 1990.
35. N. Chetty, K. W. Jacobsen, and J. K. Norskov. Optimized and transferable densities from first-principles local density calculations. *Journal of Physics: Condensed Matter*, 3(28):5437, 1991.
36. B. Farid et al. Extremal properties of the Harris-Foulkes functional and an improved screening calculation for the electron gas. *Phys. Rev. B*, 48:11602–11621, Oct 1993.
37. G. D. Bellchambers and F. R. Manby. An approximate density-functional method using the Harris-Foulkes functional. *The Journal of Chemical Physics*, 135(8):084105, 2011.
38. J. Hartford, L. B. Hansen, and B. I. Lundqvist. Harris functional densities: from solid to atom. *Journal of Physics: Condensed Matter*, 8(40):7379, 1996.
39. M. Mezbahul-Islam, A. Mostafa, and M. Medraj. Essential magnesium alloys binary phase diagrams. *Journal of Materials*, April 2014.
40. A. A. Nayeb-Hashemi, J. B. Clark, and A. D. Pelton. The Li–Mg (lithium-magnesium) system. *Bulletin of Alloy Phase Diagrams*, 5(4):365–374, Aug 1984.
41. A. A. Nayeb-Hashemi and J. B. Clark. The CaMg (calcium–magnesium) system. *Bulletin of Alloy Phase Diagrams*, 8(1): 58–65, Feb 1987.
42. N. Chetty and M. Weinert. Stacking faults in magnesium. *Phys. Rev. B*, 56:10844–10851, Nov 1997.
43. A. E. Smith. Surface, interface and stacking fault energies of magnesium from first principles calculations. *Surface Science*, 601:5762–5765, 2007.
44. A. Datta, U.V. Waghmare, and U. Ramamurty. Structure and stacking faults in layered MgZnY alloys: A first-principles study. *Acta Materialia*, 56(11):2531–2539, 2008.
45. J. A. Yasi et al. Basal and prism dislocation cores in magnesium: comparison of first-principles and embedded-atom-potential methods predictions. *Modelling and Simulation in Materials Science and Engineering*, 17(5):055012, 2009.
46. Q. Zhang et al. Ab initio study of the effect of solute atoms Zn and Y on stacking faults in Mg solid solution. *Physica B: Condensed Matter*, 416:39–44, 2013.
47. Z. Pei et al. From generalized stacking fault energies to dislocation properties: Five-energy-point approach and solid solution effects in magnesium. *Phys. Rev. B*, 92:064107, Aug 2015.
48. T. Nogaret et al. Atomistic study of edge and screw $\langle c+a \rangle$ dislocations in magnesium. *Acta Materialia*, 58(13):4332–4343, 2010.
49. J. A. Yasi, L. G. Hector, and D. R. Trinkle. First-principles data for solid-solution strengthening of magnesium: From geometry and chemistry to properties. *Acta Materialia*, 58(17):5704–5713, 2010.
50. M. Muzyk, Z. Pakiela, and K.J. Kurzydlowski. Generalized stacking fault energy in magnesium alloys: Density functional theory calculations. *Scripta Materialia*, 66(5):219–222, 2012.
51. S. L. Shang et al. Generalized stacking fault energy, ideal strength and twinnability of dilute Mg-based alloys: A first-principles study of shear deformation. *Acta Materialia*, 67:168–180, 2014.
52. M. Yuasa et al. Improved plastic anisotropy of MgZnCa alloys exhibiting high-stretch formability: A first-principles study. *Acta Materialia*, 65:207–214, 2014.
53. J. D. Robson. Effect of Rare-Earth additions on the texture of wrought magnesium alloys: The role of grain boundary segregation. *Metallurgical and Materials Transactions A*, 45(8):3205–3212, Jul 2014.
54. J. Han et al. Basal-plane stacking-fault energies of Mg: A first-principles study of Li- and Al-alloying effects. *Scripta Materialia*, 64(8):693–696, 2011.
55. Z. Wu and W. A. Curtin. The origins of high hardening and low ductility in magnesium. *Nature*, 526(7571):6267, October 2015.

# Investigation of Grain Refinement Method for AZ91 Alloy Using Carbide Inoculation

Jun Ho Bae, Young Min Kim, Ha Sik Kim, and Bong Sun You

**Abstract**

Grain refinement is an effective method to increase both the strength and toughness of structural materials. Among the various grain refinement methods, carbon inoculation is an effective method for Mg alloys containing Al. SiC is well-known as a carbon inoculation agent; however, conventional methods for using SiC are suitable only for lab-scale experiments. This study investigates the inoculation methods with regard to the use of SiC effectively and economically for grain refinement of Mg alloys, especially for application in large-scale casting processes. Al-SiC refiner was fabricated by extrusion; it demonstrated a uniform structure with well distributed SiC particles on the Al metal matrix. After adding 0.7 wt% extruded SiC refiner, the average grain size of AZ91 alloy decreased from 550 to 90 μm. It demonstrated an effective refining ability in large-scale casting facilities.

**Keywords**

Magnesium • Grain refinement • Carbon inoculation

## Introduction

Generally, the types of Mg alloys can be defined based on the presence or absence of the Al element. This classification is associated with the usage of Zr. Addition of Zr has been known as the most effective method of grain refinement for Mg alloys owing to the crystallographic similarities with α-Mg and the crystal growth inhibitory effect during the solidification [1, 2]. However, Al atoms in Mg alloys offset the refinement ability of Zr owing to the formation of stable compound, $Al_3Zr$; thus, Zr cannot be used in the Mg alloys containing Al. Mg alloys containing Al, such as AZ and AM alloy series, are the most widely used in industries because of their low price, superior castability, and relatively good mechanical properties [3]. To improve their mechanical properties in the cast state and reduce casting defects, numerous studies of grain refinement have been conducted to develop a method with a similar effect as that of Zr addition [4–9]. Various methods such as superheating process, Elfinal process, carbon inoculation, ultrasonic vibration, and melt conditioning have been developed for several decades, but none is commercially available yet due to reasons such as increased processing cost and safety concerns.

Carbon inoculation is one of the most widely studied methods for Mg alloys containing Al because of its low material cost and exceptional refining efficiency even with small amount of addition [10]. Although the grain refinement mechanism of carbon inoculation is still controversial [10–15], recent researches have focused on its application in large-scale industries efficiently. Carbon inoculation can simply be conducted by the addition of carbon-containing agents such as carbon gaseous materials ($C_2Cl_6$, $MgCO_3$, $MnCO_3$), carbides ($Al_4C_3$, SiC, $CaC_2$), and graphite. Among them, the addition of carbon gaseous materials exhibits an exceptional grain refining ability because they provide fine nucleation particles for heterogeneous nucleation and generate gas agitation effect owing to the decomposition of these materials [6, 7]. However, these materials cause severe oxidation on the melt surface owing to strong and irregular gas agitation; $C_2Cl_6$ is especially not used in the industrial field due to harmful gas emission.

Recently, several researchers have reported that the addition of SiC reduces the grain size remarkably in the AZ91 alloy [14, 16]. However, conventional SiC inoculation methods such as using the SiC powder itself and using a master alloy produced via powder metallurgy have demonstrated unstable refining ability, low efficiency, and

J. H. Bae (✉) · Y. M. Kim · H. S. Kim · B. S. You
KIMS (Korea Institute of Materials Science), 797 Changwondaero, Seongsangu, Changwon, Gyeongnam 642-831, Korea
e-mail: jhbae@kims.re.kr


D. Orlov et al. (eds.), *Magnesium Technology 2018*, The Minerals, Metals & Materials Series, https://doi.org/10.1007/978-3-319-72332-7_12

non-economical processes in terms of application in the industrial fields. Therefore, it is necessary to develop an inoculation method with regard to the SiC grain refiner that can be applied to a large-scale process of an industrial field effectively. In this study, useful methods for mass production of grain refiner using SiC powder were developed by the powder extrusion method, and its microstructure features and refining effects on the AZ91 alloy were also investigated.

## Experimental Procedure

High-purity Al and SiC powders were mixed and extruded for produce a grain refiner. To confirm the grain refinement ability of the extruded SiC refiner, commercial alloy AZ91 was melted in a steel crucible using resistance furnace under the protection of $SF_6$ gas mixed with $CO_2$. Subsequently, 0.5, 0.7 and 1 wt% extruded SiC refiner was added into the molten AZ91 alloy at 750 °C, respectively. The melt was stirred and held for 10 min and then poured into a steel billet mold ($\phi$80 × 200 mm) preheated to 200 °C. Metallographic samples were cut at the height of 20 mm from the bottom of the cast billet. All samples were polished and etched with picric acid solution consisting of 4.2 g picric acid, 10 ml acetic acid, 70 ml ethyl alcohol, and 10 ml distilled water. The microstructure was observed using optical microscopy, scanning electron microscopy (SEM), and energy-dispersive X-ray spectroscopy (EDS). The average grain size of each sample was examined using polarized light in optical microscopy and measured using the linear intercept method described in ASTM Standard E112.

## Results and Discussion

Figure 1 shows the microstructures of the extruded SiC refiner cross section perpendicular to the direction of extrusion. Figure 1a shows that the refiner consists of a mixture of Al and SiC, and Al tube surrounds it. In the powder mixture area, as shown in Fig. 1b, the SiC particles are uniformly distributed in the Al matrix without changes in size and shapes compared with the original powder. Further, in the case of Al powder, all the initial powder forms disappeared and transformed into the same metal matrix as the Al tube as shown in Fig. 1c. Although the process temperature is under the melting temperature of Al, relatively high temperature of approximately 400 °C with large strain during extrusion produces a similar effect to casting. This microstructure is nearly same as the Al-SiC composite produced by the stir casting process [17], however, usage of extrusion has several advantages such as increase in the SiC volume fraction and short process time compared to the stir casting process.

Figure 2 illustrates the XRD pattern of the extruded SiC refiner. It can be observed that the grain refiner is composed of the SiC and Al phase. Generally, commercial Al composites containing SiC have an $Al_4C_3$ phase formed at the interface during the fabrication process and it can be a factor that deteriorates the properties of the composite. To suppress the formation of the $Al_4C_3$ phase, an extra amount of Si element may be added; thus, commercial Al–SiC composite contains approximately 7 to 9 wt% Si [17]. In the case of extruded SiC refiner, there are no secondary phases including the $Al_4C_3$ phase. This is because that the process temperature and time during the extrusion are not sufficient to decompose the SiC and react with Al and carbon.

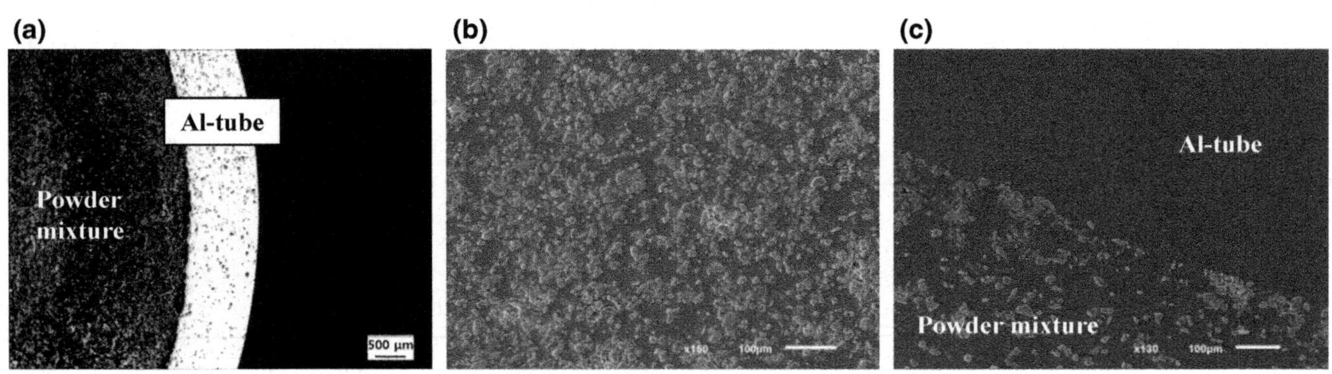


**Fig. 1** Optical (**a**) and SEM (**b**, **c**) micrographs of perpendicular sections to the extrusion direction of extruded SiC refiner

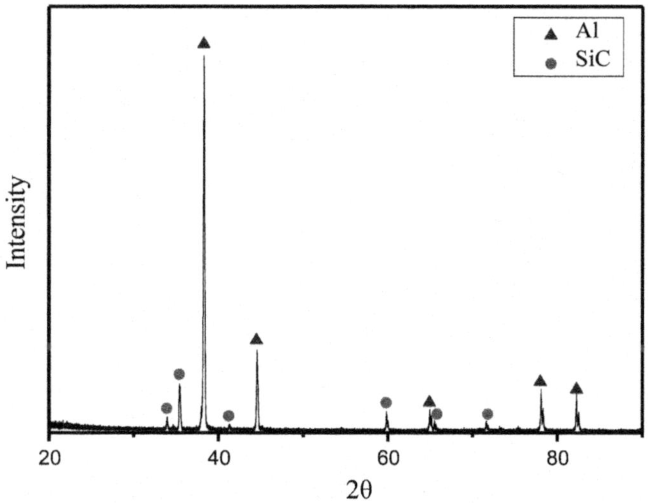


**Fig. 2** XRD pattern of extruded SiC refiner

Figure 3a shows the microstructure of as-cast commercial AZ91 alloy and its average grain size is approximately 550 μm. It exhibits a large and irregular grain structure, but the 0.5 wt% addition of the extruded SiC refiners reduce the grain size to approximately 120 μm and renders the microstructure homogeneous as shown in Fig. 3b. Further grain size reduction of approximately 97 and 90 μm was obtained on addition of 0.7 and 1 wt% refiner, respectively (Fig. 3c, d). It is noted that beyond 0.7 wt% refiner addition, the average grain size was not significantly reduced as shown in Fig. 3e.

Figure 4 illustrates the SEM microstructures of the grain refined AZ91 alloy according to the additional amount of the extruded SiC grain refiner. Typically, as-cast AZ91 alloy has $Mg_{17}Al_{12}$ and Al–Mn phases ($Al_{11}Mn$, $Al_4Mn$, $Al_8Mn_5$), but grain refined AZ91 alloy by SiC addition has an additional phase particle identified as $Mg_2Si$. The formation of the $Mg_2Si$ phase is indirect evidence that the SiC is decomposed into Si and C in the melt [14, 16]. Therefore, as the amount of SiC addition increases, the fraction of $Mg_2Si$ phase increases simultaneously as shown in Fig. 4. The $Mg_2Si$ phases are generally present at the grain boundaries with eutectic $Mg_{17}Al_{12}$ but it cannot disappear easily via the heat treatment owing to the high thermal stability; thus, it is desirable to minimize the fraction of formation because they can adversely affect the mechanical properties of the alloy. Consequently, from the results in Figs. 3 and 4, the optimum amount of the added extruded SiC refiner is obtained as 0.7 wt% and further addition would likely lead to the deterioration of the mechanical properties without significant grain refinement.

In the case of carbon decomposed from SiC, it was known that the carbide is formed by combining with the Al atoms in the alloy melt and acts as a nucleation site for α-Mg during the solidification. However, there has still been controversy as to whether $Al_4C_3$ or $Al_2MgC_2$ is a potent nucleation particle [10–16]. Nevertheless, it is evident from the results of this study that SiC is an effective grain refinement agent with regard to carbon inoculation methods for Mg alloys containing Al.

The extruded SiC refiner can be mass-produced according to the capability of the extrusion equipment and is easy to apply to commercial casting process. Actually, to confirm whether the refining efficiency is similar to the lab-scale process regarding the large-scale casting process, it was applied to commercialized ingot casting facilities with capacity of 500 kg. Figure 5 illustrates the microstructures of the commercial AZ91 and grain refined AZ91 ingot. These samples were heat treated at 400 °C for 15 h to delineate the grain boundaries. The average grain size of the unrefined and grain refined AZ91 ingots are 670 and 160 μm, respectively. It was confirmed that the extruded SiC refiner is also effective with regard to large-scale casting facilities. However, it exhibits a relatively low refining efficiency compared to the lab-scale experiments. Generally, the grain size is also affected by cooling rate related to the mold size, cooling system [9, 16]. Ingot mold size is approximately 4 times larger than the billet mold and the top surface is exposed to the atmosphere as shown in Fig. 6. Therefore, it is considered that the differences of the casting conditions may lead to the difference of grain refining efficiency.

To fabricate the grain refined final product using the grain refined ingot, it has to be re-melted and casted again with regard to the part form. Therefore, in this case, it is most important that the refinement phenomenon is maintained after re-melting and holding. Figure 7 illustrates the grain size change after re-melting of the grain refined ingot and with holding time at 700 °C. After re-melting, the grain size is considerably decreased to about 110 μm compared to the ingot due to the solidification rate related to the mold size and the cooling rate. Grain refinement efficiency according to re-melting and holding time is an exceedingly important factor in terms of industry application of grain refiner. Because inoculation fading can control as using the inoculation time, but the case of refined ingot re-melting, fading of refinement is directly affected by the holding time. In this research, it is noted that the grain refinement ability is maintained even 8 h holding after re-melting. The nucleating particles in the refined ingot are thermally stable, therefore, are not dissolved during re-melting and holding. And the

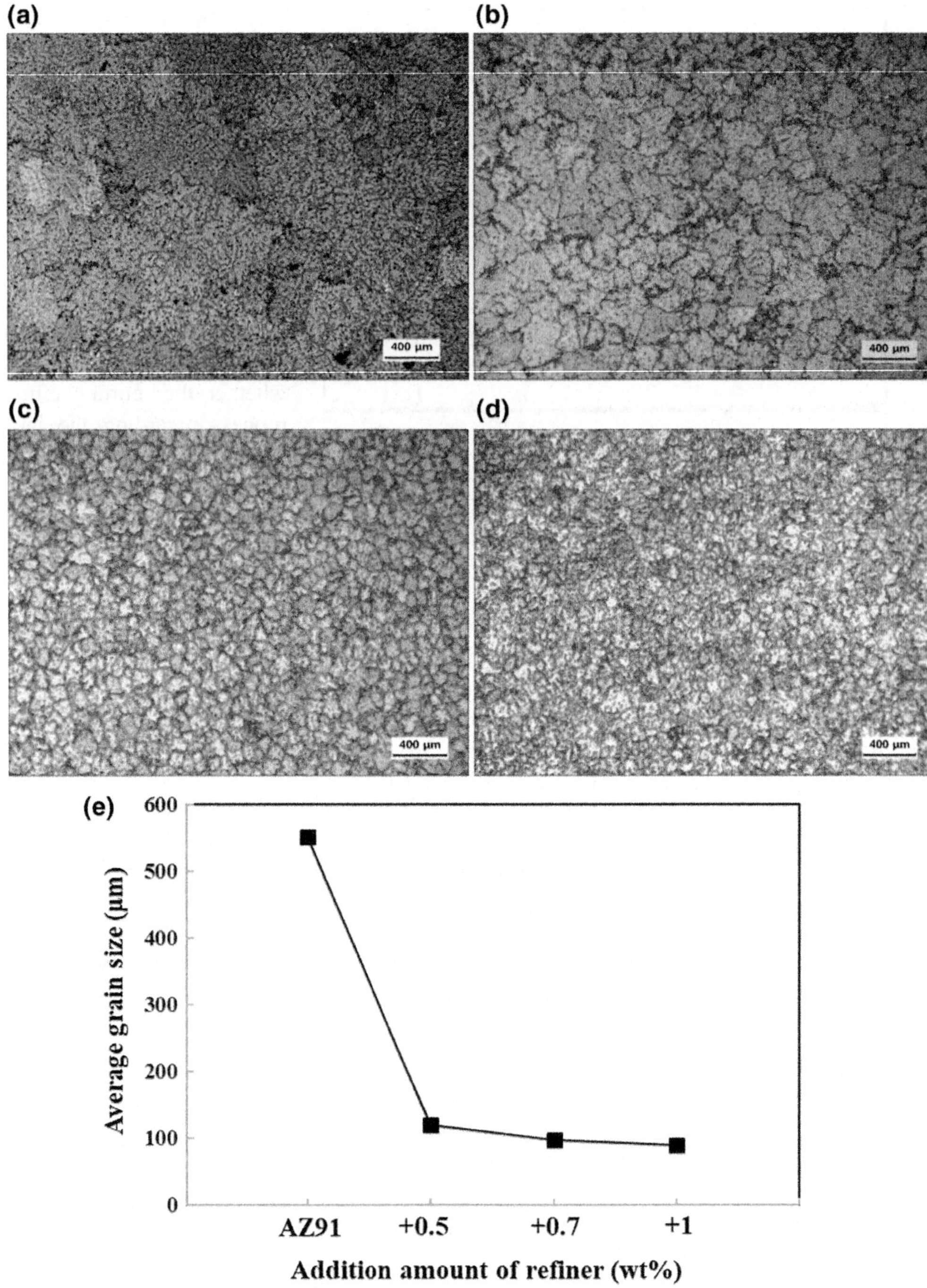


**Fig. 3** Microstructures of AZ91 alloy with the amount addition of extruded SiC refiner; **a** unrefined, **b** 0.5 wt%, **c** 0.7 wt%, **d** 1 wt% addition, and the changes in average grain size (**e**)

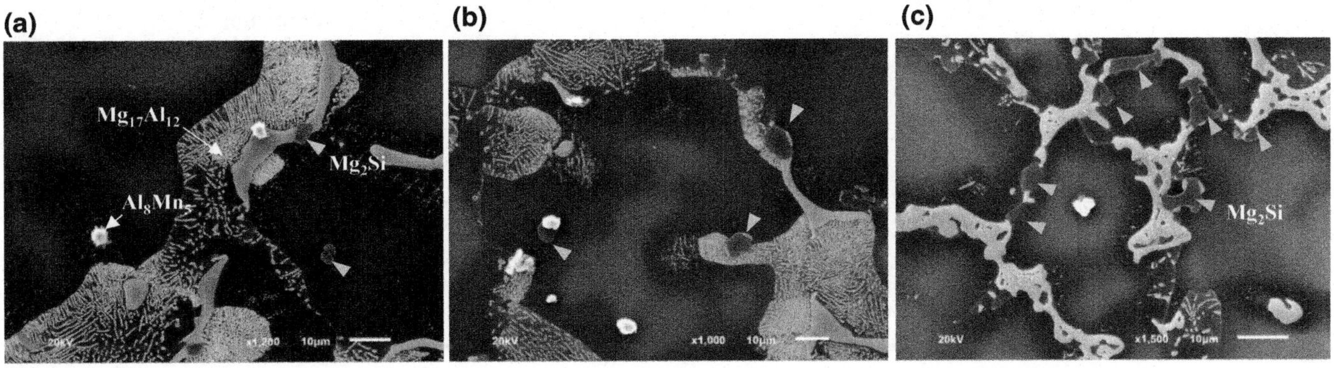


**Fig. 4** SEM micrographs of grain refined AZ91 alloy; **a** 0.5 wt%, **b** 0.7 wt%, **c** 1 wt% extruded SiC refiner added

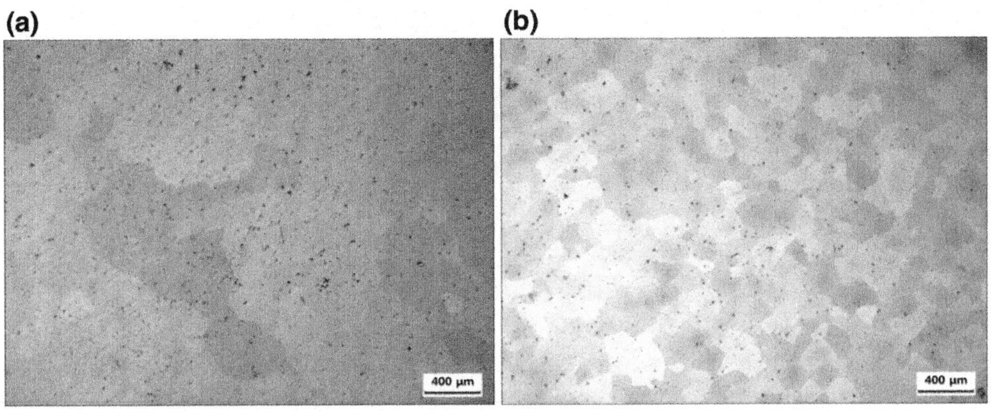


**Fig. 5** Optical micrographs of homogenized **a** commercial AZ91 ingot and **b** grain refined AZ91 ingot made in industrial casting facility

**Fig. 6** Picture of grain refined ingot casting in company

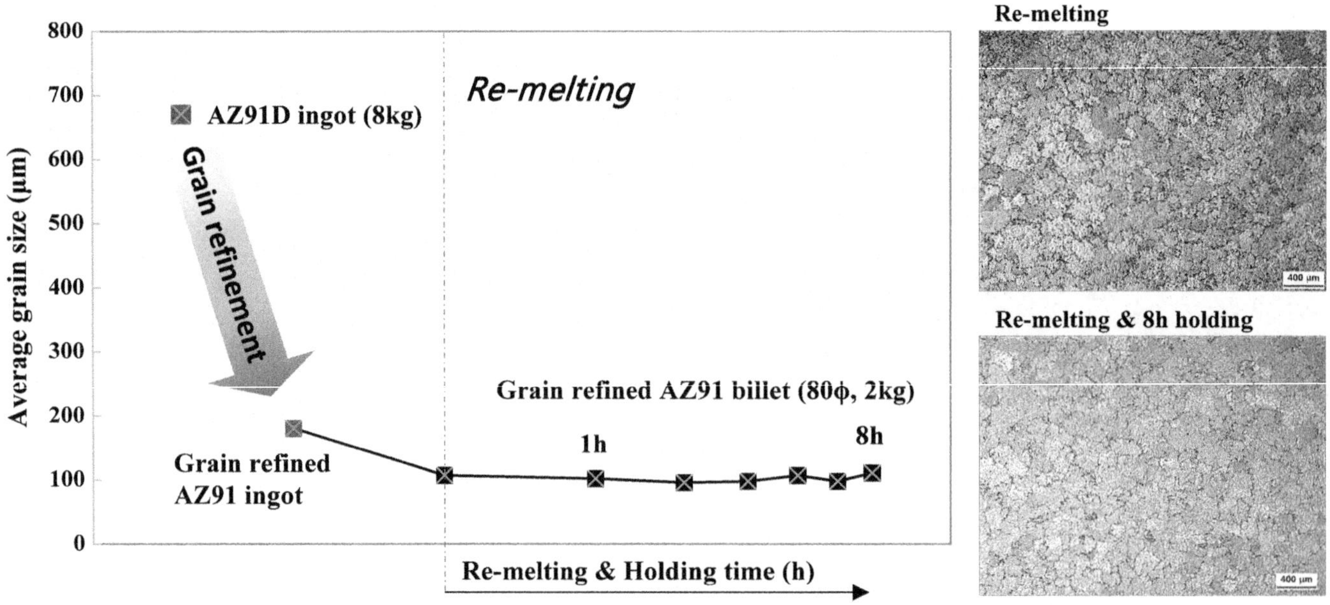


**Fig. 7** The changes in average grain size and microstructures of grain refined AZ91 ingot with re-melting and holding time

particles in the melt have better wettability, interface properties and well distribution compared to inoculant from outside of melt, therefore, it have less chance of sedimentation, floating and agglomeration. For this reasons, grain refining efficiency may be maintained for a relatively long holding time after re-melting.

## Conclusion

In the present research, the effective grain refining method of using SiC powders for AZ91 alloy was investigated. The grain refiner extruded mixed with 70 wt% Al and 30 wt% SiC powders has uniformly distributed SiC particles on the Al metal matrix that is exceedingly similar to the Al–SiC composite produced by the stir casting process. Additionally, the refiner produced by the extrusion method can further improve the efficiency of the SiC addition because no other phases are formed during the process. The average grain size of AZ91 alloy decreases from 550 μm to approximately 90 μm owing to the addition of 0.7 wt% extruded SiC refiner. It was also effectively applied to large-scale commercialized ingot casting process, and reduced the grain size of the bulk ingot substantially from 670 to 160 μm. Moreover, the ingot refined by the extruded SiC refiner has a long fading time of over 8 h after re-melting; it can be applied to other processes for final casting products such as low-pressure die casting, sand casting, etc.

**Acknowledgement** This research was supported by the Fundamental Research Program of the Korea Institute of Materials Science (KIMS).

## References


1. David H. StJohn, et al. (2005) Grain refinement of Magnesium Alloys. Metal. Mater. Trans. A. 36A. 1669–1679.
2. Ma Qian, A. Das (2005) Grain refinement of magnesium alloys by zirconium: Formation of equiaxed grains, Scrip. Mater. 54(5):881–886.
3. Yahia Ali, Dong Qiu, Bin Jiang, Fusheng Pan, Ming-Xing Zhang (2015) Current research progress in grain refinement of cast magnesium alloys: A review article. J. Alloys Comd. 619:639–651. http://doi.org/10.1016/j.jallcom.2014.09.061.
4. Z. Fan, Y. Wang. M. Xia, S. Arumuganathar (2009) Enhanced heterogeneous nucleation in AZ91D alloy by intensive melt sharing. Acta. Mater. 57:4891–4901. http://doi.org/10.1016/j.actmat.2009.06.052.
5. Gaowu W. Qin, Yuping Ren, Wei Huang, Song Li, Wenli Pei (2010) Grain refining mechanism of Al-containing Mg alloys with the addition of Mn-Al alloys. J. Alloys Comd. 507:410–413. http://doi.org/10.1016/j.jallcom.2010.07.164.
6. Young Min Kim, Ling Wang, Bong Sun You (2010) Grain refinement of Mg–Al cast alloy by the addition of manganese carbonate. J. Alloys Comd. 490:695–699. http://doi.org/10.1016/j.jallcom.2009.10.141.
7. Ling Wang, Young Min Kim, JeHyun Lee, Bong Sun You (2011) Effect of magnesium carbonate on microstructure and rolling behavior of AZ31 alloy. Mater. Sci. Eng. A. 528:1485–1490. http://doi.org/10.1016/j.msea.2010.10.053.
8. Mark Alan Easton, Ma Qian (2011) Grain refinement in Alloys: Novel Approaches. Encyc. Mater.:Sci. Tech. 1–7. http://doi.org/10.1016/B978-0-08-043152-9.02259-4.
9. T.J. Chen, R.Q. Wang, Y. Ma, Y. Hao (2012) Grain refinement of AZ91D magnesium alloy by Al–Ti–B master alloy and its effect on mechanical properties. Mater. Design 34:637–648. http://doi.org/10.1016/j.matdes.2011.05.020.
10. Ma. Qian, P. Cao (2005) Discussions on grain refinement of magnesium alloys by carbon inoculation. Scripta Mater. 52:415–419. http://doi.org/10.1016/j.scriptamat.2004.10.014.

11. Young Min Kim, Chang Dong Yim and Bong Sun You (2007) Grain refinement mechanism in Mg–Al base alloys with carbon addition. Scripta. Mater. 57:691–694. http://doi.org/10.1016/j.scriptamat.2007.06.044.
12. Guang Han, Xiangfa Liu (2009) Duplex nucleation in Mg–Al–Zn–Mn alloys with carbon inoculation. J. Alloys Comd. 487:194–197. http://doi.org/10.1016/j.jallcom.2009.08.037.
13. Sunya Nimityongskul, et al. (2010) Grain refining mechanisms in Mg-Al alloys with $Al_4C_3$ micorparticles. Mater. Sci. Eng. A. 527:2104–2111. http://doi.org/10.1016/j.msea.2009.12.030.
14. Yuanding Huang, Karl Ulrich Kainer, Norbert Hort (2011) Mechanism of grain refinement of Ma–Al alloys by SiC inoculation. Scripta Mater. 64:793–796. http://doi.org/10.1016/j.scriptamat.2011.01.005.
15. Y. Wang, M. Xia, Z. Fan, X. Zhou, G.E. Thompson (2010) The effect of $Al_8Mn_5$ intermetallic particles on grain size of as-cast Mg–Al–Zn AZ91D alloy. Intermet. 18:1683–1689. http://doi.org/10.1016/j.intermet.2010.05.004.
16. T.J. Chen, X. D. Jiang, Y. Ma, Y. D. Li Y. Hao (2010) Grain refinement of AZ91D magnesium alloy by SiC. J. Alloys Comd. 496:218–225. http://doi.org/10.1016/j.jallcom.2010.03.002.
17. D. G. Lee, C. S. Lee, K. J. Kim, S. S. Kim, S. H. Lee (2002) Fabrication of A356 Al–SiC composites by stir-casting method. J. Kor. Inst. Met. & Mater. 40(7):757–764.

# Experimental Study of the Solidification Microstructure in the Mg-Rich Corner of Mg–Al–Ce System

Charlotte Wong, Mark J. Styles, Suming Zhu, Trevor Abbott, Kazuhiro Nogita, Stuart D. McDonald, David H. StJohn, Mark A. Gibson, and Mark A. Easton

**Abstract**

The current lack of comprehensive understanding of the microstructure evolution in Mg–Al–Ce alloys hinders the accuracy of thermodynamic predictions. Our investigations have identified shortcomings within the published literature for the Mg-rich end of the Mg–Al–Ce phase diagram. In this study, the microstructure evolution in Mg–Al–Ce alloys has been studied by X-ray diffraction, scanning and transmission electron microscopy. The experimental results are compared with the Scheil-Gulliver prediction calculated using the CALPHAD method. The observed microstructure contains both the binary Al–Ce and Mg–Ce intermetallic phases in these alloys. The solidification sequence, invariant point and the phase boundaries in the liquidus projection of the Mg–Al–Ce phase diagrams that have been reported previously are inconsistent with this study. The hypoeutectic region is smaller compared to the current Mg–Al–Ce phase diagram. In addition, a hexagonal $Al_5Ce_2$ phase which is isostructural with $Al_5La_2$ has been identified in these alloys. The research addresses some of the current limitations in understanding the effect of Ce, when added in isolation, on microstructure development in Mg–Al based alloys.

**Keywords**

Magnesium alloys • Microstructure • Phase diagram

C. Wong · S. Zhu · T. Abbott · M. A. Gibson · M. A. Easton (✉)
School of Engineering, RMIT University, Carlton, VIC 3053, Australia
e-mail: mark.easton@rmit.edu.au

M. J. Styles · M. A. Gibson
CSIRO Manufacturing, Clayton, VIC 3168, Australia

K. Nogita · S. D. McDonald · D. H. StJohn
School of Mechanical and Mining Engineering, The University of Queensland, Brisbane, QLD 4072, Australia

T. Abbott
Magontec Limited, Sydney, NSW 2000, Australia

## Introduction

In recent years, there has been increasing interest in magnesium alloys due to their low density and high specific strength [1], which are beneficial for vehicle weight reductions and hence improved fuel efficiency. In addition to the commonly used AZ91 and AM50/60 alloy grades, magnesium alloys containing aluminium and rare earth (RE) elements, known as the AE series, have been developed for elevated temperature applications, such as automotive powertrain components, where creep resistance is a major concern. Two notable AE alloys are AE42 (Mg–4Al–2RE) [2] and AE44 (Mg–4Al–4RE) [3]. AE42 exhibits better creep resistance than AZ91 and AM50/60, but it tends to be susceptible to hot tearing during casting [4] and the creep resistance deteriorates rapidly at temperatures above 150 °C [5]. AE44 has improved hot tearing resistance and creep resistance over that of AE42, and has been used for a number of powertrain applications since its first use in a corvette engine cradle [6].

The RE elements used in commercial AE alloys commonly comes from mischmetal, which typically comprises 52–55% cerium (Ce), 23–25% lanthanum (La), 16–20% neodymium (Nd) and 5–6% praseodymium (Pr) [7]. Traditionally, it has been considered that all individual rare earth elements behave similarly in microstructure evolution in AE alloys [8]. In recent years, Nd and Pr prices have increased sharply due to the demand for these elements in magnetic applications. Hence, the abundant La and Ce from mischmetal has become less expensive [9]. As a result, efforts have been made to investigate the effect of individual RE elements in AE alloys. In order to better understand their attributes, such as mechanical properties, castability and creep performance, it is necessary to understand the microstructure and its evolution. However, the observed microstructures are difficult to reconcile with the published phase diagrams. For this reason, the investigation in the present study is focused on the microstructure evolution in


D. Orlov et al. (eds.), *Magnesium Technology 2018*, The Minerals, Metals & Materials Series, https://doi.org/10.1007/978-3-319-72332-7_13

Mg–Al–Ce alloys, and comparisons with the currently available thermodynamic descriptions of the Mg–Al–Ce system.

In this work, the as-cast microstructure of Mg–3Al–10Ce and Mg–4Al–12Ce (all compositions in weight percent unless otherwise specified) alloys were examined and compared with thermodynamic predictions made using the CALPHAD software package Pandat®. The two compositions were chosen because they are located on either side of the α-Mg/$Al_2Ce$ phase boundary, as shown in Fig. 1. The aim of the present study is to investigate the accuracy of the existing Mg–Al–Ce phase diagram, particularly the intermetallic phases and the solidification sequence in the current thermodynamic database. These investigations contribute to a better understanding of the microstructure evolution in AE alloys, which may lead to improvements in the thermodynamic database for the Mg–Al–Ce system.

## Materials and Methods

The Mg–3Al–10Ce (ACe310) and Mg–4Al–12Ce (ACe412) alloys were prepared from pure elemental magnesium (99.95%), aluminium (99.9%) and cerium (99.5%). The molten magnesium alloys were cast into a wedge mould consisting of two parts: permanent-mould and sand-mould. Molten magnesium and its alloys have a strong tendency to oxidize in air. Hence, the alloys were melted and degased using nitrogen mixed with H134a refrigerant gas. The pouring temperature of the molten metal into the mould was 720 °C and the preheat temperature of the mould was 95 °C. The chemical compositions of the alloys were determined by inductively coupled plasma atomic emission spectroscopy (ICP-OES). The analysed compositions are Mg–3.04Al–10.10Ce for the ACe310 alloy and Mg–3.87Al–12.40Ce for ACe412 alloy.

For microstructural examination of as-cast material, samples were cut from a wedge cast sample, as shown in Fig. 2. The samples were ground to approximately 15 μm using silicon carbide paper with grit size of 500. After grinding, the samples were then polished using cloth with diamond suspension to approximately 0.04 μm particle size. The samples were ultrasonically cleaned for 180 s after each step of grinding and polishing. The solidification microstructure of the prepared samples were analysed using a FEI Quanta 200 scanning electron microscope (SEM), equipped with an Oxford Instruments X-Max$^N$ 20 energy dispersive X-ray (EDX) spectrometer.

Phase identification was performed via X-ray powder diffraction (XRD), using a Bruker D8 diffractometer fitted with $Cu_{K\alpha}$ X-ray radiation operated at 40 kV and 40 mA. Data were collected in Bragg-Brentano geometry over the angular range 5–130° 2θ with a scan rate of 2°/min and a step size of 0.02°. The samples were continuously rotated at ~2 Hz during the measurements to improve the particle statistics as much as practical. Whilst this produced accurate relative peak intensities for the intermetallic phases observed in these samples, it did not fully mitigate the effects of the large (for XRD) α-Mg grain size. Therefore, during data analysis the α-Mg phase was modelled using the Pawley method [11] whilst the remaining intermetallic phases were modelled (concurrently) using the Rietveld method [12], as implemented in the Topas software package (version 5, Bruker).

For confirmation, the intermetallic phases in the alloys were also identified using a JEOL 2100F transmission electron microscope (TEM), equipped with an Oxford X-MaxN

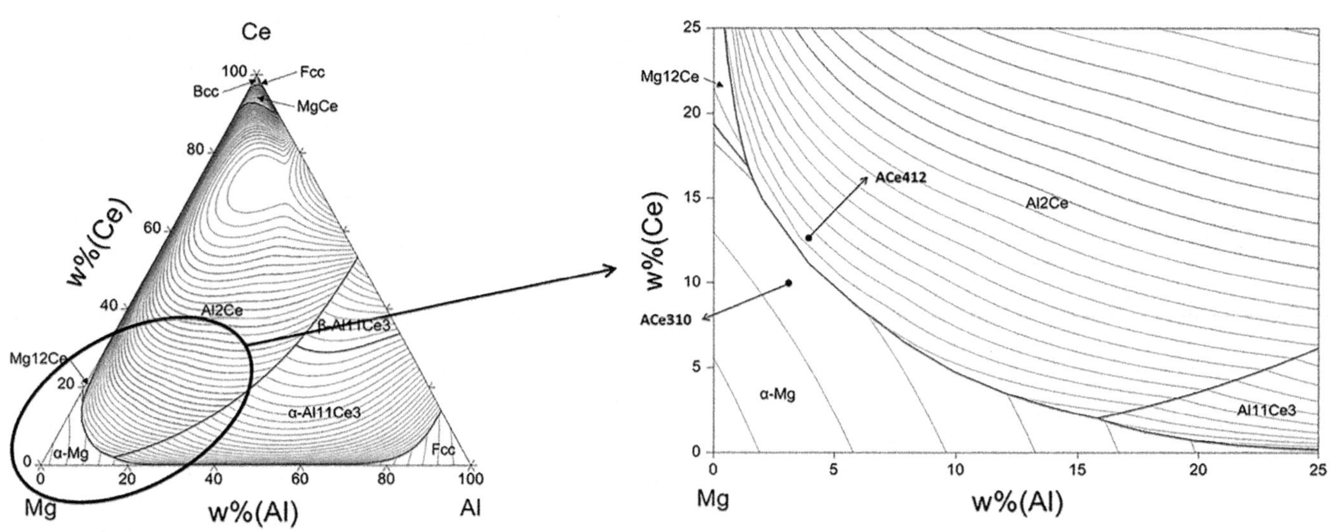


**Fig. 1** Phase diagram (left) and magnified phase diagram (right) of Mg–Al–Ce alloy system generated by Pandat PanMagnesium 2017 database [10]

**Fig. 2** Alloy was cast into two-part wedge shaped moulds consisting of a permanent metal mould (left) and a sand mould (right)

80T EDX spectrometer. The TEM foils were prepared by low-angle ion milling using a Gatan Precision Ion Polishing System (PIPS) at 4 keV with an incident angle of 4°. The phases observed by XRD and TEM were then compared with those predicted by the Scheil-Gulliver equation using PANDAT, PanMagnesium 2017 database [10].

## Results

The microstructure of the as-cast ACe310 and ACe412 alloys are shown in Fig. 3b and d respectively. Two eutectic phases are located at the interdendritic and grain boundary regions. These SEM images suggest that the intermetallic phases can be classified into three types: fine lamellar eutectic, acicular eutectic and irregular particulate-shaped particles. The SEM image of the ACe310 alloy suggests that the volume fraction of the particulate intermetallic phase is lower compared to those in ACe412 alloy, whereas the volume fraction of the acicular phase in the ACe310 alloy is higher compared to the ACe412 alloy. In addition, the particulate-shaped intermetallic particles in both the ACe310 and ACe412 alloys appear to be located near the centre of magnesium grains.

XRD and TEM were used to identify phases in the ACe310 and ACe412 alloys. Four distinct phases were identified in both alloys: α-Mg (*P6₃/mmc*), $Mg_{12}Ce$ (*I4/mmm*), $Al_2Ce$ (*Fd-3 m*) and a hexagonal phase which is isostructural with $Al_5La_2$ (*P6/mmm*) [13–15], which we have termed $Al_5Ce_2$. To our knowledge, the $Al_5Ce_2$ phase has not previously been reported and is not listed in the ICDD crystallographic database (PDF 4+) [16]. Whilst it is difficult to confirm the precise composition of this phase using the XRD data, the $Al_5Ce_2$ composition is in good agreement with the TEM data presented below. The XRD fittings of calculated diffraction patterns to the experimental diffraction patterns of the ACe310 and ACe412 alloys are shown in Fig. 3a and c respectively. There is good agreement between the calculated and observed diffraction patterns, as highlighted by the smooth difference curve and low $R_{wp}$ values. During the fitting, several small peaks were left unidentified, however, the very low intensity of these peaks suggest that they are from a minor phase which may be related to the surface oxide layer of the casting.

The identified intermetallic phases, which can be seen in Fig. 3, have also been further verified by TEM as $Al_5Ce_2$, $Al_2Ce$ and $Mg_{12}Ce$. The TEM images are shown in Fig. 4. The TEM analysis confirms that the intermetallic phase with acicular morphology is $Al_5Ce_2$; the particulate-shaped intermetallic phase is $Al_2Ce$; the eutectic phase with fine lamellar-like morphology is $Mg_{12}Ce$.

## Discussion

### Microstructure Analysis

The lattice parameters of α-Mg were refined to $a = 3.210$ Å and $c = 5.211$ Å in the XRD analysis, which is in excellent agreement with those commonly reported for pure Mg. The lattice parameters of the newly identified $Al_5Ce_2$ phase were refined to $a = 4.605$ Å and $c = 3.826$ Å, as shown in Table 1. The refined lattice parameters of this phase are significantly different from the lattice parameters of $Al_5La_2$ ($a = 4.478$ Å and $c = 4.347$ Å) reported in the literature [13–15]. Intermetallic phases such as $Al_2Ce$ and $Mg_{12}Ce$ often have reciprocal substitution of RE elements. This is because the atomic radii, valency and electronegativity of the RE elements in the mischmetal mixture (Ce, La, Pr, Nd) are similar. Therefore, the substitution of lanthanum with cerium, neodymium or praseodymium in the crystal matrix of the $Al_5Ce_2$, $Al_2Ce$ and $Mg_{12}Ce$ phases is possible based on Hume-Rothery Rules [17, 18]. Rzychon and Kielbus [14] indicated that the lattice parameters of $Al_5La_2$ ($Al_{2.12}La_{0.88}$) phase decrease slightly as small amounts of Ce substitute for La. However, the lattice parameters observed for the pure $Al_5Ce_2$ in this study are substantially smaller than those reported for $Al_5La_2$ [15]. The reason for this large difference is unclear and is worthy of further investigation.

Previous studies reported that the $Al_5La_2$ phase has a similar morphology to the $Al_2La$ phase [14, 17, 19]. In the present study, $Al_5Ce_2$ and $Al_2Ce$ have different morphologies; $Al_5Ce_2$ appears in an acicular morphology and $Al_2Ce$ as a particulate morphology. According to the binary Al–La thermodynamic database, $Al_5La_2$ is not an equilibrium phase. Rzychon and Kielbus [14] mentioned that the presence of $Al_5La_2$ phase in the Mg–Al–La alloys is due to two factors: rapid crystallisation and macrosegragation of RE alloying elements. The literature indicates that the metastable $Al_5La_2$ phase does not develop in slowly cooled gravity cast AE alloys, but develops in fast cooled HPDC AE alloys [17, 19]. Interestingly, this phase has been identified in the present study, in which samples were made by gravity die-casting. This is because the geometry of the casting mould also plays a role in developing the cooling profiles.

Another microstructure observation in the SEM images, as shown in Fig. 3b and d, is that some of the particulates are

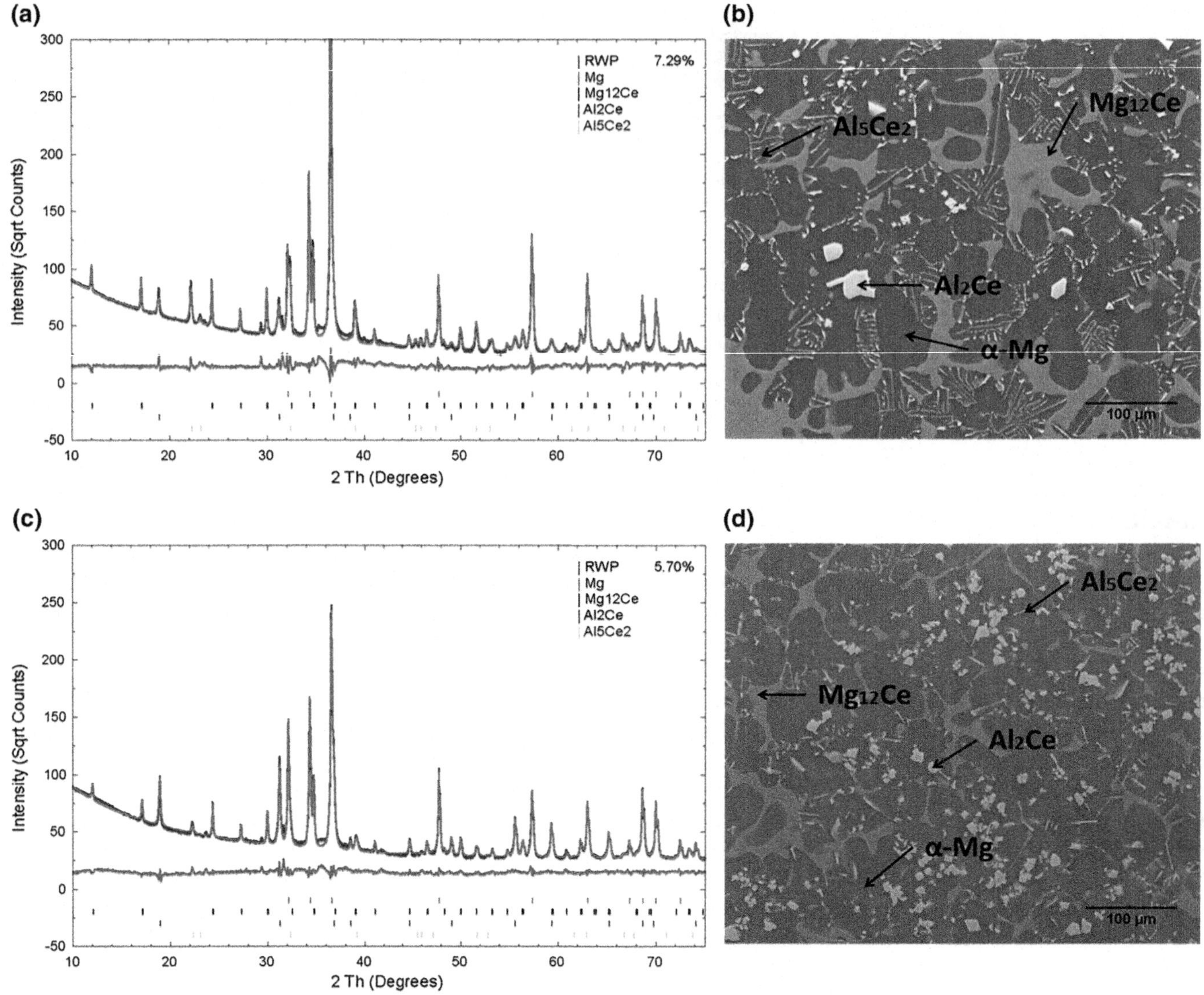


**Fig. 3** XRD analysis on **a** ACe310 alloy and **c** ACe412 alloy, the calculated patterns are represented in red and the experimental patterns represented in blue. The difference between the experimental and calculated patterns is represented in grey, SEM image on **b** ACe310 alloy and **d** ACe412 alloy

located within the magnesium grains. EDS and TEM analysis reveals that this particle is $Al_2Ce$ phase. The observation of $Al_2Ce$ within α-Mg grains suggests that $Al_2Ce$ may act as a nucleant during solidification, which is in agreement with the results of previous studies [20–22].

## Phase Diagram Evaluation

Computational thermodynamics of magnesium alloy systems have been utilised in industry to accelerate design and optimisation of alloys. In the present study, the thermodynamic database of the Mg–Al–Ce system was taken from the PanMg 2017 [10] database using Pandat. In this study, the comparison between the identified phases of the investigated alloys using ex situ characterisation methods with the existing Mg–Al–Ce phase diagram can be used to verify and improve the database for this system.

XRD and TEM analysis reveals that the phases present in ACe310 and ACe412 alloys are α-Mg, $Al_5Ce_2$, $Al_2Ce$ and $Mg_{12}Ce$. By comparing with the Mg–Al–Ce phase diagram and the Scheil-Gulliver solidification simulation, shown in Figs. 1 and 5 respectively, the phases identified in both the ACe310 and ACe412 alloys do not match the phases predicted in the thermodynamic database. Only the identified $Al_2Ce$ phase matches with the Scheil-Gulliver prediction, but not $Mg_{12}Ce$ and $Al_5Ce$. In addition, there is no $Al_{11}Ce_3$ phase observed in ACe412 alloy in the present study. It is worth mentioning that the $Al_5Ce_2$ phase is not an equilibrium phase based on the binary Al-RE thermodynamic

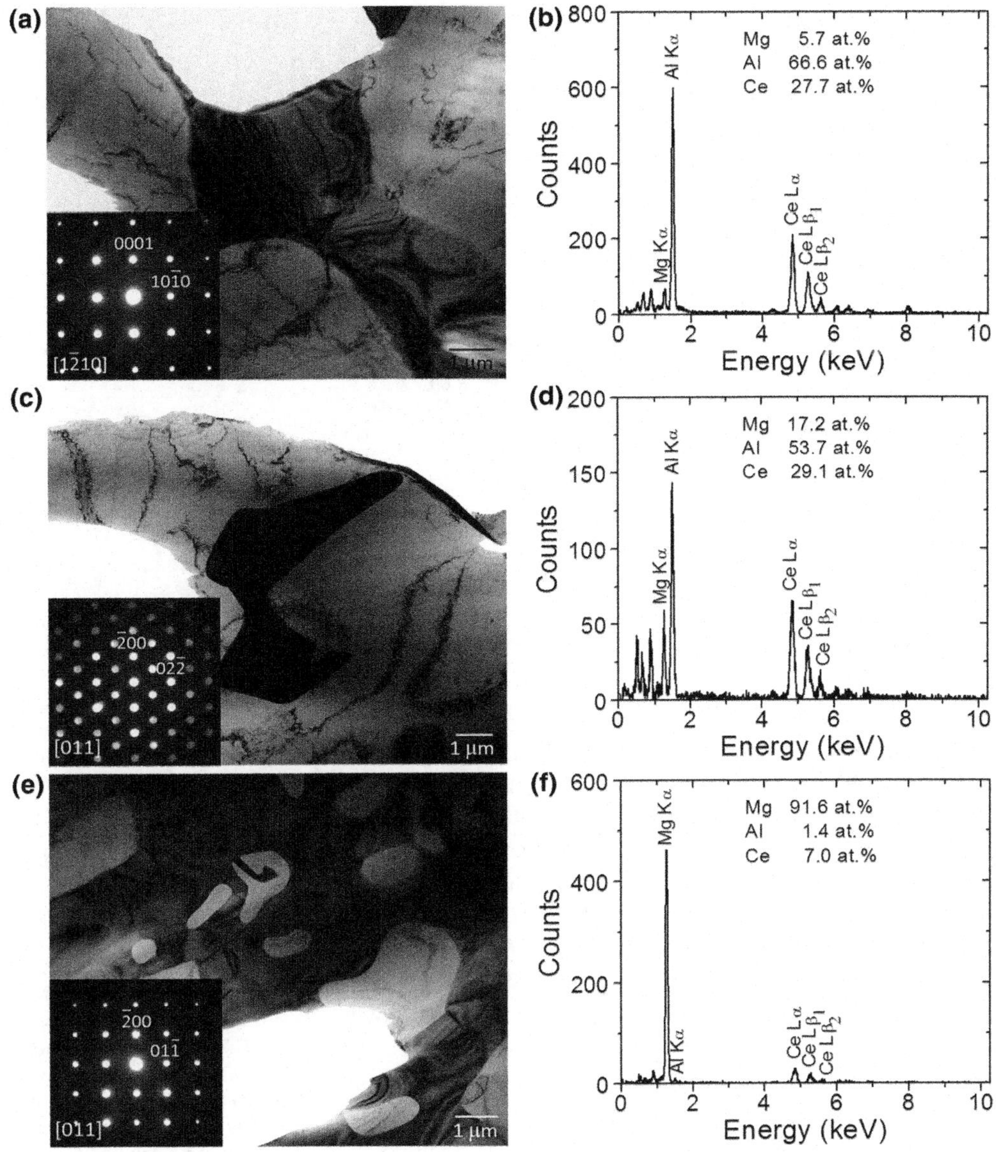


**Fig. 4** TEM image, micro beam electron diffractions and EDX spectra showing the microstructure and identification of intermetallic phases in as-cast ACe310 alloy. (**a**) and (**b**) the acicular intermetallic phase was identified as $Al_5Ce_2$, (**c**) and (**d**) the particulate-shaped intermetallic phase was identified as $Al_2Ce$, (**e**) and (**f**) the lamellar-like intermetallic was identified as $Mg_{12}Ce$

**Table 1** Space group and refined lattice parameters for phases identified in ACe310 alloy (the lattice parameters for the ACe412 alloy refined to similar values)

| Phase | Space group | Refined lattice parameters (Å) | |
|---|---|---|---|
| | | a | c |
| α-Mg | *P6₃/mmc* | 3.210 | 5.211 |
| $Mg_{12}Ce$ | *I4/mmm* | 10.316 | 5.943 |
| $Al_2Ce$ | *Fd-3 m* | 8.086 | – |
| $Al_5Ce_2$ | *P6/mmm* | 4.605 | 3.826 |

database. However, it is interesting to note that the metastable $Al_5La_2$ phase can be shown in the equivalent Mg–Al–La phase diagram by suppressing the thermodynamically stable $Al_2La$ phase in Pandat.

It is interesting that the $Al_2Ce$ particles in both the ACe310 and ACe412 alloys are located within the magnesium grains, as shown in Fig. 3b and d. This indicates that the $Al_2Ce$ phase is probably the primary phase developed in these alloys, even though the predicted primary phase in ACe310 alloy is the α-Mg phase, as shown in Fig. 1. In this case, the invariant point, phase boundaries and the predicted solidification pathway projected onto the liquidus phase diagram, as shown in Fig. 1, are inconsistent with the experimental results suggesting that the thermodynamic database for Mg–Al–Ce needs to be further refined.

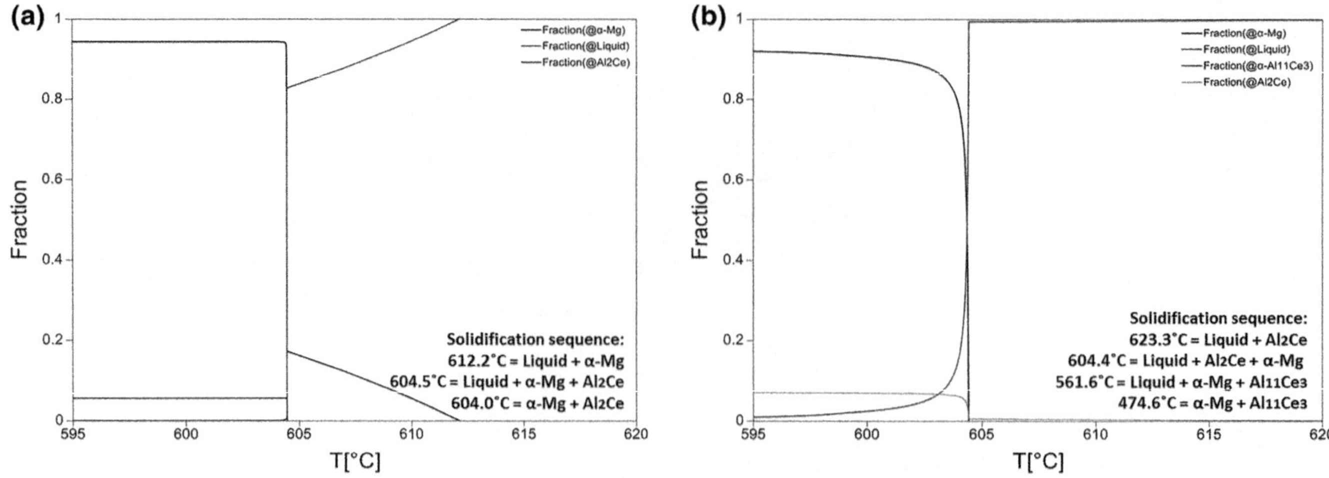


**Fig. 5** **a** ACe310 alloy and **b** ACe412 alloy solidification sequence predicted by Scheil-Gulliver equation using Pandat PanMagnesium 2017 database [10]

## Conclusions

In present study, the identified phases of ACe310 and ACe412 alloys are α-Mg, $Al_5Ce_2$, $Al_2Ce$ and $Mg_{12}Ce$. The intermetallic phase, $Al_5Ce_2$, develops an acicular morphology. The lattice parameters of this phase are significantly different from the $Al_5La_2$ phase reported in the literature. In addition, $Al_5Ce_2$ phase has not been reported in the open literature and it is currently not listed in the crystallographic database. Lastly, the phases identified in the ACe310 alloy are inconsistent with the Scheil-Gulliver predictions from PanMg 2017 thermodynamic database. The present study shows that ACe310 alloy is a hyper-eutectic alloy. The magnesium database and Pandat simulation are indeed very useful for both research and industrial purposes. However, further research is certainly required to improve the database for greater accuracy of thermodynamic predictions in the Mg–Al–Ce system.

**Acknowledgements** This study was supported by the Australian Research Council (Grant number LP130100828). RMIT Microscopy and Microanalysis Facility (RMMF) is acknowledged for access to experimental facilities. The authors would like to thank Dr. Aaron Seeber (CSIRO Manufacturing) for assistance with XRD data collection.

## References

1. Luo, A.A., *Magnesium casting technology for structural applications.* Journal of Magnesium and Alloys, 2013. 1(1): p. 2–22.
2. Aune, T.K. and T.J. Ruden, *High temperature properties of magnesium die casting alloys.* 1992, SAE Technical Paper.
3. Bakke, P., et al., *Magnesium Technology 2005.* TMS, Warrendale, PA, 2005: p. 291.
4. Bichler, L., C. Ravindran, and D. Sediako, *Onset of hot tearing in AE42 magnesium alloy.* Canadian Metallurgical Quarterly, 2009. 48(1): p. 81–89.
5. Sieracki, E.G., J.J. Velazquez, and K. Kabiri, *Compressive stress retention characteristics of high pressure die casting magnesium alloys.* 1996, SAE Technical Paper.
6. Li, N., et al., *Magnesium Engine Cradle-The USCAR Structural Cast Magnesium Development Project.* 2005, SAE Technical Paper.
7. Powell, B.R., et al., *Microstructure and creep behavior in AE42 magnesium die-casting alloy.* JOM, 2002. 54(8): p. 34–38.
8. Gröbner, J., et al., *Thermodynamic analysis of as-cast and heat-treated microstructures of Mg–Ce–Nd alloys.* Acta Materialia, 2011. 59(2): p. 613–622.
9. Zhu, S., et al., *The Influence of Individual Rare Earth Elements (La, Ce, or Nd) on Creep Resistance of Die-Cast Magnesium Alloy AE44.* Advanced Engineering Materials, 2016.
10. LLC, C., *Pandat software and PanMagnesium 2017 thermodynamic database*, Madison, Editor.
11. Pawley, G., *Unit-cell refinement from powder diffraction scans.* Journal of Applied Crystallography, 1981. 14(6): p. 357–361.
12. Rietveld, H., *A profile refinement method for nuclear and magnetic structures.* Journal of Applied Crystallography, 1969. 2(2): p. 65–71.
13. Buschow, K., *The lanthanum-aluminium system.* Philips Res. Rept., 1965. 20.
14. Rzychoń, T. and A. Kiełbus, *Microstructure and tensile properties of sand cast and die cast AE44 magnesium alloy.* Archives of Metallurgy and Materials, 2008. 53(3): p. 901–907.
15. Villars, P., L. Calvert, and W. Pearson, *Pearson's Handbook of Crystallographic Data for Intermetallic Phases. Volumes 1, 2, 3.* American Society for Metals, 1985, p. 3258.
16. ICDD, P.D.F., *International Centre for Diffraction Data (ICDD) PDF-4 + 2016.* Powder Diffraction File, Newtown Square, Pennsylvania, USA, 2016.
17. Rzychoń, T., A. Kiełbus, and L. Lityńska-Dobrzyńska, *Microstructure, microstructural stability and mechanical properties of sand-cast Mg–4Al–4RE alloy.* Materials Characterization, 2013. 83: p. 21–34.
18. Kleber, W., W.A. Wooster, and A. Wooster, *An introduction to crystallography.* 1970: VEB Verlag Technik.

19. Rzychoń, T., et al., *Microstructural stability and creep properties of die casting Mg–4Al–4RE magnesium alloy.* Materials Characterization, 2009. 60(10): p. 1107–1113.
20. Jiang, Z., et al., *Role of Al modification on the microstructure and mechanical properties of as-cast Mg–6Ce alloys.* Materials Science and Engineering: A, 2015. 645: p. 57–64.
21. Qiu, D. and M.-X. Zhang, *The nucleation crystallography and wettability of Mg grains on active Al2Y inoculants in an Mg–10 wt% Y Alloy.* Journal of Alloys and Compounds, 2014. 586: p. 39–44.
22. Hu, X., et al., *On grain coarsening and refining of the Mg–3Al alloy by Sm.* Journal of Alloys and Compounds, 2016. 663: p. 387–394.

# Material Design for Enhancing Toughness of Mg Alloy and Application for Biodegradable Devices

Toshiji Mukai

**Abstract**

Recently, Mg and its alloys have attracted much attention because of their excellent biocompatibility and biodegradability. High anisotropy of Mg crystal structure, however, limits the movement of some slip systems; therefore, pure magnesium possesses poor ductility and/or toughness. A number of studies have revealed that alloying with solute elements and modification of grain structure improved these drawbacks. In this study, to clarify the effect of adding solute elements, e.g., calcium and zinc, impact toughness testing and first-principles calculations of generalized stacking fault energy and grain boundary cohesive energy were conducted. For example, alloying magnesium with calcium and zinc and controlling the microstructure produced a Mg alloy with a high compressive fracture strain of 0.40, which was greater than the estimated maximum strain for fastening a surgical clip. This high fracture strain arose from the enhanced grain boundary cohesive energy and reduced anisotropy of slip systems by solute segregation. As a result, the alloy successfully occluded blood vessels.

**Keywords**

Mg alloy • Generalized stacking fault energy
Grain boundary cohesive energy • Toughness
Biodegradable implant

## Introduction

In laparoscopic surgery, vessels are usually occluded by surgical clips or staples instead of sutures owing to the restricted operating area. Currently, surgical clips and staples are made of pure Ti or Ti alloys since they are both strong and ductile enough to occlude blood vessels in soft tissue by forceps. In addition, it possesses excellent biocompatibility, which prevents inflammation. However, because Ti has a high corrosion resistance, its clips remain in the body permanently. The X-ray absorption coefficient of Ti is much higher than human tissue resulting in computed tomography artifacts, which hinder accurate diagnosis around the area where surgery has been performed. Therefore, biodegradable clips are required for surgery. Recently, Mg and its alloys have attracted much attention because of their excellent biocompatibility and biodegradability [1–3]. However, the high anisotropy of Mg crystal structure limits the movement of some slip systems. Therefore, pure Mg possesses poor ductility, which is confirmed by early fracturing during tensile tests and lower fracture toughness in three-point bending tests. Many studies have aimed to develop ductile Mg by controlling the microstructure and it was revealed that the basal texture affects the ductility of alloys [4, 5]. This research aimed to fabricate ductile Mg alloys containing calcium and zinc because these elements are present in the body. In addition, this research investigated the in vivo degradation behavior of the developed ductile magnesium alloy by implantation around the extraperitoneal tissue of mouse and suitability for occluding blood vessels of rat [6].

## Material Design

To clarify the effect of adding calcium and zinc to magnesium, impact toughness testing and first-principles calculations for grain boundary cohesive energy and generalized stacking fault energy (GSFE) in Mg, Mg–Ca, Mg–Ca–Zn alloys were conducted [7]. The impact toughness test demonstrated that energy absorption was improved by adding calcium and zinc. For crack initiation, the absorbed energy was higher in the ternary Mg alloy than in Mg.

T. Mukai (✉)
Department of Mechanical Engineering, Graduate School of Engineering, Kobe University, 1-1 Rokkodai, Nada, Kobe, Hyogo 657-8571, Japan
e-mail: mukai@mech.kobe-u.ac.jp


D. Orlov et al. (eds.), *Magnesium Technology 2018*, The Minerals, Metals & Materials Series, https://doi.org/10.1007/978-3-319-72332-7_14

The amount of energy absorption from the crack initiation by crack propagation below 0.4 mm was the higher in Mg–Ca–Zn than in Mg.

The Rice–Wang model proposes a theoretical relationship between the change in the microscale electronic structures and the macroscale grain boundary strengthening or decohesion [8]. In the model, the ideal work of interfacial separation ($2\gamma_{int}$) is the final important factor in grain boundary strengthening. $\gamma_{int}$ can be calculated as the difference between the two fracture surface energy ($2\gamma_s$) and grain boundary energy ($\gamma_{GB}$). It corresponds to the minimum required energy to form two fracture surfaces by dividing the grain boundary. Yamaguchi et al. referred to $2\gamma_{int}$ as the "grain boundary cohesive energy", and presented a mechanism of grain boundary decohesion caused by the grain boundary segregation of impurity atoms in bcc Fe and fcc Ni systems [9]. In the present study, $2\gamma_{int}$ was calculated for Mg, Mg–Ca, and Mg–Ca–Zn, and the effects of the addition of Ca and Zn on grain boundary strength were estimated. Actual materials contain mainly random grain boundaries in which atoms are randomly arranged. However, it is difficult to model the structure of random grain boundaries. Instead, a calculation model with a {1 1 −2 1} symmetrically tilted grain boundary, which has a relatively high grain boundary energy [10] and approximates the random grain boundary of Mg for calculating $2\gamma_{int}$, was used. Calculations of the grain boundary cohesive energy, $2\gamma_{int}$, showed that adding calcium and zinc strengthened the grain boundary in Mg alloys. The magnitude relation of the grain boundary cohesive energy corresponded to that of the energy absorbed in crack initiation by crack propagation below 0.4 mm during impact toughness testing. These results therefore suggest that the energy absorption during crack propagation is increased in the ternary Mg alloys by the addition of calcium and zinc strengthening the grain boundaries.

Calculations of GSFE for the basal and prismatic slip demonstrated that the addition of calcium and zinc to magnesium caused solid solution softening in the prismatic slip. The ratio of unstable SFE $\gamma_{us\ (basal)}/\gamma_{us\ (prism)}$ indicated that the addition of Ca and Zn decreased CRSS of the prismatic slip and made the deformation behavior of Mg isotropic. However, measurements of the V-notch opening displacement in impact toughness testing indicated that plastic blunting was increased in the ternary Mg alloy. Therefore, it is suggested that reducing the plastic anisotropy alleviated the stress concentration and increased the plastic deformability in the ternary Mg alloy [7].

## Application for Biodegradable Implant Devices

To develop a biodegradable clip, the equivalent plastic strain distribution during occlusion was evaluated by the finite element analysis (FEA) using the material data of pure Mg and Mg–Ca–Zn alloy [6]. Since the FEA suggested that a maximum plastic strain of 0.40 is required to allow the Mg clips, the alloying of magnesium with the essential elements and the control of microstructure by hot extrusion and annealing were conducted. Mechanical characterization revealed that the Mg–Ca–Zn alloy obtained by extrusion followed by annealing possessed a fracture strain over 0.40. FEA using the material data for the ductile Mg–Ca–Zn alloy also showed that the clip could occlude the simulated vessel without fracture.

The biocompatibility of the alloy was confirmed by investigating its degradation behavior and the response of extraperitoneal tissue around the Mg–Ca–Zn alloy [6]. Little gas generation was observed following implantation of the developed Mg–Ca–Zn clip by in vivo micro-CT. Histological analysis, minimal observed inflammation, and an only small decease in the volume of the implanted Mg–Ca–Zn clip confirmed its excellent biocompatibility. Microstructural observations using electron backscattering diffraction confirmed that dynamic recovery occurred during the later stage of plastic deformation of the ductile Mg–Ca–Zn alloy.

To evaluate the stability and clinical feasibility of the clip fabricated from the ductile Mg–Ca–Zn alloy bar, short-term animal experiment on Wistar rat were performed [6]. After laparotomy, left renal vein were exposed from surrounding connective tissue, occluded twice using the clips, and then cut in-between. After transection, vein stumps were observed for 10 min to evaluate the hemostatic performance. No crack was confirmable at the site of the highest equivalent plastic strain, even after the clip was fully fastened. Though the blood circulation was maintained, no drop of the blood was monitored after the excision. The rat survived the excision owing to the successful clipping, demonstrating that the ductile Mg alloy clip possessed sufficient hemostatic properties.

## Summary

First principles calculations demonstrate that alloying magnesium with calcium and zinc enhances the grain boundary cohesive energy and reduces the plastic anisotropy. Fracture behavior in a ternary Mg–Ca–Zn alloy supports the calculation results with increased crack opening displacement and enhanced absorption energy under dynamic loading.

Controlling the microstructure produced the ternary Mg–Ca–Zn alloy with a high compressive fracture strain of 0.40, which was greater than the estimated maximum strain for fastening clip. This high fracture strain arose from the dynamic recovery during plastic deformation of the alloy. The alloy successfully occluded blood vessels. Micro CT images showed that the Mg–Ca–Zn clips degraded homogeneously, resulting in gradual gas generation and producing no inflammation of the tissue around the magnesium clips.

**Acknowledgements** The author is grateful to Prof. Takumi Fukumoto, Dr. Naoko Ikeo and Mr. Takayuki Hase at Kobe University, Dr. Masatake Yamaguchi at Japan Atomic Energy Agency and Dr. Alok Singh at National Institute for Materials Science for their collaborative works. This work was financially supported by JSPS Grant-in-Aid for Scientific Research in No. 25246012, No. 17H01327 and Hyogo COE Program Promotion Project FY2016.

## References

1. F. Witte, V. Kaese, H. Haferkamp, E. Switzer, A. Meyer-Lindenberg, CJ Wirth, H. Windhagen, In vivo corrosion of four magnesium alloys and the associated bone response, Biomaterials 26 (2005) 3557–3563.
2. F. Witte, The history of biodegradable magnesium implants: A review, Acta Biomaterialia, 6 (2010), 1680–1692.
3. Z. Li, X. Gu, S. Lou, Y. Zheng; The development of binary Mg–Ca alloys for use as biodegradable materials within bone, Biomaterials, 29 (2008), 1329–1344.
4. T. Mukai, M. Yamanoi, H. Watanabe, K. Higashi, Ductility enhancement in AZ31 magnesium alloy by controlling its grain structure, Scripta Materialia, 45 (2001), 89–94.
5. H. Somekawa, M. Yamaguchi, Y. Osawa, A. Singh, M. Itakura, T. Tsuru, T. Mukai, Material design for magnesium alloys with high deformability, Philosophical Magazine, 95 (2015) 869–885.
6. N. Ikeo, R. Nakamura, K. Naka, T. Hashimoto, T. Yoshida, T. Urade, K. Fukushima, H. Yabuuchi, T. Fukumoto, Y. Ku, T. Mukai, Fabrication of a magnesium alloy with excellent ductility for biodegradable clips, Acta Biomaterialia., 29 (2016) 468–476.
7. T. Hase, T. Ohtagaki, M. Yamaguchi, N. Ikeo, T. Mukai, Effect of aluminum or zinc solute addition on enhancing impact fracture toughness in Mg-Ca alloys, Acta Materialia. 104 (2016) 283–294.
8. J. Rice and J. Wang, *Mater. Sci. Eng. A* 107, 23 (1989).
9. M. Yamaguchi, M. Shiga, and H. Kaburaki, Science 307, 393 (2005).
10. J.T. Chou, K. Ikeda, H. Nakashima, Energy and structure of [1−100] symmetric tilt grain boundaries in magnesium, J. Japan Institute Metals, 69 (2005) 303–307.

# Influences of Yttrium Content on Microstructure and Mechanical Properties of as-cast Mg–Ca–Y–Zr Alloys

Sihang You, Yuanding Huang, Karl Ulrich Kainer, and Norbert Hort

**Abstract**

The microstructure and mechanical properties of as-cast Mg–Ca–Y–Zr alloys with different Y contents were investigated. The alloy containing 0.5 wt% Y exhibited finer grains compared to the alloys with higher Y content. All alloys had a dendritic microstructure with eutectics composed of α-Mg and Ca-rich intermetallic phases. Few Mg–Y-rich intermetallic particles were also found along grain boundaries. EDS analysis showed that the solute Y segregated at dendritic and grain boundaries. The amount of Y contained in eutectics remarkably increased with increasing Y. In addition, the eutectics volume fractions of all alloys were comparable but the morphology became less continuous at higher Y contents. Both the room temperature tensile and compressive strengths were largely improved with increasing Y content. Moreover, the elevated temperature compression tests showed that the compressive yield strength first decreased slightly when the temperature rose to 175 °C, but then remained stable as the temperature increased.

**Keywords**

Mg–Ca–Y–Zr alloy • Mechanical properties
Microstructure

## Introduction

For the past few years, there is increased interest in developing Mg–Ca based alloys as Ca improves the ignition resistance, creep resistance and weakens the texture of wrought Mg alloys [1–3]. A recent study reported the addition of In to Mg–Ca alloy resulted in the formation of fine scale prismatic plate which enhanced the precipitation hardening behavior of Mg–Ca–In alloy [4]. This shows that the microstructure and mechanical properties of Mg–Ca alloys can be modified with ternary addition. RE (rare earth) elements are one of the most attractive additions due to their significant effect on enhancing the mechanical properties of Mg alloys at both room and elevated temperatures [5–7]. Among all RE elements, Y (yttrium) with a relatively high solid solubility in Mg was reported to have a remarkable solid solution strengthening effect [8–10]. In a recent work, Mg–0.3Ca–2.4Y alloy was shown to have a relatively random orientation of grains after extrusion with a high elongation to failure (37%) at room temperature [11]. In this case, the maximum yield strength was only 117 MPa [11]. Moreover, Li et al. [12] investigated Mg–1Ca–1Y alloy as a potential biomaterial. They reported that the Y addition could enhance the ductility and lower the corrosion resistance of Mg–Ca alloy. The Mg–Ca–Y alloys would not only enlarge the scope of structural application but also the biodegradable implant applications. However, the effect of Y addition on the microstructure and mechanical properties of Mg–Ca–Y–Zr alloy has not been investigated systematically.

This research aims to characterize the influence of Y addition on microstructure and mechanical properties of Mg–Ca–Y–Zr alloys. Scanning electron microscopy and optical microscope were used to characterize the microstructure of the alloys. Further, the tensile test of as-cast alloys was tested at room temperature and the compressive properties was tested at both room and elevated temperatures.

## Experimental

A series of Mg–Ca–Y–Zr alloys with different content of Y were prepared by permanent mould direct chill casting [13]. High-purity Mg was molten under Ar + 2% $SF_6$ protective atmosphere. Pure Y, pure Ca and Mg-33.3 wt% Zr master alloy were then added into the melt at 750 °C. After homogenizing by mechanical stirring under 200 rpm for

S. You (✉) · Y. Huang · K. U. Kainer · N. Hort
Magnesium Innovation Centre, Helmholtz-Zentrum Geesthacht, Max-Planck-Strasse 1, 21502 Geesthacht, Germany
e-mail: sihang.you@hzg.de


D. Orlov et al. (eds.), *Magnesium Technology 2018*, The Minerals, Metals & Materials Series, https://doi.org/10.1007/978-3-319-72332-7_15

20 min, the melt was poured into a steel crucible preheated to 680 °C and then held at 680 °C for 10 min with gas protection. Finally, the melt was solidified by lowering the crucible into cooling water with a rate of 10 mm/s. The ingot was a cylinder with a size of Φ70 mm × 180 mm. The chemical compositions of all alloys were analyzed, where Ca was analyzed by atomic absorption spectroscopy (AAS) and other elements were analyzed by X-ray fluorescence (XRF).

The as-cast samples for microstructural analysis were etched in a solution of 8 g picric acid, 6 ml acetic acid, 20 ml distilled water and 100 ml ethanol after grinding and mechanical polishing. Metallographic structures were characterized by Reichert-Jung MeF3 optical microscope with a digital camera. The average grain size was measured by the linear intercept method. The microstructure were further analyzed by a VEGA3 TESCAN scanning electron microscope (SEM) equipped with energy dispersive spectroscopy (EDS) at an accelerative voltage of 20 kV.

Tensile specimens with a gauge length of 30 mm and a diameter of 6 mm were tested at room temperature according to DIN EN 10002. Compressive specimens with a size of Φ11 × 17 mm were tested at room temperature, 175, 200, 225 and 250 °C. All the tests were performed under a strain rate of $1 \times 10^{-3}\ s^{-1}$ using a Zwick 050 machine. At least three specimens were tested under each condition.

## Results and Discussion

### Effect of Y Addition on Microstructure

The chemical composition and average grain size of the investigated alloys was listed in Table 1. The alloy containing 0.5 wt% Y exhibits a more homogeneous grain structure and the finest grains with an average grain size of 63.1 ± 25.3 μm. When the content of Y addition increases (to more than 1 wt%), the average grain size first increases to 91.3 ± 40.2 μm and then tends to be stable. However, the overall grain structure becomes inhomogeneous (Fig. 1).

**Table 1** Chemical compositions (wt%) and grain sizes of the investigated alloys

| Alloys | Y | Ca | Zr | Mg | Grain size (μm) |
|---|---|---|---|---|---|
| Mg–0.3Ca–0.5Zr–0.5Y | 0.51 | 0.26 | 0.32 | Balance | 63.1 ± 25.3 |
| Mg–0.3Ca–0.5Zr–1Y | 1.22 | 0.27 | 0.30 | Balance | 91.3 ± 40.2 |
| Mg–0.3Ca–0.5Zr–3Y | 3.21 | 0.28 | 0.39 | Balance | 89.7 ± 39.1 |
| Mg–0.3Ca–0.5Zr–5Y | 4.36 | 0.29 | 0.28 | Balance | 92.2 ± 38.4 |

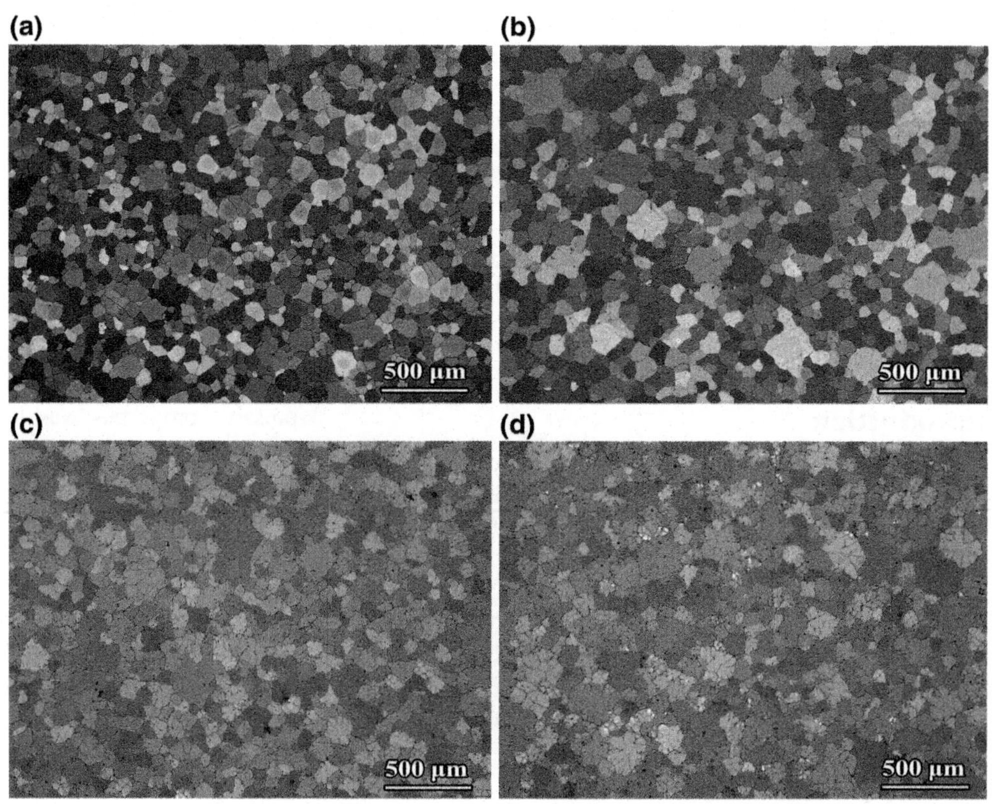


**Fig. 1** Optical microstructure of as-cast Mg–Ca–Y–Zr alloys. **a** Mg–0.3Ca–0.5Zr–0.5Y; **b** Mg–0.3Ca–0.5Zr–1Y; **c** Mg–0.3Ca–0.5Zr–3Y; **d** Mg–0.3Ca–0.5Zr–5Y

Figure 2 shows the SEM BSE (back scattered electron) micrographs of the investigated alloys. The microstructure of all the alloys consists of α-Mg matrix with dendritic morphology and Mg–Ca-rich eutectics distributed at dendrites and grain boundaries. Few Y-rich intermetallic particles at dendrites and grain boundaries are also found in the solidification microstructure. The amount of these Y-rich intermetallics slightly increases with increasing the content of Y.

The quantitative EDS analysis of the positions presented in Fig. 2 is listed in Table 2. It indicates that the eutectic contains high concentrations of Ca, which is attributed to the low solid solubility of Ca in Mg. Because there is only trace amount of Mg–Ca-rich phases existing in the microstructure, XRD (X-ray diffraction) could not be used to identify the phases. Further TEM (transmission electron microscopy) analysis will be performed in the future. According to previous investigations [14], the Mg–Ca-rich intermetallic phase is assumed to be $Mg_2Ca$. As Y is has a high solubility in Mg, it can easily segregate at the front liquid-solid interface during solidification. Therefore, most of Y atoms are soluble in α-Mg forming the segregation area along the dendrites. Less Y is presented in the Mg–Y-rich intermetallics. The content of Y in the segregation areas increases remarkably with increasing Y addition. More interestingly, Y was detected in $Mg_2Ca$ phases after its addition to Mg–Ca alloys. Its amount rapidly increases when the content of Y increases from 0.5 to 5 wt%. This could be attributed to the formation of substitutional solid solution of $Mg_2$(Ca, Y) according to the research reported by Murthy et al. [15]. They found the solubility limit of Y in $Mg_2Ca$ phase was 6.9 at.% where Ca was substituted by Y.

The volume fraction of the eutectic was determined from BSE micrograph analysis, which is listed in Table 3. It was found that the eutectic volume fraction slightly increases with the increasing Y content. According to EDS analysis, the eutectic was composed of α-Mg and $Mg_2Ca$ with Y dissolved in the intermetallic. The content of Y does not significantly change the average volume fraction of the eutectic, which can be attributed to the existence of Y in $Mg_2Ca$ phases. In addition, the morphology of the eutectic changes with different Y additions. At low contents (0.5 and 1 wt%) of Y, the eutectic mainly exhibit the continuous or quasi-continuous distribution. After adding more Y with 3 or 5 wt%, the morphology changes from initial continuous to disconnected shapes.

## Effect of Y Addition on Mechanical Properties

Figure 3 shows the typical tensile stress-strain curves of the investigated alloys at room temperature. The tensile yield strength (TYS), ultimate tensile strength (UTS) and elongation to failure (El) are presented in Table 4. At 0.5 wt% Y addition, the Mg–Ca–Y–Zr alloy has the lowest TYS and the

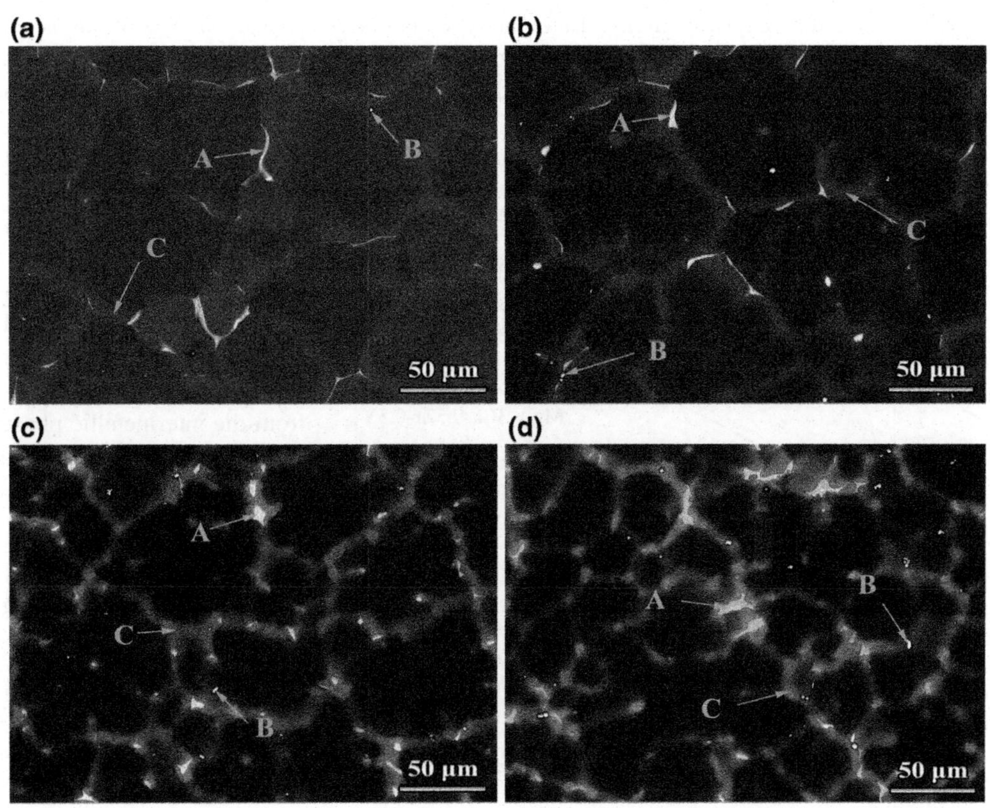


**Fig. 2** SEM BSE micrographs of as-cast Mg–Ca–Y–Zr alloys. **a** Mg–0.3Ca–0.5Zr–0.5Y; **b** Mg–0.3Ca–0.5Zr–1Y; **c** Mg–0.3Ca–0.5Zr–3Y; **d** Mg–0.3Ca–0.5Zr–5Y. Positions (**a**), (**b**), (**c**) presented in each micrographs are analyzed by EDS, which are shown in Table 2

**Table 2** Quantitative EDS analysis of the positions indicated in Fig. 2

| Areas | Positions | Composition (at.%) | | | |
|---|---|---|---|---|---|
| | | Mg | Ca | Y | Zr |
| Ca-rich phases | Figure 2a—A | 90.28 | 9.14 | 0.55 | 0.03 |
| | Figure 2b—A | 92.02 | 7.04 | 0.91 | 0.03 |
| | Figure 2c—A | 88.41 | 8.56 | 2.94 | 0.09 |
| | Figure 2d—A | 88.83 | 6.84 | 4.21 | 0.12 |
| Y-rich phases | Figure 2a—B | 80.97 | 0.84 | 16.87 | 1.32 |
| | Figure 2b—B | 79.46 | 0.65 | 18.97 | 0.92 |
| | Figure 2c—B | 76.01 | 0.77 | 21.71 | 1.51 |
| | Figure 2d—B | 78.24 | 0.53 | 20.21 | 1.02 |
| Segregation area | Figure 2a—C | 99.22 | 0.35 | 0.41 | 0.02 |
| | Figure 2b—C | 99.02 | 0.33 | 0.62 | 0.03 |
| | Figure 2c—C | 97.64 | 0.38 | 1.86 | 0.12 |
| | Figure 2d—C | 95.90 | 0.36 | 3.61 | 0.13 |

**Table 3** Average eutectic volume fraction of as-cast Mg–Ca–Y–Zr alloys

| Alloys | Volume fraction of eutectic (%) |
|---|---|
| Mg–0.3Ca–0.5Zr–0.5Y | 0.79 ± 0.12 |
| Mg–0.3Ca–0.5Zr–1Y | 0.74 ± 0.10 |
| Mg–0.3Ca–0.5Zr–3Y | 1.02 ± 0.11 |
| Mg–0.3Ca–0.5Zr–5Y | 1.17 ± 0.06 |

largest El. The alloy containing 1 wt% Y exhibits a slightly higher TYS but lower El and UTS compared to the alloy with 0.5 wt% Y. When Y content is more than 1 wt%, both the YTS and UTS increase significantly with increasing Y addition.

Obviously, the relatively higher El of alloy containing 0.5 wt% can be easily explained by its finer grain size and more homogeneous grain structure compared to other three alloys. It is known that a finer grain size is beneficial to achieve higher ductility for Mg alloys. According to the previous work, the coarse and continuous second phase formed along grain boundaries deteriorate the ductility of Mg alloys [16]. Apparently, the lowest El and UTS obtained in the alloy containing 1 wt% Y could be attributed to its coarse and continuous eutectic distributed along grain boundaries and to the coarser grain size compared to the alloy with 0.5 wt% Y. As the grain sizes of the alloys with 1, 3 and 5 wt% Y are comparable, their El are similar.

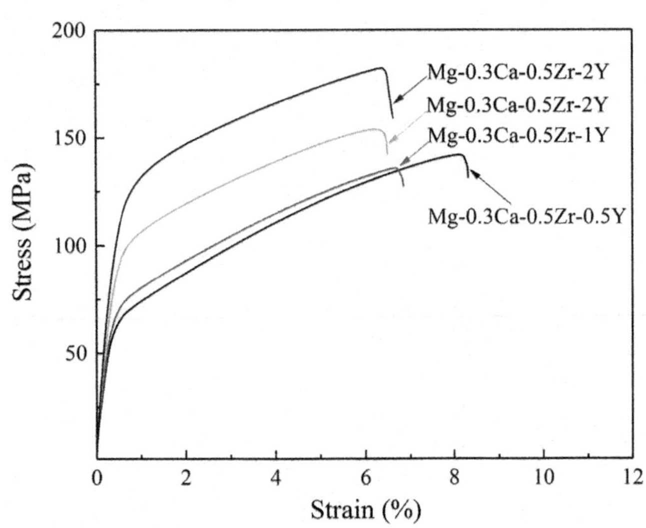


**Fig. 3** Tensile stress-strain curves of the as-cast Mg–Ca–Y–Zr alloys at room temperature

The relationship between the tensile yield strength and microstructure for Mg alloys was modelled by Cáceres et al. [17]. The contribution to yield strength consisted five key strengthening components, including Peierls stress $\sigma_0$ which is the force needed to move a dislocation within a plane of atoms in the unit cell, solid solution strengthening $\Delta\sigma_{ss}$, grain boundary strengthening $\Delta\sigma_{gb}$ which can be described as Hall-Petch relationship, grain boundary reinforcement from the intermetallic phases distributed at grain boundaries $\Delta\sigma_{gbr}$, and dislocations strengthening $\Delta\sigma_{dis}$. The overall yield stress can be described as:

$$\sigma_{0.2} = \sigma_0 + \Delta\sigma_{ss} + \Delta\sigma_{gb} + \Delta\sigma_{gbr} + \Delta\sigma_{dis} \quad (1)$$

As to the current investigation, all the alloys are at as-cast condition, the contribution dislocations strengthening $\Delta\sigma_{dis}$ can be neglected. The yield strength can then be modelled as Eq. (2),

$$\sigma_{0.2} = \sigma_0 + \Delta\sigma_{ss} + \Delta\sigma_{gb} + \Delta\sigma_{gbr} \quad (2)$$

**Table 4** Tensile properties of the as-cast Mg–Ca–Y–Zr alloys at room temperature

| Alloys | TYS (MPa) | UTS (MPa) | El (%) |
|---|---|---|---|
| Mg–0.3Ca–0.5Zr–0.5Y | 66 ± 1 | 145 ± 5 | 8.3 ± 1.2 |
| Mg–0.3Ca–0.5Zr–1Y | 72 ± 1 | 136 ± 2 | 6.3 ± 0.5 |
| Mg–0.3Ca–0.5Zr–3Y | 97 ± 1 | 150 ± 5 | 5.4 ± 1.1 |
| Mg–0.3Ca–0.5Zr–5Y | 120 ± 1 | 191 ± 7 | 6.0 ± 1.6 |

The grain boundary strengthening contribution can be estimated by Hall-Petch relationship:

$$\Delta\sigma_{gb} = kd^{-1/2} \quad (3)$$

where $k$ is a constant determined for polycrystalline materials and $d$ is the grain size. The value $k$ for Mg alloy is between 280–320 MPa $\mu m^{1/2}$ [18]. In the present case, the value $k$ = 300 MPa $\mu m^{1/2}$ will be used in the following analysis. With $\sigma_0 \approx 11$ MPa for pure Mg [19] and grain size of the experimental alloys listed in Table 1, the grain boundary strengthening contribution can be calculated as 48 MPa for alloy containing 0.5 wt% of Y and 42 MPa for the alloys containing 1, 3 and 5 wt% of Y. Obviously, the grain boundary strengthening contribution of the alloys containing higher level of Y is less than that of the alloy with 0.5 wt%, which means the significant improvement in TYS by adding more Y is possible related to other mechanisms.

It is well known that a large amount of Y can be dissolved in Mg. As a result, Y has a significant solution strengthening effect in Mg alloys. The solution strengthening component $\Delta\sigma_{ss}$ of solution heat-treated alloys is given by

$$\Delta\sigma_{ss} \cong k_s c^{2/3} \quad (4)$$

where $k_s$ is a constant and $c$ is the atom concentration. The $k_s$ value of solution treated Mg–Y alloys has been determined by Gao et al. [9] as $k_s$ = 1249 MPa (at.%)$^{-2/3}$. The $\Delta\sigma_{ss}$ of alloys containing 0.5, 1, 3 and 5 wt% calculated by Eq. (4) are 13, 23, 46 and 58 MPa. For the as-cast Mg–Ca–Y–Zr alloys investigated in this work, most of the Y atoms segregates at the dendrites but not dissolved homogeneously inside α-Mg matrix. Therefore, the solid solution contribution to the strength of as-cast alloys is expected to be less than the values estimated by Eq. (4).

As shown in Fig. 2, the size of the eutectic along the grain boundaries in the alloys with higher content of Y (3 and 5 wt%) is obviously smaller than that in the alloys with lower content of Y (0.5 and 1 wt%). It can also be observed that the morphology of eutectic becomes less continuous and increasingly disconnect with the increasing of Y content from 0.5 to 5 wt%. Since finer second phases distributed along grain boundaries are more effective to act as barriers for dislocation motion, thus the mechanical properties are improved [20]. It can be inferred that the relatively higher TYS of the alloys with higher content of Y is also related to the different size and morphology of $Mg_2Ca$ phases, although their eutectic volume fractions are relatively similar.

Table 5 shows the compressive properties of the experimental alloys at both the room and elevated temperatures, including compressive yield strength (CYS), ultimate compressive strength (UCS) and elongation to failure (El). The Y additions improve both the CYS and UCS at all temperatures.

**Table 5** Compressive properties of as-cast Mg–Ca–Y–Zr alloys at different temperatures

| Alloys | | Mg–0.3Ca–0.5Zr–0.5Y | Mg–0.3Ca–0.5Zr–1Y | Mg–0.3Ca–0.5Zr–3Y | Mg–0.3Ca–0.5Zr–5Y |
|---|---|---|---|---|---|
| CYS (MPa) | RT | 62 ± 1 | 71 ± 1 | 100 ± 2 | 126 ± 2 |
| | 175 °C | 60 ± 1 | 66 ± 1 | 85 ± 3 | 103 ± 1 |
| | 200 °C | 57 ± 1 | 68 ± 2 | 83 ± 1 | 100 ± 1 |
| | 225 °C | 59 ± 1 | 65 ± 2 | 81 ± 2 | 101 ± 2 |
| | 250 °C | 56 ± 1 | 65 ± 1 | 79 ± 2 | 96 ± 1 |
| UCS (MPa) | RT | 300 ± 4 | 312 ± 3 | 323 ± 2 | 352 ± 3 |
| | 175 °C | 253 ± 4 | 272 ± 1 | 288 ± 9 | 323 ± 4 |
| | 200 °C | 251 ± 5 | 275 ± 4 | 289 ± 2 | 318 ± 3 |
| | 225 °C | 246 ± 7 | 262 ± 2 | 292 ± 5 | 319 ± 4 |
| | 250 °C | 255 ± 9 | 254 ± 9 | 282 ± 6 | 311 ± 9 |
| El. (%) | RT | 28.2 ± 1.9 | 28.2 ± 1.2 | 26.3 ± 2.3 | 22.2 ± 2.1 |
| | 175 °C | 26.3 ± 1.3 | 24.3 ± 1.1 | 20.5 ± 1.8 | 20.3 ± 1.2 |
| | 200 °C | 31.1 ± 1.5 | 26.3 ± 0.9 | 21.4 ± 1.1 | 22.1 ± 1.4 |
| | 225 °C | 40.2 ± 1.3 | 35.3 ± 2.1 | 28.2 ± 2.3 | 25.2 ± 1.1 |
| | 250 °C | 51.2 ± 2.1 | 45.2 ± 3.2 | 38.1 ± 2.4 | 34.1 ± 3.1 |

The as-cast Mg–0.3Ca–0.5Zr–5Y alloy exhibits the highest CYS and UCS at all temperatures. For the alloys containing higher amount of Y (3, 5 wt%), their CYS decreases more significantly than that with lower Y content (0.5, 1 wt%) when the temperature increases from the room temperature to 175 °C. But then the CYS of all alloys remains stable when further increasing the test temperature to 250 °C. Similarly, the UCS of all alloys first decreases dramatically with increasing the temperature to 175 °C, but then decreases slightly when the temperature increases to 250 °C. In contrast, the elongation to failure of all alloys decreases with increasing the Y content. It increases continuously with increasing the temperature from the room temperature to 250 °C.

The good elevated temperature strength and thermal stability of the alloy can be attributed to several aspects. Firstly, the solid solubility of Y increases with increasing the test temperature, which may result in the improvement of solid solution strengthening effect. In addition, the thermal-stable $Mg_2Ca$ phases formed at the dendrites and grain boundaries could effectively hinder the grain boundary motion, especially at the elevated temperatures. The increased elongation to failure at the elevated temperatures should be attributed to the activation of non-basal slip systems.

## Conclusions

The effects of Y addition on the microstructure and mechanical properties of as-cast Mg–0.3Ca–0.5Zr–xY (x = 0.5, 1, 3, 5 wt%) were investigated. The alloy containing 0.5 wt % Y exhibited finer grains compared to the other alloys with higher Y content. The microstructure of all alloys had a dendritic morphology and consisted of α-Mg matrix, $Mg_2Ca$ phases and few Mg–Y-rich intermetallic phases. The amount of Y addition had a significant effect on the morphology of eutectics, but no obvious effect on the eutectic volume fraction. The alloys with 0.5 and 1 wt% of Y had eutectics with continuous or quasi-continuous distribution, but the eutectic morphology changed to disconnected shapes after adding more Y with 3 and 5 wt%. Both the tensile and compressive strength were largely improved with the increasing addition of Y at all temperatures, which can be attributed to the grain boundary strengthening, solid solution strengthening and grain boundary reinforcement effect. Moreover, the CYS and UCS of all alloys first decreased with increasing the temperature to 175 °C and then remained stable when the temperature further increased to 250 °C. The good mechanical property and thermal-stability of Mg–Ca–Y–Zr alloys at elevated temperatures were attributed to the thermally stable $Mg_2Ca$ phases distributed along grain boundaries and the outstanding solid solution strengthening effect of Y in Mg alloy.

**Acknowledgements** The authors are thankful to Mr. G. Meister of MagIC at Helmholtz-Zentrum Geesthacht for technical support. One of the authors (S. You) acknowledges the financial support from China Scholarship Council (CSC).

## References

1. M. Sakamoto, S. Akiyama, K. Ogi, Suppression of Ignition and Burning of Molten Mg Alloys by Ca Bearing Stable Oxide Film, Journal of Materials Science Letters 16 (1997) 1048–1050.
2. B. Jing, S. Yangshan, X. Shan, X. Feng, Z. Tianbai, Microstructure and Tensile Creep Behavior of Mg–4Al Based Magnesium Alloys with Alkaline-Earth Elements Sr and Ca Additions, Materials Science and Engineering: A 419 (2006) 181–188.
3. N. Stanford, The Effect of Calcium on the Texture, Microstructure and Mechanical Properties of Extruded Mg–Mn–Ca Alloys, Materials Science and Engineering: A 528 (2010) 314–322.
4. C. Mendis, K. Oh-ishi, T. Ohkubo, K. Hono, Precipitation of Prismatic Plates in Mg–0.3Ca Alloys with In Additions, Scripta Materialia 64 (2011) 137–140.
5. B. Mordike, T. Ebert, Magnesium: Properties—Applications—Potential, Materials Science and Engineering: A 302 (2001) 37–45.
6. M. Suzuki, H. Sato, K. Maruyama, H. Oikawa, Creep Behavior and Deformation Microstructures of Mg–Y Alloys at 550 K, Materials Science and Engineering: A 252 (1998) 248–255.
7. H. Wang, Q.D. Wang, C.J. Boehlert, D.D. Yin, J. Yuan, Tensile and Compressive Creep Behavior of Extruded Mg–10Gd–3Y–0.5Zr (Wt.%) Alloy, Materials Characterization 99 (2015) 25–37.
8. N. Stanford, R. Cottam, B. Davis, J. Robson, Evaluating the Effect of Yttrium as a Solute Strengthener in Magnesium Using in Situ Neutron Diffraction, Acta Materialia 78 (2014) 1–13.
9. L. Gao, R.S. Chen, E.H. Han, Solid Solution Strengthening Behaviors in Binary Mg–Y Single Phase Alloys, Journal of Alloys and Compounds 472 (2009) 234–240.
10. L. Gao, R.S. Chen, E.H. Han, Effects of Rare-Earth Elements Gd and Y on the Solid Solution Strengthening of Mg Alloys, Journal of Alloys and Compounds 481 (2009) 379–384.
11. N. Zhou, Z.Y. Zhang, J. Dong, L. Jin, W.J. Ding, High Ductility of a Mg–Y–Ca Alloy Via Extrusion, Materials Science and Engineering: A 560 (2013) 103–110.
12. Y. Li, P.D. Hodgson, C.e. Wen, The Effects of Calcium and Yttrium Additions on the Microstructure, Mechanical Properties and Biocompatibility of Biodegradable Magnesium Alloys, Journal of materials science 46 (2011) 365–371.
13. F.R. Elsayed, N. Hort, M.A. Salgado Ordorica, K.U. Kainer, Magnesium Permanent Mold Castings Optimization, in: (Ed.), Materials Science Forum, Trans Tech Publ, 2011, pp. 65–68.
14. H.J. Fei, G.L. Xu, L.B. Liu, H. Bo, L.J. Zeng, C.P. Chen, Phase Equilibria in Mg-Rich Corner of Mg–Ca–RE (Re = Gd, Nd) Systems at 400 Degrees C, Transactions of Nonferrous Metals Society of China 23 (2013) 881–888.
15. Y. Murthy, Y. Zhang, M. Medraj, Experimental Determination of the Phase Equilibrium in the Mg–Ca–Y System, in: (Ed.), Materials Science & Technology 2013, pp. 1555–1563.

16. B. Kim, J. Jeon, K. Park, B. Park, Y. Park, I. Park, Microstructural Characterisation and Mechanical Properties of Mg–xSn–5Al–1Zn Alloys, International Journal of Cast Metals Research 21 (2008) 186–192.
17. C.H. Cáceres, W.J. Poole, A.L. Bowles, C.J. Davidson, Section Thickness, Macrohardness and Yield Strength in High-Pressure Diecast Magnesium Alloy AZ91, Materials Science and Engineering: A 402 (2005) 269–277.
18. Z. Yang, J. Li, Y. Guo, T. Liu, F. Xia, Z. Zeng, M. Liang, Precipitation Process and Effect on Mechanical Properties of Mg–9Gd–3Y–0.6Zn–0.5Zr Alloy, Materials Science and Engineering: A 454 (2007) 274–280.
19. N. Ono, R. Nowak, S. Miura, Effect of Deformation Temperature on Hall–Petch Relationship Registered for Polycrystalline Magnesium, Materials Letters 58 (2004) 39–43.
20. N. Balasubramani, U.T.S. Pillai, B.C. Pai, Optimization of Heat Treatment Parameters in ZA84 Magnesium Alloy, Journal of Alloys and Compounds 457 (2008) 118–123.

# Strengthening and Toughening Behaviors of the Mg–9Al Alloy Containing Oxygen Atoms

Seung Won Kang and Dong Hyun Bae

**Abstract**

New Mg–O–9Al alloy has been developed by the dispersion of oxygen atoms in the Mg–9Al alloy in which oxygen atoms are supplied from the decomposition of $TiO_2$ nanoparticles in the Mg–9Al alloy melt. The dissolved oxygen atoms expand the lattice structures of both α-Mg and β-phase, inducing the reduced mismatch distance between α-Mg and β-phase. Therefore, yield stress of the Mg–O–9Al alloy is 143 MPa, much higher than 110 MPa in the Mg–9Al alloy. Fracture toughness values of 10.39 MPa $m^{1/2}$ and 12.86 MPa $m^{1/2}$ for both the Mg–9Al and Mg–O–9Al alloys are also respectively obtained. The crack propagates along the weak interface of β-phase in the Mg–9Al alloy. On the other hand, the β-phase disturbs the crack propagation route in the Mg–O–9Al alloy, showing many broken β-phases. Therefore, the improved interfacial feature with an addition of dissolved oxygen atoms in the Mg–O–9Al alloy results in much enhanced mechanical properties.

**Keywords**

Mg alloy • Oxygen atoms • Mechanical properties
Fracture toughness • Interface

## Introduction

In recent years, the needs for improved fuel efficiency in transportation industry have greatly driven the development of lightweight Mg alloys with the good specific strength and reasonable the unit cost of production [1].

S. W. Kang · D. H. Bae (✉)
Department of Materials Science and Engineering, Yonsei University, 50 Yonsei-ro Seodaemun-gu, Seoul 03722, Republic of Korea
e-mail: donghyun@yonsei.ac.kr

S. W. Kang
e-mail: vkswv@yonsei.ac.kr

Most commercial alloys (e.g., Mg–Al–Zn, and Mg–Al–Mn series) contain Al as a major alloying element because Al element enhances the high room-temperature strength and excellent castability of Mg alloys [2]. The mechanical behavior of the Mg–Al alloy system depends on the volume fraction, morphology, and size of the β-phase [3, 4]. The addition of Al atoms into the Mg alloys causes the formation of a β-$Mg_{17}Al_{12}$ phase [5] with a body-centered cubic (BCC) structure, which forms an incoherent interface with the α-Mg phase with a hexagonal closed packed (HCP) structure, thereby initiating cracks at the interface and limiting ductility of the alloys [6].

The interfacial properties of the Mg alloys can be improved by the addition of elements such as Ca, Si, and Y with the improvements of the thermal stability or morphology or interfacial stability of eutectic phase [7–10]. In the present study, we investigate the interfacial properties modified using O atoms in the Mg–9Al alloy.

## Experimental

To produce the Mg alloys containing O atoms using $TiO_2$ nanoparticles (Mg–O alloy), pure Mg was melted in a boron-nitride-coated low-carbon steel crucible. The crucible with Mg melting was protected by $SF_6 + CO_2$ mixed gas for suppressing oxidation at 720 °C. $TiO_2$ nanoparticles (approximately 50 nm in diameter) were added into the Mg melt and the melt was then held for 30 min to allow the $TiO_2$ particles to decompose. The Mg–9Al alloy containing O atoms (Mg–O–9Al) was produced by adding Al atoms into the above melt. The alloys were cast in a rectangular steel mold, to produce a specimen that was 10 mm thick. We also produced a Mg–9Al alloy specimen for comparison.

The basic characteristics, for crystallographic phase identification, were analyzed using X-ray diffraction (XRD, Rigaku, CN2301) with a Cu $K_\alpha$ radiation source (λ = 1.5405 Å). The XRD patterns were analyzed in the range from 20 to 80° with a scan rate of 0.02°/s and


D. Orlov et al. (eds.), *Magnesium Technology 2018*, The Minerals, Metals & Materials Series, https://doi.org/10.1007/978-3-319-72332-7_16

high-resolution XRD was used in the range of 30–40° with a scan rate of 0.002°/s. The interface between the α-Mg and β-phase was observed by a high-resolution transmission electron microscopy (HRTEM, JEOL, 2100FX). The composition of phases was analyzed by energy dispersive X-ray (EDX) spectroscopy equipped with HRTEM. The fracture surfaces of the samples were observed using scanning electron microscopy (SEM, JEOL JSM-7001F) at an accelerating voltage of 15 kV.

A uniaxial tension test was conducted using an Instron-type machine at a constant cross-head speed with an initial strain rate of $1 \times 10^{-4}\ s^{-1}$ at room temperature. Tensile specimens with a gage length of 10 mm, width of 6 mm, and thickness of 3 mm were machined and then finely polished. A single edge notched tensile (SENT) test was conducted at a constant cross-head speed with an initial strain rate of $1 \times 10^{-4}\ s^{-1}$ at room temperature.

## Results and Discussion

### Decomposing of $TiO_2$ Nanoparticles

Figure 1 shows a schematic image of a fabrication process of the Mg alloys containing O atoms. $TiO_2$ nanoparticles are added into liquid Mg alloy, and they are decomposed by high chemical potential energy. Most of the titanium slag sink and separate from the liquid due to the difference of density, which enables the O atoms to be dissolved in Mg alloy. Because the amount of O atoms is not satisfying to form magnesium oxides, they become supersaturated into the interstitial sites of Mg after solidification [11].

### Position of O Atoms in Mg Lattice Structure

To determine the position of O atoms in a Mg–O alloy, the microstructures of the as-cast Mg–O alloy are observed by TEM with inverse fast Fourier transform (IFFT) images as shown in Fig. 2. The α-Mg contains relatively amounts of Mg (97.3 at.%) and O (2.7 at.%); the measurement was conducted in the marked rectangle and the spectra are shown in the inset of Fig. 2a. Moiré fringes are observed in Figs. 2a and b due to the overlap of pure Mg and distorted Mg lattice by the O atoms. The satellite points which are obtained by O atoms in IFFT images (inset of Figs. 2a and b) are observed both $[\bar{1}2\bar{1}0]$ and [0001] directions at the Moiré fringes. The positions of O atoms in the Mg lattice structure are calculated by the IFFT images and described as the schematic images in Fig. 2c. To detail the positions of O atoms, the calculated positions of O atoms are combined. These results indicate the hypothesis that O atoms occupy octahedral sites within the Mg lattice rather than forming oxides.

### Microstructure of the Mg–O–9Al Alloy

Figure 3 shows SEM images of (a) the as-cast Mg–9Al and (b) Mg–O–9Al alloys. To observe the morphologies of the β-phase, both specimens were chemically etched in an acetic-picral etching solution (a mixture of 5 g picric acid, 6 ml acetic acid, 10 ml distilled water, and 100 ml ethanol). Both specimens consist of α-Mg and β-phase. The massive β-phase and (α + β) lamellar near the massive β-phase are observed in the as-cast Mg–9Al alloy. However, the as-cast Mg–O–9Al alloy only has the massive β-phase, which is relatively short because the O atoms act as nucleation sites of α-Mg during solidification [12].

### Distributed O Atoms

The lattice distortion and the composition of different phases was investigated using TEM images and EDX spectra, as shown in Fig. 4 and Table 1. The β-phase in the [011] direction, analyzed by the selective area electron diffraction (SAED) pattern as shown in Fig. 4c, to be parallel with the α-Mg in the $[\bar{1}2\bar{1}6]$ direction at the phase boundary. The α-Mg contains relatively small amounts of Al (7.7 at.%) and O (0.7 at.%) atoms compared to the β-phase which contains large amounts of Al (37.3 at.%) and O (3.8 at.%) atoms that are summarized in Table 1. The HRTEM image shows that the lattice plane spacing of $[10\bar{1}0]$ plane is increased from 2.764 to 2.777 Å in the α-Mg and that of $[2\bar{1}\bar{1}]$ plane 4.290–4.318 Å in the β-phase by dissolved O atoms as shown in Fig. 4b. The expended lattice structure of the Mg–O–9Al alloy can reduce the mismatch distance between the α-Mg and β-phase at the interface. The smaller mismatch distance between two phases among the possible combination of

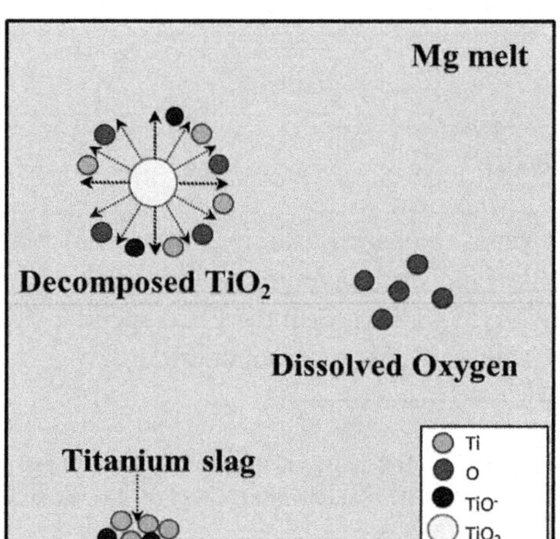


**Fig. 1** A schematic image of a fabrication process of the Mg alloy containing O atoms

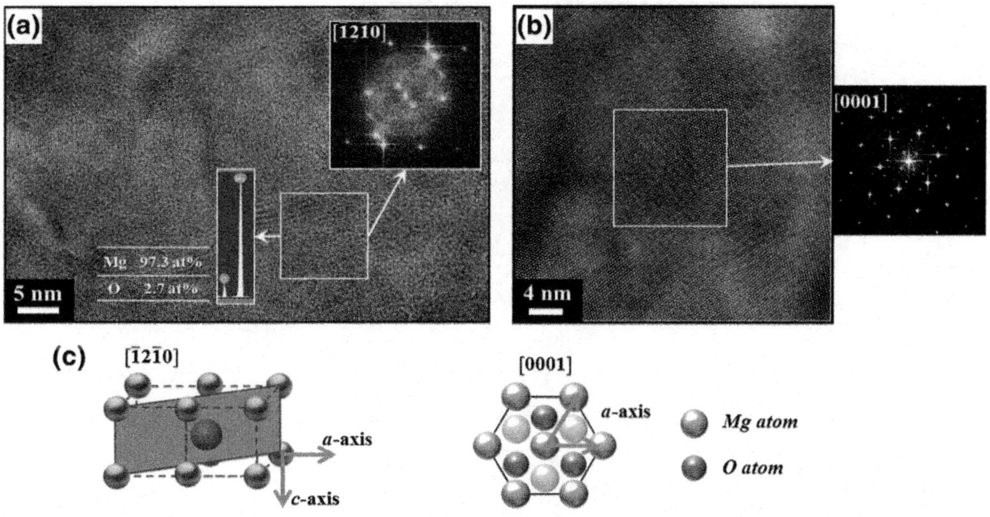


**Fig. 2** TEM images of the Mg–O alloy observed from **a** [$\bar{1}2\bar{1}0$] and **b** [0001] directions with IFFT images. **c** Schematic images show that the O atom occupy the octahedral site of Mg lattice structure

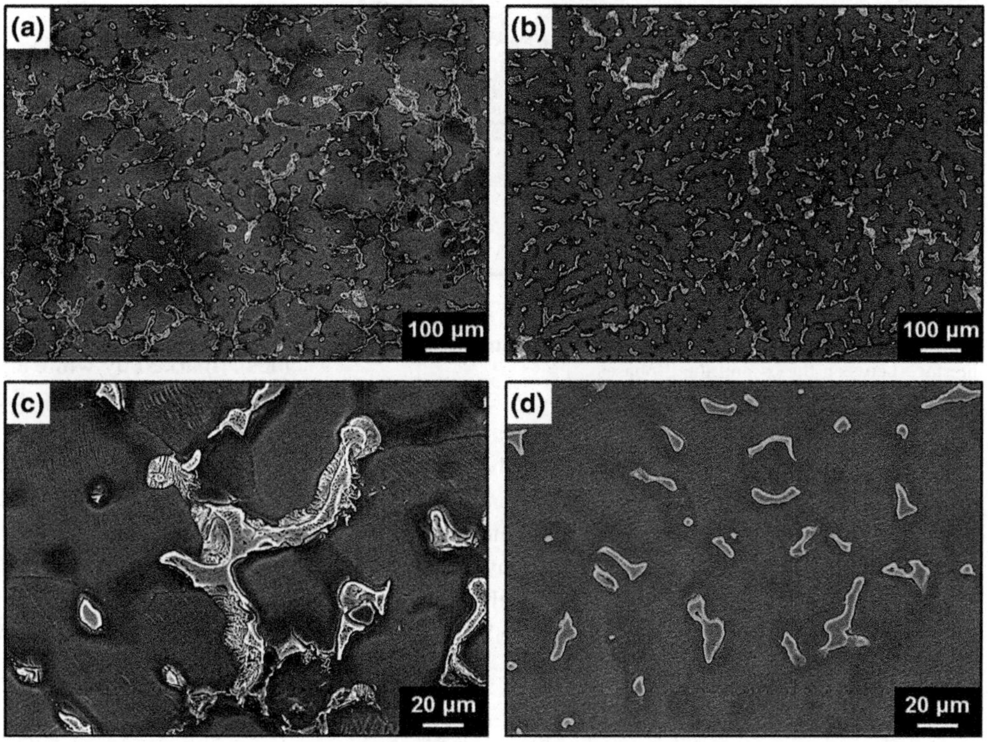


**Fig. 3** SEM images of as-cast **a** the Mg–9Al and **b** the Mg–O–9Al alloys in which the α-Mg and the β-phase are observed. Enlarged images of (**a**) and (**b**) are shown in (**c**) and (**d**), respectively, in which the massive β-phase and (α + β) lamellar are observed in (**c**) the as-cast Mg–9Al alloy and the many fine β-phases are observed in Mg–O–9Al alloy

interfacial bonding can reduce the internal energy at the interface of two phases [12].

## Mechanical Properties of the Mg–O–9Al Alloy

To evaluate the mechanical properties of the Mg–O–9Al alloys, engineering stress versus engineering strain curves obtained from uniaxial tension tests for both alloys are shown in Fig. 5a and summarized in Table 2. Much enhanced mechanical properties are observed at the Mg–O–9Al alloy compared to the Mg–9Al alloy. The yield stress and elongation values to failure are 109 MPa and 3%, respectively, for the Mg–9Al alloy and 143 MPa and 7%, respectively, for the Mg–O–9Al alloy. The yield stress and elongation values of the Mg–O–9Al alloy are respectively increased by 31 and 133% compared to the Mg–9Al alloy.

The increased elongation value to failure leads to an increase in the ultimate tensile stress of the Mg–O–9Al alloy from 148 to 211 MPa. Thus the supersaturated O atoms in the Mg–O–9Al alloy develops the finer β-phase by the increased nucleation sites of the α-Mg during solidification

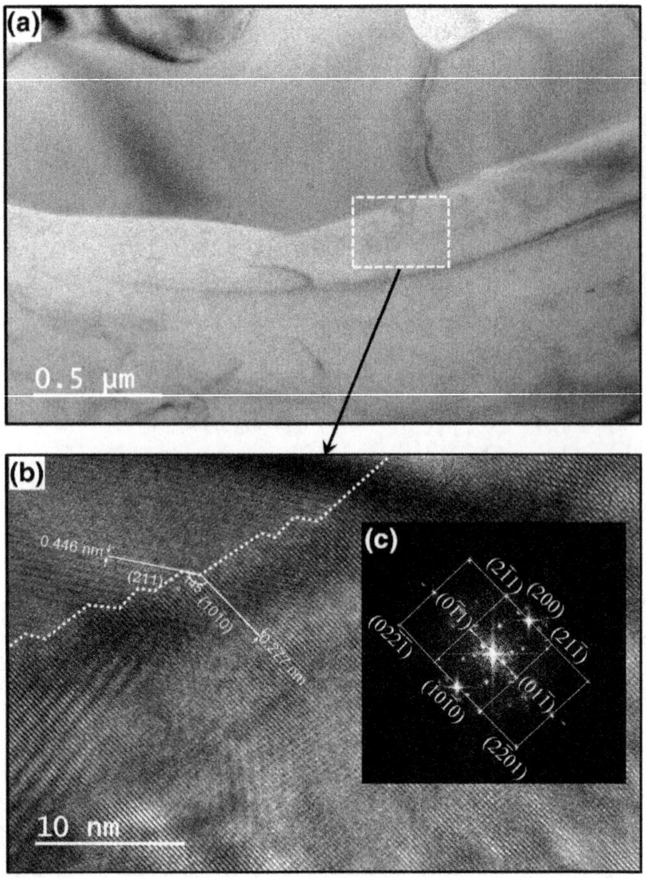


**Fig. 4** **a** TEM image of the α- and the β-phases in the Mg–O–9Al alloy. **b** HRTEM image and **c** SAED pattern of area located at the interface between the α- and the β-phases

and improves the interface features between the α-Mg and the β-phase that significantly induces higher mechanical properties in terms of strength and elongation.

To evaluate the effect of the improved interface features between α- and β-phases on the fracture behavior, the SENT tests are conducted on the Mg–9Al and the Mg–O–9Al alloys as shown in Fig. 5b. The results of the SENT tests are summarized in Table 2, indicating the enhanced fracture behavior of the Mg–O–9Al alloy. The peak load of the Mg–O–9Al alloy (1.097 MPa) is higher than that of the Mg–9Al alloy (0.93 MPa). Fracture toughness ($K_I$) can be calculated using the fracture load ($P$) with an assumption of the Mode I crack as follows [13]:

$$K_I = \frac{P}{B\sqrt{W}} * f\left(\frac{a}{W}\right) \quad (1)$$

where

$$f\left(\frac{a}{W}\right) = \frac{\sqrt{2\tan\frac{\pi a}{2W}}}{\cos\frac{\pi a}{2W}}\left[0.752 + 2.02\left(\frac{a}{W}\right) + 0.37\left(1 - \sin\left(\frac{\pi a}{2W}\right)\right)^3\right] \quad (2)$$

For the case of a linear elastic material, energy release rate ($J$) is as follows [13]:

$$J = K_I^2\left(\frac{1 - v^2}{E}\right) \quad (3)$$

where $E$ is Young's modulus and $v$ is Poisson's ratio. $K_I$ values are 10.39 and 12.86 MPa $m^{1/2}$ for the Mg–9Al and the Mg–O–9Al alloys, respectively. $J$ values are 2.37 and 3.37 N/mm for the Mg–9Al and the Mg–O–9Al alloys.

The fracture surfaces of the SENT tested specimens are observed by SEM as shown in Fig. 6. After SENT test, long secondary cracks (marked by white arrows) are formed in the Mg–9Al alloy as shown in Fig. 6a, whereas secondary cracks (marked by white arrows) of the Mg–O–9Al alloy are relatively shorter than the Mg–9Al alloy as shown in Fig. 6b. Especially, large fractured eutectic phases with debris are observed in the Mg–9Al alloy as shown in Fig. 6c. However, fine β-phase in the Mg–O–9Al alloy has many short cracks in phase interior compared to the cracks formed at the phase boundary as shown in Fig. 6d. That is, the crack is hard to form at the phase boundaries in the Mg–O–9Al alloy because of the enhanced interface features

**Table 1** The composition of α-Mg and the β-phase

| | α-Mg | | β-phase | |
|---|---|---|---|---|
| Atom | Atomic % | Weight % | Atomic % | Weight % |
| Mg | 91.6 | 91.1 | 58.9 | 57.4 |
| Al | 7.7 | 8.4 | 37.3 | 40.2 |
| O | 0.7 | 0.5 | 3.8 | 2.4 |

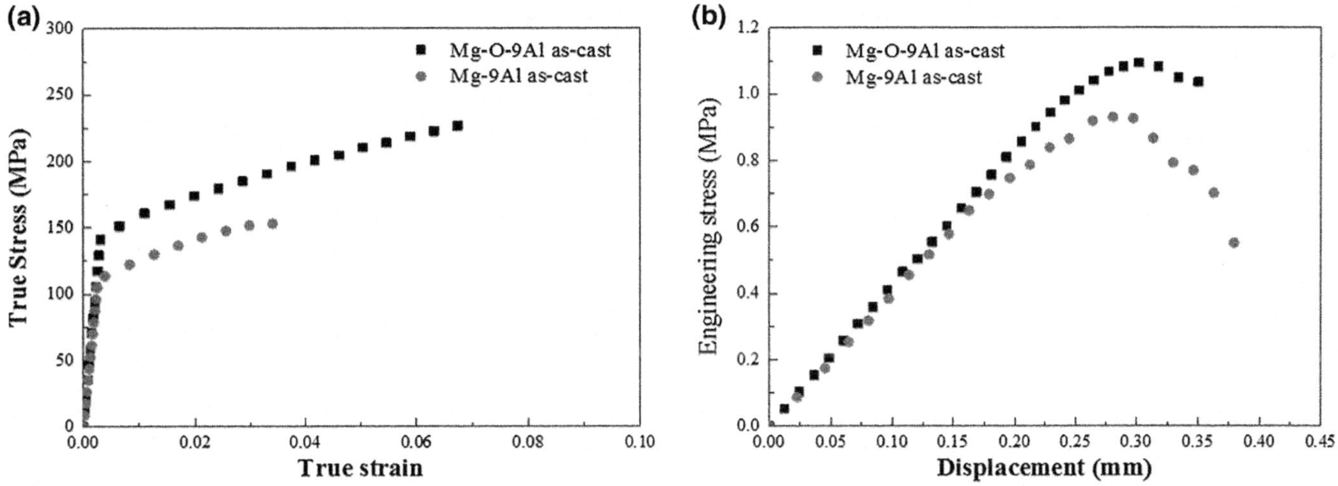


**Fig. 5** **a** Engineering stress versus engineering strain curves of the as-cast Mg–9Al and the as-cast Mg–O–9Al alloys with an initial strain rate of $10^{-4}$/s at room temperature. **b** Engineering stress versus displacement curves of the notched Mg–9Al and Mg–O–9Al alloys specimens with initial strain rate of $10^{-4}$/s at room temperature

**Table 2** Mechanical properties of the Mg–9Al and the Mg–O–9Al alloys

| Materials | Properties | | | |
|---|---|---|---|---|
| | Young's modulus (GPa) | Vickers hardness (Hv) | Yield stress (MPa) | Ultimate tensile stress (MPa) |
| Mg–9Al | 41 | 62 | 109 | 154 |
| Mg–O–9Al | 44 | 77 | 141 | 226 |
| Materials | Elongation (%) | Engineering stress (MPa) | Fracture toughness ($K_I$ = MPa × $\sqrt{m}$) | Energy release rate (J = N/mm) |
| Mg–9Al | 3 | 0.93 | 10.39 | 2.37 |
| Mg–O–9Al | 7 | 1.097 | 12.86 | 3.37 |

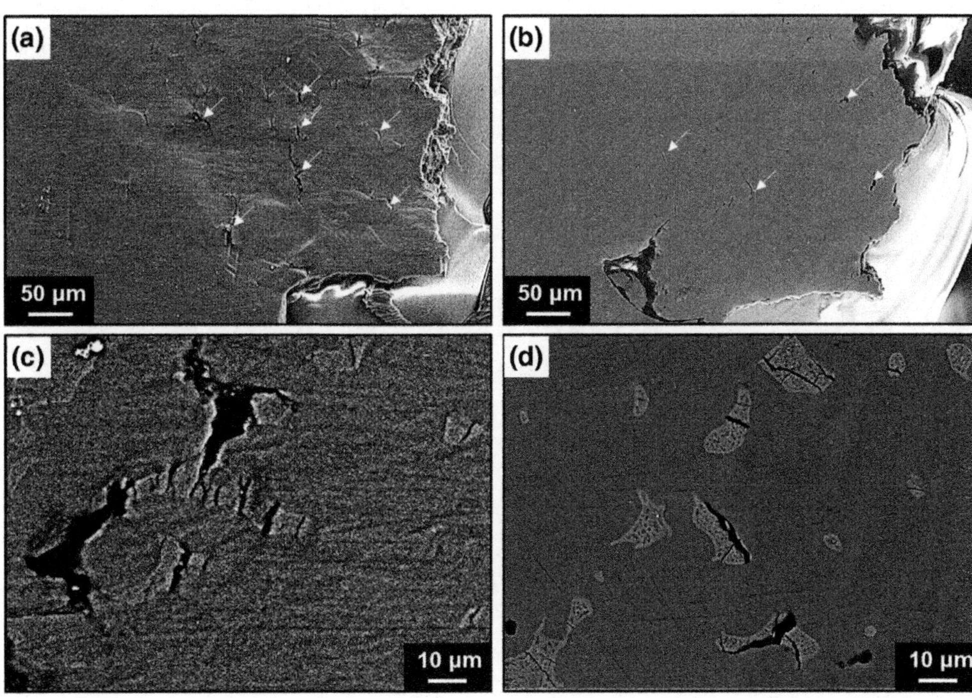


**Fig. 6** SEM images taken near notch crack after SENT test. **a** The Mg–9Al alloy with long secondary cracks marked by arrows. **b** The Mg–O–9Al alloy with short secondary cracks. The magnified fracture surfaces of **c** the Mg–9Al and **d** the Mg–O–9Al alloys show the fractured β-phase and secondary cracks

between α-phase and β-phase by the supersaturated O atoms and the modified β-phase morphology.

## Conclusion

In this work, the Mg alloys containing O atoms are produced by dispersing $TiO_2$ nanoparticles in the Mg melt and investigate the effects on the strengthening and toughening of the Mg–O–9Al alloy. The O atoms supplied by the inserted $TiO_2$ nanoparticles are dissolved into the Mg melt and supersaturated in α-Mg and β-phase. The supersaturated O atoms occupy the octahedral sites of the Mg lattice structure instead of forming oxide. During the solidification, the O atoms act as nucleation sites of α-Mg, and then fine β-phases are formed in the Mg–O–9Al alloy. The supersaturated O atoms are more concentrated in the fine β-phase in the Mg–O–9Al alloy because the solid solubility limit of O atoms is higher in liquid phase than in α-Mg, O atoms tend to move toward the β-phase solidified after α-Mg during solidification. The expanded lattice structures of the α- and β-phases in the Mg–O–9Al alloy by the O atoms decrease the mismatch distance between the α- and β-phases and enhance the interface features.

The increment of the stress and toughness for the Mg–O–9Al alloy and the suppressed crack propagation during the fracture process are due to the enhanced interface features and the modified morphology of the fine β-phase. Thereby, the secondary cracks are shortly formed by the fine β-phase and interior phase cracks. Therefore, the modified β-phase morphology with the enhanced interfacial feature by the supersaturated O atoms is effective reinforcement of mechanical properties in the Mg–Al alloy.

**Acknowledgements** The authors would like to take this opportunity to thank the National Research Foundation of Korea (NRF) (No. 2017R 1A2B2007062).

## References

1. M. Mondet, E. Barraud, S. Lemonnier, J. Guyon, N. Allain, T Grosdidier (2016) Microstructure and mechanical properties of AZ91 magnesium alloy developed by Spark Plasma Sintering. Acta. Mater. 119:55–67.
2. G. Han, X. Liu (2016) Phase control and formation mechanism of Al–Mn(–Fe) intermetallic particles in Mg–Al-based alloys with $FeCl_3$ addition or melt superheating. Acta Mater. 114:54–66.
3. C.H. Cáceres, C.J. Davidson, J.R. Griffiths, C.L. Newton (2002) Effects of solidification rate and ageing on the microstructure and mechanical properties of AZ91 alloy. Mater. Sci. Eng. A 325:344–355.
4. J. Wang, P. Song, S. Huang, F. Pan (2013) Effects of heat treatment on the morphology of long-period stacking ordered phase and the corresponding mechanical properties of Mg-9Gd-xEr-1.6Zn-0.6Zr magnesium alloys. Mater Sci. Eng. A 536:36–45.
5. D. Amberger, P. Eisenlohr, M. Göken (2012) On the importance of a connected hard-phase skeleton for the creep resistance of Mg alloys. Acta Mater. 60:2277–2289.
6. D. Duly, J.P. Simon, Y. Brechet (1995) On the competition between continuous and discontinuous precipitations in binary Mg–Al alloys. Acta Metall. Mater. 43:101–106.
7. X.G. Min, W.W. Du, F. Xue, Y.S. Sun (2002) Analysis of EET on Ca increasing the melting point of $Mg_{17}Al_{12}$ phase. Chin. Sci. Bull. 47:1082–1086.
8. A. Srinivasan, J. Swaminathan, M.K. Gunjan, U.T.S. Pillai, B.C. Paia (2010) Effect of intermetallic phases on the creep behavior of AZ91 magnesium alloy. Mater. Sci. Eng. A 527:1395–1403.
9. X.D. Li, H.T. Ma, Z.H. Dai, T.C. Qian, L.J. Hu, Y.P. Xie (2017) First-principles study of coherent interfaces of laves phase $MgZn_2$ and stability of thin $MgZn_2$ layers in Mg–Zn alloys. J. Alloy Compd. 696:109–117.
10. D.H. Bae, S.H. Kim, D.H. Kim, W.T. Kim (2002) Deformation behavior of Mg–Zn–Y alloys reinforced by icosahedral quasicrystalline particles. Acta Mater. 50:2343–2356.
11. H. Kang, H. J. Choi, S. W. Kang, S. E. Shin, G. S. Choi, D. H. Bae (2016) Multi-functional magnesium alloys containing interstitial oxygen atoms. Sci. Rep. 6:1–8.
12. S. W. Kang, H. Kang, D. H. Bae (2017) Strengthening and toughening behaviors of a Mg–9Al alloy containing oxygen atoms. J. Alloys Comp. 725:441–448.
13. T. L. Anderson (second ed) Fracture Mechanics Fundamentals and Applications. CRC press, 1994, p 604.

# Investigations on Microstructure and Mechanical Properties of Non-flammable Mg–Al–Zn–Ca–Y Alloys

Stefan Gneiger, Nikolaus Papenberg, Simon Frank, and Rudolf Gradinger

**Abstract**

Among commercial structural metals, magnesium alloys possess the lowest absolute density, featuring specific strength values superior to other structural materials such as aluminium alloys. Nevertheless, magnesium still shows some major drawbacks e.g. high oxidation tendency and more expensive processing compared to aluminium alloys. By adding calcium and yttrium to the commercially dominating AZ alloys, the oxidation behavior can be significantly improved and processing costs can be reduced. In this work, four Mg–Al–Zn–Ca–Y alloys with Al contents ranging from 3 to 9wt% were produced and compared to two standard AZ-type alloys (AZ31 and AZ91). Mechanical properties of as-extruded specimen as well as processability (casting and extrusion) and the resulting microstructures were investigated. The results show that modifying magnesium–aluminium alloys with small amounts of calcium and yttrium improves the materials' oxidation resistance and flammability behavior without deteriorating mechanical properties and processability.

**Keywords**

Magnesium • Non-flammable magnesium
Calcium • Extrusion

S. Gneiger (✉) · N. Papenberg · S. Frank · R. Gradinger
LKR Leichtmetallkompetenzzentrum Ranshofen GmbH,
Austrian Institute of Technology, 5282 Ranshofen, Austria
e-mail: stefan.gneiger@ait.ac.at

N. Papenberg
e-mail: nikolaus.papenberg@ait.ac.at

S. Frank
e-mail: simon.frank@ait.ac.at

R. Gradinger
e-mail: rudolf.gradinger@ait.ac.at

## Introduction

The global imperative for reduced ecological footprints of transportation means, which is inevitably bound to advances in weight reduction, drives aircraft and automobile manufacturers looking for alternative materials in their products. While the use of aluminium is established for a long time, the search of alternative materials with even higher specific strengths goes on. Using materials with high specific strength offers the opportunity to decrease components' weight without compromising its overall strength. Magnesium is one of the most promising materials for this application due to its low density. Compared to aluminium it is 35% lighter and possesses advantages in strength values when normalized with density. Additionally, magnesium parts are recyclable and can be easily machined, making the material more advantageous for general purpose products, especially when compared to fiber reinforced plastics.

Whilst magnesium offers a large variety of positive properties, its use is limited due to high chemical reactivity which causes low corrosion resistance and high oxidation tendency. Especially its low flammability performance is a major concern both during processing (e.g. casting and extrusion) and during operation for example in applications for commercial aviation. The use of magnesium in passenger airplane cabins was prohibited for many years due to the risk of self-ignition at high temperatures and problematic fire extinguishment once ignited. Yearlong efforts of industry together with the Federal Aviation Administration (FAA) led to the development of a laboratory-scale flammability test in order to obtain significant and repeatable results on the flammability behavior of magnesium [1]. This test is now included and described in detail in the FAA Aircraft Materials Fire Test Handbook, which describes different fire test methods [2]. Following this, the Society of Automotive Engineers (SAE) reworked their performance standard, which defines the requirements for seats in civil aircrafts [3]. This standard states that magnesium alloys may be used,


D. Orlov et al. (eds.), *Magnesium Technology 2018*, The Minerals, Metals & Materials Series, https://doi.org/10.1007/978-3-319-72332-7_17

if they meet the requirements described in the Materials fire test Handbook.

During development of the laboratory-scale flammability test, standard alloys (AZ31 and AZ91) were compared to alloys containing several weight percent rare earth elements, like WE43 and Elektron 21 [4]. In the laboratory-scale tests, the rare-earth containing alloys showed self-extinguishing behavior while the standard alloys continued to burn [1, 4].

Since then, efforts have been made to find alternatives to the expensive alloying with rare earths. Investigations concentrated on the flame retardant effect of calcium and CaO additions led to the development of Eco-Mg—alloys well known for their superior oxidation resistance [5–8]. Small additions of yttrium or other rare earth elements (with levels <1wt%) can enhance this effect [9–11]. It was discovered that calcium and yttrium form additional oxide layers on the melt surface, preventing the magnesium from further oxidation [12]. An example for results of the intentions is the Mg–Al–Ca–Y system, which is currently under development [13–19]. In the present work, the processability (casting and extrusion), the mechanical properties in the extruded state and the flammability resistance of four “non-flammable” Mg–Al–Ca–Y–(Zn, Mn) alloys were investigated and compared to two commercial available Mg–Al–Zn alloys, namely AZ31 and AZ91.

## Methodology

Six alloys with their nominal compositions given in Table 1 were produced. Two alloys representing commercially available AZ alloys while four alloys based on the same system were modified with small quantities of Ca and Y. The manganese content was adjusted in the Y containing alloys in order to decrease the formation of undesired ternary Al–Mn–Y phases which reduce the amount of Y available for the formation of other phases. While the alloys containing 9wt% Al are preferential used for cast applications, the alloys with lower Al contents can be considered as wrought alloys.

The alloys were prepared using commercial pure Mg ingots (99.8wt%), Al ingots (99.9wt%), Zn ingots (99.8wt%), calcium granules (99.6wt%), manganese chloride and a magnesium—30wt% yttrium master alloy. The melt was prepared in a resistance heated furnace using mild steel crucibles. For casting, the melt temperature was held at 720 °C and the melt was covered with a protective gas atmosphere (Ar + 1% $SF_6$). Billets with a diameter of 63 mm and a length of 240 mm were cast into a preheated steel mold. The mold is represented in Fig. 1. For controlling the flowability of the melt, AZ91 and AZXW9100 alloys were cast into tempered, spiral shaped steel molds.

After casting, the alloys were homogenized at 425 °C (AZ91D, AZXW8000 and AZXW9100) and 440 °C (AZ31B, AZXW3100 and AZXW6000) for 24 h under protective gas atmosphere (Ar + 1% $SF_6$).

Cylindrical dilatometer test samples with a diameter of 6 mm and a length of 10 mm were machined out of the homogenized billets. The samples have been tested in a Bähr 805 A/D deformation dilatometer with varied speed (0.1 $s^{-1}$, 1 $s^{-1}$, 5 $s^{-1}$) and temperature (300 °C, 350 °C, 400 °C).

After homogenization, the billets were turned down to a diameter of 58.5 mm, suitable for extrusion pressing. Direct extrusion pressing was done in a laboratory-scale extrusion press with a maximum extrusion force of 1.5 MN. The billets were inductively heated to 390 °C before extrusion and the extrusion speed was chosen to be 1.0 mm/s (ram speed). Rectangular shaped profiles (43 mm × 7.4 mm) were extruded for investigations of the mechanical properties and the flammability resistance.

Flat tensile test samples according to EN ISO 6892 with a gauge length of 50 mm and a cross section of 2.4 mm 12.5 mm were machined out of the extruded profiles. Flammability test samples with a dimension of 6.35 mm 38.1 mm × 508 mm were fabricated.

For the flammability tests, the test setup shown in Fig. 2 was used. The samples were mounted horizontally and exposed to a gas flame for 120 s. The burner positioning as well as the gas flow was kept constant during the tests. The test procedure and setup is following those prescribed in the FAA flammability test with some minor differences: While the FAA test uses kerosene-type fuel, the burner presented here is fired by propane gas. Additionally, due to the different flamer setup and resulting flame temperatures, the test duration of the present test was limited to 120 s, in contrast

**Table 1** Nominal chemical compositions of the investigated alloys

| Alloy | Al | Zn | Mn | Ca | Y | Mg |
|---|---|---|---|---|---|---|
| AZ31B | 2.5–3.5 | 0.7–1.3 | >0.2 | 0.0 | 0.0 | Bal. |
| AZ91D | 8.5–9.5 | 0.45–0.9 | >0.17 | 0.0 | 0.0 | Bal. |
| AZXW3100 | 3.0 | 0.8 | 0.1 | 0.5 | 0.2 | Bal. |
| AZXW6000 | 6.0 | 0.3 | 0.1 | 0.5 | 0.2 | Bal. |
| AZXW8000 | 8.0 | 0.3 | 0.1 | 0.3 | 0.2 | Bal. |
| AZXW9100 | 9.0 | 0.8 | 0.1 | 0.3 | 0.2 | Bal. |

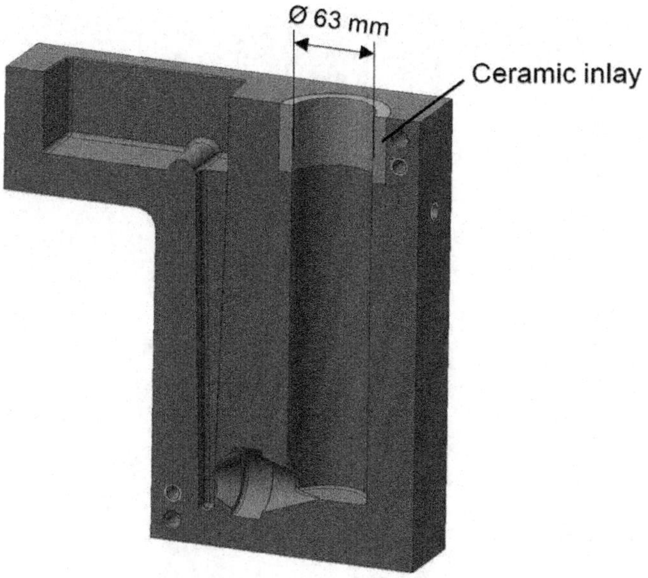


**Fig. 1** Steel mold

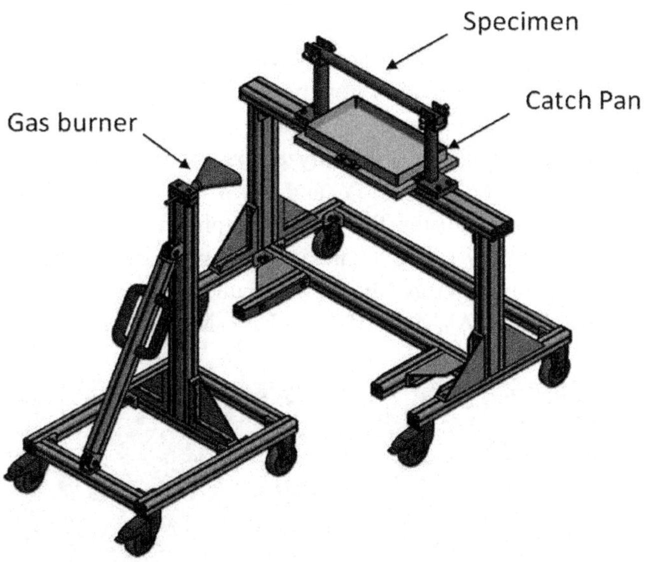


**Fig. 2** Flammability test setup

to 240 s applied in the FAA test. Nevertheless, this laboratory scale test is able to reveal the specific flammability resistance of alloys.

## Results and Discussion

### Melt Preparation/Casting

Due to the additions of Ca and Y, melt preparation of AZXW alloys is more complex compared to the standard AZ alloys. Nevertheless, when using a Mg–Y master alloy, the yield is very high and alloying is simple. Alloys containing Ca show reduced reactivity and the amount of cover gas needed for melting and melt handling can be reduced by a great amount and under certain circumstances melt covering is not necessary at all e.g. in a properly closed furnace. Additionally, the surface quality of the cast parts is better for Ca containing alloys due to the protective effect of Ca.

The flowability of AZXW9100 is superior compared to conventional AZ91 alloy, which leads to advantages in casting of parts with small wall thickness. Figure 3 shows the spirals of AZXW9100 and AZ91 in direct comparison. On average, the flow length of AZXW9100 is 35% higher than that of AZ91. Apparent porosity in the cast billets and spirals was equal for both alloys. According to H. S. Kim et al., the flow length is reduced when 1.31wt% Ca is added to Mg due to increasing solidification range [20]. In contrast, Fu et al. noticed an increase of melt fluidity caused by Ca in Mg–Zn–Ce alloys and suggested grain refinement and absence of oxide inclusions as the main effects [21]. Further investigations on the grain structure, the melt purity and the solidification behavior are recommended for clarifying the role of Ca on the flowability of Mg–Al melts. Also it should be noted that measuring the flowability by cast spirals is strongly dependent on temperatures (melt and mold) and on casting procedure (speed, melt volume in ladle etc.) and is therefore only meaningful for direct comparisons of casts done in quick succession.

### Microstructure

Figure 4 shows the as-cast microstructure of AZ91. Blocky or needle-shaped Al–Mn-phases as well as β-phase ($Mg_{17}Al_{12}$) can be found. After homogenization, the β-phase was

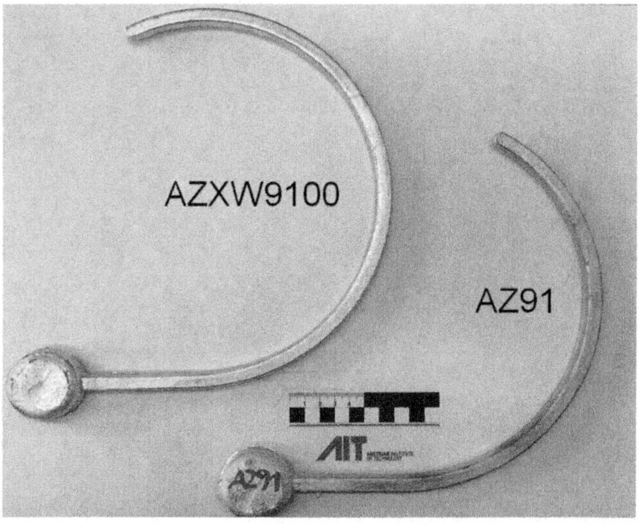


**Fig. 3** Cast spirals showing the flowability of AZ91 and AZXW9100

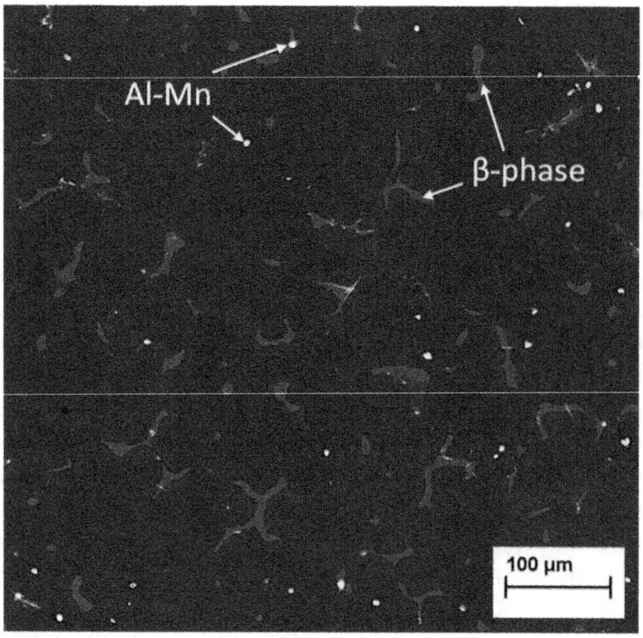


Fig. 4 SEM micrograph of AZ91 (as-cast)

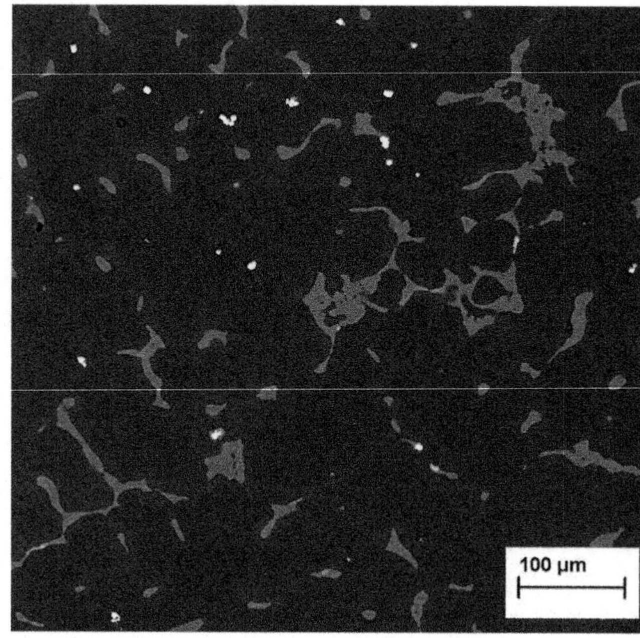


Fig. 6 SEM micrograph of AZXW9100 (as-cast)

completely dissolved whereas the Al–Mn-phases remained unchanged (Fig. 5).

Figure 6 shows the microstructure of AZXW9100 (as-cast). EDS mappings showed that Ca can be found preferably in the β-phase and in small quantities also in an Al–Ca phase. The overall fraction of β-phase is higher compared to AZ91. After homogenization, the β-phase was partially dissolved while all other phases remained unchanged (Fig. 7). Ca is known to increase the thermal stability of $Mg_{17}Al_{12}$ [22], an effect considered as main reason for less effective homogenization treatment. The brightest phases in Fig. 7 were identified as Al–Mn, Al–Y and Al–Mn–Y, where the majority of the analyzed phases were Al–Mn–Y. As can be seen, the Al–Y and Al–Mn–Y phases appear almost identical in the SEM micrographs. Therefore, comprehensive EDS analyzes are important for differentiation. Inquiring the literature, the phases found in the samples can be specified as $Al_8Mn_5$, $Al_2Y$ and $Al_{10}Mn_2Y$ [23–26].

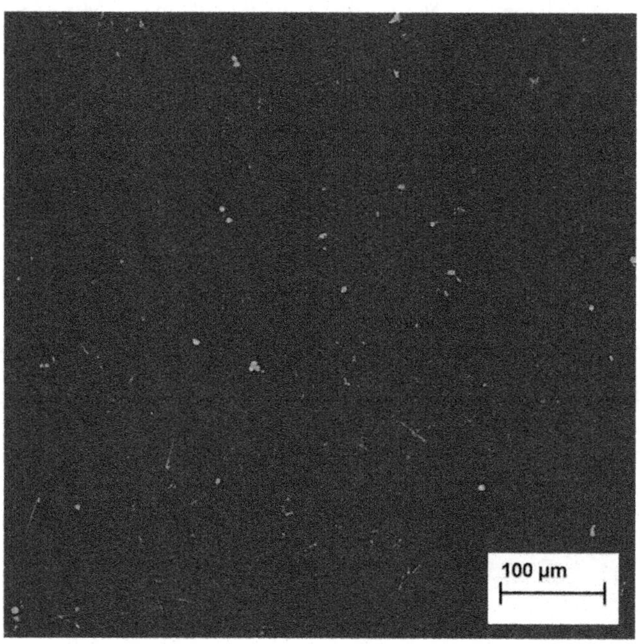


Fig. 5 SEM micrograph of homogenized AZ91

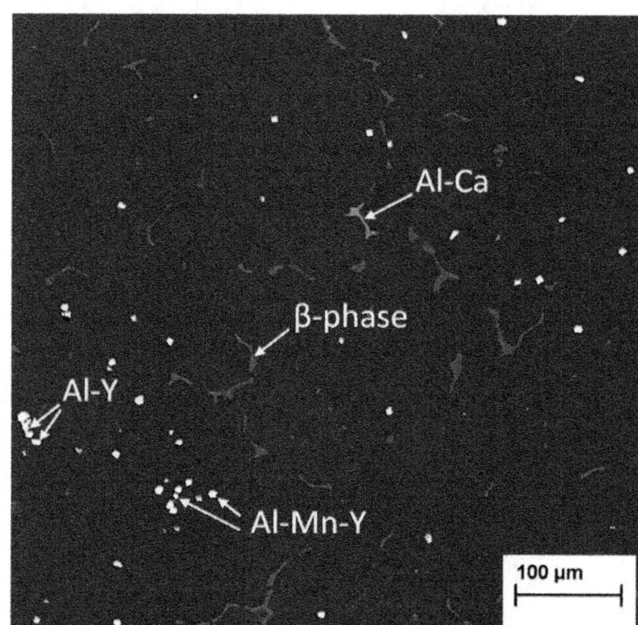


Fig. 7 SEM micrograph of homogenized AZXW9100

## Dilatometer Tests

Compression testing with varying temperatures and strain rates are widely used as assessment tool for forming processes. Thereby a rough estimation of process flow stresses and resulting microstructure can be made. The process microstructure depends on the stock material as well as on various processing parameters e.g. temperature, strain rate and degree of deformation. While the temperature and strain rate can be kept near constant throughout the testing, the degree of deformation varies within the sample. The highest effective strain can be found in the center of the sample while the outer edge experiences hardly any deformation. This is also easily visible in the resulting microstructure which is depending on the local effective strain.

All samples were tested without failures or cracks at 400 °C and a strain rate of 0.1 $s^{-1}$ ($\sim$1 mm/s), the results are presented in Fig. 8. The material shows typical dynamic recrystallization behavior as a rapid softening after the peak stress has been reached [27]. As expected, the curves show a general increase in peak stress with increasing Al content, whereas the flow stress of all alloys at high strains (steady stage) is similar. The flow stress curves of AZ31 and AZXW3100 are nearly identical, which implies that the processability of both alloys is equal.

Figure 9 shows the center of a tested sample (AZXW6000) with a fine recrystallized grain structure. The grains on sample rim (Fig. 10) on the other hand are interspersed with various twins, which shows that the temperature of 400 °C and the low amount of deformation was not sufficient to start dynamic recrystallization.

Figure 11 shows the microstructure of as extruded AZXW6000. Compared to Fig. 9 one can see that the microstructure is not fully recrystallized as twins and twin bands are still visible. Also the grain size of the extruded sample is substantially bigger. Additionally, trails of particles can be easily recognized which is typical for extruded material.

## Extrusion Pressing

For evaluating the extrusion process, diagrams of every single pass were recorded. Measured data are the extrusion pressure, ram- and die exit speeds and the ram- and container temperatures. One example diagram is shown in Fig. 12. Additionally, the surface quality of the profiles was evaluated.

The extrusion pressures and speeds of all investigated alloys were within common deviances and no significant differences between the materials were observed. In contrast to the experimental data obtained in [16], differences were less significant as the extrusion ratio was substantially smaller and obviously not sufficient to work out details. The surface quality of all profiles was good which was anticipated at the applied speeds.

## Tensile Tests

Figure 13 shows the tensile test results of the extruded samples. As anticipated, the strength values increase with increasing Al content for all alloys and for AZ alloys the elongation decreased. The addition of Ca and Y to AZ alloys has only a marginal influence on the strength values. The elongation decreased from AZ31 to AZXW3100 whereas the elongation from AZ91 to AZXW9100 increased. Compared to the results from [19], the results presented here are

**Fig. 8** Flow curves of homogenized AZ31 and AZXW alloys, T = 400 °C, $\dot{\varepsilon} = 0.1\ s^{-1}$

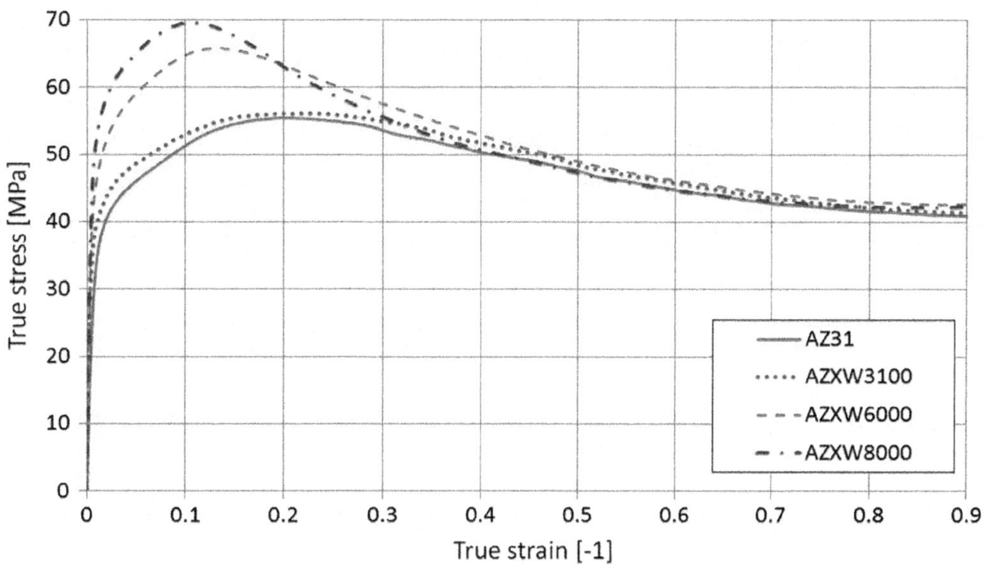

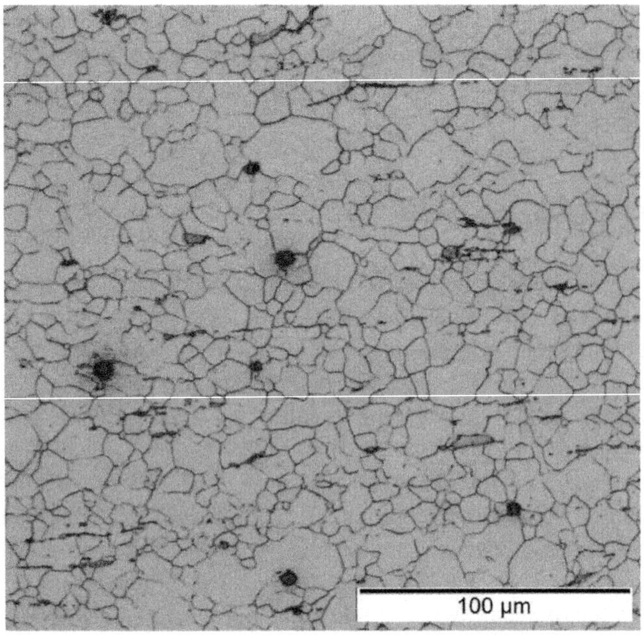


**Fig. 9** AZXW6000 dilatometer sample, center

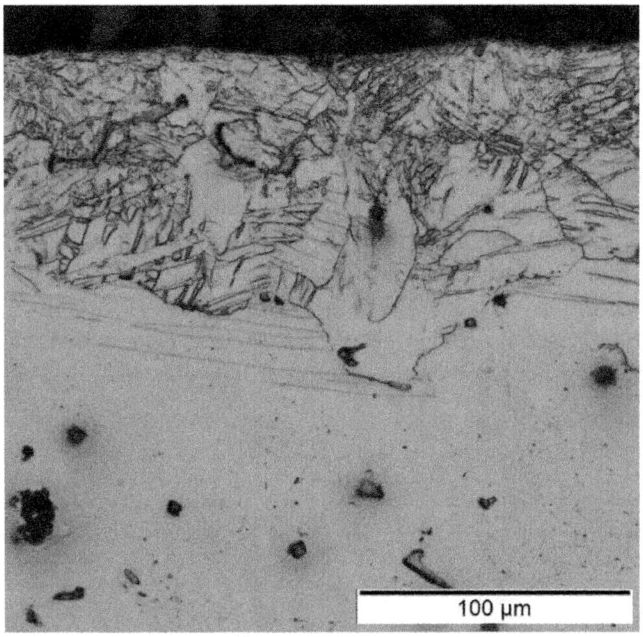


**Fig. 10** AZXW6000 dilatometer sample, rim

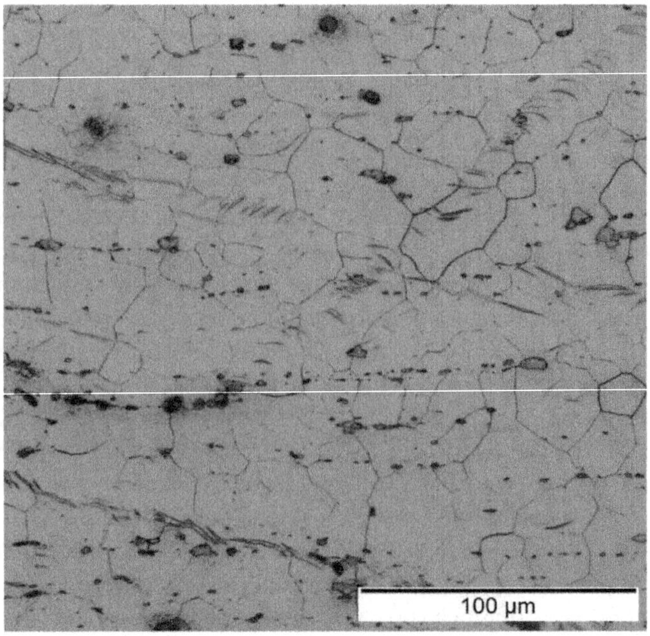


**Fig. 11** AZXW6000 extruded sample, center

inferior. Reason for this could be the longer gauge length of the flat test samples used in the present investigations and the substantially lower extrusion ratio (1:9 compared to 1:25). Further investigations are necessary to fully understand the particular decrease in elongation of alloy AZXW8000.

## Flammability Tests

In the flammability tests AZ31 as well as AZ91 started to ignite in the given circumstances while all AZXW alloys passed without ignition. Main reason for ignition is the formation of a melt droplet, dripping off the material. Droplets remain on the specimen endings which are still directly exposed to the flame and therefore prone to overheating and ignition. Figures 14 and 15 shows specimen of AZ31 and AZXW3100 during the test, 60 s after the burner was removed. AZ31 started to ignite on the remaining ends close to the burner impingement area and continued burning after dripping while AZXW3100 showed no ignition at all.

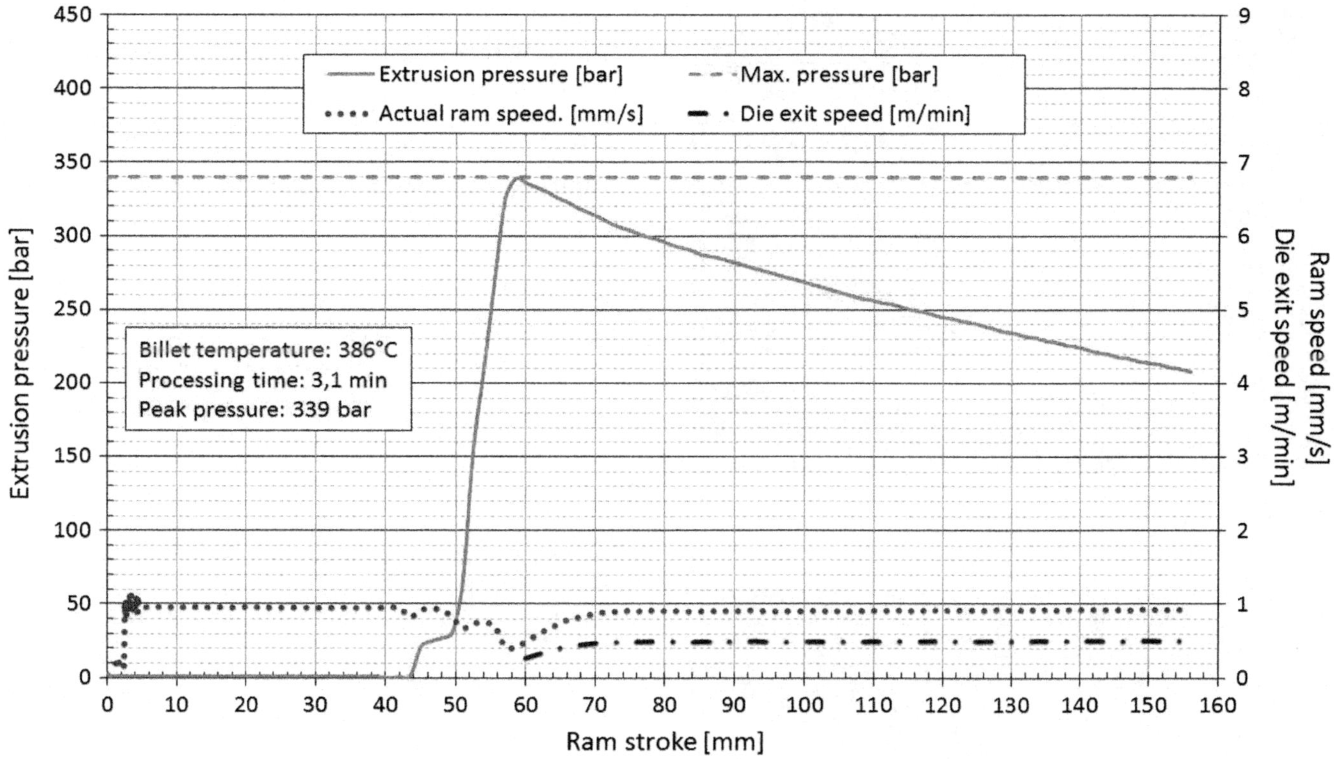


**Fig. 12** Experimental data of the extrusion of AZXW3100

**Fig. 13** Mechanical properties (tensile) of the extruded alloys

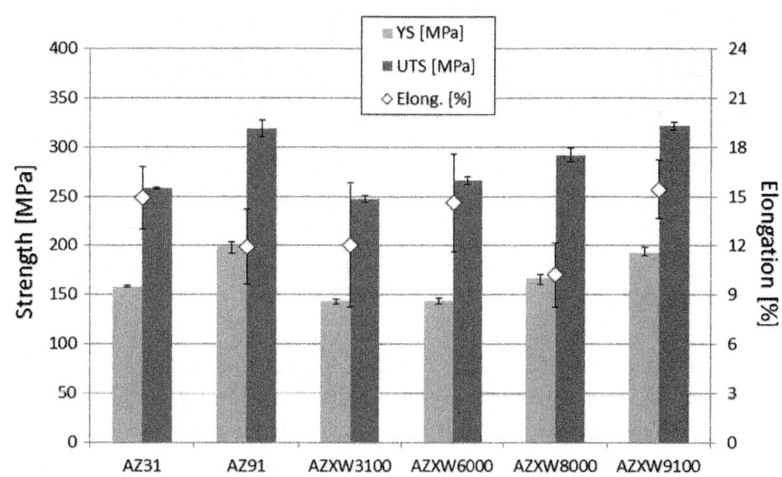

**Fig. 14** AZ31 bar during flammability test

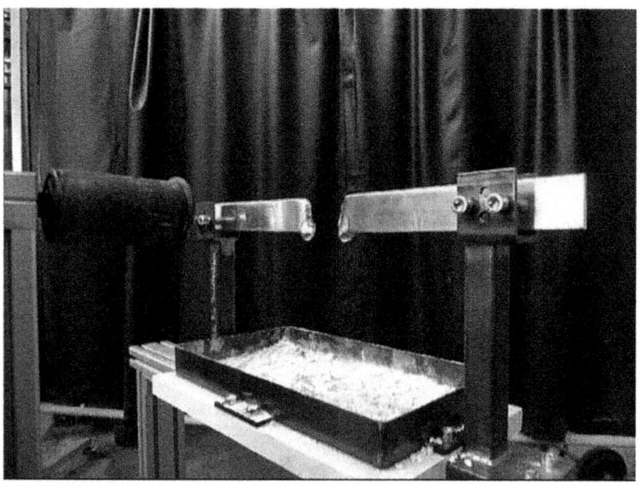

**Fig. 15** AZXW3100 bar during flammability test

## Conclusions

1. Small amounts of Ca and Y added to AZ-type magnesium alloys improve their castability (flowability and oxidation behavior) without showing a detrimental effect on the mechanical properties of the alloys in the extruded state.
2. As known, higher Al content in Mg–Al alloys leads to increased strength and flow stress. This behavior can also be confirmed for AZXW alloys in all tested temperature ranges. The small amount of additional alloying elements show no significant change in flow stress in comparison to the original AZ-type alloys.
3. The melt reactivity and flammability of AZ alloys can be reduced effectively by addition of Ca and Y. This allows a reduction of protective gas usage throughout casting and heat treatment processes.
4. Altogether, AZXW alloys can be an alternative to flammability resistant alloys containing large amounts of rare earths with a significant reduction in both cost and environmental footprint. It has also been shown, that AZXW alloys can be used as an alternative to classical highly reactive magnesium alloys.

**Acknowledgements** This work has been supported by the European Regional Development Fund (EFRE) in the framework of the EU-program "IWB Investition in Wachstum und Beschäftigung Österreich 2014–2020", and the federal state Upper Austria.

## References

1. T. R. Marker, "Development of a Laboratory—Scale Flammability Test for Magnesium Alloys Used in Aircraft Seat Construction," Federal Aviation Administration, DOT/FAA/TC-13/52, Feb. 2014.
2. "Aircraft Materials Fire Test Handbook," Federal Aviation Administration, DOT/FAA/AR-00/12.
3. "AS8049C: Performance Standard for Seats in Civil Rotorcraft, Transport Aircraft, and General Aviation Aircraft—SAE International." *[Online]*. Available: http://standards.sae.org/as8049c/.
4. T. Marker, "Evaluating the Flammability of Various Magnesium Alloys During Laboratory- and Full-Scale Aircraft Fire Test," US Department of Transportation, Federal Aviation Administration, Atlantic City, New Jersey, DOT/FAA/AR-11/3, 2013.
5. S. Cheng, G. Yang, J. Fan, Y. Li, and Y. Zhou, "Effect of Ca and Y additions on oxidation behavior of AZ91 alloy at elevated temperatures," *Trans. Nonferrous Met. Soc. China*, vol. 19, no. 2, pp. 299–304, Apr. 2009.
6. S.K. Kim, J.K. Lee, J. Cho, H.H. Jo, W. Ha, and Y.J. Kim, "The behavior of CaO in magnesium alloys", *Magnesium Technology 2005*, R.N. Neale, pp. 285–289, Feb. 2005.
7. S. K. Kim, "Design and Development of High-Performance Eco-Mg Alloys", *Magnesium Alloys - Design, Processing and Properties*, F. Czerwinski (Ed.), InTech, 2011, http://doi.org/10.5772/13836.
8. J.K. Lee, and S.K. Kim, "Effect of CaO Addition on the Ignition Resistance of Mg–Al Alloys", *Mat. Trans.*, vol 52, No. 7, 2011, pp. 1483–1488.
9. N. V. Ravi Kumar, J. J. Blandin, M. Suéry, and E. Grosjean, "Effect of alloying elements on the ignition resistance of magnesium alloys," *Scr. Mater.*, vol. 49, no. 3, pp. 225–230, Aug. 2003.
10. J. F. Fan, C. L. Yang, G. Han, S. Fang, W. D. Yang, and B. S. Xu, "Oxidation behavior of ignition-proof magnesium alloys with rare earth addition," *J. Alloys Compd.*, vol. 509, no. 5, pp. 2137–2142, Feb. 2011.
11. A. Prasad, Z. Shi, and A. Atrens, "Influence of Al and Y on the ignition and flammability of Mg alloys" *Corros. Sci.*, vol. 55, pp. 153–163, Feb. 2012.
12. N. Zhou, Z. Zhang, J. Dong, L. Jin, and W. Ding, "Selective oxidation behavior of an ignition-proof Mg–Y–Ca–Ce alloy", *J. Rare Earths*, vol. 31, no. 10, pp. 1003–1008, Oktober 2013.
13. B.S. You, W.W. Park, and I.S. Chung, "Effect of calcium additions on the oxidation behavior in magnesium alloys," *Scr. Mater. - Scr. MATER*, vol. 42, no. 11, pp. 1089–1094, 2000.
14. S. Gneiger, S. Frank, and A. Betz, "Untersuchung des Elementeinflusses von Ca und Y auf Magnesiumlegierungen" presented at the *9. Ranshofener Leichtmetalltage*, Bad Ischl, 2016.

15. R. Gradinger, S. Gneiger, and A. Betz, "Investigations on Fire-Resistant Magnesium Alloys for Aerospace Applications" in *7th Light Metals Technology Conference 2015*, Port Elizabeth - South Africa, 2015.
16. S. Frank and S. Gneiger, "Development of non-flammable magnesium alloys," presented at the *ISDM 2017*, Leoben, 2017.
17. US Patent Application for "Magnesium Alloy with Excellent Ignition Resistance and Mechanical Properties, and Method of Manufacturing the same", Patent Application (Application 20130280121, October 24, 2013).
18. Y. M. Kim, B. S. You, M. -S. Shim, and N. J. Kim, "Mechanical Properties and High-temperature Oxidation Behavior of Mg–Al–Zn–Ca–Y Magnesium Alloys" in *Magnesium Technology 2012*, S. N. Mathaudhu, W. H. Sillekens, N. R. Neelameggham, and N. Hort, Eds. Wiley, 2012, pp. 217–219.
19. S. Gneiger, R. Gradinger, C. Simson, Y.M. Kim, and B.S. You, "Investigations on microstructure and mechanical properties of non-flammable Mg–Al–Zn–Ca–Y extruded alloys," presented at the *7th European Conference For Aeronautics And Space Sciences (EUCASS)*, Milan, 2017.
20. H. S. Kim, and T. G. Kim, "The Effect of Ca and Mn Addition on Ignition Temperature, Castability, and Electrochemical Properties of Mg-Alloy by Casting", *Sol. St. Phen.*, Vol. 118, pp. 485–490, 2006.
21. Y. Fu, H. Wang, X.T. Liu, and H. Hao, "Effect of calcium addition on microstructure, casting fluidity and mechanical properties of Mg–Zn–Ce–Zr magnesium alloy", *J. Rare Earth.*, vol. 35, no. 5, pp. 503–509, 2017.
22. X. Min, W. Du, F. Xue, and Y. Sun, "Analysis of EET on Ca increasing the melting point of $Mg_{17}Al_{12}$ phase," Chin. Sci. Bull., vol. 47, no. 13, pp. 1082–1086, Jul. 2002.
23. Y. Wang, M. Xia, Z. Fan, X. Zhou, and G. E. Thompson, "The effect of Al8Mn5 intermetallic particles on grain size of as-cast Mg–Al–Zn AZ91D alloy," *Intermetallics*, vol. 18, no. 8, pp. 1683–1689, Aug. 2010.
24. A. Suzuki, N. D. Saddock, J. W. Jones, and T. M. Pollock, "Solidification paths and eutectic intermetallic phases in Mg–Al–Ca ternary alloys," *Acta Mater.*, vol. 53, no. 9, pp. 2823–2834, May 2005.
25. L. Zhang, R. Jia, D. Li, W. Zhang, and F. Guo, "Effect of Intermetallic Phases on Corrosion Initiation of AZ91 Alloy With Rare Earth Y Addition," *J. Mater. Sci. Technol.*, vol. 31, no. 5, pp. 504–511, Mai 2015.
26. B. Nami, S. G. Shabestari, H. Razavi, S. Mirdamadi, and S. M. Miresmaeili, "Effect of Ca, RE elements and semi-solid processing on the microstructure and creep properties of AZ91 alloy," *Mater. Sci. Eng. A*, vol. 528, no. 3, pp. 1261–1267, Jan. 2011.
27. T. Al-Samman and G. Gottstein, "Dynamic recrystallization during high temperature deformation of magnesium," *Mater. Sci. Eng. A*, vol. 490, no. 1, pp. 411–420, Aug. 2008.

# Development of BioMg® 250 Bioabsorbable Implant Alloy

R. Decker, S. LeBeau, D. LaCroix, S. Makiheni, and J. Allison

**Abstract**

The alloy development of bioabsorbable BioMg® 250 is described in terms of design of mechanical properties, biocorrosion rate and biocompatibility. The basic mechanistic role of microalloyig elements Zn, Ca and Mn is discussed as related to microstructures. In vitro corrosion and in vivo animal studies are reported. Finally, we list potential orthopedic applications in bone fixation devices fabricated from BioMg 250.

**Keywords**

Magnesium • Bioabsorbable • Implant • Alloy

## Introduction

Bone trauma, including fracture and osteoporosis, is a major challenge for biomedical engineering. There are about 6 million bone fractures reported in the US annually with a significant number requiring some type of fixation device to facilitate healing. The majority of fixation devices are made of non-bioabsorbable conventional implants such as titanium (Ti) or stainless steel. However, a significant portion of these non-degradable implants require secondary removal surgeries which are painful and costly; but also risk infections and further fractures. In addition to the implant being permanent, post-surgical complications as a result of local tissue reaction and protrusion have been recorded.

Devices constructed with bioabsorbable polymers, e.g. PLLA/PLGA, have been developed to overcome the current concerns and complications with stainless and Ti. Due to their low strength of 80–100 MPa, these polymer-based devices have limited utility in load-bearing sites and in self-tapping screws. In addition, some of the polymer devices have shown some long term foreign body reactions.

There is a need to develop alternate material-based implants for orthopaedic applications. Magnesium (Mg) and its alloys are such candidates due to their biodegradability, non-toxicity and excellent mechanical properties. Compared to Ti or stainless steel, Mg alloys have physical and mechanical properties that more closely match those of bone. Several Al-containing Mg alloys have been evaluated. A rare earth–containing Mg alloy, Magnezix® MgYRE, is in commercial use for compression screws. Mg is the 4th most abundant element present in the body. $Mg^{+2}$ is biocompatible and osteoconductive in promoting new bone growth [1–5].

NanoMAG has developed a proprietary Al- and RE-free Mg alloy, BioMg® 250, leveraging microalloying with small multiple alloying additions of zinc (Zn), calcium (Ca) and manganese (Mn)- in preference to larger singular or binary additions [6]. This approach provides: (a) improved strength, (b) lower potential toxicity (c) improved ductility, formability and toughness. Microalloying affords less corrosion compared to excessive alloying additions and, therefore, controlled release of hydrogen which the human body can accommodate. The design for mechanical properties, bioabsorbing corrosion rate and biocompatibility are described below—along with in vivo animal tests.

## Alloy Design

### Alloying for Mechanical Properties

Mechanical properties double those of bioabsorbable polymers are required for structural bone fixation; both for surgical insertion and also to insure rigid fixation of the repaired bone array. Such >200 MPa strength levels are seen in Mg alloys that are hardened by Al; but that element is suspect in

R. Decker (✉) · S. LeBeau · D. LaCroix
NanoMAG, LLC, Livonia, MI, USA
e-mail: rdecker@Nanomag.us

R. Decker · S. Makiheni · J. Allison
The University of Michigan, Ann Arbor, MI, USA

D. LaCroix
Michigan Technological University, Houghton, MI, USA


D. Orlov et al. (eds.), *Magnesium Technology 2018*, The Minerals, Metals & Materials Series, https://doi.org/10.1007/978-3-319-72332-7_18

dementia, Alzheimer's disease and bone dissolution. Therefore an alternate strengthening mechanism is demanded for bioabsorbable Mg alloys.

Therefore, nanoMAG selected a new combination of alloying elements to fortify the mechanical properties of the Mg base. Narrowing the search to elements that are established nutrients to the body, nanoMAG utilized Quantum Mechanics First Principles to select and optimize ternary additions. At the same time, the principles of microalloying were practiced—to capture synergisms amongst the alloying elements at low levels while avoiding the detrimental effects on corrosion and ductility of phases introduced by excessive alloying.

Thus the selected ternary alloying elements were Zn, Ca and Mn, The solid solution is strengthened, especially with Mn [7] (Table 1); but also Ca [7, 8]. The mixing enthalpy between Ca and Zn is negatively large at −22 kJ/mol, an order of magnitude larger than Mg–Ca. Thus there is a strong attractive interaction between Ca and Zn in the Mg matrix [9]. The free energy of the alloy system can then be lowered during processing by the segregation of alloying elements—(a) to cluster on dislocations and grain boundaries and (b) to nucleate nano-sized phases (**nanoMAG**) [10]. Thus, the alloy is strengthened. Ca and Zn lower the stacking fault energy of Mg, thus resulting in activation of non-basal slip [11]. Grain boundary cohesive energy, $2\gamma_{int}$, is increased to the benefit of ductility and toughness (**nanoMAG**) [10, 12, 13].

Indeed, that is the case wherein clusters or short range ordered zones (known as Guinier-Preston (GP) zones) form on the basal {0001} planes of Mg. These GP zones are one atomic layer thick (<0.5 nm) and about 15 nm in diameter (see Fig. 1). The array of Zn and Ca atoms in these ordered zones is pictured in Fig. 2. The Zn to Ca ratio is 2:1 and interparticle distance is 10 nm on the basal plane. The population is about $10^{22}$–$10^{23}$/m$^3$. These zones resist dislocation deformation on the basal planes, shunting this deformation to pyramidal planes for higher strength and formability.

In addition, the alloy and processing were designed for fine grains. Zn decreases grain size of Mg, as cast and as thermomechanically processed. Likewise, Mn serves the same benefit; but due to spherical α Mn particles of 10–120 nm size (see Fig. 3). Thus the Hall-Petch effect is utilized to boost strength combined with ductility.

The end result is tensile strengths of 250–320 MPa and elongations of 12–20%, depending on the process steps (see Tables 2, 3). The excellent ductility and formability are demonstrated in Fig. 4.

## Alloying for Low Texture and Good Formability and Ductility

The Ca atoms and Zn atoms co-segregate to grain boundaries in a strong interaction—both elements minimizing the

**Table 1** Alloying elements in BioMg® 250 and strengthening potential

| Alloying element in Mg base | Solid solution strengthening potency (MPa/Atomic %) [7] |
|---|---|
| Zn | 33 |
| Ca | 84 |
| Mn | 121 |

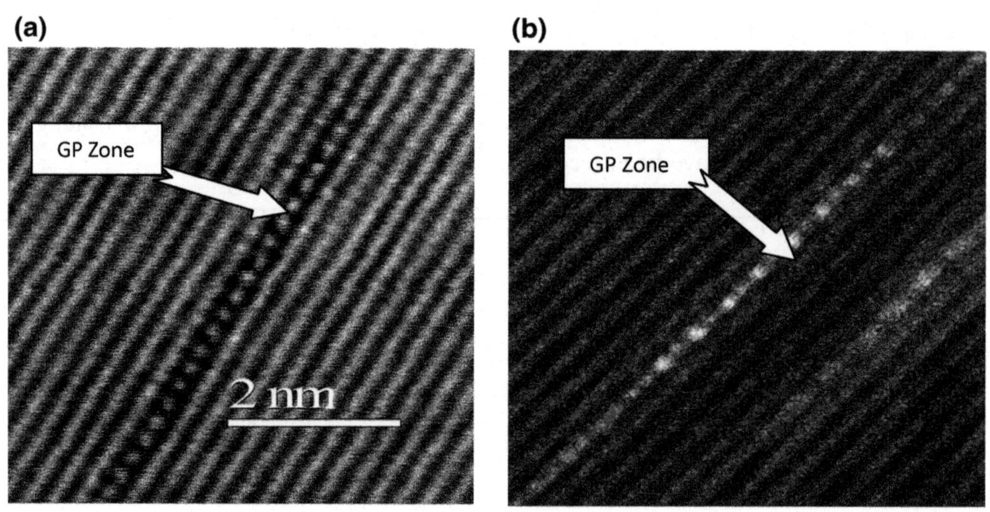


**Fig. 1** Electron micrograph images of GP zone in BioMg 250, **a** brightfield image, **b** darkfield image in which bright atoms are Zn and Ca (Allison, Makiheni)

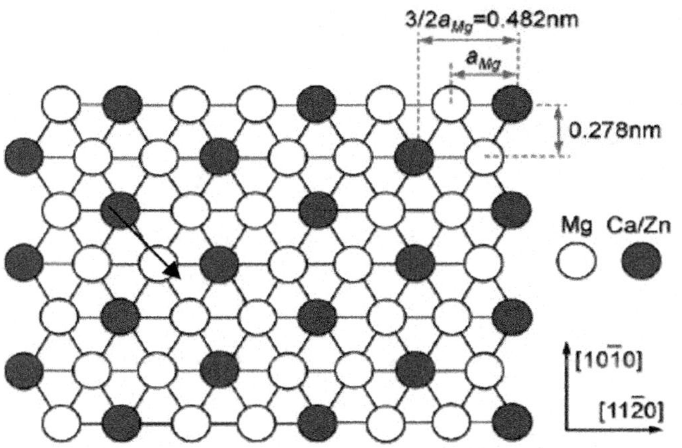


**Fig. 2** Array of Zn and Ca atoms in GP zone on basal plane of Mg matrix of BioMg 250

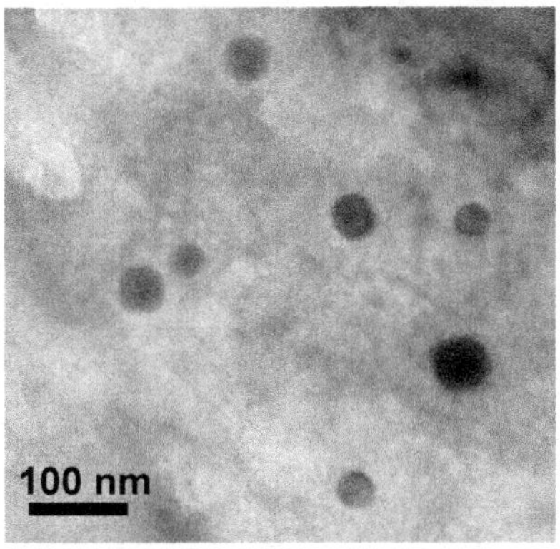


**Fig. 3** TEM of spherical α Mn particles, 10–120 nm diameter, for grain refinement (Allison, Makiheni)

elastic strains of the dislocations in the grain boundaries. By this mechanism, grain growth of highly oriented <1120> grains is inhibited; thus randomizing the growth of grains with other orientations (**nanoMAG**) [12]. Also contributing is the presence of second phase Mn particles during thermomechanical processing. Low texture of MRD 3.396 was produced in the stock of BioMg 250 used for plate implants as Treatment A in Table 2 (see Fig. 5). Rather than domination of basal slip, deformation is shared on pyramidal planes. This makes for superior formability compared to conventional Mg alloys which exhibit higher textures of about Multiple of Random Distribution = 8–10.

## Alloying for Biocorrosion

Small alloying additions of Zn, Ca and Mn were all beneficial to corrosion resistance in Synthetic Body Fluids (SBF) at 37 °C. However, with each element, excessive additions introduced extra phases that led to excessive degradation rate. Thus microalloying with the three elements resulted in a rate intermediate between ZK60 (too fast) and AZ91D (too slow) for bioabsorption in 1 year (Fig. 6).

**Zn** Optimum small additions to the Mg base reduce corrosion and $H_2$ evolution. The oxidized Zn ions report to Mg $(OH_2)$ replacing $Mg^{2+}$ cations, thus suppressing penetration of $Cl^-$ anions through the $Mg(OH_2)$ corrosion layer. Zn forms $Mg_2Zn$ particles that are anodic to the Mg matrix and $Ca_3Mg_6Zn_2$ particles that are cathodic, thus affording control of corrosion rates. Excessive Zn additions lead to excessive phases, promoting excessive corrosion and $H_2$ evolution.

**Ca** Likewise, small additions reduce corrosion of Mg; but excessive additions lead to high corrosion and $H_2$ release—notably due to $Ca_3Mg_6Zn_3$ and $Mg_2Ca$ phases at grain boundaries.

## Stress Corrosion

BioMg 250 demonstrated good resistance to stress corrosion cracking during in vitro exposure in SBF for 50 days. As seen in Fig. 7 and Table 4, 4-point loaded bend samples stressed to 100, 150, 200, 250, 300, 350 and 400 MPa survived without cracking, The residual bend stresses and bending strains after SBF exposure were independent of exposure stress level.

## Alloying for Biocompatibility—Primary Criteria Being Selection of Elements that are Nutrients and Promote Bone Growth (Osteoconductivity)

The base **Magnesium (Mg)** is a bone promoting, biocompatible and biodegradable material.

Magnesium is an essential element in the body. $Mg^{+2}$ is the fourth most abundant cation in the human body and is a co-factor to many enzymes and metabolic processes [1–5]. Yoshizawa and Sfeir et al. [14] identified extracellular matrix proteins and transcription factors positively affected by $Mg^{+2}$ that are responsible for the enhanced bone generation observed around degradable Mg orthopaedic/craniofacial devices. Cheng [15] found that Mg interference screws for anterior cruciate ligament (ACL) fixation in vivo promoted the expression of bone morphogenic protein-2 (BMP-2) and vascular endothelial growth factor (VEGF) and fibrocartilaginous enthesis regeneration. In vitro and in vivo studies indicated that Mg-based implants have good biocompatibility without inflammatory reactions [16, 17]. Mg alloys better match bone strength and elastic modulus than Ti implants, thus mitigating stress shielding tendencies. Push-out testing revealed significantly greater

**Table 2** Mechanical properties of BioMg 250 plates and screws

| Process | YS (MPa) | UTS (MPa) | Elong. (%) | Grain size (μm) |
|---|---|---|---|---|
| A for medium strength | 150 | 260 | 20 | 15 |
| B for high strength | 200 | 260 | 20 | 5 |
| C for highest strength | 300 | 320 | 12 | 3 |

**Table 3** Mechanical properties of BioMg 250 wire of 0.3 mm diameter

| Condition | YS (MPa) | UTS (MPa) | Elong. (%) | Bendability |
|---|---|---|---|---|
| Hard | 400 | 400 | 3 | <1.2 mm radius |
| Anneal C | 352 | 352 | 11 | |
| Anneal D | 228 | 269 | 19 | Tight knot (Fig. 4) |

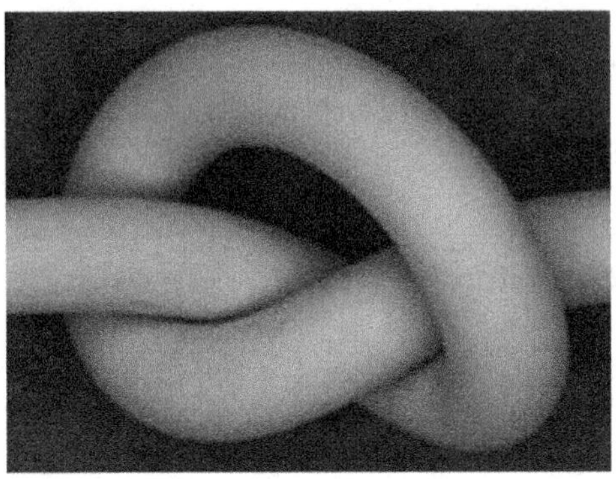

**Fig. 4** Knot tied at room temperature in 0.3 mm BioMg 250 wire, 2 μm grain size

pull-out force, ultimate shear strength and energy absorption in Mg implants than in Ti implants—with significantly higher bone-implant contact and bone volume with the Mg implants [4]. Upon absorption, the volume surrounding the Mg implant becomes alkaline as an advantage over the harmful acidity around certain polymer implants. The rapidly growing body of evidence for Mg's osteogenic enhancing effects suggests it could be harnessed to improve bone defect regeneration. In a 48 patient clinical study, Dewei-Zhao [18] found that biodegradable Mg screws provided successful fixation of vascularized bone graft in treating osteonecrosis of the femoral head (ONFH). Mg had antibacterial effects, as enhanced by Zn [19].

**Calcium (Ca)** is known to be an essential nutrient element and the most abundant mineral in the body, about 1000–1300 g for a healthy adult. It can regulate the normal physiological function of the organs, tissues and systems. Ca is the major mineral component of bone and teeth, playing a crucial role in the formation of bone. Ca can also promote the activity of enzymes and the activity of many enzymes involved in cell metabolism requires the activation of Ca ion, such as lipase and amylase [12, 20].

**Zinc (Zn)** is a critical alloying addition to BioMg 250. Its benefits are many, as follows:

**Bone Growth**—Enhanced

1. Zn enhances bone formation with mineralization and is an essential element of osteoblastic proliferation [19, 21].
2. Zn has been implicated in the prevention and reversal of the osteoporosis process [22].

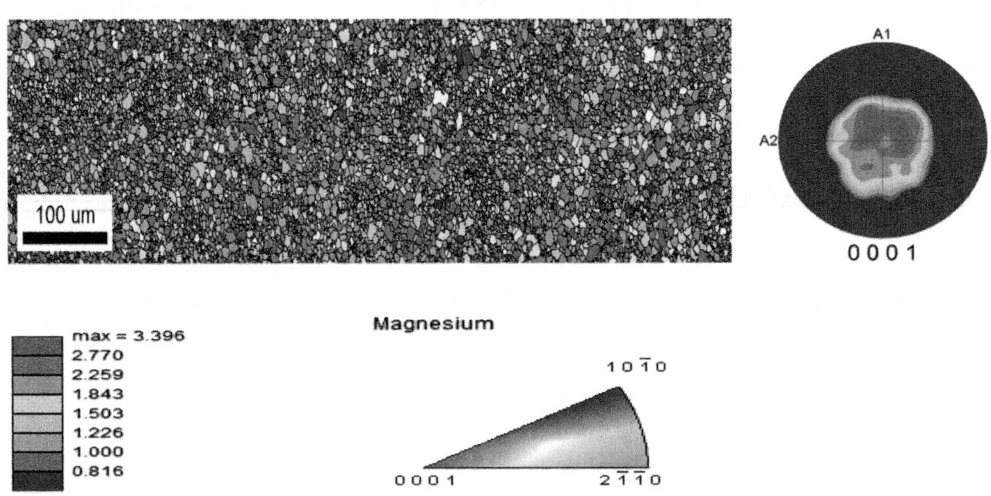


**Fig. 5** Grain structure and texture of BioMg 250 with treatment A of Table 2 (Allison)

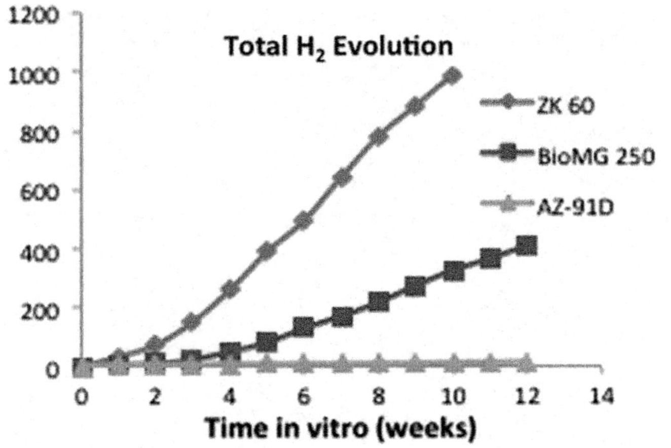


**Fig. 6** Corrosion of BioMg 250 in SBF, compared to ZK60 and AZ91D

3. $Zn^{+2}$ has anti-inflammatory/antibacterial effect [19, 23].
4. Zn was found at 0.3% in new bone grown in vivo on BioMg 250 in 52 weeks (nanoMAG data).

**Bone Growth Mechanisms**

1. Zn has a significant role in the metabolism of proteins, carbohydrates and lipids [24].
2. Zn release during skeletal breakdown can inhibit osteoclastic bone resorption [22].
3. The antibacterial effect of $Zn^{+2}$ induces faster bone healing by preventing bacterial colonization and degradation of premature membrane [25].
4. Zn stimulates cellular proliferation and differentiation of osteoblastic cells and inhibits the activity and differentiation of osteoclastic cells [26].
5. $Zn^{+2}$ increases ATPase activity and regulates transcription of osteoblastic differentiation genes such as collogen, osteopontin and osteocalcin [27].

## Alloying Effects In Vivo

The reaction of BioMg 250 implants and its alloying elements was measured during 52 week in vivo tests in rabbits at North Carolina State A&T. By Scanning Electron Microscope (SEM) images and linear scans, the alloying elements were located in (a) the residual implant, (b) a transition corrosion layer and (c) in new bone that encapsulated and replaced the implant (Fig. 8). The SEM images for Mg, Zn, Ca and Mn are seen in Figs. 9, 10, 11 and 12 along with P (Fig. 13). Quantitative measurements of alloying contents in the transition corrosion layer and new bone are included in Table 5.

In concert with the literature and the role in corrosion, Zn and Mn were found in the transition corrosion layer; in the case of Zn segregated in particles (Fig. 12). All the alloying

**Fig. 7** Unbroken BioMg 250 stress corrosion sample exposed at 400 MPa bend stress after 50 days in SBF

**Table 4** Post bend test results on stress corrosion samples of BioMg 250 exposed 50 days in SBF at 37 °C and pH 7.8

| SCC stress level (MPa) | 0 | 100 | 150 | 200 | 250 | 300 | 350 | 400 |
|---|---|---|---|---|---|---|---|---|
| Residual strength (MPa) | 175 | 230 | 228 | 226 | 168 | 250 | 193 | 274 |
| Bending strain (mm) | 0.14 | 0.25 | 0.21 | 0.19 | 0.19 | 0.27 | 0.16 | 0.36 |

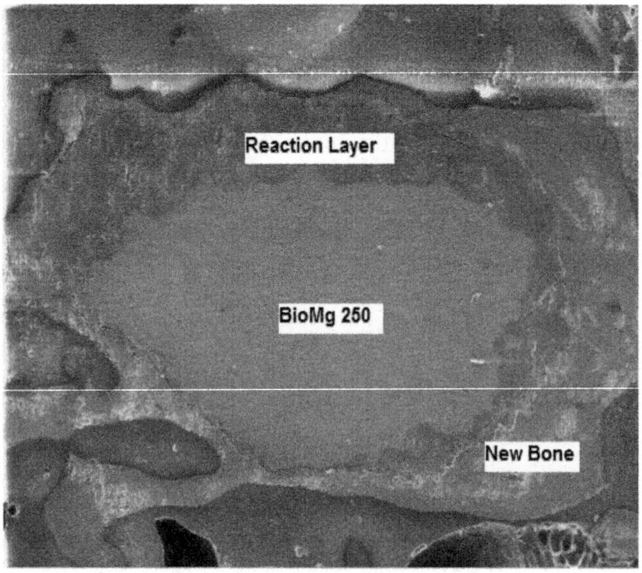


**Fig. 8** SEM of residual implant after 52 weeks in vivo, with corrosion reaction layer and encapsulating new bone

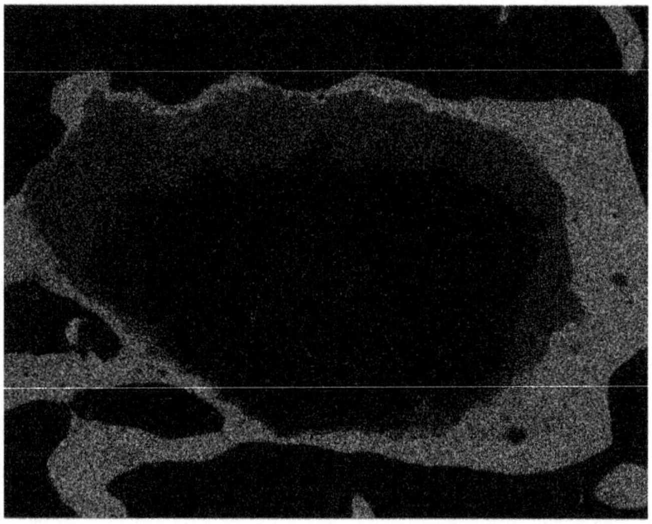

**Fig. 10** Ca map

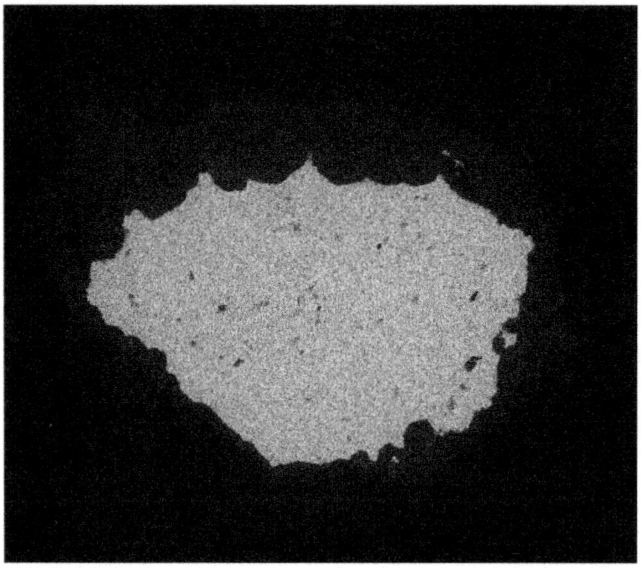

**Fig. 9** Mg map

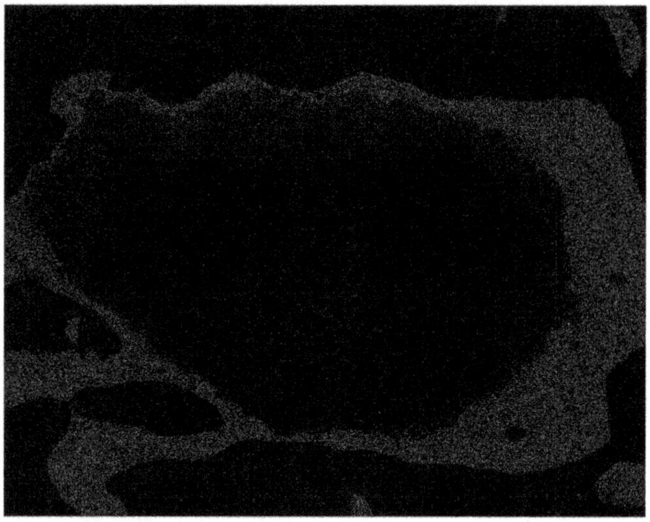

**Fig. 11** P map

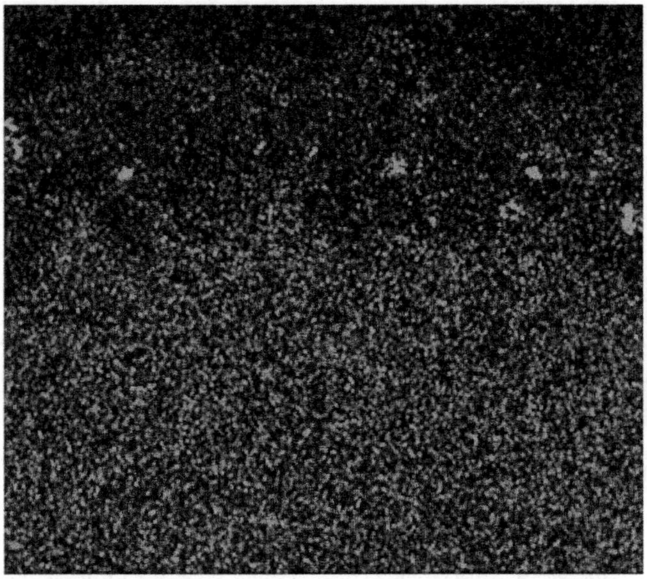

**Fig. 12** Zn map

**Fig. 13** Mn map

**Table 5** SEM linear scan results on BioMg 250 implant after 52 weeks in vivo in rabbit

| Element | % in transition layer | % in new encapsulating bone |
|---|---|---|
| Mg | 23–30 | 1.4–6.6 |
| Zn | 1.0–1.7 | 0.22–0.31 |
| Ca | 11–15 | 38–41 |
| Mn | 0.2–0.5 | 0.08–0.14 |
| P | 4–10 | 18–19 |
| O | 49–53 | 37–40 |

elements were found in the new bone, probably related to their osteoconductivity. Ca, P and O were derived from the body.

## Biocompatibility Testing

BioMg 250 was subjected to ISO 10993 tests, with results as follows:

1. Cytotoxicity (ISO10993-5)—no evidence of lysis or cell toxicity.
2. Irritation or intracutaneous reactivity (ISO 10993-10)—met requirements.
3. Muscle implantation (ISO 10993-6)—classified as a non-irritant.
4. Pyrogenicity (ISO10993-11)—non-pyrogenic.
5. Genotoxicity (ISO 10933-3)—non-mutagenic to 5 strains
6. Risk Assessment of Toxicology (ISO 10993-12)—no risk to patients from extractable materials.

## Potential Biomedical Applications

Potential Biomedical Applications for BioMg 250 are in ACL fixation, foot and ankle, hand and wrist, spine, craniofacial and dental locations—in the form of screws, plates,

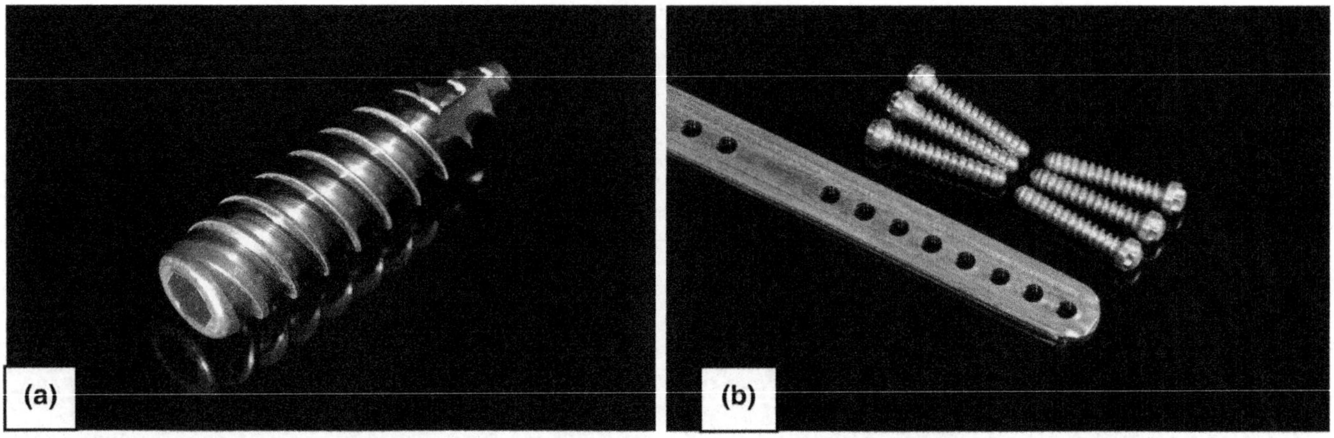


**Fig. 14** BioMg 250 implants **a** intererence screw for ACL fixation (courtesy MDC), **b** screws and plate for craniofacial fixation

wire, membranes and K wire. Some of these devices are seen in Fig. 14.

## Conclusions

1. Microalloying the Mg base with Zn, Ca and Mn produces the mechanical properties, biocorrosion rates and biocompatibility needed for orthopedic bone fixation.
2. Such Mg alloy fixation offers relief from the cost, pain and infection of secondary removal operations that might be required for non-absorbable implants.
3. BioMg 250 is expanding in vivo trials for several of the above-mentioned implant applications.

**Acknowledgements** NanoMag appreciates strong support under NSF Contract 0847198 and NIH Grant 4R44DEO24919-02.

## References

1. Xin, X., Hu, T., Chu, P., (2011), *In Vitro* studies of biomedical magnesium alloys in a simulated physiological environment: a review, Acta Biomatrialia, 7, p. 1452.
2. Staiger, M., Pietak, A., Huadmai, J., Dias,G., (2006), Magnesium and its alloys as orthopedic materials: a review, Biomaterials, 27, p. 1728.
3. Hartwig, A., (2001), Role of magnesium in genomic stability, Mutat Res/Fund Mol Mech Mutagen, 457, p. 13.
4. Wolf, F., Cittadini, A., (2003), Chemistry and biochemistry of magnesium, Mol Aspects Med, 24, p. 3.
5. Vormann, J., (2003), Magnesium: nutrition and metabolism, Mol Aspects Med, 24, p. 27.
6. Decker, R., LeBeau, S., Young, S., Patent WO201414145672AI, September 18, 2014.
7. Yasi, J., Hector, L., Trinkle, D., (2010), First-principles for solid-solution strengthening of Magnesium: From geometry and chemistry to properties, Acta Materialia, 58, p. 5704.
8. Saal, J., Wolverton, C., (2014), Thermodynamic stability of Mg-based ternary long-period stacking ordered structures, Acta Materialia, 68, p. 325.
9. Oh-ishi, K., Watanabe, R., Mendis, C., Hono, K., (2009), Age-hardening response of Mg-0.3 at% Ca alloys with different Zn contents, Materials Science and Engineering A, 526, p. 177.
10. Zeng, Z., Zhu, Y., Xu, S., Bian, M., Davies, C., Birbilis, N. Nie, J., (2016), Texture evolution during static recrystallization of cold-rolled magnesium alloys, Acta Materialia, 105, p. 479.
11. Yuasa, M., Hayashi, M., Mabuchi, M., Chino, Y., (2014), Improve plastic anisotropy of Mg–Zn–Ca alloys exhibiting high-stretch formability—a first principles study, Acta Materialia, 65, p. 207.
12. Zeng, Z., Xu, S., Bian, M., Davies, C., Birbilis, N. Nie, J., (2016) Effects of Dilute Additions of Zn and Ca on Ductility of Magnesium Alloys, Materials and Science & Engineering A, 674, p. 459.
13. Hase, T., Ohhtagaki, T., Yamaguchi, M., Ikeo, N., (2016), Effect of aluminum or zinc solute addition on enhancing impact toughness in Mg–Ca alloys, Acta Materialia, 104, p. 283.
14. Yoshizawa S, Brown A, Barchowsky A, Sfeir C. (2014), Magnesium ion stimulation of bone marrow stromal cells enhances osteogenic activity, simulating the effect of magnesium alloy degradation. Acta Biomaterialia, 10(6), p. 2834.
15. Cheng, P., Han, P., Zhao, C., Zhang, S., Wu, H., Ni, J., Peng, J., Zhang, Y., Liu, J., Xu, H., Liu, S., Zhang, X., Zheng, Y., (2016), High Purity magnesium interference screws promote fibrocartilaginous entheses regeneration in the anterior cruciate ligament reconstruction rabbit model via accumulation of BMP-2 and VEGF, Biomaterials, 81, p. 14.
16. Witte, F., Kaese, V., Haferkamp, H., Switzer, E., Meyer-Lindenberg, A., Wirth, C., (2005), *In vivo* corrosion of four magnesium alloys and the associated bone response, Biomaterials, 26, p. 3557.
17. Castellani, C., Lindtner, R., Hausbrandt, P., Tschegg, E., Stanzl-Tschegg, S., Zanoni, G., Beck, S., Weinberg, A., (2011), Bone-implant interface strength and osseointegration: Biodegradable magnesium alloy versus standard titanium control, Acta Biomaterialia, 7, p. 432.
18. Dewei-Zhao, M., et al., (2016), Vascularized bone grafting fixed by biodegradable magnesium screw for treating osteonecrosis of the femoral head, Biomaterials, 81, p. 84.
19. Yu, Y., et al., Multifunctions of dual Zn/Mg ion-implanted titanium on osteogenesis, angiogenesis and bacterial inhibition for dental implants, Acta Biomaterialia, article in press, 2016.11.067.

20. Zheng, Y., Magnesium Alloys as Degradable Biomaterials, CRC Press (2016), ISBN-13: 978-1-4665-9804-1.
21. Kallyanashis, P., Lee, B., Abueva, C., Kim, B., Jun Choi, J., Bae, S., Byong Taek Lee, B., (2017) In vivo evaluation of injectable calcium phosphate cement composed of Zn-and Si-incorporated β-tricalcium phosphate and monocalcium phosphate monohydrate for a critical sized defect of the rabbit femoral condyle, J. Biomedical Materials Research B. Applied Biomaterials, 105, p. 260.
22. Fielding, G., Bandyopadhyay, A., Bose, S., (2012), Effects of silica and zinc oxide doping on mechanical and biological properties of 3-D printed tricalcium phosphate tissue engineering scaffolds, Dental Materials, 2012, 28, p. 113.
23. Moreno-Eutimio, M., Nieto-Velazquez, N., Espinosa-Monroy, L., Torres-Ramos, Y., Montoya-Estrada, A., Cueto, J.,,Hicks, J., Acosta-Altamirano, G., (2014), Potent Anti-Inflammatory Activity of Carbohydrate Polymer with Oxide of Zinc, Biomed Res Int, 2014, p. 8.
24. Molokwu, C., Li, Y., (Fall 2006), Zinc Homeostasis and Bone Density, Ohio Research and Clinical Review, 15, p. 7.
25. Chou, A., et al, Implant Dentistry, 2007, 16, p. 89.
26. Yang, F., et al, Oral Surg Oral Med Oral Pathol Oral Radiol, 2012, 113, p. 313.
27. Ghorbani, F., Kaffashi, B., Shokrollahi, P., Seyedjafari, E., Ardeshirlylajima, A., (2015), PCL/chitisan/Zn-doped nHA electrospun nanocomposite scaffold promotes adipose derived stem cells adhesion and proliferation, Carbohydr Polym,118, p. 133.

# The Electrolytic Production of Magnesium from MgO

James C. Withers, John Laughlin, and Jeffery Babis

**Abstract**

There is only one primary producer of magnesium (Mg) metal in the U.S. utilizing an electrolytic process with feed from the Great Salt Lake. While electrolytic extraction of Mg from anhydrous $MgCl_2$ is 1.3× more energy efficient than the ferrosilicon reduction of dolomite, this Pidgeon process consumes over 2.3× the energy and produces 5× the $CO_2$ of the electrolytic $MgCl_2$ process. However, direct electrolytic reduction of MgO provides an opportunity to produce Mg at 20% less cost than the Pigeon process, conserve energy, and reduce $CO_2$ emissions. It has been demonstrated Mg can be electrolytically produced just above its melting point utilizing a composite anode of MgO+C as well as in a vapor state at approximately 1200 °C when MgO is dissolved in a select all fluoride fused salt. The high temperature processing is being bench pilot scale demonstrated to confirm the lower energy requirement and cost savings for full commercial scale demonstration.

**Keywords**

Electrolytic magnesium • MgO electrolytic-magnesium Mg $2/kg

## Introduction

Magnesium has one of the highest strength to weight ratios of frequently used structural metals at 158 kNm/kg compared to aluminum at 130 kNm/kg and steel at 38 kNm/kg. Magnesium is widely viewed as essential to lightweighting for both automotive and aircraft applications to reduce carbon emissions. There are a plethora of commercial and consumer applications for magnesium that will significantly expand with the availability of lower cost magnesium. However, greater adoption of magnesium will only be realized when it can be produced at a lower cost with energy savings and reduced carbon emissions. The world's largest producer of magnesium is China using the Pidgeon process which is the reduction of dolomite (MgO) with ferrosilicon (FeSi). The cost of producing Chinese-Pidgeon magnesium is $2.50/kg [1, 2]. Recently CSIRO of Australia has claimed the ever elusive carbothermic reduction of MgO utilizing a supersonic nozzle to separate the magnesium vapor from the back reaction from the gaseous CO which, if such a process can be commercialized, has the potential to be economical.

The sole producer of magnesium in the U.S. is U.S. Magnesium, LLC who utilize magnesium chloride from the Great Salt Lake to electrolytically produce magnesium which although more environmentally friendly than the thermal Pidgeon process costs approximately $3.31/kg [2]. A comparison of the energy efficiencies, carbon emissions, cost and strength ratio of magnesium and the comparative structural metals of aluminum and steel is shown in Table 1 [2–5].

The cost of the electrolytic magnesium process has been estimated as shown in Fig. 1 [2–6].

A goal for magnesium to expand its use domestically is that its production reach parity with steel on cost, energy consumption, and $CO_2$ emissions. Table 2 shows the necessary energy, emissions and cost requirements for magnesium to equal parity with steel. As shown in Table 2, the cost requirement is approximately 20% lower than the Chinese-Pidgeon process and 40% lower than the current magnesium chloride based electrolytic process with approximately 38% reduction in energy requirements.

Korenko, et al. [7] reported a process of 0.3 mol% MgO dissolved in 52.4 mol% $MgF_2$-49.6 mol% $CaF_2$ operated at

J. C. Withers (✉) · J. Laughlin · J. Babis
ATS-MER, LLC, Tucson, USA
e-mail: jwithers@atsmer.com

J. Laughlin
e-mail: jlaughlin@atsmer.com

J. Babis
e-mail: jbabis@atsmer.com


D. Orlov et al. (eds.), *Magnesium Technology 2018*, The Minerals, Metals & Materials Series, https://doi.org/10.1007/978-3-319-72332-7_19

**Table 1** Strength, energy, emissions, cost, and density for aluminum, magnesium, and steel

| | Aluminum | Magnesium | Steel |
|---|---|---|---|
| Strength-to-weight ratio (kNm/kg) | 130 | 158 | 38 |
| Processing energy (kWh/kg) | *Hall-Heroult*: 56 | *Western electrolytic*: 43.6<br>*Pidgeon process*: 102 | 6.4 |
| Theoretical minimum energy (kWh/kg) | 7.5 | 5.8 | 2.4 |
| Emissions ($kgCO_2/kg$) | *Hall-Heroult*: 22 | *Western electrolytic*: 6.9<br>*Pidgeon process*: 37 | 2.3 |
| Domestic production cost ($/kg) | 2.00 | 3.31 | 0.47 |
| Density ($kg/m^3$) | 2700 | 1800 | 7870 |

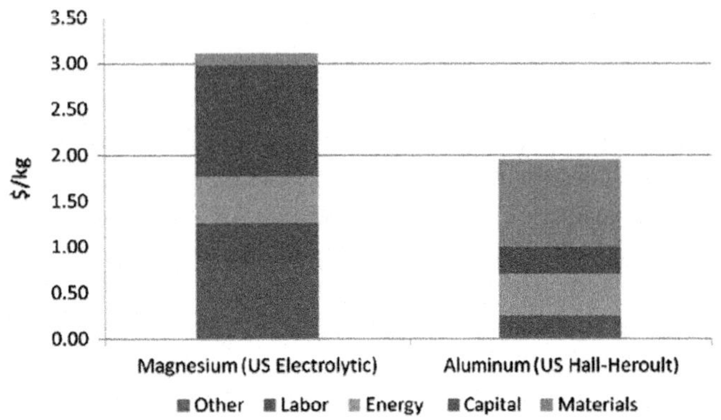


**Fig. 1** The estimated cost profile for U.S. magnesium

**Table 2** Energy, emissions and cost requirements for parity with steel

| | Magnesium | |
|---|---|---|
| | Current | Steel parity |
| Energy (kWh/kg) | 43.6 | 27.3 |
| Emissions ($kgCO_2/kg$) | 6.9 | 9.8 |
| Cost ($/kg) | 3.31 | 1.98 |

1250 K electrolytically produced magnesium at a cathode and $CO_2$ at a graphite anode. The liquid magnesium at the cathode was transported with flowing argon and condensed outside the electrolysis cell then refined from the salt electrolyte and cast into solid blocks. A 20 A cell was operated which produced a 90% cathodic efficiency resulting in a projected cost of $2.46/kg of magnesium produced and 4.78 kg $CO_2$/kg Mg. The experimental projected cost is less than the magnesium chloride electrolytic process and comparable to the mature Chinese-Pidgeon process. The $CO_2$ emissions are lower than either of the other magnesium processes. Although this [7] electrolytic process uses the desirable lower cost MgO feed with low $CO_2$ emissions, it does not meet the parity cost with steel requiring $1.98/kg.

The author [8] patented the composite anode process consisting of stoichiometric MgO–C in a densely pressed anode. Many fused salt electrolytes can be used that include all chlorides, all fluorides, and mixed chlorides and fluorides operated at just above the melting point of magnesium such as 650+ °C–700 °C. This lower operating temperature is an advantage to harvest the magnesium. In the mid-1980s this composite anode process was scaled to large bench cells which projected it would be possible to produce magnesium for approximately $2.00/kg or possibly less. If this can be confirmed, the goal of magnesium parity with steel as presented in Table 2 would be achieved establishing the U.S. as the low cost producer and significantly expand the magnesium market.

## Electrolytic Processes to Produce Mg Metal

Under ARPA-E sponsorship, the high temperature process of dissolving MgO in $MgF_2$–$CaF_2$ developed at Valpariso University under the direction of Professor Palumbo [7] is being scaled-up to a 500 A size. It is expected that electrolysis will begin in early 2018. This scale from 20 to 500 A shall confirm the operation of the system with the potential to achieve a lower cost than the current projection of $2.46/kg for the MgO in $MgCF_2$–$CaF_2$ process.

The metal oxide-carbon composite anode process has been demonstrated for several metals including magnesium, aluminum, titanium, the rare earths, beryllium, boron, chromium, zirconium, hafnium, vanadium, as well as producing alloys in some cases by combining two oxides with the carbon in the same anode. The author [8] has bench scale fused salt cells available at ATS-MER and is seeking partners to further investigate the MgO–C composite anode process to operate in the 670–700 °C range to verify that magnesium can be produced for $2.00/kg or less. The goal is to utilize magnesium hydroxide or carbonate as a feed as these compounds are the most economical source of magnesium which converts to MgO when pressed in an anode and fired with the carbon precursor to produce a composite anode. The large volume scale price of $Mg(OH)_2$ or Mg $(CO_3)_2$ is approximately $0.20/kg providing a low cost feed that helps in the potential to achieve the $2.00/kg goal. The electrolysis step at the lower temperature (690–700 °C) is less expensive than aluminum production in the Hall-Heroult processing which is $2.00/kg or less in the most efficient plant. Producing the MgO–C composite anode is also less expensive than producing the carbon anodes for the Hall-Heroult cells. At the production scale in comparison to aluminum it can be projected that the MgO–C composite anode process can be produced for approximately \$2.00/kg.

## Discussion

To meet parity of producing Mg with steel cost which is the key to Mg's widespread use, the production process must be less than the Chinese-Pidgeon process or laboratory feasibility demonstrated process at Valpariso [2] which is about the same as the Pidgeon process. However, the Valparaiso process meets the $CO_2$ emissions goal. Although scale-up often opens avenues to reduce cost, a reduction of approximately 20% will be challenging but may be possible using lower cost feed material. Innovative transporting and harvesting of the Mg vapor may further help in reducing the cost to \$2.00/kg. It appears the lower temperature (670–700 °C) composite anode process has the most potential to achieve the parity cost of magnesium with steel at \$2.00/kg.

## References


1. Das S (2010) Life Cycle Energy and $CO_2$ Analysis of Solid Oxide Membrane Primary Magnesium Production Technology, U.S. DOE Report No. DE-AC5-00OR22725.
2. Das S (2008) Primary Magnesium Production Costs for Automotive Applications, JOM 60(11):66.
3. Hamilton B (2013) Aluminum Cost Profiles. Unpublished Report, March 2013. Analyses are based on costs reported by Hale W (2000) The Global Light Metals Sector Outlook: A Primary Aluminum Perspective. Industrial Insight. 26–30. Nov. 2000.
4. Electric Arc Furnace Steelmaking Costs 2012, Metals Consulting International, 2012.
5. Wulandari W, Brooks G, Rhamdhani M, Monaghan B (2010) Magnesium: current and alternative production routes. Paper presented at 40$^{th}$ Annual Chemeca: Australasian Conference on Chemical Engineering, Adelaide, 26–29 September 2010.
6. Choragudi A, Kuttolamadom M, Jones J, Mears M, Kurfess T (2010) Investigation of the Machining of Titanium Components in Lightweight Vehicles. Paper presented at the SAE 2010 World Congress, Detroit, 13–15 April 2010.
7. Korenko M, Larson C, Blood K, Palumbo R, Venstrom L (2017) Technical and economic evaluation of a solar thermal MgO electrolysis process for magnesium production. Energy 135:182–194.
8. Withers, JC (1983) Cell with Composite Anode for Electrolytic Production of Magnesium, US. Patent 4,409,083. 11 October 1983.

# Empirical Examination of the Formation of Mechanical Properties of Heated Twin-Roll-Cast Magnesium Strips

Claudia Kawalla, Marie Teuber, and Michael Höck

**Abstract**

Magnesium components using strips produced by the Twin-Roll casting technology are of high interest for the automotive industry. Depending on the components application Original Equipment Manufactures define rigid mechanical property standards for magnesium sheets to ensure a high finished product quality. To meet these requirements continuously in a large-scale production via quality assurance the highly complex relationships during Twin-Roll casting need to be investigated holistically. This paper proposes a structural equation model for the assessment of the interrelationships of Twin-Roll casting process parameters and mechanical properties in transverse direction of heat-treated AZ31 Twin-Roll-Cast (TRC) strips using partial least square structural equation modeling (PLS-SEM). Within this context, the impact of the casting parameters, thickness profile and segregation formation on mechanical properties of magnesium strips after heating and before hot rolling will be approximated. It will be shown that the TRC thickness profile indicates mechanical properties irregularities due to local segregations.

**Keywords**

Twin-Roll casting • Magnesium alloy AZ31
Thickness profile • Mechanical properties
Structural equation modeling • Partial least square

C. Kawalla (✉) · M. Höck
Institute of Industrial Management, Operations and Logistics, TU Bergakademie Freiberg, Freiberg, Germany
e-mail: claudia.kawalla@bwl.tu-freiberg.de

M. Teuber
Institute of Metal Forming, TU Bergakademie Freiberg, Freiberg, Germany


D. Orlov et al. (eds.), *Magnesium Technology 2018*, The Minerals, Metals & Materials Series, https://doi.org/10.1007/978-3-319-72332-7_20

## Introduction

Twin-Roll casting is an energy-efficient process to produce Twin-Roll-Cast (TRC) strips, especially in the field of light metals [1, 2]. This technology was further developed for the TRC strip production of various magnesium alloys at the TU Bergakademie Freiberg [3, 4]. Moreover, hot rolled TRC strips were successfully applied as prototypes in the automotive industry [5]. The pilot plant of the university enables the production of TRC strips in a thickness range of 3.0–7.0 mm and a width up to 780 mm. The hot rolling of the TRC strip occurs either continuously or discontinuously after an intermediate annealing and enables the manufacturing of hot rolled strips with a thickness up to 0.8 mm. After hot rolling the strips are subjected to a final annealing to set the required final properties [6, 7].

Final properties of hot rolled strips result, inter alia, from the quality of TRC strips, which strongly depends on the individual process conditions. The Twin-Roll casting process is a highly complex process and depends on many interactive and partially unknown parameters which are not directly derivable from measurement data. To identify the impact factors on the thickness profile quality of TRC strips, a structural equation model was developed. Evaluated by the partial least square structural equation modeling (PLS-SEM) approach [8], a multivariate analysis method, the interrelationship between the TRC width, length of contact arc, deformation resistance, strip forming force and the thickness profile was demonstrated [6]. The model was expanded by the level of segregations [9]. Based on the research results [6, 9] further casting parameters were deliberately varied to determine the interrelationship between thickness profile, macrosegregations and mechanical properties of annealed TRC strips in transverse direction. Thus, TRC strips should be already selected for further production during processing to produce hot rolled strips with optimal properties.

In this paper, we propose a structural equation model for the determination of the impact of various process conditions

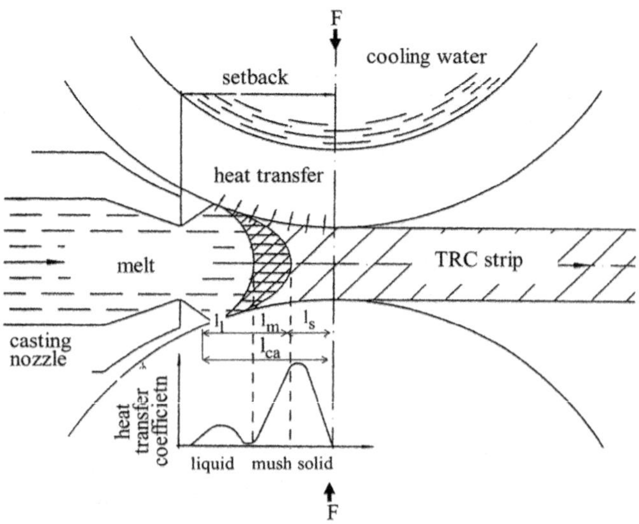


**Fig. 1** Twin-Roll casting process according to [12, 13]

on the formation of the thickness profile and the associated microstructure evolution as well as the mechanical properties after annealing in transverse direction, which was approximated by the predictive modeling approach PLS-SEM.

## Twin-Roll Casting Including Strip Forming and Heat Treatment Processes

During Twin-Roll casting, the melted magnesium alloy AZ31 is poured into the roll gap of two rotating work rolls by a special cast system via nozzle. The melt solidifies by contact with the water-cooled work rolls within the roll gap, and the solid phase undergoes a plastic deformation [10, 11]. The area between the work rolls can be divided into several sections: the liquid phase with solidified shells at the work rolls $l_l$, partially solidified phase $l_m$ (thixoforming) and solid phase $l_s$ (see Fig. 1).

The total distance between the nozzle entry and the narrowest point between the work rolls is called a setback [14]. The strip forming force (F) is the sum of all forces that are needed for merging the solidified strip shells and forming the solid phase [6, 15]. The TRC strip formation process and the resulting thickness profile are described in detail in [6] and [9]. Depending on the TRC conditions segregations may occur during TRC strip forming and the related thickness profile formation [9]. Micro- and macrosegregations can be distinguished. Microsegregations arise between secondary dendrite arms, whereas macrosegregations are formed in the center of the joint shells, i.e. center segregations. They occur depending on the nozzle position, but commonly in the center of the strip and indicate a potential defective position of the casting nozzle. Figure 2 depicts the microstructure before and after homogenization treatment along cast strip thickness.

Another type of macrosegregations are inverse segregations (see Fig. 3). They are caused by the disturbances of the laminar melt flow emitted from the nozzle. They spread out linearly along the cast/TRC direction (see Fig. 1). Inverse segregations can negatively influence the forming behavior of the TRC strip. They are characterized by the enrichment of aluminum and can pass up to the surface of the TRC strip. In this case, aluminum reacts during the heat treatment with the oxygen of the furnace atmosphere. Thus, it leads to the material discontinuity, which is characterized by the

**Fig. 2** Twin-Roll-Cast strip before (**a**) and after homogenizing treatment (**b**) [16]

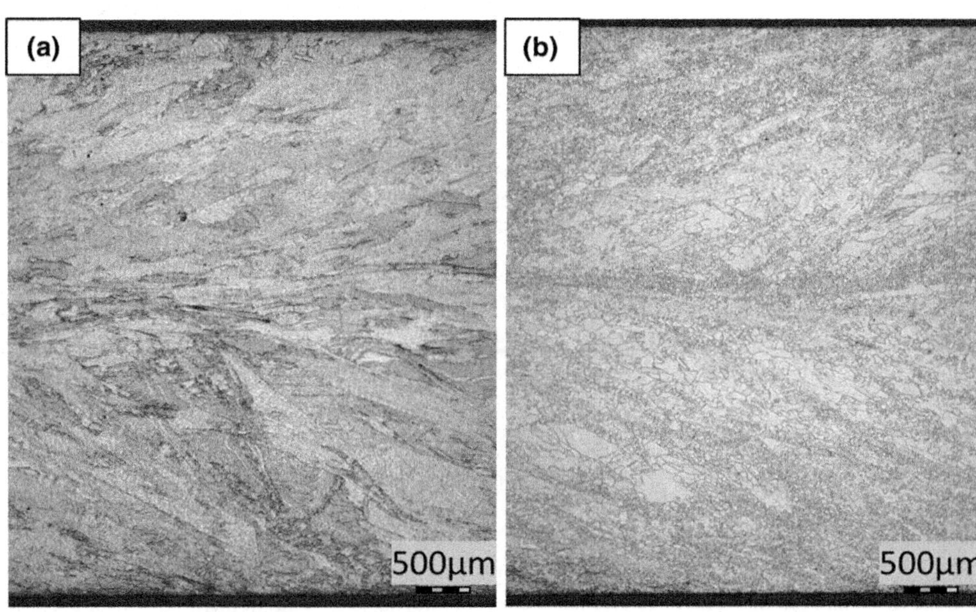

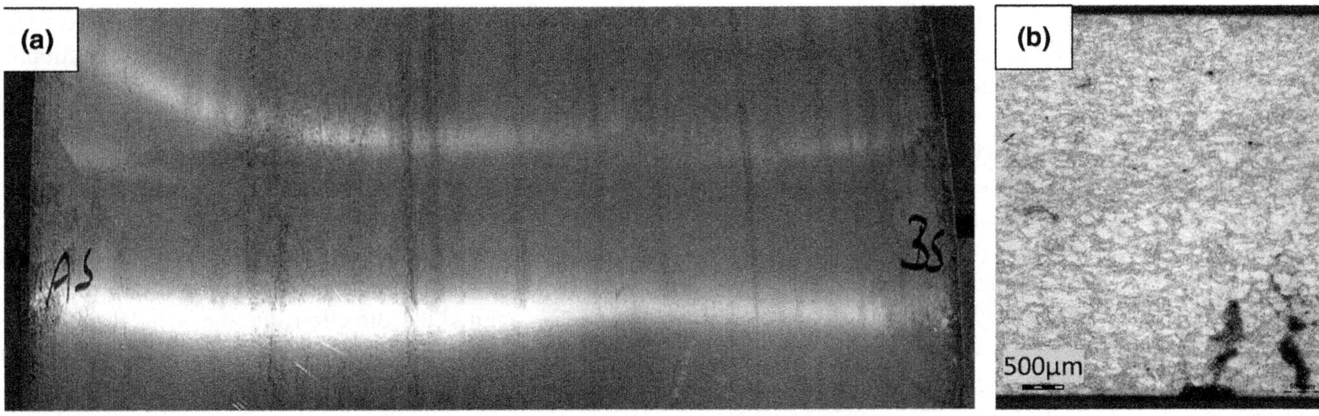


**Fig. 3** Heat-treated TRC strip with inverse segregations (**a**) [17], microstructure including inverse segregations (**b**) [18]

notch effect during the deformation. As a result, the mechanical properties of the TRC strip are adversely affected in transverse direction to the casting/rolling direction.

Both types of macrosegregations can be qualitatively evaluated by metallographic analysis. The first type of macrosegregation is determined, as standard, by metallographic specimens taken along the casting direction and regularly from the same location of the TRC strip [19].

Within this investigation, the influence of macrosegregations on mechanical properties in transverse direction is analyzed. Moreover, the relationship between thickness profile and mechanical properties in transverse direction is examined by means of the expanded structural equation model of [6, 9].

Based on theory and previous research [6, 9] the following hypotheses can be derived:

- **Hypothesis 1–3**: Width, length of contact arc and deformation resistance have a positive interrelationship with strip forming force [6]
- **Hypothesis 4**: Strip forming force has a positive impact on the thickness profile [6]
- **Hypothesis 5**: Length of contact arc influences segregations positively [9]
- **Hypothesis 6**: Deformation resistance is negatively related to segregations [9]
- **Hypothesis 7**: Thickness profile in transverse direction has a negative impact on segregations [9]
- **Hypothesis 8**: Thickness profile is positively related to mechanical properties
- **Hypothesis 9**: Segregations have a negative impact on mechanical properties

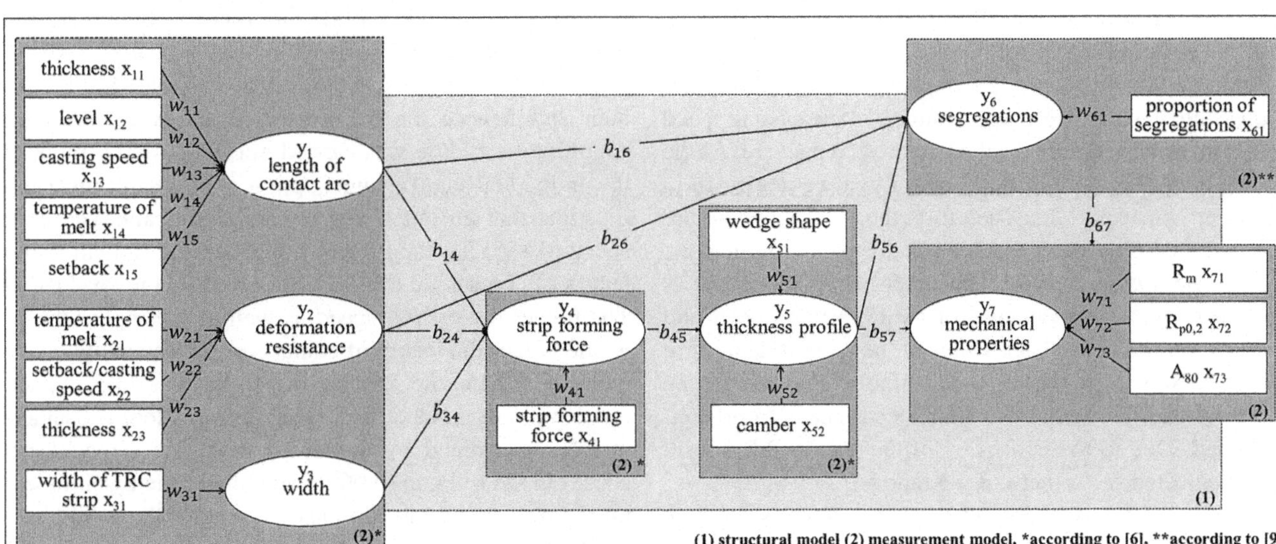


**Fig. 4** Structural equation model for the assessment of the interrelationships of the TRC process parameters and the mechanical properties in transverse direction of heated AZ31 TRC strips

The corresponding structural equation model is shown in Fig. 4 and consists of the structural model and the measurement model. The structural model illustrates the presumed hypotheses between unobservable and not measurable variables, also called latent variables ($y_1$ to $y_7$), whereas the measurement model operationalises the latent variables with the help of indicators ($x_{11}$ to $x_{71}$).

## Data

The Institute of Metal Forming at the TU Bergakademie Freiberg provided 43 coils from 13 casting campaigns of magnesium alloy AZ31 and an overall strip length of 500 m. For the calculation of the structural equation model, 30 coils were used. These were strips whose thickness profiles could be evaluated in terms of camber [20] and wedge shape [21]. The strip thickness which varies slightly between 5.2 and 5.4 mm, was not considered in the following investigation. The strip width of the coils, which were produced under stationary conditions, differed between 600 and 750 mm. Furthermore, the mechanical properties of the TRC strips after heat treatment at temperatures above 400 °C were obtained by tensile tests. The used tensile test samples were taken from the TRC strip in longitudinal and transverse direction to the casting direction. In transverse direction, the samples were taken interleaved across the overall strip width. The obtained mechanical properties were yield strength ($R_{p0.2}$), tensile strength ($R_m$) and elongation ($A_{80}$).

Based on previous research, the provided data of the pilot plant and the experimental design, the indicators of the latent variables were defined as

(1) *Width*: The width of the TRC strip is used for the spread of the material in the roll gap [6].
(2) *Length of contact arc*: The thickness of the TRC strip, pressure of melt (level), temperature of melt, casting speed [6] and setback determine the length of contact arc. As the strip thickness is almost constant, it will be neglected in further considerations. Therefore, the level, temperature of melt, casting speed and setback are used as indicators.
(3) *Deformation resistance*: The deformation resistance is defined by the temperature of melt, casting speed and thickness [6]. As setback varies in the current data set in comparison to [6, 9], the dwell time of the melt in the rolling gap, i.e. setback/casting speed, needs to be considered. Due to the constancy of the strip width, it will be neglected in further considerations.
(4) *Strip forming force*: The strip forming force is the sum of all forces in the roll gap [6].
(5) *Thickness profile*: It is characterized by the camber and wedge shape [6].
(6) *Segregations*: The area of segregations was measured by metallographic specimens along the rolling direction of the respective strip according to [19].
(7) *Mechanical properties*: The average mean of the yield strength ($R_{p0.2}$), tensile strength ($R_m$) and elongation ($A_{80}$) of all samples taken from the respective TRC strip (5–7 specimens), were used to evaluate the mechanical properties. Because of collinearity between $R_m$ and $A_{80}$, only $A_{80}$ and $R_{p0.2}$ are considered further on. In contrast to the longitudinal direction, the mechanical properties in transverse direction, which were evaluated, vary widely.

## Model Estimation and Evaluation

The structural equation model was estimated by the PLS-SEM approach [22, 23] using the software SmartPLS 3.2.6 [24]. The prediction-oriented PLS-SEM approach is a statistical method based on a series of ordinary least squares regressions fitting networks of latent variables to empirical data. In the following section, the PLS results concerning the measurement model and the structural model are evaluated and reported according to the procedures and criteria of [8].

### Formative Measurement Model Assessment

Table 1 depicts the estimation results of the formative measurement model. The variance inflation factor (VIF) values, which are considered to check the indicators for critical collinearity, are uniformly below the recommended threshold of 5 [25]. The outer weights give information on the contribution of the indicators and their relevance for their respective latent variable [8]. To check their significance the bias-corrected and accelerated bootstrapping procedure was carried out with 5000 samples and the individual sign change option. The results in Table 1 illustrate that all outer weights are significant. The casting speed (0.885) has the biggest impact on the formation of the length of contact arc ($l_m + l_s$), followed by the level (0.460). The higher the casting speed as well as the bigger the level is, the bigger the length of contact arc is. The temperature of melt (0.258) and the setback (0.199) have a lower impact on the length of contact arc. The deformation resistance is mainly determined by the dwell time (i.e. setback/casting speed) in the roll gap (0.969). The higher the dwell time, the more time for solidification and cooling of the solid phase is available and the higher the deformation resistance is. The indicators of the thickness profile show outcomes of the improvement process based on the research results of [6].

**Table 1** Evaluation of formative measurement model

| Latent variable[a] | Indicator | Outer weight | VIF | 90% bias-corrected bootstrap confidence interval[b] | Significant? |
|---|---|---|---|---|---|
| Length of contact arc | Temperature of melt | 0.258 | 1.243 | [0.003, 0.375] | Yes |
| | Casting speed | 0.885 | 2.715 | [0.919, 1.170] | Yes |
| | Level | 0.460 | 1.149 | [0.252, 0.767] | Yes |
| | Setback | 0.199 | 3.249 | [0.010, 0.343] | Yes |
| Deformation resistance | Temperature of melt | −0.305 | 1.003 | [−0.443, −0.031] | Yes |
| | Setback/casting speed | 0.969 | 1.003 | [0.876, 1.034] | Yes |
| Thickness profile | Wedge shape | −0.325 | 1.003 | [−0.572, −0.072] | Yes |
| | Camber | 0.964 | 1.003 | [0.811, 1.026] | Yes |
| Mechanical properties | $A_{80}$ | 0.546 | 1.148 | [0.399, 0.876] | Yes |
| | $R_{p0,2}$ | 0.664 | 1.148 | [0.465, 0.841] | Yes |

[a]Single-item constructs (width, strip forming force, segregations) are excluded in this table (i.e. their outer relationship is 1.0)
[b]*Note* Bias-accelerated and corrected bootstrapping (BCa) procedure; 5000 bootstrapping subsamples; individual sign change option

The included thickness profiles mainly consist of symmetric thickness profiles with a camber (0.964).

## Structural Model Assessment

An important criterion for evaluating the structural model is the amount of explained variance of the latent variable illustrated for the respective latent variables of the model in Table 2 (i.e. their $R^2$-value). In case of the strip forming force, the coefficient of determination has a value of 0.975 and can be classified as substantial. Based on the improved production conditions and the optimized thickness profile, the thickness profile has only an $R^2$-value of 0.286, which is caused by the low data variation. Concerning macrosegregations, the $R^2$-value is relative high (0.559). This value can be considered very well because the determined segregations are allocated locally in a sample. Thus, they can only be regarded as an indicator of a possible impact of dissolved precipitations for the entire TRC strip. The $R^2$-value of the mechanical properties after annealing is 0.314. This is bound up with the fact that the macrosegregations, which consist mainly of gamma phase, are largely dissolved during heating.

All path relationships of the structural model are significant with the exception of "segregations → mechanical properties", which were explained above. In contrast to hypothesis 1 a negative interrelationship exists between width and strip forming force (−0.898) in the present data set. It depends on the variation of the process conditions and especially on the selection of the setback and casting speed. Moreover, it clarifies the interaction of the process parameters. According to hypotheses 2 and 3, the deformation resistance (0.244) and the length of contact arc (0.146) are positively related to the strip forming force. The value of the path coefficient between the strip forming force and the thickness profile is 0.535. Furthermore, it can be concluded with a path coefficient of 0.510 that the bigger the length of contact arc is, the more segregations are formed. The deformation resistance has, in contrast to [9], a positive relation to segregations (0.646). This will be clarified in subsequent investigations in the near future. Despite local capturing of segregations, the thickness profile has a negative relationship (−0.730) with the segregations. The more asymmetric the thickness profile is, the higher the segregation proportion is. The thickness profile is not only indicator for the segregation formation, but also for the mechanical properties of the TRC strip after annealing. The better the thickness profile (symmetric) is, the better the mechanical properties are. This is related to the formation of the inverse segregations. Detailed explanations of the appeared phenomenon will follow in a future analysis.

The calculation of the bias-corrected and accelerated (BCa) confidence interval (see Table 2) reflects the partly heterogeneous data structure, which can be expected of research with the aim to capture different impacts.

The relevance of the path coefficient is substantiated by the $f^2$ effect size values.

In addition, the blindfolding procedure [26] was carried out to assess the predictive relevance of the model by the Stone-Geissers criterion ($Q^2$) [27, 28]. At an omission distance of seven, all values are well above zero. In accordance with the evaluation, a valid model was developed.

**Table 2** Structural model evaluation results

| Endogenous latent variable | $R^2$ value | | $Q^2$ value | |
|---|---|---|---|---|
| Strip forming force | 0.975 | | 0.854 | |
| Thickness profile | 0.286 | | 0.112 | |
| Segregations | 0.559 | | 0.369 | |
| Mechanical properties | 0.314 | | 0.150 | |
| **Relation** | **Path coefficient** | **90% bias-corrected bootstrap confidence interval**[a] | **Significant?** | $f^2$ **value** |
| Length of contact arc → strip forming force | 0.146 | [0.076, 0.369] | Yes | 0.347 |
| Length of contact arc → segregations | 0.510 | [0.135, 0.787] | Yes | 0.246 |
| Deformation resistance → strip forming force | 0.244 | [0.167, 0.429] | Yes | 0.817 |
| Deformation resistance → segregations | 0.646 | [0.511, 1.410] | Yes | 0.379 |
| Width → strip forming force | −0.898 | [−0.982, −0.841] | Yes | 15.288 |
| Strip forming force → thickness profile | 0.535 | [0.292, 0.715] | Yes | 0.400 |
| Thickness profile → segregations | −0.730 | [−0.933, −0.586] | Yes | 1.129 |
| Thickness profile → mechanical properties | 0.393 | [0.072, 0.672] | Yes | 0.137 |
| Segregations → mechanical properties | −0.223 | [−0.500, −0.012] | No | 0.044 |

[a]*Note* Bias-accelerated and corrected bootstrapping (BCa) procedure, 5000 bootstrapping subsamples; individual sign change option

## Conclusion

The Twin-Roll casting process for the magnesium TRC strip production was introduced. Achievable mechanical properties in transverse direction to the casting/rolling direction of annealed AZ31 magnesium alloy TRC strips and their impact factors were investigated, which have an influence on the property level. Impacts of process conditions were evaluated by the approximation of the structural equation model using the PLS-SEM approach.

Based on the analysis, most important influencing factors were identified. The main factor affecting the thickness profile formation is the strip forming force with respect to the special process conditions. The deformation resistance and the length of contact arc, as well as their influencing process parameters determine the strip forming force. In case of the current data set, the impact of the strip width must be regarded separately. It is an evidence of the interactions between the process parameters themselves.

Major finding of the empirical examination is the identified relationship between the thickness profile and the mechanical properties. The better the thickness profile (more symmetric) is, the better the mechanical properties are and especially the higher the elongation ($A_{80}$) transverse to rolling/casting direction are.

The PLS method has proved in research studies with partly restricted variable data sets and various process parameters. In conclusion, it has provided a qualitative relationship between the strip thickness and the mechanical properties. Based on these results clues could be derived in respect of further required data, which should be additionally captured to reveal some vague relationships.

**Acknowledgements** The authors are grateful to the financial support by the European Union (European Social Fund) and Saxonian Government under grant 100270111.

## References

1. Hadadzadeh A, Wells MA (2013) Mathematical modeling of thermo-mechanical behavior of strip during twin roll casting of an AZ31 magnesium alloy. Journal of Magnesium and Alloys 1(2): 101–114. https://doi.org/10.1016/j.jma.2013.04.001
2. Li BQ (1995) Producing thin strips by twin-roll casting—part I: Process aspects and quality issues. Journal of the Minerals, Metals and Materials Society 47(5): 29–33

3. Kawalla R, Ullmann M, Schmidt C et al. (2013) Casting and Rolling of Magnesium Alloys. VII. International Youth Scientific and Practical Conference 2013
4. Neh K, Ullmann M, Kawalla R (2014) Twin-roll-casting and hot rolling of magnesium alloy WE43. Procedia Engineering 81: 1553–1558. https://doi.org/10.1016/j.proeng.2014.10.189
5. Ullmann M, Berge F, Neh K et al. (2014) Application of magnesium sheets and strips in vehicle construction. In: Kawalla R (ed) AutoMetForm - SFU 2014: New materials for vehicle components; 3rd to 5th November 2014. Inst. für Metallformung, Freiberg, pp 37–45
6. Kawalla C, Höck M, Ullmann M et al. (2018) An empirical examination of the thickness profile formation of twin-roll-cast magnesium strips. Archives of Civil and Mechanical Engineering 18(1): 227–234. https://doi.org/10.1016/j.acme.2017.06.006
7. Kawalla R, Ullmann M, Schmidt C et al. (2011) Properties of Magnesium Strips Produced by Twin-Roll-Casting and Hot Rolling. MSF 690: 21–24. https://doi.org/10.4028/www.scientific.net/MSF.690.21
8. Hair JF, Hult GTM, Ringle CM et al. (2017) A primer on partial least squares structural equation modeling (PLS-SEM) (2. ed.). Sage Publications, Thousand Oaks
9. Kawalla C, Ullmann M, Hoeck M (2017) Procedure to Analyze the Formation of Segregations Using the PLS-SEM Approach. KEM 746: 75–83. https://doi.org/10.4028/www.scientific.net/KEM.746.75
10. Miehe A, Gross U (2012) Modelling of heat transfer and solidification processes in horizontal twin-roll casting of magnesium AZ31. In: IOP Conference Series 33: Materials Science and Engineering
11. Kawalla C, Hoeck M (2017) Quality assurance system for hot rolled TRC magnesium strips. Scientific Papers of Silesian University of Technology. Organization and Management Series 100: 171–179
12. Frischknecht B, Thun G (2001) Bandgießtechnologien - Stand der Technik und Trends. Aluminium 77(10): 746–751
13. Strid J (1992) Twin roll casting, a technological over view. Proceedings of the 3rd International 3: 321–356
14. Hadadzadeh A, Wells M, Essadiqi E (2012) Mathematical Modeling of the Twin Roll Casting Process for AZ31 Magnesium Alloy - Effect of Set-Back Distance. In: Mathaudhu SN,Sillekens WH, Neelameggham NR, Hort N (eds) Magnesium Technology 2012. Springer, Cham
15. Hoffmann H, Neugebauer R, Spur G (eds) (2012) Handbuch Umformen. Carl Hanser Verlag GmbH & Co. KG
16. Institut für Metallformung (2014) Bericht Bandqualität. Intern (413–6)
17. Institut für Metallformung (2011) Bericht Bandqualität. Intern (324–3)
18. Institut für Metallformung (2012) Bericht Bandqualität. Intern (360–3)
19. Institut für Metallformung Interne Prüfvorgaben. Intern
20. Tautz W (10.06.92) Verfahren zur Regelung totzeitbehafteter Regelstrecken (EP 0 575 636 AI)
21. Degner M, Neuschütz E (1998) Walzen von gleichmäßigem (keilfreiem) Warmbanddickenprofil: Europäische Kommission, Generaldirektion Forschung und Innovation. EUR (Luxembourg)
22. Wold HOA (1982) Soft modelling: the basic design and some extensions. In: Jöreskog KG & Wold HOA (eds) Systems under Indirect Observations, Part II, North Holland, Amsterdam
23. Lohmöller JB (1989) Latent Variable Path Modeling with Partial Least Squares, Physica, Heidelberg
24. Ringle CM, Wende S, Becker JM SmartPLS 3. Boenningstedt: SmartPLS GmbH, http://www.smartpls.com
25. Diamantopoulos A, Riefler P (2008) Formative Indikatoren: Einige Anmerkungen zu ihrer Art, Validität und Multikollinearität. Zeitschrift für Betriebswirtschaft 78(11): 1183–1196
26. Tenenhaus M, Vinzi VE, Chatelin Y-M et al. (2005) PLS path modeling. Computational statistics & data analysis 48(1): 159–205
27. Geisser S (1974) A predictive approach to the random effect model. Biometrika: 101–107
28. Stone M (1974) Cross-validatory choice and assessment of statistical predictions. Journal of the royal statistical society. Series B (Methodological): 111–147

# The Morphology and Distribution of $Al_8Mn_5$ in High Pressure Die Cast AM50 and AZ91

G. Zeng, X. Zhu, S. Ji, and C. M. Gourlay

**Abstract**

The morphology and distribution of $Al_8Mn_5$ is studied in AM50 and AZ91 produced by hot and cold chamber high pressure die casting (HPDC). It is found that, in HPDC, primary $Al_8Mn_5$ particles take a wide range of morphologies within the same casting spanning from faceted polyhedra to weakly-faceted dendrites. These different morphologies exist across the whole cross-section without any clear trend in morphology versus radial position. A comparison with $Al_8Mn_5$ in samples solidified at low cooling rate suggests that the larger polyhedral particles are externally solidified crystals (ESCs) that nucleate and grow in the shot chamber analogous to αMg ESCs, and that the dendritic $Al_8Mn_5$ nucleated and grew at high cooling rate in the die cavity.

**Keywords**

HPDC • Externally solidified crystals • Porosity $Al_8Mn_5$

## Introduction

In recent years there has been a drive to use high pressure die casting (HPDC) for structural and crash-worthy automotive applications which has required improved HPDC quality through optimised melt handling, die design, process parameters and vacuum systems [1–3].

HPDC Mg components typically contain externally solidified αMg crystals (ESCs) which nucleate and grow in the shot chamber before being injected into the die cavity [4–6]. The volume fraction of αMg ESCs is commonly 10–30 vol.% and depends on the melt superheat and shot chamber characteristics including the fill fraction and the temperature of the sleeve walls and plunger tip [7, 8]. αMg ESCs are typically ~100 μm and are significantly larger than the in-cavity solidified αMg grain size of ~5–30 μm, leading to a bimodal grain size distribution [9]. The presence of larger αMg ESCs has been linked to a decrease in the mean elongation to fracture [10] and to an increased variability in ductility [11].

Most Mg–Al-based alloys contain a small Mn addition to ameliorate against the negative effects of Fe on corrosion resistance [12, 13]. Generally, AM, AZ, AS, and AE series alloys contain sufficient Al and Mn that $Al_8Mn_5$ is a primary phase (i.e. $Al_8Mn_5$ forms before αMg during solidification) as shown in Fig. 1 [14] for the alloys used in this work (Table 1). A consequence of this in HPDC is that $Al_8Mn_5$ can form and settle in the holding pot/crucible (enhanced by Fe pick-up from the pot/crucible), leading to die casting sludge [15]. Furthermore, since $Al_8Mn_5$ forms at higher temperature than αMg, it might be expected that $Al_8Mn_5$ begins to form in the shot chamber along with the externally solidified αMg crystals, and that the ESCs injected into the cavity are a mixture of $Al_8Mn_5$ and αMg.

This work was conducted to explore whether $Al_8Mn_5$ crystals form in the shot chamber and are injected into the cavity, and to understand how they are distributed in the final casting. Furthermore, this behaviour is compared in hot and cold chamber HPDC where the thermal conditions in the shot chamber/gooseneck are different and, therefore, the formation of ESCs is likely to be different.

## Methods

Mg alloys AM50 and AZ91 with compositions in Table 1 were used in this work. The AM50 steering wheel in Fig. 2a

G. Zeng (✉) · C. M. Gourlay (✉)
Department of Materials, Imperial College London, London, SW7 2AZ, UK
e-mail: g.zeng@imperial.ac.uk

C. M. Gourlay
e-mail: c.gourlay@imperial.ac.uk

X. Zhu · S. Ji
Brunel Centre for Advanced Solidification Technology (BCAST), Institute of Materials & Manufacturing, Brunel University London, Uxbridge, UB8 3PH, UK


D. Orlov et al. (eds.), *Magnesium Technology 2018*, The Minerals, Metals & Materials Series, https://doi.org/10.1007/978-3-319-72332-7_21

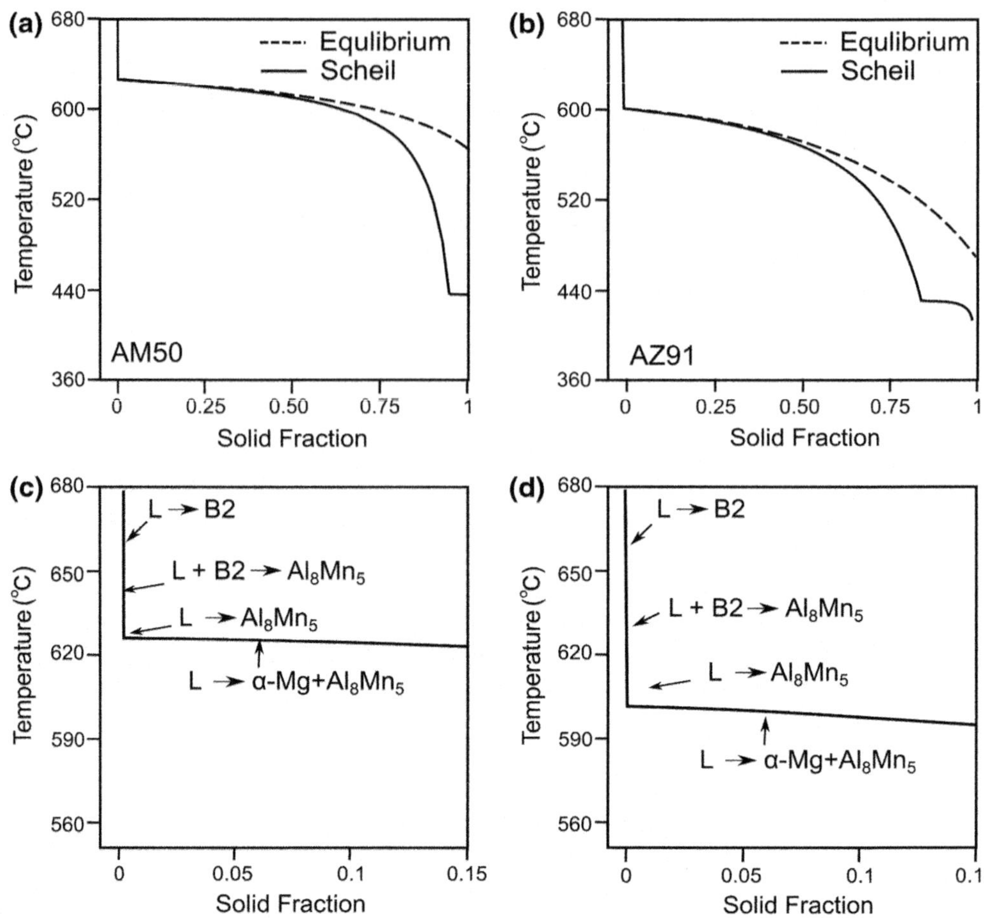


**Fig. 1** Scheil and equilibrium solidification paths of AM50 (**a**, **c**) and AZ91 (**b**, **d**) calculated with the Thermo-Calc TCMG4 database [14]. The top row shows the whole solidification path. The bottom row is a zoom-in on the beginning of solidification up to 15% solid fraction

was produced industrially via the hot chamber HPDC process from the same die as in Ref. [16]. The AZ91 tensile bar castings were produced by cold chamber HPDC on a 5 MN locking force Frech DAK 450-54. The multicavity die shown in Fig. 2b was preheated to 150 °C. Liquid alloy at 675 °C (~75 °C superheat) was ladled into a preheated shot chamber. The plunger velocity was set to 0.3 $ms^{-1}$ for the first phase and 4 $ms^{-1}$ for the filling stage, and the intensification pressure was 36 MPa.

Samples for microstructural analysis were cut from the positions marked on Fig. 1 into slices of 10 mm × 10 mm × 0.5 mm. Metallographic polishing was carried out down to 0.05 μm colloidal silica by standard preparation methods. Samples were etched in a solution of 200 ml ethylene glycol, 68 ml distilled water, 4 ml nitric acid and 80 ml acetic acid. A Zeiss AURIGA field emission gun SEM (FEG-SEM) with an Oxford Instruments INCA x-sight energy dispersive X-ray spectroscopy (EDX) detector and a BRUKER e-FlashHR electron backscatter diffraction (EBSD) detector were used. Bruker ESPRIT 2.1 software was used to analyse the obtained EBSD patterns.

## Results and Discussion

$Al_8Mn_5$ particles were identified by combining EDX with EBSD. A typical example is shown in Fig. 3 for the polyhedral particle marked with a cross in Fig. 3a. The EDX spectrum in Fig. 3b contains no Mg peaks, indicating that the interaction volume did not contain any αMg matrix, and the composition 60.9Al–38.5Mn–0.6Fe (at.%) is consistent with $Al_8(Mn,Fe)_5$. Figure 3c shows an EBSD pattern collected from the same particle. The next image in Fig. 3c shows the pattern indexed as rhombohedral $Al_8Mn_5$ [17] (with Strukturbericht designation $D8_{10}$) using the standard Hough transform method in BRUKER ESPRIT 2.2. To further confirm the phase, dynamical simulations were conducted in BRUKER DynamicS as shown in the right-most image of Fig. 3c. It can be seen that there is good agreement in the band positions and band intensities between the experimental and dynamically simulated patterns. Thus, the particle is identified as $Al_8Mn_5$ with $D8_{10}$ structure by EDX and EBSD.

Typical microstructures from the AM50 steering wheel are shown in Fig. 4 where the classical HPDC

**Table 1** AM50 and AZ91 alloy compositions

| | Mg | Al | Zn | Mn | Si | Fe | Cu | Ni | ppm Be |
|---|---|---|---|---|---|---|---|---|---|
| AZ91E | Bal. | 8.95 | 0.72 | 0.19 | 0.039 | <0.001 | 0.001 | <0.001 | 0.007 |
| AM50 | Bal. | 4.79 | 0.007 | 0.23 | 0.011 | <0.001 | <0.002 | 0.001 | \ |

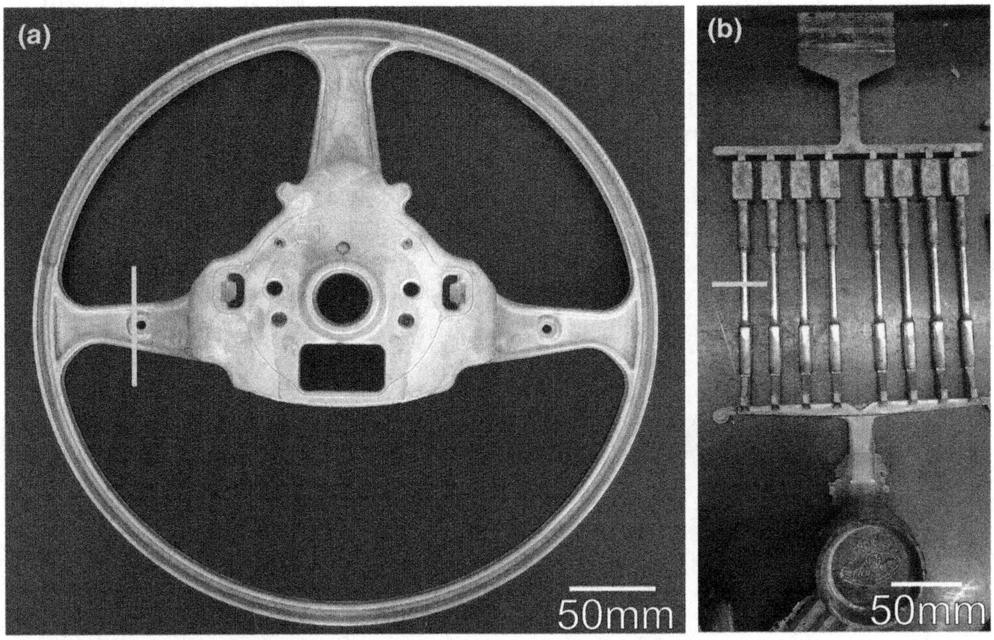


**Fig. 2** Photographs of the two castings. The position of the sectioning planes is indicated by the superimposed lines

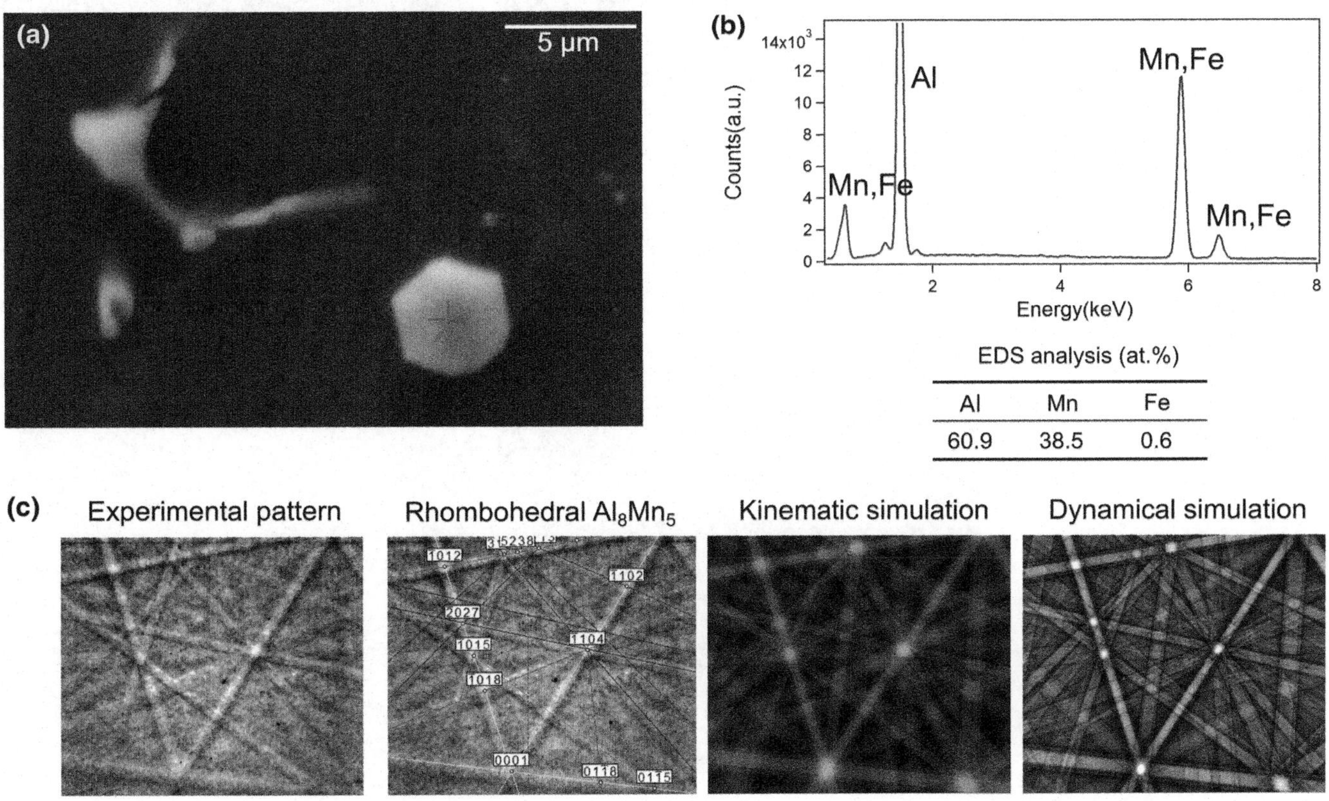


**Fig. 3** Phase analysis on Al–Mn particles by EDS and EBSD

microstructural features are evident: The polished section in Fig. 4a, c contains two pore bands following the contour of the casting, similar to those discussed in past work [18, 19]. After etching, Fig. 4b, d show αMg ESCs as well as the pore bands which appear as dark bands. The area fraction of αMg ESCs in this sample is ~5%, which is a low value for Mg HPDC. Higher magnification imaging of the pore bands in Fig. 4f shows that they contain a higher fraction of $Mg_{17}Al_{12}$ (i.e. more eutectic) than the surrounding regions, similar to past work [6]. In most regions of the casting, the $Al_8Mn_5$ appear as small particles that can be easily identified in backscattered electron (BSE) mode due to the much higher atomic-number of Mn compared with Mg and Al. For example, in Fig. 4e, the numerous bright particles are $Al_8Mn_5$ and the lighter grey particles are $Mg_{17}Al_{12}$.

$Al_8Mn_5$ particles in the AM50 steering wheel are examined in more detail in Fig. 5. In low magnification BSE images (e.g. Fig. 5a) $Al_8Mn_5$ particles can be seen that have much larger size than the small $Al_8Mn_5$ particles present at higher magnification (e.g. in Fig. 4e). In Fig. 5a, these larger $Al_8Mn_5$ particles are indicated with arrows. Figure 5b examines a large $Al_8Mn_5$ particle at higher magnification. The $Al_8Mn_5$ particle is ~20 μm long and has a faceted morphology. Adjacent, is a smaller particle (~2 μm across) with a similar composition that has a dendritic morphology and is, therefore, also expected to be primary $Al_8Mn_5$. It is likely that the smaller $Al_8Mn_5$ with dendritic morphology formed at significantly higher cooling rate than the large particle. To compare the two sizes of $Al_8Mn_5$ particle with the bimodal distribution of αMg grains, Fig. 5c shows that the αMg ESCs are an order or magnitude larger than the in-cavity solidified αMg grains. Based on the significantly different sizes of primary $Al_8Mn_5$ in Fig. 5b and the presence of αMg ESCs, it seems that the large $Al_8Mn_5$ particles

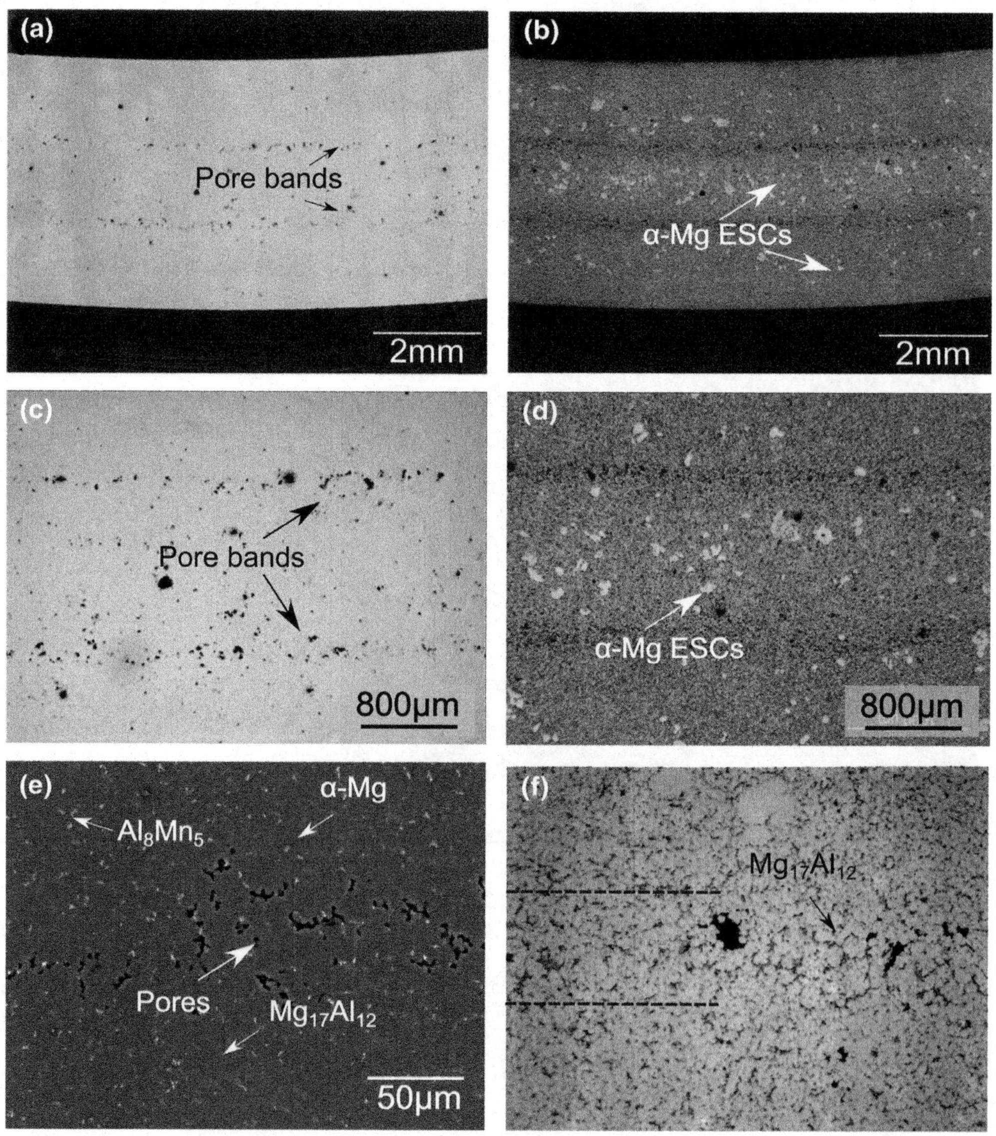


**Fig. 4** Macro- and microstructures of AM50 steering wheel. **a** and **c** as-polished sample optical micrograph showing the distribution of porosity. The bulk flow direction during HPDC was into the page(X). **b** and **d** etched-sample micrograph revealing the ESCs of α-Mg (white grains) and defects bands of macrosegregation. **e** BSE and **f** optical images of defect bands showing concentrated porosity and also fine eutectic $Mg_{17}Al_{12}$ network

in Fig. 5a, b are $Al_8Mn_5$ ESCs that nucleated and grew in the shot chamber or goose neck at low cooling rate before being injected into the die cavity, whereas the smaller primary $Al_8Mn_5$ nucleated and grew dendritically at higher cooling rate. This interpretation will be further supported at the end of this paper when the size of large $Al_8Mn_5$ particles are shown to be similar to primary $Al_8Mn_5$ formed at low cooling rate.

The distribution of large $Al_8Mn_5$ particles across the cross-section (i.e. vertically across Fig. 4b) is quantified in Fig. 5d and a similar quantification is shown for αMg ESCs in Fig. 5e. It can be seen that there are more αMg ESCs near the centre and fewer near the edge of the cross-section similar to past work [5], while the large $Al_8Mn_5$ particles are distributed more uniformly across the cross-section. However, the total number of $Al_8Mn_5$ ESCs is low and the distribution in Fig. 5d should be treated with caution.

Typical microstructures from the AZ91 tensile bars are shown in Fig. 6. Similar to the steering wheel, there is a pore band following the contour of the casting (Fig. 6a).

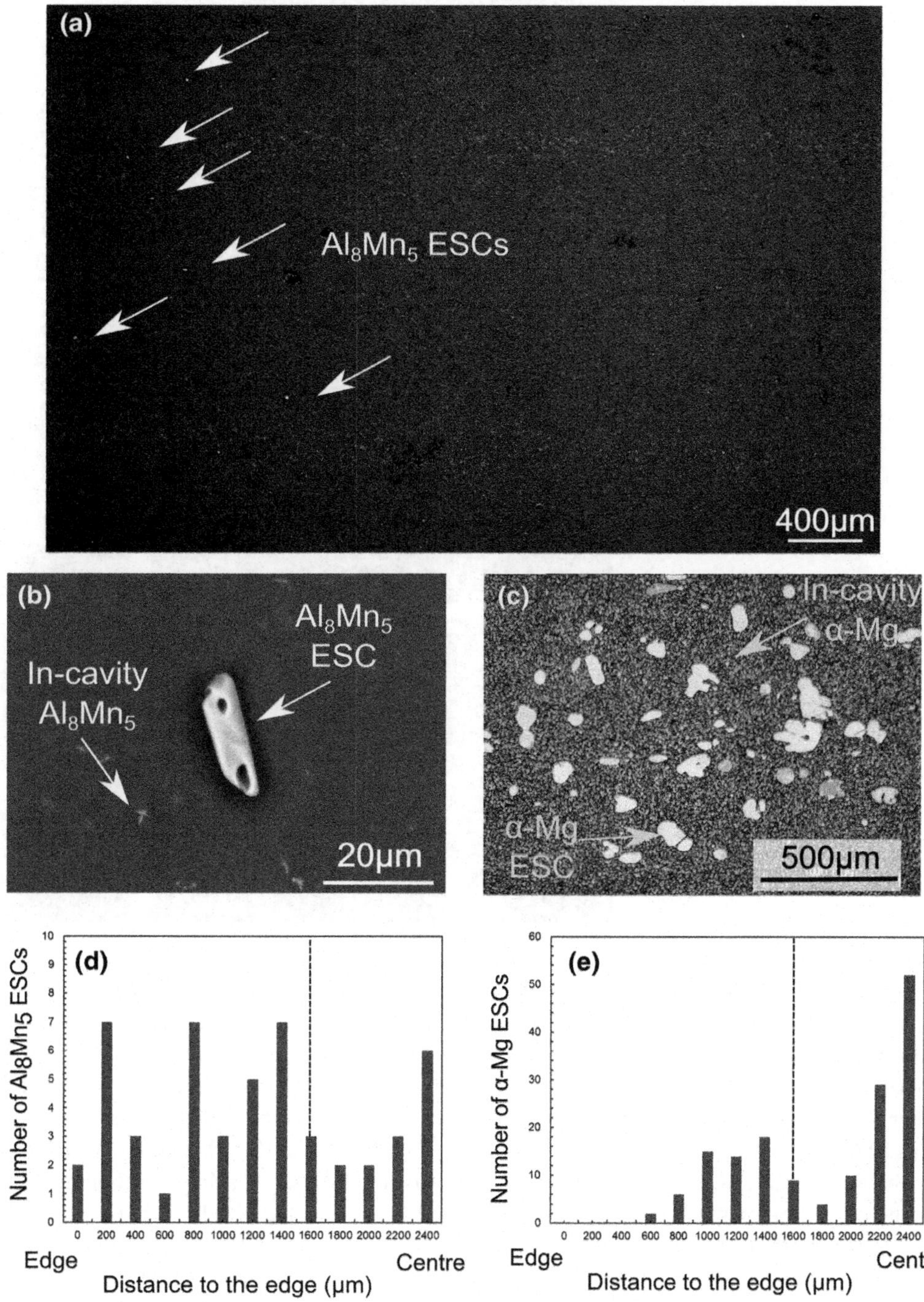


**Fig. 5** Distribution of ESCs of $Al_8Mn_5$ and α-Mg **a** BSE image at low magnification. **b** ESCs of $Al_8Mn_5$ and $Al_8Mn_5$ solidified in die cavity. Note the size different of two type of $Al_8Mn_5$ **c** ESCs of α-Mg (white grains). **e**

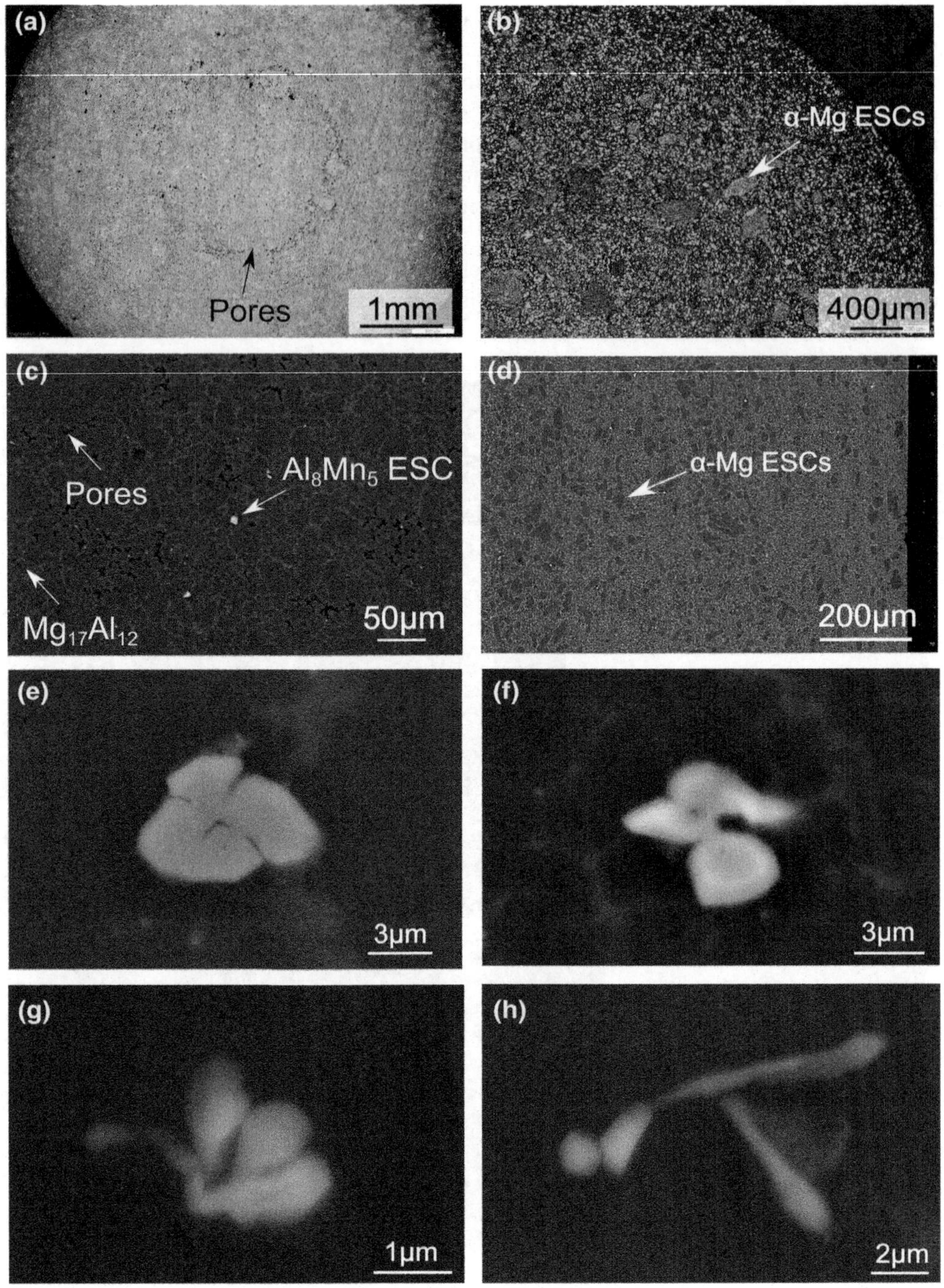


**Fig. 6** Transverse view of HPDC AZ91 tensile bar sample. **a** macrostructure showing defect bands. **b** Etched microstructure and **d** BSE image shows macro-segregation and the ECSs of α-Mg grains. **c** BSE image at higher magnification shows pores, $Al_8Mn_5$ ECSs and eutectic $Mg_{17}Al_{12}$

However, a significant difference is the much higher fraction of αMg ESCs in the cold chamber HPDC AZ91 tensile bars, estimated to be >30 vol.%. This can be seen in Fig. 6b, d where αMg ESCs exist across most of the cross-section, either as large dendrites (Fig. 7b), coarsened flakes (Fig. 7b) or as ESC fragments (Fig. 6d and 7b). Examining Fig. 6d, it can be seen that ESC fragments exist all the way to the casting edge. Fragmented and partially-globularised ESCs in cold chamber HPDC have been linked to the high shear rate during the filling stage in past work [20]. The high αMg ESC fraction and mixture of full ESC dendrites, ESC fragments and in-cavity solidified grains leads to a complex microstructure throughout the cross-section; for example, in Fig. 6c, many ~50 μm ESC αMg fragments can be seen surrounded by the smaller in-cavity solidified αMg grains.

Similarly, the AZ91 tensile bars contained a wide range of $Al_8Mn_5$ morphologies. A representative range is shown in Fig. 6e–h: Fig. 6e has a faceted polyhedral morphology, Fig. 6f–g have complex branched morphologies with weaker faceting, and Fig. 6h are complex flakes. While the $Al_8Mn_5$

in Fig. 6h probably formed during a eutectic reaction, the faceted polyhedral particle in Fig. 6(a) and branched crystals in Fig. 6f–g are significantly larger than the surrounding eutectic intermetallics and are likely to be primary $Al_8Mn_5$. Some of the polyhedral $Al_8Mn_5$ particles were significantly larger than surrounding primary $Al_8Mn_5$ and are likely to be $Al_8Mn_5$ ESCs. Examples are given in Fig. 6c and Fig. 7a of polyhedral $Al_8Mn_5$ particles >5 μm in diameter.

To further confirm that the larger polygonal $Al_8Mn_5$ are ESCs (i.e. formed in the gooseneck or shot sleeve/chamber before being injected into the cavity), Fig. 8 compares the large $Al_8Mn_5$ from HPDC with $Al_8Mn_5$ in AZ91 solidified at low cooling rate. Figure 8a are $Al_8Mn_5$ from the original AZ91 ingot (i.e. prior to melting and HPDC) and Fig. 8c are $Al_8Mn_5$ from AZ91 cooled at 1 K/s from the work in Ref [21]. In both slow-cooled cases the primary $Al_8Mn_5$ have an equiaxed-polyhedral faceted morphology and a narrow size range [21] of ~5–15 μm. In Fig. 8b, d it can be seen that the large $Al_8Mn_5$ from the HPDC samples have a similar morphology and a similar size to those formed at low cooling rate in the laboratory. This, combined with the presence of a wide range of primary $Al_8Mn_5$ particles in HPDC including $Al_8Mn_5$ dendrites, shows that the larger polyhedral $Al_8Mn_5$ formed at lower cooling rate.

The observation of both αMg and $Al_8Mn_5$ ESCs in AM50 and AZ91 high pressure die castings is consistent with the solidification paths predicted by Thermo-Calc with the TCMG4 database in Fig. 1, where B2–Al(Mn,Fe) and $D8_{10}$–$Al_8Mn_5$ are expected to form before αMg in the solidification sequence. Combining the microstructural observations and thermodynamic calculations, it can be concluded that the largest polyhedral $Al_8Mn_5$ particles originally form in the shot chamber and are injected into the die cavity as a mixture of $Al_8Mn_5$, αMg and liquid.

The size of the largest $Al_8Mn_5$ particles in the casting is expected to be determined by the same factors that govern the size and volume fraction of αMg ESCs (melt superheat, fill fraction, thermal profile in the shot chamber, dwell time

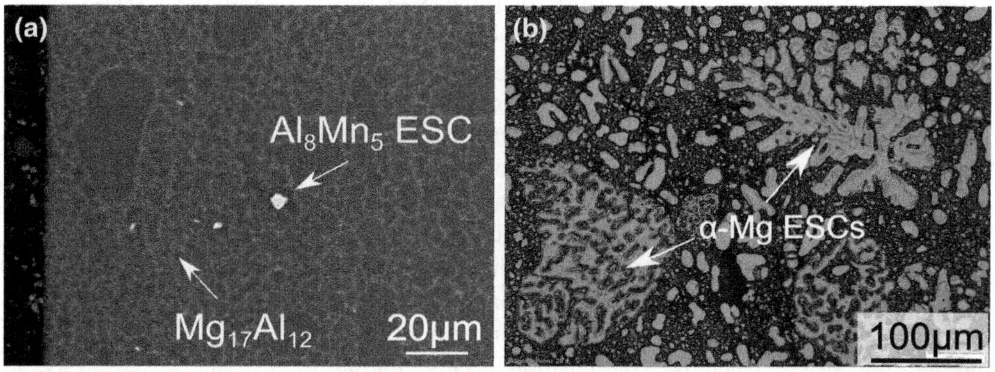


**Fig. 7** ESCs of **a** $Al_8Mn_5$ and **b** α-Mg grains surrounded by eutectic $Mg_{17}Al_{12}$/α-Mg

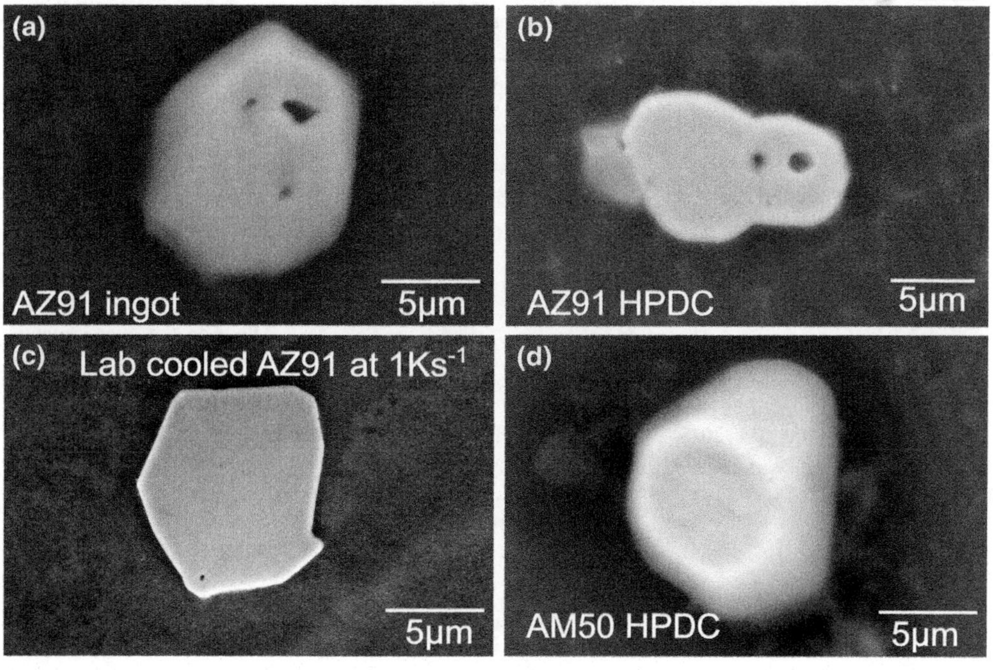


**Fig. 8** Faceted $Al_8Mn_5$ crystals in **a** AZ91 ingot, **b** HPDC AZ91, **c** AZ91 solidified at 1 K/s and **d** AM50 steering wheel by HPDC

etc. [5]). Furthermore, since most commercial Mg–Al alloys (AM, AZ, AE, AJ etc.) contain sufficient Al and Mn for $Al_8Mn_5$ to be a primary phase, it is likely that $Al_8Mn_5$ ESCs will also form in the HPDC of these alloys. $Al_8Mn_5$ ESCs seem to be an integral part of the HPDC process of Mg–Al-based alloys and result in a small population of 5–20 μm equiaxed-faceted $Al_8Mn_5$ crystals that are significantly larger than the in-cavity solidified $Al_8Mn_5$. However, Al–Mn particles form a very low volume fraction ($\sim$0.6 vol.% [14]) and are rarely larger than 20 μm even at low cooling rate.

## Conclusions

$Al_8Mn_5$ particles have been studied in a hot chamber HPDC AM50 steering wheel and a cold-chamber HPDC AZ91 tensile bar casting. In both castings, polyhedral $Al_8Mn_5$ particles have been found to exist in the castings that have a similar morphology and size as primary $Al_8Mn_5$ formed in AZ91 at low cooling rate ($\sim$1 K $s^{-1}$). These polyhedral $Al_8Mn_5$ particles are significantly larger than the branched primary $Al_8Mn_5$ in the same HPDC cross-sections. It is concluded that the large polyhedral $Al_8Mn_5$ particles nucleated and grew in the shot chamber along with the αMg externally solidified crystals (ESCs) and were injected into the die cavity as a mixture of $Al_8Mn_5$, αMg and liquid. This interpretation is supported by thermodynamic calculations with the Thermo-Calc TCMG4 database which predicts that primary $Al_8Mn_5$ should begin to form at higher temperature than αMg and, therefore, that a mixture of primary $Al_8Mn_5$ and αMg externally solidified crystals (ESCs) should exist in the shot sleeve when pre-solidification occurs. Partial solidification in the shot chamber followed by in-cavity solidification at higher cooling rate leads to a wide range of primary $Al_8Mn_5$ morphologies in the same cross section, spanning from large ($\leq$20 μm) polyhedral particles to significantly smaller branched primary crystals.

**Acknowledgements** Financial support from EPSRC (UK) under grant number EP/N007638/1 is gratefully acknowledged.

## References

1. Luo AA (2013) Magnesium casting technology for structural applications. Journal of Magnesium and Alloys 1(1):2–22.
2. Wang QL, Xiong SM (2015) Effect of multi-step slow shot speed on microstructure of vacuum die cast AZ91D magnesium alloy. Transactions of Nonferrous Metals Society of China 25(2):375–380.
3. Ji S, Yang W, Jiang B, Patel J, Fan Z (2013) Weibull statistical analysis of the effect of melt conditioning on the mechanical properties of AM60 alloy. Mater. Sci. Eng., A 566:119–125.
4. Bowles A, Griffiths J, Davidson C (2001) Ductility and the Skin Effect in High Pressure Die Cast Mg-Al Alloys. Magnesium Technology 2001:161–168.
5. Laukli H, Lohne O, Sannes S, Gjestland H, Arnberg L (2003) Grain size distribution in a complex AM60 magnesium alloy die casting. Int. J. Cast Met. Res. 16(6):515–521.
6. Gourlay C, Laukli H, Dahle A (2007) Defect band characteristics in Mg-Al and Al-Si high-pressure die castings. Metall. Trans. A 38 (8):1833–1844.
7. Laukli H, Graciotti A, Lohne O, Gjestland H, Sannes S (2002) The effect of solidification of metal prior to injection in HPDC on the grain size distribution in a complex die casting. NADCA Transactions 21(T02-035):1–4.
8. Bi C, Xiong S, Li X, Guo Z (2016) Development of a Fluid-Particle Model in Simulating the Motion of External Solidified Crystals and the Evolution of Defect Bands in High-Pressure Die Casting. Metall. Trans. B 47(2):939–947.
9. Bowles A, Nogita K, Dargusch M, Davidson C, Griffiths J (2004) Grain size measurements in Mg-Al high pressure die castings using electron back-scattered diffraction (EBSD). Mater. Trans., JIM 45(11):3114–3119.
10. Easton M, Abbott TB, Cáceres CH, The effect of microstructural features and defects on the ductility of high pressure die cast AS21, AM60 and AZ91. Materials Science Forum 147–152.
11. Gjestland H, Sannes S, Svalestuen J, Westengen H, Optimizing the magnesium die casting process to achieve reliability in automotive applications. SAE Technical Paper, 2005.
12. Hanawalt JD, Holdeman GE, Nelson CE, Removal of iron from magnesium base alloys, US. Patent 2.267.862 December 1941.
13. Liu M, Uggowitzer PJ, Nagasekhar AV, Schmutz P, Easton M, Song G-L, Atrens A (2009) Calculated phase diagrams and the corrosion of die-cast Mg–Al alloys. Corros. Sci. 51(3):602–619.
14. Thermo-Calc. TCMG Database version 4.0, 2015.
15. Corby C, Ricketts N, Qian M (2004), Investigation of intermetallics in magnesium die-casting sludge, Magnesium Technology 2004: 209–214.
16. Otarawanna S, Gourlay C, Laukli H, Dahle A (2009) The thickness of defect bands in high-pressure die castings. Mater. Charact. 60(12):1432–1441.
17. Ellner M (1990) The structure of the high-temperature phase MnAl(h) and the displacive transformation from MnAl(h) into Mn5Al8. Metall. Trans. A 21:1669–1672.
18. Gourlay CM, Meylan B, Dahle AK (2008) Shear mechanisms at 0-50% solid during equiaxed dendritic solidification of an AZ91 magnesium alloy. Acta Mater. 56:3403–3413.
19. Gourlay CM, Dahle AK, Laukli HI (2004) Segregation band formation in Al-Si die castings. Metall. Trans. A 35(9):2881–2891.
20. Ghomashchi M (1995) High-pressure die casting: effect of fluid flow on the microstructure of LM24 die-casting alloy. J. Mater. Process. Technol. 52(2–4):193–206.
21. Zeng G, Xian JW, Gourlay CM (2017) Growth of $Al_8Mn_5$ Intermetallic in AZ91, Magnesium Technology 2017:85–92.

# Study on the Production of Metallic Magnesium from Nickel-Containing Serpentine

Huimin Lu and Guangzhi Wu

**Abstract**

About 1.4 billion tons of nickel-containing serpentine are stored in the territory of Tuquan County, China's Inner Mongolia Autonomous Region. In this paper, serpentine as raw material, silicon aluminum alloy as a reducing agent, the use of the corresponding additives and catalyst, the first were mixed and then dry pressed into the ball and placed in the furnace under the appropriate vacuum to heat the balls, the magnesium steam was collected to get crystalline magnesium, and the mixed residue crushed, the nickel-iron alloy was separated from the tailings by magnetic separation. The tailings were used in the manufacture of cement. The process has no $CO_2$ emissions, low-carbon environmental clean production, and the laboratory using a new type of continuous production of heating equipment for the reduction reaction to get magnesium steam, can achieve continuous production, expand production scale and improve production efficiency.

**Keywords**

Aluminum-silicon thermal reduction • Magnesium
Aluminum-silicon alloy • Nickel-containing serpentine
Nickel-iron alloy

## Introduction

Magnesium is the lightest metal of the commonly used metals and can be applied widely. For example, there are metallurgical applications, chemical applications, structural applications, and automobile parts production [1, 2]. In 2016 the global production of magnesium reached 900 thousand tons. The production of magnesium is based on Pidgeon process and electrolytic process, and more than 90% production of magnesium was produced by Pidgeon process from China, and all magnesium producers use dolomite as magnesium reduction feedstock.

Magnesium production in the world is currently dominated by the Pidgeon process which uses silicon, in the form of ferrosilicon, to reduce magnesia from calcined dolomite under vacuum. The overall reaction of the process can be written as follows [3]:

$$\begin{aligned} &2MgO(s) + 2CaO(s) + (xFe)Si(s) \\ &= 2Mg(g) + Ca_2SiO_4(s) + xFe(s) \end{aligned} \quad (1)$$

The reaction is performed in batch mode within steel retorts that operate around 1200 °C and a vacuum of 10–20 Pa to produce approximately 20 kg of Mg over an eight to ten hour period. The process suffers from high energy usage and low productivity.

In Tuquan County, Inner Mongolia, China, magnesium resources are not the traditional dolomite but serpentine, huge reserves (about 1.4 billion tons), extraction of magnesium with the type of serpentine ore, there is no precedent all over the world, it is a great challenge. Therefore, that the study of extraction technology of metal magnesium from serpentine ore (magnesium silicate type) and the development of new type of vertical vacuum semi-continuous magnesium reduction equipment, to replace the traditional the Pidgeon process, it is of great significance.

In this paper, serpentine as raw material, silicon aluminum alloy as a reducing agent, the use of the corresponding additives and catalyst, the first were mixed and then dry pressed into the ball and placed in the furnace under the appropriate vacuum to heat the balls, the magnesium steam was collected to get crystalline magnesium, and the mixed residue crushed, the nickel-iron alloy was separated from the tailings by magnetic separation. The tailings were

H. Lu (✉)
School of Materials Science and Engineering, Beihang University, 37 Xueyuan Road, 100191 Beijing, China
e-mail: lhm0862002@aliyun.com

G. Wu
Inner Mongolia Xintai Construction and Installation Group Co., Ltd, Huafeng Street, 137599 Inner Mongolia, China
e-mail: 13910518379@126.com


D. Orlov et al. (eds.), *Magnesium Technology 2018*, The Minerals, Metals & Materials Series, https://doi.org/10.1007/978-3-319-72332-7_22

used in the manufacture of cement. The process has no $CO_2$ emissions, low-carbon environmental clean production, and the laboratory using a new type of continuous production of heating equipment for the reduction reaction to get magnesium steam, can achieve continuous production, expand production scale and improve production efficiency.

## Serpentine Ore Process Mineralogy and Reduction Principle

### The Structure and Characteristics of Serpentine Ore

Through the microscopic observation of serpentine lithofacies, the structure type of serpentine is mainly composed of scaly crystal structure and fiber crystal structure.

(1) Scale crystal structure: most of the serpentine crystals under the polarizing microscope are tabular and arranged in a directional orientation (Fig. 1a).
(2) Fiber crystal structure: visible fibrous serpentine crystal almost aligned or arranged in a row (Fig. 1b).
(3) Substituted structure: visible hematite along the magnetite octahedron cleft, or fissure account, hematite formation was significantly later than magnetite. See also serpentine exchange olivine.
(4) Substituted residual structure: visible serpentine exchange olivine, olivine was account for the remnants.
(5) Self-shaped and semi-self-shaped grain structure: chromite, magnetite often from the self shaped and semi-self-shaped particles distributed in the serpentine granules.

### Mineral Composition of Serpentine Ore

The results show that the mineral species in the serpentine sample is relatively simple, and the main ore minerals are serpentine. Also found are olivine, hematite, magnetite, chromite, chrome spinel and nickel pyrite. From the structure of the rock and the composition of the mineral can be seen, the original rock should be super-basic rock (peridotite).

### Identification of Rock and Mineral Facies

In the microscopic observation of the serpentine sample, there are many hematite veins inserted in the broken magnetite, in which the main metal minerals are magnetite and

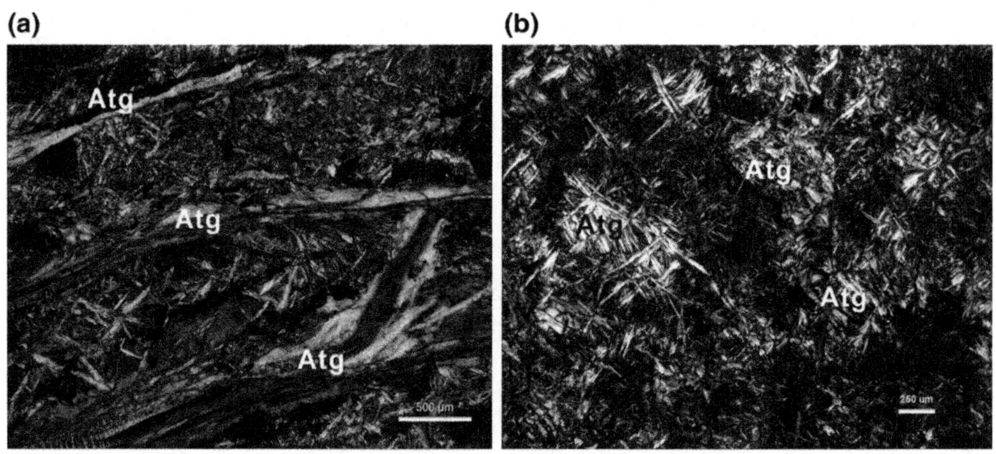


**Fig. 1** The main structure of the serpentine type under the microscope. **a** serpentine in the rock mass; **b** serpentine intertwined with fiber crystal structure; Atg-serpentine

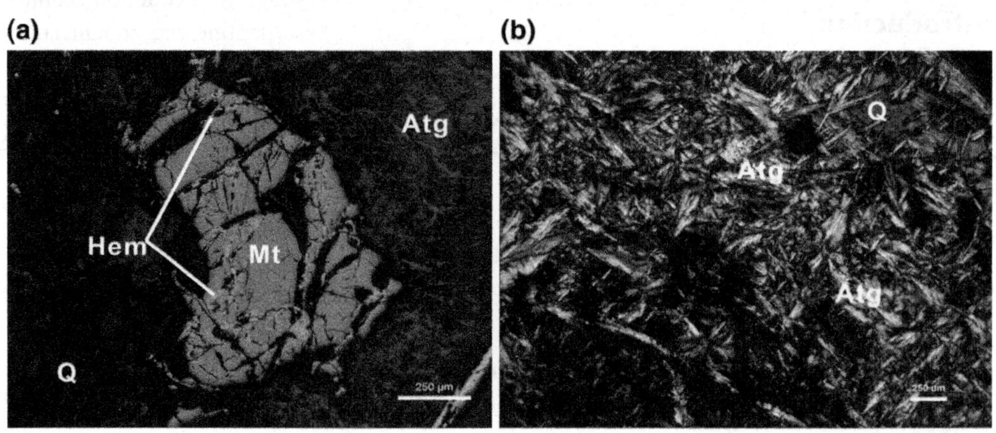


**Fig. 2** Micrograph of minerals in serpentine ore. **a** fragmented magnetite interspersed with a number of hematite vein; **b** serpentine was intertwined structure distributed in the rock Hem-hematite; Mt-magnetite; Atg-serpentine

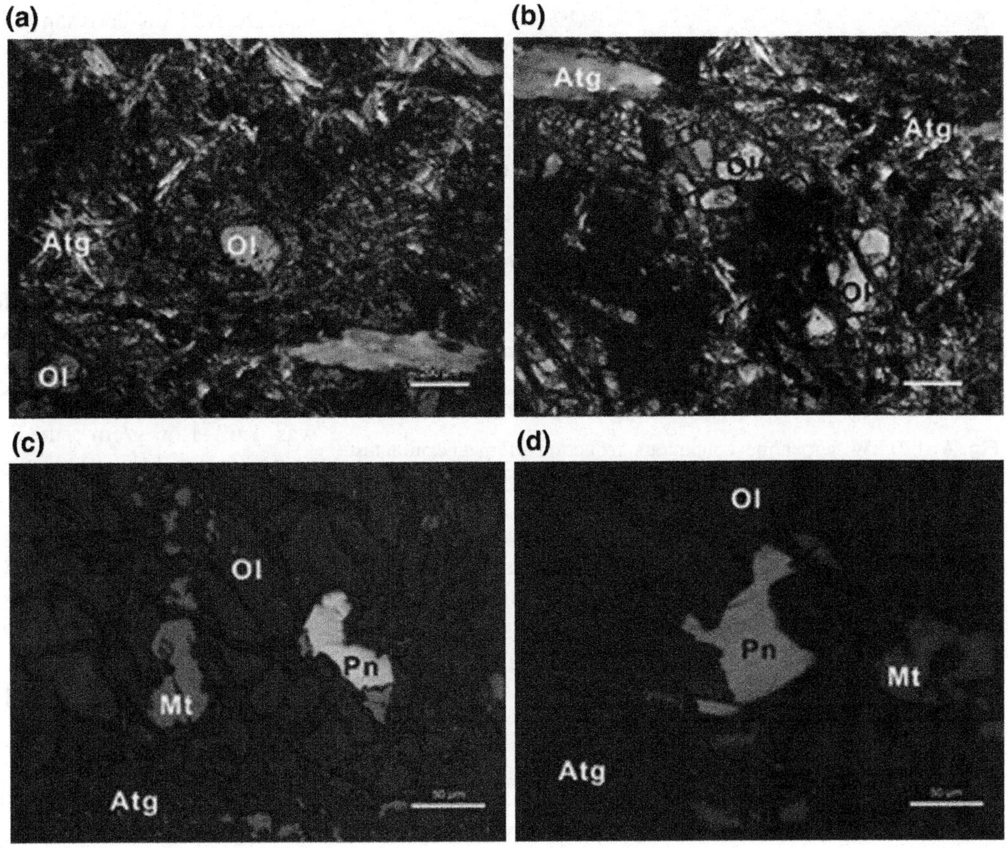


**Fig. 3** Microscope photographs of mineral minerals in serpentine. **a** serpentine and metasomatous residual olivine; **b** debris structure olivine and serpentine formed along the edge; **c**, **d** magnetite and nickel pyrite in algal-alteration; Ol-olivine; Mt-magnetite; Atg-serpentine; Pn-nickel pyrite

hematite (Fig. 2a), chromite, nickel pyrite (Fig. 3c, d) and a small amount of chalcopyrite, non-metallic minerals mainly serpentine, olivine (Fig. 3a, b), serpentine in the mirror fibrous, interlaced structure distribution (Fig. 2b), or metasomatic olivine.

## The Nature of Serpentine Ore

The main mineralogical characteristics of the ore by mineralization, SEM /EDS and XRD analysis of serpentine ore are as follows:

(1) Serpentine is the main ore minerals in the serpentine ore, and with a small amount of olivine and chrome spinel, metal minerals, including magnetite, hematite, chromite, nickel pyrite and chalcopyrite. Serpentine ore is fiber crystal, granular crystal, fiber interwoven structure, substituted structure and substituted residual structure, block structure.
(2) Serpentine fiber length is generally in the tens and hundreds of microns, fiber length changes in a large number of individual mm.

## Reduction Principle

Magnesium production in the laboratory is carried on which uses aluminum-silicon alloy as reduction agent, to reduce magnesium silicate from serpentine under vacuum. In Al–Si alloy reduction process of magnesium, due to the presence of calcium oxide, resulting in a more stable $12CaO \cdot 7Al_2O_3$, making the theoretical minimum temperature of the reaction reduced by 249.2 °C, showing the presence of calcium oxide makes the reaction easier. The overall reaction of the process can be written as follows:

$$\begin{aligned}&54CaO(s) + 14[2MgO \cdot SiO_2(s)] + 7[MgO \cdot SiO_2(s)]\\&\quad + 14Al(s) + 7Si = 12CaO \cdot 7Al_2O_3(s)\\&\quad + 14[2CaO \cdot SiO_2(s)] + 35[Mg](g) + 14[CaO(s) \cdot SiO_2(s)]\end{aligned} \tag{2}$$

## Experimental Procedure

Laboratory experiments for producing magnesium were conducted in a 120 kW serpentine continuous reduction of magnesium test furnace, as pictured in Fig. 4. Process

**Fig. 4** 120 kW serpentine continuous reduction of magnesium test furnace

conditions were as follows: the reaction was performed in semi-continuous feeding charge 30 kg each within magnesium test furnace that operate around 1200 °C and a vacuum of 10–20 Pa over an two to four hour period.

## Raw Materials

***Serpentine***. Serpentine ore powder was taken from Inner Mongolia Xintai Construction and Installation Group Co., Ltd, the size was smaller than 0.5 mm. The main minerals in the serpentine ore are 96 mass% serpentine, and with a small amount of olivine, chrome spinel, and metal minerals including magnetite, hematite, chromite, nickel pyrite and chalcopyrite. The assay of the serpentine ore is shown in Table 1.

## Test Methods

The raw materials for producing magnesium were mixed uniformly in the given proportions in a stirring machine, dry pressure briquetted into ovals with the major axis 40 mm and the minor axis 20 mm. The average density of briquettes was 1.62–1.66 g/cm$^3$, the porosity was 44–47%. The briquettes were charged into the 120 kW serpentine continuous reduction of magnesium test furnace with temperature 1150–1200 °C, vacuum of 10–20 Pa and time 2–4 h. The Mg was discharged. The samples were analysed by XRD, SEM analysis and chemical analysis.

## Comprehensive Experiment for Producing Magnesium

According to the condition test parameters, the comprehensive experiment was carried out on a 120 kW new serpentine semi-continuous magnesium reduction furnace. Each filling

**Table 1** Analysis of serpentine ore used in this work/mass%

| | | | | | | |
|---|---|---|---|---|---|---|
| MgO | $Al_2O_3$ | CaO | $Fe_2O_3$ | FeO | $SiO_2$ | $Na_2O$ |
| 35.54 | 0.64 | 3.36 | 5.23 | 1.47 | 37.17 | 0.0097 |
| $K_2O$ | Mn | Ni | Cr | $H_2O$ | P | S |
| 0.055 | 0.048 | 0.22 | 0.12 | 0.48 | 0.0022 | 0.0009 |

**Fig. 5** Comprehensive experimental raw materials and products left upper: pellet; left lower: crystalline magnesium; right upper: nickel-iron alloy; right lower: slag

**Table 2** The chemical analysis results of the crude magnesium/mass%

| Si | Cr | Ni | Cu | Zn | Ca | Fe | Al | Mg |
|---|---|---|---|---|---|---|---|---|
| 0.001 | 0.002 | 0.002 | 0.002 | 0.002 | 0.003 | 0.005 | 0.002 | 99.58 |

35.8 kg, of which dry serpentine 20 kg, lime 12 kg, Al–Si alloy 3 kg, fluorite 0.8 kg, mixed even dry powder ball into the furnace, keeping vacuum 10 Pa, temperature 1200 °C, to maintain the reduction time 3 h. Comprehensive experimental raw materials and products taken from the serpentine semi-continuous magnesium reduction furnace were pictured in Fig. 5. The recoveries of magnesium extracted from serpentine by Al–Si alloy can reach 85%. The chemical analysis results of the crude magnesium obtained in the test is shown in Table 2. Visible, as crude magnesium, its quality is very good. Further refined, high quality Mg products can be obtained. The reduction slag can be used as a cement raw material due to the transformation of calcium disilicate crystal in form of expansion and self-crushing loose white powder, the size smaller than 50 μm, the chemical analysis results were as follows: 61.69 mass%CaO, 13.45 mass% $Fe_2O_3$, 17.44 mass%$SiO_2$, 4.39 mass%$Al_2O_3$, and 3.03 mass%MgO. The obtained residue was subjected to magnetic separation to obtain a nickel iron block. The chemical analysis results were as follows: 2.32mass%Ni, 97.68mass% Fe. Nickel iron recovery rate was about 85%.

## Conclusions

1. The process of extracting magnesium by the aluminum-silicon alloy heat-reducing from serpentine is reasonable. The Al–Si alloy can extract the metallic magnesium from the serpentine at 10–20 Pa of vacuum and from 1100 to 1200 °C. In the process of magnesium extraction, at the same time these nickel-iron alloy and dicalcium silicate and aluminum silicate slag used as raw materials for the preparation of cement can be obtained.
2. Laboratory scale of comprehensive experimental studies has shown that magnesium metal extraction from serpentine is feasible, and magnesium recovery rate reaches 80%. This is a cleaning process with no carbon dioxide emissions.
3. Laboratory semi-continuous reduction furnace improves magnesium production efficiency and reduces costs, it is worth further study, and it has a bright future. Raw materials for production of Al–Si alloy are obtained easily, such as coal gangue, fly ash and so on. Semi-continuous process with aluminum-silicon alloy from the serpentine extraction of metal magnesium is simple, clean production, and low cost.

## References


1. D. Eliezer, et al., "Magnesium Science, Technology and Applications," Advanced Performance Materials, Vol. 5(1998), 201–212.
2. M. Halmann, et al., "Magnesium Production by Pidgeon Process Involving Dolomite Calcinations and MgO Silicothermic Reduction: Thermodynamic and Environmental Analyses," Industrial & Engineering Chemistry Research, Vol. 47(2008), 2146–2154.
3. G. Hanko, H. Antrekowitsch, P. Ebner, "Recycling Automotive Magnesium Scrap", JOM, 2 (2002), 51–54.

# Fabrication of $Mg(OH)_2$ by Electrolysis Using $MgCl_2$ Aqueous Solution

Xijuan Pan, Zhihe Dou, Ting-an Zhang, Yukun Ren, Guozhi Lyu, Junjie Zhang, and Xiuxiu Han

**Abstract**

$Mg(OH)_2$ is usually used as a flame retardant agent, a heavy metal removing agent and a flue gas desulfurization agent in the field of environmental protection. In this work, $Mg(OH)_2$ was prepared by electrolytic method using $MgCl_2$ aqueous solution as the main raw materials. And the X-ray diffraction results proved that pure $Mg(OH)_2$ crystals can be fabricated by the electrolysis method. The $Mg(OH)_2$ prepared has two microscopic forms, shaped in block and elongated fibrous. The effects of concentration of $MgCl_2$ solution, current density, temperature and time of electrolysis on the current efficiency and energy consumption in the electrolysis process were studied in this paper. Within the experimental range, current efficiency increased but energy consumption decreased with the increase of $MgCl_2$ solution concentration, electrolysis temperature and time. The effects of current density on the current efficiency and energy consumption rested on the concentration of the $MgCl_2$ solution and electrolysis temperature.

**Keywords**

$Mg(OH)_2$ • $MgCl_2$ aqueous solution • Electrolysis Current efficiency • Energy consumption

X. Pan · Z. Dou · T. Zhang (✉) · Y. Ren · G. Lyu · J. Zhang · X. Han
Key Laboratory of Ecological Metallurgy of Multi-Metal Intergrown Ores of Ministry of Education, Special Metallurgy and Process Engineering Institute, Northeastern University, Shenyang, 110819, Liaoning, China
e-mail: zta2000@163.net

X. Pan
e-mail: 1248374494@qq.com

Z. Dou
e-mail: douzh@smm.neu.edu.cn

Y. Ren
e-mail: 416372442@qq.com

G. Lyu
e-mail: lvgz@smm.neu.edu.cn

J. Zhang
e-mail: 1316656838@qq.com

X. Han
e-mail: 974219685@qq.com

## Introduction

In recent years, $Mg(OH)_2$ has attracted the attention of many experts and scholars for its excellent properties. As a chemical product and intermediate, $Mg(OH)_2$ has excellent properties such as flame resistance, smoke suppression and fallibility. Therefore, $Mg(OH)_2$ has been widely used in the preparation of flame retardant, acidic wastewater treatment agent, heavy metal removal agent, flue gas desulfurization agent and other products [1–3]. There are many methods for the production of $Mg(OH)_2$, such as chemical precipitation, hydrothermal, sol-gel, microemulsion, ammonia bubble reaction, ultrasound assisted, microwave synthesis and electrochemical methods, etc. Because the electrochemical method has many advantages, for instance, less investment, high purity and easy to control crystal shape, the electrochemical method is easier to produce on a large scale [4–6].

The principle of preparing $Mg(OH)_2$ by electrolysis is similar to that of chlor alkali industry. $MgCl_2$ solution is used as raw material to carry out electrolysis of solution in a cation membrane electrolyzer. The chloride ion loses electrons into chlorine on the anode plate, and the water gets electrons into hydrogen and hydroxide on the cathode plate. Magnesium ions under the effect of direct current pass through the cation exchange membrane, and then they are combined with the hydroxyl in the cathode chamber, becoming the production of $Mg(OH)_2$. Because the solubility product constant (Ksp) of $Mg(OH)_2$ is very small, which is $1.2 \times 10^{-11}$, so it is easy to get the precipitate $Mg(OH)_2$ [6]. The electrode reactions and total reaction are as follows [7]:


D. Orlov et al. (eds.), *Magnesium Technology 2018*, The Minerals, Metals & Materials Series, https://doi.org/10.1007/978-3-319-72332-7_23

The anode reaction:

$$2Cl^- - 2e^- \rightarrow Cl_2 \quad (1)$$

The cathode reaction:

$$2H_2O + 2e^- \rightarrow 2OH^- + H_2 \quad (2)$$

$$Mg^{2+} + 2OH^- = Mg(OH)_2\downarrow \quad (3)$$

The total reaction: $MgCl_2 + 2H_2O \overset{electrify}{\longrightarrow} Mg(OH)_2 \downarrow + H_2\uparrow + Cl_2\uparrow \quad (4)$

In this work, the X-ray diffraction results proved that pure $Mg(OH)_2$ crystals can be fabricated by the electrolysis method. And the effects of concentration of $MgCl_2$ solution, current density, temperature and time of electrolysis on the current efficiency and energy consumption in the electrolysis process were studied in this paper. The experimental results provide the basic theory evidence for large-scale preparation of $Mg(OH)_2$ by electrochemical method, and lay the foundation for further experimental and theoretical research.

## Experimental

### Materials and $Mg(OH)_2$ Synthesis

The $MgCl_2 \cdot 6H_2O$ chemicals used in this experiment were purchased from Tianjin Kermel Chemical Reagent Co., Ltd, AR, the purity is 97%. The $CO_2$ gas (99.95%) was bought from Shenyang Industrial Gas Co., Ltd. Deionized water disposed by ourselves was used in the preparation of all aqueous solutions.

The cation membrane electrolytic cell used in this work was made by ourselves, which is separated by a cation exchange membrane into two electrode chambers, in the right one fixed with a titanium anode plate coated of ruthenium, indoor for titanium cathode plate in the left one. The volumes 100 ml of different concentration of $MgCl_2$ solution were added into the two chambers, with a current density as a fixed value for experiments. When the electrolytic time reached to a certain period of time, the experiment would be stopped. After $Mg(OH)_2$ in the cathode liquid was filtered and washed, the dried particles is of high purity $Mg(OH)_2$ powders.

### Characterization of $Mg(OH)_2$ Particles

The X-ray diffract meter used in this work was D8 Advance X Bruker ray analyzer produced by German Bruker Company. Light tube type was Cu target, ceramic X light tube. $\lambda$ = 0.15406 nm, scan range was 10°–90°, the scanning speed was 2 degrees/min. The micro morphology was observed by a scanning electron microscope. All samples were prepared for SEM examination by pasting on the conductive adhesive. SEM images were collected on a HITACHI SU8010 SEM at 3.0 kV.

### Calculation of Current Efficiency and Energy Consumption

In the electrolysis transformation process, the products are often less than the theoretical amount. The causes of this phenomenon include physical factors such as convection, diffusion and carry-under and chemical factors such as side reaction, decomposition, recombination, etc. In addition, the reverse shift of the $OH^-$ will also lead to a reduction in the current efficiency. At the same time, the chlorine produced by the anode will be hydrolyzed to form HClO and $ClO^-$, and further react to $ClO^{3-}$, which will result in the reduction of the current efficiency. As a result, the amount of electricity used is less than the actual amount supplied. Therefore, current efficiency is used to measure the utilization of electricity [8, 9].

Since the target product in this work is the cathode chamber product $Mg(OH)_2$, the current efficiency is calculated at the final yield of $Mg(OH)_2$. In other words, the cathode current efficiency represents that of the electro conversion reaction. The cathode current efficiency is the ratio of the actual $Mg(OH)_2$ mass produced by the electrolysis to the theoretical mass. The solid content of the Mg $(OH)_2$ produced in the cathode chamber can be obtained by weighing, and the theoretical quality can be calculated by the law of Faraday electric transformation. The equation is as follows [10–15]:

$$\eta = \frac{n_{Mg(OH)_2}}{\frac{I\Delta t}{F}} \quad (5)$$

where $\eta$ is current efficiency (%), $n_{Mg(OH)_2}$ is the actual Mg $(OH)_2$ amount of substance produced by the electrolysis (mol), I is current (A), $\Delta t$ is reaction time (s), F is the Faraday constant (about 96,500 C/mol).

DC power consumption is an important test index. For the reaction (4) of $Mg(OH)_2$ by elecrolysis from $MgCl_2$ aqueous solution, 0.5 mol $Mg(OH)_2$ can be produced per 1F of electricity. Based on the calculation of 1 kg $Mg(OH)_2$ converted by electrolysis, the equation DC power consumption is as follows [10–15]:

$$W = \frac{2UF}{3600 \times M_{Mg(OH)_2} \cdot \eta} \quad (6)$$

where W is DC power consumption (kWh/kg), U is average operating voltage (V), $M_{Mg(OH)_2}$ is relative molecular mass of $Mg(OH)_2$, η is current efficiency (%).

## Results and Discussion

### X-Ray Diffraction of $Mg(OH)_2$ Particles

As shown in Fig. 1, the XRD analysis results of $Mg(OH)_2$ prepared under different conditions are presented. Figure 1a is the contrastive XRD result of original $MgCl_2$ concentration from 50 to 250 g/L. Figure 1b is the contrastive XRD result of current density from 67 to 335 mA/cm$^2$. Figure 1c is the contrastive XRD result of react temperature from 25 to 65 °C. Figure 1d is the contrastive XRD result of react time from 60 to 180 min. As can be seen from Fig. 1, the peak height increases with the increase of the current density, temperature and react time, but the concentration of $MgCl_2$ has little influence on the peak height.

### SEM Analysis of $Mg(OH)_2$ Particles

Some typical SEM pictures of $Mg(OH)_2$ particles prepared by electrolysis are given in Fig. 2. Figure 2a is a typical bulk structure and Fig. 2b, c have both a block structure and a fibrous structure. As we can see from the top right corner enlargement of Fig. 2b, c, the surface of the fiber structure is similar to that of the block structure, and the surface is rough.

### Effect of Different Electrolysis Condition for Current Efficiency and Energy Consumption

Figure 3a is the current efficiency and the unit energy consumption of electrolysis process when the concentration of $MgCl_2$ solution increased from 50 to 250 g/L. And the Fig. 3b–d represent the changes of the current efficiency and the unit energy consumption when current density, react temperature and react time changed respectively.

As can be seen from Fig. 3a, the current efficiency increased from 46.39 to 48.78% with the concentration of $MgCl_2$ solution from 50 to 250 g/L, and the slope of <150 g/L section became larger, meaning the increasing speed of current efficiency increased. The unit energy consumption decreased with the increase of $MgCl_2$ concentration, and decreased from 23.88 to 14.53 kWh/kg. This is because the increasing concentration of $MgCl_2$ aqueous solution leaded to the number of ions in contact with the electrode increasing, which accelerated the reaction and the occurrence of $Mg(OH)_2$ precipitation.

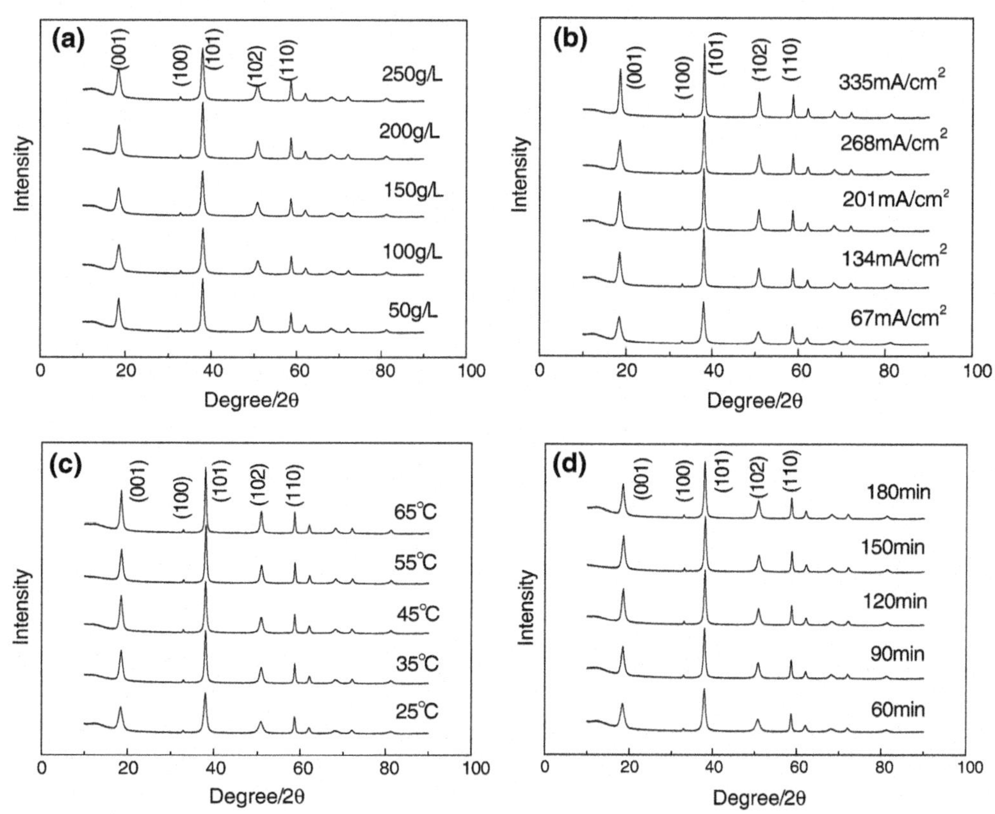


**Fig. 1** The X-ray diffraction of $Mg(OH)_2$ powders prepared under different conditions

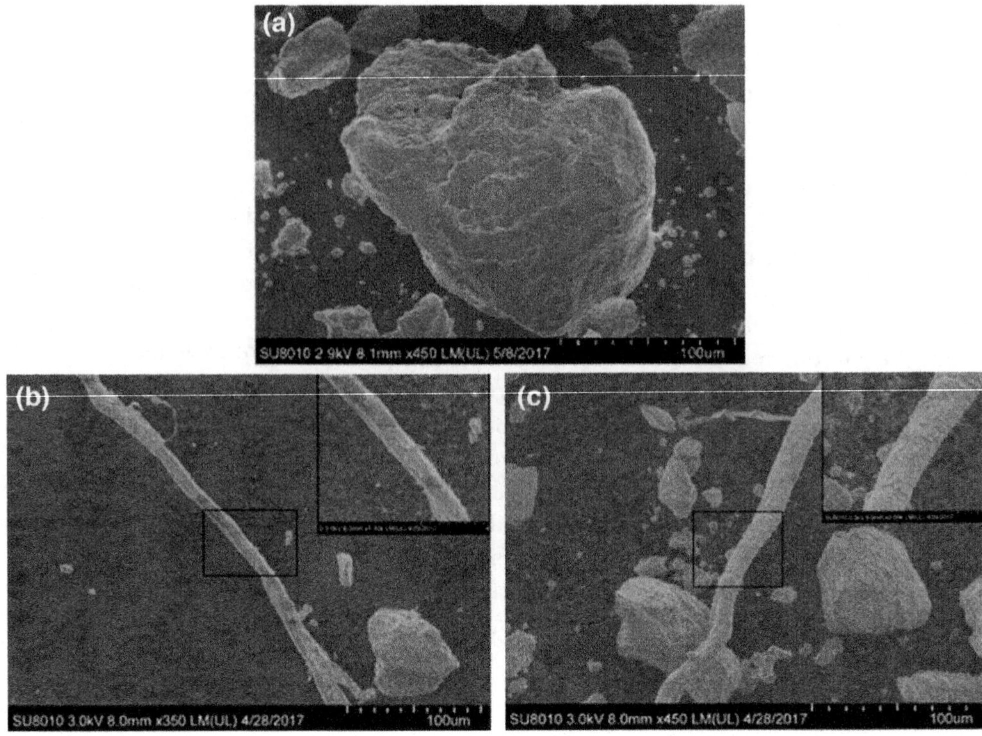


**Fig. 2** The typical SEM pictures

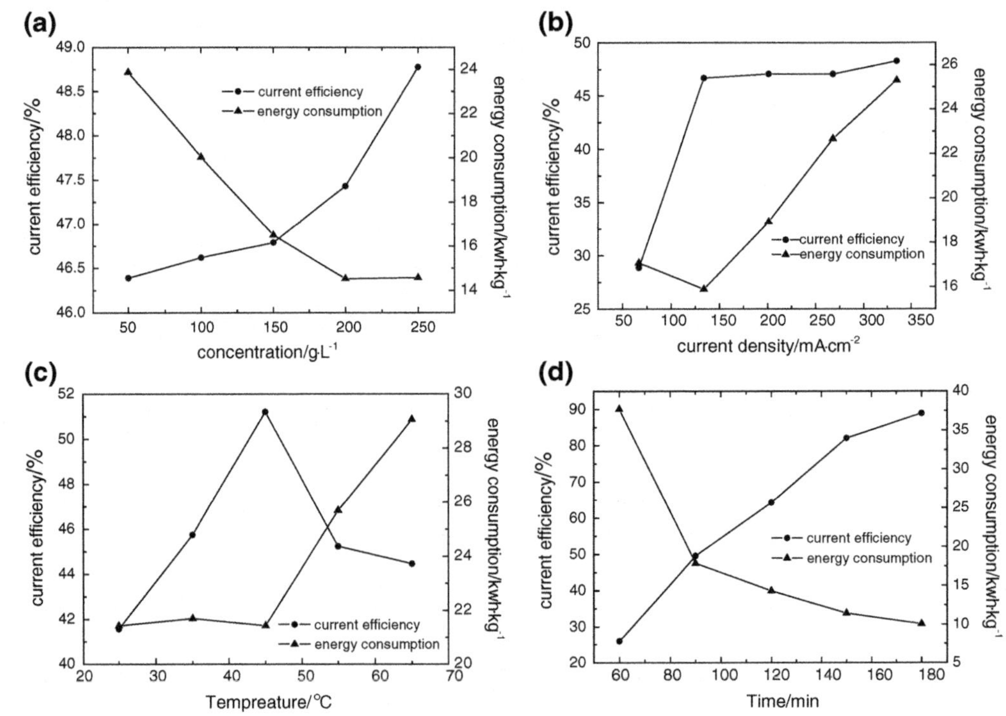


**Fig. 3** Variations of current efficiency and energy consumption as changing electrolysis condition

Figure 3b presents that there is a turning point at 134 mA/cm$^2$ with the current density increasing from 67 to 335 mA/cm$^2$. Current efficiency quickly increased from 28.86 to 46.69% from the current density increasing from 67 to 134 mA/cm$^2$and then maintain stable. Accordingly, the unit energy consumption reduced from 17.07 to 15.89 kWh/kg, and then increased to 25.32 kWh/kg.

There is also a turning point at 45 °C as can be found from Fig. 3c. With the react temperature increasing from 25 to 45 °C, current efficiency quickly increased from 41.57 to

51.21% and unit energy consumption maintained at about 25 kWh/kg. Then current efficiency decreased 25.06% and unit energy consumption increased to 29.06 kWh/kg with temperature increasing to 65 °C. This indicated that the ion activity was higher and the membrane permeability was better and the precipitation efficiency was higher at 45 °C.

Figure 3d shows that with the increase of the react time, current efficiency increased from 25.92 to 88.93% and unit energy consumption decreased from 37.70 to 10.05 kWh/kg.

## Conclusions

In this work, $Mg(OH)_2$ was prepared by electrolytic method using $MgCl_2$ aqueous solution as the main raw materials. The effects of concentration of $MgCl_2$ solution, current density, temperature and time of electrolysis on the current efficiency and energy consumption in the electrolysis process were studied in this paper.

(1) The X-ray diffraction results proved that pure $Mg(OH)_2$ crystals can be fabricated by the electrolysis method. The $Mg(OH)_2$ prepared has two microscopic forms, shaped in block and elongated fibrous.
(2) Within the experimental range, current efficiency increased but energy consumption decreased with the increase of $MgCl_2$ solution concentration, electrolysis temperature and react time. The effects of current density on the current efficiency and energy consumption rested on the concentration of the $MgCl_2$ solution and electrolysis temperature.

**Acknowledgements** This research was supported by the National Natural Science Foundation of China (U1508217), the Fundamental Research Funds for the Central Universities (N162505002).

## References

1. X. Feng, Y. Z. Liu, H. Y. Shen, J. H. Bai. Research status of superfine magnesium hydroxide flame retardant [J]. China Plastics Industry, 2013, 41(02):7–10 + 69.
2. Z. L. Zu, Q. Sun, H. Li, P. Ye. Research progress in preparation and application of magnesium hydroxide flame retardant material [J]. China Powder Science and Technology, 2013, 19(06):37–40.
3. L. M. Ji, L. J. Li, F. Nie, Z. M. Zeng. Research status of magnesium hydroxide flame retardant in China [J]. Journal of Salt Lake Research, 2007, (02):62–72.
4. Z. Y. Zhao, X. G. Li, S. R. Wang, Y. Xiao. Study on the priority growth of (001) plane of hexagonal magnesium hydroxide [J]. Journal of Synthetic Crystals, 2014, 43(07):1611–1619. https://doi.org/10.16553/j.cnki.issn1000-985x.2014.07.014.
5. M. Y. Su, H. J. Wang, L. S. Tang. Application situation and research progress of magnesium hydroxide flame retardants [J]. Contemporary Chemical Industry, 2015, 44(01):114–116+119.
6. H. Shen. The research on multiplexing of modified magnesium hydroxide/expandable graphite [D]. Chengdu University of Technology, 2013.
7. Northeastern University. A method for converting magnesium chloride into high-purity Magnesium Oxide [P]. China patent, 201710324299.X.
8. Y. Y. Wang, W. J. Peng. A study of the cathodic current efficiency in zincate zinc plating [J]. Electroplating & Pollution Control, 2003, (06):12–16.
9. Chun Rong Wang, Sha Chang, Min Ye, Qin Yi Ren. Current Efficiency and Energy Consumption of Electrochemical Oxidation for Ammonia Removal from Coking Wastewater Using Boron-Doped Diamond Electrodes [J]. Applied Mechanics and Materials, 2013, 2301(295).
10. M. G. Pavlovic, N. D. Nikolic, K. I. Popov. The current efficiency during the cathodic period of reversing current in copper powder deposition and the overall current efficiency [J]. Journal of the Serbian Chemical Society, 2003, 68(8–9):.
11. Geir Martin Haarberg. The Current Efficiency for Aluminium Deposition from Molten Fluoride Electrolytes with Dissolved Alumina [M]. Springer International Publishing: 2016.
12. V.M. Volgin, V.V. Lyubimov, I.V. Gnidina, A.D. Davydov, T.B. Kabanova. Effect of Current Efficiency on Electrochemical Micromachining by Moving Electrode [J]. Procedia CIRP, 2016, 55:.
13. Élen Rufino, Mario Santana, Luiz Faria, Leonardo Silva. Influence of lead dioxide electrodes morphology on kinetics and current efficiency of oxygen-ozone evolution reactions [J]. Chemical Papers, 2010, 64(6).
14. X. Z. Deng, Y. W. Wang, J. P. Peng, K. J. Liu, N. X. Feng and Y. Z. Di. Surface area control of nanocomposites Mg(OH)2/graphene using a cathodic electrodeposition process: high adsorption capability of methyl orange [J]. RSC Advances, 2016, 91 (6):88315–88320. https://doi.org/10.1039/c6ra15804h.
15. Q. X. Tan, S. L. Liu. Current efficiencies of electroplating bath and measurement method thereof [J]. Plating & Finishing, 2008, 30 (4):27–29.

# Update on Ballistic Characterization of the Scalability of Magnesium Alloy AMX602

Tyrone L. Jones

**Abstract**

The US Army Research Laboratory (ARL) and the Osaka University Joining and Welding Research Institute (JWRI) continued a collaborative partnership with Taber Extrusions, Epson, Pacific Sowa, Kurimoto, and National Material LP to reproduce and scale-up military grade magnesium alloy AMX602 at the Taber Extrusions manufacturing facility in Russellville, AR. The latest effort was to extrude 304.8-mm (12-in) wide AMX602 plate, exploring 5 extrusion scenarios, and dynamically characterizing the penetration resistance of these plates against two projectiles with different nose shapes. The results were parametrically analyzed and compared to conventionally processed magnesium alloy AZ31B-H24 and aluminum alloy AA5083-H131. None of the 304.8-mm wide AMX602 plates could meet or exceed the penetration threshold of AA5083 against the fragment simulating projectile. Future research is required to define the manufacturing variables that correlate to maintaining the quality of the defeat mechanisms in the AMX602 plate as the width increases.

**Keywords**

Armor • AMX602 • AA5083 • AZ31B • Ballistic characterization • Manufacturing processes Scalability

## Introduction

The US Army Research Laboratory's (ARL) ballistic analysis of magnesium (Mg) alloys over the last 11 years has led to an increased understanding of the material's failure mechanisms and relationship between Mg alloy strength and ductility requirements for lightweight armor applications [1]. While Mg alloys have been used for military structural applications since WWII, very little research has been done to improve its mediocre ballistic performance [2]. The highest strength commercial Mg alloy available in plate form, AZ31B (1.78 g/cm$^3$), may be an effective substitute armor material for AA5083 (2.66 g/cm$^3$) against armor-piercing projectiles at an equal weight basis that is dependent on the projectile diameter [3]. Mg alloy AZ31B can substitute for AA5083 against fragment simulating projectiles (FSPs) within a limited areal density range [4]. However, shearing and scabbing are the dominant failure modes during ballistic impact. The ballistic data generated by ARL was used to develop the first set of Mg alloy acceptance standards incorporated into MIL-DTL-32333 (MR), "Armor Plate, Magnesium Alloy, AZ31B, Applique" [5]. Ultimate tensile strength (UTS), tensile yield strength (YTS), ductility, and grain size are all important parameters in determining the ballistic performance of metals. The bulk material properties of AZ31B are shown in Table 1. Figure 1 correlates Charpy impact energy absorption (J) versus Mg armor alloy AZ31B grain size (μm) [6].

In 2009, ARL collaborated with the Joining and Welding Research Institute (JWRI) of Osaka University under contract through the International Technology Center-Pacific to develop and evaluate high-strength, high-ductility Mg alloy plate for structural applications. Initial evaluation showed grain refinement alone did not clearly improve the ballistic performance of Mg armor alloy AZ31B [7]. The results showed a research gap that needed to be filled to make Mg alloys a viable armor material that could compete with current aluminum (Al) armor alloy solutions [4]. Based on the preliminary material and ballistic analysis, the ARL/JWRI program set goals to develop Mg alloys with the mechanical properties shown in Table 1.

T. L. Jones (✉)
US Army Research Laboratory, Adelphi, USA
e-mail: tyrone.l.jones20.civ@mail.mil


D. Orlov et al. (eds.), *Magnesium Technology 2018*, The Minerals, Metals & Materials Series, https://doi.org/10.1007/978-3-319-72332-7_24

**Table 1** Objective mechanical properties of Mg alloy [8]

| Alloy | Ultimate tensile strength (MPa) | Tensile yield strength (MPa) | Elongation to failure (%) |
|---|---|---|---|
| AZ31B | 245 | 150 | 7 |
| Proposed Mg alloy | 400 | 350 | 20 |

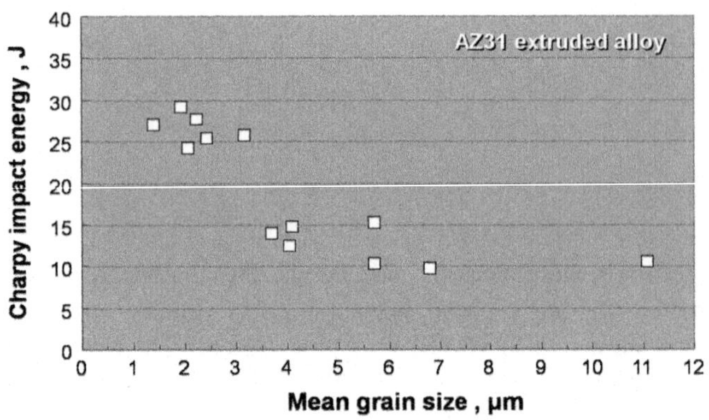


**Fig. 1** AZ31B grain size versus charpy impact energy absorption

Clearly, there were two potential paths forward towards achieving these set goals:

1. Discover new chemical compositions to create high strength, high ductility Mg alloys, while not compromising the desirably low density of 1.78 g/cm$^3$.
2. Improve grain refinement through novel processing techniques to produce high-strength, high-ductility Mg alloys.

As a result, ARL and JWRI collaboratively created two new experimental Mg alloys, AMX602 and ZAXE1711, in extruded 40-mm wide bars starting with an advanced metallurgical powder process. The successful ballistic and corrosion evaluation of each material evolved into a coalition to reproduce and scale up the lateral dimensions of AMX602 bars into 305-mm × 305-mm (12-inch × 12-inch) plates for commercial production in the United States. Cost was the driver in our selection of AMX602 over ZAXE1711. However, once scaling up of AMX602 is achieved, the same processing methodology can be applied to reproduce ZAXE1711. Details of the scale up process and ballistic evaluation of AMX602 are discussed in the next sections.

## Material Exploration

AMX602 (Mg-6Al-0.5Mn-2Ca/mass%) Mg alloy powders produced by the Spinning Water Atomization Process (SWAP) were used as raw input materials [9, 10]. The coarse Mg alloy powders had a particle dimension of 1–5-mm. It was previously verified that the coarse Mg powders of these sizes were of low explosion risk. The α-Mg grain size of the raw powders was less than 0.5 μm. Powder compaction and hot extrusion were applied to these raw powders to fabricate the extruded bars. The bar had dimensions of 24.5 mm 40 mm × 1000 mm. Tensile test specimens machined from these bars were evaluated at room temperature. The material microstructures were observed using an optical microscope.

## Experimental Evaluation of Raw Materials

In SWAP powder preparation, schematically illustrated in Fig. 2a, AMX602 magnesium alloy ingots were melted at 1053 K in a ceramic crucible covered by a protective inert gas. The molten metals were directly streamed inside the spinning water chamber from a crucible nozzle. Table 2 shows the chemical composition in weight percent of AMX602 alloy powders prepared by SWAP. The calcium is necessary because it reduces the combustive properties of the magnesium alloys. The impurity content of Fe and Cu is controlled to less than 0.005% because they are known corrosive elements in magnesium alloys. As shown in Fig. 2b, the length of the coarse AMX602 powders prepared by SWAP is ~1–4 mm, and they are of irregular shape.

## Powder Metallurgy

The powder was consolidated at room temperature using a 2000 kN hydraulic press machine to fabricate the green compact. The green compact had a relative density of 85% and 42 mm diameter. The columnar compact and cast ingot were heated at ~573–673 K for 180 s in an argon gas atmosphere, then immediately consolidated into full density material by hot extrusion. An extrusion ratio of 37 and an extrusion speed 1 m/s were used in this study.

Taber Extrusions manufactured the 304.8-mm wide 25.4-mm thick plate of AMX602 in five tempers. The method of fabrication process and details of the mechanical property measurements are detailed in the fabrication report available from Taber Extrusions.

The mechanical properties of the specimens from each temper widely varied. None of the specimens met the mechanical property objectives. In some cases, the specimens were up to 90% lower than the 20% elongation objective.

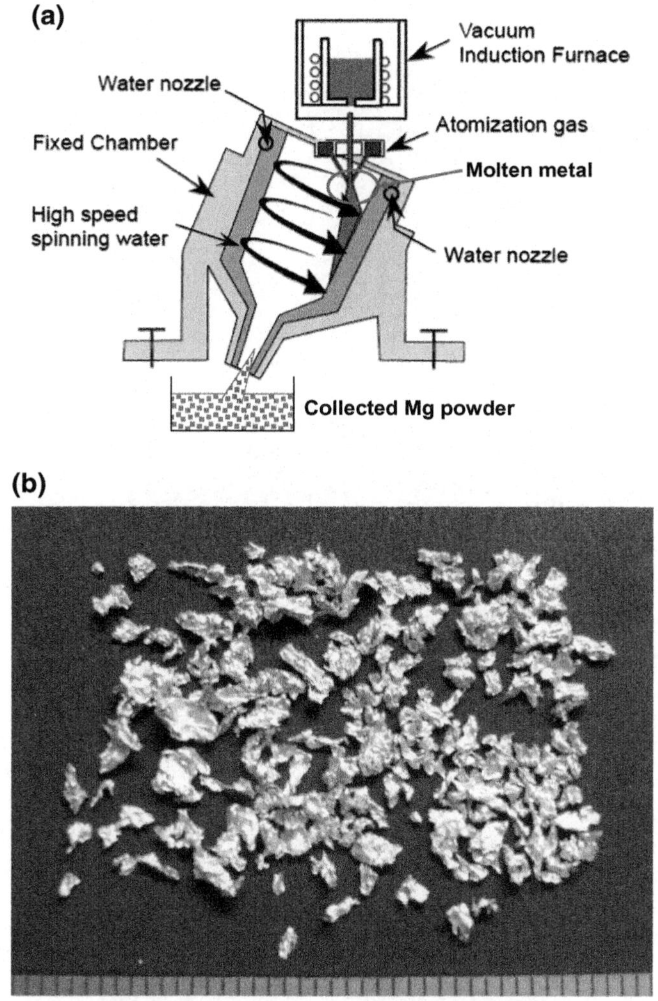


**Fig. 2** **a** Schematic illustration of SWAP equipment to produce rapidly solidified Mg alloy powders; **b** Morphology of coarse magnesium alloy powder prepared by SWAP

## Ballistic Experimental Procedure

The ballistic projectiles were selected in a manner that would allow for direct comparison to other metal alloy armor standards, particularly to the aluminum alloy 5083 (AA5083) armor plate standard. The weldable, AA5083 is currently specified for use in many vehicle armor systems. The experimental projectiles were selected based on the required projectile in accordance to the magnesium armor standard, MIL-DTL-32333 and the aluminum armor standard, MIL-DTL-46027 K at 45.21 kg/m$^2$ (9.26 lb per square foot) areal density [5, 11]. At 45.21 kg/m$^2$ or 25.4-mm thick, the magnesium alloy AZ31B standard requires the 0.30-cal. Fragment Simulating Projectile (FSP) at 0° obliquity at impact. At 45.21 kg/m$^2$ or 17.0-mm thick, the aluminum alloy 5083 standard requires the 0.30-cal. Armor Piercing M2 (APM2) at 30° obliquity at impact. The experimental setup of target at 0° obliquity at impact is illustrated in Fig. 3. The experiment setup for the 30° obliquity at impact is illustrated in Fig. 4.

Ballistic testing of all magnesium alloy plate samples was performed by ARL at Aberdeen Proving Grounds, Maryland in accordance with MIL-STD-662F, issued 18 December 1997 [12]. Ballistic results were characterized using the standard $V_{50}$ test methodology, also documented in MIL-STD-662F.

## Experimental Projectiles

The first ballistic projectile used to test the magnesium and aluminum alloy targets was the 0.30-cal FSP designed in accordance with MIL-DTL-46593B (MR), issued 6 July 2006, as depicted below in Fig. 5 [13]. The weight and hardness specifications are shown in Table 3. In accordance with MIL-STD-662F, the maximum allowable yaw was 5° for an acceptable FSP ballistic impact test.

The second ballistic threat used to test the magnesium and aluminum alloy plate samples was the 0.30-cal. APM2. The cross-section of this projectile is shown in Fig. 6 [14]. The APM2 projectile has a steel core with a hardness in the range of 60–65 Rockwell C. The physical characteristics of the core of the 0.30-cal. APM2 projectile are included in Table 4 [14]. The maximum allowable yaw was 3° for an acceptable APM2 ballistic impact test.

## Ballistic Experimental Results

The $V_{50}$ ballistic limit of the 304.8-mm wide AA5083 plate was measured and provided for comparison. The $V_{50}$ ballistic limit of the 304.8-mm wide AZ31B plate, manufactured by Magnesium Elektron North America, was also measured and provided for comparison. All materials were nominally at an equivalent areal density (i.e., mass per unit surface area) of 45.21 kg/m$^2$. The actual plate thicknesses for each target was measured. The hardness of the Mg plates was measured on a Brinell 500-kg scale, while the aluminum plate was measured on the 3000-kg scale. Table 5 compares the $V_{50}$ ballistic limits of the AMX602

**Table 2** Chemical composition of AMX602 magnesium alloy powders

| Al | Zn | Mn | Fe | Si | Cu | Ca | Mg |
|---|---|---|---|---|---|---|---|
| 6.01 | 0.007 | 0.26 | 0.002 | 0.038 | 0.004 | 2.09 | Bal. |

(a) Front View

(b) Profile View

(c) Rear View

**Fig. 3** **a** Front view of 0° obliquity experimental setup; **b** Profile view of experimental setup; rear view of experimental setup

(a) Front View

(b) Profile View

(c) Rear View

**Fig. 4** **a** Front view of 30° obliquity experimental setup; **b** Profile view of experimental setup; rear view of experimental setup

plate of 304.8-mm width to previously manufactured AMX602 against the FSP [15].

Table 5 illustrates the trend of a reduction in penetration resistance as the width of the AMX602 plate increases. The AZ31B-H24 was the threshold material and is highlighted in red. The AA5083 was the objective material and is highlighted in green. The material that did not exceed the objective threshold was highlighted in blue. The AMX602-K1 yielded the highest $V_{50}$ ballistic limit of 1014 m/s for the 304.8-mm wide AMX602 plates. None of the 304.8-mm wide AMX602 plates met or exceeded the ballistic limit of the AA5083. The radius of the scabbing and subsequent spallation off the rear of the AMX602 plates against the 0.30-cal. FSP was larger than the scabbing and spallation from the lower width plates. This is indicative of the lower % elongation in the 304.8-mm wide plates. As previously documented in prior research papers [15], ductility is equal if not more critical as strength properties in the penetrative resistance of magnesium alloys. Figure 7 illustrates the damage of the 25.4-mm × 101.6-mm (1-inch × 4-inch) AMX602-Plate B plate, the 25.4 mm 152.4 mm (1-inch × 6-inch) AMX602-Plate E plate, and the 25.4 mm × 304.8-mm (1-inch × 12-inch) AMX602-K1 plate.

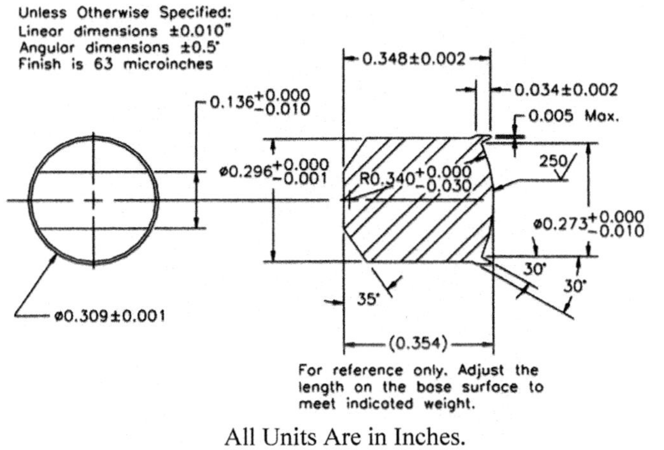


**Fig. 5** The 0.30-cal. FSP schematic diagram. All units are in inches

**Table 3** 0.30-cal. FSP weight and hardness

| Weight (g) | Rockwell hardness C |
|---|---|
| 2.85 ± 0.03 | 30 ± 2 |

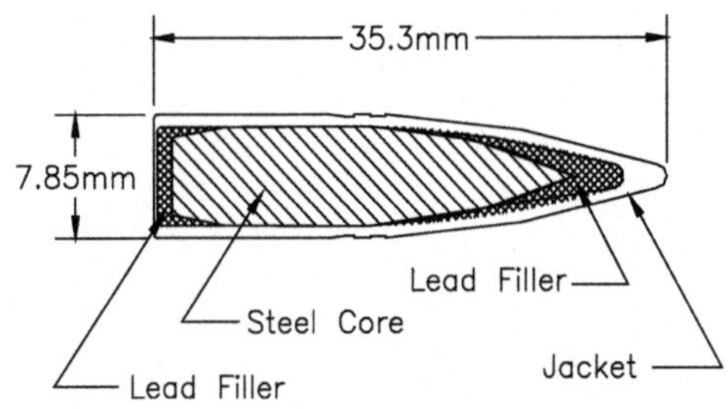


**Fig. 6** The 0.30-cal. APM2 cross-section

**Table 4** 0.30-cal. APM2 projectile dimensions and weight

| Projectile length (mm) | Projectile diameter (mm) | Projectile weight (g) | Core length (mm) | Core diameter (mm) | Core weight (g) |
|---|---|---|---|---|---|
| 35.3 | 7.85 | 10.8 | 27.4 | 6.2 | 5.3 |

Table 6 compares the $V_{50}$ ballistic limits of the AMX602 plate of 304.8-mm width to previously manufactured AMX602 against the 0.30-cal. APM2. The $V_{50}$ ballistic limit of the standard production 304.8-mm wide AZ31B plate and the 304.8-mm wide AA5083 plate was measured and included for comparison. No data exists for the thinner width AMX602 plates against the 0.30-cal. APM2.

The T1 and the GP-T tempers yielded the highest $V_{50}$ ballistic limits for the 304.8-mm wide AMX602 plates. In contrast to the FSP results, all of the 304.8-mm wide AMX602 plates exceeded the ballistic limit of the AA5083 plate. The AA5083 plate was not thick enough for the 0.30-cal. APM2 projectile to be decelerated by the AA5083 tensile strength properties as efficiently as the magnesium plates at impact. The radius of the scabbing and subsequent spallation off the rear of the AMX602 plate against the 0.30-cal. APM2 was smaller than scabbing and subsequent spallation from the 0.30-cal. FSP plates. Figure 8 illustrates the damage of the 25.4 mm × 304.8-mm (1-inch × 12-inch) AMX602-T1 plate.

**Table 5** V50 ballistic limits at 0° obliquity versus the 0.30-cal. FSP

| Metal Alloys | Manufacturer | Plate Thickess | Plate Areal Density | Plate Width | *Hardness | Ballistic Limit | Standard Deviation |
|---|---|---|---|---|---|---|---|
| | | mm | kg/m² | mm | HBN | m/s | m/s |
| AA5083 (2016) | -- | 16.834 | 44.78 | 304.8 | 80 | 1020 | 7 |
| AZ31B-H24 (2016) | Mg-Elektron | 26.626 | 47.39 | 304.8 | 52 | 943 | 6 |
| AMX602-1 (JWRI) | JWRI/KURIMOTO | 25.190/25.235 | 44.84/44.92 | 40 | 80 | 1061 | 10 |
| AMX602-2 (JWRI) | JWRI/KURIMOTO | 25.171/25.210/25.197 | 44.80/44.87/44.85 | 40 | 80/80/83 | 1092 | 6 |
| AMX602-3 (JWRI) | JWRI/KURIMOTO | 25.171/25.178 | 44.80/44.82 | 40 | 80/80 | 1105 | 19 |
| AMX602-B1 | Taber Extrusions | 25.489 | 45.37 | 38.1 | 99 | ~1122 | ~54 |
| AMX602-B2 | Taber Extrusions | 25.565 | 45.51 | 38.1 | 86 | ~1070 | ~2 |
| AMX602-Plate A | Taber Extrusions | 25.286 | 45.01 | 101.6 | 77 | 1103 | 12 |
| AMX602-Plate B | Taber Extrusions | 25.317 | 45.06 | 101.6 | 83 | 1105 | 18 |
| AMX602-Plate C | Taber Extrusions | 25.4 | 45.21 | 101.6 | 77 | 1037 | 35 |
| AMX602-Plate D | Taber Extrusions | 25.267 | 44.98 | 101.6 | 72 | 998 | 30 |
| AMX602-Plate E | Taber Extrusions | 25.229 | 44.91 | 152.4 | 83 | 1048 | 10 |
| AMX602-T1 | Taber Extrusions | 25.121 | 44.72 | 304.8 | 72 | 1006 | 6 |
| AMX602-GP-T | Taber Extrusions | 25.343 | 45.11 | 304.8 | 77 | 1008 | 10 |
| AMX602-T2 | Taber Extrusions | 24.613 | 43.81 | 304.8 | 80 | 992 | 10 |
| AMX602-K2 | Taber Extrusions | 24.841 | 44.22 | 304.8 | 74/83 | 971 | 6 |
| AMX602-K1 | Taber Extrusions | 24.994 | 44.49 | 304.8 | 77 | 1014 | 8 |
| The ballistic limits of B1 & B2 are estimated; ran out of material. | | | | | | | |
| *Mg Alloys measured on 500-kg scale; AA5083 meaured on 3000-kg scale | | | | | | | |

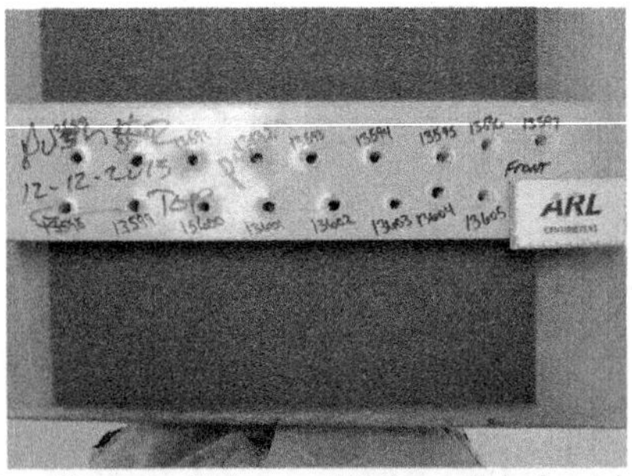

Entry of Impact - 25.4-mm x 101.6-mm (1-inch x 4-inch)

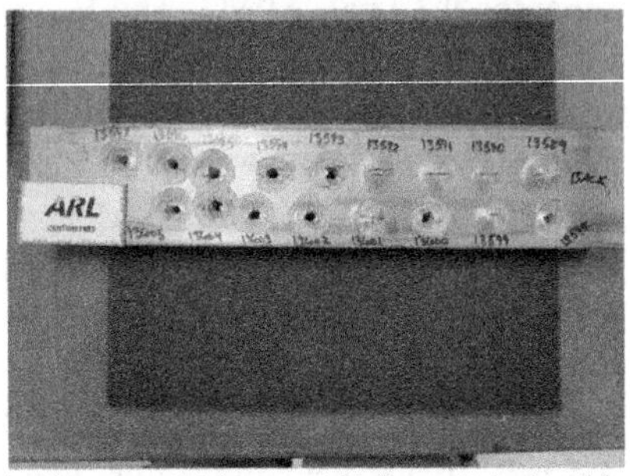

Exit of Impact - 25.4-mm x 101.6-mm (1-inch x 4-inch)

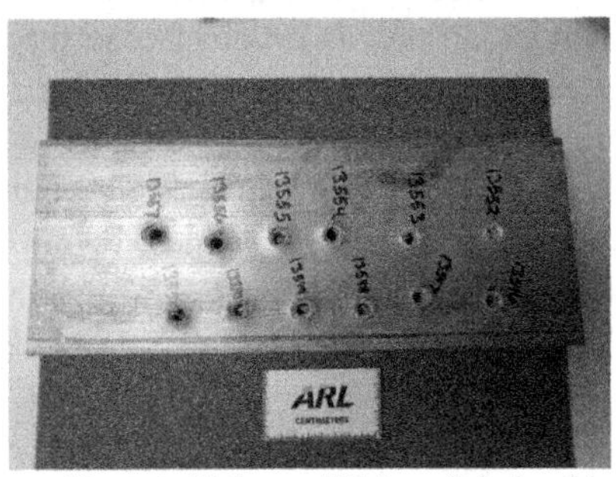

Entry of Impact - 25.4mm x 152.4mm (1-inch x 6-inch)

**Fig. 7** Post-ballistic images of the AMX602 plates against the 0.30-FSP

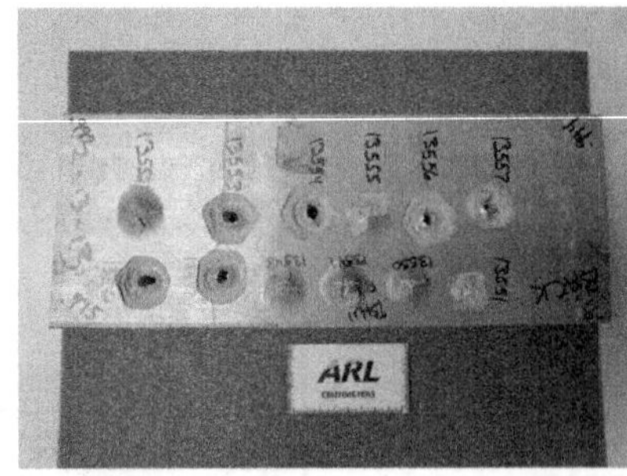

Exit of Impact - 25.4mm x 152.4mm (1-inch x 6-inch)

Entry of Impact - 25.4mm x 304.8-mm (1-inch x 12-inch)

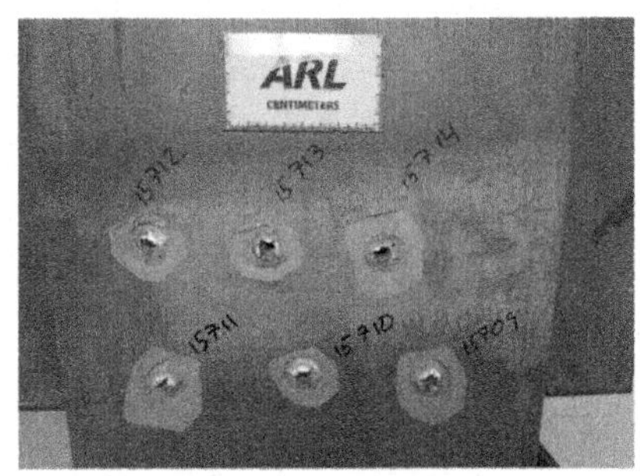

Exit of Impact - 25.4mm x 304.8-mm (1-inch x 12-inch)

**Fig. 7** (continued)

**Table 6** V50 ballistic limits at 30° obliquity versus the 0.30-cal. APM2

| Metal Alloys | Manufacturer | Plate Thickess | Plate Areal Density | Plate Width | *Hardness | Ballistic Limit | Standard Deviation |
|---|---|---|---|---|---|---|---|
| | | mm | kg/m$^2$ | mm | HBN | m/s | m/s |
| | | | | | | | |
| **AA5083 (2016)** | -- | 16.834 | 44.78 | 304.8 | 80 | 499 | 7 |
| **AZ31B-H24 (2016)** | Mg-Elektron | 26.626 | 47.39 | 304.8 | 52 | 539 | 7 |
| **AMX602-T1** | Taber Extrusions | 25.121 | 44.72 | 304.8 | 72 | 556 | 11 |
| **AMX602-GP-T** | Taber Extrusions | 25.343 | 45.11 | 304.8 | 77 | 555 | 8 |
| **AMX602-T2** | Taber Extrusions | 24.613 | 43.81 | 304.8 | 80 | 546 | 4 |
| **AMX602-K2** | Taber Extrusions | 24.841 | 44.22 | 304.8 | 74 | 538 | 7 |
| **AMX602-K1** | Taber Extrusions | 24.994 | 44.49 | 304.8 | 77 | 544 | 10 |
| ***Mg Alloys measured on 500-kg scale; AA5083 meaured on 3000-kg scale** | | | | | | | |

Entry of Impact - 25.4-mm x 304.8-mm (1-inch x 12-inch) AMX602-T1 plate

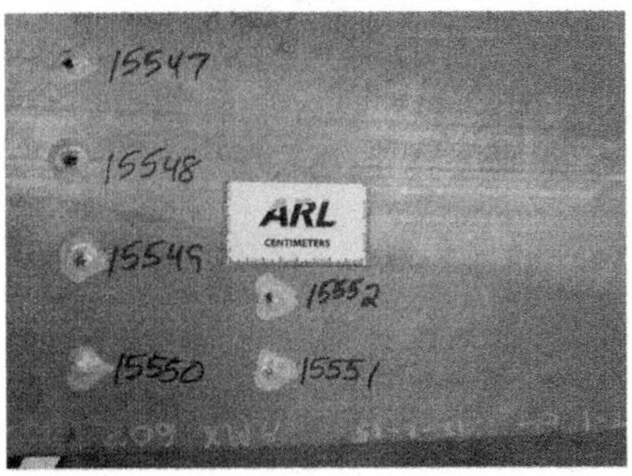


Exit of Impact - 25.4-mm x 304.8-mm (1-inch x 12-inch) AMX602-T1 plate

**Fig. 8** Post-ballistic images of the AMX602 plates against the 0.30-cal. APM2

## Conclusion

The 304.8-mm scaled-up magnesium alloy AMX602 plates showed limited ballistic resistance improvements over the ballistic performance of AA5083. Five tempers of 304.8-mm wide AMX602 plate were fabricated but these 304.8-mm wide plates did not meet objective mechanical properties. The % elongations of these plates were significantly lower than the % elongations of lower width AMX602 plates from prior work. This resulted in the 304.8-mm width AMX602 plates failing to match or exceed the ballistic performance of AA5083 against the 0.30-cal. FSP. The 304.8-mm width plates did not exhibit good scab containment at the back of the material against the 0.30-cal. FSP. As expected, the 304.8-mm width AMX602 plate exceeded the ballistic performance of AA5083 against the 0.30-cal. APM2 projectile. Future research should focus on understanding how to improve ultimate and yield strength (transverse and longitudinal) and % elongation of the AMX602 plate as the width of the plate increases during the manufacturing process. This will be pivotal in boosting the ballistic performance of large scale AMX602 plate under dynamic impact loading from kinetic energy penetrators.

## References


1. Jones TL, DeLorme R, Burkins M, Gooch W (2007) Ballistic Evaluation of Magnesium Alloy AZ31B. ARL-TR-4077, US Army Research Laboratory: Aberdeen Proving Ground, MD, April 2007.
2. Mathaudhu, S, Nyberg, E (2010) Magnesium Alloys in Army Applications: Past, Current and Future Solutions. Paper presented at the 139th TMS Annual Meeting, Seattle, WA, February 2010.
3. Jones, T, DeLorme, R (2008) Development of a Ballistic Specification for Magnesium Alloy AZ31B. ARL-TR-4664, US Army Research Laboratory: Aberdeen Proving Ground, MD; December 2008.
4. Jones, T, DeLorme, R (2008) A Comparison of the Ballistic Performance Between Rolled Plate in AZ31B-H24 Magnesium and 5083-H131 Aluminum. Paper presented at the 24th International Symposium on Ballistics, New Orleans, LA, September 2008.
5. MIL-DTL-32333 (2009) Armor Plate, Magnesium Alloy, AZ31B, Applique, US Army Research Laboratory: Aberdeen Proving Ground, MD.
6. Liao, J, Hotta, M, Kaneko, K, Kondoh, K (2009) Enhanced Impact Toughness of Magnesium Alloy by Grain Refinement. Scripta Materialia 61:208–211.
7. Jones, T, Kondoh, K (2009) Initial Evaluation of Advanced Powder Metallurgy Magnesium Alloys for Dynamic Applications. ARL-TR-4828, US Army Research Laboratory, Aberdeen Proving Ground, MD, May 2009.
8. Jones, T et al. (2012) Ballistic and Corrosion Analysis of New Military-Grade Magnesium Alloys AMX602 and ZAXE1711 for Armor Applications. ARL-TR-5931, US Army Research Laboratory: Aberdeen Proving Ground, MD, February 2012.
9. Sakamoto, M, Akiyama, S, Hagio, T, Ogi, K (1997) Control of Oxidation Surface Film and Suppression of Ignition of Molten Mg-Ca Alloy by Ca Addition. Journal of Japan Foundry Engineering Society, 69:227–233.
10. Nishida, S, Motomura, I (2008) Estimation of Heat Transfer Coefficient and Temperature Transition on Melt Drag Process of AZ31 Magnesium Alloy by Heat Transfer and Solidification Analysis. Journal of Japan Institute of Light Metals, 58:439–442.
11. MIL-DTL-46027 K (2013) Armor Plate, Aluminum Alloy, AA5083, Weldable. US Army Research Laboratory: Aberdeen Proving Ground MD.
12. MIL-STD-662F (1997) $V_{50}$ Ballistic Test for Armor, US Army Research Laboratory: Aberdeen Proving Ground MD.
13. MIL-DTL-46593B (2006) Projectile, Calibers .22, .30, .50, and 20MM Fragment-Simulating. US Army Research Laboratory: Aberdeen Proving Ground MD.
14. Gallardy, D (2012) Ballistic Evaluation of 7085 Aluminum. ARL-TR-5952, US Army Research Laboratory, Aberdeen Proving Ground, MD, March 2012.
15. Jones, T et al. (2015) Ballistic Characterization of the Scalability of Magnesium Alloy. ARL-TR-7341, US Army Research Laboratory, Aberdeen Proving Ground, MD, July 2015.

# Experimental Study on the Reversion Reaction Between Magnesium and CO Vapor in the Carbothermic Reduction of Magnesia Under Vacuum

Yang Tian, Bao-qiang Xu, Bin Yang, Da-chun Liu, Tao Qu, Hai Liu, and Yong-nian Dai

**Abstract**

The mechanism of magnesia production was investigated experimentally in reversion reaction process in vacuum. Condensation temperature and temperature gradient which effected on the condensation of the magnesium vapor produced by magnesia carbothermic reduction in vacuum have been investigated by Mg recovery efficiency, XRD, SEM and EDS. The results show that the higher recovery efficiency was obtained when the condensing temperature which is closer to the dew point of magnesium in the constant temperature gradient. Under the condition of appropriate condensation temperature, the lower the temperature gradient is, the better the crystallization of magnesium vapor is. The XRD patterns of the profile and undersurface of the condensation product of 1873 K show that the profile of the condensation contains Mg only and the purity of the metal magnesium is high. But the undersurface of the condensation contains Mg and MgO. The SEM and EDS images of the profile and undersurface of condensation indicate that the microstructure of the undersurface of condensation is largely flocculent structure and irregular arrangement and the crystal morphology was poor and particles were also tiny. This is due to magnesium vapor reacted with CO vapor at cooling phase, MgO and C obtained covered the undersurface and stopped magnesium vapor condensing.

**Keywords**

Condensation temperature • Carbothermic reduction
Temperature gradient • Vacuum

## Introduction

The carbothermic reduction of magnesia to produce magnesium has attracted extensive attention in the last decades due to its advantages of low reductant cost, high equipment utilization rate, and environment-friendliness compared to the silicothermic and electrolytic processes for the production of magnesium [1–3]. However, the reversion reactions between Mg vapor and CO can affect the condensation of magnesium vapor, and become to the major technical challenges [4, 5]. Most of the researchers focused on the condensation of magnesium produced by carbothermic reduction and concluded that there are two routes for condensing magnesium vapor: the "quench" route and the "solvent" route [6]. The CRIRO is currently studying the carbothermic reduction of magnesia with graphite at 1600 °C using a Laval nozzle to quench the resultant magnesia vapor [7–11]. Recent years, the carbothermic reduction of magnesia under vacuum conditions had been studied at KMUST [12–14] Previous studies found that condensing temperature and temperature gradients have a significant impact on the condensation process of magnesium vapor [15]. But it is not clear that how do the reversion reaction impact on the condensation of magnesium vapor. Therefore, through the experiment studying, we try to reveal the impact of reversion reaction on condensation of magnesium vapor in order to provide the basic theory for the mechanism of condensation of magnesium vapor.

Y. Tian · B. Xu · B. Yang (✉) · D. Liu · T. Qu · H. Liu · Y. Dai
State Key Laboratory of Complex Nonferrous Metal Resources Clear Utilization, Kunming University of Science and Technology, Kunming, 650093, China
e-mail: kgyb2005@126.com

Y. Tian · B. Xu · B. Yang · D. Liu · T. Qu · H. Liu · Y. Dai
National Engineering Laboratory for Vacuum Metallurgy, Kunming University of Science and Technology, Kunming, 650093, China

Y. Tian · B. Xu · B. Yang · D. Liu · T. Qu · Y. Dai
Key Laboratory of Non-Ferrous Metals Vacuum Metallurgy of Yunnan Province, Kunming University of Science and Technology, Kunming, 650093, China

Y. Tian · B. Xu · B. Yang · D. Liu · T. Qu · Y. Dai
College of Metallurgy and Energy Engineering, Kunming University of Science and Technology, Kunming, 650093, China


D. Orlov et al. (eds.), *Magnesium Technology 2018*, The Minerals, Metals & Materials Series, https://doi.org/10.1007/978-3-319-72332-7_25

## Experimental

### Raw Material

The chemical compositions of the reducing agents and of the magnesium oxide are shown in Tables 1 and 2, respectively.

### Experimental Apparatus

The experiment was done by using an internally heated vacuum furnace (Fig. 1).

### Experiment Method

Coking coal and magnesium oxide were used as raw materials. These materials were pulverized into 200 mesh particles, uniformly mixed, and then placed in a graphite crucible after compression molding. The crucible was then placed in a vacuum furnace, and the cooling water valve is then opened. The furnace was heated when the pressure decreased to below 60 Pa. The heating rate was kept constant at approximately to 11 K/min. The furnace was initially heated to 1073 K (which is the optimal coking temperature of coal) and was then insulated for 30 min [12]. The temperature was then increased to reaction temperature (1573–1973 K). After the reaction time (3 h), the reaction product was cooled to room temperature and took out.

In the experimental process, the condensing zone temperature changed by changing the reduction temperature, and the temperature gradient changed by changing the condenser pipe length. The reduction temperatures are 1573 K, 1673 K, 1773 K, 1873 K, and the temperature gradients are $T_1'$, $T_2'$, $T_3'$, $T_4'$, ($T_1' > T_2' > T_3' > T_4'$). We studied the impact of temperature and temperature gradient on reversion reaction and the impact of reversion reaction on condensation of magnesium vapor. Figure 2 shows the experimental flow chart.

## Analysis Methods

The crystalline phase of the products was identified by X-ray diffraction (XRD) instrument (D/max-3B) using Cu $K_\alpha$ radiation with a scanning rate of 2(°)/min which was made by Rigaku Corporation of Japan. Surface morphology and chemical composition of the condensing product were characterized by scanning electron microscopy and EDAX pattern, respectively, which were made by Philips of Holland. The below formula means the direct recovery of the magnesium metal obtained, the obtained value was then used as the technological index of magnesium vapor condensation.

$$\eta = \frac{M}{m} \times 100\%$$

$$m = n \times \varepsilon\% \times \frac{3}{5}$$

Where η is the direct recovery of the magnesium metal, M is the mass of the obtained magnesium metal in the condensing zone, m is the mass of the magnesium mate reduced from the materials, n is the magnesium oxide content of the materials and ε% is the reduction rate of magnesium oxide.

## Results and Discussion

### Influence of Reaction Temperature and the Temperature Gradient on the Direct Recovery

Figure 3 shows the relationships of the reduction temperature and recovery efficiency, at the constant temperature gradient and the average pressure of 60 Pa. Heat preservation was performed for 3 h. We can see from Fig. 3 that the direct recovery increased with reaction temperature increasing when the temperature gradient was $T_1'$, $T_2'$, $T_3'$.

**Table 1** Chemical compositions of reducing agent (mass fraction, wt%)

| Sample | Fixed carbon | Volatile | Water | Ash (30.65) | | |
|---|---|---|---|---|---|---|
| | | | | $SiO_2$ | $Al_2O_3$ | $TiO_2$ |
| Coking coal | 47.14 | 21 | 1.21 | 60.75 | 26.3 | 2.47 |

**Table 2** Chemical compositions of magnesium oxide (mass fraction, wt%)

| Ingredient | MgO | $SiO_2$ | $Al_2O_3$ | CaO |
|---|---|---|---|---|
| Content | ≥95 | 0.15 | <0.10 | <0.50 |

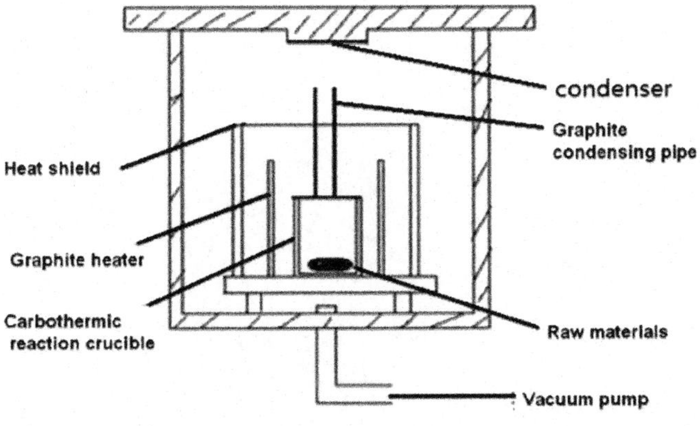


**Fig. 1** Schematic diagram of vacuum furnace

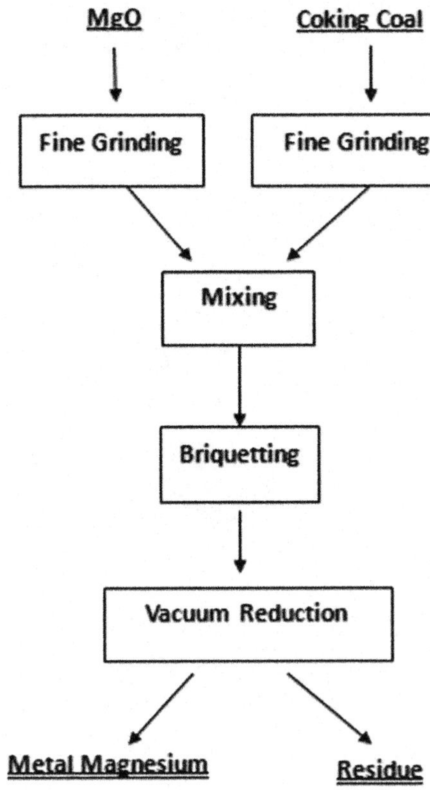


**Fig. 2** The experimental flow chart

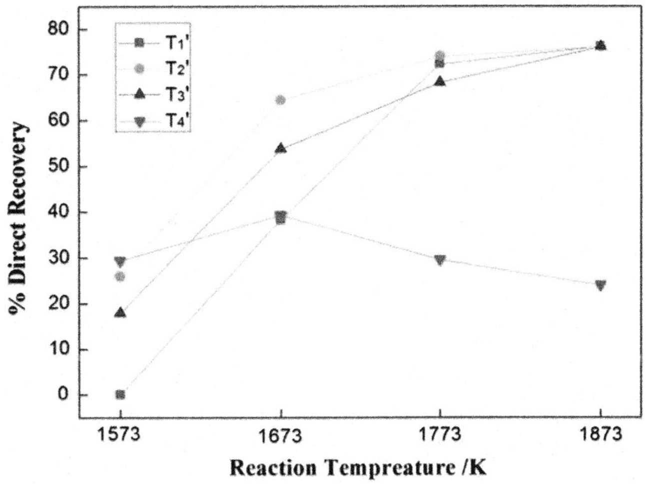


**Fig. 3** Relationships of the reduction temperature and direct recovery

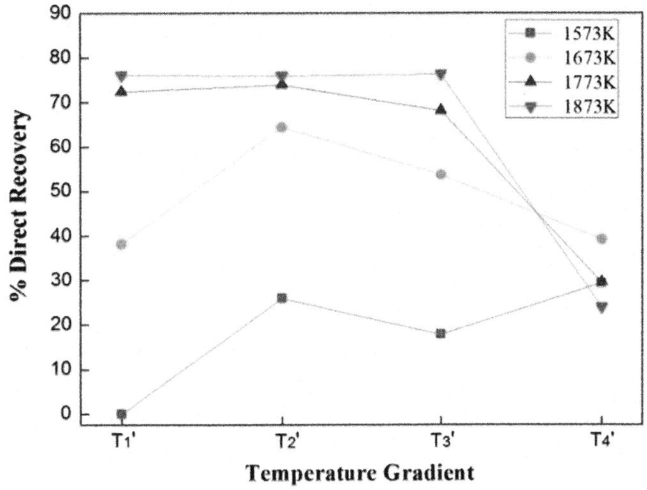


**Fig. 4** Relationships of the temperature gradient and direct recovery

This is due to that increasing reaction temperature will make condensation temperature increase and magnesium vapor is easier to condense and crystallize when the condensation temperature is closer to the dew point temperature of the metal magnesium. However, the direct recovery essentially unchanged with reaction temperature increasing when the temperature gradient was $T_4'$. This is because that the condensation temperature is higher than the dew point temperature when the temperature gradient was $T_4'$ and it is difficult magnesium vapor condenses and crystallizes. In conclusion, under the constant temperature gradient, magnesium vapor is easier to condense and crystallize when the condensation temperature is closer to the dew point temperature of the metal magnesium, and the direct recovery also is higher.

Figure 4 shows the relationships of the temperature gradient and recovery efficiency, at the constant reaction temperature and the average pressure of 60 Pa. Heat preservation was performed for 3 h. We can see from Fig. 4 that the direct recovery increased when the temperature gradient decreased from $T_1'$ to $T_2'$ under the condition that the reaction temperature is 1573 and 1673 K. However, the direct recovery decreased when the temperature gradient continued decrease to $T_3'$ and $T_4'$. This is because that the condensation temperature is higher than the dew point temperature the temperature gradient was $T_3'$ and $T_4'$ and it is difficult that magnesium vapor condenses and crystallizes. When the reaction temperature were 1773 and 1873 K, heat from the reaction zone made the condensation zone temperature too high and so the temperature gradient had a little effect on the direct recovery. In conclusion, the small temperature gradient can improve the direct recovery under the appropriate condensation temperature.

## Macroscopic Analysis of Metal Magnesium Condensation

At an average pressure 60 Pa, when the reduction temperature is 1573 and 1873 K, the temperature gradient is $T_3'$, the physical map of the condensation is shown by the Figs. 5 and 6, respectively. It can be seen from Fig. 5 that the condensation product of 1573 K was small pieces and it's structure was loose and frangibility. Figure 6 shows that the condensation product of 1873 K was a big size and it's structure was compact and better hardness.

The condensation product of 1873 K is shown in the Fig. 7. A and B are the profile and undersurface, respectively. It can be seen from Fig. 7 that the profile of the condensation is needle-like and has metal luster, macroscopically. However, Fig. 6 shows that the undersurface of the condensation is granular and ash black.

**Fig. 6** Condensation product of 1873 K

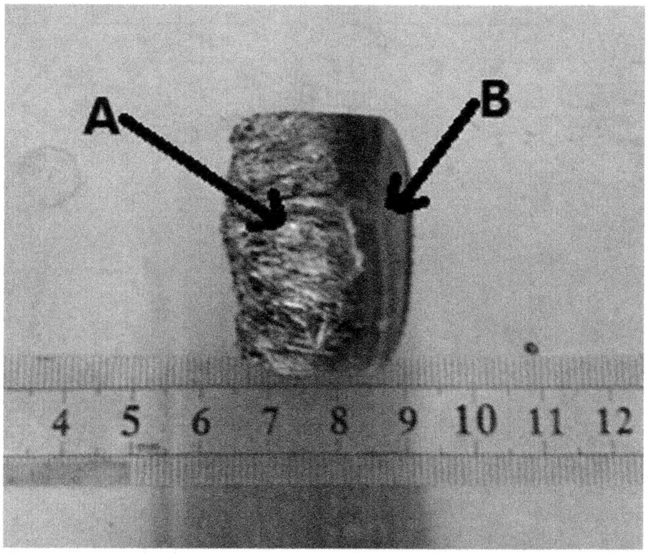


**Fig. 7** The picture of the condensation

## Microcosmic Analysis of Metal Magnesium Condensation

Figures 8 and 9 show the XRD patterns of the profile and undersurface of the condensation product of 1873 K, respectively. It can be seen from Fig. 8 that the profile of the condensation contains Mg only and all peaks are sharp and well-defined, which indicated the high purity of the metal magnesium. As can be seen from the Fig. 9 that the undersurface of the condensation contain Mg and MgO.

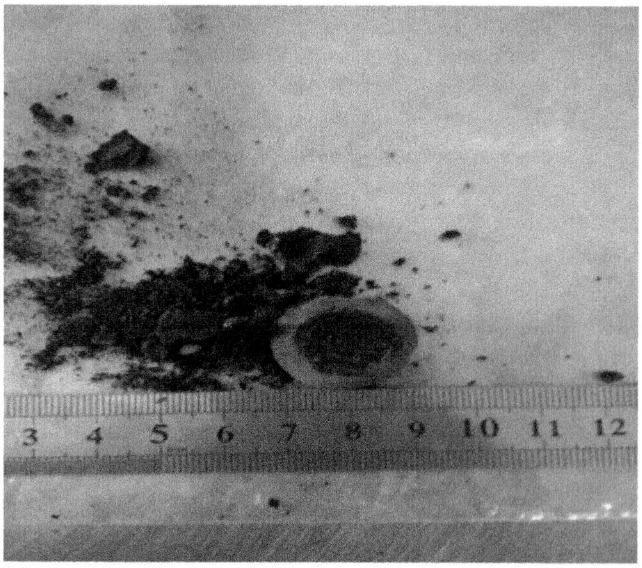


**Fig. 5** Condensation product of 1573 K

Figure 10 shows the scanning electron microscopy (SEM) and energy dispersive spectrometer (EDS) images of the profile of condensation. We can see from Fig. 10 that magnesium atoms grow like branch and condensation have a linear connecting and regularity structure. The EDS shows that the content of magnesium is 96.93 wt%, oxygen is 3.03 wt%.

Figure 11 shows the SEM and EDS images of the undersurface of condensation. As was shown in the Fig. 11, the microstructure of the undersurface of condensation is largely flocculent structure and irregular arrangement. The EDS

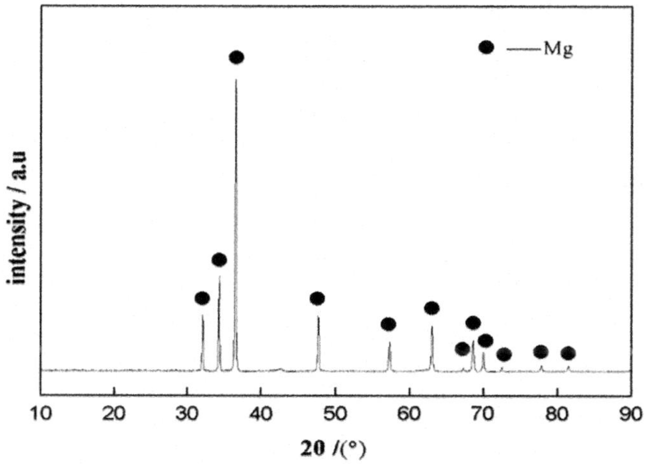


**Fig. 8** The XRD of the profile of condensation

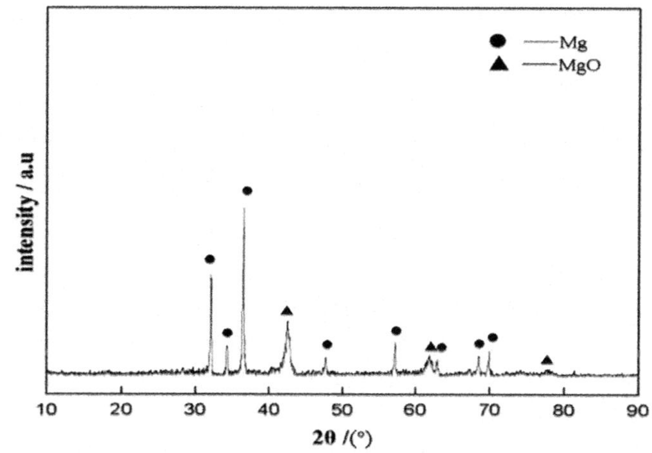


**Fig. 9** The XRD of the undersurface of condensation

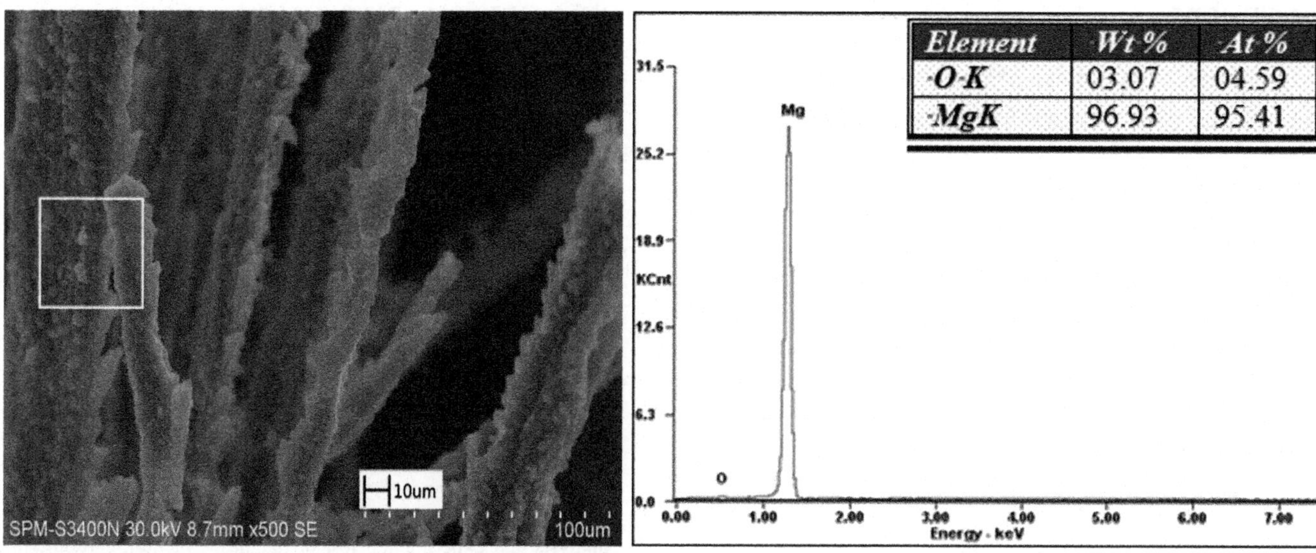


**Fig. 10** The SEM and EDS images of the profile of condensation

**Fig. 11** The SEM and EDS images of the undersurface of condensation

shows that the content of magnesium is 75.43 wt%, oxygen is 12.39 wt%, and carbon is 12.18%. Compared to Fig. 10, the crystal morphology was poor in Fig. 11 and particles were also tiny. Loosely crystallized magnesium in Fig. 11 was due to the carbon and magnesia produced by the reverse reaction. The carbon and magnesia obtained by the reverse reaction block the aggregation of nucleating vapors and prevent coalescence, thereby decreasing the rates of magnesium nucleation and crystal growth. This is due to that magnesium vapor reacted with CO vapor at cooling phase, MgO and C obtained covered the undersurface and stopped magnesium vapor condensing. When the temperature is lower than the gas-liquid transition temperature, the magnesium vapor would transform from gaseous to solid state rapidly, and then the magnesium powder would reacted with CO vapor directly.

## Conclusions

1. Under the constant temperature gradient, magnesium vapor is easier to condense and crystallize when the condensation temperature is closer to the dew point temperature of the metal magnesium, and the direct recovery also is higher. Under the appropriate condensation temperature, the small temperature gradient can improve the direct recovery under the appropriate condensation temperature.
2. When the average pressure is 60 Pa and the temperature gradient is $T_3'$, the condensation product of 1573 K was small pieces and it's structure was loose and frangibility; and the condensation product of 1873 K was a big size and it's structure was compact and better hardness. The profile of the condensation product of 1873 K is needle-like and has metal luster, macroscopically. However, the undersurface of the condensation product is granular and ash black.
3. The XRD patterns of the profile and undersurface of the condensation product of 1873 K show that the profile of the condensation contains Mg only and the purity of the metal magnesium is high. But the undersurface of the condensation contains Mg and MgO. The SEM and EDS images of the profile and undersurface of condensation indicate that the microstructure of the undersurface of condensation is largely flocculent structure and irregular arrangement and the crystal morphology was poor and particles were also tiny. This is due to magnesium vapor reacted with CO vapor at cooling phase, MgO and C obtained covered the undersurface and stopped magnesium vapor condensing.

**Funding** Project supported by fund of free inquiry of chinese academician (NO. 2017HA006); Fund of Yunnan Collaborative Innovation Center of Complex Nonferrous Metal Resources Clear Utilization (NO. 2014XTZS005).

## References

1. L.I. Rong-ti, P.A.N Wei, Masamichi Sano, Metal. Mater. Trans. 8 (2003):433–437.
2. Geoffrey Brooks, Simon Trang, Peter Witt, etc. The Carbothermic route to Magnesium [J]. JOM, 2006, 58, 5:51–55.
3. Tassios Steven, Production Process, WO, 2020012042A1 [P], 2010.
4. Yang Tian, Tao Qu, Bin Yang, etc. Magnesium Technology: Mechanism of Carbothermic Reduction of Magnesia and Reversion Reaction [C]. In: TMS2012, 2012, pp. 511–516.
5. Q.C. Yu, B. Yang, Xu B.Q, Liu D.C, Li Z.H, Dai Y.N, in: Proceedings of the Ninth Vacuum Metallurgy and Surface Engineering Conference, 2009, pp. 428–433.
6. H.P. Leon, W.N. Michael, R.D. Timothy, T. Steven, T. Benny, J. Peter, K. Keri, in: TMS2012, 2012, pp, 31–34.
7. PRENTICE L, NAGLE M, CONSTANTI-CAREY K. Impurities in the carbothermal production of magnesium: To 1500 °C, in: High temperature processing symposium [C]//Swinburne University, Hawthorn, Australia, 2009: 365.
8. PRENTICE L, NAGLE M. Mechanism and kinetics of reduction of magnesium oxide with carbon, in: Magnesium technology [C]// The Minerals, Metals, and Materials Society, 2009: 35 39.
9. PRENTICE L, WAI POI N, HAQUE N. Life cycle assessment of carbothermal production of magnesium in australia, in: ima 67th annual world magnesium conference [C]//International Magnesium Association: Hong Kong, PRC, 2010: 77 82.
10. PRENTICE L. Psuedo-steady-state control of high temperature gas-solid reaction, in: Chemeca2011 [C]//Sydney, Australia: Engineers Australia, 2011: 332.
11. KHAN H, WITT P, BROOKS G, BARTON T, of ultra-rapid quenching of metallic vapors, in: Proceedings of materials processing fundamentals: process modelling [C]. TMS2006. Warrendale Pa, 2006: 699.
12. LI Zhi-hua, DAI Yong-nian, XUE Huai-sheng. Thermodynamical analysis and experimental test of magnesia vacuum carbothermic reduction [J]. Nonferrous metals, 2005, 1(57):56–59.
13. Tian Yang, Liu Hong-xiang, Yang Bin, Qu Tao, Dai Yong-nian, Peng Peng. Magnesium Extraction from Magnesia by Carbothermic Reduction in Vacuum [J]. Chinese Journal of Vacuum Science and Technology, 2012, 32(4):43–48.
14. Tian Yang, Liu Hong-xiang, Yang Bin, Qu Tao, Dai Yong-nian, Xu Bao-qiang, Geng Seng, Behavior analysis of CaF2 in magnesia carbothermic reduction process in vacuum [J]. Metallurgical and Materials Transaction B 2012, 43:657–661.
15. Tian Yang, Cheng-bo Yang, Qu Tao, Yang Bin, Qi Luo, Hong-xiang Liu, Dai Yong-nian. Magnesium Technology: Study on Mechanism of Magnesium Production by Reversion Reaction Process in Vacuum [C]. In: TMS2012, 2012, pp, 511–516.

# Study on Metal Smelting Process Under Microwave Irradiation

Satoshi Fujii, Eiichi Suzuki, Naomi Inazu, Shuntarou Tsubaki, Masahiko Maeda, and Yuji Wada

**Abstract**

Chemical reactions carried out under microwave irradiation often have high reaction rates and high selectivities, which enable compact reactor sizes and energy-conservation processes. Thus, microwave chemical processing and chemical synthesis have attracted considerable interest, as they will be employed for greatly improving process efficiencies and conserving energy for realizing "Green Chemistry" or "Green Engineering". We have applied this technology to smelting process of magnesium metal. But Oxide (dolomite) does not absorb microwave energy well and does not generate heat. This time, when electrically conductivity ferrosilicon used as a reducing agent was mixed with the raw dolomite material and made into an antenna structure, it became easier to absorb the microwave energy and reduce the temperature. We have successfully obtained small amount of magnesium metal using a microwave irradiation with high yield of 71%, and also showed quarter of energy consumption in comparison with conventional process, which is called Pidgeon process.

**Keywords**

Microwave irradiation • Pidgeon method • Smelting magnesium

## Introduction

Chemical reactions performed under microwave irradiation often demonstrate high reaction rates and high selectivities, which allows for more compact reactors and more energy-efficient processes than conventional heating methods. Microwave chemical processing and chemical synthesis have therefore attracted significant attention as means of improving process efficiencies and conserving energy for realizing "green chemistry" or "green engineering" [1]. Currently, the smelting of magnesium metal is mainly performed using the Pidgeon method (thermal reduction method) where the material temperature is raised using a large amount of coal as the heat source. About 80% of magnesium metal is produced in China. A large amount of coal is consumed for smelting, resulting in the generation of the air pollutant PM 2.5 (fine particulate matter) and the release of carbon dioxide to the atmosphere, which are major problems. The Pidgeon method is a technique for heating dolomite ore and silicon iron to high temperatures and then cooling the evaporated magnesium to obtain magnesium metal.

$$2MgO(s) + 2CaO(s) + (Fe)Si(s) \rightarrow 2Mg(g) + Ca_2SiO_4(s) + Fe(s)$$

* s: Solid, g: Gas

Dolomite mineral: MgO, CaO; Ferrosilicon: FeSi, Heat source: Coal.

In this study, we have investigated a new Pidgeon process, using microwaves instead of coal as the heat source, for producing Mg with lower energy inputs and a decrease in greenhouse gas emissions [2].

S. Fujii (✉) · E. Suzuki · N. Inazu · S. Tsubaki · Y. Wada
Department of Applied Chemistry, Graduate School of Science and Engineering, Tokyo Institute of Technology, 2-12-1 Ookyama, Meguro-ku, Tokyo, 152-8550, Japan
e-mail: fujii.s.ap@m.titech.ac.jp

S. Fujii
Department of Information and Communication System Engineering, National Institute of Technology, Okinawa College, 980 Henoko, Nago-shi, Okinawa, 905-2192, Japan

M. Maeda
Oricon Inc., 6-8-10 Roppongi, Minato-ku, Tokyo, 106-0032, Japan


D. Orlov et al. (eds.), *Magnesium Technology 2018*, The Minerals, Metals & Materials Series, https://doi.org/10.1007/978-3-319-72332-7_26

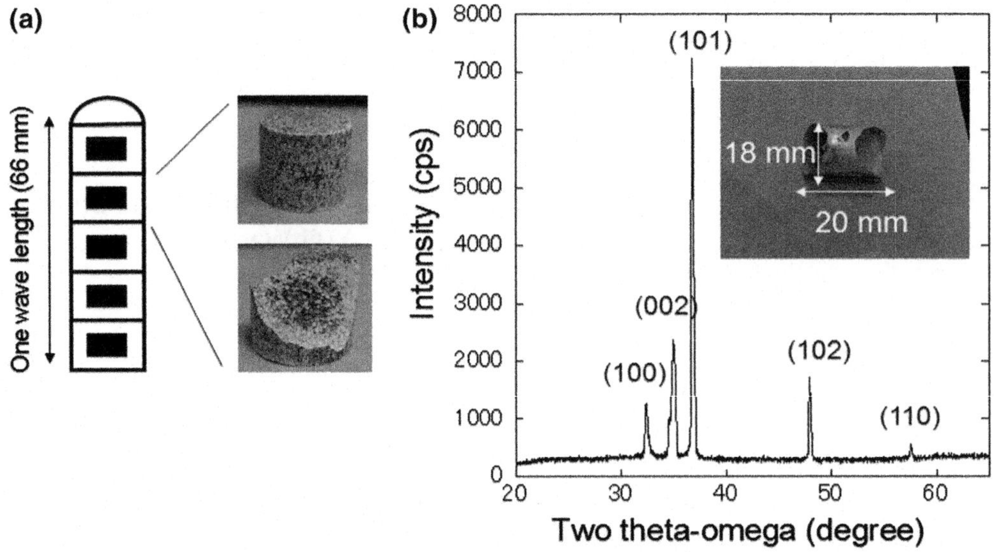


**Fig. 1** **a** Schematic of the antenna structure of briquettes, and photographs of diagonal and cross-sectional views of briquettes **b** X-ray diffraction pattern of the obtained Mg and a photograph of the metal sample

## Experimental

Normally, dolomite is a poor absorber of microwave energy and does not generate heat. However, by using ferrosilicon as the reducing agent, devising the shape of the raw material pellet obtained by mixing dolomite and ferrosilicon and forming it as an antenna so that it has a resonance structure of 2.45 GHz (same as the frequency for microwave ovens), it was possible to confine the microwave energy to the briquettes. The small-scale experimental reactor consists of a multi-mode applicator, a stirrer fan, and a magnetron as a microwave source and experiments were conducted to confirm the antenna effects and smelting briquettes. Then, a demonstration furnace about 5 times larger than the experimental furnace was constructed and conducted to estimate the energy consumption of smelting Mg using microwave irradiation.

## Results and Discussion

In a small-scale experimental reactor, 1 g of magnesium metal was smelted successfully. Figure 1 shows the result of X-ray diffraction and the photo of the obtained magnesium metal smelting magnesium. And in a large-scale experimental reactor, 7 g of magnesium metal was also obtained and the energy consumption of the new microwave Pidgeon method could be reasonably estimated as 58.6 GJ/t (Mg).

## Conclusion

This success in saving energy for smelting magnesium metal has led to the possibility of this technique being developed and applied to the high temperature reduction process of oxides. In the future, through further development of this research, it will be applied to the smelting of other metal materials to save energy with steel, metals, materials, and chemistry, which have not advanced, and help reduce carbon dioxide, which is one of the causes of global warming.

## References

1. Y. A. Çengel, "Green thermodynamics," *Int. J. Energy Res.*, vol. 31, pp. 1088–1104, 2007.
2. Y. Wada, S. Fujii, E. Suzuki, M. M. Maitani, S. Tsubaki, S. Chonan, M. Fukui, N. Inazu, "Smelting magnesium metal using a microwave Pidgeon method," *Sci. Rep.* 2017, 7, 46512, https://doi.org/10.1038/srep46512.

# Thermogravimetric Analysis of Simultaneous Decomposition and Formation of $MgB_2$

Muhammad A. Imam and Ramana G. Reddy

**Abstract**

This study provides the simultaneous thermogravimetric (TGA) decomposition and formation of $MgB_2$. This thermal decomposition of $MgB_2$ to $MgB_4$ was investigated to determine the kinetic barriers associated with the decomposition process. At the same time, the formation of $MgB_2$ from $MgB_4$ was also studied. A list of models available from the literature was also validated in the present study by using the Coats-Redfern equation to determine the mechanism involved in the decomposition and formation reactions. A second order reaction model was more linearly fitted with the CR equation than other available models. A computational approach was used to determine the precise reaction order (n = 2.2) for both decomposition and formation. The activation energy of decomposition was 205.81 ± 1.5 kJ/mol and formation was 241.5 ± 2.6 kJ/mol, both of which are in close agreement with the literature. The standard formation enthalpy of $MgB_2$ (−18.16 ± 1.78 kJ/mol) and $MgB_4$ (−13.86 ± 0.71 kJ/mol) was also obtained.

**Keywords**

Thermogravimetric analysis • $MgB_2$ • $MgB_4$
Activation energy

## Introduction

The $MgB_2$ superconducting phase possesses a higher critical temperature ($T_c$) of ~39 K compared to other superconducting phases [1]. After the invention of $MgB_2$, different processing routes were studied to optimize the superconducting properties [2–5]. Along with the formation of $MgB_2$, other semiconducting phases were also investigated in the Mg-B binary system [6]. The latest Mg-B binary phase diagram, reported by Liu et al., consists of $MgB_2$, $MgB_4$, and $MgB_7$ binary phases [7]. Some other intermediate phases were reported in the early literature but not experimentally confirmed. The thermodynamic studies of Mg-B system were not established yet. But the CALPHAD and ab initio based thermodynamic calculations of Mg-B binary system were performed to obtain better thermodynamic boundaries of the existing phases [8]. Cook et al. and Brutti et al. investigated the vaporization calorimetry techniques to understand the high-temperature thermodynamic properties [9–11]. Recently, Reddy et al. conducted the equilibrium studies using emf cell technique to understand the solid state high-temperature phase equilibria [12].

The kinetics and diffusion studies of Mg-B binary system are limited due to experimental difficulties. Ma et al. and Shi et al. reported the formation kinetics of sintered $MgB_2$ from an elemental Mg and B mixture [13, 14]. The decomposition of the $MgB_2$ thin film was studied by Fan et al., which is kinetically limited to the thin film application. Additionally, the reported kinetically limited Mg partial pressure did not agree with the thermodynamically predicted Mg partial pressure. On the other hand, the synthesis of bulk $MgB_4$ and $MgB_7$ has largely depended on the decomposition of $MgB_2$. It is reported in the literature that the solid phase $MgB_2$ directly decomposes to solid phase $MgB_4$ [15]. Furthermore, similar study with other Mg based alloys was also performed by Reddy et al. [16].

The reaction kinetics of solids was observed by using the experimental techniques like differential thermal analysis (DTA) [16], thermogravimetric analysis (TGA) [17, 18], and differential scanning calorimetry (DSC) [19]. The kinetic studies were performed under isothermal conditions in the early studies [20, 21]. The reaction rates in solids were determined using non-isothermal techniques. However, these results were not used for kinetic assessments until the 1930s. Therefore, the concepts of solid-state kinetics were established by experiments carried out under isothermal conditions. Initially, non-isothermal kinetic studies had been

M. A. Imam · R. G. Reddy (✉)
Department of Metallurgical and Materials Engineering, The University of Alabama, Tuscaloosa, AL 35487, USA
e-mail: rreddy@eng.ua.edu


D. Orlov et al. (eds.), *Magnesium Technology 2018*, The Minerals, Metals & Materials Series, https://doi.org/10.1007/978-3-319-72332-7_27

ignored. Freeman and Carroll studied some TGA-based weight loss/gain kinetics [22]. They stated some advantages over conventional isothermal studies. After Coats and Redfern (CR) proposed non-isothermal kinetic equations based on TGA weight loss, several studies are suggested using this approach.

In the present work, TGA weight change was evaluated using the CR equation [23].

In the reaction:

$$2MgB_2 \rightleftharpoons MgB_4 + Mg(g) \quad (1)$$

Currently accepted kinetic equation might express the rate of decomposition of $MgB_2$ as:

$$\frac{d\alpha}{dt} = kf(\alpha) \quad (2)$$

$$k = A\exp^{-E/RT} \quad (3)$$

where conversion factor, $\alpha = \frac{m_0 - m_t}{m_0 - m_\infty}$; $m_0$ = intial mass, $m_t$ = mass after time t, $m_\infty$ = mass after reaction complete, k = rate constant, A = pre exponential frequency factor, R = universal gas constant, T = absolute temperature, f(α) is a function of the reacted fraction, that depends on the reaction model [9], E = activation energy.

For a linear heating rate, $\beta = dT/dt$, Eq. (1) and (2) gives:

$$\frac{d\alpha}{dT} = \frac{A}{\beta}\exp^{-E/RT} f(\alpha) \quad (4)$$

$$\int_0^1 \frac{d\alpha}{f(\alpha)} = g(\alpha) = \frac{A}{\beta}\int_0^T \exp^{-E/RT} dT \quad (5)$$

There is no exact integral of $\int_0^T \exp^{-E/RT} dT$, but making substitution E/RT = x:

$$g(\alpha) = \frac{ART^2}{\beta E}\int_x^\infty \frac{\exp^{-x}}{x^2}dx = \frac{ART^2}{\beta E}P(x) \quad (6)$$

where P(x) has no analytical solution and using the relation:

$$\int_x^\infty \frac{e^{-x}}{x^b}dx \approx x^{1-b}e^{-x}\sum_{n=0}^{\infty}\frac{-1^n(b)n}{x^{n+1}} \quad (7)$$

$$g(\alpha) = \frac{ART^2}{\beta E}\left[1 - \frac{2RT}{E}\right]e^{\frac{-E}{RT}} \quad (8)$$

by taking natural logs:

$$\ln\frac{g(\alpha)}{T^2} = \ln\frac{AR}{\beta E}\left[1 - \frac{2RT}{E}\right] - \frac{E}{RT} \quad (9)$$

Therefore a plot of $\ln\frac{g(\alpha)}{T^2}$ vs $-\frac{1}{RT}$ should be a straight line of slope E, activation energy.

In this study, a set of different models shown in Table 1 were chosen to calculate the activation energy using Eq. 8. The best linear fitted model was chosen as the driving mechanism for this decomposition or formation of $MgB_2$. X-ray diffraction (XRD) of $MgB_2$ before and after TGA confirmed the reversible reaction. The scanning electron microscope (SEM) analysis are also showed the similar morphology before and after the TGA measurements.

## Experimental

### Materials

Magnesium diboride was synthesized using the elemental mixture of Mg and B. Pure elemental Mg (99.99% metal basis, ~325 mesh) and B (99.99% metal basis, ~325 mesh) powders were purchased from Alfa Aesar, USA. A mixture containing stoichiometric amounts of elemental Mg and B (1:2) was mixed in a glass jar. The powders were mixed for 24 h in a jar mill to get a homogenous mixture of Mg and B. The powder mixture was cold-pressed using a Carver, Inc. (Model 43-50L) to obtain a cylindrical pellet (2B + Mg) with 1.3 cm diameter. This process was performed for 20 min at 55 MPa to obtain a high-dense pellet.

**Table 1** Solid state reaction mechanisms, kinetic models, and conversion functions [24]

| Mechanism | Kinetic model | Conversion functions, g(α) |
|---|---|---|
| Nucleation model | Power law | $\alpha^{1/n}$, $n = 2$–$4$ |
| Diffusion model | 1-D diffusion | $\alpha^2$ |
| | Diffusion control (Janders) | $[1-(1-\alpha)^{1/3}]^2$ |
| | Diffusion control (Crank) | $1-(2/3)\alpha-(1-\alpha)^{2/3}$ |
| Reaction order | Mampel (first order) | $-\ln(1-\alpha)$ |
| | Second order | $(1-\alpha)^{-1}-1$ |
| Geometrical construction | Contracting cylinder | $1-(1-\alpha)^{1/2}$ |
| | Contracting sphere | $1-(1-\alpha)^{1/3}$ |

Subsequently, sintering of the pellet was carried out at 973 K for 3 h in an argon gas flowing atmosphere.

## Thermogravimetric Analysis (TGA)

The thermogravimetric analysis of freshly prepared $MgB_2$ was performed from 373 to 1250 K at 10 K/min using a Perkin Elmer TGA7 instrument according to ASTM standard E2550 [25]. About 20 mg of the $MgB_2$ was used to determine the weight change due to decomposition and formation of $MgB_2$. An inert atmosphere was maintained using the ultra-high pure (99.9%) argon gas (purchased from Airgas, USA) at a purge rate of 20 ml/min.

## Characterization of $MgB_2$ Powder

After sintering, the sample was characterized using X-ray Diffraction (XRD) and scanning electron microscopy (SEM). XRD was performed using Philips X'pert MPD instrument. The XRD data were obtained at a scan rate of 0.01°/s in the range of diffraction angle $2\theta$ = 20–80° with the Cu-K$\alpha$ ($\lambda$ = 1.5405 A°) monochromatic source. Phase analysis was conducted using the Jade 6.0 software (Materials Data, Inc.). The analysis was based on the standard ICDD diffraction data. The surface morphology was performed using FESEM (Model JEOL JSM 7000F) before and after the TGA. The images of nanoparticles were obtained at a high resolution ($\gtrsim$30000×) with an accelerating voltage of 20 kV.

## Results and Discussion

### Thermogravimetric Analysis (TGA)

Figure 1 shows the TGA analysis for the different samples. The increase in weight is $\sim 1\%$ over the temperature range of 300–1185 K due to the surface oxidation of $MgB_2$ sample. XRD results confirm the formation of $MgB_2$ after the intermediate steps.

As seen from Fig. 1, the onset decomposition temperature is observed at $1194 \pm 4$ K. The activation energy of the decomposition and formation reactions are determined from the TGA weight loss and weight gain data using the Coats-Redfern (CR) Eq. (8).

### X-Ray Diffraction of $MgB_2$

Figure 2 shows an X-ray diffraction pattern obtained (before and after TGA) from the crystalline particles (>100 μm in size), which matched with the $MgB_2$ phase (ICCD-PDF# 38-1369). The plane (101) is the characteristic peak for $MgB_2$ at Bragg's angle of 42.61° that match with the ICCD PDF file. Along with this, other peaks also matched with significant crystallographic planes, i.e. (001), (100), (101), (002), (110), (102), (111), (200), and (201) of the $MgB_2$ PDF file. Only a trace amount (<2%) of MgO was found in the XRD spectrum. The peak was observed at Bragg's angle of 62.302° for (220) plane of MgO. The trace amount of MgO was formed as a product of formation of $MgB_2$ [26].

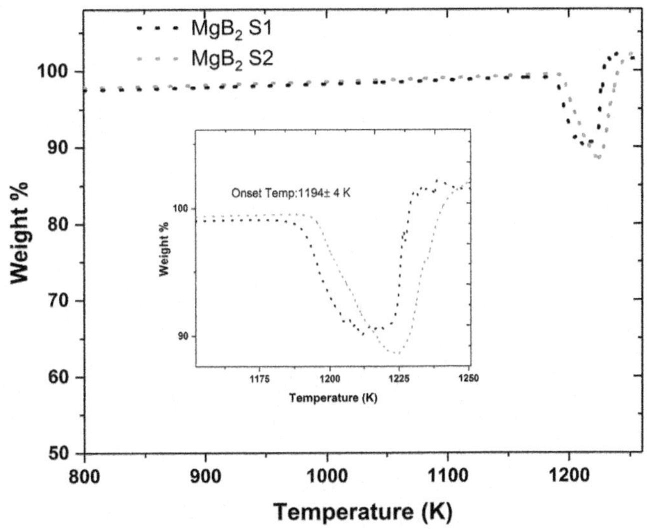


**Fig. 1** Simultaneous weight loss and gain of $MgB_2$ using TGA analysis

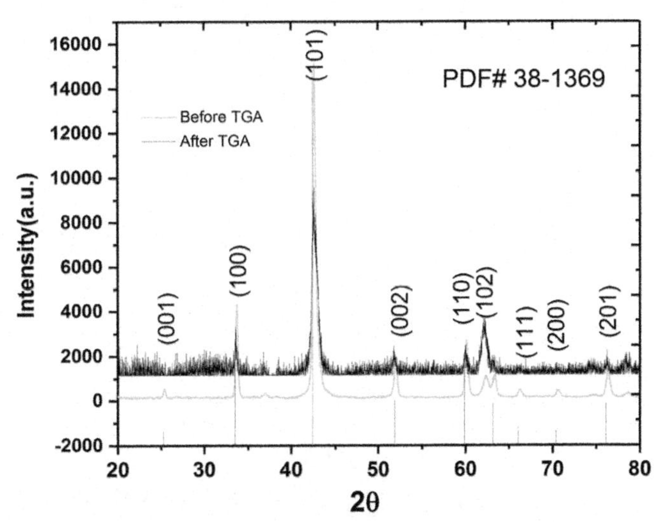


**Fig. 2** XRD of $MgB_2$ compound before and after TGA

## Scanning Electron Microscope (SEM)

Figure 3 shows the morphology of $MgB_2$ obtained from FESEM before and after TGA. Figure 3a shows the morphology of $MgB_2$ which was sintered at 973 K for 3 h in an argon flowing atmosphere, while Fig. 3b represents the formation of $MgB_2$ from the reversible reaction (Eq. 1). The formation of $MgB_2$ from an elemental mixture of Mg and B at 973 K was due to the solid-liquid rearrangement of particles and their subsequent reaction which led to re-precipitation and grain growth (Ostwald ripening mechanism) of $MgB_2$ [13]. Additionally, the particle sizes (Fig. 3a) are small (100–300 nm) due to the solution precipitation. On the other hand, the $MgB_2$ formed during TGA analysis is governed by the solid state reversible reaction. The particle sizes (Fig. 3b) are much larger compared to the sintered $MgB_2$. That is might be due to the difference in the formation mechanism.

## Activation Energy Calculation by Model Validation

Various kinetic models mentioned in Table 1 were used with the CR equation to confirm the mechanism. The corresponding ln $[g(\alpha)/T^2]$ versus 1/RT plot is presented in Fig. 4a, b for decomposition (weight loss) and formation (weight gain) respectively. All the linear fit kinetics data are also tabulated in Tables 2 and 3. As seen from Fig. 4a, b, the most linear fitted curve for decomposition and formation is a second order reaction model mechanism. In both cases, the coefficient of determination ($R^2$) is 0.92 and 0.93 for decomposition and formation respectively. It needs to be mentioned that all the other kinetics models were not showing a linear relationship ($R^2 < 0.86$). However, the accurate model or reaction mechanism could not be determined by these ($R^2 = 0.92$ or 0.93) coefficients of determination.

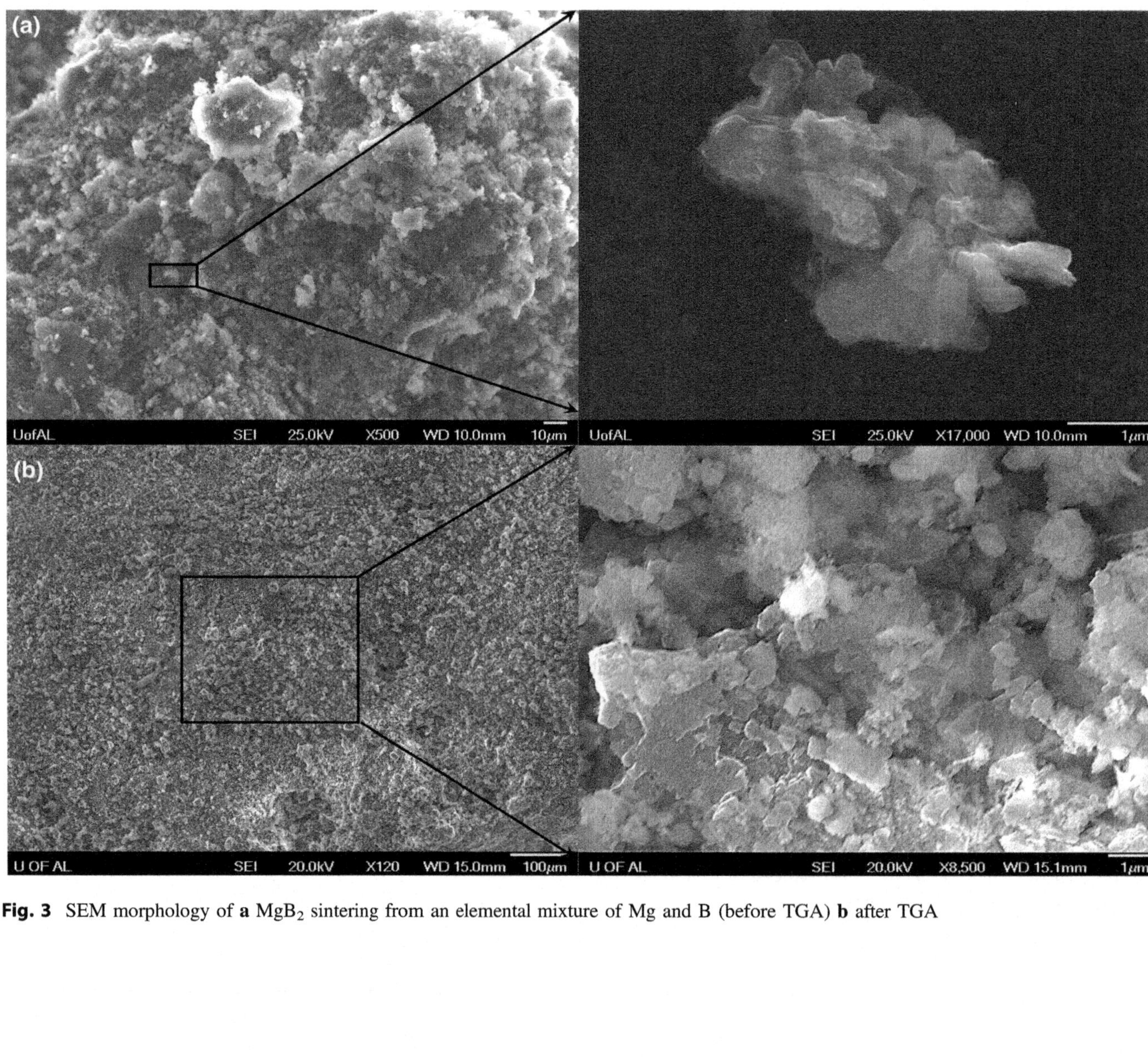


**Fig. 3** SEM morphology of **a** $MgB_2$ sintering from an elemental mixture of Mg and B (before TGA) **b** after TGA

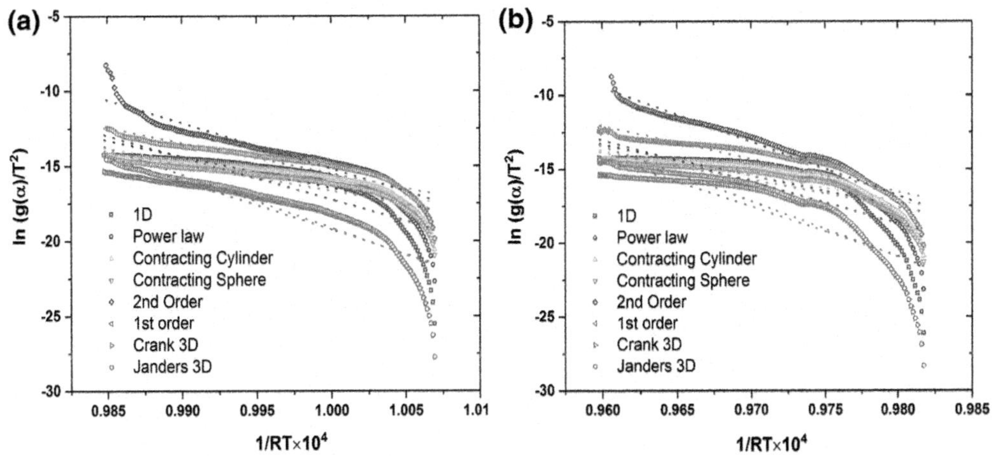


**Fig. 4** Plots of ln [g(α)/T$^2$] versus 1/RT for different models using values of α for $MgB_2$ **a** decomposition and **b** formation

**Table 2** Model verification for TGA weight loss [$0 \leq \alpha \leq 1$]

| Model | 1D | Power law | Contracting cylinder | Contracting sphere | 2nd order | 1st order | Crank 3D | Janders 3D |
|---|---|---|---|---|---|---|---|---|
| Intercept | 266.55 ± 14.08 | 195.86 ± 10.56 | 150.17 ± 6.76 | 160.58 ± 6.61 | 289.62 ± 7.06 | 186.75 ± 6.35 | 307.68 ± 13.65 | 337.35 ± 13.21 |
| Slope | −283.78 ± 14.14 | −212.33 ± 10.60 | −166.46 ± 6.78 | −177.22 ± 6.63 | −304.79 ± 7.09 | −202.21 ± 6.38 | −326.91 ± 13.70 | −356.46 ± 13.26 |
| $R^2$ | 0.71 | 0.71 | 0.78 | 0.81 | 0.92 | 0.86 | 0.77 | 0.81 |
| Adj. $R^2$ | 0.71 | 0.71 | 0.78 | 0.81 | 0.92 | 0.86 | 0.77 | 0.81 |

**Table 3** Model verification for TGA weight gain [$0 \leq \alpha \leq 1$]

| Model | 1D | Power law | Contracting cylinder | Contracting sphere | 2nd order | 1st order | Crank 3D | Janders 3D |
|---|---|---|---|---|---|---|---|---|
| Intercept | 259.17 ± 15.83 | 190.82 ± 11.87 | 152.52 ± 7.80 | 166.19 ± 7.65 | 341.89 ± 6.72 | 203.50 ± 7.16 | 307.75 ± 15.73 | 346.61 ± 15.31 |
| Slope | −283.52 ± 16.33 | −212.64 ± 12.24 | −173.13 ± 8.04 | −187.51 ± 7.89 | −366.01 ± 6.92 | −224.52 ± 7.38 | −335.32 ± 16.22 | −375.02 ± 15.78 |
| $R^2$ | 0.65 | 0.65 | 0.74 | 0.77 | 0.93 | 0.85 | 0.72 | 0.77 |
| Adj. $R^2$ | 0.64 | 0.64 | 0.74 | 0.77 | 0.93 | 0.85 | 0.72 | 0.77 |

A computational approach (inset of Fig. 5a, b) was used to select the value of *n* (*n = 2.2*), giving the best linear fitting $R^2 = 0.98$ and $R^2 = 0.97$ for decomposition and formation respectively. From this analysis, it could be inferred that the reaction order (*n = 2.2*) mechanism was maintained for decomposition and formation.

The slope of the $ln\,[g(\alpha)/T^2]$ versus $-1/RT$ gives the activation energy (*E*) after normalizing the units. The activation energy of decomposition is 205.65 ± 1.5 kJ/mole and formation is 241.4 ± 4.75 kJ/mole, which is in close agreement with the published literature, $\Delta H_{vap.} = 238.1 \pm 2.6$ kJ/mol [11]. The activation energy was summarized in Table 4.

Additionally, this activation energy $(E = \Delta_r G = \Delta_r H + T\,\Delta_r S)$ was used to derive the standard formation of enthalpy $(\Delta_f H^o)$ assuming that $(\Delta_r S)$ is negligible. The standard formation of enthalpy of formation $(\Delta_f H^o)$ was calculated using the following relationship for Eq. $1(2\mathrm{MgB_2} \rightleftharpoons \mathrm{MgB_4} + \mathrm{Mg}(g))$:

$$\Delta_r H(\mathrm{Mg}B_2) = \Delta_f H^0_{298.15}(\mathrm{MgB_4}) + \Delta_f H^0_{298.15}(\mathrm{Mg}) - 2\,\Delta_f H^0(\mathrm{MgB_2});$$

where the sublimation enthalpy $\Delta_f H^0_{298.15}(\mathrm{Mg}) = 146.4\,\mathrm{kJ/mol}$, $\Delta_r H(\mathrm{MgB_2}) = E$ (TGA formation) $= 241.5\,\mathrm{kJ/mol}$ and $\Delta_f H^0_{298.15}(\mathrm{MgB_4})$ was calculated from the following equation:

$$7/3\mathrm{MgB_4}(s) = 4/3\mathrm{MgB_7}(s) + \mathrm{Mg}(g)$$

$$\Delta_r H(\mathrm{MgB_4}) = 4/3 * \Delta_f H^0_{298.15}(\mathrm{MgB_7}) + \Delta_f H^0_{298.15}(\mathrm{Mg}) - 7/3 * \Delta_f H^0(\mathrm{MgB_4})$$

It needs to be mentioned that the formation enthalpy of $MgB_7$ was taken from literature [10]. Table 5 summarizes the formation enthalpy from this work and literature.

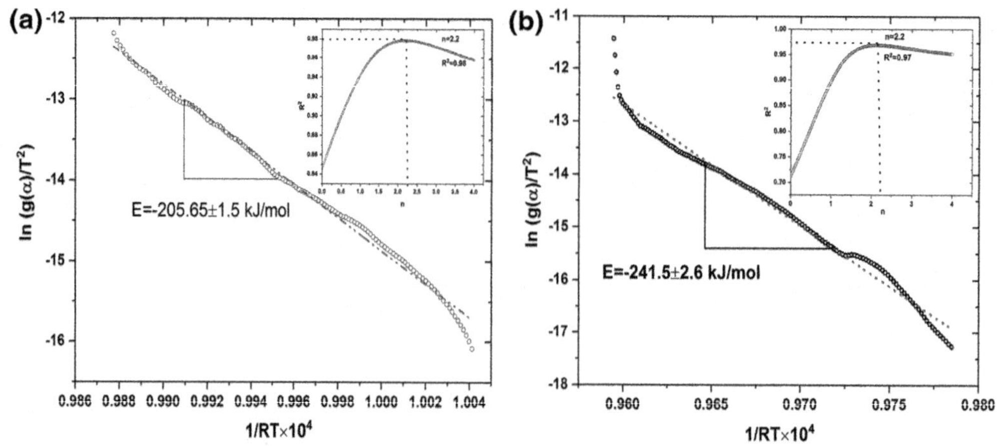


**Fig. 5** Reaction order kinetics ($n = 2.2$) for **a** decomposition ($R^2 = 0.98$) and **b** formation ($R^2 = 0.97$) using a computational approach (MATLAB)

**Table 4** Comparison of activation energy from this study and literature

| Activation energy/$\Delta H_{vap}$(kJ/mole) | Study |
|---|---|
| 205.81 ± 1.5 | Decomposition (this study) |
| 241.5 ± 2.6 | Formation (this study) |
| 238.1 ± 2.6 | Brutti et al. [11] |
| 152.18 | Y. Guo et al. [27] |
| 252.76 ± 6.3 ($\Delta_r H_{298}$) | JANAF [28] |
| 252.2 | Yan et al. (DSC $MgB_2$ decomposition) [29] |

**Table 5** Comparison between $\Delta_f H^0_{298.15}$(kJ/mol) of Mg-B from this work and literature

| $\Delta_f H^0_{298.15}$ | $MgB_4$ | $MgB_2$ |
|---|---|---|
| TGA | −13.86 ± 0.71 | −18.16 ± 1.78 |
| Rybakova et al. [30] (calorimetry) | −14.73 ± 0.5 | −18.55 ± 1.12 |
| Ariya et al. [31] (calorimetry) | −14.39 ± 0.75 | −17.17 ± 1.39 |

## Conclusion

The thermal decomposition of $MgB_2$ to $MgB_4$ was studied to determine the kinetic barriers associated with the decomposition process. At the same time, the formation of $MgB_2$ from $MgB_4$ was investigated. XRD confirms the formation of $MgB_2$ after TGA. The morphology study was performed using FESEM. Several models available from the literature were also validated in the present study using the Coats-Redfern equation to determine the mechanism involved in the decomposition and formation reaction. A second order reaction model was linearly fitted with the CR equation. A computational approach was used to determine the precise reaction order (n = 2.2) for both decomposition and formation. The activation energy of decomposition is 205.81 ± 1.5 kJ/mol and formation is 241.5 ± 2.6 kJ/mol, which is in close agreement with the published literature, $\Delta H_{vap.}$ = 238.1 ± 2.6 kJ/mol. The standard formation enthalpy of $MgB_2$ and $MgB_4$ was also derived.

**Acknowledgements** The authors gratefully acknowledge the financial support of the National Science Foundation (NSF, Grant No. DMR-1310072). The authors thank Jacob Young for proofreading the manuscript.

## References

1. J. Nagamatsu, N. Nakagawa, T. Muranaka, Y. Zenitani, and J. Akimitsu, Superconductivity at 39 K in magnesium diboride, Nature, **410**, 63 (2001).
2. D. H. A. Blank, H. Hilgenkamp, A. Brinkman, D. Mijatovic, G. Rijnders, and H. Rogalla, Superconducting Mg–B films by pulsed-laser deposition in an in situ two-step process using multicomponent targets, Applied Physics Letters, **79**, 394–396 (2001).
3. K. Ueda and M. Naito, As-grown superconducting MgB 2 thin films prepared by molecular beam epitaxy, Applied Physics Letters, **79**, 2046–2048 (2001).
4. W. Jo, J. U. Huh, T. Ohnishi, A. F. Marshall, M. R. Beasley, and R. H. Hammond, In situ growth of superconducting $MgB_2$ thin films with preferential orientation by molecular-beam epitaxy, Applied Physics Letters, **80**, 3563–3565 (2002).

5. S.-D. Bu, D. M. Kim, J. H. Choi, J. Giencke, E. E. Hellstrom, D. C. Larbalestier, *et al.*, Synthesis and properties of c-axis oriented epitaxial MgB 2 thin films, Applied Physics Letters, **81**, 1851–1853 (2002).
6. M. E. Yakıncı, Y. Balcı, M. A. Aksan, H. I. Adigüzel, and A. Gencer, Degradation of superconducting properties in MgB 2 by formation of the $MgB_4$ phase, Journal of superconductivity, **15**, 607–611 (2002).
7. Z.-K. Liu, D. G. Schlom, Q. Li, and X. X. Xi, Thermodynamics of the Mg–B system: implications for the deposition of $MgB_2$ thin films, Applied Physics Letters, **78**, 3678–3680 (2001).
8. S. Kim, D. S. Stone, J.-I. Cho, C.-Y. Jeong, C.-S. Kang, and J.-C. Bae, Phase stability determination of the Mg–B binary system using the CALPHAD method and ab initio calculations, Journal of Alloys and Compounds, **470**, 85–89 (2009).
9. L. P. Cook, R. Klein, W. Wong-Ng, Q. Huang, R. A. Ribeiro, and P. C. Canfield, Thermodynamics of $MgB_2$ by calorimetry and Knudsen thermogravimetry, IEEE transactions on applied superconductivity, **15**, 3227–3229, (2005).
10. G. Balducci, S. Brutti, A. Ciccioli, G. Gigli, P. Manfrinetti, A. Palenzona, *et al.*, Thermodynamics of the intermediate phases in the Mg–B system, Journal of Physics and Chemistry of Solids, **66**, 292–297 (2005).
11. S. Brutti, A. Ciccioli, G. Balducci, and G. Gigli, Vaporization thermodynamics of $MgB_2$ and $MgB_4$, Applied Physics Letters, **80**, 2892–2894 (2002).
12. M. A. Imam and R. G. Reddy, Thermodynamic Studies on the Mg-B System Using Solid State Electrochemical Cells, in *Applications of Process Engineering Principles in Materials Processing, Energy and Environmental Technologies*, 457–464 (Springer, 2017).
13. Q. Z. Shi, Y. C. Liu, Q. Zhao, and Z. Q. Ma, Phase formation process of bulk $MgB_2$ analyzed by differential thermal analysis during sintering, Journal of alloys and compounds, **458**, 553–557 (2008).
14. Z. Ma and Y. Liu, The varied kinetics mechanisms in the synthesis of $MgB_2$ from elemental powders by low-temperature sintering, Materials Chemistry and Physics, **126**, 114–117 (2011).
15. Z. Y. Fan, D. G. Hinks, N. Newman, and J. M. Rowell, Experimental study of $MgB_2$ decomposition, Applied Physics Letters, **79**, 87–89 (2001).
16. M. R. Bogala and R. G. Reddy, Reaction kinetic studies of metal-doped magnesium silicides, Journal of Materials Science, **52**, 11962–11976 (2017).
17. M. Ramachandran, D. Mantha, C. Williams, and R. G. Reddy, Oxidation and Diffusion in Ti-Al-(Mo, Nb) Intermetallics, Metallurgical and Materials Transactions A, **42**, 202–210 (2011).
18. I. C. I. Okafor, X. Wen, and R. G. Reddy, Interdiffusion in the $TiO_2$ oxidation product of $Ti_3Al$, Metallurgical and Materials Transactions A, **32**, 491–495 (2001).
19. M. A. Imam, S. Jeelani, V. K. Rangari, M. G. Gome, and E. A. B. Moura, Electron-Beam Irradiation Effect on Thermal and Mechanical Properties of Nylon-6 Nanocomposite Fibers Infused with Diamond and Diamond Coated Carbon Nanotubes, International Journal of Nanoscience, **15**, 1650004 (2016).
20. G. N. Lewis, Autocatalytic decomposition of silver oxide, 719–733 (1905).
21. P. Vallet, Theoretical study of the decomposition of bodies in linearly increasing temperature, Comptes Rendus, **200**, 315–17 (1935).
22. E. S. Freeman and B. Carroll, The application of thermoanalytical techniques to reaction kinetics: the thermogravimetric evaluation of the kinetics of the decomposition of calcium oxalate monohydrate, The Journal of Physical Chemistry, **62**, 394–397 (1958).
23. A. W. Coats and J. P. Redfern, Kinetic Parameters from Thermogravimetric Data, Nature, **201**, 68–69 (1964).
24. A. Khawam and D. R. Flanagan, Role of isoconversional methods in varying activation energies of solid-state kinetics: II. Non-isothermal kinetic studies, Thermochimica Acta, **436**, 101–112 (2005).
25. Standard Test Method for Thermal Stability by Thermogravimetry, (ASTM International, 2011).
26. P. Toulemonde, N. Musolino, and R. Flükiger, High-pressure synthesis of pure and doped superconducting $MgB_2$ compounds, Superconductor Science and Technology, **16**, 231 (2003).
27. Y. Guo, W. Zhang, X. Zhou, and T. Bao, Magnesium boride sintered as high-energy fuel, Journal of thermal analysis and calorimetry, **113**,787–791 (2013).
28. M. W. Chase, J. L. Curnutt, J. R. Downey, R. A. McDonald, A. N. Syverud, and E. A. Valenzuela, JANAF Thermochemical Tables, 1985 Supplement, Journal of Physical and Chemical Reference Data, **1**, 226 (1985).
29. S. C. Yan, G. Yan, C. F. Liu, Y. F. Lu, and L. Zhou, Experimental study on phase transformation between $MgB_2$ and $MgB_4$, Journal of the American Ceramic Society, **90**, 2184–2188 (2007).
30. G. A. Rybakova, L. A. Pavlinova, M. P. Morozova, and D. V. Korol'kov, Enthalpy of formation and nature of the phases in the beryllium-boron, magnesium-boron, iron-sulfur, cobalt-sulfur, and nickel-sulfur systems, Probl. Sov. Khim. Koordinates Doedin. 146 (1978).
31. S. M. Ariya, M. P. Morozova, G. A. Semenov, and G. A. Ryabakova, Magnesium-boron system and determination of the enthalpy of formation of magnesium borides, Zhur. Fiz. Khim., **45**, 181 (1971).

# Dislocations in Mg Alloys with Rare-Earth Element Addition

Zhiqing Yang and Hengqiang Ye

**Abstract**

Improvement of ductility is important for applications of Mg alloys. Basal dislocation motion and twinning are the two major deformation modes in Mg alloys. However, the basal slip system cannot support homogeneous plastic deformation of Mg alloys. Twin-boundaries in Mg alloys are potential sites for cracking. Therefore, activation of non-basal slip is expected to play an important role in improving ductility of Mg alloys. We have studied both **<a>** and **<c+a>** dislocations, as well as their interactions with solute atoms, stacking-faults and grain-boundaries in RE-containing Mg alloys. Cottrell atmospheres along dislocations in deformed strengthening phases of a Mg–Zn–Y alloys were observed and quantified. Based on atomic resolution characterizations, a binding energy of about 0.05 eV is deduced between basal dislocations and surrounding solute atmospheres. Besides providing enough independent slip systems, **<c+a>** dislocations can cut and react with basal stacking-faults, and react with grain-boundary dislocations, and drive migration of grain-boundaries, benefiting both strength and ductility of Mg alloys.

**Keywords**

Dislocation • Cottrell atmosphere • Suzuki segregation Grain boundary • Strength • Ductility

## Introduction

Mg alloys have tremendous potential to achieve energy efficiency in many industries, since they are the lightest metallic materials for structural applications [1, 2]. Proper plastic formability is one of the important properties for most industrial applications of metallic materials. Dislocation motion and twinning are the two main plastic deformation modes in alloys [3]. Basal **<a>** slip is usually activated first at the early stage of plastic deformation of Mg alloys [4]. With further increase of flow stress, deformation twinning may start to accommodate the plastic deformation of Mg together with basal slip [5]. Therefore, motion of dislocations and their interactions with other crystal defects play a critical role in ductility of Mg alloys. However, Mg alloys usually exhibit ductility not high enough for making structural components through plastic processing, because their easiest slip system, $(0001)<11\bar{2}0>$ in the hexagonal close-packed (HCP) structure, cannot supply five independent slip systems to achieve homogeneous plastic deformation in polycrystalline materials [6, 7]. Moreover, twin boundaries (TBs) were potential sites for nucleation and propagation of microcracks [8]. Cracks often occur in many Mg alloys (such as AZ31, AZ31B) during hot processing [9–11]. Therefore, poor formability is one of the primary barriers to industrial applications of Mg alloys.

Activation of pyramidal **<c+a>** dislocations which can provide five independent slip systems is thus a potential way to improve the ductility of HCP alloys. Therefore, **<c+a>** dislocations in HCP metals and alloys have been extensively investigated for decades [12–14]. But, molecular dynamics simulations showed that **<c+a>** dislocations showed high tendency to transform from dissociated geometry on pyramidal slip planes into sessile locks on basal planes in Mg, preventing their long-range glide on pyramidal planes, which might be one of the reasons for the poor ductility of Mg alloys [15, 16]. So, it is necessary to prevent **<c+a>** dislocations from dissociating into sessile locks, in order to achieve their long-range pyramidal glide. Interestingly, there have been studies showing that pyramidal slip of **<c+a>** dislocations occurred in Mg alloys with addition of rare-earth elements (RE), such as Ce and Y [17–20]. The improvement of ductility of Mg alloys with the addition of RE was also confirmed by V-bending tests at room

Z. Yang (✉) · H. Ye
Shenyang National Laboratory for Materials Science,
Institute of Metal Research, Chinese Academy of Science,
110016 Shenyang, China
e-mail: yangzq@imr.ac.cn


D. Orlov et al. (eds.), *Magnesium Technology 2018*, The Minerals, Metals & Materials Series, https://doi.org/10.1007/978-3-319-72332-7_28

temperature in a Mg–Zn–Y alloy without cracking with a minimum bending radius per thickness of 3.3, while a large crack appeared in a Mg–Al–Zn sample [21].

Addition of RE in Mg alloys may lead to the formation of long-period stacking ordered (LPSO) strengthening phases. And it has been revealed that deformation twinning was large prohibited in Mg–Zn–Y alloys containing LPSO [22, 23]. Therefore, dislocation motion should play an important role in plastic deformation of Mg alloys with RE addition. In the present study, dislocations and their interaction with other crystal defects, such as solute atoms, stacking faults (SFs) and grain boundaries (GBs), in Mg–Zn–Y and Mg–Zn–Gd alloys were investigated using aberration-corrected scanning-transmission electron microscopy (STEM), in order to obtain some in-depth understanding of the influence of dislocations on strength and ductility of Mg alloys with RE addition.

## Experiments

A $Mg_{97}Zn_1Y_2$ (at.%) alloy and a $Mg_{97}Zn_1Gd_2$ (at.%) alloy were prepared using a high frequency induction furnace under argon atmosphere protection. The as-cast $Mg_{97}Zn_1Y_2$ (at.%) alloy has a duplex microstructure with LPSO phases and Mg matrix containing a high density of SFs [22]. The as-cast $Mg_{97}Zn_1Gd_2$ alloy has no LPSO phase, so it was heated at 500 °C for 10 h, in order to obtain LPSO strengthening phases. Besides formation of LPSO phases, a high density of SFs was also formed in the heat-treated $Mg_{97}Zn_1Gd_2$ alloy, according to STEM observations, as shown in Fig. 1. Microstructure of the $Mg_{97}Zn_1Gd_2$ alloy treated at 500 °C is quite similar to that of the as-cast $Mg_{97}Zn_1Y_2$ alloy. LPSO phase in the $Mg_{97}Zn_1Gd_2$ alloy has a 14H structure, which is different from 18R LPSO observed in as-cast Mg–Zn–Y alloys [22].

Compression tests were performed at 300 °C with strain rates ranging from $2 \times 10^{-4}$ to $1 \times 10^{-3}$ $s^{-1}$, in order to study dislocations and their interactions with other crystal defects. Samples for microstructural investigations were prepared by standard ion milling techniques. High-angular annular dark field (HAADF) STEM studies were performed on a Titan 60–300 microscope operating at 300 kV. The beam convergence half-angle was set as 25 mrad, and the collection half-angles of the HAADF detector ranged from about 60 mrad to about 290 mrad for atomic-resolution observations, which can basically eliminate the influence of strain contrast around dislocations [24, 25].

## Results and Discussion

### Basal Slip and Cottrell Atmospheres in LPSO Phases

Plastic deformation of LPSO phases in Mg alloys has been found dominated by basal slip [22]. Basal <**a**> dislocations may rearrange to form tilt GBs which are often referred to as

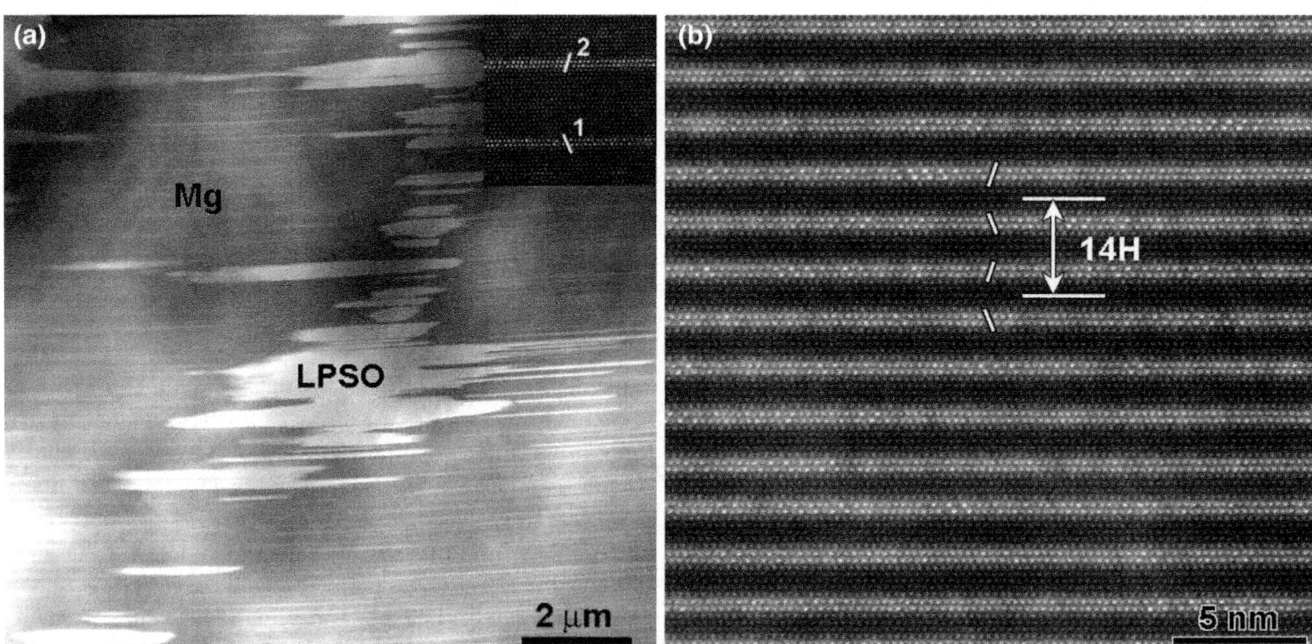


**Fig. 1** **a** Low-magnification HAADF-STEM image showing microstructure in the $Mg_{97}Zn_1Gd_2$ alloy heated at 500 °C for 10 h, **b** atomic resolution HAADF-STEM image of LPSO phase. The upper right inset is an atomic resolution HAADF-STEM image recorded from Mg matrix containing a high density of SFs with solute segregation

kink boundaries in deformed LPSO phases. Figure 2a shows an atomic resolution HAADF-STEM image recorded from a LPSO plate in a $Mg_{97}Zn_1Y_2$ sample deformed at 300 °C to a strain of about 20% at a strain rate of $5 \times 10^{-3}$ $s^{-1}$. It is seen that a kink boundary with a tilt angle of about 6.5° was formed in the deformed LPSO plate, according to direct measurement on the image. There should be six basal 60° <a> dislocations within the region indicated by the Burgers circuit, according to its closure failure. Interestingly, the kink boundary shows brightness higher than the LPSO matrix, demonstrating the occurrence of solute segregation at the boundary [24–26]. Moreover, the enhanced brightness is not homogenous along the boundary, but the centers of tiny regions with enhanced brightness are basically coincident with dislocation cores, suggesting attraction of solutes towards dislocations (Fig. 2a).

Figure 2b is an atomic resolution HAADF-STEM image recorded from a region far away from kink boundaries in deformed LPSO phase in a $Mg_{97}Zn_1Y_2$ sample. The closure failure of the Burgers circuit indicates presence of an basal 60° <a> dislocation. There are also atomic columns showing enhanced brightness due to solute enrichment around this isolated dislocation line, demonstrating the formation of Cottrell atmosphere. Isolated dislocations in the bulk lattice of LPSO like the one shown in Fig. 2b should be moving either toward a nearby kink boundary or somewhere to form a new kink boundary together with other dislocations of the same sign. Therefore, the observed Cottrell atmosphere shown in Fig. 2b should be associated with a moving dislocation during plastic deformation of the LPSO strengthening phase. It may be expected that a fraction of the solute atoms within the Cottrell atmospheres along dislocations in the LPSO matrix could be brought to kink boundaries together with the moving dislocations. The solute concentration within the Cottrell atmospheres in deformed LPSO phase is estimated to be about 2.5 times of that in the bulk LPSO, based on quantitative analyses of the image brightness in combination with image simulations [26]. Furthermore, a binding energy of about 0.05 eV is derived between extra solutes in the Cottrell atmosphere and the associated dislocation, based on the following equation, $C = C_0\exp(u/kT)$ [27], where $C$ is the local solute concentration in atmospheres, $C_0$ is the solute concentration in bulk LPSO, $k$ is Boltzmann's constant, $T$ is the deformation temperature, and $u$ is the binding energy between the dislocation and the excess solute atom. The formation of Cottrell atmospheres along dislocations in LPSO phases should play a role in strengthening the alloy, since it increases the energy barrier for dislocation motion in LPSO during deformation.

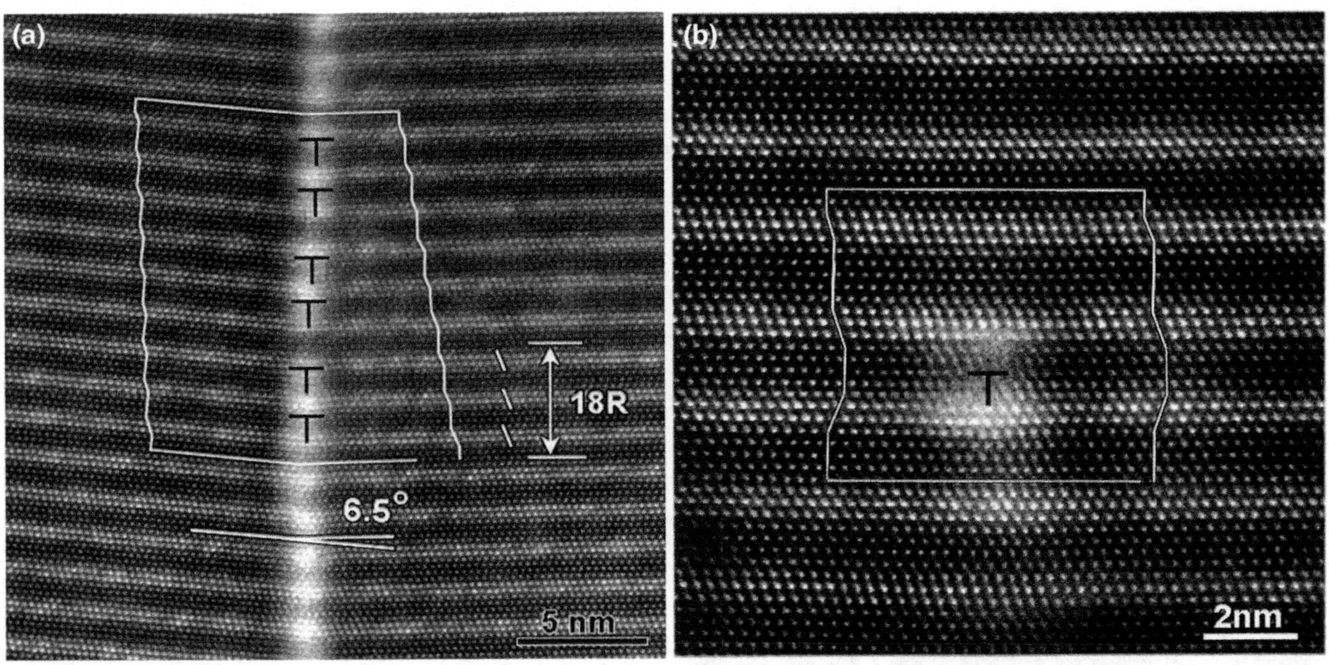


**Fig. 2** **a** Atomic resolution HAADF-STEM image showing enhanced brightness at a kink boundary with a tilt angle of about 6.5° in a deformed $Mg_{97}Zn_1Y_2$ sample. **b** One isolated basal 60° <a> dislocation observed in deformed LPSO phase within the $Mg_{97}Zn_1Y_2$ alloy. The images were recorded with the electron beam parallel to the zone axis of $<11\bar{2}0>$. Symbols '⊥' indicate basal 60° <a> dislocations

## Basal Slip, Suzuki Segregation and Cottrell Atmospheres in Mg Matrix

Plastic deformation through dislocation motion also took place in the Mg matrix in both alloys, besides activation of basal slip systems in LPSO strengthening phases, since the Mg matrix is softer [22]. Figure 3 is a typical high resolution HAADF-STEM image recorded from the Mg matrix in a deformed $Mg_{97}Zn_1Gd_2$ sample with a strain of 8.0%, showing a dissociated basal **<a>** dislocation and occurrence of Suzuki segregation at the SF [28]. This kind of nanometer-sized SFs are present widely in hot deformed samples of $Mg_{97}Zn_1Y_2$ and $Mg_{97}Zn_1Gd_2$. Nanometer-sized SFs were not observed in both as-cast $Mg_{97}Zn_1Y_2$ and $Mg_{97}Zn_1Gd_2$ treated at 500 °C, since most SFs spread over whole grains, as shown in Fig. 1a. Therefore, nanometer-sized SFs like that shown in Fig. 3 were produced by dissociation of moving dislocations during hot compression of both alloys. Occurrence of Suzuki segregation increases resistance to glide, and hence has contribution to strengthening the alloys [28]. In addition, it is seen that there are some atomic columns within in the tension region of the 30° partial showing brightness higher than those in the surrounding matrix, demonstrating the formation of Cottrell atmospheres.

## Pyramidal Slip and Formation of GBs Composed of <c+a> Dislocations in Mg Matrix

Isolated **<c+a>** dislocations in Mg matrix of Mg–Zn–RE alloys were reported previously [17–20]. Interestingly, we also observed GBs formed by **<c+a>** dislocations with either extended or relatively compact cores in plastically deformed samples of $Mg_{97}Zn_1Y_2$, as shown in Fig. 4. Shown in Fig. 4a is a GB with a tilt misorientation angle of about 9.5° between basal planes of Mg lattice on both sides. Close analyses of closure failure of Burgers circuits indicate presence of **<c+a$_{60}$>** dislocations with slightly dissociated core structures at this low angle GB, as shown in Fig. 4a. These **<c+a$_{60}$>** dislocations dissociated slightly on non-basal planes, forming pairs of partials.

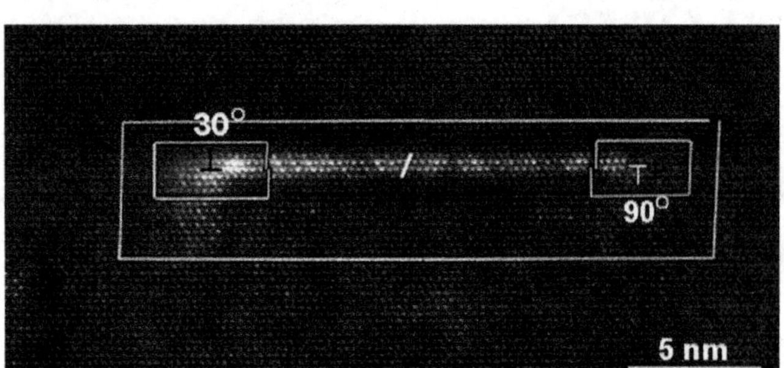


**Fig. 3** High resolution HAADF-STEM image showing dissociation of a basal 60° **<a>** dislocation and associated solute segregation observed widely in Mg grains within the $Mg_{97}Zn_1Gd_2$ alloy deformed to a final strain of 8.0% at a strain rate of $2 \times 10^{-4}$ $s^{-1}$ at 300 °C

Figure 4b shows an atomic resolution HAADF-STEM image recorded from another region in Mg grains within the deformed $Mg_{97}Zn_1Y_2$ sample. These defects are narrow five-layer basal SFs, as indicated by a white bar. The closure failure indicates that these five-layer basal SFs are associated with dislocations with a **<c>** component, as indicated by the arrow in Fig. 4b. The formation of five-layer deformation SFs on basal planes in Mg is due to dissociation of **<c+a>** dislocations with a screw **<a>** component, according to our previous studies [28]. The observation of GBs composed of **<c+a>** dislocations is a clear evidence for the activation of pyramidal slip and long-range motion of of **<c+a>** dislocations. Additionally, many SFs in Mg grains were found being cut by **<c+a>** dislocations. The activation and long range motion of **<c+a>** dislocations, on the one hand, requires higher applied stress, on the other hand, provides enough slip systems for homogeneous plastic deformation of the alloy. Therefore, it is believed that **<c+a>** dislocations play an important role in excellent mechanical properties of Mg–Zn–RE alloys.

## Sliding and Migration of GBs

Rearrangement of dislocations in deformed alloys may lead to the formation of GBs, as shown in Fig. 2a. Those newly formed GBs may serve as obstacles for dislocation motion, strengthening the materials. With further increase in plastic strain, it can be expected that newly formed GBs may be activated to move driven by the applied stress or through interaction with dislocations arriving at the GBs. Figure 5 shows images of tilt GBs formed during hot deformation, and their sliding and migration observed in $Mg_{97}Zn_1Y_2$ samples with a plastic strain of 40%. Solute-rich sub-layers on both sides of the tilt GB in Fig. 5a have a rigid-body translation along the GB, as indicated by the white steps, demonstrating occurrence of GB sliding during plastic deformation. The steps in the upper region are 2-basal-layer high, but the ones in the bottom region contain 3 basal layers; and there is an extra basal plane in the left grain, as indicated by an arrow. So, the GB sliding should be achieved through movement of non-basal dislocations with a $1/2\langle 0001\rangle$ component [29]. GB sliding can thus provide deformation along c-axis, which is beneficial for homogeneous plastic deformation of polycrystalline HCP alloys.

Figure 5b shows a high resolution HAADF-STEM image for a tilt GB with a deflection, indicating the occurrence of GB migration. No change in misorientation angle is observed along the GB. There is a relative translation of

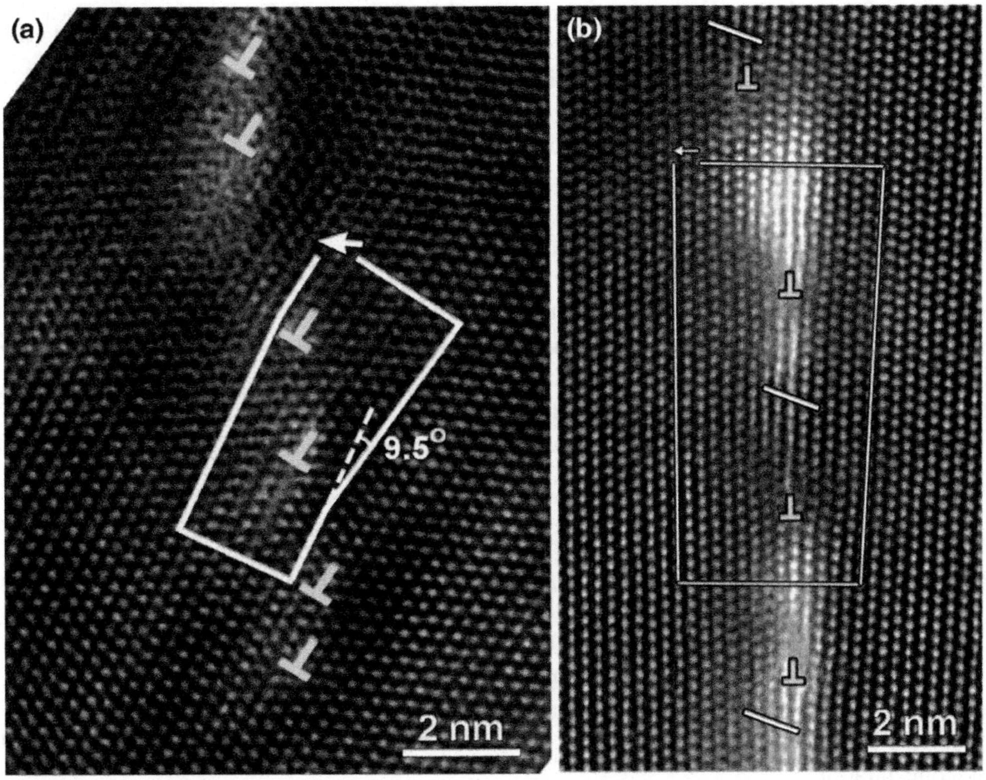


**Fig. 4** High resolution HAADF-STEM images showing formation of low-angle GBs with dissociated <c+a> dislocations in deformed $Mg_{97}Zn_1Y_2$ samples. The <a> components are 60° basal dislocations and a screw dislocations in (**a**) and (**b**), respectively

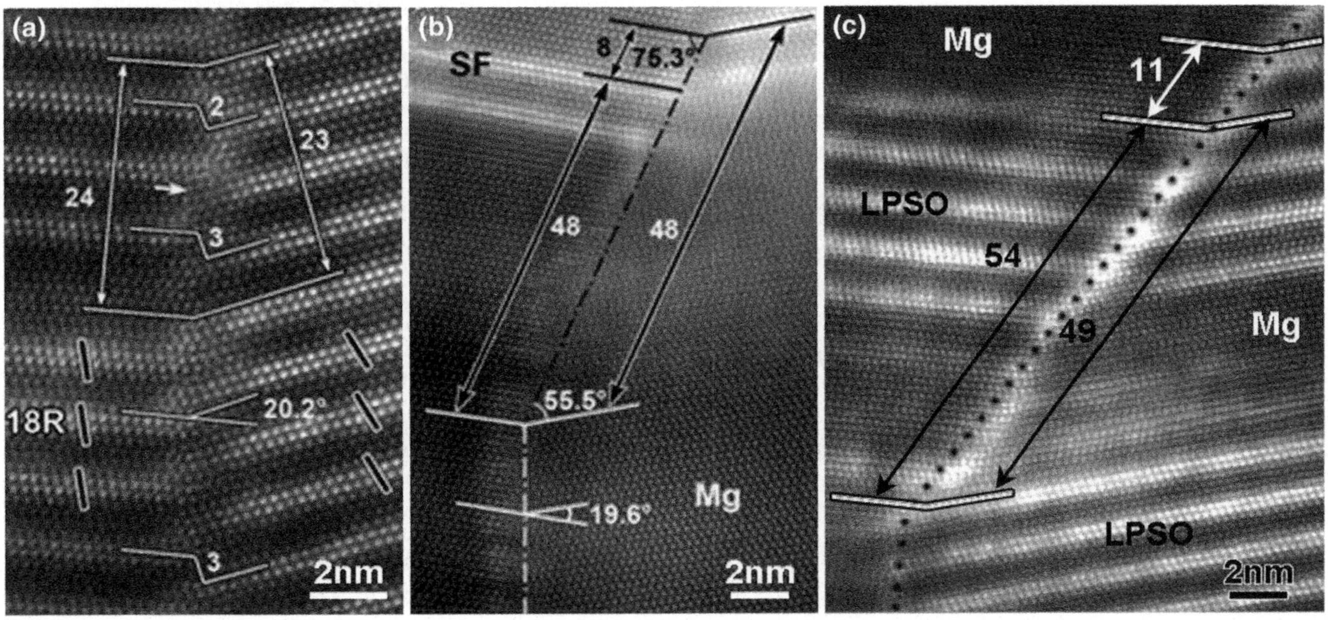


**Fig. 5** High resolution HAADF-STEM images recorded along $\langle 11\bar{2}0\rangle$ direction, showing GBs in $Mg_{97}Zn_1Y_2$ samples deformed to a strain of about 40%. **a** GB sliding, **b** GB migration, **c** simultaneous sliding and migration of GB

eight basal layers between SFs on both sides of the deflected part of the GB. However, there are both 48 basal planes between the deflection point and the upper SFs on both sides of the GB, as indicated by the double-arrowed lines, so no rigid-body GB sliding occurred. Migration of this GB should be achieved through glide of GB dislocations under the applied stress. Migration of GBs could also be associated with interactions between the GBs and **<c+a>** dislocations [29, 30].

Figure 5c is an atomic resolution HAADF-STEM image for another GB with a deflection in deformed $Mg_{97}Zn_1Y_2$ samples. The amounts of basal layers on both sided of the migrated GB are not equal within the region indicated by double-arrowed region, which is different from the case of simple GB migration shown in Fig. 5b. Therefore, it is most likely that GB sliding also occurred, besides migration. GB sliding itself would lead to local stress concentration at triple points or other obstructions along the GB [31]. GB migration also plays a role in eliminating the stress concentration caused by GB sliding at triple points as well as other obstructions along the GB, which may retard cracking at GBs [31]. Therefore, the observed GB migration is also believed important for the ductility of Mg alloys, since it can reduce local stress concentration.

## Conclusion

In summary, we studied dislocations and their interactions with other crystal defects in two Mg alloys with RE addition. **<a>** and **<c+a>** dislocations were activated during hot deformation in both alloys. Besides tilt GBs formed through rearrangement of basal **<a>** dislocations, low-angle GBs composed of **<c+a>** dislocations with either extended or relatively compacted cores were observed in deformed samples. The occurrence of long-range motion of **<c+a>** dislocations should play an important role in ductility of this kind of Mg alloys. Local solute segregation, in the form of Cottrell atmospheres or/and Suzuki segregation, was observed in LPSO strengthening phases or Mg matrix of deformed samples. The solute atmospheres may have higher pinning effects on dislocations compared with randomly distributed individual solute atoms, which can play a role in improving the strength of the alloys. Additionally, both sliding and migration of newly formed GBs occurred, which should also have contribution to ductility of the alloys.

**Acknowledgements** This work was supported by the National Natural Science Foundation of China (51371178, 51390473 and 51771202), Key Research Program of Frontier Sciences, CAS (QYZDY-SSW-JSC027).

## References


1. Luo AA, Recent magnesium alloy development for automotive powertrain applications, in: Kojima Y, Aizawa T, Higashi K, Kamado S (eds.), Magnesium Alloys 2003, Pts 1 and 2, vol. 419-4, Zurich-Uetikon: Trans Tech Publications Ltd, 2003, pp. 57–65.
2. Pollock TM, Science 328 (2010) 986–987.
3. Yoo MH, Metall. Mater. Trans. A 12 (1981) 409–418.
4. Agnew. SR, Yoo. MH, Tome CN, Acta Mater. 49 (2001) 4277–4289.
5. Koike J, Metall. Mater. Trans. a-Phys. Metall. Mater. Sci. 36A (2005) 1689–1696.
6. von Mises R. Z. Angew, Math. Mech. 8 (1928) 161–185.
7. Hirth JP, Lothe J, Theory of Dislocations, John Wiley & Sons, New York, 1982.
8. Somekawa H, Singh A, Mukai T. Philos. Mag. Lett. 89 (2009) 2–10.
9. Guo F, Zhang D, Yang X, Jiang L, Chai S, Pan F, Mater. Sci. Eng. A 607 (2014) 383–389.
10. Kuang J, Li X, Zhang R, Ye Y, Luo AA, Tang G, Mater. Des. 100 (2016) 204–216.
11. Onuki Y, Hara K, Utsunomiya H, Szpunar JA, J. Mater. Eng. Performance 24 (2015) 972–985.
12. Price PB, J. Appl. Phys. 32 (1961) 1750–1758.
13. Groves GW, Kelly A, Philos. Mag. 8 (1963) 877–887.
14. Feaugas X, Clavel M, Acta Mater. 45 (1997) 2685–2701.
15. Wu Z, Curtin WA, Scripta Mater. 116 (2016) 104–107.
16. Wu Z, Curtin WA, Nature 526 (2015) 62–67.
17. Chino Y, Kado M, Mabuchi M, Acta Mater. 56 (2008) 387–394.
18. Agnew SR, Horton JA, Yoo MH, Metall. Mater. Trans. a-Phys. Metall. Mater. Sci. 33 (2002) 851–858.
19. Agnew. SR, Yoo. MH, Tome CN, Acta Mater. 49 (2001) 4277–4289.
20. Sandlöbes S, Zaefferer S, Schestakow I, Yi S, Gonzalez-Martinez R, Acta Mater. 59 (2011) 429–439.
21. Itoi T, Inazawa T, Kuroda Y, Yamasaki M, Kawamura Y, Hirohashi M, Mater. Lett. 64 (2010) 2277–2280.
22. Shao XH, Yang ZQ, Ma XL, Philos. Mag. Lett. 94 (2014) 150–156.
23. Shao XH, Yang ZQ, Ma XL, Acta Mater. 58 (2010) 4760–4771.
24. Pennycook SJ, Annu. Rev. Mater. Sci. 22 (1992) 171–195.
25. Pennycook SJ, Ultramicroscopy 30 (1989) 58–69.
26. Hu WW, Yang ZQ, Ye HQ, Scripta Mater. 117 (2016) 77–80.
27. Thompson K, Flaitz PL, Ronsheim P, Larson DJ, Kelly TF, Science, 317 (2007) 1370–1374.
28. Yang ZQ, Chisholm MF, Duscher G, Ma X, Pennycook SJ, Acta Mater. 61 (2013) 350–359.
29. Hu WW, Yang ZQ, Ye HQ, Adv. Eng. Mater. DOI:https://doi.org/10.1002/adem.201700516.
30. Hu WW, Yang ZQ, Ye HQ, Acta Mater. 124 (2017) 372–382.
31. Sherby OD, Wadsworth J., Prog. Mater. Sci. 33 (1989) 169–221.

# Microstructure, Mechanical Properties and Deformation Behavior of Mg–Gd–Zn Alloy

K. Li, V. S. Y. Injeti, P. Trivedi, and R. D. K. Misra

**Abstract**

We describe here the microstructure, mechanical properties and deformation behavior of an ultrafine-grained (UFG) Gd and Zn-containing magnesium alloy that was characterized by high strength-high ductility combination. The deformation behavior was studied by nanoindentation and post-mortem electron microscopy analysis of the deformed region. The behavior is compared with low strength-low ductility coarse-grained (CG) counterpart. Extensive dislocation slip was an active deformation mechanism in the UFG alloy, while in contrast, mechanical twinning occurred in the CG alloy. We attribute these observed differences in the deformation mechanism to the grain size effect.

**Keywords**

Ultrafine-grained • Deformation • Structure, magnesium-rare earth alloy

## Introduction

Magnesium and its alloys are characterized by high specific strength and specific stiffness [1–3]. However, their practical usage is restricted because of their hexagonal close-packed (HCP) structure with limited slip systems. This characteristic renders Mg alloys difficult to plastically deform at ambient temperature. The strength of Mg-rare earth (RE) alloys can be enhanced via precipitation strengthening. Refinement of grain size is another method that increases the strength of metal and alloys. Recently, ultrafine-grained (UFG) structure was obtained by us in a Mg–2Zn–2Gd alloy that was intriguingly characterized by high strength-high ductility through an approach involving annealing and multiaxial forging (MAF). The processing details of the multiaxially forged alloy are described elsewhere [1, 4]. The objective here is to describe the relationship between grain size and deformation mechanism in Mg–2Zn–2Gd alloy.

## Experiments

The as-cast alloy + annealed alloy and multiaxially forged (MAF) alloy samples were cut into specimens of 5 mm 5 mm × 5 mm dimensions and metallography polished to mirror finish, followed by etching for ~2 min using a solution consisting of 8 g picric acid, 5 ml acetic acid, 10 ml distilled water, and 100 ml ethanol. Electron microscopy (Hitachi H-9500) was carried out at 300 kV using 3 mm disks, electropolished in a solution containing 3% perchloric acid in ethanol.

To study the deformation mechanism, nanoindentation experiments were carried out at a constant loading rate of 2 μN/s, with maximum load set to 0.5 mN using a Berkovich three-sided pyramidal diamond indenter that had a nominal angle of 65.3° and indenter diameter of 20 nm. An array of indents of matrix 10 × 10 was made with the indent gap of 10 μm. After the indentation experiments, the disk were electropolished from the side opposite to the indented surface. This procedure enabled the indents to be examined by TEM, as described previously [1, 5]. Tensile properties were measured at room temperature at a strain rate of $2 \times 10^{-3}\ s^{-1}$.

## Results and Discussion

The microstructure of the as-cast + annealed and as-cast + annealed + MAF alloy is presented in Fig. 1. During annealing, as expected, grain growth took place and the grain size was increased from 25 μm in the as-cast alloy to 44 ± 5 μm in the annealed alloy [1]. But after multiaxial

K. Li · V. S. Y. Injeti · P. Trivedi · R. D. K. Misra (✉)
Department of Metallurgical, Materials and Biomedical Engineering, University of Texas at El Paso, 500 W. University Avenue, El Paso, TX 79968-0521, USA
e-mail: dmisra2@utep.edu


D. Orlov et al. (eds.), *Magnesium Technology 2018*, The Minerals, Metals & Materials Series, https://doi.org/10.1007/978-3-319-72332-7_29

forging (MAF), the grain size of the annealed alloy was dramatically refined to 416 ± 140 nm. We refer as-cast + annealed alloy as coarse-grained (CG) alloy (average grain size ~44 ± 5 μm) and multiaxially forged (MAF) alloy as UFG alloy (average grain size ~416 ± 140 nm) [1]. The morphology and precipitates present in the α-Mg matrix are illustrated in Fig. 1c,d. The precipitates in the α-Mg matrix were ~50–80 nm in diameter and were characterized to be $Mg_3Zn_3Gd$ with orientation relationship of $(2\bar{2}01)[\bar{1}2\bar{1}6]_{\alpha-Mg}$ $//(220)[\bar{1}11]_{Mg_3Zn_3Gd}$, from the analysis of the diffraction pattern.

The mechanical properties of CG and UFG alloys are listed in Table 1. The UFG alloy had yield strength of 227 MPa, and elongation of 30%. These properties are excellent for the lean Mg–2Zn–2Gd alloy.

Post-mortem TEM analysis of plastically deformed region surrounding the indentation for UFG and CG Mg–2Zn–2Gd alloy are presented in Figs. 2 and 3, respectively. In the UFG alloy, there was appreciate activity in ~60% of the grains. There were high density of dislocations in the UFG alloy. The dislocation activity was characteristic of $1/3<11\bar{2}3>$ pyramidal slip system, observed in the two-beam condition of g = [0002]. $1/3 <11\bar{2}0>$ basal slip system was not observed in this condition. The observation of <c+a> dislocation slip alters the active deformation mode on the pyramidal planes in Mg–2Zn–2Gd alloy [1].

Figure 2 shows an intriguing illustration of dislocation structure associated with nanoindentation (Fig. 2a-top right corner). The indented grain was out of contrast in the TEM imaging, but the dark grain in Fig. 2a showed a very high dislocation activity (which is reverted to a bright grain in the dark-field image in Fig. 2b using the [0002] operating reflection in the SAED pattern of Fig. 2c [1]. In the bottom left grain in Fig. 2a, dislocation profiles (zones 3A, 3B) are shown being emitted from the grain boundary, while the overall dislocation density in this grain is significantly low compared to that in the central (dark) grain in Fig. 2a. We envisage that the logical sequence of events would have a very large number of dislocations emitted by the nanoindentor in Fig. 2a. This would create a high density of dislocation pile-ups against the upper grain boundary, and emitting a high density of similar dislocation profiles in the dark grain. These profiles act as pile-ups against the opposite (lower) grain boundary, and activate only two sources which emit dislocation profiles shown at 3A and 3B in Fig. 2a; as an indication of only minimal residual stress (or strain) at the lower grain boundary [1].

In striking comparison to the UFG alloy, $\{10\bar{1}1\}$ nanoscale twinning was an active deformation mechanism in the CG alloy. But the frequency of observation of twinning was less at ~15% of all grains. Examples of twinning are presented in Fig. 3 [1]. From the analysis of diffraction

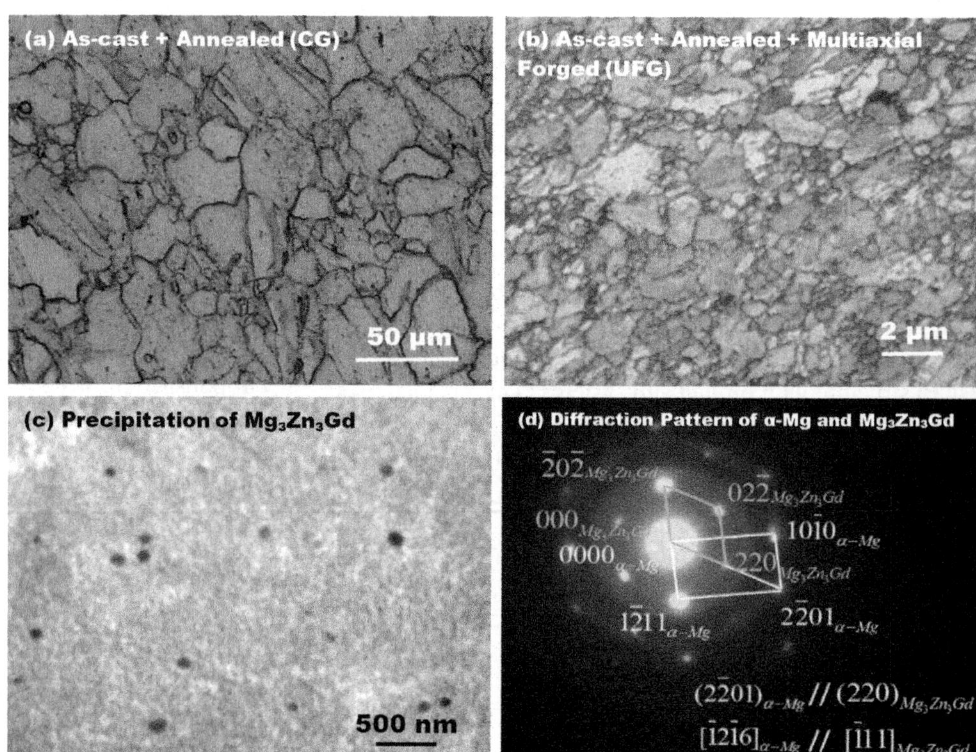


**Fig. 1** **a** Microstructure of as-cast + annealed alloy coarse-grained (CG) alloy and **b** microstructure of as-cast + annealed + multiaxially forged ultrafine-grained (UFG) Mg–2Zn–2Gd alloy (**b**). **c** TEM micrograph illustrating $Mg_3Zn_3Gd$ precipitates in α–Mg matrix that were confirmed by (**d**) electron diffraction pattern [1]

**Table 1** Average grain size and mechanical properties of CG and UFG Mg–2Zn–2Gd alloys [1]

| Property | Coarse-grained (CG) alloy | Ultrafine-grained (UFG) alloy |
|---|---|---|
| Grain size | 44 ± 5 μm | 416 ± 140 nm |
| YS (MPa) | 46 | 227 |
| UTS (MPa) | 70 | 272 |
| % Elongation | 7 | 30 |

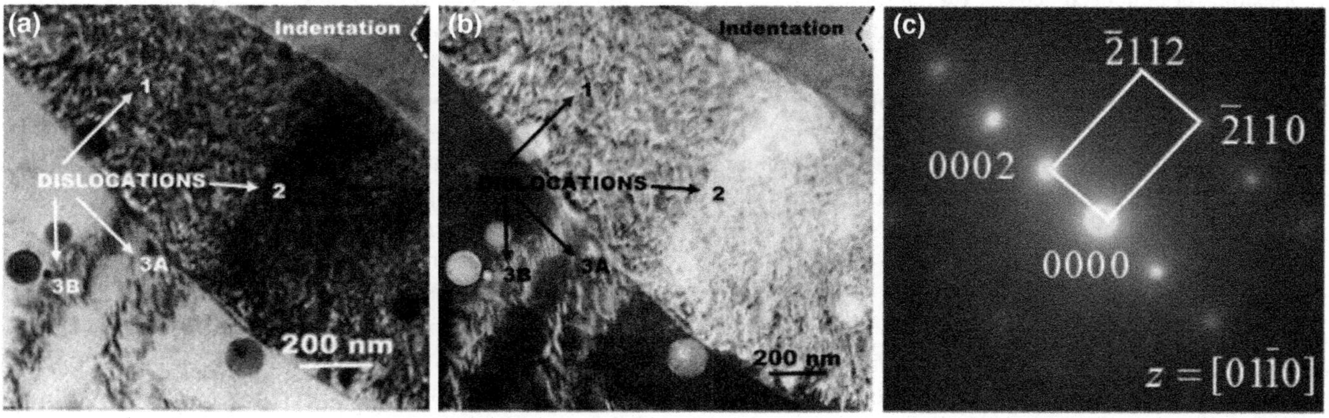


**Fig. 2** **a** Bright field and **b** dark field TEM images showing high density of dislocations in the plastic zone surrounding the indentation in ultrafine-grained (UFG) Mg–2Zn–2Gd alloy. Zones 1 and 2 have high density of dislocations and zones 3A and 3B illustrate dislocations emitted by the grain boundary. **c** Selected area electron diffraction pattern for grain in (**a**) and (**b**) showing [0002] showing operating reflection [1]

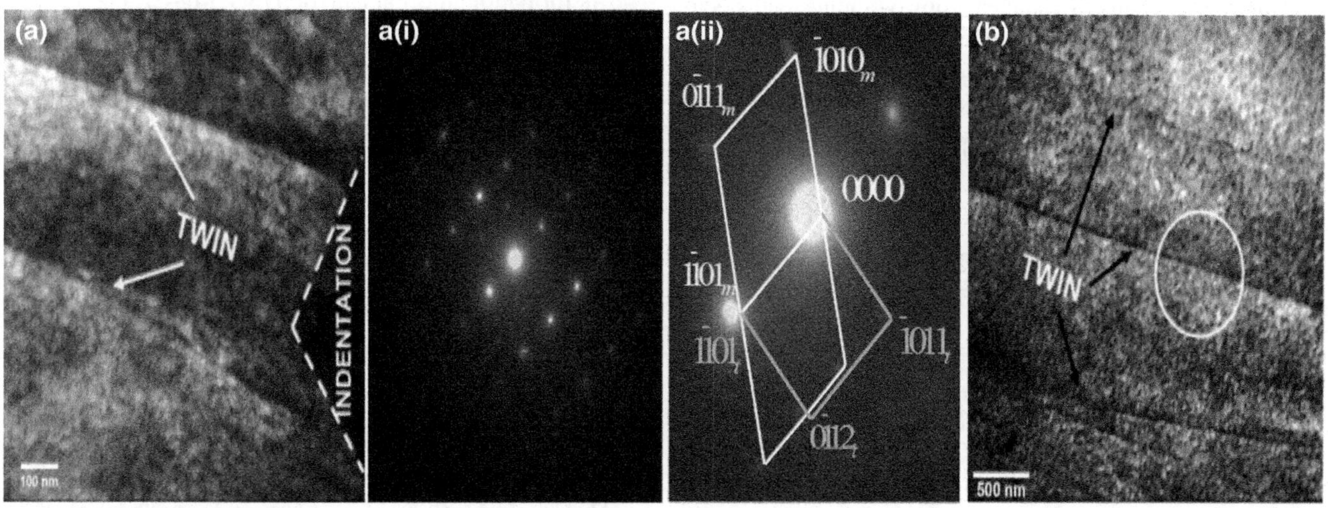


**Fig. 3** **a** TEM images of the plastically deformed zone surrounding the indentation, illustrating twinning in the CG Mg–2Zn–2Gd alloy and the diffraction pattern. **b** Second illustration of twinning in the CG alloy. From the diffraction pattern, the twin $\{1\bar{1}01\} < 0\bar{1}12 >$ belongs to the family of $\{10\bar{1}1\} < 10\bar{1}12 >$ compression twin. a(i) and a(ii) show original and enlarged (and indexed) selected area electron diffraction pattern for (**a**) [1]

pattern (Fig. 3a (i), (ii)), the twin was $\{1\bar{1}01\} < 0\bar{1}12 >$ and belongs to the family of $\{10\bar{1}1\} < 10\bar{1}2 >$ compression twinning, in line with previous work on CG Mg alloys [6, 7].

Twins during deformation accommodate the applied strain. Studies on 100 μm grain size Mg–Li, Mg–Zn, Mg–Al alloys at a strain rate of 1/s suggested that rapid strain hardening took place because of twinning. This was a consequence of activation of <c+a> dislocations and interactions between dislocations and twin boundaries [1].

We can conclude from Figs. 2 and 3 that there was a change of deformation mechanism from twinning in the CG alloy to extensive dislocation activity in the UFG alloy. This suggests grain size effect, where twinning is suppressed in the UFG Mg–RE alloy. This is in agreement with the recent study of Tsai and Chang [8] in Mg–Al–Zn alloy, and the work of Kumar et al. [9, 10] in hcp crystals. The present study underscores that grain refinement and nanometer-sized precipitates in the UFG multiaxially forged alloy contributed to high ductility in the high strength Mg–2Zn–2Gd alloy [1].

Modified Peierls potential [1, 11, 12] and/or change in stacking fault energy [1, 12–14] can be also considered to be responsible for enhanced activity on non-basal planes in the UFG alloy. The Peierls potential is defined as the energy required to move a dislocation from one stable position to the next [1, 12]. But Peierls potential is affected only to a small extent for activation of partial dislocations and twinning, hence this aspect is debatable. One can also suggest other reasons that contributed to the high ductility of UFG alloy. Literature suggests that stacking fault energy (SFE) increases with decrease in grain size because the dissociation of triple dislocation nodes diminishes with decrease in grain size [1, 15]. But SFE is only affected by the chemical composition and not by grain size, because grain boundaries have no role in *ab-initio* models for SFE. Thus, we cannot consider that SFE changes with grain size.

It is clear from the observations presented in Figs. 2 and 3 that the change in the deformation from twinning in the CG alloy to high dislocation activity in the UFG alloy, is a grain size effect and there must be a critical grain size at which the deformation mechanism changes from twinning to dislocation slip. This may involve change in activation volume [1].

The activation volume V*, given by [16, 17]:

$$V^* = \sqrt{3} * kT\left(\frac{\partial ln\dot{\varepsilon}}{\partial\sigma}\right) = \frac{3\sqrt{3} * kT}{mH} \tag{1}$$

where $k$ is Boltzmann's constant, T is the temperature and $m$ is the strain-rate sensitivity. Using Ref. [17], the activation volume of CG and UFG Mg–RE alloys is 45–105$b^3$ and 25–81$b^3$, respectively, and are different. Therefore, the activation volume V* may be a factor that impacts the deformation mechanism in Mg–2Zn–2Gd alloys with different grain size [1]. It is proposed that the deformation mechanism of Mg-rare earth alloys with different grain size might be related to the plastic zone surrounding the nanoindentation (which is ~5 or 20 μm for the quasi-static and dynamic indentations, respectively). The critical grain size, $d_{cri}$, at which the dominant deformation mechanism changes from twinning to dislocation slip [1] can be estimated by [18]:

$$d_{cri} = \left(\frac{0.15lnZ - 12.2}{73 - 3.8lnZ}\right) \tag{2}$$

where the Z-parameter equals $\dot{\varepsilon} \times \exp(Q/RT)$. Q is the apparent activation energy for lattice diffusion in Mg, and $\dot{\varepsilon}$ is expressed as a creep power-law for the low applied stress of nanoindentation [19, 20]:

$$\dot{\varepsilon} = A\left(\frac{\sigma}{G}\right)^n D_0 exp\left(\frac{-Q}{RT}\right) \tag{3}$$

where $A$ is a constant, $n$ value is the dislocation climb (=4–10 for magnesium) and G is the shear modulus. The critical grain size for Mg–RE alloy at room temperature is ~4 μm [21]. When the grain size is finer than the critical grain size, the deformation mechanism is dislocation slip, while the deformation mechanism alters to twinning, when the grain size is greater than the critical grain size [1]. The grain size of UFG alloy was 416 ± 140 nm, which is smaller than the plastic zone size, while the grain size of CG alloy was significantly larger at 44 ± 5 μm. Thus, the UFG alloy showed multiple dislocation slip because of the need for compatibility near the grain boundaries. In the CG alloy, deformation twinning compensated for the lack of dislocation slip [1]. But less frequency of twinning in the CG alloy suggests reduced tensile elongation of ~7% in comparison to the high elongation of ~30% in the UFG alloy. The UFG alloy with large number of grain boundaries promoted dislocation slip [1].

## Conclusions

The deformation mechanism of UFG Mg–2Zn–2Gd alloy with high strength-high ductility combination was different from the CG counterpart, and is related to the grain size effect. In UFG alloy, extensive dislocation slip contributed high ductility, whereas the extent of twinning dictated ductility in the CG alloy.

**Acknowledgements** The authors gratefully acknowledge S. Goel, R. Jayganthan, and A. Srinivasan for their help with processing of magnesium alloy.

## References

1. K. Li, V.S.Y. Injeti, P. Trivedi, L.E. Murr, R.D.K. Misra, J. Mater. Sci. Technol. https://doi.org/10.1016/j.jmst.2017.07.023.
2. D. Qiu, M.X. Zhang, J.A. Taylor, P.M. Kelly, Acta Mater. 57 (2009) 3052–3060.
3. M. Yamasakia, K. Hashimoto, K. Hagihara, Y. Kawamura, Acta Mater. 59 (2011) 3646–3656.
4. P. Trivedi, K.C. Nune, R.D.K. Misra, S. Goel, R. Jayganthan, A. Srinivasan, Mater. Sci. Eng. A, 668 (2016) 59–64.
5. R.D.K. Misra, V.S.A. Challa, P.K.C. Venkatsurya, Y.F. Shen, M. C. Somani, L.P. Karjalainen, Acta Mater. 84 (2015) 339–345.
6. S. Sandlöbes, S. Zaefferer, I. Schestakow, S. Yi, R. Gonzalez-Martinez, Acta Mater. 59 (2011) 429–436.
7. S. Sandlöbes, M. Friák, J. Neugebauer, D. Raabe, Mater. Sci. Eng. A 576 (2013) 61–65.
8. M.S. Tsai and C.P. Chang, Mater. Sci. Technol., 29 (2013) 759–763.

9. M.A. Kumar, I.J. Beyerlein, R.J. McCabe and C.N. Tone, Nature Communications, 7:13826 (2016) 1–8.
10. M.A. Kumar, I.J. Beyerlein, and C.N. Tone, J. Appl. Phys., 120 (2016) 155105–155111.
11. S. Ando, M, Tanaka, H. Tonda, Mater. Sci. Forum 419–422 (2003) 87–93.
12. H. Yan, R. Chen, N. Zheng, J. Luo, S. Kamado, E. Han, J. Magnesium Alloys 1 (2013) 23–29.
13. W.Y. Wang, S.L. Shang, Y. Wang, Mater. Res. Lett. 2(2014) 29–36.
14. S. Sandlöbes, M. Friák, S. Zaefferer, Acta Mater. 60 (2012) 3011–3021.
15. J.H. Dun and C.S. Choi, Mater. Sci. Eng. A, 257 (1998) 353–356.
16. H. Pan, G. Qin, Y. Huang, Q. Yang, J. Alloy Compd. 688 (2016) 149–156.
17. H. Somekawa, C.A. Schuh, Scripta Mater. 68 (2013) 416–419.
18. M.R. Barnett, Z. Keshavarz, A.G. Beer, Acta Mater. 52 (2004) 5093–5099.
19. H.J. Frost, M.F. Ashby: Deformation Mechanism Map, Pergamon press, Oxford, 1982.
20. S.X. Song, J.A. Horton, N.J. Kim, T.G. Nieh, Scripta Mater. 56 (2007) 393–396.
21. H. Somekawa, C.A. Schuh, J. Mater. Research, 27 (2012) 1295–1302.

# Twin-Slip Interaction at Low Stress Stage Deformation in an AZ31 Mg Alloy

Peng Chen, Bin Li, Duke Culbertson, and Yanyao Jiang

**Abstract**

Extruded magnesium alloys with strong basal texture present tension and compression asymmetry. Dislocation slip dominates plastic deformation during tension along the extrusion direction (ED), whereas twinning is the main contributor to plastic strain when compressed along the ED. In this work, an extruded AZ31 Mg alloy was prestrained by tension along the ED to 5 and 10% of total strain, followed by compression, in order to investigate twin-slip interaction. The results show that the yield stress in compression only slightly increases with increasing prestrain. Notably, the hardening rate at the low stress stage during compression remains almost unchanged, compared to specimens without prestrain. Our results suggest that the contribution of twin-slip interaction to hardening is negligible in deformation of Mg alloys.

**Keywords**

Magnesium alloy • Prestrain • Twin-slip interaction Work hardening

## Introduction

Deformation twinning and dislocation slip are two important modes during plastic deformation in hexagonal-close-packed (hcp) metals. Especially, the $\{10\bar{1}2\}10\bar{1}\bar{1}$ twinning mode is the most prevalent twinning mode in hcp metals [1–4]. The favorable stress state for the $\{10\bar{1}2\}$ extension twinning can be achieved by tension along c-axis or compression perpendicular to c-axis. In situ neutron diffraction [5–8], in situ X-ray diffraction [9] as well as acoustic emission [6, 10] studies on early stage of plastic deformation in Mg alloys reveal that $\{10\bar{1}2\}$ extension twinning accounts for a major portion of the plastic deformation. Recent measurement of contribution of $\{10\bar{1}2\}$ twinning to plastic strain in an extruded Mg alloy also indicated that $\sim 90\%$ of the plastic strain comes from $\{10\bar{1}2\}$ twining during low stress stage deformation [11]. Extruded Mg alloys with strong basal texture exhibit tension-compression asymmetry. Yield stress and strain hardening behavior are largely dependent on the dominant mechanism of plastic deformation. During tension along the extrusion direction (ED), dislocation slip dominates plastic deformation. However, when compressed along the ED, $\{10\bar{1}2\}$ twinning is the main contributor during early stage of plastic deformation in Mg alloys. Thus, the yield strength under ED compression is lower than that under ED tension. Also, the strain hardening rate is higher during ED tension compared with ED compression.

A number of researchers investigated the influence of prestrain on flow curve and work hardening behavior [12–16]. Slip-slip [17–20], twin-twin [21–23] and slip-twin interactions [24–27] during deformation were studied. The slip-slip interaction take place during plastic flow, especially when it is accomplished by dislocation glide on multiple slip systems. The interaction generally gives rise to a reduction in mobile dislocations, thus promoting work hardening [28]. The sub-grains resulted from twinning and twin-twin interaction also provide a strengthening mechanism, which is attributed to the dynamic Hall-Petch effect [29]. The twin-slip interaction is a more complex mechanism and still not well understood. Serra et al. [30] considered twin-dislocation interaction as an important hardening mechanism because twin boundaries hinder subsequent slip and twinning deformation. Proust et al. [31] proposed that during the preloading, <a> and <c+a> dislocations are

P. Chen (✉) · B. Li
Department of Chemical and Materials Engineering, University of Nevada, Reno, USA
e-mail: pchencor@gmail.com

P. Chen · B. Li
Nevada Institute for Sustainability, University of Nevada, Reno, USA

D. Culbertson · Y. Jiang
Department of Mechanical Engineering, University of Nevada, Reno, USA


D. Orlov et al. (eds.), *Magnesium Technology 2018*, The Minerals, Metals & Materials Series, https://doi.org/10.1007/978-3-319-72332-7_30

introduced. These dislocations may act as barriers to twin nucleation or/and twin propagation. El Kadiri and Oppedal [32] developed dislocation transmutation theory, which hypothesizes that dislocations inside parent are incorporated and transmuted into immobile defects in twins, promoting latent hardening. Some researchers suggest when lattice dislocations meet with a twin boundary, dislocation dissociation will occur, which is also a mechanism for enlarging or shrinking the twin domain [33].

The current work aims to investigate twin-slip interaction by taking the advantage that slip and twinning dominated deformation can be activated separately by controlling the loading direction on extruded specimens. By pre-tension of extruded specimens to various levels of plastic strain, dislocations of different density are first introduced. Then the deformation is changed to compression such that twinning is activated. As such, twin-slip interaction and its contribution to strain hardening can be studied without ambiguity.

## Experimental Method

An extruded, commercially obtained AZ31B Mg alloy was used for the current investigation. The circular extruded bar had a diameter of 38.1 mm. Dog-bone specimens were machined from the extruded bar with the gage section along the extrusion direction. The specimens had a gage length of 13.0 mm and gage diameter of 9.0 mm. An Instron load frame with a loading capacity of 25 kN was used to conduct the experiments. The specimens were carefully aligned to avoid triggering early buckling. Lubrication was applied between the moving surfaces. The extruded AZ31 Mg alloy was first prestrained by tension along the ED to 5 and 10% of total strain, respectively, followed by consecutive compression along the ED. The reason we did not pre-tensile the specimen to higher strain levels lies in that the fracture strain is approximately 12% for the material under tension. The mechanical experiments were conducted in strain-control mode at a strain rate of $8 \times 10^{-4}$ at ambient temperature.

For electron backscatter diffraction (EBSD) scans, cylindrical specimens with a thickness of 5 mm were sectioned in the middle of the gage section perpendicular to the ED. The specimens were ground mechanically down to 1200/4000 grit number on SiC sand papers, followed by electrochemical polishing with a solution of 5% nitric acid, 0.5% perchloric acid and 94.5% ethanol at 20 V for ~20 s. EBSD scans were conducted on a JEOL 7100F field emission scanning electron microscope (SEM) with an Oxford HKL Channel 5 instrument. An acceleration voltage of 20 kV and a working distance of 25 mm were used. For better statistics and analysis of the twinning behavior, at least three different regions were scanned by EBSD with a step size ranging from 0.5 to 1 μm for each companion specimen. All the EBSD scans were performed on a cross-section perpendicular to the ED or the loading axis.

## Results and Discussion

The initial grain structure and the texture of the as-extruded specimens are shown in Fig. 1a, b, respectively. The as-received AZ31B presents a typical rod-texture in which the (0001) basal pole is nearly perpendicular to the ED, and the $\{10\bar{1}0\}$ pole presents a strong intensity along the ED. Intensity peaks appear at ~4 and 10 o'clock positions in (0001) PF in Fig. 1b, which is likely due to the inhomogeneity in microstructure and the region for EBSD scan off the center of the extruded bar.

The stress-strain curves for 5% prestrain and 10% prestrain are displayed in Fig. 2. Firstly, we prestrained the specimens to 5% and 10% total strain in tension, then subjected the specimens to compression. The red dots denote strain levels at which the compression was interrupted and EBSD scans were performed on the specimens. The

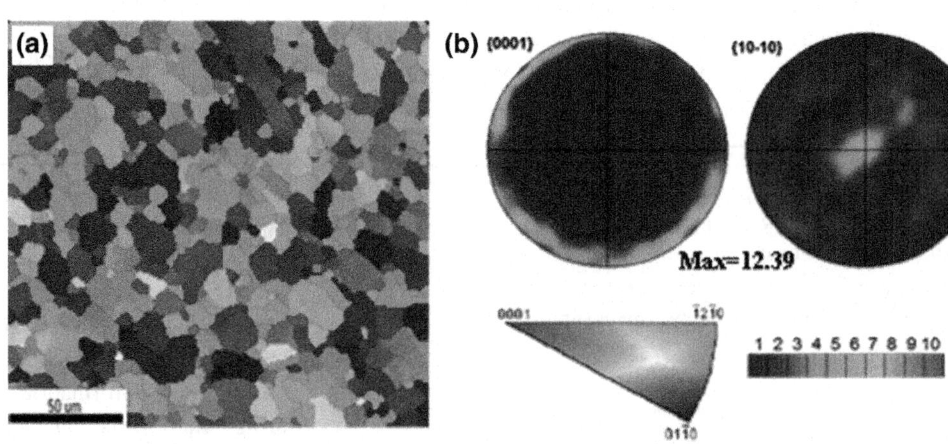


**Fig. 1** Initial texture of as-extruded AZ31 alloy. The scanned surface is perpendicular to the ED. **a** Inverse pole figure (IPF) map. **b** Pole figures. The basal plane is mostly parallel to the ED, which is a typical rod-texture in extruded Mg alloys

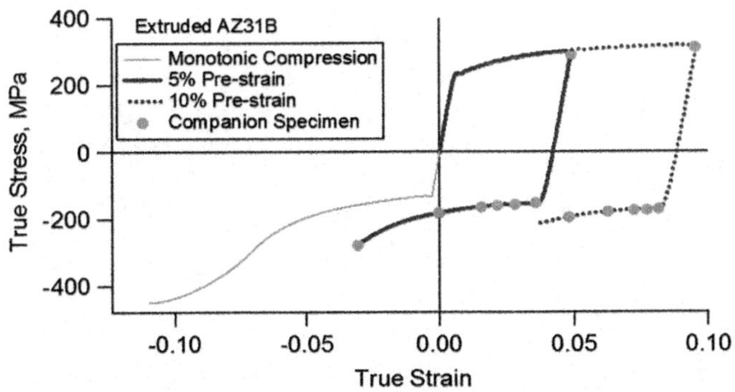


**Fig. 2** Stress-strain curves of monotonic compression (thin solid line), 5% pre-strained (solid line) specimen and 10% pre-strained (dash line) specimen. The red dots denote strain levels at which the compression was interrupted for EBSD analysis

influence of prestrain on the compression behavior is shown in Fig. 3 which compares the stress-strain curves for 0% (i.e. monotonic compression), 5 and 10% prestrains.

Several important features can be noted in Fig. 3. First, the yield stresses in compression of the prestrained specimens are slightly higher than the specimen without prestrain. As the prestrain increases, the yield stress in compression slightly increases. In our most recent measurements of twin volume fraction at low stress stage compression of extruded AZ31 specimens [11], we observed that at plastic strain ~0.25%, deformation twinning has already been activated but with a low volume fraction. The main contributor to plastic strain is dislocation slip. But the twin volume fraction rapidly increases as the strain increases. During the low stress stage deformation, twinning accounts for 80–90% contribution to the plastic strain. Thus, the increase in yield stress in Fig. 3 can be attributed to the hardening effect of the pre-existing dislocations generated in prestraining.

Second and most interestingly, it can be observed that the hardening behavior during the low stress stage deformation remains almost identical. Figure 4 plots the hardening rate in the three scenarios. The difference between the zero-prestrain, 5% prestrain and 10% prestrain specimens is obvious. The prestrained specimens present slightly lower hardening rates if compared to the un-prestrained specimen. This raises a question as to what a role twin-slip interaction plays in the hardening. Twin-slip interaction has been considered as an important contributor to the increase in hardening rate during twinning and after twinning [13, 30, 34, 35]. Dynamic Hall-Petch effect [36–38] has been proposed as one of the mechanisms that contributes to strain hardening when twinning is activated along with dislocation slip. Twin boundaries act as new grain boundaries that reduce the effective grain size and the mean free path of dislocations. However, our experimental results indicate that the contribution of twin-slip interaction to the hardening is negligible.

The EBSD results in Fig. 5 show the evolution of extension twins of the specimens with and without prestrain. In the zero-prestrain specimen (Fig. 5a), which is under monotonic compression at low stress stage of plastic deformation in an extruded Mg alloy [11], $\{10\bar{1}2\}$ twinning is activated at very early stage of plastic deformation, the density of twins increases rapidly with increasing strain. The specimen with 5% prestrain (Fig. 5b) presents a similar trend: as the plastic strain increases, the twin volume fraction increases. However, for the 10% prestrained specimen, substantial difference in twinning behavior upon subsequent tension was observed, that is, twin nucleation is retarded by the pre-introduced dislocations. It can be observed that at $\varepsilon_p = 1.22\%$ (Fig. 5c), twins with a very low density are activated. However, after nucleation, the twin volume

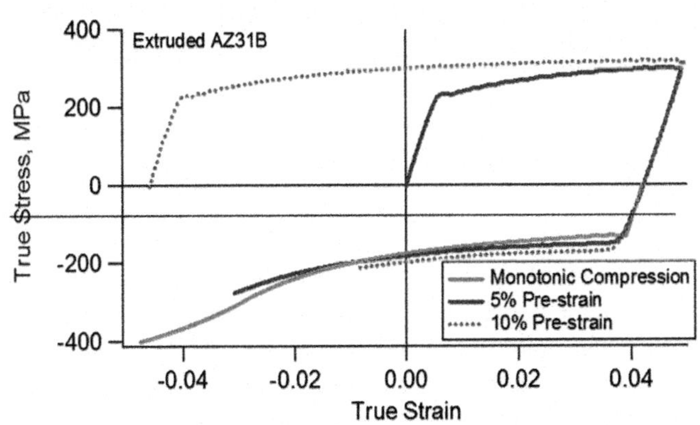


**Fig. 3** Comparison of stress-strain curves for 0% (monotonic compression), 5 and 10% prestrain. The yield stress under compression only slightly increases with increasing pre-strain. However, the hardening rate at the low stress stage during compression remains almost the same

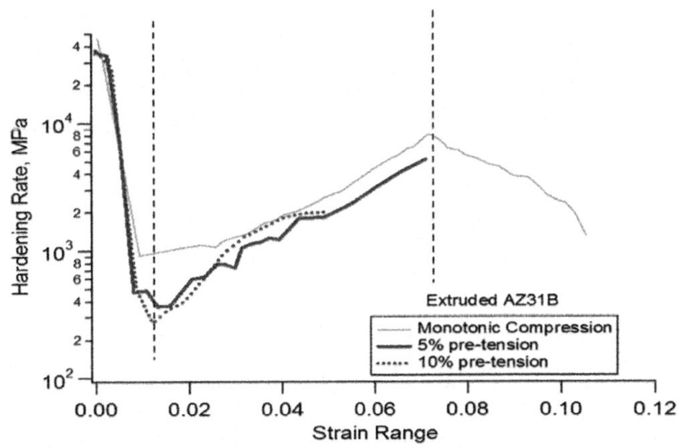


**Fig. 4** The strain hardening rate as a function of true strain during compression along the ED. After yielding, the strain hardening rates present the lowest value which is attributed to the twin nucleation. It is worth noting that the prestrained specimens present lower hardening rate than monotonic compression (0% prestrain)

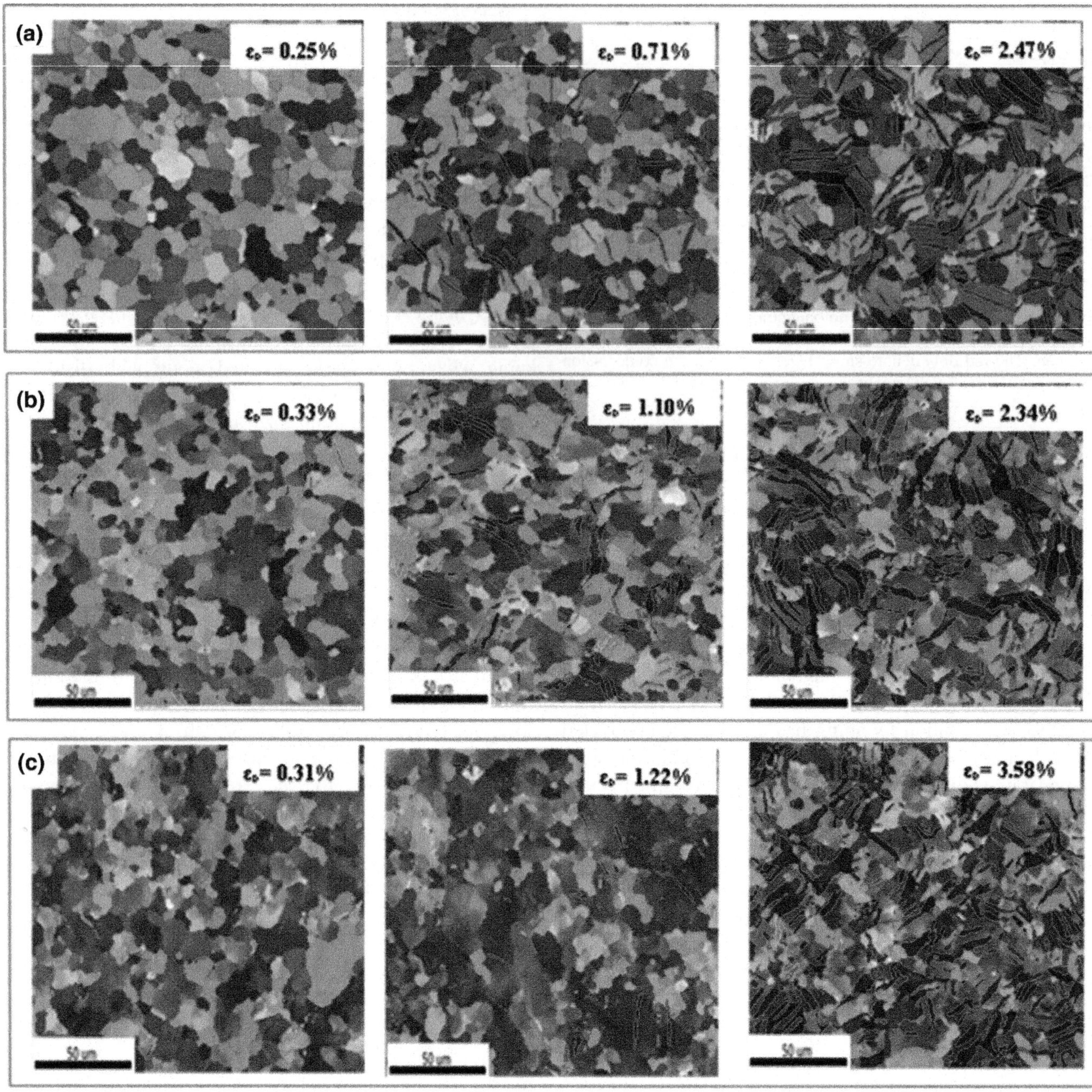


**Fig. 5** Inverse pole figure (IPF) maps at selected plastic strain levels for: **a** 0% pre-strain (monotonic compression); **b** 5% pre-strain; **c** 10% pre-strain. Compared with 5% pre-strain and monotonic compression, the number of twins is less at the early stage ($\varepsilon_p < 2.18\%$). Obviously, twin nucleation is impeded under 10% pre-strain

fraction rapidly increases as does in the zero-prestrain and the 5% prestrain specimens. Mahajan [39] reported that prestraining of iron prior to shock-loading inhibits the formation of shock twins, there are a lot of mobile dislocations, fewer twins are required to accommodate plastic strain. Boucher and Christian [40] studied the influence of prestrain on deformation twinning in niobium single crystals, it is found that prestrain is effective in suppressing twinning at temperatures down to 77 K. These dislocations induced by prestrain may increase the energy barriers for twin nucleation. This would explain the increase of the stress corresponding to the onset of twinning with increasing prestrain. Since conventional EBSD is unable to resolve dislocation type and density, transmission electron microscopy (TEM) and computer simulations are needed to reveal how twinning interacts with slip in our future work.

Skrotzki [41] reported that in γ-TiAl, one part of the possible glide dislocations remains completely unaffected by the twin interface when these dislocations come across twin boundary. Zhu et al. [42] found that there is no energy change (barrier) when 90° partial pass the twin boundary by cross-slip in bcc metals. Hence, the twin boundary will not necessarily act barriers for dislocation slip. Recently, Li and Ma [43] showed that $\{10\bar{1}2\}$ twinning in Mg alloy is actually mediated by atomic shuffling. A large deviation between the actual twin boundaries and the theoretical $\{10\bar{1}2\}$ twinning plane was also observed [44]. More recently, Li and Zhang [45, 46] showed that the twinning shear of $\{10\bar{1}2\}10\bar{1}\bar{1}$ mode can only be zero because the twinning plane is not an invariant plane and thus no shear deformation mediated by twinning dislocations can occur on the twinning plane. These new findings, along with our experimental observations, call for reconsideration of the nature of twin-slip interaction in Mg alloys and other HCP metals. Computer simulations may be needed to resolve the mechanisms.

## Conclusion

By prestraining in tension of extruded AZ31B specimens, we investigated twin-slip interaction during plastic deformation. The results show that the yield stress in compression only slightly increases with increasing prestrain. However, the hardening rate at the low stress stage during compression remains almost unchanged, indicating that the contribution of twin-slip interaction to hardening is negligible in deformation of Mg alloys.

**Acknowledgements** Bin Li gratefully thanks support from the U.S. National Science Foundation (NSF) (CMMI-1635088). Yanyao Jiang acknowledges support from the NSF (CMMI-1462885).

## References


1. Christian JW, Mahajan S (1995) Deformation twinning. Prog Mater Sci 39:1–157.
2. Yoo MH, Lee JK (1991) Deformation twinning in hcp metals and alloys. Philos Mag A 63:987–1000.
3. Yoo MH (1981) Slip, twinning, and fracture in hexagonal close-packed metals. Metall Trans A 12:409–418.
4. Barnett MR (2007) Twinning and the ductility of magnesium alloys. Mater Sci Eng A 464:8–16. https://doi.org/10.1016/j.msea.2007.02.109.
5. Agnew SR, Brown DW, Tomé CN (2006) Validating a polycrystal model for the elastoplastic response of magnesium alloy AZ31 using in situ neutron diffraction. Acta Mater 54:4841–4852. https://doi.org/10.1016/j.actamat.2006.06.020.
6. Muránsky O, Barnett MR, Carr DG, et al (2010) Investigation of deformation twinning in a fine-grained and coarse-grained ZM20 Mg alloy: Combined in situ neutron diffraction and acoustic emission. Acta Mater 58:1503–1517. https://doi.org/10.1016/j.actamat.2009.10.057.
7. Muránsky O, Carr DG, Šittner P, Oliver EC (2009) In situ neutron diffraction investigation of deformation twinning and pseudoelastic-like behaviour of extruded AZ31 magnesium alloy. Int J Plast 25:1107–1127. https://doi.org/10.1016/j.ijplas.2008.08.002.
8. Agnew SR, Tomé CN, Brown DW, et al (2003) Study of slip mechanisms in a magnesium alloy by neutron diffraction and modeling. Scr Mater 48:1003–1008. https://doi.org/10.1016/S1359-6462(02)00591-2.
9. Lynch PA, Kunz M, Tamura N, Barnett MR (2014) Time and spatial resolution of slip and twinning in a grain embedded within a magnesium polycrystal. Acta Mater 78:203–212. https://doi.org/10.1016/j.actamat.2014.06.030.
10. Lou XY, Li M, Boger RK, et al (2007) Hardening evolution of AZ31B Mg sheet. Int J Plast 23:44–86. https://doi.org/10.1016/j.ijplas.2006.03.005.
11. Chen P, Li B, Culbertson D, Jiang Y, Contribution of extension twinning to plastic strain at low stress stage deformation of a Mg-3Al-1Zn alloy, Mater. Sci. Eng. A. 709 (2018) 40–45.
12. Zhao S, Guo E, Wang L, et al (2015) Effect of pre-compressive strain on microstructure and mechanical properties of Mg–2.7Nd–0.4Zn–0.5Zr alloy. Mater Sci Eng A 647:28–33. https://doi.org/10.1016/j.msea.2015.08.092.
13. Xin Y, Lv L, Chen H, et al (2016) Effect of dislocation-twin boundary interaction on deformation by twin boundary migration. Mater Sci Eng A 662:95–99. https://doi.org/10.1016/j.msea.2016.03.061.
14. Hama T, Nagao H, Kuchinomachi Y, Takuda H (2014) Effect of pre-strain on work-hardening behavior of magnesium alloy sheets upon cyclic loading. Mater Sci Eng A 591:69–77. https://doi.org/10.1016/j.msea.2013.10.083.
15. Hama T, Kariyazaki Y, Hosokawa N, et al (2012) Work-hardening behaviors of magnesium alloy sheet during in-plane cyclic loading. Mater Sci Eng A 551:209–217. https://doi.org/10.1016/j.msea.2012.05.009.
16. Zhang J, Xi G, Wan X, Fang C (2017) The dislocation-twin interaction and evolution of twin boundary in AZ31 Mg alloy. Acta Mater 133:208–216. https://doi.org/10.1016/j.actamat.2017.05.034.
17. Nogaret T, Curtin WA, Yasi JA, et al (2010) Atomistic study of edge and screw <c+a> dislocations in magnesium. Acta Mater 58:4332–4343. https://doi.org/10.1016/j.actamat.2010.04.022.
18. Li B, Zhang QW, Mathaudhu SN (2017) Basal-pyramidal dislocation lock in deformed magnesium. Scr Mater 134:37–41. https://doi.org/10.1016/j.scriptamat.2017.02.040.
19. Geng J, Chisholm MF, Mishra RK, Kumar KS (2014) The structure of <c+a> type dislocation loops in magnesium. Philos Mag Lett 94:377–386. https://doi.org/10.1080/09500839.2014.916423.
20. Lavrentev FF (1980) The type of dislocation interaction as the factor determining work hardening. Mater Sci Eng 46:191–208. https://doi.org/10.1016/0025-5416(80)90175-5.
21. Yu Q, Wang J, Jiang Y, et al (2014) Co-zone 1¯012 Twin Interaction in Magnesium Single Crystal. Mater Res Lett 2:82–88. https://doi.org/10.1080/21663831.2013.867291.
22. El Kadiri H, Kapil J, Oppedal AL, et al (2013) The effect of twin–twin interactions on the nucleation and propagation of {101¯2} twinning in magnesium. Acta Mater 61:3549–3563. https://doi.org/10.1016/j.actamat.2013.02.030.
23. Yu Q, Wang J, Jiang Y, et al (2014) Twin–twin interactions in magnesium. Acta Mater 77:28–42. https://doi.org/10.1016/j.actamat.2014.05.030.

24. Lv L, Xin Y, Yu H, et al (2015) The role of dislocations in strain hardening of an extension twinning predominant deformation. Mater Sci Eng A 636:389–395. https://doi.org/10.1016/j.msea.2015.04.007.
25. Zhou N, Zhang Z, Jin L, et al (2014) Ductility improvement by twinning and twin–slip interaction in a Mg-Y alloy. Mater Des 1980-2015 56:966–974. https://doi.org/10.1016/j.matdes.2013.12.014.
26. Cepeda-Jiménez CM, Molina-Aldareguia JM, Pérez-Prado MT (2015) Origin of the twinning to slip transition with grain size refinement, with decreasing strain rate and with increasing temperature in magnesium. Acta Mater 88:232–244. https://doi.org/10.1016/j.actamat.2015.01.032.
27. Ezaz T, Sangid MD, Sehitoglu H (2011) Energy barriers associated with slip–twin interactions. Philos Mag 91:1464–1488. https://doi.org/10.1080/14786435.2010.541166.
28. Madec R, Devincre B, Kubin L, et al (2003) The Role of Collinear Interaction in Dislocation-Induced Hardening. Science 301:1879–1882. https://doi.org/10.1126/science.1085477.
29. Kalidindi S r., Salem A a., Doherty R d. (2003) Role of Deformation Twinning on Strain Hardening in Cubic and Hexagonal Polycrystalline Metals. Adv Eng Mater 5:229–232. https://doi.org/10.1002/adem.200300320.
30. Serra A, Bacon DJ, Pond RC (2002) Twins as barriers to basal slip in hexagonal-close-packed metals. Metall Mater Trans A 33:809–812. https://doi.org/10.1007/s11661-002-0149-7.
31. Proust G, Tomé CN, Jain A, Agnew SR (2009) Modeling the effect of twinning and detwinning during strain-path changes of magnesium alloy AZ31. Int J Plast 25:861–880. https://doi.org/10.1016/j.ijplas.2008.05.005.
32. El Kadiri H, Oppedal AL (2010) A crystal plasticity theory for latent hardening by glide twinning through dislocation transmutation and twin accommodation effects. J Mech Phys Solids 58:613–624. https://doi.org/10.1016/j.jmps.2009.12.004.
33. El Kadiri H, Barrett CD, Wang J, Tomé CN (2015) Why are twins profuse in magnesium? Acta Mater 85:354–361. https://doi.org/10.1016/j.actamat.2014.11.033.
34. Wang F, Agnew SR (2016) Dislocation transmutation by tension twinning in magnesium alloy AZ31. Int J Plast 81:63–86. https://doi.org/10.1016/j.ijplas.2016.01.012.
35. Xu J, Guan B, Yu H, et al (2016) Effect of Twin Boundary–Dislocation–Solute Interaction on Detwinning in a Mg–3Al–1Zn Alloy. J Mater Sci Technol 32:1239–1244. https://doi.org/10.1016/j.jmst.2016.08.023.
36. Park SH, Hong S-G, Lee JH, Lee CS (2012) Multiple twinning modes in rolled Mg–3Al–1Zn alloy and their selection mechanism. Mater Sci Eng A 532:401–406. https://doi.org/10.1016/j.msea.2011.11.003.
37. Oppedal AL, El Kadiri H, Tomé CN, et al (2012) Effect of dislocation transmutation on modeling hardening mechanisms by twinning in magnesium. Int J Plast 30:41–61. https://doi.org/10.1016/j.ijplas.2011.09.002.
38. Ma Q, El Kadiri H, Oppedal AL, et al (2012) Twinning effects in a rod-textured AM30 Magnesium alloy. Int J Plast 29:60–76. https://doi.org/10.1016/j.ijplas.2011.08.001.
39. Mahajan S (1970) Effects of existing substructure on shock-twinning behaviour of iron. Phys Status Solidi A 2:217–223. https://doi.org/10.1002/pssa.19700020205.
40. Boucher NA, Christian JW (1972) The influence of pre-strain on deformation twinning in niobium single crystals. Acta Metall 20:581–591. https://doi.org/10.1016/0001-6160(72)90013-2.
41. Skrotzki B (2000) Crystallographic aspects of deformation twinning and consequences for plastic deformation processes in γ-TiAl. Acta Mater 48:851–862. https://doi.org/10.1016/S1359-6454(99)00385-7.
42. Zhu YT, Wu XL, Liao XZ, et al (2011) Dislocation–twin interactions in nanocrystalline fcc metals. Acta Mater 59:812–821. https://doi.org/10.1016/j.actamat.2010.10.028.
43. Li B, Ma E (2009) Atomic Shuffling Dominated Mechanism for Deformation Twinning in Magnesium. Phys Rev Lett 103:035503. https://doi.org/10.1103/PhysRevLett.103.035503.
44. Zhang XY, Li B, Wu XL, et al (2012) Twin boundaries showing very large deviations from the twinning plane. Scr Mater 67:862–865. https://doi.org/10.1016/j.scriptamat.2012.08.012.
45. Li B, Zhang XY (2014) Global strain generated by shuffling-dominated {101-2}<10-1-1>twinning. Scr Mater 71:45–48. https://doi.org/10.1016/j.scriptamat.2013.10.002.
46. Li B, Zhang XY (2016) Twinning with zero twinning shear. Scr Mater 125:73–79. https://doi.org/10.1016/j.scriptamat.2016.07.004.

# In Situ Neutron Diffraction and Acoustic Emission During the Biaxial Loading of AZ31 Alloy

Jan Čapek, Tobias Panzner, Karl Sofinowski, Daria Drozdenko, and Kristián Máthis

**Abstract**

The evolution of twinning in randomly textured magnesium alloy and rolled AZ31 alloy during biaxial mechanical tests has been monitored using concurrent application of acoustic emission and neutron diffraction methods. The influence of the loading path on twinning is discussed in detail. It is shown that the twinning is strongly sensitive to the load path.

**Keywords**

Twinning • Biaxial testing • In situ methods

## Introduction

Deformation mechanisms of magnesium alloys are an often discussed topic of the scientific community in the last two decades. It is a widely established fact that in addition to basal slip, deformation twinning and non-basal slip play an important role in plastic deformation. The contribution of these mechanisms to plasticity significantly depends on materials' characteristics, such as the initial texture [1] and grain size [2] as well as on the loading path. Among different experimental techniques, the in-situ neutron diffraction (ND) method has been successfully utilized for characterization of deformation mechanisms in hexagonal materials. It has been shown that the integrated intensity of particular reflections (parent and twin grain families) is sensitive to the reorientation of the crystal lattice (e.g. due to twinning) during the loading. Further, the activity of different slip systems can be deduced from the evolution of the lattice strains with the applied stress. Recently, another in-situ method, the acoustic emission (AE) technique has been used concurrently with neutron diffraction measurements [3, 4]. AE technique yields information on the dynamic processes involved in plastic deformation of alloys, as e.g. deformation twinning and dislocation glide. The advantage of both methods is to obtain information from entire volume in real time. Further, in the case of magnesium alloys they provide complementary information: the AE is sensitive only to twin nucleation, whereas neutron diffraction characterizes their growth as well.

The combination of AE and ND measurement was used by several authors and brought completely new information about the deformation behavior of magnesium alloys during the uniaxial loading. However, the multi-axial testing is also important, since this method better simulate the in-service conditions.

## Experimental Methods

Rolled AZ31 with a grain size of $(25 \pm 10)$ μm and strong basal texture (Fig. 1) was used for the experiments.

The biaxial tests were carried out at POLDI instrument using cruciform specimens. The arms of the samples were 55 mm. The specimen was thinned in the middle in order to ensure uniform deformation in this area. The deformation tests were carried out at room temperature using a load frame at a force control with a rate of 20 N/sec. In order to collect the ND data, the tests were stopped for approx. 45 min. at

J. Čapek (✉) · D. Drozdenko · K. Máthis
Department of Physics of Materials, Charles University,
Ke Karlovu 5, 12116 Prague 2, Czech Republic
e-mail: jan.capek89@gmail.com

J. Čapek
Nuclear Physics Institute,
The Czech Academy of Sciences, Řež 130,
250 68 Řež, Czech Republic

T. Panzner
Laboratory for Neutron Scattering, NUM, Paul Scherrer Institute,
5232 Villigen, Switzerland

K. Sofinowski
Swiss Light Source, Paul Scherrer Institute,
5232 Villigen, Switzerland

K. Sofinowski
Neutrons & Xrays Mech Mat, IMX, Ecole Polytech Fed Lausanne,
1012 Lausanne, Switzerland


D. Orlov et al. (eds.), *Magnesium Technology 2018*, The Minerals, Metals & Materials Series, https://doi.org/10.1007/978-3-319-72332-7_31

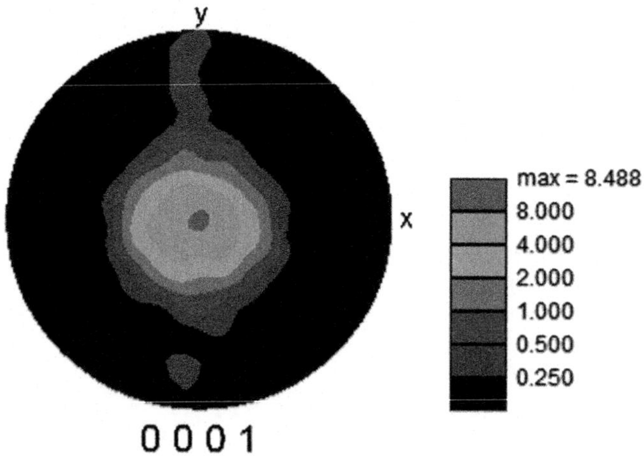


**Fig. 1** Pole figure of texture of rolled AZ31 sheet

predefined force levels. The schematic illustration of the experimental setup is shown at Fig. 2.

The AE testing was performed using a Physical Acoustics PCI-2 acquisition board and a broadband AE sensor from the same company was mounted on the outside the gauge length using vacuum grease and a pin. The AE was amplified by 40 dB in the frequency range 100–1200 kHz. The tests were repeated in continuous mode without the ND acquisition in order to collect the continuous AE data.

There are several challenges connected to the biaxial testing. The evaluation of the stress state at the center of a sample is not straightforward. The sample does not have well-defined cross-section [5]. Moreover, the circular thickness reduction at the center of specimen causes a so-called "ring-effect"—by compressing the sample along axis *x* we induce the tensile stress along the axis *y* [6]. Due to this fact the "uniaxial" deformation does not have the radial symmetry along the loading axis. Digital image correlation (DIC) was used to measure the strain field of the sample.

## Results

AZ31 samples had texture which enhanced twinning during the compression along the *x* or *y*-axis. The Strain-Force curves, measured during uniaxial compression along the *x*-axis and equibiaxial compression, respectively are shown in Fig. 3.

Detailed view on the deformation can be obtained by separation of the strain into individual components (Fig. 4). Figure 4a shows the results for the equibiaxial loading. The dominant deformation component in this case is the elongation along the *z*-axis. At the same time the compressive strain has arisen along the *x* and and *y* axes. Thus, the deformation has similar character as a simple uniaxial tension along *z*-axis.

The results for uniaxial compression are shown at Fig. 4b. The dominant deformation component is along *x*-axis, i.e. it has the same sense as the loading axis. Nevertheless, the ring effect causes the tensile stress along the *y*-axis. Moreover, the largest elongation is observed along the *z*-axis. This effect can be elucidated by extension twinning. During this mechanism the grains undergoing twinning elongates along their crystallographic *c*-axis. In the particular case of our specimens, the majority of grains have their *c*-axis parallel to the *z*-axis (Fig. 1). Thus, the observed elongation along *z*-axis is comprehensible.

$\varepsilon_x$ strain maps of the center area of the samples at the same load level, 5.5 kN for the horizontal arms are shown in Fig. 5. It is obvious that the deformation in the middle part, where the sample was examined by the neutron beam, was uniform.

In order to reveal the active deformation mechanisms, AE response was analyzed (Fig. 6). The count rate as a function of applied force is shown at Fig. 6a. It is possible to observe, that the AE activity is stronger for equibiaxial loading up to the yield point than for the uniaxial loading. The equibiaxial loading causes higher local stresses which leads to microyielding. Around and after the yield point the AE is

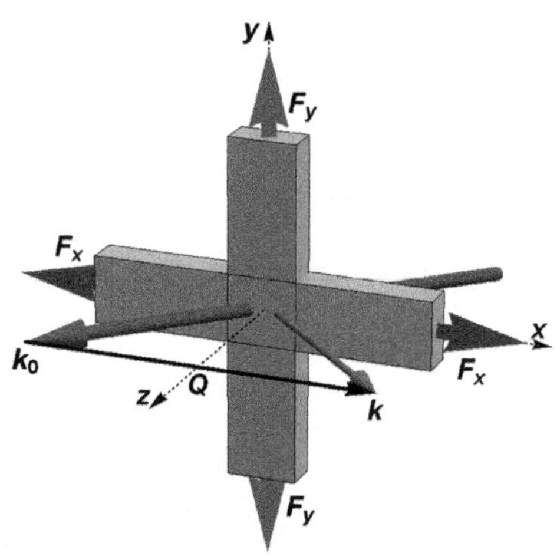


**Fig. 2** Schematic illustration of the experimental setup

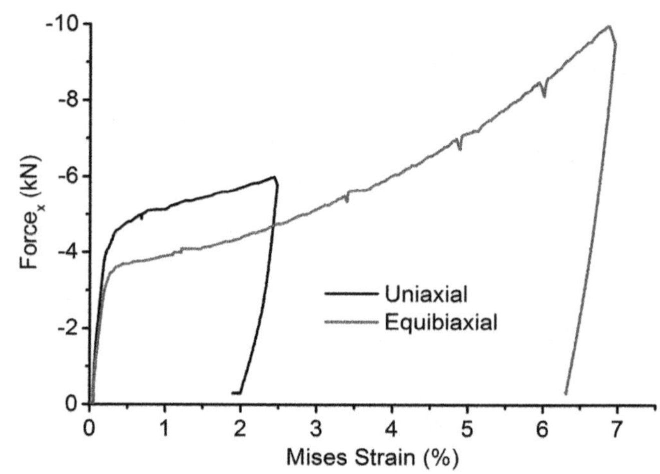


**Fig. 3** Strain Force curve for uniaxial and equibiaxial loading of AZ31

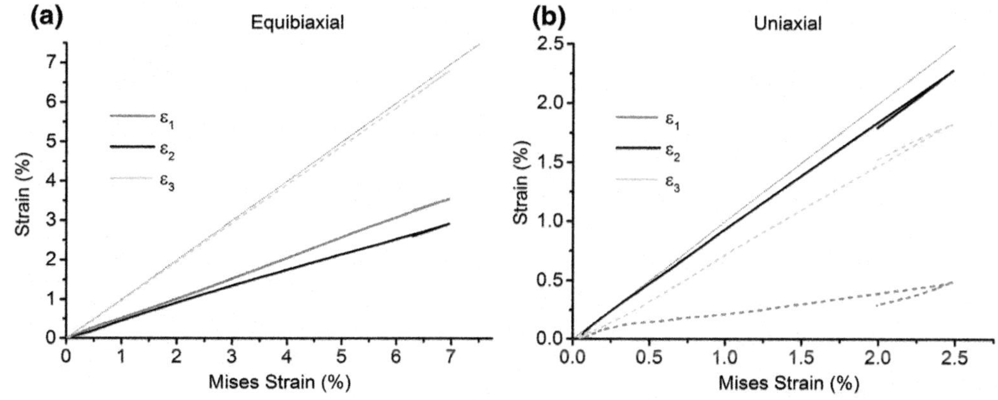


**Fig. 4** Comparison of the individual strain components to the Mises strain. Red curve is deformation along axis 1, Black curve along the axis 2, Green along the axis 3. Solid line represent the compressive strain, dashed line tensile strain. Gray line symbolize 1:1 ratio

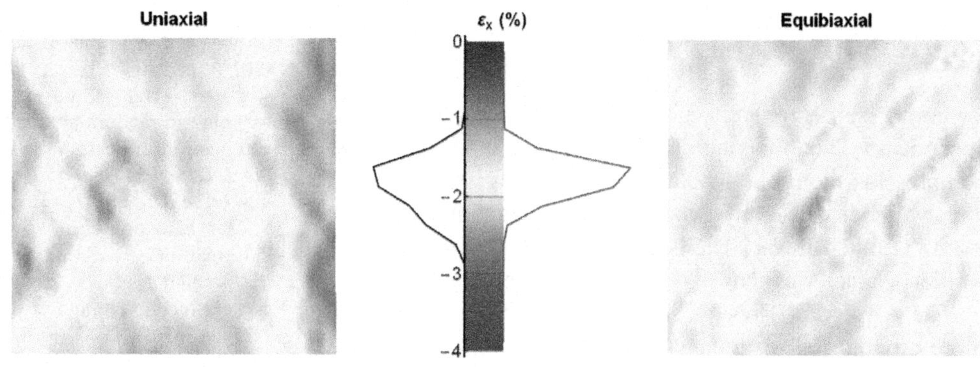


**Fig. 5** DIC strain maps in the horizontal direction recorded at 5.5 kN load on horizontal arm for uniaxial and equibiaxial mode

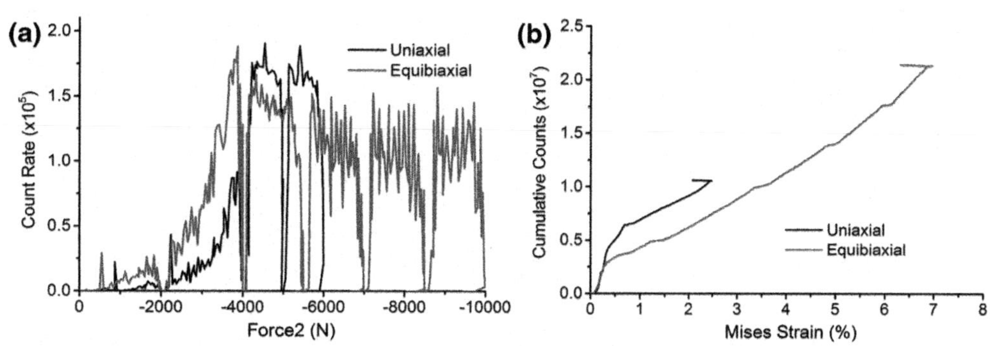


**Fig. 6** **a** count rate as a function of applied force; **b** cumulative counts as a function of Mises Strain for uniaxial and equibiaxial deformation

slightly higher for uniaxial deformation. The strain evolution of cumulative counts is shown in Fig. 6b. It is obvious that at the beginning of the deformation the cumulative counts are higher for uniaxial deformation. Since the twin nucleation is the main source of AE in magnesium, the results suggest that twin nucleation is more pronounced for uniaxial deformation and the nucleated twins are larger.

Since the ND data were collected from the direction of *x*-axis, the change of the relative intensities of peaks related to parent grains has to be evaluated with respect to the $\varepsilon_x$ (Fig. 7) [7]. The results show, that the twinned volume is slightly higher for equibiaxial deformation. In our previous work we highlighted, that during compression deformation fewer twin variants are nucleated than that for tensile deformation [8]. Since, as discussed above, tensile strains have arisen during equibiaxial deformation, this results indicate that the number of the nucleated twin variants are also higher for this deformation path.

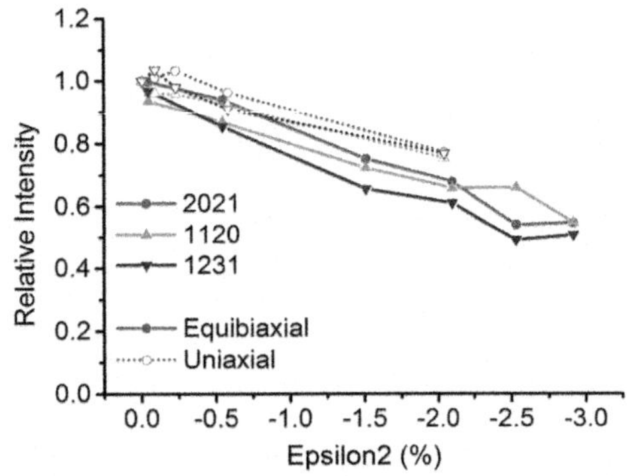


**Fig. 7** Change of the ralative intensity for grains favorable oriented for twinning during the compression deformation

## Conclusions

Twinning has a strong effect on the deformation of the AZ31 specimens. The uniaxial deformation has a character of compression along the loading axis and elongation along $z$-axis, which is caused by twinning. The elongation along the $y$-axis is much lower. The equibiaxial deformation has the similar character as tensile deformation along $z$-axis.

The AE and ND results suggests that twin nucleation is more pronounced for uniaxial deformation, but the twinned volume is higher for equibiaxial loading.

**Acknowledgements** The authors are grateful for the financial support of the Czech Science Foundation under the contract 14-36566G. JČ acknowledges the support by the Charles University in Prague, Faculty of Mathematics and Physics, project GA UK Nr. 251715. This work was supported by the project "Nanomaterials centre for advanced applications", Project No. CZ.02.1.01/0.0/0.0/15_003/0000485, financed by ERDF. KAS and TP thank the European Research Council for the financial support within the advanced grant MULTIAX (339245).

## References

1. E.W. Kelly and W.F. Hosford, "DEFORMATION CHARACTERISTICS OF TEXTURED MAGNESIUM". *Transactions of the Metallurgical Society of AIME*. **242**(4) (1968): p. 654–&.
2. E.O. Hall, "The deformation and ageing of mild steel.3. Discussion of results". *Proceedings of the Physical Society of London Section B*. **64**(381) (1951): p. 747–753.
3. O. Muransky, et al., "Investigation of deformation twinning in a fine-grained and coarse-grained ZM20 Mg alloy: Combined in situ neutron diffraction and acoustic emission". *Acta Materialia*. **58**(5) (2010): p. 1503–1517.
4. J. Capek, et al., "Study of the loading mode dependence of the twinning in random textured cast magnesium by acoustic emission and neutron diffraction methods". *Materials Science and Engineering A-Structural Materials Properties Microstructure and Processing*. **602** (2014): p. 25–32.
5. S. Van Petegem, et al., "In-situ neutron diffraction during biaxial deformation". *Acta Materialia*. **105** (2016): p. 404–416.
6. V. Bonnand, et al., "Investigation of multiaxial fatigue in the context of turboengine disc applications". *International Journal of Fatigue*. **33**(8) (2011): p. 1006–1016.
7. M.A. Gharghouri, et al., "Study of the mechanical properties of Mg-7.7at.% Al by in-situ neutron diffraction". *Philosophical Magazine a-Physics of Condensed Matter Structure Defects and Mechanical Properties*. **79**(7) (1999): p. 1671–1695.
8. J. Čapek, et al., "Dependence of twinned volume fraction on loading mode and Schmid factor in randomly textured magnesium". *Acta Materialia*. **130** (2017): p. 319–328.

# Acoustic Emission Study of High Temperature Deformation of Mg–Zn–Y Alloys with LPSO Phase

Klaudia Horváth, Daria Drozdenko, Kristián Máthis, Gerardo Garcés, and Patrik Dobroň

**Abstract**

Magnesium alloys with different content of zinc (Zn) and yttrium (Y) were extruded at an extrusion ratio of 18:1 at 350 °C. The alloying elements in both Mg alloys formed a long period stacking ordered (LPSO) phase, which during the extrusion process was elongated along the extrusion direction (ED). The magnesium matrix has bimodal character composed by fine dynamically recrystallized (DRX-ed) grains and initial coarse grains elongated along ED. Compression tests with concurrent acoustic emission (AE) measurements were performed along ED at 200, 300, and 400 °C. The deformation mechanisms and the mechanical properties at 200 °C are very similar to those obtained at ambient temperatures, i.e. in the alloy with low volume fraction of the LPSO phase (<10%) twinning controls the yielding, while in the alloy with high volume fraction of the LPSO phase (around 35%) dislocation slip and kink formation are dominant. At 300 °C the reinforcing effect of the LPSO phase is reduced and at 400 °C it is not effective anymore.

**Keywords**

Mg alloys • LPSO phase • High temperature
Mechanical properties • Acoustic emission

## Introduction

Thanks to low density, high specific strength, and recyclability of magnesium (Mg) alloys, they have a great potential in replacing aluminum alloys and steel in engineering systems [1]. The main limitation for their use is their poor corrosion resistance, poor ductility at room temperature and degradation of their mechanical properties with increasing temperature (for example lower strength at temperatures above 200 °C [2]). Recently, Mg alloys containing long-period stacking ordered (LPSO) phase received a lot of interest due to their excellent mechanical properties [3–5]. The LPSO phase is formed in the Mg–Y–Zn alloy when the atomic ratio of Y/Zn is 2:1. The Y and Zn atoms are placed periodically in the Mg basal planes and they form an ordered structure [6].

It was shown that the presence of the LPSO phase increases the critical resolved shear stress (CRSS) of the basal slip [7]. Thus, activation of non-basal slip systems in the Mg matrix is required. Usually, during the extrusion process, texture with basal planes oriented parallel to the extrusion direction (ED) is formed. During compression along ED, such a texture promotes the nucleation of (10–12) extension twins [8–10]. On the other hand, it was shown in [11, 12] that in the regions with high density of the LPSO phase the twin nucleation and growth is suppressed. The main deformation mode of the LPSO phase when the loading direction is aligned with the LPSO fibers is kinking [4]. Since the LPSO phase has higher hardness and Young's modulus than the Mg matrix, the presence of the LPSO phase reinforces the alloy [13, 14]. According to Hagihara, the LPSO phase strengthens the material via short-fiber strengthening, similarly to composite materials [15].

The influence of the LPSO phase on the dynamic recrystallization (DRX) was reported in [16, 17]. It seems that the presence of 18R type LPSO phase increases the temperature for the DRX [18]. Garcés et al. show that the hardening due to the occurrence of the LPSO phase during tensile tests is effective up to 523 K (250 °C) [19].

K. Horváth (✉) · D. Drozdenko · K. Máthis · P. Dobroň
Department of Physics of Materials, Faculty of Mathematics and Physics, Charles University, Ke Karlovu 5, 121 16 Prague, Czech Republic
e-mail: horvathk@karlov.mff.cuni.c

K. Horváth
Nuclear Physics Institute of the CAS, 250 68 Řež, Prague, Czech Republic

G. Garcés
Department of Physical Metallurgy, CENIM-CSIC, Avenida Gregorio Del Amo 8, 28040 Madrid, Spain


D. Orlov et al. (eds.), *Magnesium Technology 2018*, The Minerals, Metals & Materials Series, https://doi.org/10.1007/978-3-319-72332-7_32

The high temperature compressive behavior of Mg alloys with the LPSO phase is not clear yet. In the present paper, we applied acoustic emission (AE) technique in order to monitor active deformation mechanisms. AE monitoring gives real time information about the dynamic processes during loading of the material. Its main advantage is an excellent temporal resolution (few microseconds) and the ability to characterize the entire sample volume. The study focuses on the effect of the alloying elements content on the high temperature deformation of Mg alloys with the LPSO phase.

## Experimental Methods

The WZ42 (Mg + 3.5 wt% Y + 1.6 wt% Zn, $Mg_{98.5}Y_1Zn_{0.5}$ in at.%) and the WZ104 (Mg + 10 wt% Y + 3.7 wt% Zn, $Mg_{95.5}Y_3Zn_{1.5}$) magnesium alloys were cast in KITECH (Korea). The cast billets were extruded with an extrusion speed of 0.5 mm/s and an extrusion ratio of 18:1 at 350 °C in Centro Nacional de Investigaciones Metalúrgicas (CENIM) Madrid.

The deformation tests were carried out using a universal testing machine INSTRON® 5882 with a constant strain rate of $10^{-3}$ $s^{-1}$. The samples with a length of 12 mm and a diameter of 8 mm were compressed along ED at 200, 300, and 400 °C. Before deformation the samples were heated up to the test temperature and held for 300 s to obtain an even temperature distribution within the gauge volume.

The AE activity during deformation tests was monitored by a computer-controlled MICRO-II system developed by Physical Acoustics Corporation (PAC), which allows a continuous storage of AE signals with a sampling frequency of 2 MHz. A piezoelectric S9215 sensor (PAC) with a maximum operating temperature of 540 °C was used. High signal/noise ratio was ensured by using a 60 dB preamplifier. A threshold-level detection of the recorded AE signal was performed to achieve a comprehensive set of AE parameters. The threshold level was set as 24 dB.

Samples for microstructure investigation were cut from the middle part of the longitudinal section of the extruded bars. The sample surfaces were ground on SiC papers and subsequently polished by diamond paste with particle size decreasing to 0.25 μm. For the light microscopy observation, the polished surfaces were first etched with 5% Nital for 5 s and then with a solution of 50 ml of ethanol, 9 ml of water, 4 ml of acetic acid and 6 g of picric acid (98%) for 10 s.

## Results

The as-extruded microstructures of the WZ42 and the WZ104 alloys are presented in Fig. 1. In both alloys, the microstructure consists of the initial coarse Mg grains, DRX-ed Mg grains, and the LPSO phase. The coarse Mg grains and the LPSO phase are elongated along ED. In the magnified microstructure (Fig. 1b, d), the coarse grains are marked by black arrows while the LPSO phase is indicated by red arrows. The volume fraction of the LPSO phase is significantly higher in the WZ104 alloy (35 ± 5)% compared to the WZ42 alloy (7 ± 2)% [20]. The size of the DRX-ed grains is (4.3 ± 0.3) μm and (2.4 ± 0.2) μm for the WZ42 and the WZ104 alloys, respectively. For the calculation of the grain size of the DRXed grains, grains smaller than 10 μm were considered.

The true stress versus true strain compression curves at 200, 300, and 400 °C are shown in Fig. 2. With increasing temperature, a gradual decrease in the yield stress and ultimate compression stress (UCS) values can be observed. Both the yield stress and the UCS are higher for the WZ104 alloy compared to the WZ42 alloy at 200 and 300 °C. The differences in mechanical properties for the two alloys are negligible at 400 °C. The yield stress of the WZ104 alloy decreases from 320 MPa at 200 °C to 35 MPa at 400 °C. These values for the WZ42 alloy are 212 and 44 MPa for 200 and 400 °C, respectively. The deformation curves at 200 °C have very different character. The WZ104 alloy shows a typical dislocation dominated convex curve while the shape of the WZ42 alloy has a concave S-shape. Nevertheless, both alloys undergo a significant work hardening. The WZ104 alloy deformed at 300 °C shows more or less a steady state flow after the yield point around 240 MPa. The WZ42 alloy at 300 °C still exhibits hardening but it is less significant than at 200 °C. The softening starts around 20 and 18% of deformation at 300 °C for the WZ104 alloy and for the WZ42 alloy, respectively. The deformation curves are very similar to each other at 400 °C and they exhibit a steady state flow at 36 MPa.

Figure 3 shows the AE response at 200 and 300 °C for both alloys. During deformation at 400 °C, no AE signals were detected. At 200 and 300 °C, AE signals appear from the beginning of the test and the maximum of the AE count rate in all cases corresponds to the yield point. Despite the similar general trend, there are several differences in the AE response of the specimens. At 200 °C, AE persists until the end of the test for both alloys, but broader maximum is

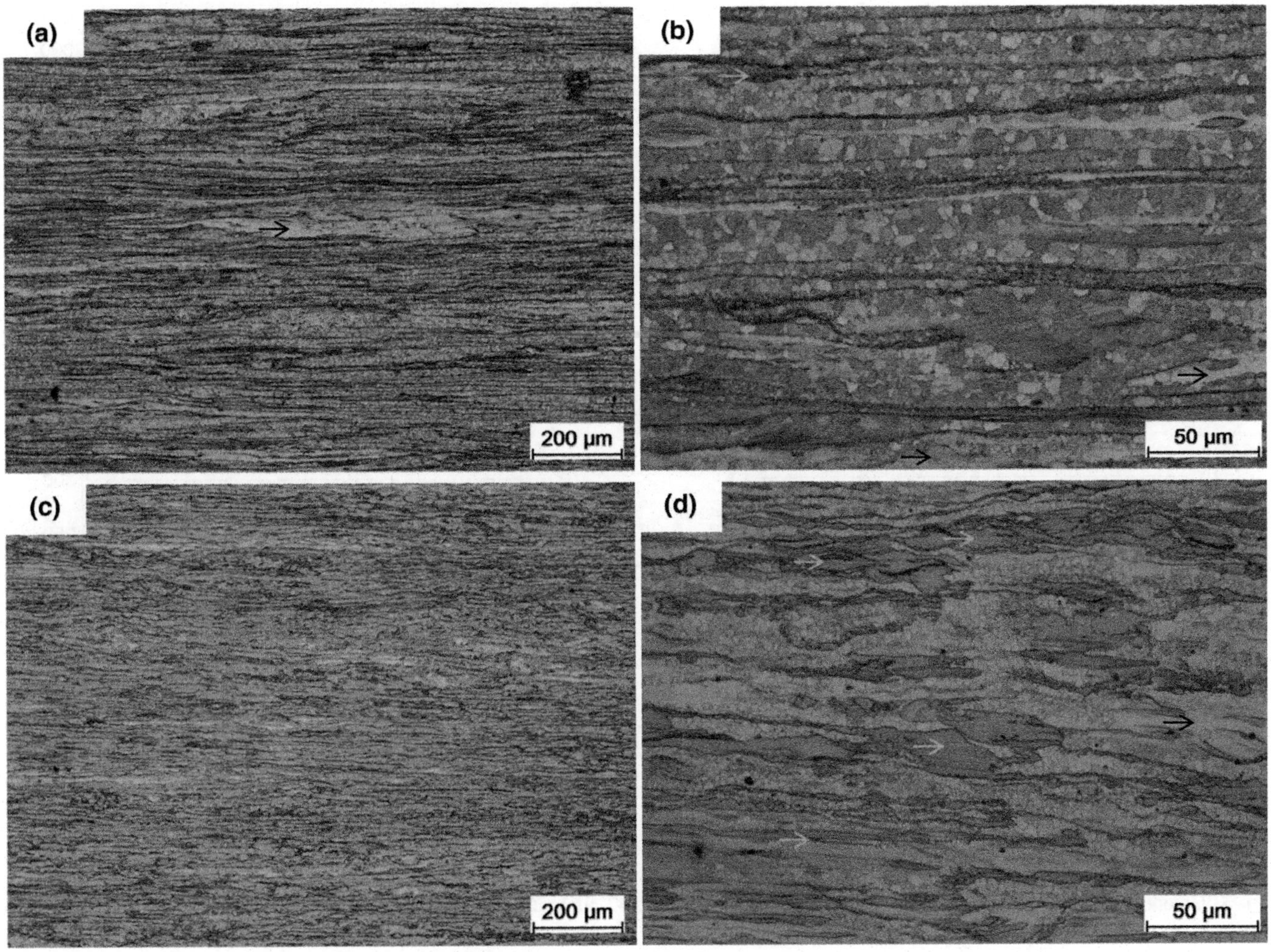


**Fig. 1** The initial microstructure of the as-extruded Mg alloys: **a** and **b** microstructure of the WZ42 alloy; **c** and **d** microstructure of the WZ104 alloy

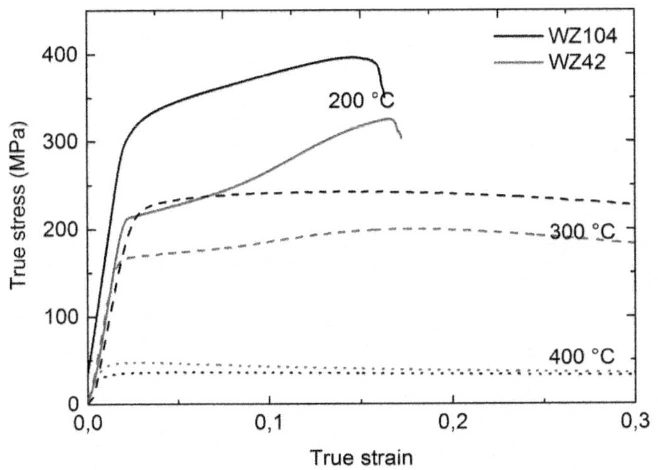


**Fig. 2** True stress true strain curves for the WZ42 and the WZ104 alloys deformed at 200, 300, and 400 °C

observed for the WZ104 alloy than for the WZ42 alloy. At 300 °C the maximum of AE count rate is reduced by two orders of magnitude and fewer signals are detected.

## Discussion

Previous studies already reported the effect of extrusion speed and extrusion ratio on the microstructure of Mg alloys with the LPSO phase [20, 21]. The bimodal microstructure observed in both alloys is very probably the result of low extrusion speed of 0.5 mm/s. As it was shown previously [20], the increasing content of Y and Zn leads to an increase in the volume fraction of the LPSO phase and reduces the volume fraction of original coarse grains and the grain size of the DRX-ed grains.

The reinforcing effect of the LPSO phase is clearly seen on the deformation behavior of the WZ42 and WZ104 alloys

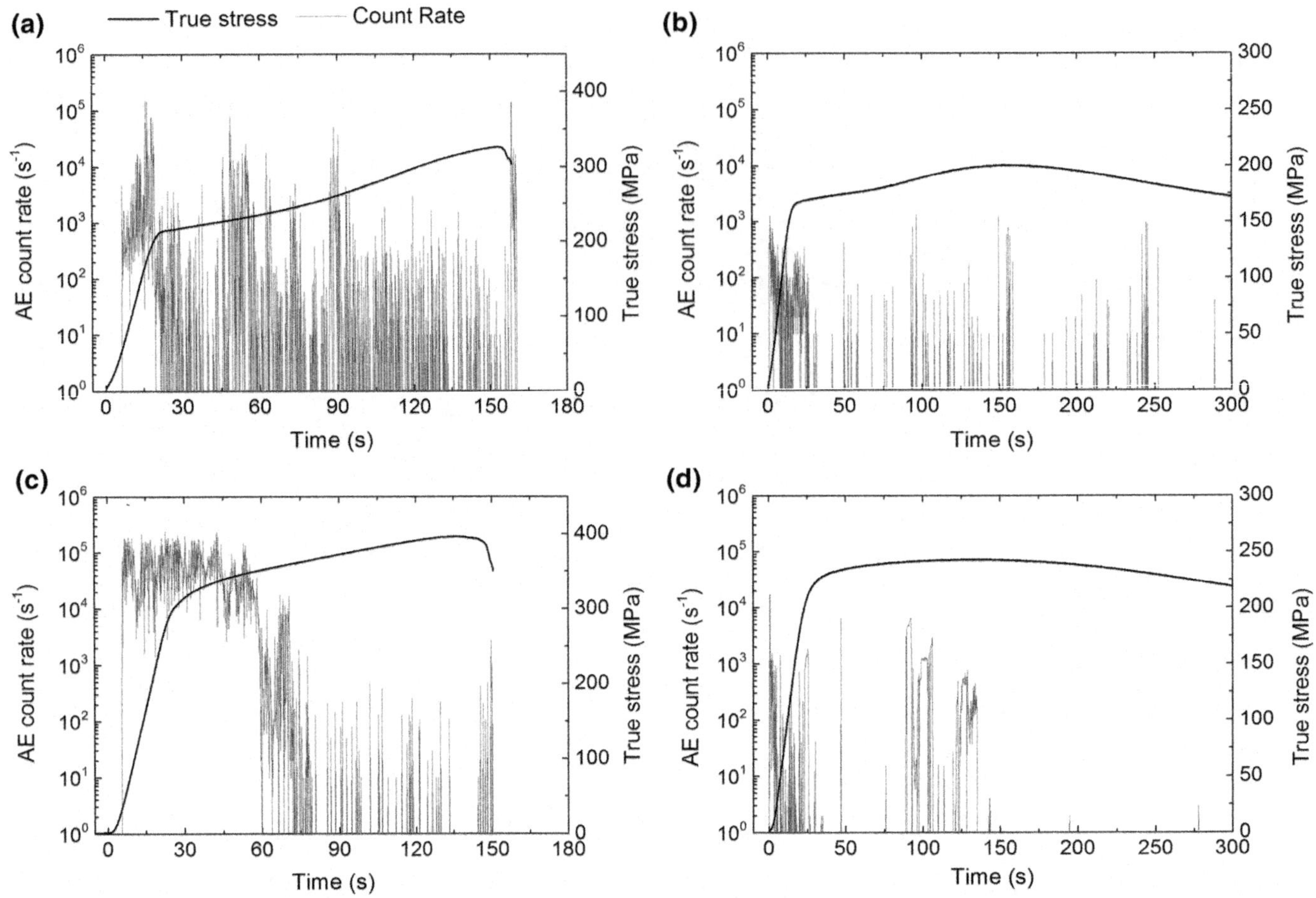


**Fig. 3** Compression test with a concurrent AE measurement of the WZ42 alloy at **a** 200 °C and at **b** 300 °C and that of the WZ104 alloy at **c** 200 °C and at **d** 300 °C

at 200 and 300 °C. It was previously reported that the addition of Y and the formation of the LPSO phase could effectively increase the difficulty of hot deformation [22]. At 200 °C, the deformation curves are very similar to those obtained at room temperature [11]. The yield stress at 200 °C for both alloys decreases only by 40 MPa if compared to the values measured at room temperature [11]. The S-shaped deformation curve of the WZ42 alloy is a typical sign of twinning activity. However, the presence of the LPSO phase suppresses the twinning activity [11, 12]. This reflects in the convex shape of the deformation curve of the WZ104 alloy. Both alloys show a significant hardening after the yield point which is the result of increasing number of barriers for dislocation motion.

At 300 °C, the difference in the yield stress is lower compared to that at 200 °C. It was reported that in the case of tensile tests the LPSO phase loses its reinforcing effect at 523 K (250 °C) [19]. It is consistent with our result and suggests that the difference in the yield stress could be linked to different deformation modes. The deformation curve of the WZ42 alloy still has a concave part which proposes that twins could be activated. It was reported that above 200 °C the CRSS for non-basal slip systems is reduced, what decreases the importance of twinning [23, 24]. However, the increase of the CRSS of basal slip in the presence of the LPSO phase was also documented [7] and some twins can possibly nucleate in the large initial grains well oriented for extension twinning. The nucleation of twins in these grains could be the reason for the lower yield stress of the WZ42 alloy than of the WZ104 alloy and be responsible for the different work hardening. After reaching the peak stress of the WZ42 alloy, a dynamical softening occurs. The deformation curve of the WZ104 alloy reflects the mechanism of hot deformation to some point. In the initial stage the flow stress increased with increasing strain and then this increasing trend decreases gradually. In contrast to the WZ42 alloy, no significant softening occurred what can be the result of higher Y content which causes a delay of the DRX [22].

At 400 °C, the two alloys have very similar mechanical properties. It seems that the volume fraction of the LPSO phase has no effect on the deformation behavior of the alloys at this temperature. After reaching the yield stress both materials have a steady-state flow. The high elongation obtained at this temperature could be linked to grain boundary sliding [19, 25].

The different deformation behavior of the alloys at 200 °C is reflected in the AE response, too. The maximum in the AE count rate is very narrow in the WZ42 alloy, while in the WZ104 alloy the count rate remains very high even after the yield point. The sharp decrease of the AE count rate after the yield point is usually linked to twinning activity. Twin nucleation is an excellent source of AE, emitting large amplitude signals, but their growth does not produce detectable AE signal [26–28]. It means that the twin nucleation at the yield point of the WZ42 alloy results in high AE count rate and due to their growth the count rate is reduced above the yield point. Since it does not drop to zero a massive dislocation motion should be active [26, 27]. The AE signals persist till the end of the test which supposes that the deformation is accumulated by a slip of hundreds of dislocation and by the kinking of the LPSO phase. Contrary, in the WZ104 alloy the AE count rate does not fall even after the yield point what proposes that the deformation is controlled by dislocation slip, not by twinning. After the decrease of the first maximum, another local peak in the AE count rate is observed. These signals could be related to the kink formation in the LPSO phase what is realized by an avalanche-like motion of (0001) basal dislocations.

The amplitude of the AE count rate at 300 °C is reduced by two orders of magnitude for both alloys compared to the measurement at 200 °C. It is reasonable since the collective dislocation motion is not very likely due to the easy activation of non-basal slip systems [23, 24].

## Conclusions

The effect of the volume fraction of the LPSO phase on the high temperature mechanical properties has been studied. The following conclusion could be drawn:

- At 200 °C the LPSO phase successfully strengthens both the WZ42 and the WZ104 alloy, and their mechanical properties are comparable with those obtained at ambient temperatures.
- At temperatures below 300 °C twinning controlled the yield behavior in the WZ42 alloy with a low volume fraction of the LPSO phase (<10%). In the WZ104 alloy with high volume fraction of the LPSO phase (around 35%) dislocation slip and kink formation are dominant.
- With increasing temperature, the reinforcing effect of the LPSO phase decreased. At 400 °C mechanical properties of two alloys are similar.

**Acknowledgements** The authors are grateful for support from the Czech Science Foundation under grant Nr. 16-12075S; the Grant Agency of the Charles University under grant Nr. 1262217; the grant SVV-2017-260442; the Operational Programme Research, Development and Education, The Ministry of Education, Youth and Sports (OP RDE, MEYS) under the grant CZ.02.1.01/0.0/0.0/16_013/0001794 and the Spanish Ministry of Economy and Competitiveness under the grant number MAT2012-34135.

## References

1. Pan H, Ren Y, Fu H, Zhao H, Wang L, Meng X, Quin G (2016) Recent developments in rare-earth free wrought magnesium alloys having high strength: a review. J Alloys Compd. 663: 146–156.
2. Kainer Ku (2000) Magnesium alloys and their applications. Wiley-VCH, Weinheim, Germany.
3. Yamasaki M, Hashimoto K, Hagihara K, Kawamura Y (2011) Effect of multimodal microstructure evolution on mechanical properties of Mg–Zn–Y extruded alloy. Acta Mater. 59: 3646–3658.
4. Hagihara K, Yokotani N, Umakoshi Y (2010) Plastic deformation behavior of Mg12YZn with 18R long-period stacking ordered structure. Intermetallics 18: 267–276.
5. Hagihara K, Sugino Y, Fukusumi Y, Umakoshi Y, NakanoT (2011) Plastic Deformation Behavior of Mg12ZnY LPSO-Phase with 14H-Typed Structure. Mater. Trans. 52: 1096–1103.
6. Abe E, Kawamura Y, Hayashi K, Inoue A (2002) Long-period ordered structure in a high-strength nanocrystalline Mg-1at% Zn-2at% Y alloy studied by atomic-resolution Z-contrast STEM. Acta Mater 50: 3845–3857.
7. Matsuda M, Ando S, Nishida M (2005) Dislocation structure in rapidly solidified $Mg_{97}Zn_1Y_2$ alloy with long period stacking order phase. Mater Trans 46: 361–364.
8. Bohlen J, Yi S, Letzig D, Kainer KU (2010) Effect of rare earth elements on the microstructure and texture development in magnesium–manganese alloys during extrusion. Mater. Sci. Eng. A 527: 7092–7098.
9. Dobroň P, Chmelík F, Yi S, Parfenenko K, Letzig D, Bohlen J (2011) Grain size effects on deformation twinning in an extruded magnesium alloy tested in compression. Scripta Mater. 65: 424–427.
10. Victoria-Hernandez J, Yi S, Letzig D, Hernandez-Silva D, Bohlen J (2013) Microstructure and texture development in hydrostatically extruded MgAlZn alloys during tensile testing at intermediate temperatures. Acta Mater. 61: 2179–2193.
11. Garces G, Perez P, Cabeza S, Lin HK, Kim S, Gan W, Adeva P (2015) Reverse tension/compression asymmetry of a Mg-Y-Zn alloys containing LPSO phases. Mater. Sci. Eng. A 647: 287–293.
12. Shao XH, Yang ZQ, Ma XL (2010) Strengthening and toughening mechanisms in Mg–Zn–Y alloy with a long period stacking ordered structure. Acta Mater. 58: 4760–4771.
13. Itoi T, Seimiya T, Kawamura Y, Hirohashi M (2004) Long period stacking structures observed in $Mg_{97}Zn_1Y_2$ alloy. Scr. Mater. 51: 107–111.
14. Garces G, Morris DG, Munoz-Morris MA, Perez P, Tolnai D, Mendis C, Stark A, Lim HK, Kim S, Shell N, Adeva P (2015) Plasticity analysis by synchrotron radiation in a $Mg_{97}Y_2Zn_1$ alloy with bimodal grain structure and containing LPSO phase. Acta Mater. 94: 78–86.
15. Hagihara K, Kinoshita A, Sugino Y, Yamasaki M, Kawamura Y, Yasuda HY, Umakoshi Y (2010) Effect of long-period stacking ordered phase on mechanical properties of $Mg_{97}Zn_1Y_2$ extruded alloy. Acta Mater. 58: 6282–6293.

16. Zhou X, Liu C, Gao Y, Jiang S, Han X, Chen Z (2017) Evolution of LPSO phases and their effect on dynamic recrystallization in a Mg-Gd-Y-Yn-Yr alloy. Metall. Mater. Trans. A 48: 3060–3072.
17. Liu H, Ju J, Yang X, Yan J, Song D, Jiang J, Ma A (2017) A two-step dynamic recrystallization by LPSO phases and its impact on mechanical property of severe plastic deformation processed $Mg_{97}Y_2Zn_1$ alloy. J. Alloys Compd. 704: 509-517.
18. Lv BJ, Peng J, Peng Y, Tang AT, Pan FS (2013) The effect of LPSO phase on hot deformation behavior and hot workability of Mg-2.0Zn-0.3Zr-5.8Y alloy. Mater. Sci. Eng. A 579: 209–216.
19. Onorbe E, Garcés G, Perez P, Adeva P (2012) Effect of the LPSO volume fraction on the microstructure and mechanical properties of Mg-$Y_{2x}$-$Zn_x$ alloys. J Mater Sci 47: 1085–1093.
20. Horváth K, Drozdenko D, Daniš S, Garcés G, Máthis K, Kim S, Dobroň P (2017) Characterization of Microstructure and Mechanical Properties of Mg-Y-Zn Alloys with Respect to Different Content of LPSO Phase. Adv. Eng. Mater. DOI:https://doi.org/10.1002/adem.201700396.
21. Horváth K, Drozdenko D, Garcés G, Máthis K, Dobroň P (2017) Effect of Extrusion Ratio on Microstructure and Resulting Mechanical Properties of Mg Alloys with LPSO Phase. Magnesium Technology 2017: 29–34.
22. Tahreen N, Zhang DF, Pan FS, Jiang XQ, Li DY, Chen DL (2015) Hot deformation and work hardening behavior of an extruded Mg-Zn-Mn-Y alloy. J Mater Sci Technol 31: 1161–1170.
23. Ion SE, Humphreys FJ, White SH (1982) Dynamic recrystallization and the development of microstructure during the high temperature deformation of magnesium. Acta Metall. 30: 1909–1919.
24. Kocks UF, Westlake DG (1967) The importance of twinning for the ductility of CPH polycrystals. Trans. Metall. Soc. 239: 1107–1109.
25. Perez P, Eddahbi M, González S, Garces G, Adeva P (2011) Refinement of the microstructure during superplastic deformation of extruded $Mg_{94}Ni_3Y_{1.5}CeMM_{1.5}$ alloy. Scr. Mater. 64: 33–36.
26. Heiple CR, Carpenter SH (1987) Acoustic Emission Produced by Deformation of Metals and Alloys – A Review: Part I. J. Acoust. Emission 6: 177–204.
27. Heiple CR, Carpenter SH (1987) Acoustic Emission Produced by Deformation of Metals and Alloys – A Review: Part II. J. Acoust. Emission 6: 215–237.
28. Drozdenko D, Bohlen J, Chmelík F, Lukáč P, Dobroň P (2016) Acoustic emission study on the activity of slip and twin mechanisms during compression testing of magnesium single crystals. Mater. Sci. Eng. A 650: 20–27.

# Deformation and Recrystallization Mechanisms and Their Influence on the Microstructure Development of Rare Earth Containing Magnesium Sheets

Changwan Ha, Sangbong Yi, Jan Bohlen, Xiaohua Zhou, Heinz-Günter Brokmeier, Norbert Schell, Dietmar Letzig, and Karl Ulrich Kainer

**Abstract**

Many studies have shown that textures with less distinct alignment of basal planes and the related improvement of formability are found in alloys that contain rare-earth (RE) elements and zinc. However, the effect of the combination of these additional elements on the texture modification has not been yet clearly understood. In this work, sheet samples from Mg–Zn–RE alloys rolled at 400 °C were used for in situ synchrotron X-rays diffraction measurements under tensile loading at different temperatures, in order to track the development of diffraction profiles and textures during deformation. In Mg–Zn–RE alloys, a significantly retardation of recovery and dynamic recrystallization during the high temperature deformation is observed in comparison to the RE-free Mg–Zn alloy. The differences in the active deformation mechanisms as well as the dynamic recrystallization mechanisms are reviewed with respect to the texture alteration. For discussion of the impact of different mechanisms, EBSD observations reveal the microstructure development.

**Keywords**

Mg–Zn–RE alloy • Deformation • Recrystallization mechanisms • In situ measurement • EBSD

## Introduction

Conventional Mg-based sheet alloys such as AZ31, with strong basal type textures, have been investigated in many studies in order to evolve potential for the limited sheet formability by restricted activities of slip systems. To improve the formability of Mg sheet, it is required to develop weaker textures, especially a less distinct alignment of basal planes parallel to the sheet plane. Texture weakening has been revealed in Mg alloy containing yttrium (Y), cerium (Ce), or neodymium (Nd), i.e. rare earth (RE) elements, during rolling. It is reported that their addition in binary Mg–RE or ternary Mg–Zn–RE alloys contribute to higher accommodation of deformation by active basal <a> slip [1–4]. Furthermore, several studies have shown that texture weakening contributes to the activation of various deformation mechanisms, e.g. the formation of different twin types, shear bands, and the activation of <c+a> slip, which also leads to an improvement of the formability [1, 5, 6]. However, it is also noted that such a texture modification is caused by the same various slip modes, shear bands, and different twin types during sheet rolling [1, 2]. In the case of alloy ZE10, varied texture developments could be achieved by only small changes in the alloy composition and the application of different processing methods [3, 7]. Such effects still lack fundamental understanding and the effect of element concentration changes needs to be related to the activity of deformation and recrystallization mechanisms. Synchrotron X-ray radiation with short exposure time by transmission technique allows to analyze the properties of materials, as well as microstructure information [8]. Electron backscatter diffraction (EBSD) measurement allows the analysis of the microstructure development [2, 5, 6, 9]. In this work, both techniques are used to reveal the differences in the deformation and recrystallization behavior on microstructure development of two magnesium sheet alloys during tensile deformation.

C. Ha (✉) · S. Yi · J. Bohlen · D. Letzig · K. U. Kainer
Magesium Innovation Centre, Helmholtz-Zentrum Geesthacht, Max-Planck-Str. 1, D-21502 Geesthacht, Germany
e-mail: changwan.ha@hzg.de

X. Zhou · H.-G. Brokmeier
Institute of Materials Science, TU Clausthal Engineering, Agricolastr. 6, 38678 Clausthal-Zellerfeld, Germany

N. Schell
HZG Outstation at DESY, Helmholtz-Zentrum Geesthacht, Notkestr. 85, 22607 Hamburg, Germany


D. Orlov et al. (eds.), *Magnesium Technology 2018*, The Minerals, Metals & Materials Series, https://doi.org/10.1007/978-3-319-72332-7_33

## Experimental Procedures

Two Mg alloys, Z2 and ZE21 as a RE containing counterpart, were machined in form of slab for rolling experiment after gravity casting of rectangular samples. The alloy composition and the homogenization heat treatment prior to rolling are listed in Table 1. A 50 ton laboratory rolling stand with cold rolls was used for rolling plates. The geometry of rolling slabs was 200 × 100 × 20 $mm^3$. The rolling experiments were conducted with increasing degree of deformation per pass, 3 passes of 0.1 followed by 3 passes of 0.2 on Z2 and 4 passes of 0.1 followed by 3 passes 0.15 on ZE21. Different rolling temperatures were applied, 400 °C for Z2 and 450 °C for ZE21. After each rolling pass, the sheets were re-heated to the rolling temperature for 15 min. The final gauge of the sheets was comparable at 7.95 and 8.25 mm, respectively. After rolling, the sheets were annealed with different annealing conditions, in order to receive similar microstructures (Table 1), i.e. comparable grain sizes. The texture of the annealed sheets was measured using a Panalytical X-ray diffractometer and Cu $K_\alpha$ radiation. The (10.0) and (00.2) pole figures were recalculated using a computer code MTEX [10]. The microstructure was observed by using standard metallographic sample preparation techniques and an etchant based on picric acid [11].

A universal testing machine was installed at the High Energy Material Science beamline, HEMS, P07B, at Petra III (DESY, Hamburg). In this work, a hard X-ray beam (87 keV, λ = 0.1420 Å) was chosen with beam size of 0.5 × 0.5 $mm^2$ and a sample detector distance of 1201 mm to investigate the texture development and the diffraction profiles during tensile testing. $LaB_6$ powder has been used as standardization material. Tensile tests were performed using round tensile samples with 4 mm in diameter and 24 mm gauge length along the rolling direction (RD), at 3 temperatures, room temperature (R.T.), 100, and 200 °C. The initial strain rate was $1 \times 10^{-4}$/s. Texture measurement were repeated by using a fast PerkinElmer XRD 1621 detector during tensile loading at selected 5 points, 0.02, 0.05, 0.10, 0.15, and 0.20 strain. During texture measurements at R.T., the loading was stopped at each measuring point for texture measurement. In the case of measurement at 100 and 200 °C, the strain rate was reduced to $2.9 \times 10^{-5}$/s to minimize stress relaxation. Sample rotation was applied for 105° in 3° steps for the texture measurements. Debye-Scherrer patterns at each measuring point were collected to reveal the (10.0), (00.2), (10.1), (10.2), (11.0), and (10.3) reflections from the area detector using software Fit2D [12]. The pole figures and orientation distribution function (ODF) were recalculated using open source software MTEX [10]. Furthermore, a separate continuous experiment without stop or change in the strain rate has been carried out in order to collect the development of the diffraction profiles from in situ diffraction profiles measurement. The initial strain rate was the same and no sample rotation was applied, keeping the normal direction (ND) of the sample parallel to the beam during loading. In order to evaluate diffraction profiles, the collected Debye-Scherrer patterns were integrated along loading direction (LD) with a sector of 10°, using Fit2D. Figure 1 shows an exemplary Debye-Scherrer pattern. The integrated Debye-Scherrer pattern allowed to reveal peak properties. The measured Full-width at half maximum (FWHM) was evaluated by using a modified Williamson-Hall plot (WH-plot) after instrumental and thickness correction using $LaB_6$ powder [8]. The first 15 reflections were considered for calculating the dislocation density. The modified WH-plot uses the FWHM variation assuming that strain broadening is caused by dislocation [13]. In the present investigation, the following expression was used [8, 13]:

$$\Delta K \cong \gamma/D + \left(\pi M^2 b^2/2\right)^{1/2} \rho^{1/2} K\sqrt{\overline{C}} + O(K^2\overline{C}),$$

where D is the coherently diffraction domain size, γ equals to 0.9 when using FWHM in the analysis, ρ and b are the average dislocation density and the length of the Burgers vector of dislocations, respectively. $\Delta K = \cos\theta[\Delta(2\theta)]/\lambda$, where $\Delta(2\theta)$ equals FWHM of the diffraction peak. M is a dislocation arrangement parameter depending on the effective outer cut-off radius of dislocations. O is for a higher order term of $K_2\ \overline{C}$. $\overline{C}$ is a contrast factor of dislocations [8, 13]. A more detailed introduction to the analysis can be found elsewhere [8, 13, 14]. In the case of hexagonal crystals, the contrast factor should be applied differently from cubic crystals because the hexagonal crystal has different types of slip system with different Burgers vectors. For the contrast factor of dislocation for hexagonal crystals, see details in [15, 16].

EBSD observations for tensile sample were prepared by electrochemical polishing using a Struers™ AC2 solution

**Table 1** Chemical composition in wt% and experimental information of the examined alloys

| Alloy | Zn (wt %) | Ce | Mg | Homo-genization | Final thickness (mm) | Annealing condition | Average grain size (μm) |
|---|---|---|---|---|---|---|---|
| Z2 | 1.87 | – | Bal. | 20 h at 400 °C | 8.25 | 10 min. at 385 °C | 33 |
| ZE21 | 1.97 | 1.13 | Bal. | 16 h at 450 °C | 7.95 | 20 min. at 350 °C | 30 |

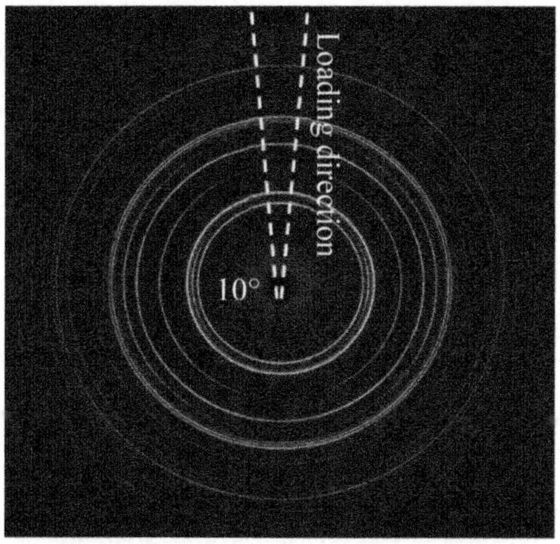


**Fig. 1** Debye-Scherrer pattern of the tensile sample and sketch of the integrated sector along the LD

at −25 °C. EBSD measurements were performed on the deformed area after 0.20 strain of tensile samples using a FE-SEM (Zeiss, Ultra 55), equipped with an EDAX™/TSL EBSD system with a Hikari detector. An accelerating voltage of 15 kV was applied to scan an area of 250 × 600 $\mu m^2$ with a step size of 0.6 μm. Evaluation was carried out using the TSL-OIM 7 analysis software. A clean-up procedure consisting of grain confidence index (CI) standardization and a neighbor CI correlation was applied.

## Results and Discussion

The annealed microstructure and texture are shown in Fig. 2. A similar grain size of the 2 alloys was adjusted by different parameters of rolling and annealing. The average grain size of Z2 and ZE21 are 33 and 30 μm, respectively (Table 1). However, the textures of the annealed sheets are different concurrent with the addition of Ce as a RE element [3]. After annealing, Z2 has a strong basal-type texture with $P_{m.r.d.} = 9.2$, comparable to AZ31 as a conventional Mg alloy, while ZE21 has a weak basal texture ($P_{m.r.d.} = 3.8$) and a spread of basal planes toward the transverse direction.

Stress-strain curves collected during the in situ texture and diffraction profiles measurements are shown in Fig. 3. The stress relaxation is clearly visible at the texture measuring points. However, the stress-strain curves of the measurements for diffraction profiles of the same alloy are very similar so that the impact of stress relaxation on the texture development is neglected at this point. Z2 shows relatively low yield strength along RD at all temperatures in comparison with ZE21. Higher strain hardening is reached in Z2 than in ZE21 up to the ultimate tensile strength (UTS). Such a behavior has also been found in earlier work, corresponding to changed critically resolved shear stresses (CRSS) for individual slip system [3, 7].

Figure 4 shows the pole figure after the tests ($\varepsilon = 0.20$). The (10.0) pole strengthens toward LD, while the (00.2) pole develops a transverse spread perpendicular to LD. The intensities of the (00.2) pole figure does not exhibit a clear tendency after the tests at 3 temperatures. The intensities of the (10.0) pole developed toward LD in both alloys and shows an increase after tests regardless on testing temperature. The significant differences between the 2 alloys are that the (00.2) pole of ZE21 alloy broadens clearly faster towards the TD than of Z2. Besides, the intensity of the (10.0) pole figure shows a different behavior in ZE21 compared to Z2 with increasing temperature. In Z2, there is no clear trend revealed with temperature after deformation. In ZE21, intensities are increasing with temperature, respectively. The broadening of

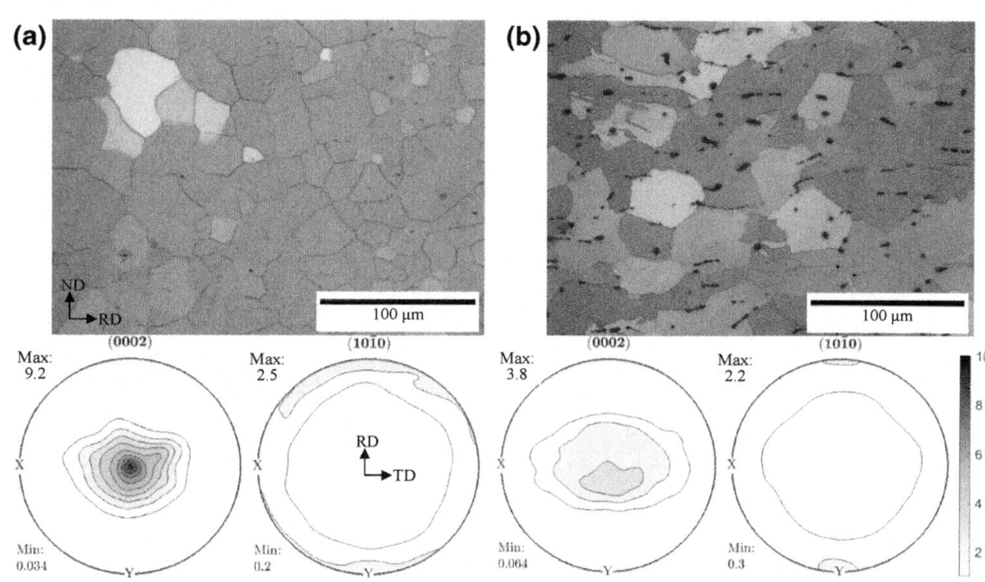


**Fig. 2** Annealed microstructure and texture of the 2 sheets, **a** Z2 and **b** ZE21 alloy, prior to mechanical testing

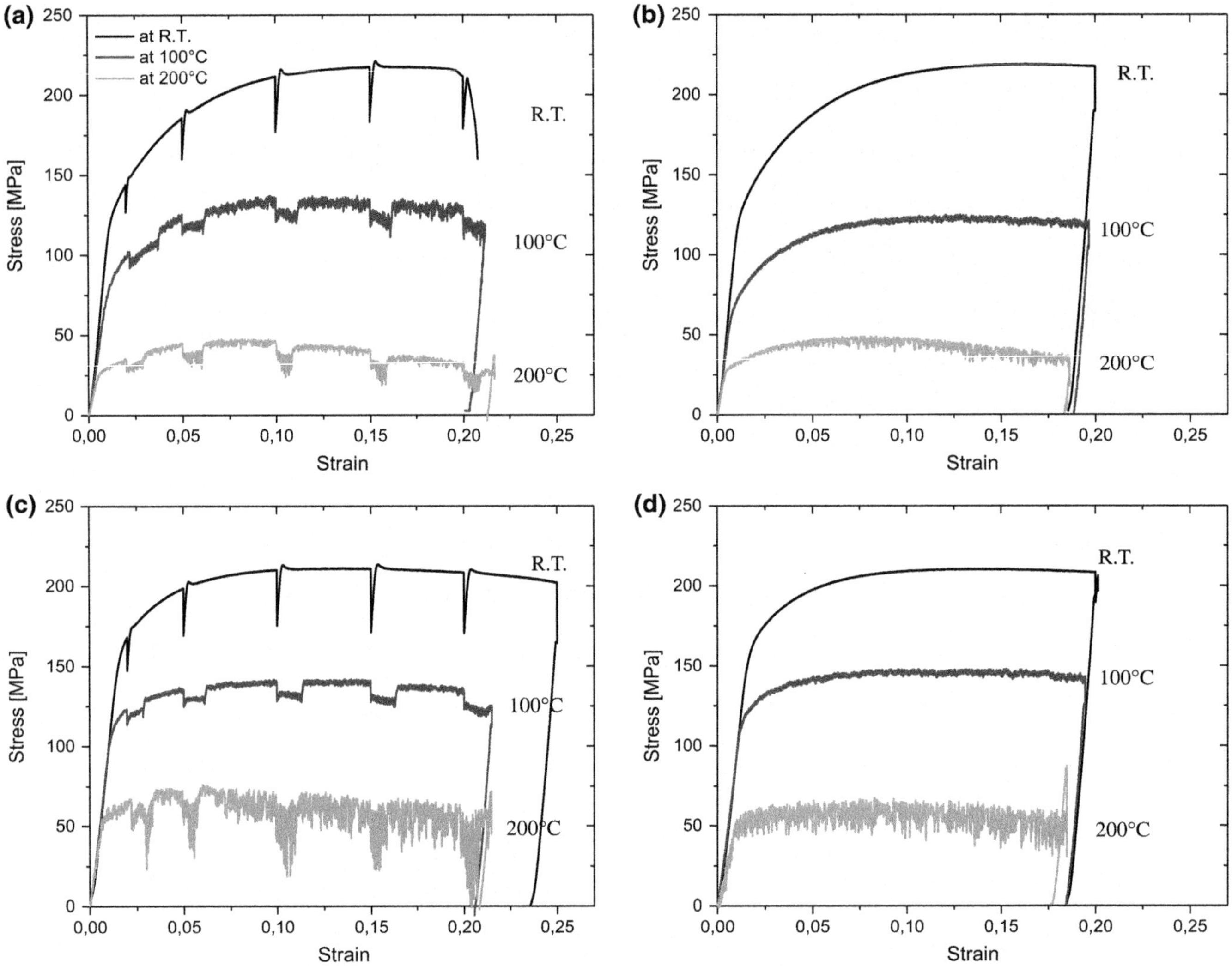


**Fig. 3** Stress-strain curves for **a**, **c** in situ texture and **b**, **d** diffraction measurements during tensile test along the RD in **a**, **b** Z2 and **c**, **d** ZE21 sheets

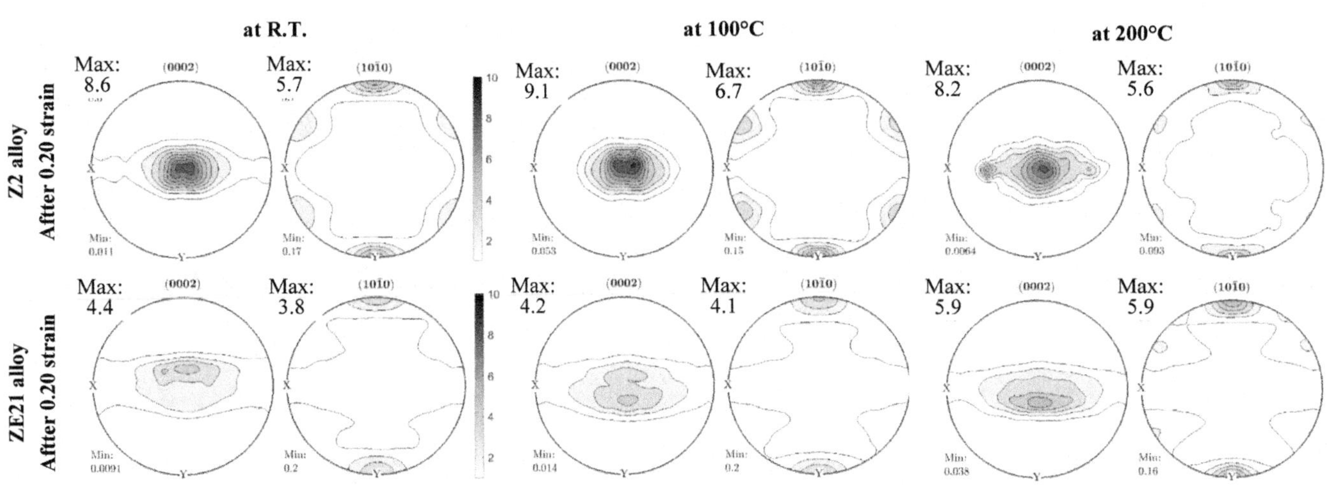


**Fig. 4** Calculated pole figures of in situ measurement after deformation ($\varepsilon = 0.20$) at 3 temperatures in the 2 alloys

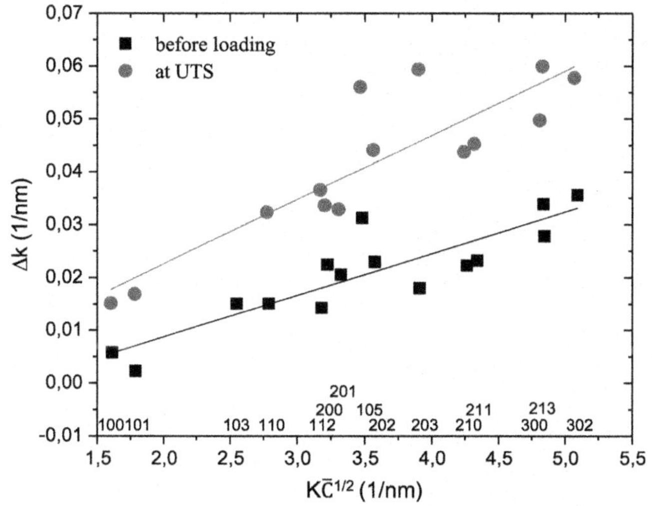


**Fig. 5** An example of modified WH-plot before loading and at UTS on the ZE21 sheet

the (00.2) pole as well as the intensity variation of the (10.0) pole can be related to a higher activation of prismatic $\langle a \rangle$ slip in ZE21 than in Z2 with increasing temperature.

The diffraction profiles were evaluated by the slope of the modified WH-plot, in order to reveal the dislocation density changing through changes on individual property of peaks at selected strain points, i.e. FWHM variation. Figure 5 shows an example of the modified WH-plot for the sample before loading and at UTS. The evaluation of the dislocation density of both alloys during tensile loading is shown in Fig. 6. The dislocation densities increase during tensile loading at RT in both alloys. However, Z2 shows faster relaxation of dislocation at 100 and 200 °C than ZE21. In other words, dislocation densities in Z2 seem to retain from the beginning in overall tendency, while dislocation densities in ZE21 increased tendencially during deformation at 100 and 200 °C. The decreased dislocation density can be related to recovery (RV) and dynamic recrystallization (DRX) during microstructure development [17]. To compare the microstructure development after deformation, EBSD measurements were carried out using the same samples. The inverse pole figure (IPF) map, the grains with lower interior orientation spread (GOS) map, and misorientation histogram on the center of the samples after deformation ($\varepsilon = 0.20$) are presented in Fig. 7 for the experiments at 3 temperatures. In Z2, the IPF map at R.T. shows deformed grain and twins. Deformed grains are still visible after deformation at 100 °C, whereas the twins disappear. At 200 °C, locally deformed band or shear band are revealed, as indicated by white lines. The IPF map of ZE21 at R.T. shows also deformed grains and twins. However, at 100 °C more distortional grains are found, corresponding to higher stored strain in the grains compared to Z2. At 200 °C, no shear bands are observed in ZE21. Such shear bands have been observed to accommodate the deformation in RE- free Mg alloy, e.g. AZ31. Such locally concentrated deformation leads to inhomogeneous deformation [18]. Thus, there is a significant difference in the deformation and recrystallization behavior during tensile testing. To identify recrystallized grains, GOS maps are considered, assuming that recrystallized grains will appear with low orientation spread whereas non-recrystallized deformed grains do not [19, 20]. Grains with GOS < 2° and high angle grain boundaries (HAGB, >15°) are considered. At R.T. and 100 °C, a very small amount of recrystallized grains is identified in both alloys.

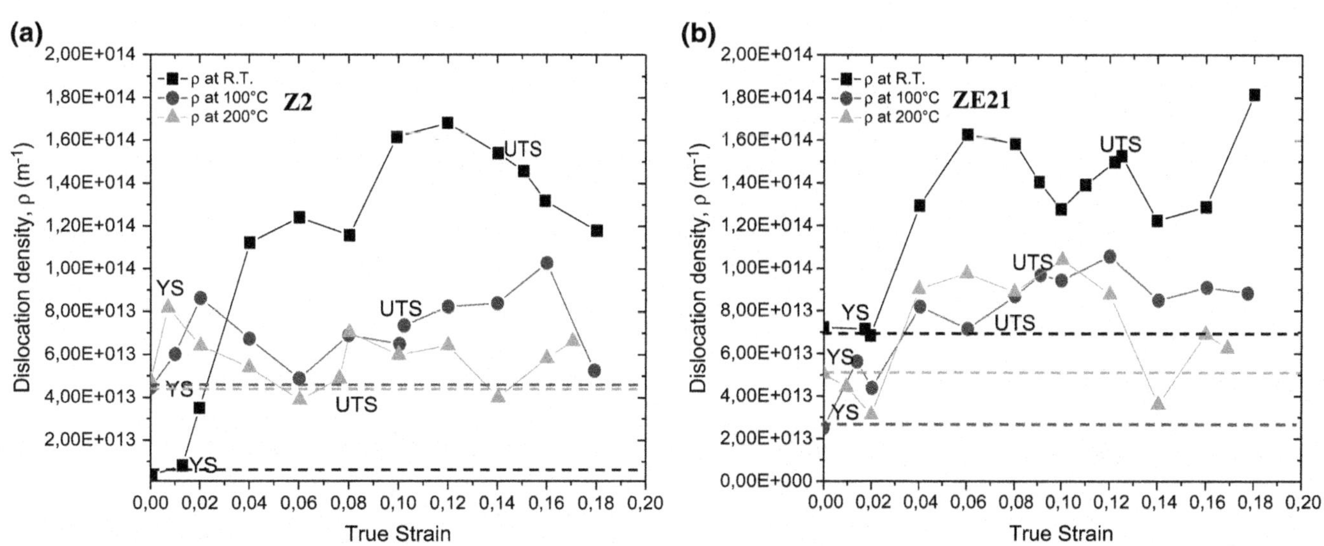


**Fig. 6** Dislocation density variations of the **a** Z2 and **b** ZE21 at 3 temperatures during deformation

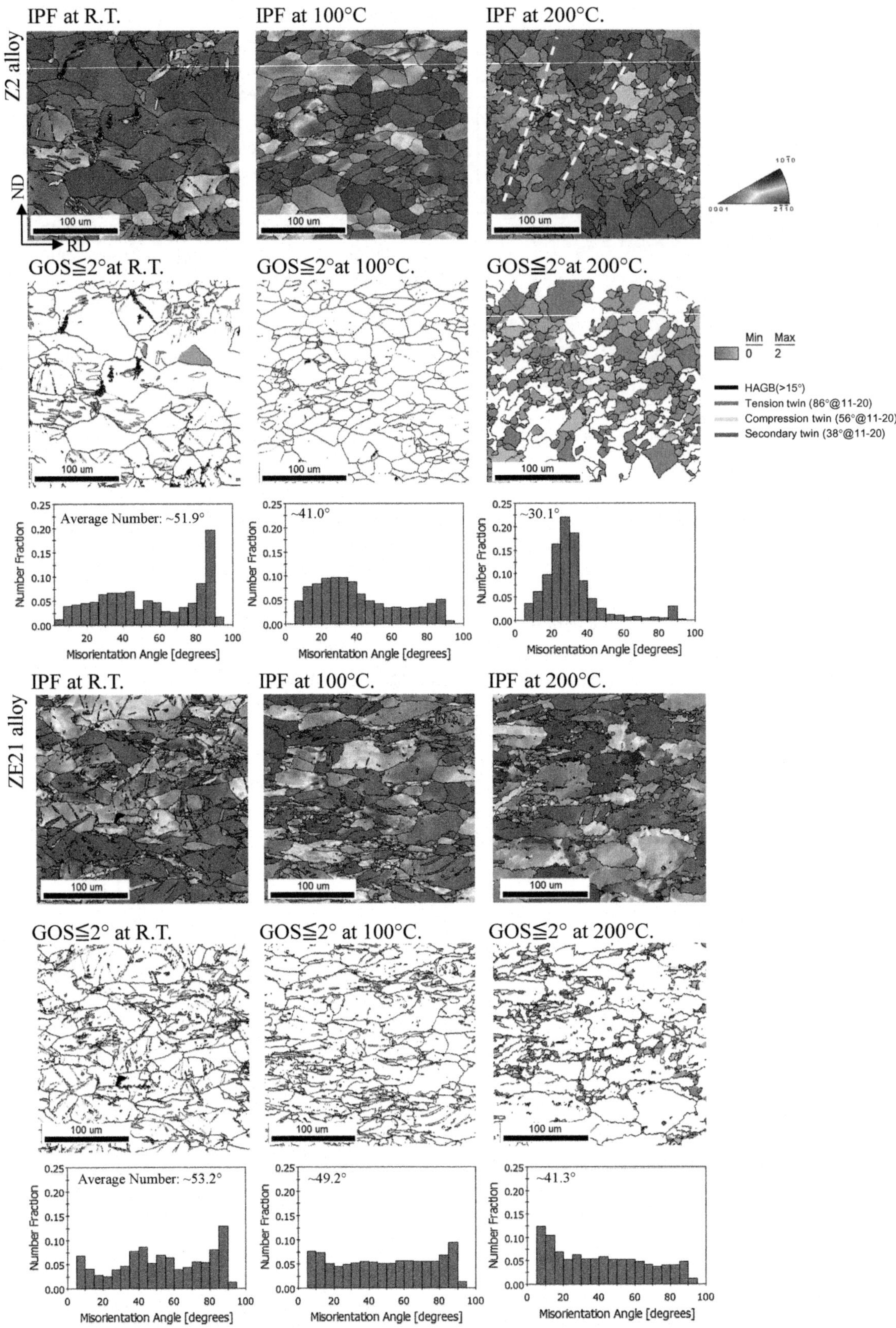


**Fig. 7** Sections of IPF maps, GOS maps with HAGB(>15 °), and respective misorientation histograms of EBSD measurement after deformation ($\varepsilon = 0.20$) at 3 temperatures

Small grains recrystallized are only observed at 200 °C. In comparison, there are concentrated in the above mentioned bands in Z2 whereas in ZE21 they are preferentially found at grain boundaries. In Z2 the selected grains are larger compared to ZE21 indicating a growth restriction during recrystallization. In summary, only Z2 after deformation at 200 °C exhibits a distinct fraction of recrystallized grains. Furthermore, misorientation angle histograms are shown which also help to reveal the deformation and recrystallization behavior [5, 9]. In case of Z2 at R.T., a pronounced peak is observed around 86°, resulting from a high amount of {1012} tension twin boundaries after deformation. With increasing test temperature, recrystallized grains with peaks around 30° in misorientation increase. Such peaks are related with recrystallization, with preferential rotation of 30° around c-axis [21]. The increase of the fraction of such orientations seems to indicate the beginning of recrystallization between R.T. and 100 °C, which is not fully consistent with the finding of recrystallized grains in the GOS map. It is hypothesized that this is related to recovery of grains, which is also consistent with the decrease of the dislocation density in the WH-plot during deformation in Fig. 6. At 200 °C, recrystallized grains and misorientation peaks around 30° can be related to DRX, with new grains concentrated along shear bands [17]. However, in case of ZE21, a clearly lower fraction of recrystallized grains is found. No evidence of the shear bands that played a role during recrystallization in Z2 is found in this alloy. Recrystallized grains also remain smaller after testing at 200 °C compared to Z2. The retarded recrystallization is also consistent with the misorientation histogram, where no distinct peak formation at 30° misorientation is observed. However, peaks at misorientation lower than 15° occur with increasing temperature. Such peaks are consistent with higher stored energy [9], associated with higher dislocation density or low angle grain boundaries (LAGB). This finding is also consistent with the result of modified WH-plot in Fig. 6. Thus, in ZE21 recrystallization appears retarded and leads to continuous increase of internal strain as well as the dislocation densities during testing even at 200 °C.

For DRXed grains in GOS map, in Z2, the new recrystallized grains are created along shear band [22], while they are created around grain boundary in ZE21. This difference in the nucleation of new grains may have an impact on the recrystallization kinetics as well. In summary, the addition of RE elements leads to a retardation of DRX, which in the RE-free alloy Z2 has been enhanced in shear band [1, 3, 7]. Also the decreasing intensity of the (10.0) pole between the tensile tests at 100 and 200 °C can be explained by a proposed enhancement of recrystallization along shear band and a resulting higher fraction of recrystallized grains at 200 °C (Fig. 4).

## Summary

The microstructure development using in situ Synchrotron radiation measurement and EBSD measurement at 3 different temperatures of RE free- and RE elements containing Mg sheets was analyzed in the present study. In comparison with the Z2 alloy, a high activity of prismatic <a> slip has been revealed with increasing testing temperature, especially in case of ZE21, resulting in an intensity increase of the (10.0) component toward LD and the (00.2) spread perpendicular to LD. The visibility of this increasing activity of prismatic <a> slip was reduced with the reformation of the microstructure due to enhanced recrystallization associated with a preferred shear band nucleation of grains in Z2 at 200 °C. The dislocation densities have been revealed by using a modified WH-plot which show a clear increase during deformation at R.T.. The dislocation densities in Z2 at 100 and 200 °C decreased due to RV and DRX during deformation.

**Acknowledgements** The authors appreciate Mr. B. Schwebke and Mr. A. Reichart for technical support during experiments as well as the DESY for the provision of beamtime and the facilities of Petra III (proposal. I-20150212 and I-20160228). This work has been financially supported by the Deutsche Forschungsgemeinschaft (Grants BO 2461/4-1, Yi 103/2-1, and BR 961/7-1).

## References

1. K. Hantzsche et al. (2010) Effect of rare earth additions on microstructure and texture development of magnesium alloy sheets. Scripta Materialia 63:725–730.
2. J. Bohlen et al. (2015) The Influence of the Combination of Alloying Elements on the Microstructure and Texture Development During Rolling and Annealing of Magnesium. Paper presented at the 10th International Conference, Mg 2015, Jeju, 11–16 October 2015.
3. J. Bohlen et al. (2007) The texture and anisotropy of magnesium–zinc–rare earth alloy sheets. Acta Materialia 55:2101–2112.
4. Y. Chino et al. (2011) Effects of Ca on Tensile Properties and Stretch Formability at Room Temperature in Mg-Zn and Mg-Al Alloys. Materials Transactions 52(7):1477–1482.
5. D.W. Kim et al. (2013) Texture Evolution in Mg-Zn-Ca Alloy Sheets. Metallurgical and Materials transactions A 44A:2950–2961.
6. S.B. Yi et al. (2010) Mechanical anisotropy and deep drawing behaviour of AZ31and ZE10 magnesium alloy sheets. Acta Materialia 58:592–605.
7. L. Stutz et al. (2011) Influence of the processing of magnesium alloys AZ31 and ZE10 on the sheet formability at elevated temperature. Key Engineering Materials 473:335–342.
8. Z.Y. Zhong et al. (2015) Dislocation density evolution of AA 7020-T6 investigated by in-situ synchrotron diffraction under tensile load. Materials Characterization 108:124–131.
9. J. Victoria-Hernández et al. (2016) Strain-induced selective grain growth in AZ31 Mg alloy sheet deformed by equal channel angular pressing. Materials Characterization 113:98–107.

10. F. Bachmann et al. (2010) Texture Analysis with MTEX – Free and Open Source Software Toolbox. Solid State Phenom. 160:63–68.
11. V. Kree et al. (2004) The Metallographical Examination of Magnesium Alloys, Practival Metallography 41:233–245.
12. A.P. Hammersley (1998) FIT2D V12.012 Reference Manual V6.0 ESRF98HA01T. ESRF Internal Report. Accessed 8 August 2017.
13. T. Ungár et al. (2001) Densities and character of dislocations and size-distribution of subgrains in deformed metals by X-ray diffraction profile analysis. Mater. Sci. Eng. A 319–321:274–278.
14. T. Ungár et al. (2001) Crystallite size distribution and dislocation structure determined by diffraction profile analysis: principles and practical application to cubic and hexagonal crystals. J. Appl. Crystallogr. 34:298–310.
15. K. Máthis et al. (2004) The evolution of non-basal dislocations as a function of deformation temperature in pure magnesium determined by X-ray diffraction. Acta Materialia 52:2889–2894.
16. I.C. Dragomir et al. (2002) Contrast factors of dislocations in the hexagonal crystal system. J. Appl. Cryst. 35:556–564.
17. T. Ungár (2004) Microstructural parameters from X-ray diffraction peak broadening. Scripta Materialia 51:777–781.
18. M.R. Barnett et al. (2007) Influence of microstructure on strain distribution in Mg-3Al-Zn. Scripta Materialia 57(12):1125–1128.
19. J. Victoria-Hernandez et al. (2014) The influence of the recrystallization mechanisms and grain growth on the texture of a hot rolled AZ31 sheet during subsequent isochronal annealing. Journal of Alloys and Compounds 616:189–197.
20. J. Bohlen et al. (2015) Calcium and zirconium as texture modifiers during rolling and annealing of magnesium–zinc alloys. Materials Characterization 101:144–152.
21. N. Bozzolo et al. (2002) Recrystallization Textures in some Hexagonal Alloys. Materials Science Forum 408–412:901–906.
22. A. Galiyev et al. (2001) Correlation of plastic deformation and dynamic recrystallization in magnesium alloy ZK60. Acta mater. 49:1199–1207.

# Thermo-Mechanical Treatment of Extruded Mg–1Zn Alloy: Cluster Analysis of AE Signals

Patrik Dobroň, Daria Drozdenko, Marius Hegedűs, Juraj Olejňák, Klaudia Horváth, and Jan Bohlen

**Abstract**

The proper thermo-mechanical treatment can improve mechanical properties of extruded Mg alloys through a solute segregation and precipitation along twin boundaries. The effect of heat treatment on mobility of twin boundaries with respect to applied loading direction was studied in extruded Mg–1Zn alloy using the acoustic emission (AE) technique. The adaptive sequential k-means clustering (ASKC) was applied to analyze the AE data in order to determine the dominant deformation mechanism in a given time period. The AE energy, median frequency and the number of elements in individual AE clusters are the main parameters of presented clustering analysis. Active deformation mechanisms are discussed with respect to mutual orientation of grains and loading direction.

**Keywords**

Magnesium • Twinning • Annealing • Acoustic emission Cluster analysis

P. Dobroň (✉) · D. Drozdenko · M. Hegedűs · J. Olejňák · K. Horváth
Department of Physics of Materials, Faculty of Mathematics and Physics, Charles University, Ke Karlovu 5, 12116 Prague 2, Czech Republic
e-mail: dobronp@karlov.mff.cuni.cz

K. Horváth
Nuclear Physics Institute, Czech Academy of Sciences, Řež 130, 250 68 Řež, Prague, Czech Republic

J. Bohlen
Magnesium Innovation Centre, Helmholtz-Zentrum-Geesthacht, Max-Planck-Straße 1, D21502 Geesthacht, Germany


D. Orlov et al. (eds.), *Magnesium Technology 2018*, The Minerals, Metals & Materials Series, https://doi.org/10.1007/978-3-319-72332-7_34

## Introduction

One of the ways to improve mechanical properties of wrought magnesium alloys is an application of thermo-mechanical treatment (TMT). A strong basal texture in extruded Mg alloys promotes the formation of extension twins during compression along the extrusion direction (ED) [1]. Thus, twinning becomes a key mechanism of plastic deformation in the Mg alloys. Precipitation at twin boundaries can cause the pinning effect and leads to higher compression yield strength. Thus, a reduction of the compression-tension yield asymmetry can be achieved.

Tensile or compressive loading of the pre-compressed and annealed Mg alloy may influence their deformation behavior. Therefore, it is essential to get more insight into individual mechanisms of deformation (dislocation slip, twinning, twinning—detwinning) of the TMT Mg alloys with respect to applied loading direction.

The acoustic emission (AE) technique provides real time information about the collective processes involved in plastic deformation of crystalline materials. It is based on sensitive detection of the transient elastic waves generated within the material during deformation due to sudden localized structure changes [2, 3]. A direct correlation of the AE parameters with the stress-strain curve yields information on the dynamic changes indicating e.g. activation of various deformation mechanisms during plastic deformation of Mg alloys [4–6]. To determine the dominant active deformation mechanism at the various stages of deformation in Mg alloys, advanced statistical methods for analysis of AE response can be applied. The adaptive sequential k-means clustering (ASKC) method was found to be useful for such analysis [7–9].

The aim of the present work is to examine an influence of TMT on deformation behavior of extruded binary Mg–1Zn alloy using the clustering analysis of AE signals. A special attention is paid to characterization of active deformation mechanisms with respect to the loading direction.

## Experimental

Specimens with a diameter of 11 mm and a gauge length of 17 mm were machined from the extruded round bar of binary Mg-1 wt%Zn (Mg–1Zn) alloy parallel to ED. Deformation tests were performed in a universal testing machine Instron 5882 at room temperature and at a constant strain rate of $10^{-3}$ $s^{-1}$.

Influence of TMT on mechanical behavior of Mg–1Zn alloy was studied with respect to loading direction (tension or compression). Firstly, all samples were pre-compressed up to 75 MPa (shortly after the yield point—YP) to introduce deformation twins into the structure and then they were subjected to:

(a) annealing at 200 °C for 8 h and subsequent compression up to 100 MPa (TMT-c sample)
(b) annealing at 200 °C for 8 h and subsequent tension up to 75 MPa (TMT-t sample)

AE during deformation tests was monitored by computer-controlled PCI-2 device (Physical Acoustic Corporation) using a piezoelectric transducer (micro 30S - PAC) and a2/4/6-preamplifier giving a gain of 60 dB. The sensor was attached to the specimen surface using a clamp. The AE waveform streaming (2 MHz) provides the AE data for a cluster analysis. In this study, the adaptive sequential k-mean clustering (ASKC) developed by Pomponi and Vinogradov [7] was used to determine the dominant deformation mechanism in a given time period. More details about the procedure and an application of ASKC to the AE data recorded during loading of Mg alloys can be found in [8, 9].

Global texture information about the alloy in the as-received condition was obtained using the X-ray diffraction. A Panalytical X-ray diffractometer setup using CuKα radiation was employed to measure pole figures on polished samples in reflection geometry to a sample tilt of 70°.

Standard metallography procedures on longitudinal sections by using an etchant based on picric acid [10] were applied to reveal the grain structure.

## Results

The Mg–1Zn alloy has almost homogeneous microstructure with the average grain size of (50 ± 2) μm (Fig. 1). The initial basal texture is shown in Fig. 2.

Results of the ASKC, represented by energy, median frequency and the number of elements in individual AE clusters, for samples subjected to different procedures are presented in Figs. 3, 4 and 5. Deformation curves are correlated with the number of elements in individual AE clusters on a time scale (Figs. 3, 4 and 5b) to provide information about development of the dominant deformation mechanism. While the ASKC analysis determine only the dominant mechanisms, the concurrent activity of other deformation mechanisms during plastic deformation cannot be excluded.

**Fig. 1** Microstructure of extruded Mg–1Zn alloy (↑ ED)

For all samples, preference ranges in the energy-median frequency dependence for individual clusters can be seen. Cluster 1 is found at low energies and in a wide range of median frequencies (Figs. 3, 4 and 5a). It is observed before the deformation experiment (Figs. 3, 4 and 5b) and therefore is associated with noise signals. Cluster 2 is found at medium energies in a wide range of median frequencies and it occurs short after the beginning of the deformation test. Thus, this cluster is linked to activity of basal slip. Cluster 3 associated with twinning consists of events with higher energies and high median frequencies. Cluster 4 appears only after the yield point (YP) and is related to non-basal slip (Figs. 3, 4 and 5b).

For an assignment of events into clusters the ASK algorithm includes more parameters than the energy and the median frequency presented in Fig. 3, 4 and 5. Therefore, an overlap in the energy-frequency plots for different clusters is essential (Figs. 3, 4 and 5a).

All deformation curves exhibit S-shape (Figs. 3, 4 and 5b) which is typically seen in extruded Mg alloys compressed along ED. The lowest yield strength (57 ± 1 MPa) is observed in the compressed sample, whereas both TMT samples have similar yield strength (65 ± 1 MPa). For all the three samples, a slight difference in the yield behavior can be seen. This behavior is also reflected in the AE results.

Deformation is firstly realized by basal slip, and then twinning and non-basal slip are activated. The highest AE activity represented by energy and number of elements was observed in the compressed sample (Fig. 3). Basal slip

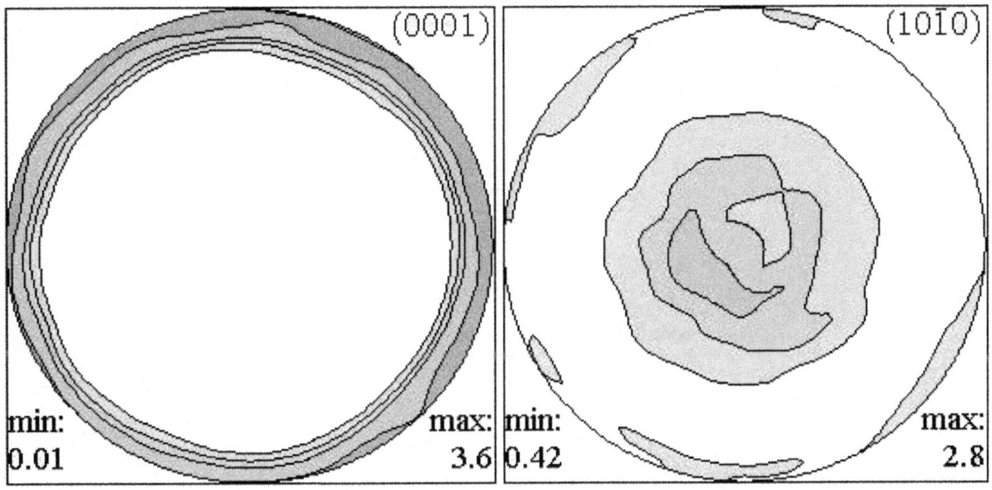


**Fig. 2** Initial texture of extruded Mg–1Zn alloy (× ED)

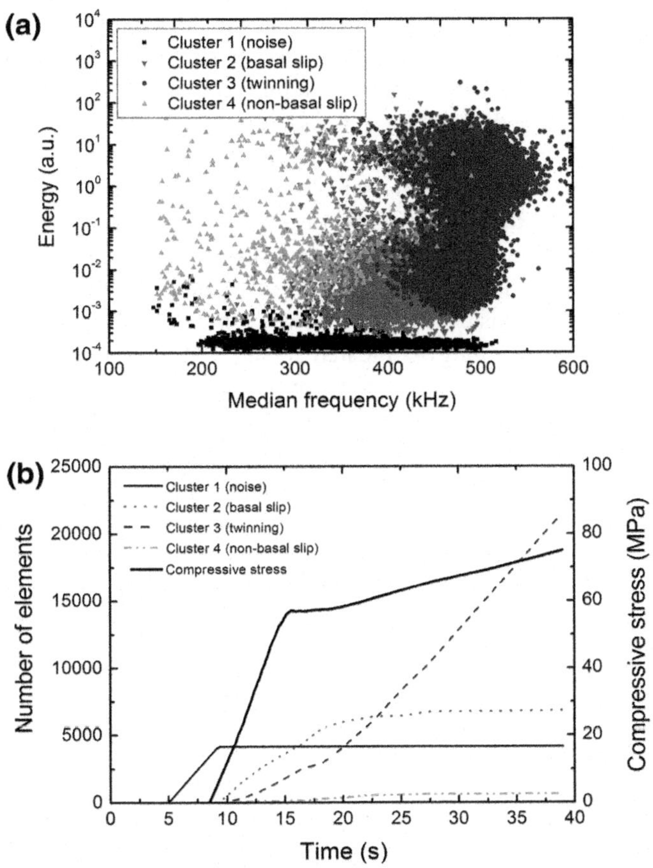


**Fig. 3** Dependence of energy on the median frequency (**a**), and the number of elements and stress vs. time in the AE clusters (**b**) for the sample compressed up to 75 MPa

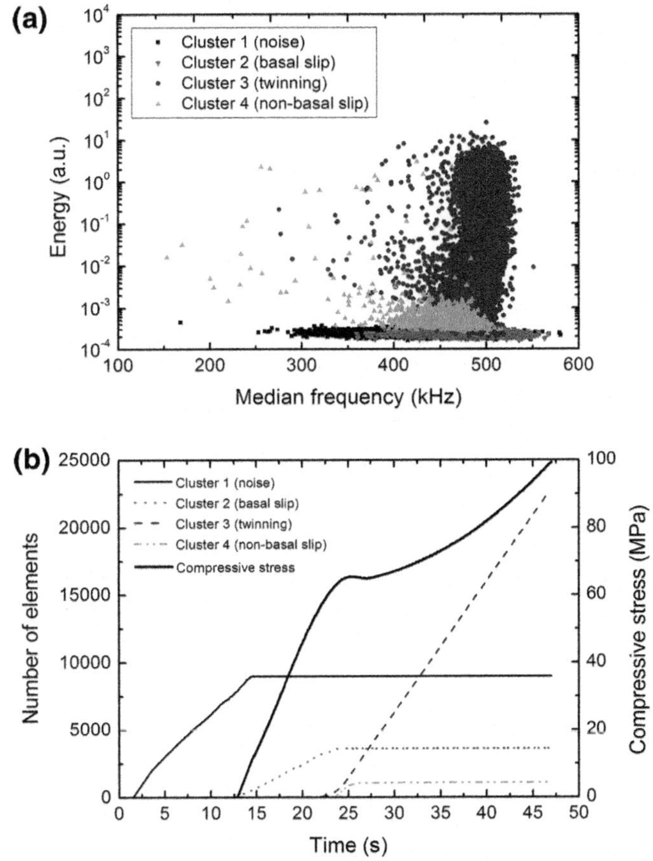


**Fig. 4** Dependence of energy on the median frequency (**a**), and the number of elements and stress versus time in the AE clusters (**b**) for the pre-compressed sample up to 75 MPa, annealed at 200 °C for 8 h and subsequently compressed up to 100 MPa (the TMT-c sample)

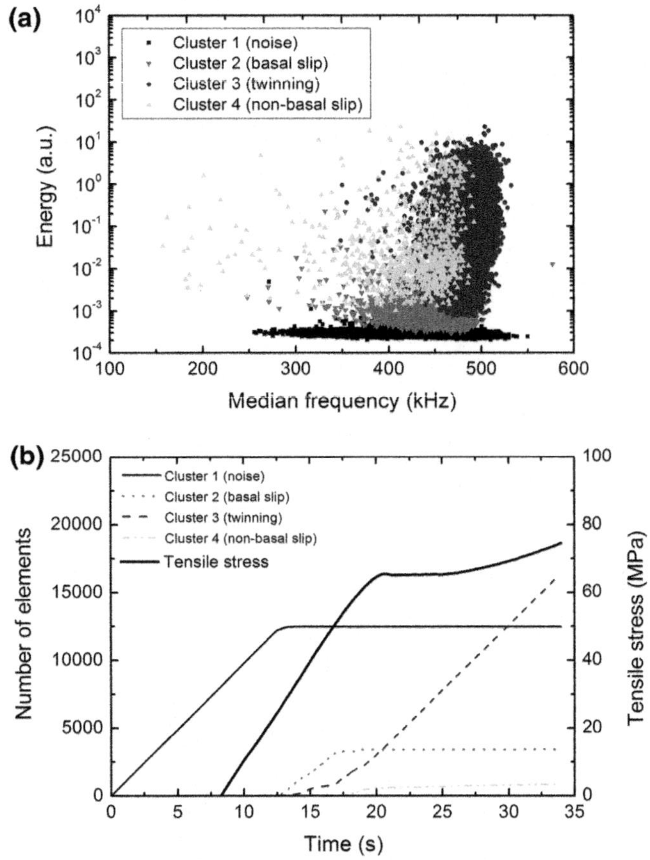


**Fig. 5** Dependence of energy on the median frequency (**a**), and the number of elements and stress versus time in the AE clusters (**b**) for the pre-compressed sample up to 75 MPa, annealed at 200 °C for 8 h and subsequently loaded in tension up to 75 MPa (the TMT-t sample)

(cluster 2) is less active in the TMT samples compared to its activity in the compressed sample (Figs. 3, 4 and 5). Twinning (cluster 3) starts to be dominant in the TMT-c sample at the macroscopic YP whereas in the TMT-t sample twinning occurs before YP. Furthermore, higher twinning activity can be seen in the TMT-c sample.

## Discussion

The Mg–1Zn alloy has fully recrystallized and homogeneous grain structure exhibiting a basal texture. Microstructure observation showed that sample after pre-compression up to 75 MPa (shortly after YP) contains {10-12} <10-11> extension twins. This is a typical deformation behavior in extruded Mg alloys. The ASKC data for the compressed sample were used as a reference set to ASKC data for deformation tests on the TMT samples. The effect of TMT with respect to applied loading direction is not pronounced and it can be manifested only in the yield behavior without a change in the yield strength. The AE technique is very sensitive to collective dislocation processes and to nucleation of twins. The twin growth does not produce detectable AE [11] and therefore, observed AE is a result of collective dislocation motion and/or twin nucleation. Assignment of clusters 2–4 to individual deformation mechanisms is based on previous results [7–9] and with respect to possible deformation mechanisms in the deformation mode.

The highest AE activity in the compressed sample (Fig. 3a) compared to the TMT samples can be explained by a higher mean free path of moving dislocation. In the TMT samples, a large number of obstacles in a form of twin boundaries and precipitates reduce the mean free path, and thus the AE energy of dislocation slip is lower (clusters 2 and 4, Figs. 3, 4 and 5a). The nucleation of twins (cluster 3, Figs. 3, 4 and 5a) is an excellent AE source and therefore the reduction of the AE energy in this cluster by an occurrence of obstacles is very low.

The reasons for the difference in the AE energy for dislocation clusters in the TMT samples (clusters 2 and 4, Figs. 4 and 5a) lies in applied loading direction. In the TMT-t sample, the detwinning process plays an important role. According to Christian and Mahajan [12], a higher stress is required for nucleation than for the propagation of twins. Thus, the extension twins produced during pre-compression become thinner and the mean free path of moving dislocation increases. This is also reflected in higher AE energy for dislocation clusters and in a lower number of elements in cluster 3 (twinning) compared to the TMT-c sample. The tensile loading reopens dislocation sources, so collective dislocation motion produces detectable AE signals with a relative high AE energy

The pinning effect of twin boundaries can be seen in the activity of cluster 3 (twinning, Fig. 4b). Plastic deformation is firstly realized mainly by basal slip and twinning occurs later at the yield strength, whereas in the sample without TMT twinning occurs almost concurrently with the basal slip (Fig. 3b).

## Conclusions

The influence of thermo-mechanical treatment consisting of pre-compression, annealing at 200 °C for 8 h of extruded binary Mg–1Zn alloy on deformation behavior was studied during either tension or compression loading. The application of ASKC to the AE data revealed that the AE energy originated from twinning is almost not dependent on loading direction. On the contrary, the AE energy from dislocation (basal, non-basal) slip is strongly affected by loading direction, where higher AE energy was observed during tension compared to the compressive loading. In tension, lower number of elements in cluster assigned to twinning can be explained by the fact that detwinning reduces the number of newly created twins. Increasing detwinned area promotes collective dislocation motion, what results in the strong AE energy of both basal and non-basal slip.

**Acknowledgements** This work received support from the Czech Science Foundation under grant No. 17 - 21855S; the Grant Agency of the Charles University under grant Nr. 1262217; the Operational Programme Research, Development and Education, The Ministry of Education, Youth and Sports (OP RDE, MEYS) under the grant CZ.02.1.01/0.0/0.0/16_013/0001794.

## References

1. C.J. Bettles, M.A. Gibson, "Current wrought magnesium alloys: strengths and weakness," *JOM*, 57 (2005), 46–49.
2. C.R. Heiple, S.H. Carpenter, "Acoustic emission produced by deformation of metals and alloys – A review: Part I.," *J. Acoustic Emission*, 6 (1987), 177–204.
3. C.R. Heiple, S.H. Carpenter, "Acoustic emission produced by deformation of metals and alloys – A review: Part II.," J. Acoustic Emission 6 (1987) 215–237.
4. A. Vinogradov, D. Orlov, A. Danyuk, Y. Estrin, "Effect of grain size on the mechanisms of plastic deformation in wrought Mg–Zn–Zr alloy revealed by acoustic emission measurements," *Acta Mater.*, 61 (2013), 2044–2056.
5. J. Čapek, K. Máthis, B. Clausen, J. Stráská, P. Beran, P. Lukáš, Study of the loading mode dependence of the twinning in random textured cast magnesium by acoustic emission and neutron diffraction methods, *Mat. Sci. Eng. a-Struct. 602 (2014), 25–32.*
6. D. Drozdenko, J. Bohlen, F. Chmelík, P. Lukáč, P. Dobroň, "Acoustic emission study on the activity of slip and twin mechanisms during compression testing of magnesium single crystals", *Mater. Sci. Eng. A 650 (2016), 20–27.*
7. E. Pomponi, A. Vinogradov, "A real-time approach to acoustic emission clustering" *Mech. Syst. Signal Process. 40 (2013) 791–804.*
8. K. Máthis, G. Csiszár, J. Čapek, J. Gubicza, B. Clausen, P. Lukáš, A. Vinogradov, S.R. Agnew, "Effect of the loading mode on the evolution of the deformation mechanisms in randomly textured magnesium polycrystals - Comparison of experimental and modeling results", *Int. J. Plast. 72 (2015) 127–150.*
9. K. Horváth, D. Drozdenko, K. Máthis, J. Bohlen, P. Dobroň " Deformation behavior and acoustic emission response on uniaxial compression of extruded rectangular profile of Mg-Zn-Zr alloy", *J. Alloys and Compds. 680 (2016) 623–632.*
10. V. Kree, J. Bohlen, D. Letzig, K.U. Kainer, *Practical Metallography, 41/5 (2004), pp. 233–246.*
11. J.P. Toronchuk, "Acoustic emission during twinning of zinc single crystals," *Mater. Eval.,* (1977), 51–53.
12. J.W. Christian, S. Mahajan, "Deformation twinning", Prog. Mater. Sci. 39 (1995), 1–157.

# The Effect of Initial Texture on Deformation Behaviors of Mg Alloys Under Erichsen Test

Jaiveer Singh, Min-Seong Kim, and Shi-Hoon Choi

**Abstract**

Deformation and fracture behaviors of AZ31 and E-form Mg alloys sheets were investigated during Erichsen test. Formability of Mg alloys was discussed in terms of Erichsen index (IE) and tests were conducted at room temperature using conventional Erichsen tester. The role of difference in initial textures and grain sizes in Mg alloys sheets was investigated to understand the deformation and fracture mechanisms in Mg alloys during Erichsen test. The evolution of the microstructure and microtexture of the deformed Mg alloys was analyzed via an electron back-scattered diffraction (EBSD) technique. Crystal plasticity finite element method (CPFEM) was used to predict the micromechanical deformation behavior of Mg alloys during Erichsen test. EBSD analysis revealed that deformation twins along with shear localization by dislocation slip were the main deformation mechanisms during Erichsen test. E-form Mg alloys with a weaker basal texture show higher IE compared to AZ31 alloy with a stronger basal texture.

**Keywords**

Mg alloys • Erichsen test • Fracture • Formability CPFEM

## Introduction

Magnesium (Mg) alloys are attractive for commercial applications in aerospace/automotive industry and known to be one of the lightest structural alloys [1]. However, commercial usage of Mg alloys is limited because of its low formability at room temperature (RT). The critical resolved shear stress (CRSS) for the basal slip is much smaller than that of non-basal slips and the activation of non-basal slips are hardly occurs at RT in the Mg alloys [2]. Therefore, Mg alloys exhibit poor plastic formability at RT. Although, the grain reorientation and refinement via severe plastic deformation have been proposed to enhance the RT ductility [3]. The addition of rare earth (RE) and Ca elements in Mg alloys have succeeded in improving the formability, good creep resistance and higher strength among the lightest structural alloys [4, 5]. The addition of specific elements is attributed to a reduction in basal texture intensity which results in enhanced stretch formability [6].

The deformation mechanisms and failure behaviors of Mg alloys could be significantly affected by complex stress states during Erichsen test. Thus far, many studies were mainly focused on the enhancement of stretch formability in Mg alloys by cross-wavy bending [7], pre-stretching and annealing treatment [8], grain boundary sliding [9], the addition of Ca and Ce [10, 11], twinning-mediated formability [12] and pre-twinning [13]. Huang et al. [14] studied the effects of initial microstructure on the microstructural evolution and stretch formability of warm rolled AZ31 Mg alloy sheets. Park et al. [15] reported the remarkable improvement of the stretch formability of rolled AZ31 Mg alloy at RT which was attributed by introducing initial $\{10\bar{1}2\}$ twins. However, many efforts have been made in understanding the effects of initial microstructures/textures on deformation and fracture behaviors in Mg alloys under the Erichsen test at RT [14, 16–21]. Therefore, the present work was motivated to study the effects of initial texture on Erichsen index (IE) and deformation behaviors under the complex stress states during Erichsen tests at RT. The IE is defined as the distance traveled by punch onto the sheet specimen till cracks begin to appear on the surface and is a measure for the formability of the sheet specimens during stretch forming.

Simulation of forming process such as Erichsen test for Mg alloys via finite element analysis (FEA) remains a challenge [7, 13, 22, 23]. During the sheet metal forming for automotive applications, heterogeneous deformation occurs

J. Singh · M.-S. Kim · S.-H. Choi (✉)
Department of Printed Electronics Engineering, Sunchon National University, Suncheon, Jeonnam 57922, Republic of Korea
e-mail: shihoon@scnu.ac.kr


D. Orlov et al. (eds.), *Magnesium Technology 2018*, The Minerals, Metals & Materials Series, https://doi.org/10.1007/978-3-319-72332-7_35

through the thickness direction [24]. The polycrystal plasticity modeling of hexagonal close-packed (hcp) materials can be used to analyze the mechanical behavior, deformation mechanisms, and microstructure/texture evolution. Many studies based on the crystal plasticity finite element method (CPFEM) have been conducted to simulate the plastic deformation behaviors and texture evolution in Mg alloys [25–28]. In the previous study, Choi et al. [25] used the CPFEM model, considering both slip and twin as deformation modes to simulate the texture evolution and macroscopic properties of AZ31 Mg alloy. In another study [26], CPFEM model was also used to predict the spatial stress concentration during in-plane compression in polycrystalline AZ31 Mg alloy. Staroselsky and Anand [28] used the crystal-mechanics-based constitutive model, which also accounts both crystallographic slip and deformation twinning to predict the mechanical response and texture evolution in AZ31B Mg alloy.

In the present work, the effects of initial texture on Erichsen index (IE) and deformation behavior during Erichsen test at RT were performed on the AZ31 and E-form Mg alloys. EBSD technique was used to investigate the microstructural and microtextural evolution through the thickness direction during Erichsen test. CPFEM was used to analyze the micromechanical deformation and fracture behaviors during Erichsen test.

## Experimental and Simulation Details

Hot-rolled AZ31 and E-form® (Mg–Al based Easy-formable alloy) Mg alloys sheet of 0.5 mm initial thickness were used in the present study. The microstructure and microtexture of the as-rolled specimen and as-deformed specimens by Erichsen test were characterized using electron back-scattered diffraction (EBSD) technique. The samples were polished through standard metallography polishing by mechanical grinding using abrasive papers (#800, #1200, and #2400) followed by ethanol-based 3–1 μm diamond suspension and final polishing using colloidal silica. All the microstructural measurements were made on a plane containing rolling (RD) and normal (ND) directions (RD-ND section). The EBSD scans were measured on a JEOL (JSM-7100F) by scanning an area of 100 μm × 100 μm at a step size of 0.25 μm. The EBSD data were processed with the condition of confidence index (CI) >0.1. The Erichsen test of the sheets with 0.5 mm thickness was performed at RT using conventional Erichsen tester. The dimensions of the specimens were 55 mm × 55 mm and punch speed was kept at 1 mm/min.

For simulation, the commercial software, ABAQUS v6.10 was used to understand the macroscopic deformation behavior during Erichsen test. For a detailed description of the crystal plasticity model, readers are suggested to refer the previous papers [25–27]. In the crystal plasticity model, four slip and two twin systems were considered. The set of parameters for AZ31 and E-form were determined by representative volume element (RVE) based on the results of uniaxial tension and compression tests. For defining the hardening parameters, a simple 3-D mesh ($20 \times 20 \times 10 = 4000$ elements) and an ODF (orientation distribution function) measured by X-ray diffraction were used. To understand the micromechanical deformation behavior during Erichsen test, the initial geometry of tools and sheet was considered as the same dimension with experiment. A 3-D C3D8R (in ABAQUS) element type with eight nodes and one integration point was used and a total number of elements was 43,899. To impose initial orientations of the elements, 40 orientations randomly selected from the 4000-grain orientations were mapped onto each integration point in the finite element mesh. It was assumed that the stress state at each integration point would be determined by a volume-average over the total number of grains comprising the respective material point. This scheme is called random mapping (RM) as given in Ref. [29]. For Erichsen test modeling, the friction coefficient $\mu = 0.05$ was imposed between the blank and tools. The IE of 2.8 and 4.5 mm were used for AZ31 and E-form Mg alloys for predictions, respectively.

## Results and Discussion

In order to compare the stretching formability of the AZ31 and E-form Mg alloys, Erichsen tests were carried out at RT. Figure 1a shows the top view of deformed specimens after Erichsen test. The failure in AZ31 by cracking occurred after the punch displacement of 2.63 mm, while in E-form, 4.84 mm. The black arrow indicates the macroscopic cracks on the deformed specimens. Erichsen test indicates that E-form shows much higher formability in terms of IE at RT than AZ31 Mg alloy. Figure 1b, c show the punch displacement-force curves and comparison of IE for AZ31 and E-form Mg alloys, as determined via conventional Erichsen tester at a displacement rate of 1 mm/min at RT, respectively. However, it should be noted that cracks in AZ31 occurred along the transverse direction (TD), while in E-form, along the rolling direction (RD).

Figure 2 collates the initial microstructure and microtexture of AZ31 and E-form Mg alloys measured through the thickness direction of the as-rolled sheet. Figure 2a, b show inverse pole figure (IPF), image quality (IQ) maps and pole figures of the RD-ND plane for the as-rolled AZ31 and E-form Mg alloys, respectively. The average grain sizes of AZ31 and E-form Mg alloys were 8.6 and 11.9 μm, respectively. The morphology of grains was almost equiaxed for both AZ31 and E-form Mg alloys. The maximum intensities of AZ31 and E-form alloys were 15.5 and 12.4,

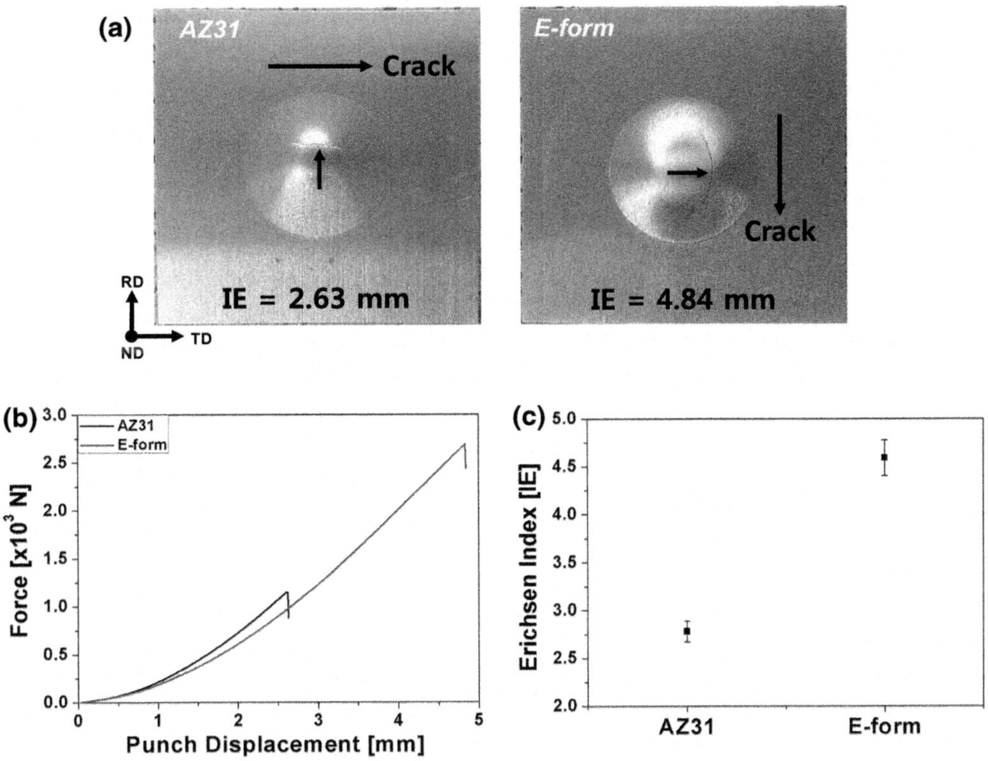


**Fig. 1** **a** Top view of AZ31 and E-form Mg alloys after Erichsen tests at room temperature. **b** Punch displacement-force curves and **c** Erichsen index of AZ31 and E-form Mg alloys were 2.63 and 4.84 mm, respectively

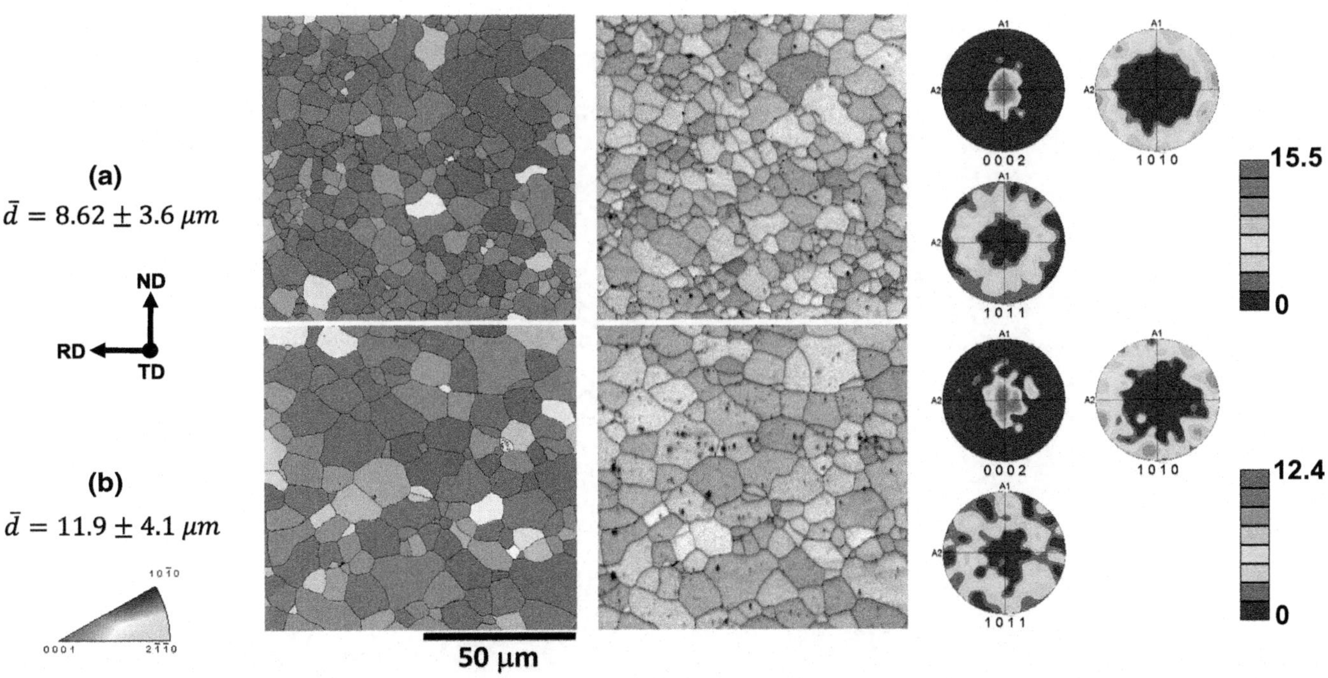


**Fig. 2** ND-IPF (normal direction—inverse pole figure), IQ (image quality) and pole figure (PF) maps of **a** AZ31 and **b** E-form Mg alloys measured through the thickness direction of as-received rolled sheets

respectively. Therefore, it is evident that the higher IE of the E-form Mg alloy is closely related to the weaker basal texture compared to AZ31 Mg alloy. The IE of the AZ31 specimen having strong basal texture was only 2.63 mm, which means poor formability. The strong basal texture places most grains with *c*-axis parallel to ND, which are difficult to deform since the RSS (resolved shear stress) in the basal plane is equal to zero under in-plane tension or compression, which results in stress localization and shear failure. However, in the case of E-form Mg alloy with lower basal texture intensity, a large number of grains will be favorably orientated for deformation, which contributed to higher IE and improvement in the formability. To examine the microstructure and microtexture evolution during Erichsen test at RT, Erichsen tested specimens have been analyzed by EBSD. Complex stress states can occur throughout the whole specimens during Erichsen testing. Figure 3a, b shows the IPF, IQ maps and corresponding pole figures of the AZ31 and E-form Mg alloys after Erichsen test at RT, respectively. Analysis was carried out near the cracks present in the deformed specimen. The maximum intensities of the AZ31 and E-form specimens after Erichsen test were 18.7 and 20.1, respectively. These results suggest the enhancement in the microtexture after Erichsen test at RT. We can also observe the deformation twins in the E-form Mg alloy, while AZ31 indicates mostly localized shear regions. Although, localized deformation zones were mostly confined to the grain boundaries during Erichsen test. The above-mentioned results show that there is a significant activity of deformation twins in the E-form Mg alloy during Erichsen test which results in the accommodation of high deformation. Hence, E-form Mg alloy had higher IE as compared to AZ31 Mg alloy.

In this study, CPFEM was used to understand the micromechanical deformation behavior of Mg alloys under Erichsen test at RT. However, the simulation using macroscopic FEA demonstrated the evolution of effective stress and strain distribution in any region of the deformed blank of Mg alloys after Erichsen test. Figure 4 shows the effective stress and strain distributions in the blank (top and bottom view) of AZ31 and E-form Mg alloys after Erichsen tests at RT. The color contour represents the magnitude of the effective stress and strain distributions on the blank deformed to the punch displacement of 2.8 and 4.5 mm for the AZ31 and E-form Mg alloys, respectively. The orientation option in ABAQUS was used to specify local axis systems (1, 2, and 3) for material that is parallel to RD, TD, and ND, respectively, with respect to a deformed specimen. The effective stress and strain distribution also indicate probable sites for the high-stress concentration which could result in a crack. Similarly, Fig. 5 shows the effective stress and strain distribution on cross-sectional planes. The strain and stress distributions for both AZ31 and E-form Mg alloys were concentrated at the center of the sheets and E-form alloy had a higher level of strain and stress compared to AZ31 Mg alloy. The effective strain distribution was

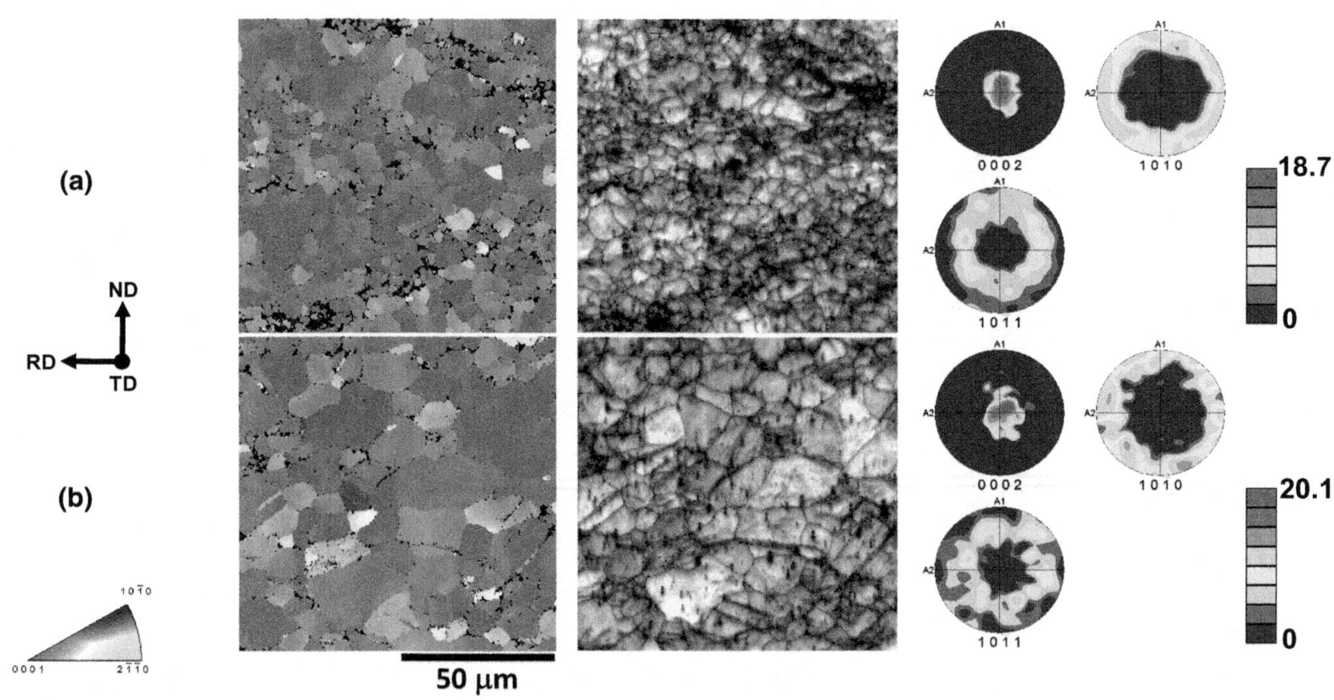


**Fig. 3** ND-IPF (normal direction—inverse pole figure), IQ (image quality) and pole figures (PF) maps of **a** AZ31 and **b** E-form Mg alloys. The measurement was done near the crack position through thickness direction after Erichsen tests at room temperature

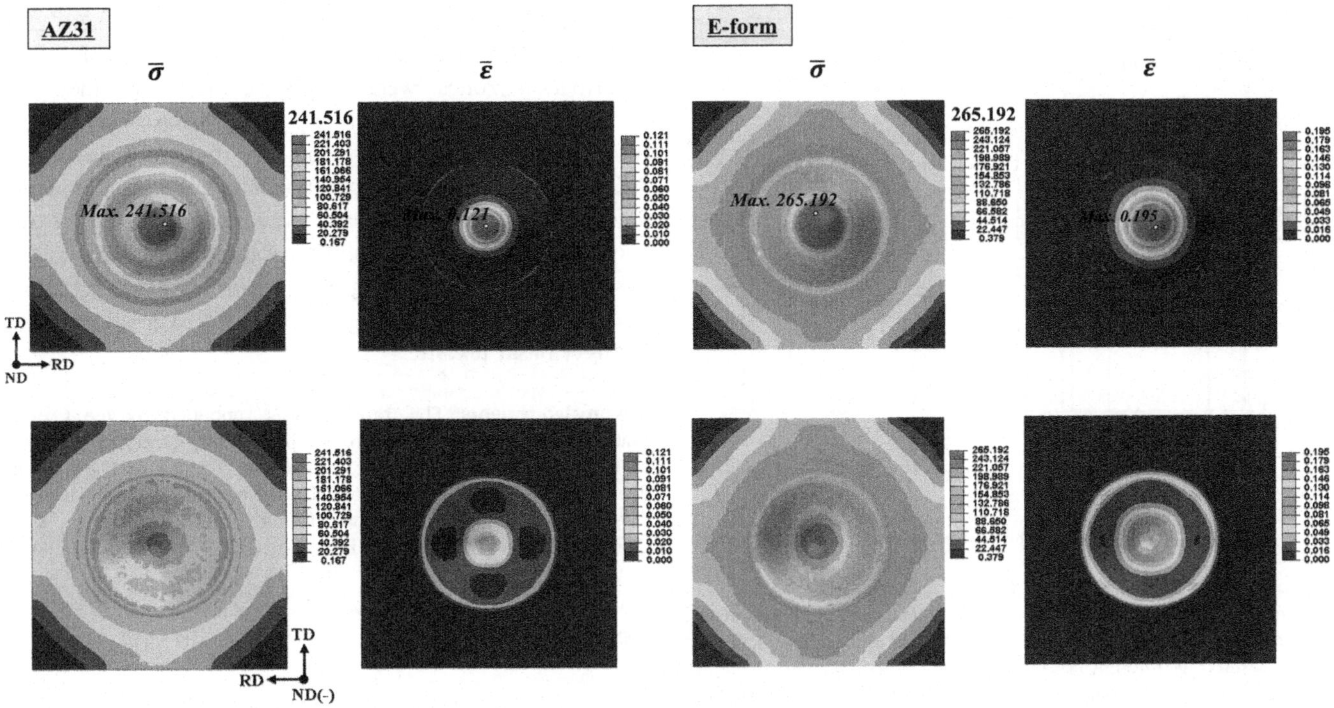


**Fig. 4** Effective stress and strain distribution (top and bottom view) in the AZ31 and E-form Mg alloys after Erichsen tests at room temperature

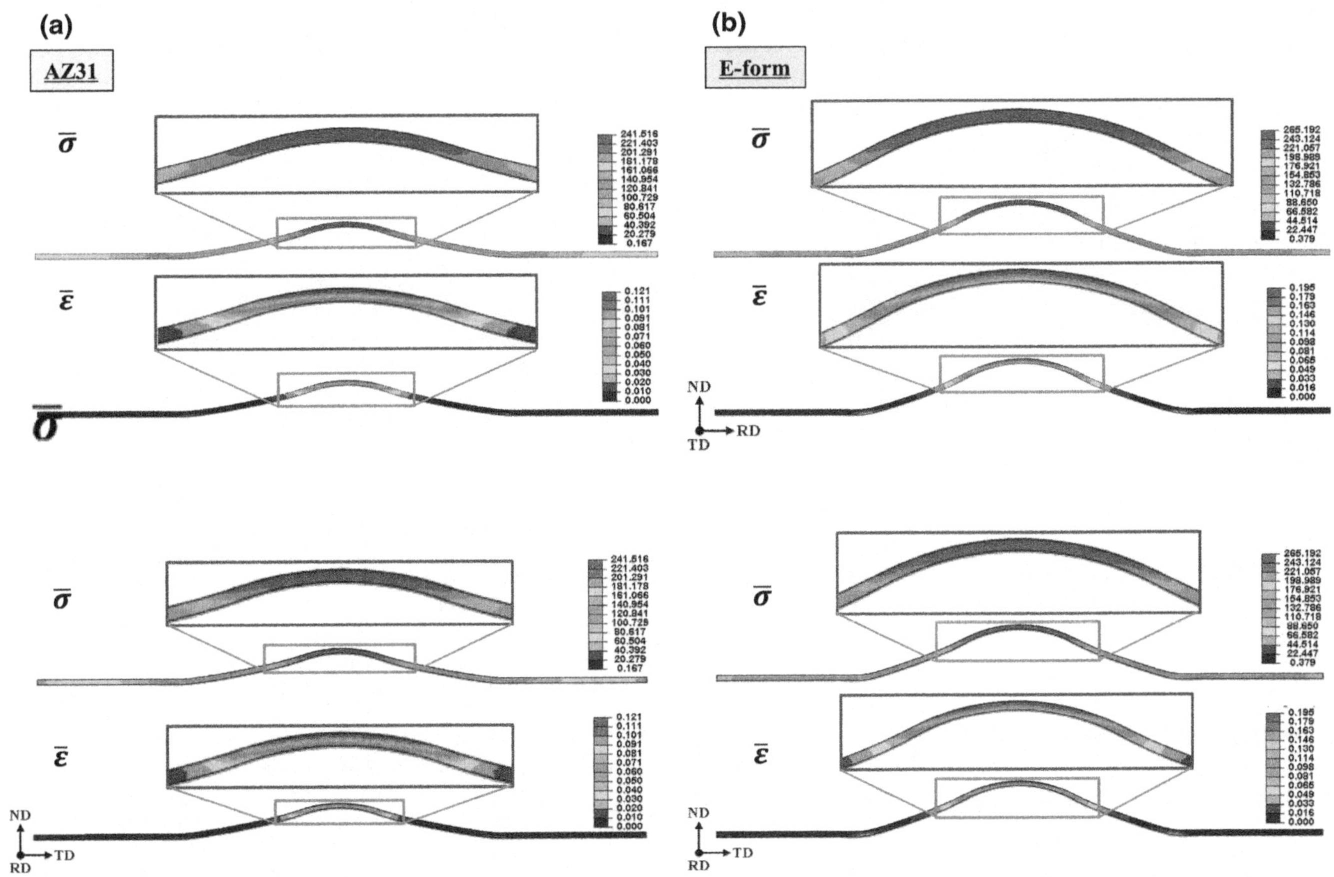


**Fig. 5** Effective stress and strain distribution on cross-sectional planes **a** AZ31 and **b** E-form Mg alloys after Erichsen tests at room temperature

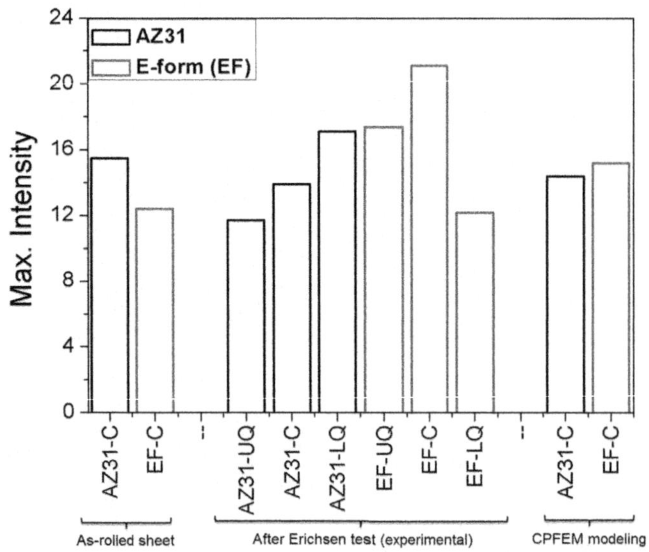


**Fig. 6** Maximum intensities of (0002) pole figures of AZ31 and E-form Mg alloys before and after Erichsen tests at room temperature. The microtextures were obtained from as-rolled sheets, after Erichsen test (measured at different positions viz. UQ: upper quarter, C: center and LQ: lower quarter) and microstructure based CPFEM simulation

heterogeneous through the thickness direction during Erichsen test. It should be noted that E-form Mg alloys underwent the highest level of effective stress and strain in the hot spots while AZ31 Mg alloys experienced a slightly lower level of effective stress and strain which could be observed in the through thickness direction.

Figure 6 shows the maximum intensities of (0002) pole figure of AZ31 and E-form Mg alloys before and after Erichsen tests at RT. The microtexture intensities were obtained from the as-rolled sheets, Erichsen tested specimens (measured at different positions in the through thickness direction viz. UQ: upper quarter, C: center and LQ: lower quarter) and microstructure based CPFEM for AZ31 and E-form Mg alloys. This result indicates the microtexture strengthening after Erichsen tests at RT. Therefore, it should be noted that microstructure based CPFEM model successfully captured the microtexture enhancement as observed in the experimental work. The detailed results obtained from the microstructure based CPFEM model will be published separately. However, shear localized deformation along with deformation twins was the main contributions of microtexture evolution through the thickness direction during Erichsen test.

## Conclusions

Evolution of microstructure and microtexture in the AZ31 and E-form Mg alloys through the thickness direction was examined using EBSD technique under Erichsen test at RT. The effects of the initial crystallographic texture and grain sizes which exerted on the IE were studied. The localized deformation zones were mostly confined to the grain boundaries during Erichsen test. CPFEM has been successfully captured the deformation and failure behaviors of AZ31 and E-form Mg alloys under Erichsen test and predicted the effective stress and strain distributions. Texture evolution was also successfully captured via microstructure based CPFEM. E-form Mg alloys with a weaker basal texture show higher IE compared to AZ31 Mg alloy with a stronger basal texture.

**Acknowledgements** This research was supported by the Creative Materials Discovery Program through the National Research Foundation of Korea (NRF) funded by the Ministry of Science, ICT and Future Planning (NRF-2015M3D1A1069710) and the Basic Science Research Program through the National Research Foundation of Korea (NRF) funded by the Ministry of Education (NRF-2014R1A6A1030419).

## References


1. A. Stalmann, W. Sebastian, H. Friedrich, S. Schumann, and K. Dröder, "Properties and processing of magnesium wrought products for automotive applications," Adv. Eng. Mater., 3 (2001), 969–974.
2. S.R. Agnew, and Ö. Duygulu, "Plastic anisotropy and the role of non-basal slip in magnesium alloy AZ31B," Int. J. Plast., 21 (2005), 1161–1193.
3. S. Suwas, G. Gottstein, and R. Kumar, "Evolution of crystallographic texture during equal channel angular extrusion (ECAE) and its effects on secondary processing of magnesium," Mater. Sci. Eng. A., 471 (2007), 1–14.
4. N. Stanford, and M.R. Barnett, "The origin of "rare earth" texture development in extruded Mg-based alloys and its effect on tensile ductility," Mater. Sci. Eng. A., 496 (2008), 399–408.
5. J.H. Kim, N.E. Kang, C.D. Yim, and B.K. Kim, "Effect of calcium content on the microstructural evolution and mechanical properties of wrought Mg-3Al-1Zn alloy," Mater. Sci. Eng. A., 525 (2009), 18–29.
6. Y. Chino, T. Ueda, Y. Otomatsu, K. Sassa, X. Huang, K. Suzuki, and M. Mabuchi, "Effects of Ca on Tensile Properties and Stretch Formability at Room Temperature in Mg-Zn and Mg-Al Alloys," Mater. Trans., 52 (2011), 1477–1482.
7. Q. Huo, X. Yang, H. Sun, B. Li, J. Qin, J. Wang, and J. Ma, "Enhancement of tensile ductility and stretch formability of AZ31 magnesium alloy sheet processed by cross-wavy bending," J. Alloys Compd., 581 (2013), 230–235.
8. H. Zhang, G. Huang, L. Wang, and J. Li, "Improved formability of Mg-3Al-1Zn alloy by pre-stretching and annealing," Scr. Mater., 67 (2012), 495–498.
9. H. Somekawa, A. Kinoshita, K. Washio, and A. Kato, "Enhancement of room temperature stretch formability via grain boundary sliding in magnesium alloy," Mater. Sci. Eng. A., 676 (2016), 427–433.
10. Y. Chino, X. Huang, K. Suzuki, and M. Mabuchi, "Enhancement of Stretch Formability at Room Temperature by Addition of Ca in Mg-Zn Alloy," Mater. Trans., 51 (2010), 818–821.
11. Y. Chino, M. Kado, and M. Mabuchi, "Enhancement of tensile ductility and stretch formability of magnesium by addition of 0.2 wt%(0.035 at%)Ce," Mater. Sci. Eng. A., 494 (2008), 343–349.

12. B.-C. Suh, J.H. Kim, J.H. Hwang, M.S. Shim, and N.J. Kim, "Twinning-mediated formability in Mg alloys," Sci. Rep., 6 (2016), 22364.
13. W. He, Q. Zeng, H. Yu, Y. Xin, B. Luan, and Q. Liu, "Improving the room temperature stretch formability of a Mg alloy thin sheet by pre-twinning," Mater. Sci. Eng. A., 655 (2016), 1–8.
14. X. Huang, K. Suzuki, M. Yuasa, and Y. Chino, "Effects of initial microstructure on the microstructural evolution and stretch formability of warm rolled Mg-3Al-1Zn alloy sheets," Mater. Sci. Eng. A., 587 (2013), 150–160.
15. S.H. Park, S.G. Hong, and C.S. Lee, "Enhanced stretch formability of rolled Mg-3Al-1Zn alloy at room temperature by initial {10-12} twins," Mater. Sci. Eng. A., 578 (2013), 271–276.
16. H.L. Kim, W.K. Bang, and Y.W. Chang, "Deformation behavior of as-rolled and strip-cast AZ31 magnesium alloy sheets," Mater. Sci. Eng. A., 528 (2011), 5356–5365.
17. J. Dong, D. Zhang, Y. Dong, S. Chai, and F. Pan, Microstructure evolution and mechanical response of extruded AZ31B magnesium alloy sheet at large strains followed by annealing treatment," Mater. Sci. Eng. A., 618 (2014), 262–270.
18. X. Huang, K. Suzuki, and N. Saito, "Microstructure and mechanical properties of AZ80 magnesium alloy sheet processed by differential speed rolling," Mater. Sci. Eng. A., 508 (2009), 226–233.
19. J.-Y. Lee, Y.-S. Yun, B.-C. Suh, N.-J. Kim, W.-T. Kim, and D.-H. Kim, "Comparison of static recrystallization behavior in hot rolled Mg-3Al-1Zn and Mg-3Zn-0.5Ca sheets," J. Alloys Compd., 589 (2014), 240–246.
20. H. Zhang, G. Huang, D. Kong, G. Sang, and B. Song, "Influence of initial texture on formability of AZ31B magnesium alloy sheets at different temperatures," J. Mater. Process. Technol., 211 (2011), 1575–1580.
21. H.L. Kim, W.K. Bang, and Y.W. Chang, "Effect of initial texture on deformation behavior of AZ31 magnesium alloy sheets under biaxial loading," Mater. Sci. Eng. A., 552 (2012), 245–251.
22. M.. Mekonen, D. Steglich, J. Bohlen, L. Stutz, D. Letzig, and J. Mosler, "Experimental and numerical investigation of Mg alloy sheet formability," Mater. Sci. Eng. A., 586 (2013), 204–214.
23. J. Yoon, and Y. Lee, "Fracture mechanism of Mg-3Al-1Zn sheet at the biaxial state with respect to forming temperatures," Mater. Des., 55 (2014), 43–49.
24. J. Singh, M.-S. Kim, and S.-H. Choi, "The effect of strain heterogeneity on the deformation and failure behaviors of E-form Mg alloy sheets during a mini-V-bending test," J. Alloys Compd., 708 (2017), 694–705.
25. S.-H. Choi, D.H. Kim, H.W. Lee, and E.J. Shin, "Simulation of texture evolution and macroscopic properties in Mg alloys using the crystal plasticity finite element method," Mater. Sci. Eng. A., 527 (2010), 1151–1159.
26. S.-H. Choi, D.H. Kim, S.S. Park, and B.S. You, "Simulation of stress concentration in Mg alloys using the crystal plasticity finite element method," Acta Mater., 58 (2010), 320–329.
27. S.-H. Choi, D.W. Kim, B.S. Seong, and A.D. Rollett, "3-D simulation of spatial stress distribution in an AZ31 Mg alloy sheet under in-plane compression," Int. J. Plast., 27 (2011), 1702–1720.
28. A. Staroselsky, and L. Anand, "A constitutive model for hcp materials deforming by slip and twinning: Application to magnesium alloy AZ31B," Int. J. Plast., 19 (2003), 1843–1864.
29. E.-Y. Kim, H.S. Yang, S.H. Han, J.H. Kwak, and S.-H. Choi, "Effect of initial microstructure on strain-stress partitioning and void formation in DP980 steel during uniaxial tension," Met. Mater. Int., 18 (2012), 573–582.

# Measurement of Twin Formation Energy Barriers Using Nudged Elastic Band Molecular Statics

Deepesh Giri, Christopher Barrett, and Haitham El Kadiri

**Abstract**

The nudged elastic band (NEB) method [1, 2] is used to find the energy barrier separating stable twinned and untwinned states in Magnesium. This technique enables identification of the minimum energy path between the two stable states. The effects of dislocations, grain boundaries, and other defects which enhance twin nucleation produce a measurable effect on the minimum energy path [3, 4] and energy barrier. Thus, the NEB technique enables direct comparisons of the material twin-ability under a variety of boundary conditions. By simulating twinning in a variety of boundary conditions, NEB calculations provide key insight into the twin nucleation process by demonstrating both what defect arrangements encourage twin nucleation, and how likely twinning is to occur under given boundary conditions. This data is crucial for incorporating twin nucleation accurately into higher scale modeling.

**Keywords**

NEB • Energy • Twinning

## Introduction

One of the greatest challenges in materials modeling is the effective upscaling of relevant material properties while the description of material behavior is continually simplified from that which is used at lower scales. In particular, in moving from the atomistic modeling to higher scales such as dislocation dynamics, phase field modeling, and crystal plasticity, the material description changes from describing individual atoms which have their own freedom to move to treating a continuous material which experiences stress and deformation as a solid body. This drastic simplification of the material description must be coupled with the introduction of many new material properties which are automatically present in the discrete framework. The material response to plasticity involves slip, twinning, grain boundary migration, grain boundary sliding, recrystallization, faceting [5] and many other effects which are by-products of the discrete nature of atoms, but can be described in a continuous material by introducing new laws of behavior. A particular challenge in this regard is how best to capture the role of twinning in higher scale plasticity.

Twinning drastically alters the positions of atoms in its wake, inducing a transformation strain in a particular region which has a newly oriented structure and is surrounded by new interfaces, shuffling [6] all of the atoms inside this region. This complex process occurs very quickly as twins can shoot from one side of a grain to the other and in magnesium [7, 8] such twins have been known to grow until the entire parent grain they occur in has been consumed. Such an unusual process requires dedicated treatment in higher scale modeling, and often can only roughly approximate experimental data. In this work, we focus on the difficult question of how such twins originate.

Twin nucleation is of great concern because the ensuing growth of twins is highly dependent on the points of their origin. Identification of origin points in an initial microstructure undergoing deformation is still at infantile stages, making it a ripe area for dedicated study. To quantify the points where twins prefer to nucleate and fully describe the nucleation process up to the point that a stable embryo is present and ready to begin rapid growth, we turn the nudged elastic band (NEB) method.

The NEB method is an algorithm which is an approximation defined to find the minimum energy path (MEP). The MEP is a curve through phase space that finds its way from wells over saddle points [9] to connect the wells by

D. Giri (✉) · C. Barrett · H. E. Kadiri
Department of Mechanical Engineering, Mississippi State University, Starkville, MS 39762, USA
e-mail: dg1295@msstate.edu

D. Giri · C. Barrett · H. E. Kadiri
Center for Advanced Vehicular System, Mississippi State University, Starkville, MS 39762, USA


D. Orlov et al. (eds.), *Magnesium Technology 2018*, The Minerals, Metals & Materials Series, https://doi.org/10.1007/978-3-319-72332-7_36

using the lowest energies possible to create that curve. In other words, the NEB method searches for how to move all of the atoms synchronously to get from the state with no twin to the state with the twinned embryo while needing the least amount of energy possible. The algorithm guesses initial path for all the atoms, iterates several times to relax these paths and finally gives the MEP along with the height of the activation barrier [10]. To govern the identification of the MEP, NEB first identifies a linear path traversed by all the atoms from the initial to final states. Then it defines several new states along that path and allows the atoms in those states to relax toward a lower energy configuration while being constrained with a spring force that prevents their distance from neighboring states from growing too large. This enables all of the states to relax while remaining continuous. The phase space MEB path is then defined as the combined motions of all the atoms as they gradually move from the initial to the final state.

The MEP is a path for a rearrangement of a group of atoms from one stable configuration to another. Although other methods like path of slowest ascent, chain of states etc. were earlier used to find the MEP, they had the shortcomings such as inaccuracies resulting from the MEP path being much longer than a linear path, and higher uncertainty near the saddle point [3, 4], which is the most crucially needed measurement in many cases. [1]. The NEB method has been proven ideal for measuring the MEP path in comparison to these other algorithms, and therefore, minimization of the elastic band is carried out to solve this problem [1].

## Methodology

For our simulations, we used single crystals and bicrystals of Magnesium and Titanium. Both are HCP metals and undergo twinning to accommodate c-axis deformation [9, 19, 20]. At first, we relaxed **a** and **c** parameters for Mg by minimizing 2-atom unit cell in Lammps [11]. At first, we relaxed **a** and **c** parameters for Mg by minimizing 2-atom unit cell in Lammps [11]. Using these values, we created a Mg single crystal in Matlab with almost 125,000 atoms. This crystal is a totally pristine crystal and doesn't contain any kind of defects in it. We took the same crystal and inserted a {10-12} twin embryo in it using a Matlab code which solves for the displacement field due to a series of twinning dislocation loops plus shuffling inside the embryo using the isotropic approximation. We used 4 dislocation loops, separated by a distance of 2*c to generate this embryo. The twin embryo consists of around 3000 atoms. Next, we minimized these two crystals—pristine and with twin embryo. Let's call these minimum files as $\text{pristine}_{\text{min}}$ and $\text{embryo}_{\text{min}}$ respectively. The Conjugate Gradient (cg) method was used for minimization. Upon minimization, we require that the twin embryo remained stable in the crystal. For this purpose, we applied a shear stress of 1.2 GPa. We considered the $\text{pristine}_{\text{min}}$ and $\text{embryo}_{\text{min}}$ as two stable states of twin embryo nucleation and obtained the intermediate states by taking a fraction of the difference between these two stable states. We used the difference between the atomic coordinates of the two stable files and created 10 files altogether to represent the transition. Then we used Lammps to calculate the energies of each state and obtained an energy graph of the process as shown in Fig. 1. This illustrates the concept of the atom paths through phase space and the evolution of the energy along the MEP.

Unfortunately, for this simplistic conceptual illustration, we were unable to get a real MEP using NEB because of the instability of the twin embryo.

For a more realistic measurement, we took the same Mg pristine crystal but inserted a twin embryo at the surface. As earlier, the twin embryo was created using 4 dislocation loops but this time the loops were much larger to generate a bigger embryo. We sliced the crystal along the basal plane of the twin using Matlab. This reduced the number of atoms in the crystal to almost 110,000. The sliced embryo contains around 3900 atoms. Accordingly, we sliced the pristine crystal as well so that both the crystals contain atoms with same atom ID. We minimized both the pristine and twin embryo crystals. This time we got stable twin embryo without using any stress. The boundary conditions used were surface(x), surface(y) and periodic(z). Then we used NEB method to simulate this transition. We used 8, 16 and 24 replicas in 3 different simulations. Figure 2 shows the two stable crystals used in this process.

In the third case, we took a Magnesium bicrystal with a <2110> asymmetric tilt grain boundary (GB) [12, 13]. This crystal contains 188760 atoms. Tension was applied on this bicrystal [14] along the c-axis of the top grain and the resulting state contained two twin embryos along with deformed interface and presence of dislocations. The initial bicrystal and the one with twin embryo were minimized under zero stress to obtain the two stable states of this transition. Let's call these stable states as $\text{GB}_{\text{inital}}$ and $\text{GB}_{\text{final}}$. Then we used the NEB method with 16 replicas to obtain the MEP of this phenomenon, as shown in Fig. 3. The boundary conditions in this case were periodic, periodic, surface in x, y and z respectively.

For both the second and third case, minimization was done using 'quickmin' [2] method in Lammps. The NEB method works only with 'quickmin' and 'fire' minimization styles [15]. These files were opened in Ovito [16] and color coding was done as per their energy values.

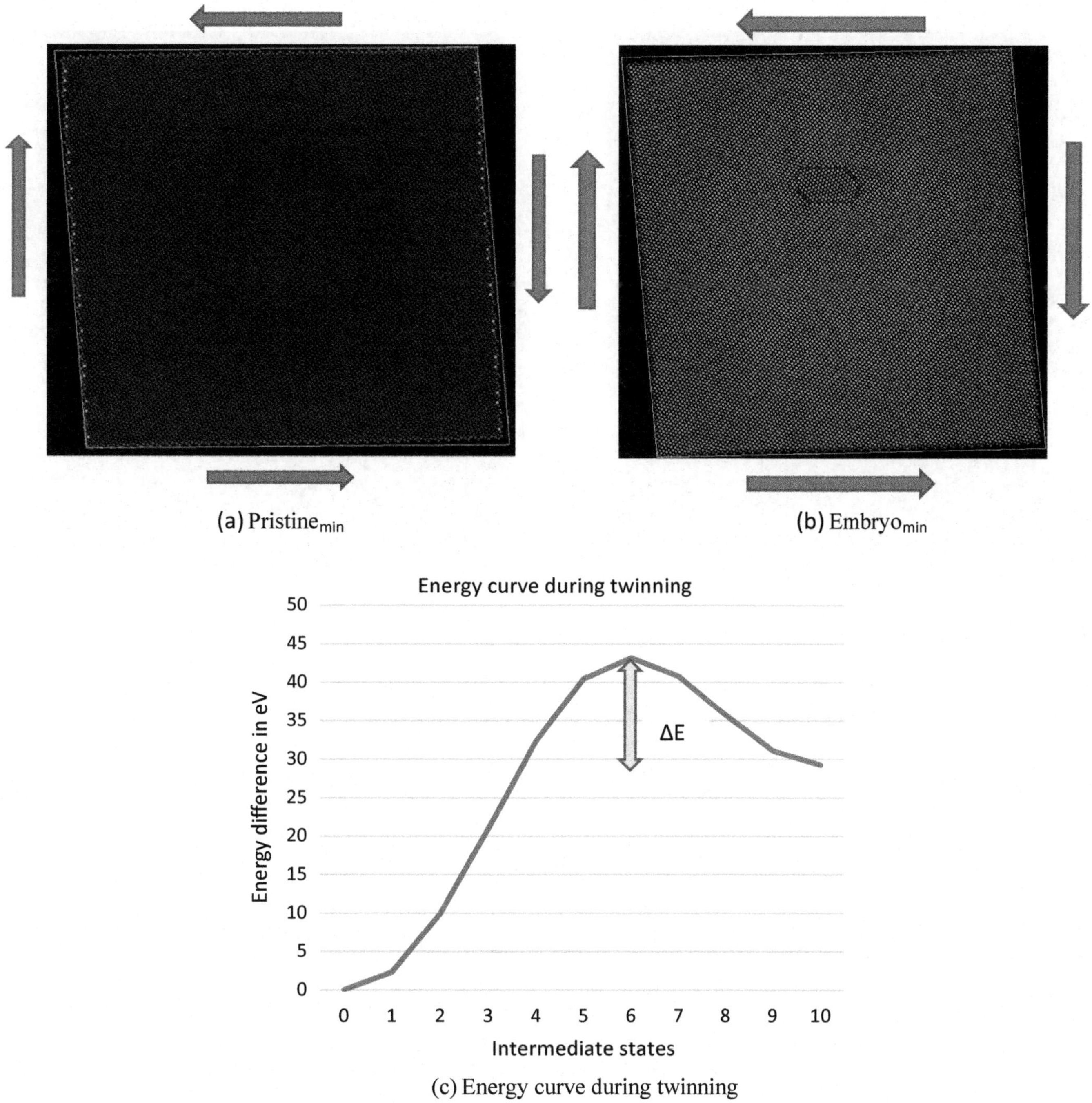


**Fig. 1** **a** Minimized pristine file. The arrows indicate the application of xy shear stress. **b** Minimized {10-12} twin embryo. **c** Energy curve of transition. The energy barrier for this case was 43.16 eV

## Results and Discussion

For the first case, although we were able to get good energy curve by calculating the energies of individual files, we were not able to get it done using NEB method. The possible reason behind this could be because we used totally an ideal crystal for the purpose. Presence of a twin embryo in the middle of a homogenous crystal is practically impossible. The crystal doesn't contain any defects or dislocations and inserting a twin embryo using Matlab doesn't resemble a real-case scenario. We think NEB method wasn't designed to simulate an ideal case and hence, probably we were not able to get good results. We included these results in our methodology to illustrate the MEP concept.

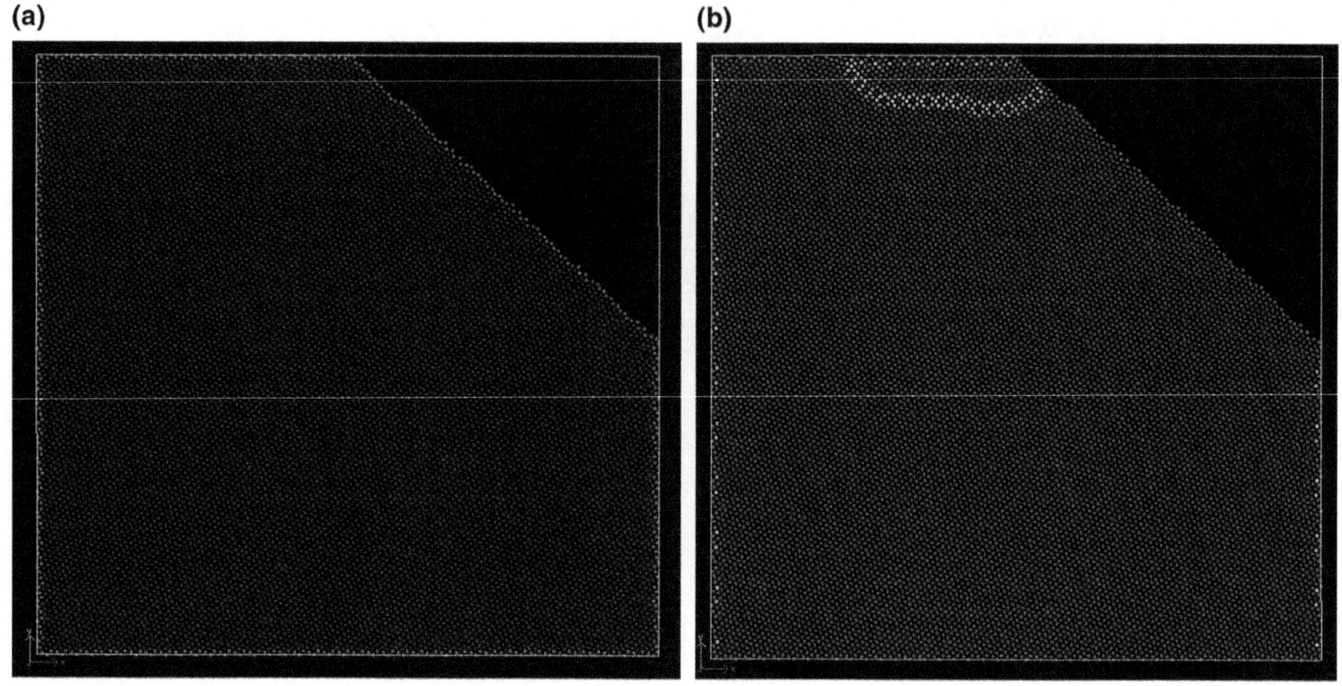

**Fig. 2** **a** Minimized Pristine crystal. **b** Minimized {10-12} twin embryo. Both are sliced along the basal plane of the twin

The second case better resembles a real case scenario. A twin embryo on the surface of a crystal is more likely to be formed than in the middle. If we look closely, the twin embryo has been sliced along the basal plane. The energy of the surface of the basal plane is lower than that of the prismatic plane and this gives it more favorability to form twin. The basal plane is closer packed than the prismatic plane and hence the atoms in basal plane have lower energy. Upon using 24 replicas to simulate this process, an activation barrier of 52.04 eV was obtained. Visualizing the dump files in Ovito, it is seen that the twin embryo is gradually growing until it reaches the last replica. The twin embryo actually starts forming from the 14th replica and the energy actually drops at that point and gradually increases until the penultimate replica. The last replica has slight reduction in energy (0.26 eV).

The third case looks more realistic than the other two since we used crystals with defects like GB and dislocations. Like the earlier cases, we minimized the crystals at first to obtain the two stable states and then used the NEB method. We used 16 replicas to simulate this transition and obtained the energy graph as shown in Fig. 3d. If we visualize the obtained dump files using Ovito, then it is seen that the twin embryo starts forming 9th replica onwards. However, there is sudden rise in energy in the 8th replica and there is some energy drop in the 9th one. This is consistent with the second case where energy dropped when twin embryo just started forming. Prior to this, there was almost a linear increase in energy. And even after the drop, the energy keeps increasing. It's natural for the energy to increase during twinning but the drop in the middle replica is a matter requiring further research.

We used a spring constant of 10 eV/Ang$^2$ in all the simulations and we think adjusting the values of spring constant and increasing the minimization of each replica would give better results which is a part of our future work. Also, we are working to compare the behavior of Titanium [17] to Magnesium twin nucleation under the same conditions. Like Magnesium, we have generated the a and c parameters for Ti and working on the implementation of NEB method for similar cases.

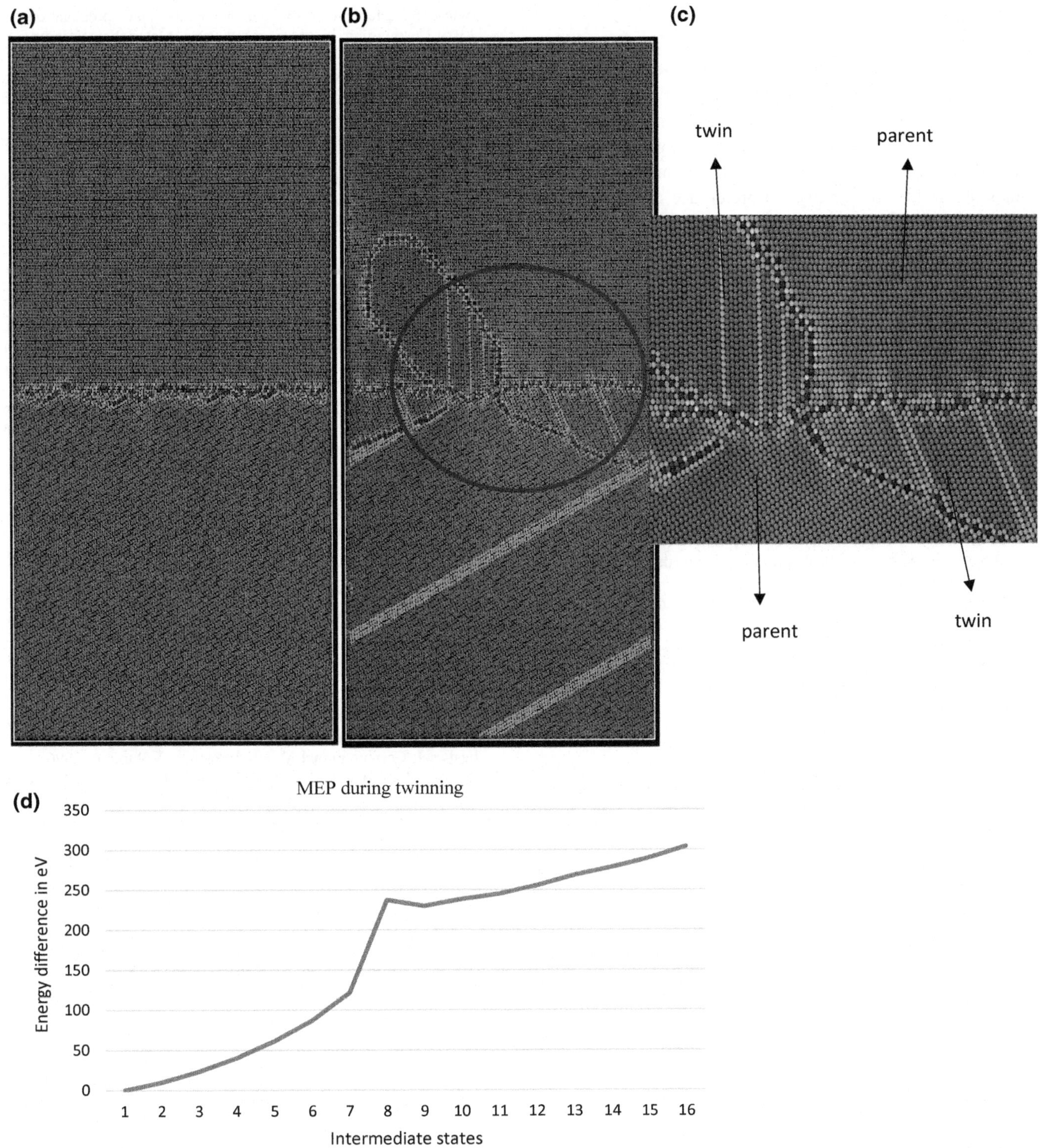


**Fig. 3** **a** Mg bicrystal. **b** {10-12} twin embryos generated in the crystal after applying sufficient tension. **c** Zoomed view of the highlighted portion in **b**. **d** The MEP of the transition

## Conclusion

From these results, we can conclude that the NEB method can be of great help to find the MEP during twinning. While this work is still in early stages, it shows great promise as more realistic scenarios are introduced and a wider variety of nucleation scenarios are explored. NEB method demonstrates the ability to quantify the preference of various sites such as low angle grain boundaries and free surfaces on various planes to activate twin embryos and also illustrates the process by which these embryos activate. Thus, critical insight for higher scale modeling can be obtained allowing better incorporation of twinning in continuum scale models.

## References


1. Christopher D. Barrett, Haitham El Kadiri; Fundamentals of mobile tilt grain boundary faceting; Scripta Materialia, Volumes 84–85, 2014
2. Ivan Milas, Berit Hinnemann, Emily A. Carter; Diffueion of Al, O, Pt, Hf and Y atoms on α-$Al_2O_3$ (0001): Implications for the role of alloying elements in thermal barrier coatings; Journal of Materials Chemistry, 2011
3. Graeme Henkelman, Blas P. Uberuaga, Hannes Jonsson; A climbing image nudged elastic band method for finding saddle points and minimum energy paths; Journal of Chemical Physics, Volume 113, 2000
4. Graeme Henkelman, Hannes Jonsson; Improved tangent estimate in the nudged elastic band method for finding saddle points and minimum energy paths; Journal of Chemical Physics, Volume 113, 2000
5. Haitham El Kadiri, Christopher D. Barrett, Mark A. Tschoop; The candidacy of shuffle and shear during compound twinning in hexagonal close packed structures; Acta Materialia, Volume 61, 2013
6. Haitham El Kadiri, Christopher D. Barrett, Jian Wang, Carlos N. Tome; Why are {10-12} twins profuse in magnesium? Acta Materialia, Volume 85, 2015
7. OVITO 2017; https://ovito.org
8. Haitham El Kadiri, J. Kapil, A. L. Oppedal, L. G. Hector Jr, Sean R. Agnew, M. Cherkaoui, S. C. Vogel; The effect of twin-twin interactions on the nucleation and propagation of {10-12} twinning in magnesium; Acta Materialia, Volume 61, 2013
9. Pierre-Antoine Geslin, Riccardo Gatti, Benoit Devincre, David Rodney; Implementation of the nudged elastic band method in a dislocation dynamics formalism: Application to dislocation nucleation; Journal of the Mechanics and Physics of Solids, Volume 108, 2017
10. Cameron Sobie, Laurent Capolungo, David L. McDowell, Enrique Martinez; Scale transition using dislocation dynamics and the nudged elastic band method; Journal of the Mechanics and Physics of Solids, Volume 105, 2017
11. Hannes Jonsson, Greg Mills and Karsten W. Jacobsen; Nudged elastic band method for finding minimum energy paths of transition; http://theory.cm.utexas.edu/henkelman/pubs/jonsson98_385.pdf
12. Hong Qin, John J. Jonas, Hongbing Yu, Nicolas Brodusch, Raynald Gauvin, Xiyan Zhang; Initiation and accommodation of primary twin in high-purity titanium; Acta Materialia, Volume 71, 2014
13. Christopher D. Barrett, Haitham El Kadiri; Impact of deformation faceting on {10-12}, {10-11} and {10-13} embryonic twin nucleation in hexagonal close-packed metals; Acta Materialia; Volume 70, 2014
14. Hong Yang, Bin Jiang, Junjie He, Zhongtao Jiang, Jianyue Zhang, Fusheng Pan; {10-12} twinning behavior in magnesium bicrystal; Journal of Alloys and Compounds, Volume 725, 2017
15. LAMMPS molecular dynamics simulator software; http://lammps.sandia.gov
16. Daniel Sheppard, Rye Terrell, Graeme Henkelman; Optimization methods for finding minimum energy paths; The journal of Chemical Physics, Volume 128, 2008
17. Christopher D. Barrett, Haitham El Kadiri; The roles of grain boundary dislocations and disclinations in the nucleation of {10-12} twinning; Acta Materialia, Volume 63, 2014
18. G. Kresse, J. Furthmuller; Efficiency of ab-initio total energy calculations for metals and semiconductors using a plane-wave basis set; Computational Material Science, Volume 6, 1996
19. M. S. Hooshmand, M. J. Mills, M. Ghazisaeid; Atomistic modeling of dislocation interactions with twin boundaries in Ti; Modeling and Simulation in Materials Science and Engineering, Volume 25, 2017
20. M. Ghazisaeid, W, A. Curtin; Analysis of dissociation of <c> and <c+a> dislocations to nucleate (10-12) twins in Mg; Modeling and Simulation in Materials Science and Engineering, Volume 21, 2013

# Microstructure and Mechanical Properties of Mg-7.71Gd-2.39Nd-0.17Zr Alloy After the Different Heat Treatments

Shifeng Luo, Guangyu Yang, Lei Xiao, and Wanqi Jie

**Abstract**

Microstructure and mechanical properties of Mg-7.71Gd-2.39Nd-0.17Zr alloy after different heat treatments were investigated. The microstructure of the as-cast alloy was composed of α-Mg matrix, bone-shaped α-Mg + β-$Mg_5$(Gd, Nd) eutectic, a little amount of small cuboid phase ($GdH_2$) and Zr-rich cluster within α-Mg matrix. The optimal solution treatment was determined to be at 515 °C for 4 h. After solution treatment, bone-shaped α-Mg + β-$Mg_5$(Gd, Nd) eutectic was almost dissolved into α-Mg matrix and the grain size increased slightly. Furthermore, a large amount of $GdH_2$ was precipitated along the grain boundaries and within α-Mg matrix. After subsequent aging treatment at 200 °C for 32 h, $Mg_5Gd$ phases were precipitated along the grain boundaries. For the peak-aged alloy, the peak hardness of 105 HV was achieved and the ultimate tensile strength, yield strength and elongation at room temperature were up to 273.7, 188.2 MPa and 4.1%, respectively, which may be mainly attributed to the β″ and β′ phase precipitated within α-Mg matrix after ageing treatment.

**Keywords**

Microstructure • Mechanical properties • Mg–Gd–Nd–Zr alloy • Solution treatments • Ageing-hardening

## Introduction

Due to their high specific strength and stiffness, low density and the excellent mechanical properties both at room temperature and elevated temperature, magnesium alloys containing heavy rare earth elements (RE) have attracted more and more attention as the structural materials in applications where weight saving is of great importance [1–3]. Among these Mg-RE alloys, Mg–Gd based alloy was considered to be one of the promising candidates for a novel Mg-based heat-resistant alloy, which has an excellent age hardening response. To date, many investigations have been conducted on the microstructure and mechanical properties of Mg–Gd based alloys [4–8]. Nodooshan et al. [4] studied the compositional dependence of the age hardening response and the high temperature tensile properties of Mg-xGd-3Y-0.5Zr (x = 3, 6, 10, and 12) (wt%, used throughout the paper unless noted) alloys and found that peak-aged Mg-12Gd-3Y-0.5Zr alloy maintained a high ultimate tensile strength of more than 300 MPa up to 250 °C, which was mainly associated with solution strengthening and precipitation hardening of the cuboid-shaped phases and β′ precipitates in Mg matrix. Peng et al. [7] investigated the age hardening behavior and mechanical properties of Mg-12Gd-4Y-2Nd-0.3Zn-0.6Zr alloy and its ultimate tensile strength, yield strength and elongation of the peak-aged alloy were 293, 261 MPa and 5.8% at 300 °C, respectively, which were mainly attributed to the fine microstructure, fine metastable precipitate and the dispersed precipitates in the matrix. However, these Mg–Gd based alloys with more than 10% Gd increase the density and the costs, it is very necessary to develop some novel kinds of Mg–Gd based alloys containing lower Gd content.

Nd, as another kind of light rare earth element, was also added into Mg alloys with the aim to modify the precipitation microstructure and thereby improve the precipitation hardening [9–15]. Improved strength of Mg-6Zn-1Mn-4Sn alloy containing 0–1.5% Nd has been obtained by Hu et al. [11], which was mainly attributed to the high number density of rod-shaped $\beta_1'$ precipitates in the Mg matrix after ageing. Microstructure and mechanical properties of Mg-11Gd-0.5Zr alloy containing 2 wt% Nd has been investigated by Zheng et al. [13], and found that the ultimate tensile strength of the peak-aged alloy was 330 and 250 MPa at 250 and 300 °C, respectively. Negishi et al. [14, 15] studied the aging

S. Luo · G. Yang (✉) · L. Xiao · W. Jie
State Key Laboratory of Solidification Processing, Northwestern Polytechnical University, No. 127 Youyi Western Road, Xi'an, 710072, People's Republic of China
e-mail: ygy@nwpu.edu.cn


D. Orlov et al. (eds.), *Magnesium Technology 2018*, The Minerals, Metals & Materials Series, https://doi.org/10.1007/978-3-319-72332-7_37

characteristics and tensile properties of Mg–Gd–Nd(–Zr) alloys. The results showed that the ultimate tensile strengths of Mg-10Gd-3Nd–Zr alloy were 310 and 300 MPa at 200 and 250 °C, respectively. Compared with Gd element, Nd element has a lower density and costs, and it may be the better substitutable element for Gd in Mg-RE alloys.

In this paper, Mg-7.71Gd-0.17Zr alloy with low addition of 2.39% Nd was prepared, and the microstructure and mechanical properties of the alloy after different heat treatment were investigated, and the strengthening mechanism was also discussed.

## Experimental Procedures

The experimental alloy was prepared from pure Mg, pure Nd, Mg-21.6Gd master alloy and Mg-33Zr master alloy in an electrical-resistant furnace under the mixed atmosphere of $CO_2$ and $SF_6$ with the ratio of 100:1. After being refined with $Cl_6C_2$ and holding at 780 °C for 30 min, the melt was cast in a steel mold at 740 °C. The real compositions of the experimental alloy were determined to be Mg-7.71Gd-2.39Nd-0.17Zr through inductively coupled plasma atomic emission spectrum (ICP-AES) apparatus. Specimens were cut from the ingot by electric spark linear cutting machine and solution treated at 485–515 °C for 2–16 h following by quenching in cold water. After solution treatment, the specimens were isothermally aged at 200 °C.

The microstructure of the experimental alloys was observed by Olympus PM-G3 optical microscopy (OM), JEOL JSM-5800 scanning electron microscopy (SEM) equipped with energy dispersive spectroscopy (EDS) and Technai 30F transmission electron microscopy (TEM). TEM specimens were prepared by twin jet electropolishing in a solution of 25% $HNO_3$ and 75% methanol cooled down to −20 °C. Further thinning was performed using low energy ion beam milling. The grain size of the experimental alloys was measured by using linear intercept method. The X-ray diffraction (XRD) was performed on X'Pert PRo MPD instrument in the scanning angle of 20°–90°, using Cu Kα (λ = 0.154 nm) as the radiation source. Vickers hardness teats were carried out using 49 N load with a holding time of 15 s, and HV value was calculated by averaging at least 10 points to reduce the error. Tensile tests were performed by using cylindrical specimens with a gauge length of 28 mm and a diameter of 5 mm at a strain rate of $6 \times 10^{-4}$ $s^{-1}$ on a Zwick 150 type universal tensile testing machine at room temperature. At least three specimens were tested at each condition to ensure the reproducibility of the data.

## Results and Discussion

### Microstructures of the As-Cast and Solution Treated Alloy

Figure 1 shows the OM and SEM images of as-cast Mg–Gd–Nd–Zr alloys. The microstructure of the experimental alloy was composed of dendritic α-Mg and semi-continuous eutectic structure distributed along grain boundaries, as shown in Fig. 1a, and the average grain size was about 120 μm. In the magnified SEM image shown in Fig. 1b, the black matrix, boned-shaped eutectic structure along the grain boundaries, cuboid phase and rod-shaped phase were observed, which are indexed by A, B, C and D, respectively. In addition, the corresponding EDS results were also listed in Table 1. According to XRD patterns of the experimental alloy in different states shown in Fig. 2, the black matrix and eutectic structure can be determined to be α-Mg and α-Mg + β-$Mg_5$(Gd, Nd) eutectic, respectively. For the rod-shaped phase, its composition was determined to be Mg-0.6Gd-7.65Zr (at. %) by EDS analysis, which was thought to be Zr-rich cluster and acted as the nucleation site for α-Mg during solidification, leading to the grain refinement [16]. The cuboid phase will be discussed below. It was worth noting that the cuboid phase and Zr-rich clusters were not detected in the XRD patterns due to the low volume fraction.

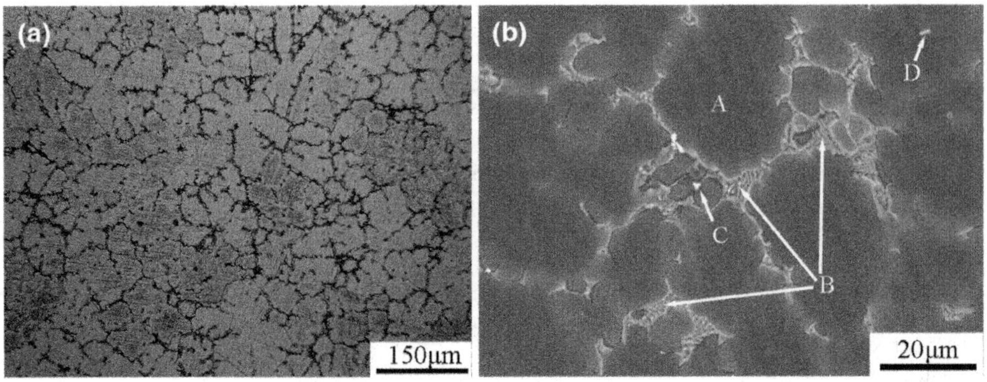


**Fig. 1** Microstructures of as-cast experimental alloy: **a** OM image and **b** SEM image

**Table 1** EDS results of as-cast experimental alloy in Fig. 1b (at.%)

| Positions | Mg | Gd | Nd | Zr | Phase |
|---|---|---|---|---|---|
| A | 99.71 | 0.29 | – | – | α-Mg |
| B | 93.77 | 2.50 | 3.73 | – | α-Mg + $Mg_5Gd$ |
| C | 88.71 | 8.81 | 2.48 | – | Cuboid phase |
| D | 91.75 | 0.60 | – | 7.65 | Zr-rich cluster |

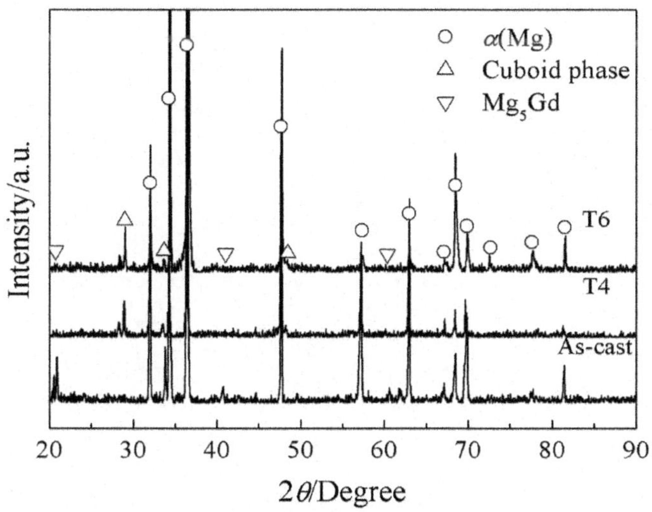


**Fig. 2** XRD patterns of the experimental alloys in different state in which, as-cast, T4: 515 °C/4 h (solution treated), T6: 200 °C/32 h (peak-aged)

Figure 3 shows the microstructure of the experimental alloys solution treated at the different heat treatment conditions. As shown in Fig. 3a, b, it can be found that the second phase was not dissolved into the matrix entirely after solution treated at 485 °C for 16 h or at 500 °C for 8 h, respectively. In contrast, after solution treated at 515 °C for 4 h shown in Fig. 3c, the second phase was almost dissolved into matrix and many small cuboid phases were precipitated within matrix and along the grain boundaries, which were confirmed by the XRD analysis, as shown in Fig. 2. Compared with the as-cast alloy, the grain size of solution treated alloy increased slightly, about 140 μm. Further prolonging the solution time to 8 h, some grains began to grow up abnormally, as shown in Fig. 3d, and the average grain size was about 224 μm. Therefore, the optimal solution treatment parameter was determined to be 515 °C for 4 h, in which the second phases were almost dissolved into the matrix and the average grain size did not increase remarkably. Furthermore, it was worth noting that the grains did not grow up distinctly when cuboid phases existed along the triple grain boundaries, as donated by black circles in Fig. 3c, d, indicating that the cuboid phases distributed along the triple grain boundaries can inhibit the grain growth at some extent.

The morphology of the cuboid phases in the alloy solution treated at 515 °C for 4 h was shown in Fig. 3e and the corresponding EDS result was shown in Fig. 3f, which revealed that these particles were RE-rich phase. Combined with the XRD pattern shown in Fig. 2, the RE-rich phase can be determined to be $GdH_2$ compound. These particles were also confirmed by selected-area diffraction patterns of the cuboid phase shown in Fig. 4b. It can be indexed to fcc $GdH_2$ precipitate with a = 0.524 nm, which is almost the same as the reported value of 0.5293 nm [17]. Yang et al. [18] identified the formation of hydride $NdH_2$ in the solution treated Mg–Nd binary alloy and proposed that the formation of hydride $NdH_2$ occurred during annealing treatment by the reaction of previously existed hydrogen with Nd. Another mechanism that the formation of hydrides was attributed to the decomposition of Mg-RE intermetallic phases by hydrogen was proposed by Zhu et al. [19]. Furthermore, Huang et al. [20] studied the influences of alloying element REs and heat treatments on the formation of hydrides in Mg-RE alloy and considered that the formation mechanism was attributed to the surface reaction of Mg-RE alloys with water. Therefore, at present, the responsible formation mechanisms for the RE-rich phases are still argued even if the identification of the RE-rich phases has been solved. Further investigations are still needed.

## Ageing-Hardening Response and Microstructure After Aged Treatment

Figure 5 shows the age-hardening curve of the experimental alloy aged at 200 °C. The hardness increased rapidly at first with prolonging the ageing time, and reached the peak of 105 HV after 32 h. Further ageing led to a slight decline in hardness. The OM image of the peak-aged alloy was shown in Fig. 6. The $GdH_2$ phases were non-uniformly distributed along the grain boundaries and within the matrix and no obvious grain growth was observed after ageing (the average grain size was about 143 μm), which was similar to the OM image of solution treated alloy. According to the XRD pattern shown in Fig. 2, the microstructure of the peak-aged alloy was composed of α-Mg, $Mg_5Gd$ and $GdH_2$.

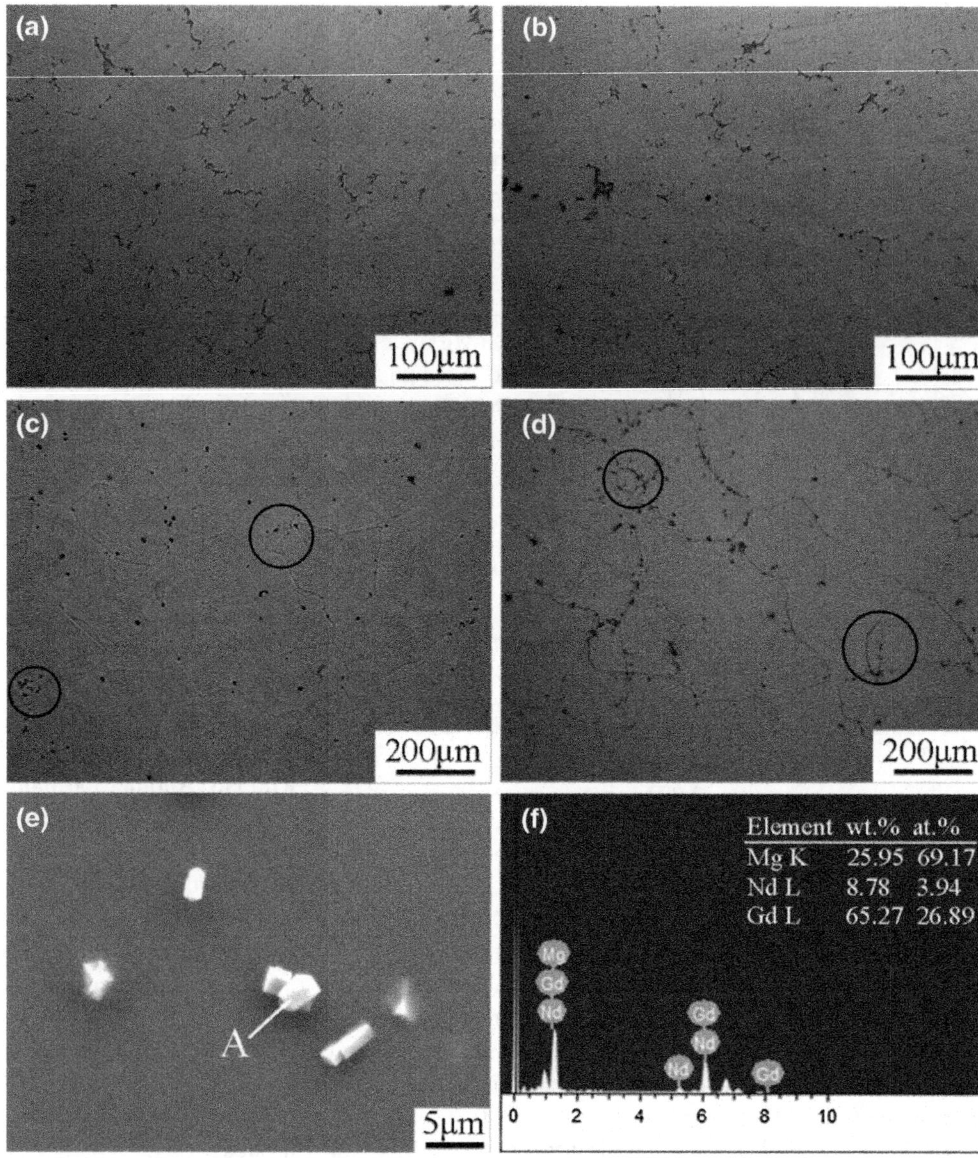


**Fig. 3** **a** OM image of the alloy solution treated at 485 °C for 16 h; **b** OM image of the alloy solution treated at 500 °C for 8 h; **c** OM image of the alloy solution treated at 515 °C for 4 h; **d** OM image of the alloy solution treated at 515 °C for 8 h; **e** SEM image of the alloy solution treated at 515 °C for 4 h; **f** EDS result of the cuboid phase indexed by A in (**e**)

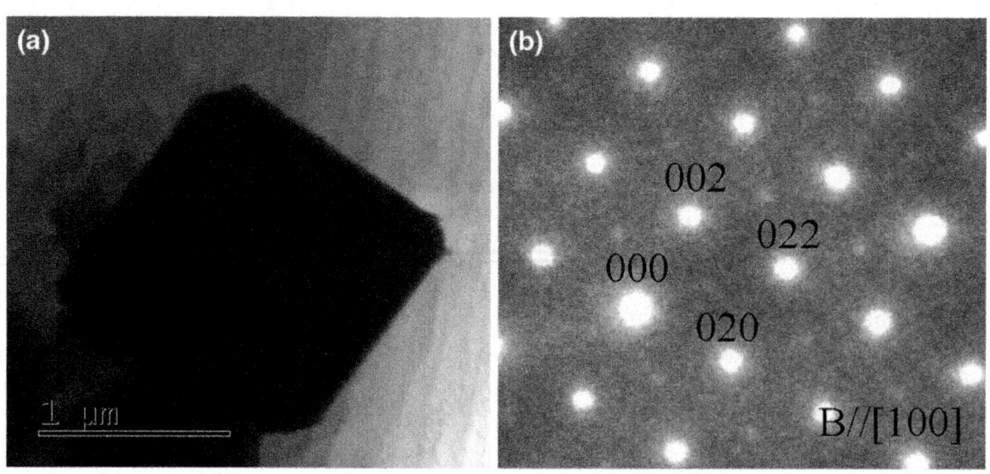


**Fig. 4** TEM image of the cuboid phase in solution treated specimen (515 °C for 4 h). **a** TEM bright field and **b** SAED pattern of the cuboid phase

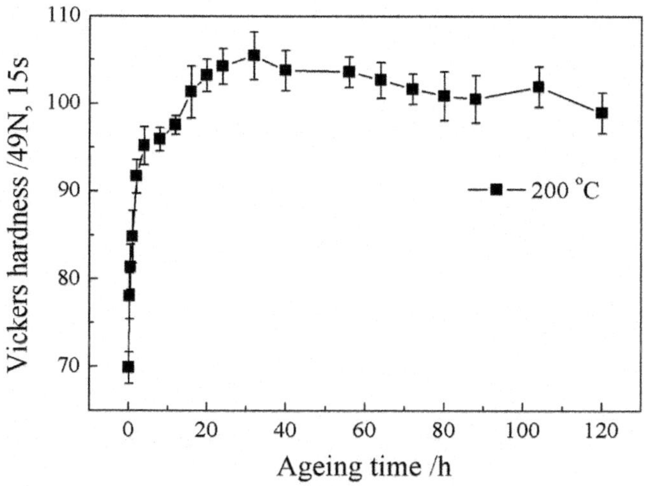


**Fig. 5** Age-hardening curve of the experimental alloy aged at 200 °C

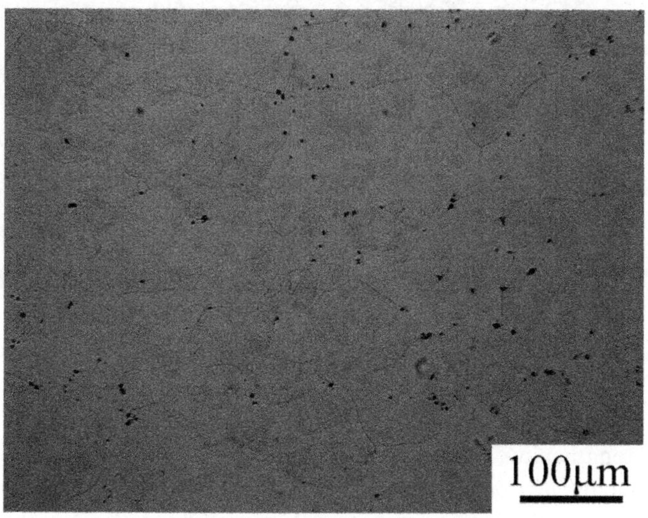


**Fig. 6** OM image of the quenched experimental alloy after aged at 200 °C for 32 h

Figure 7 shows the grain boundary structure of the alloy 200 °C for 32 h. Precipitates were formed along the grain boundaries linearly and the precipitate free zone occurred simultaneous, which was also observed by Zheng et al. [13] and the precipitates along the grain boundaries was identified as equilibrium $Mg_5Gd$ phase. Figure 8 shows the dispersive precipitates formed in experimental alloy after aged at 200 °C for 32 h. Many plate precipitates with the length of about 50 nm and the thickness of about 5 nm were observed in $[01\bar{1}0]_\alpha$ projections, as shown in Fig. 8a. Furthermore, some globular particles were also found from $B//[2\bar{1}\bar{1}0]_\alpha$ images. From the SAED pattern shown in Fig. 8b and d that the diffraction spots at $1/2\{01\bar{1}0\}$ can be clearly seen, and faint spots at $1/4\{01\bar{1}0\}$ and $3/4\{01\bar{1}0\}$ were also found, which indicated that the precipitates observed in the T6 state alloy

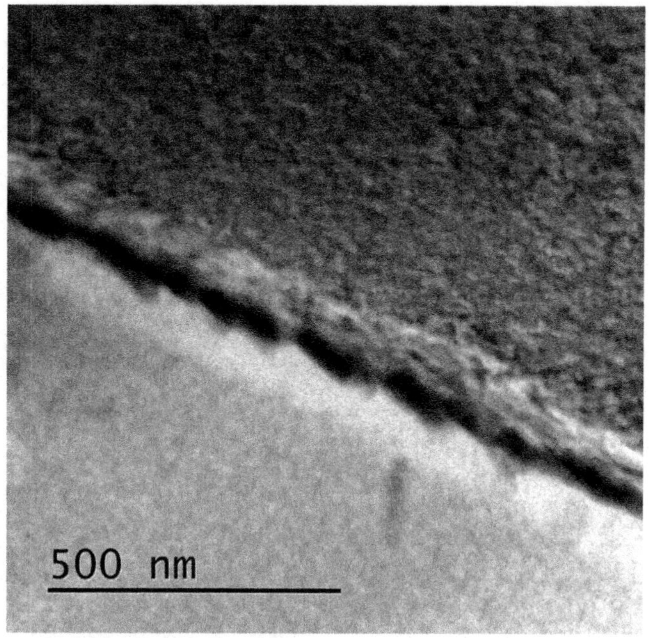


**Fig. 7** TEM image of grain boundary structure on the alloy aged at 200 °C for 32 h

were β″ with DO19 structure ($a \sim 2a_{Mg} = 0.64$ nm, $c \sim c_{Mg} = 0.52$ nm) and β′ phase with bco structure ($a \sim 2a_{Mg} = 0.64$ nm, $b \sim 2.2$ nm, $c \sim c_{Mg} = 0.52$ nm) [13, 21]. β″ and β′ precipitates can be related to the maximum hardness.

## Mechanical Properties

Figure 9 shows the room temperature mechanical properties of the experimental alloy in different states in which, as-cast, T4: 515 °C/4 h (solution treated), T6: 200 °C/32 h (peak-aged). After solution treatment, the alloy had higher ultimate tensile strength (UTS), yield strength (YS) and elongation than those of the as-cast alloy. Further ageing treatment led to an increase in UTS and YS but a decrease in elongation. The highest UTS and YS were reached when the alloy peak-aged at 200 °C for 32 h, about 273.7 and 188.2 MPa, respectively. The elongation significantly decreased from 7.5% in solution treated state to 4.1% in peak-aged state.

It is well accepted that many factors, such as grain size, morphology and distribution of the second phases, have a significant effect on the room temperature mechanical properties of an alloy [22]. It should be noted that the grain size of the experimental alloy in different heat treatment conditions did not change significantly, as discussed above, which indicated that the difference in mechanical properties were mainly attributed to the morphology and distribution of the second phases. For the as-cast alloy, coarse bone-shaped α-Mg + β-$Mg_5$(Gd, Nd) eutectic structure distributed along

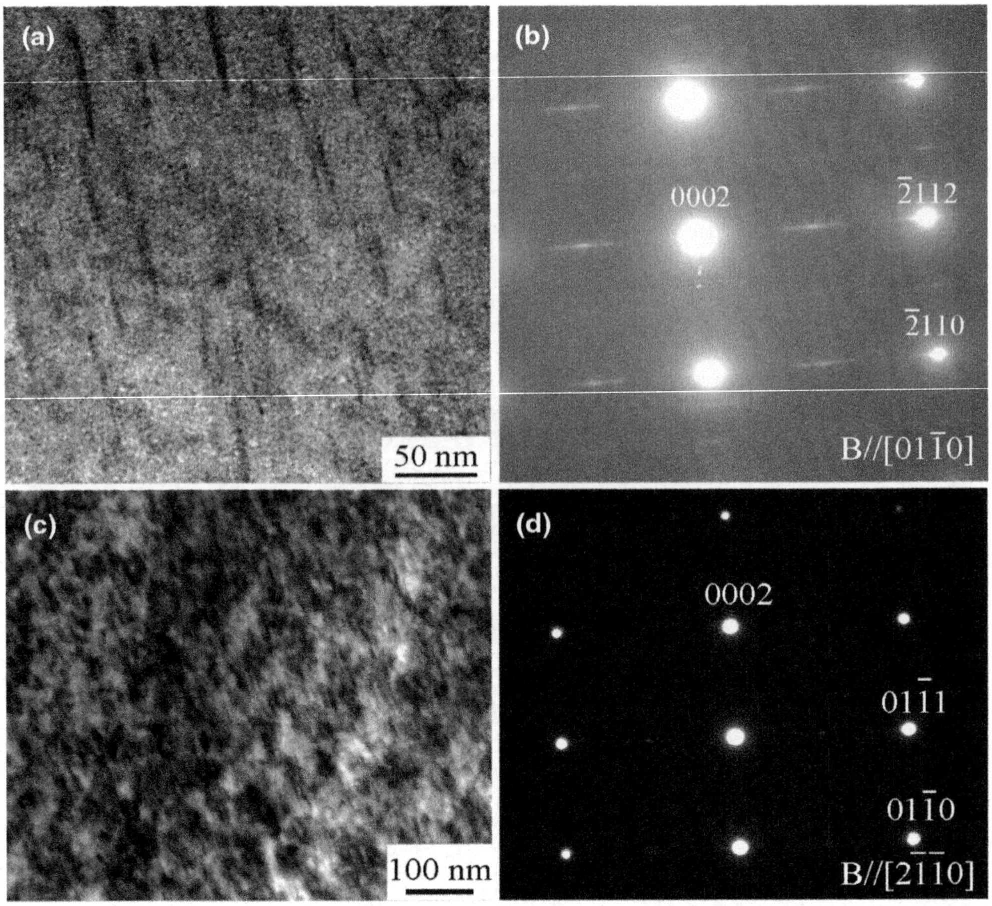


**Fig. 8** TEM images and SAED patterns for 200 °C/32 h aged alloy with B//[01$\bar{1}$0]$_\alpha$ in (**a**, **b**), and B//[2$\bar{1}\bar{1}$0]$_\alpha$ in (**c**, **d**)

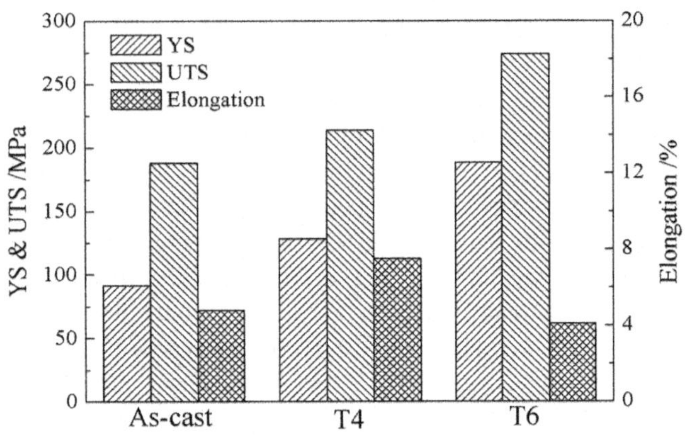


**Fig. 9** Tensile properties of the experimental alloy in different state in which, as-cast, T4: 515 °C/4 h (solution treated), T6: 200 °C/32 h (peak-aged)

the grain boundaries semi-continuously was prone to fracture due to the local stress concentration during tensile test, which was detriment to the mechanical properties, and thereby the as-cast alloy had the lowest fracture strength. In contrast, α-Mg + β-$Mg_5$(Gd, Nd) eutectic structure was dissolved into the matrix after solution treatment, which significantly reduced the regions for crack initiation [23], and many tiny $GdH_2$ were precipitated within the matrix and along the grain boundaries, which could impede the dislocation slip at some extent. The improved mechanical properties were mainly attributed to the solid solution strengthening. After ageing treatment, dispersive β″ and β′ precipitates formed within the matrix. Although solid solution strengthening contribution decreased due to the fact that the matrix was depleted of solute as the precipitation process proceeded, precipitation strengthening contribution caused by β″ and β′ precipitates increased significantly. Therefore, precipitation strengthening was the major contributor to the high strength of T6 state alloy.

## Conclusions

Microstructure and mechanical properties of Mg-7.71Gd-2.39Nd-0.17Zr alloy after heat treatment were investigated systematically. The main conclusions can be drawn as follows:

(1) The microstructure of the as-cast Mg-7.71Gd-2.39Nd-0.17Zr alloy was composed of α-Mg matrix, bone-shaped α-Mg + β-$Mg_5$(Gd, Nd) eutectic, a little mount of small cuboid phase and Zr-rich cluster within α-Mg matrix.

(2) The optimal solution treatment parameters were determined to be 515 °C for 4 h, in which bone-shaped α-Mg + β-$Mg_5$(Gd, Nd) eutectic was almost dissolved into α-Mg matrix and the grain size was increased slightly. Some tiny $GdH_2$ particles were precipitated along the grain boundaries and within α-Mg matrix.
(3) After ageing treatment at 200 °C for 32 h, the peak hardness of 105 Hv was achieved. The UTS, YS and elongation of the alloy at room temperature were 273.7, 188.2 MPa and 4.1%, respectively, which was mainly attributed to the dispersive β″ and β′ precipitates formed within the matrix.

**Acknowledgements** This work was supported by the National Natural Science Foundation of China (Grant Nos. 51771152, 51227001 and 51420105005) and the Research Fund of the State Key Laboratory of Solidification Processing (NWPU), China (Grant No. 138-QP-2015).

## References

1. Nodooshan HRJ, Liu W, Wu G, Rao Y, Zhou C, He S, Ding W, Mahmudi R (2014) Effect of Gd content on microstructure and mechanical properties of Mg–Gd–Y–Zr alloys under peak-aged condition, Mater. Sci. Eng. A. 615: 79–86
2. Huang S, Wang J, Hou F, Huang X, Pan F (2014) Effect of Gd and Y contents on the microstructural evolution of long period stacking ordered phase and the corresponding mechanical properties in Mg–Gd–Y–Zn–Mn alloys, Mater. Sci. Eng. A. 612: 363–370
3. Xu C, Zheng MY, Wu K, Wang ED, Fan GH, Xu SW, Kamado S, Liu XD, Wang GJ, Lv XY (2013) Effect of ageing treatment on the precipitation behaviour of Mg–Gd–Y–Zn–Zr alloy, J. Alloy Compd. 550: 50–56
4. Nodooshan HRJ, Wu G, Liu W, Wei G, Li Y, Zhang S (2016) Effect of Gd content on high temperature mechanical properties of Mg–Gd–Y–Zr alloy, Mater. Sci. Eng. A. 651: 840–847
5. Wu J, Chiu YL, Jones IP (2016) Effect of Gd on the microstructure of as-cast Mg-4.2Zn-0.8Y (at.%) alloys, J. Alloy Compd. 661: 455–460
6. Zheng KY, Dong J, Zeng XQ, Ding WJ (2008) Effect of precipitation aging on the fracture behavior of Mg–11Gd–2Nd–0.4Zr cast alloy, Mater. Charact. 59(7): 857–862
7. Peng Q, Hou X, Wang L, Wu Y, Cao Z, Wang L (2009) Microstructure and mechanical properties of high performance Mg–Gd based alloys, Mater. Des. 30(2): 292–296
8. Li RG, Xin RL, Liu Q, Liu JA, Fu GY, Zong L, Yu YM, Guo SG (2015) Effect of Ag addition on microstructure and mechanical properties of Mg–14Gd–0.5Zr alloy, Mater. Charact. 109: 43–49
9. Li Q, Wang Q, Wang Y, Zeng X, Ding W (2007) Effect of Nd and Y addition on microstructure and mechanical properties of as-cast Mg–Zn–Zr alloy, J. Alloy Compd. 427(1): 115–123
10. Yang J, Wang J, Wang L, Wu Y, Wang L, Zhang H (2008) Microstructure and mechanical properties of Mg–4.5Zn–xNd (x = 0, 1 and 2, wt%) alloys, Mater. Sci. Eng. A. 479(1): 339–344
11. Hu G, Zhang D, Tang T, Shen X, Jiang L, Xu J, Pan F (2015) Effects of Nd addition on microstructure and mechanical properties of Mg–6Zn–1Mn–4Sn alloy, Mater. Sci. Eng. A. 634: 5–13
12. Wan YC, Jiang SN, Liu CM, Wang BZ, Chen ZY (2015) Effect of Nd and Dy on the microstructure and mechanical property of the as extruded Mg–1Zn–0.6Zr alloy, Mater. Sci. Eng. A. 625: 158–163
13. Zheng KY, Dong J, Zeng XQ, Ding WJ (2008) Precipitation and its effect on the mechanical properties of a cast Mg–Gd–Nd–Zr alloy, Mater. Sci. Eng. A. 489(1): 44–54
14. Negishi Y, Iwasawa S, Kamado S, Kojima Y, Ninomiya R (1994) Effect of yttrium and neodymium additions on aging characteristics and high temperature tensile properties of Mg-10mass%Gd and Mg-10mass%Dy alloys, J. JPN I. Met. 44(10): 549–554
15. Negishi Y, Nishimura T, Iwasawa S, Kamado S, Kojima Y, Ninomiya R (1994) Aging characteristics and tensile properties og Mg–Gd–Nd–Zr and Mg–Dy–Nd–Zr alloys, J. JPN I. Met. 44(10): 555–561
16. Song G, StJohn D (2002) The effect of zirconium grain refinement on the corrosion behaviour of magnesium-rare earth alloy MEZ, J. Light Met. 2(1): 1–16
17. Peng Q, Huang Y, Meng J, Li Y, Kainer KU (2011) Strain induced $GdH_2$ precipitate in Mg–Gd based alloys, Intermetallics. 19(3): 382–389
18. Yang Y, Peng L, Fu P, Hu B, Ding W (2009) Identification of $NdH_2$ particles in solution-treated Mg–2.5%Nd (wt.%) alloy, J. Alloy Compd. 485(1): 245–248
19. Zhu SM, Nie JF, Gibson MA, Easton MA (2014) On the unexpected formation of rare earth hydrides in magnesium–rare earth casting alloys, Scripta Mater. 77: 21–24
20. Huang Y, Yang L, You S, Gan W, Kainer KU, Hort N (2016) Unexpected formation of hydrides in heavy rare earth containing magnesium alloys, J. Magn. Alloy. 4(3): 173–180
21. Nie JF (2012) Precipitation and Hardening in Magnesium Alloys, Metall. Mater. Trans. A. 43(11): 3891–3939
22. Liu SJ, Yang GY, Luo SF, Jie WQ (2015) Microstructure evolution during heat treatment and mechanical properties of Mg–2.49Nd–1.82Gd–0.19Zn–0.4Zr cast alloy, Mater. Charact. 107: 334–342
23. Fu P, Peng L, Jiang H, Chang J, Zhai C (2008) Effects of heat treatments on the microstructures and mechanical properties of Mg–3Nd–0.2Zn–0.4Zr (wt.%) alloy, Mater. Sci. Eng. A. 486(1): 183–192

# Superplasticity in a Chip-Consolidated $Mg_{97}Zn_1Y_2$ Alloy with LPSO Phase

Kazuha Suzawa, Shin-ichi Inoue, Michiaki Yamasaki, Yoshihito Kawamura, Michimasa Miyanaga, Katsuhito Yoshida, and Nozomu Kawabe

**Abstract**

Novel $Mg_{97}Zn_1Y_2$ alloy of equiaxed fine grains with long-period stacking ordered (LPSO) phase was prepared by consolidation of chips machined from a twin-roll-cast alloy. The chip-consolidated $Mg_{97}Zn_1Y_2$ alloy exhibited superplasticity at a wide initial strain-rate ranging from $1 \times 10^{-3}$ to $3 \times 10^{-1}$ s$^{-1}$ at temperatures of 623 and 673 K. The maximum elongation was approximately 560% at $3 \times 10^{-2}$ s$^{-1}$ at 673 K. The strain rate sensitivity index (*m*-values) was shown to be greater than 0.3 in this study; this value seems to be large enough to ensure uniform elongation during superplastic deformation. SEM/EBSD measurements revealed that the chip-consolidated $Mg_{97}Zn_1Y_2$ alloy was composed of mixed equiaxed fine-grains of α-Mg and LPSO phases of a mean size of ~1 μm. The grain boundary sliding in the equiaxed fine-grains appears to account for most of the mechanism of superplastic deformation in this alloy.

**Keywords**

Mg–Zn–Y • Long-period stacking ordered phase
Chip consolidation • Superplasticity

## Introduction

Varieties of high-strength Mg alloys of increased ductility have been extensively developed in recent years in hopes that they will find wider use in automotive, aerospace, and biomaterial applications [1, 2]. Among others, the novel $Mg_{97}Zn_1Y_2$ (at.%) alloy which was developed in 2001 by a process of rapidly solidified powder metallurgy (RS P/M) has two claims to a special mention: it not only exhibits a significantly high yield strength of 610 MPa and reasonable elongation of 5% at room temperature [3], but also has high-strain-rate superplasticity with an elongation as large as ~800%. The RS P/M $Mg_{97}Zn_1Y_2$ alloy microstructurally features nanocrystalline α-Mg grains embedding novel long-period stacking ordered (LPSO) structure [4]. The LPSO phase in the $Mg_{97}Zn_1Y_2$ alloy constitutes mainly the 18*R* structure (Ramsdell notation) with a close-packed plane stacking sequence of ABABAB|CACACA|BCBCBC. Both Zn and Y elements are enriched in four atomic fcc layers sandwiching intrinsic stacking faults in the closely packed plane at six period intervals, and in-plane long-range ordering of Zn and Y atoms occurs in the quadruple atomic layers, thereby forming $L1_2$-typed $Zn_6Y_8$ clusters, locally and periodically [5]. The LPSO phase acts as an alloy-strengthening phase owing to a unique atomic arrangement. More recently, it has been found that the LPSO phase forms in cast ingots as well as in rapidly solidified alloys. With the Mg–Zn–Y Mg/LPSO two-phase alloys prepared by conventional thermo-mechanical methods, such as extrusion and rolling, excellent mechanical properties have also been brought to light [6–8]. The extruded $Mg_{97}Zn_1Y_2$ alloy features a multimodal microstructure. The multimodally grained Mg–Zn–Y alloy consists of three regions; a dynamically recrystallized (DRXed) α-Mg fine-grain region, a hot-worked α-Mg grain region, and a hot-worked LPSO phase grain region [8]. The DRXed fine α-Mg grains are effective at increasing ductility, whereas the fiber-textured α-Mg and LPSO coarse grains strengthen the alloys. In a ward, multimodal microstructure development helps improve both the strength and ductility of the alloy, but being heterogeneous by nature, the multimodal microstructure can be detrimental to the superplasticity of the resultant alloy. In this study, we sidestepped this problem by a fortunate combination of a novel Mg/LPSO two-phase

K. Suzawa · S. Inoue · M. Yamasaki (✉) · Y. Kawamura
Magnesium Research Center, Department of Materials Science, Kumamoto University, 2-39-1, Kurokami, Chuo-Ku, Kumamoto, 860-8555, Japan
e-mail: yamasaki@gpo.kumamoto-u.ac.jp

K. Suzawa · M. Miyanaga · K. Yoshida · N. Kawabe
Magnesium Alloy Development Division, Sumitomo Electric Industries, Ltd., 1-1-1, Koyakita, Itami-shi, Hyogo 664-0016, Japan


D. Orlov et al. (eds.), *Magnesium Technology 2018*, The Minerals, Metals & Materials Series, https://doi.org/10.1007/978-3-319-72332-7_38

alloy composed of equiaxed fine-grains and a new technique of chip-consolidation combined with twin-roll casting. Twin-roll casting with high cooling rate was adopted to obtain a fine dendrite structure of α-Mg and LPSO phases and a novel $Mg_{97}Zn_1Y_2$ alloy of equiaxed fine grains was prepared by consolidation of chips machined from a twin-roll-cast master alloy. Superplastic behavior and the microstructure of the $Mg_{97}Zn_1Y_2$ alloy obtained by consolidating chips made from a twin-roll cast alloy was examined by high temperature tensile testing, SEM observation, and electron backscatter diffraction (EBSD) analysis.

## Experimental

The master alloy sheet of nominal composition, $Mg_{97}Zn_1Y_2$ (at.%), was prepared by twin roll casting in air. Alloy chips were prepared by machining the cast alloy in a milling machine. Machined chips were pressed into a copper billet, degassed for 15 min at 523 K. The extrusion was performed at an extrusion ratio of 15 at 623 K. The schematic diagram of chip-consolidation combined with twin-roll casting is shown in Fig. 1. The tensile test was carried out in an Instron-type tensile testing machine at initial strain rates of $1 \times 10^{-3}\ s^{-1}$, $1 \times 10^{-2}\ s^{-1}$, and $1 \times 10^{-1}\ s^{-1}$ at 623 K, and $3 \times 10^{-3}\ s^{-1}$, $3 \times 10^{-2}\ s^{-1}$, and $3 \times 10^{-1}\ s^{-1}$ at 673 K. The specimen had a gauge length of 10 mm and a gauge diameter of 2 mm. The tensile direction was parallel to the direction of extrusion. The microstructure of the specimens was investigated by scanning electron microscopy (SEM; JEOL JSM-7001F) before and after the tensile test. The textures of the alloys were analyzed by electron back-scatter diffraction (EBSD) analysis. SEM specimen was polished by a cross-section polisher (JEOL SM-09010).

## Results and Discussion

Figure 2 shows SEM images taken from the longitudinal section of the $Mg_{97}Zn_1Y_2$ alloy prepared by consolidation of chips machined from a twin-roll-cast (TRC) alloy. When examined by SEM back scatter electron (SEM-BSE) microscopy, LPSO phase grains seemed to be elongated in the direction of extrusion at low magnification (Fig. 2a), but at high magnification the LPSO phase grains looked fragmented during consolidation by extrusion (Fig. 2b).

Figure 3 shows an inverse pole figure (IPF) map taken from the longitudinal section of the $Mg_{97}Zn_1Y_2$ alloy prepared by consolidation of chips machined from a TRC alloy. This IPF map indicates that the chip-consolidated $Mg_{97}Zn_1Y_2$ alloys (extrusion ratio was 15, extrusion temperature was 623 K) consists of equiaxied fine grains with a mean grain size of $\sim 1\ \mu m$ with random crystallographic orientation. It was known that the LPSO phase-containing Mg–Zn-rare earth alloys developed a strong $< 10\bar{1}0 >$ fiber texture considered LPSO-stimulated during extrusion of the Mg/LPSO two-phase cast alloy [9, 10]. In the case of TRC-chip-consolidated Mg–Zn–Y alloy, however, texture evolution was not observed. Chipping of the TRC alloy may bring about LPSO grain fragmentation which in turn prevented LPSO-stimulated texture evolution.

Figure 4 shows nominal stress-strain curves for the TRC-chip-consolidated $Mg_{97}Zn_1Y_2$ alloy at several strain rates at 623 K. The stress-strain behavior of the alloy exhibited a substantial increase in flow stress with increasing strain rate, suggesting that the alloy is highly sensitive to the strain rate. Elongation at strain rates of $1 \times 10^{-3}\ s^{-1}$ and $1 \times 10^{-2}\ s^{-1}$ exceeded 200%. Figure 5 shows nominal stress-strain curves for the alloy at 673 K. The flow stress was reduced significantly when the testing temperature was

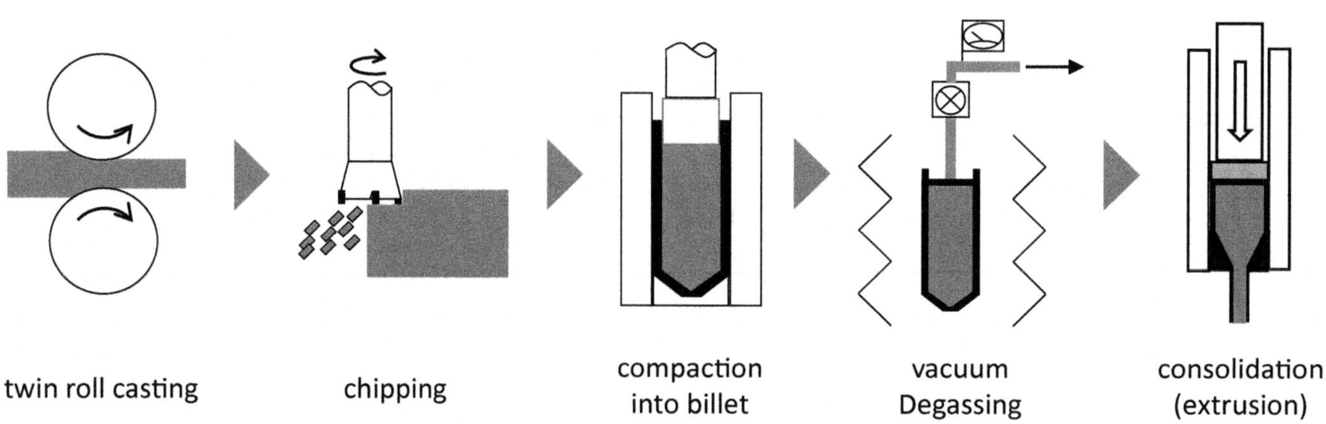


**Fig. 1** Schematic diagram for the chip-consolidation processing combined with twin-roll casting

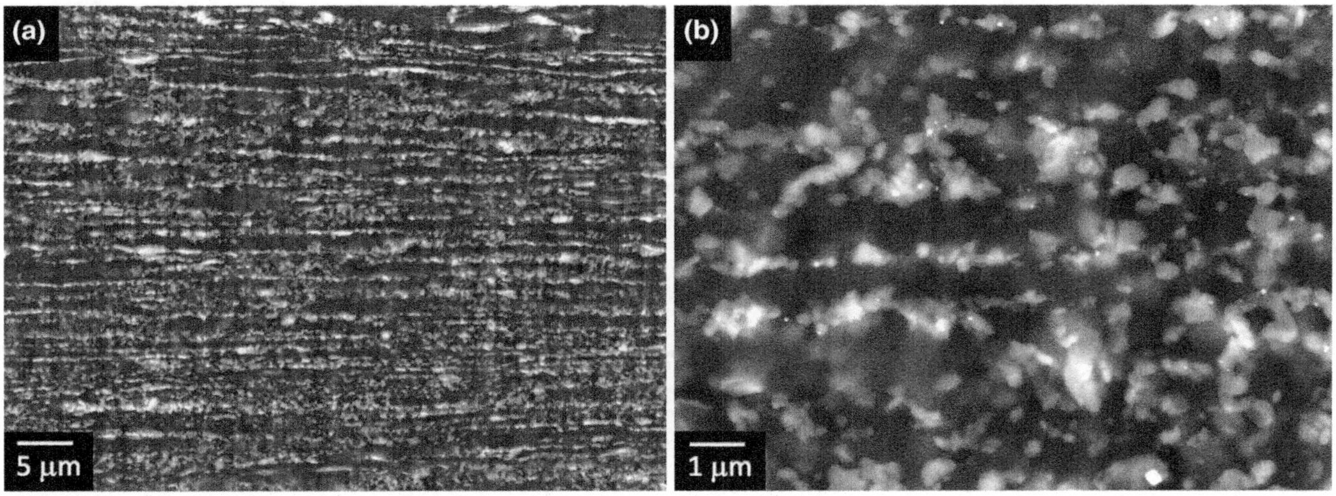


**Fig. 2** **a**, **b** SEM images taken from the longitudinal section of the $Mg_{97}Zn_1Y_2$ alloy prepared by consolidation of chips machined from a TRC alloy. Magnifications of (**a**) and (**b**) were 1400x and 7000x, respectively

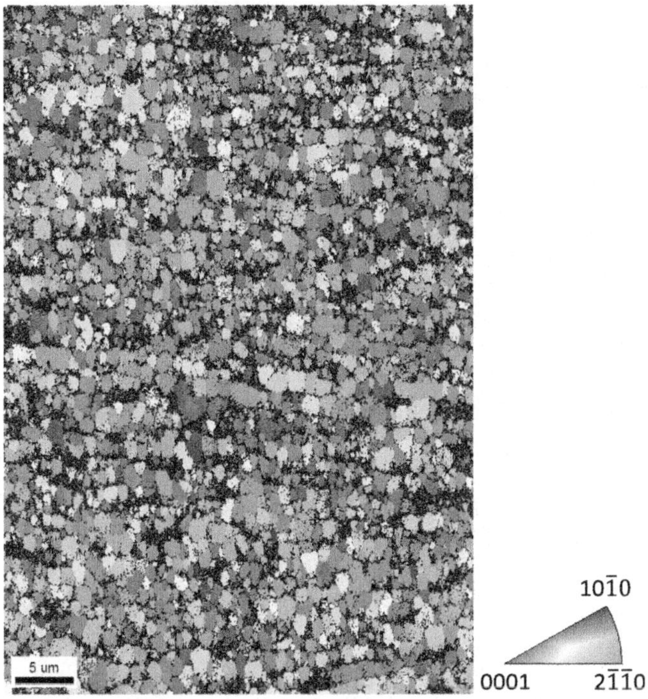


**Fig. 3** Inverse pole figure map taken from the longitudinal section of the $Mg_{97}Zn_1Y_2$ alloy prepared by consolidation of chips machined from a TRC alloy. Magnification was 1400x

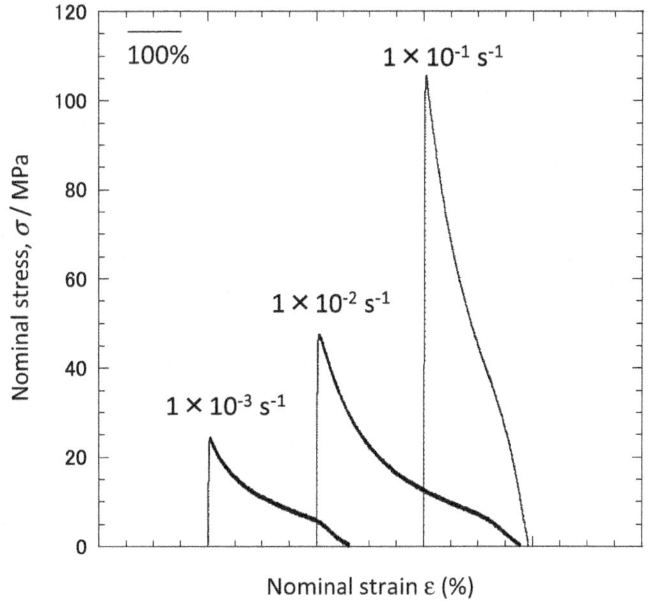


**Fig. 4** Stress-strain curves for the TRC-chip-consolidated $Mg_{97}Zn_1Y_2$ alloy at 623 K

increased from 623 to 673 K. It should be noted, however, that elongation increased with increasing testing temperature, with elongation exceeding 200% in the all specimens examined in this study and even reaching 560% at a strain rate of $3 \times 10^{-2}$ $s^{-1}$ at 673 K.

Figure 6 shows change in the elongation as a function of initial strain rate. Strain rate-dependence of elongation can be depicted as convex curves at both temperatures of 623 and 673 K. The strain rates that induced maximum elongation at 623 and 673 K were $1 \times 10^{-2}$ $s^{-1}$ and $3 \times 10^{-2}$ $s^{-1}$, respectively. Optimum strain rate moved from low strain rate to high.

Figure 7 shows the variation in flow stress as a function of strain rate at different temperatures. Flow stresses for tensile deformation at 623 and at 673 K increased with increasing strain rate. The strain rate sensitivity indices (*m*-values) for the deformation at 623 K and at 673 K were 0.32 and 0.39, respectively. Given that with conventional metals, the *m*-value does not exceed 0.3 even at elevated temperatures, it may well be greater than 0.3 for superplastic materials [11, 12].

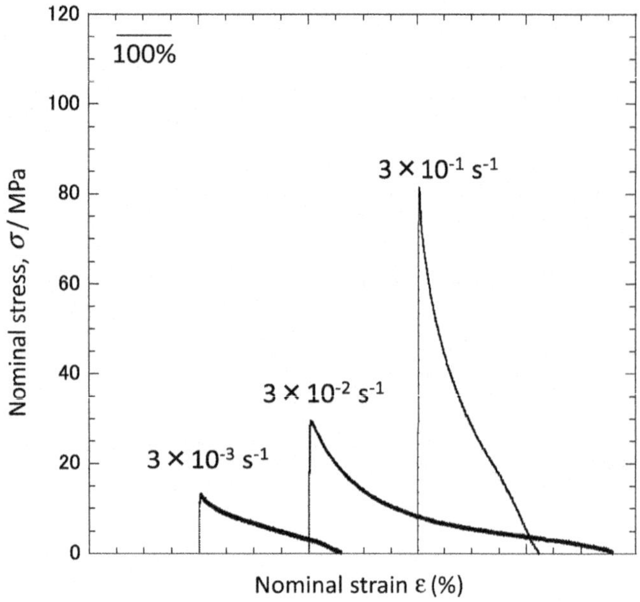


**Fig. 5** Stress-strain curves for the TRC-chip-consolidated $Mg_{97}Zn_1Y_2$ alloy at 673 K

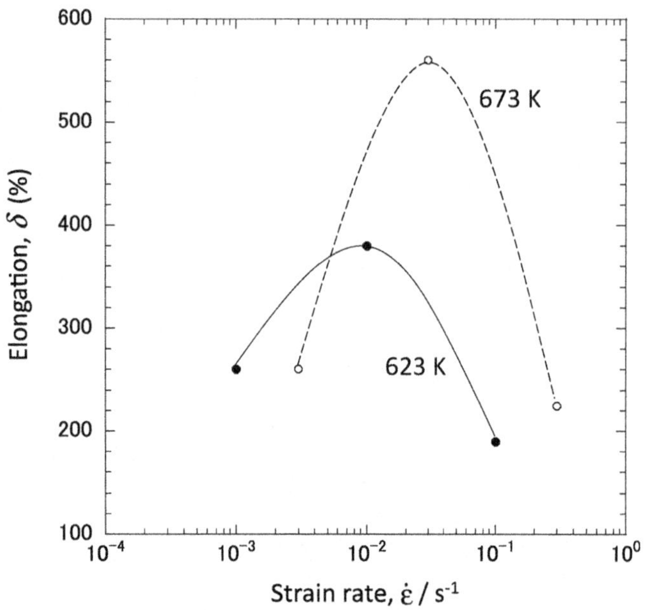


**Fig. 6** Change in the elongation as a function of strain rate in the tensile testing for the TRC-chip-consolidated $Mg_{97}Zn_1Y_2$ alloy

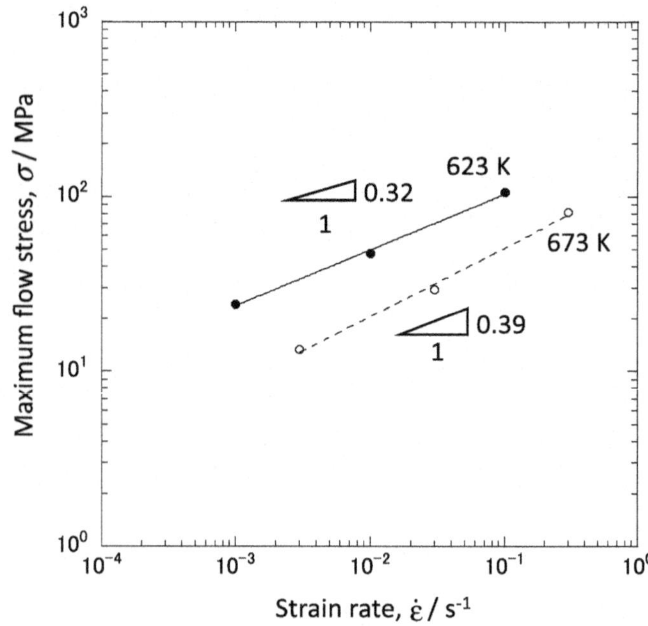


**Fig. 7** Variation in the flow stress as a function of strain rate in the tensile testing for the TRC-chip-consolidated $Mg_{97}Zn_1Y_2$ alloy

For materials to be accepted as superplastic are required to achieve at least 200% elongation in uniaxis tension. In this study, 200% or larger elongation and 0.3 or higher *m*-values were observed with the TRC-chip-consolidated $Mg_{97}Zn_1Y_2$ alloy at strain rates ranging from $1 \times 10^{-3}$ $s^{-1}$ to $1 \times 10^{-2}$ $s^{-1}$ at 623 K and from $3 \times 10^{-3}$ $s^{-1}$ to $3 \times 10^{-1}$ $s^{-1}$ at 673 K. Our results may be taken as evidence to suggest that the alloy under discussion is a high-strain-rate superplastic material. In addition, the high *m*-values suggest that the grain boundary sliding of Mg grains account for most of the mechanism of superplastic deformation in the TRC-chip-consolidated $Mg_{97}Zn_1Y_2$ alloy [13].

It is well known that grain boundary sliding takes place more easily with grain refinement. Mukai et al. reported that the grain size required to cause superplastic forming at high strain rates was estimated to be ~2 μm with commercially available Mg alloys [14]. Thermo-mechanical treatment, if adequate, enables commercially available Mg alloys to promote dynamic recrystallization whereby equiaxed fine grains of a grain size of less than 2 μm are formed. With LPSO phase-containing Mg alloy, on the other hand, LPSO grains hardly recrystallize during thermo-mechanical treatment but for kink deformation and LPSO-stimulated texture evolution that easily occur [8–10, 15]. Coarse and strong textured LPSO grains lead to low elongation of the extruded Mg/LPSO alloys. To overcome these problems we attempted in this study to establish a new technique by combining chip-consolidation processing and twin-roll casting. Chipping the twin-roll-cast alloy that possesses a fine dendrite structure of α-Mg and LPSO phases on one hand and extruding chips on the other one can form equiaxed fine grains of a mean grain size of more than 2 μm. Fragmentation of LPSO grains by chipping treatment may be effective for formation of equiaxed grain with random crystallographic orientation [16].

## Summary

Fine grained Mg/LPSO two-phase alloy was prepared by consolidation of chips machined from a twin-roll-cast $Mg_{97}Zn_1Y_2$ alloy. Its microstructure and superplastic behavior were investigated. The results are summarized as follows.

(1) The TRC-chip-consolidated $Mg_{97}Zn_1Y_2$ alloy was composed of the α-Mg and LPSO phases, and the α-Mg and LPSO grains had a mean grain size of ~1 μm. A stacking fault-like LPSO phase was frequently observed inside α-Mg grains.

(2) The TRC-chip-consolidated $Mg_{97}Zn_1Y_2$ alloy exhibited superplasticity at temperatures of 623 and 673 K. Large elongation of ~560% was observed at an initial strain rate of $3 \times 10^{-2}\ s^{-1}$ at 673 K. Given that the *m*-value showed more than 0.3 for all tensile tests in this study, the large *m*-values may be taken as evidence to support that uniform elongation occurs during superplastic deformation.

**Acknowledgements** This study is supported by grants-in-aid for Scientific Research(A) No. 16H02404 and (B) No. 17H03431 from the MEXT Japan, and by the Sumitomo Electric Industries, Ltd. Endowed Chair in Kumamoto University.

## References

1. Aghion E, Bronfin B, Von Buch F, Schumann S, Friedrich H (2003) Newly developed magnesium alloys for powertrain applications, JOM 55(11) 30–33.
2. Easton M, Beer A, Barnett M, Davies C, Dunlop G, Durandet Y, Blacket S, Hilditch T, Beggs P (2008) Magnesium alloy applications in automotive structures, JOM 60(11) 57–62.
3. Kawamura Y, Hayashi K, Inoue A, Masumoto T (2001) Rapidly solidified powder metallurgy $Mg_{97}Zn_1Y_2$ alloys with excellent tensile yield strength above 600 MPa, Mater. Trans. 42(7) 1172–1176.
4. Abe E, Kawamura Y, Hayashi K, Inoue A (2002) Acta Mater. 50 (15) 3845–3857.
5. Egusa D, Abe E (2012) The structure of long period stacking/order Mg–Zn–RE phases with extended non-stoichiometry ranges, Acta Mater. 60(1) 166–178.
6. Yoshimoto S, Yamasaki M, Kawamura Y (2006) Microstructure and Mechanical Properties of Extruded Mg-Zn-Y Alloys with 14H Long Period Ordered Structure, Mater. Trans. 47(4) 959–965.
7. Hagihara K, Kinoshita A, Sugino Y, Yamasaki M, Kawamura Y, Yasuda HY, Umakoshi Y (2010) Effect of long-period stacking ordered phase on mechanical properties of $Mg_{97}Zn_1Y_2$ extruded alloy, Acta Mater. 58(19) 6282–6293.
8. Yamasaki M, Hashimoto K, Hagihara K, Kawamura Y (2011) Effect of Multimodal Microstructure Evolution on Mechanical Properties of Mg-Zn-Y Extruded Alloy, Acta Mater. 59(9) 3646–3658.
9. Jono Y, Yamasaki M, Kawamura Y (2013) Effect of LPSO Phase-Stimulated Texture Evolution on Creep Resistance of Extruded Mg-Zn-Gd Alloys, Mater. Trans. 54(5) 703–712.
10. Jono Y, Yamasaki M, Kawamura Y (2015) Quantitative evaluation of creep strain distribution in an extruded Mg-Zn-Gd alloy of multimodal microstructure, Acta Materialia, 82(1) 198–211.
11. Pilling J, Ridley N (1989) Superplasticity in Crystalline Solids, The Institute of Metals, Maney Publishing, London.
12. Abu-Farha F, Curtis R (2011) Standard for superplastic forming of metals. in Giuliano G (ed) Superplastic forming of advanced metallic materials, Method and applications. Woodhead Publishing Limited, Cambridge, pp. 34–48.
13. Clarisse L, Bataille A, Pennec Y, Crampon J, Duclos R (1999) Investigation of grain boundary sliding during superplastic deformation of a fine-grained alumina by atomic force microscopy Ceram. Inter. 25(4) 389–394.
14. Mukai T, Watanabe H, Higashi K (2000) Application of superplasticity in commercial magnesium alloy for fabrication of structural components, Mater. Sci. Tech. 16(11–12) 1314–1319.
15. Yamasaki M, Hagihara K, Inoue S, Hadorn JP, Kawamura Y (2013) Crystallographic classification of kink bands in an extruded Mg–Zn–Y alloy using intragranular misorientation axis analysis, Acta Mater. 61(6) 2065–2076.
16. Fudetani S, Itoi T, Kubo T, Kawamura Y, Hirohashi M (2009) Microstructures and Mechanical Properties of $Mg_{96}Zn_2Y_2$ Alloy Prepared by Extrusion of Machined Chips, Mater. Trans. 50(2) 349–353.

# Technological Solutions to Apply Magnesium Bulk Materials in Dynamic Bending and Axial Compression Load Cases

Elmar Beeh, Horst E. Friedrich, Philipp Straßburger, William Altenhof, Ping Zhou, Michael Worswick, and Samuel Kim

**Abstract**

Typical magnesium bulk materials, like AZ31B, show high potential to reduce vehicle weight in automotive applications. Technical limitations are coming from the material behavior under crash loads, where a risk of catastrophic failure is given in buckling deformation modes. To potentially implement magnesium into new applications, the behavior of magnesium AZ31B structures in dynamic bending and axial compression load cases have been studied. In structures with a bending load, such as a bumper, the stabilization of the section of the profile leads to a significant improvement of specific energy absorption and to a lower risk of catastrophic failure. Rectangular section beams have been constructed and tested by using the quasi-static/dynamic three-point bending facilities at German Aerospace Centre (DLR)—Institute of Vehicle Concepts. For axial loads, cutting or peeling mode based mechanisms have been developed and investigated, which allow the use of magnesium in these challenging applications.

**Keywords**

Magnesium alloy • Three-point bending
Tension-compression asymmetry • Energy absorption
Cutting

E. Beeh (✉) · H. E. Friedrich · P. Straßburger
German Aerospace Center (DLR)—Institute of Vehicle Concepts, Pfaffenwaldring 38-40, 70569 Stuttgart, Germany
e-mail: elmar.beeh@dlr.de

H. E. Friedrich
e-mail: horst.friedrich@dlr.de

P. Straßburger
e-mail: philipp.strassburger@dlr.de

W. Altenhof
University of Windsor, Windsor, ON, Canada
e-mail: altenh1@uwindsor.ca

P. Zhou · M. Worswick · S. Kim
University of Waterloo, Waterloo, ON, Canada
e-mail: pingchou83@gmail.com

M. Worswick
e-mail: worswick@lagavulin.uwaterloo.ca

S. Kim
e-mail: samuel.kim@uwaterloo.ca

## Introduction

Driven by lightweight goals, magnesium and its alloys have attracted increasing attention in the transportation industry due to their inherent advantages such as low density, high strength-to-weight ratio, high damping resistance and easy recycling [1]. Up to now, the majority of the applications of Mg alloys are high pressure die castings (HPDC), such as transmission cases, instrument panels and door inners [2]. Recently, there is a desire to employ Mg alloys as crashworthy automotive components; therefore more attention has been paid on wrought Mg alloys, i.e. rolled sheets and extrusions, which generally exhibit higher strength and greater ductility than Magnesium HPDCs. Compared with cast and forged magnesium alloys, wrought magnesium alloys (i.e. rolled sheets and extruded profiles) exhibit better mechanical properties, including higher strength and ductility and accordingly better energy absorption performance. However, the ductility and formability of wrought magnesium alloys at room temperature are relatively poor compared with commonly-used steels and aluminium alloys, resulting in difficulties with the rolling/forming/stamping processes of magnesium sheets and their crashworthiness applications.

Crash load cases play a major role for many components in vehicle applications. In addition to the body structure, doors and tailgates, seats and other interior components must also be crashworthy. In terms of structural applications for crash related areas, much attention in the past decade have been paid to the deformation mechanisms and mechanical behavior of magnesium alloy thin-walled structures subjected to axial compressive loads. Wagner et al. [3, 4]


D. Orlov et al. (eds.), *Magnesium Technology 2018*, The Minerals, Metals & Materials Series, https://doi.org/10.1007/978-3-319-72332-7_39

studied the deformation characteristics of magnesium AM30 extruded beams and AZ31 sheet beams in axial crushing. In such cases, magnesium beams do not produce the so-called accordion shaped deformation mode [5] like steel and aluminum beams; instead, they crack and split into large fragments. Steglich et al. [6] assessed the crashworthiness of the rectangular tubes made of magnesium AZ31/ZE10 sheet and extrusion under quasi-static axial crushing. It was observed that the higher work hardening rate in uniaxial compression tests contributes to higher energy absorption by the formation of multiple material folds.

For bending load cases, where a loaded structure sees separated zones of tension and compression loads, the strong tension-compression asymmetry of typical magnesium bulk materials is the dominating material property which has to be addressed in an engineering design process. It has been extensively reported over the past years, that the tension-compression asymmetry of wrought magnesium alloys represents not only a much lower yield stress in compression than in tension, but also a distinct difference of strain hardening rate between tension and compression [7–15]. In stage II of a stress-strain curve in compression, wrought magnesium alloys show high strain hardening rates due to multiple slip and interactions among dislocations on non-parallel planes. It was considered that the strong strain hardening brings advantages on energy absorption in bending and buckling modes compared with steels and aluminium alloys because the large radius of curvature of the bent sheets may have been caused by the strong strain hardening [16–18]. Because of the low yield strength under compression compared to steel and aluminium, these theoretical potentials will only lead to lighter magnesium parts, if the structures are well designed and optimized by precise numerical simulations.

## Relevant Properties of Mg AZ31B

For the development of concept ideas and the validation with structural tests, a commercial-grade magnesium alloy AZ31B (Mg–3Al–1Zn, wt%) in the form of cast-rolled sheets with 1.80 and 2.92 mm thicknesses and extruded profiles with 3 mm wall thickness have been used. This Mg alloy has an elastic modulus of 45 GPa and a density of 1.77 g/cm$^3$.

The test results shown in Fig. 1 have been measured by quasi-static tensile and compression testing on a universal testing machine Zwick/Roell Z250. The dimensions of the specimens used for tension testing are outlined in the standard ASTM E8-11 with a gage length of 50 mm. For each specimen, the longitudinal strain along the gage length was recorded by a sensor arm extensometer. Simultaneously, the ARAMIS® two-dimensional digital image correlation (2D-DIC) system (GOM GmbH, Germany) was used to measure the longitudinal strain and transverse strain.

For uniaxial compressive testing, a new simply-constructed testing method recently developed by the current authors [20] was employed. For convenience, the new method will be briefly introduced here. As shown in Fig. 2, the thin-sheet specimen is clamped by the back anti-buckling plate and the two front anti-buckling plates. The compressive load is introduced via the both ends of the specimen. A special thin-sheet specimen with a single-sided groove in the parallel section was developed, as shown in Fig. 3a–c. The reduced thickness leads to an off-axial load which generates a bending moment $M_p$ over the parallel section, as shown in Fig. 3d. The self-balance between the bending moment $M_p$ and the normal force $N_p$ exerted by the back anti-buckling plate prevents the buckling.

## Possible Areas of Application for Magnesium Bulk Materials

To utilize magnesium bulk materials in structural applications, e.g. in vehicle structures, it's crucial to understand the functional requirements, where the material properties can lead to lightweight potential. Currently magnesium sheets are often suggested to be used in plane structures like roof panels or inner reinforcement panels for the engine hood (Fig. 4).

Even if magnesium shows only a low elastic modulus, the material can benefit from a high specific bending stiffness. As a result of the very low density of magnesium, the geometrical moment of inertia, achieved by thicker wall thickness, overcompensates the low elastic modulus.

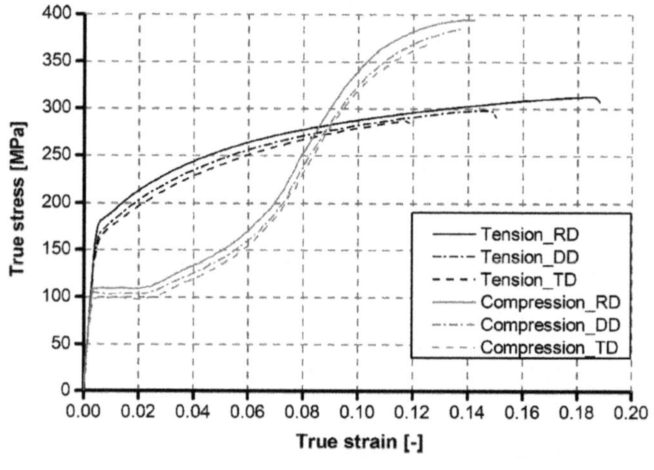


**Fig. 1** Quasi-static tensile/compressive true stress/strain responses for magnesium alloy AZ31B (RD—rolling direction, DD—diagonal direction, TD—transverse direction) [19]

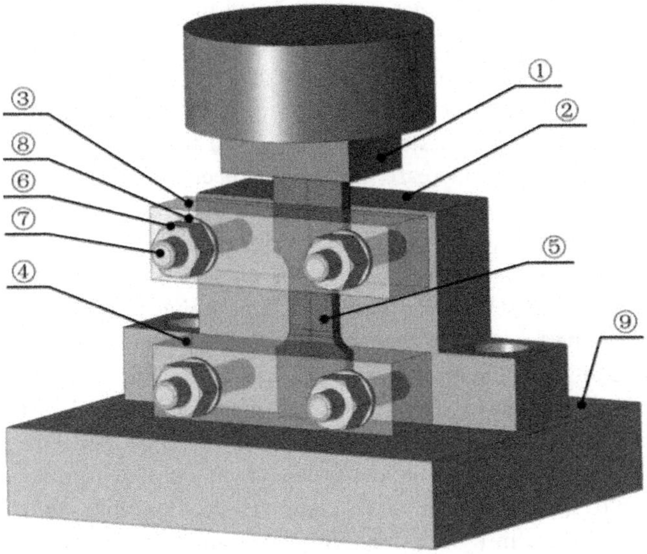


**Fig. 2** Schematic illustration of the new uniaxial compression testing apparatus, including parts: ① compression platen, ② back anti-buckling plate, ③ front upper anti-buckling plate, ④ front lower anti-buckling plate, ⑤ grooved thin-sheet specimen, ⑥ nuts, ⑦ pin bolts, ⑧ spring washers, ⑨ base support [20]

A similar effect can be used beneficially by building sandwich-like structures, with thin walled magnesium extrusions. These parts currently are limited regarding their geometrical complexity, but can be potentially implemented for flat floor structures or the back rest of the rear seat in cars.

Further potential application is in using magnesium bulk materials in crash related structures with a bending-dominant load. These kinds of structures, namely, bumpers or door reinforcements, are typically dimensioned by the crash loadcase and see only low static loads in normal use. For achieving lightweight potential in such a structural area, it is necessary to utilize tension-compression asymmetry together with the strong work hardening of magnesium bulk materials. The design of such a component requires a dynamic simulation based design process, where the tension/compression asymmetry, the energy absorption through plastic deformation, the work hardening and the rate sensitivity of material properties are considered. DLR-Institute of Vehicle Concepts has done the required research to be capable to perform these kinds of design

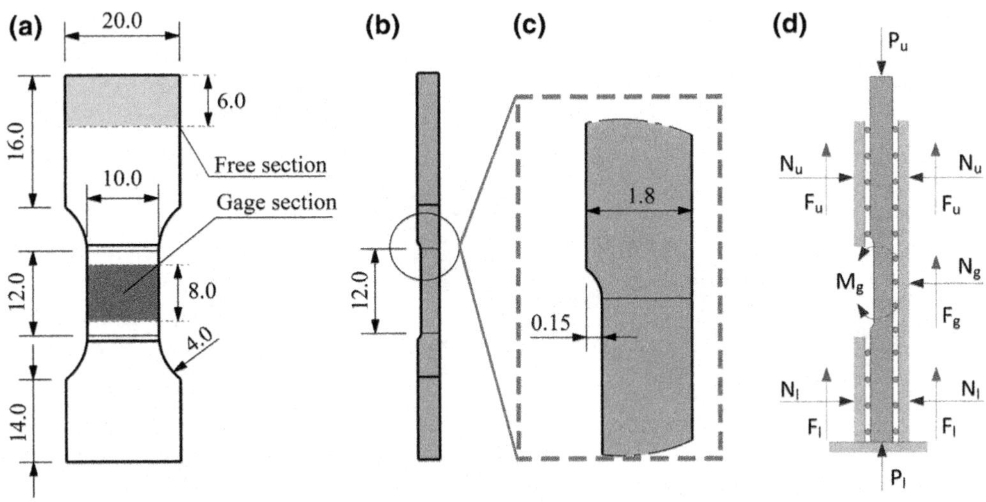


**Fig. 3** Design concept of the grooved thin-sheet specimen: **a** front view; **b** left view; **c** detail view; **d** free body force diagram [20]

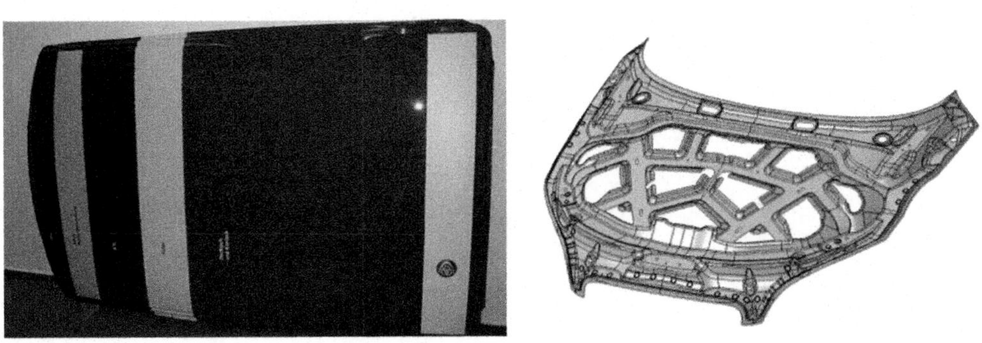

**Fig. 4** Roof structure made from (left) and inner hood designed for magnesium sheet (right) [21]

processes. The potential applications will be shown in the next section.

## Concept and Solution for Applying Magnesium in Vehicle Structures Bending Loads

The main objective for utilizing the material properties of wrought magnesium is to exploit the strong tension/compression asymmetry and the significant compression work hardening. Especially in load cases where the tension and compression zones are clearly separated within a larger deformation, the properties of magnesium can be beneficial. In components such as the rocker panel, bumper or door impact beam a profile is loaded transversely. This produces compressive stresses on the inside and tensile stresses on the outside.

One way for optimizing the design is to distribute material asymmetrically in the profile section, so that more material is located in the compressive stress area (see Fig. 5). This design leads to a compensation of the tension/compression asymmetry, but it will be hard to balance and utilize the work hardening effect in a beneficial way. Compressed areas generally tend to fail earlier due to buckling or kinking. Providing members with core structures (e.g. foam cores) is an effective method of avoiding premature buckling or kinking by reinforcing the cross-section of the member. For structures fabricated from DC04 steel it was possible to increase the weight-specific energy absorption by a factor of 3 [22], where in this case the tension side of the member was very heavily elongated.

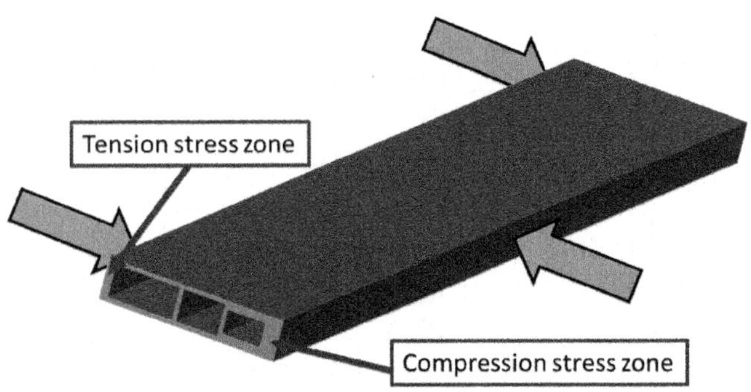


**Fig. 5** Profile with asymmetric material distribution in tension and compression loaded areas

To investigate the behavior of magnesium in this load case, quasi-static and dynamic three-point bending tests were performed to investigate the deformation modes, fracture and energy absorbing performance of rectangular tubes. For these basic tests no asymmetric material distribution was chosen. As shown in Fig. 6, Mg AZ31B sheets with 1.8 mm wall thickness were bent cold to form half shells and then welded to complete profiles with dimensions 130 mm by 88 mm using shielding gas. The bending radius used was 20 mm, because a smaller radius would have pre-damaged the part. Extrusions from the same material and a similar outer shape have been produced with a wall thickness of 3 mm. For both profile variants the length was 1650 mm. The profiles destined for foam filling were provided with a cataphoretic coating in order to guarantee a good bond between the foam and the profile. The profiles were then filled by DOW Chemical Company with BETAFOAM in three different densities (200, 300 and 400 kg/m$^3$).

The three-point bending rig consisted of two fixed lower supports having a diameter of 96 mm, and an upper indenter with a 300 mm diameter. The two fixed lower supports provided a beam span of 1368 mm. Quasi-static bending tests were conducted at a constant speed of 60 mm/s and at a maximum penetration distance of 450 mm, using an MTS horizontal servo-hydraulic testing machine with a 250 kN load cell. Dynamic bending tests were performed on the DLR-dynamic component testing system, a moving sled with a mass of 767 kg, an acceleration sensor and 3 Photron FASTCAM SA3 high-speed-cameras, which can also measure speed and displacement. The dynamic loads

**Fig. 6** Cross-sections of empty/foam-filled magnesium beams. *Note* All dimensions are in millimetres [19]

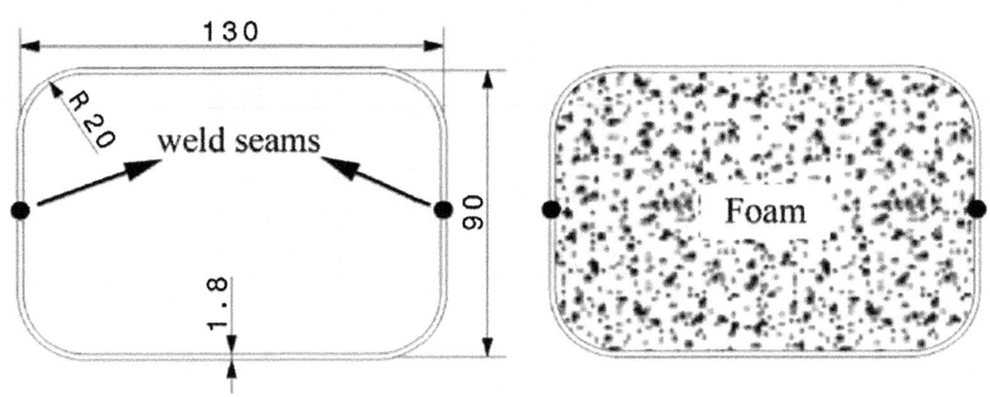

were calculated from the mass and deceleration of the moving sled mass. The dynamic tests were conducted at an initial speed ranges from 1.8 to 6 m/s.

The properties of the hollow profiles were compared to those of the foam filled profiles in quasi-static and dynamic 3-point bending tests. Figure 7 shows the results of the quasi-static test.

Both the sheet and the extruded hollow magnesium profiles outperformed the steel tube with an higher specific energy absorption but it has to be considered that hollow AZ31 profiles are effected by very early fracture of local areas and that the steel grade DC04 is normally not used for structural applications because of its low yield strength.

The reason why DC04 was used here was the very high breaking elongation of this steel material which will lead to a strong increase in the SEA for the foam filled profile. Figure 8 shows the results for SEA for the different tests. The SEA for DC04 steel increases by more than three times, while for magnesium profiles with BETAFOAM200 the SEA approximately doubles. The magnesium sheet profile achieves a value of SEA more than 100 J/kg greater than that of the steel profile. Using this option the component does not fracture in either the quasi-static or the dynamic test.

The analysis of the deformation behaviour is interesting. In the bending loadcase the steel profiles are strongly elongated on their tension side, while the Mg profiles show strong plastification on the compression side (Fig. 9). This is caused by the pronounced asymmetry in the compressive and tensile properties of magnesium, so that the plastification is concentrated on the weaker compression side there. This can be beneficially exploited in this application, because this effect, in conjunction with an optimal foam filling, means that the deformation can be distributed across a large area. Using this design it is possible to beneficially make use of tension/compression asymmetry and the work hardening effect of magnesium sheets and extrusions and to utilize this material in crash related areas with bending load cases.

## Solutions for Applying Magnesium in Axial Load Cases

In the case of an accident, front and rear crash structures and the vehicle side members typically convert energy through deformation, generally by regular buckling behavior. The deformation capacity of magnesium sheets and profiles is generally insufficient to meet such demands. Together with the University of Windsor (UW), Canada, DLR is working on concepts for utilizing magnesium as a crash absorber. The fundamental idea is to utilize cutting or peeling processes for energy absorption. These kinds of processes, which are usually used for machining, are well known in industrial production. It is well known, that many light metals, even if they have a low breaking strain, possess a good machinability. UW's concept is to longitudinally disintegrate magnesium profile tubes using cutting blades. DLR's concept is to peel the outer skin of a metallic load bearing structure, like e.g. the crashbox or the longitudinal rail.

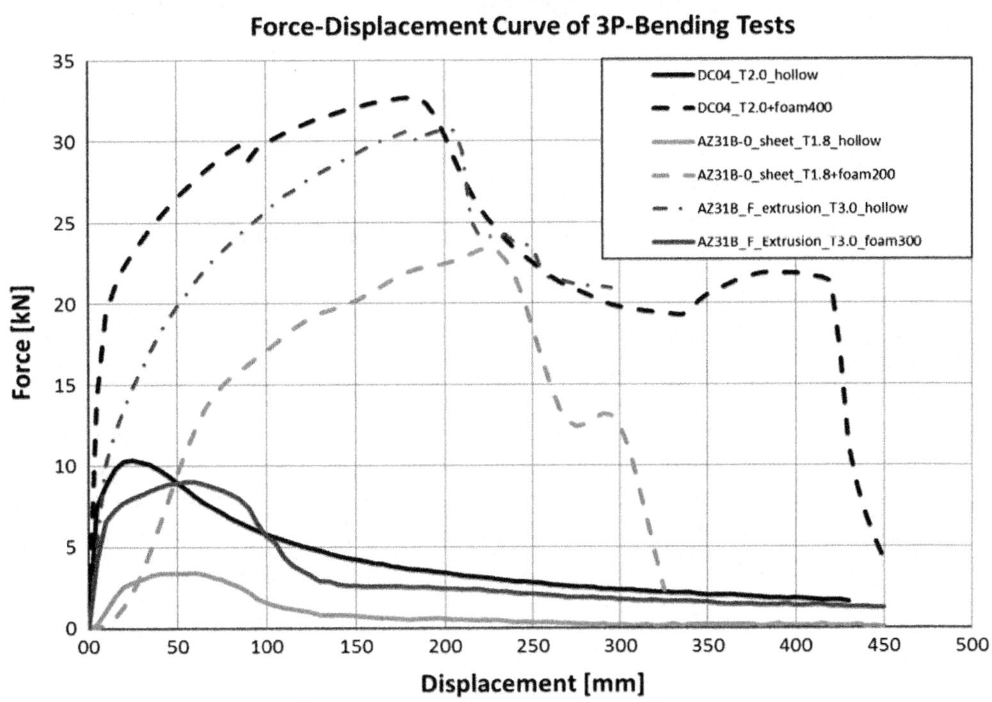


**Fig. 7** Force-displacement responses of quasi-static 3-point bending tests compared to former tests with steel DC04

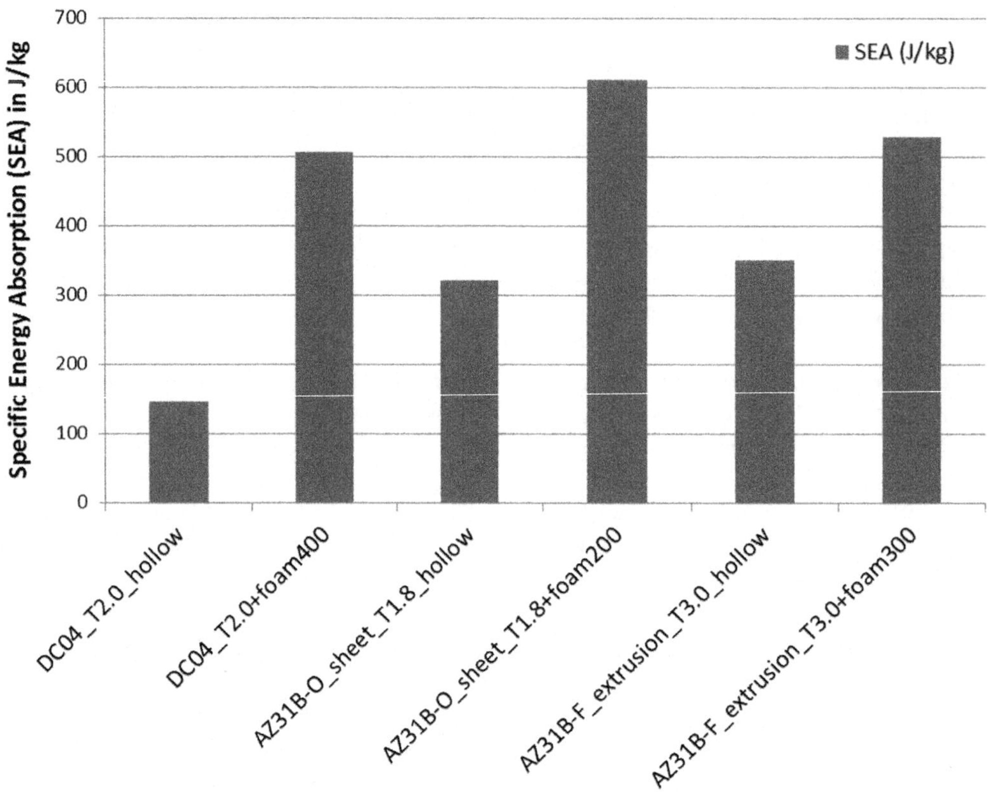


**Fig. 8** Comparison of SEA of hollow and foam filled steel and magnesium profiles

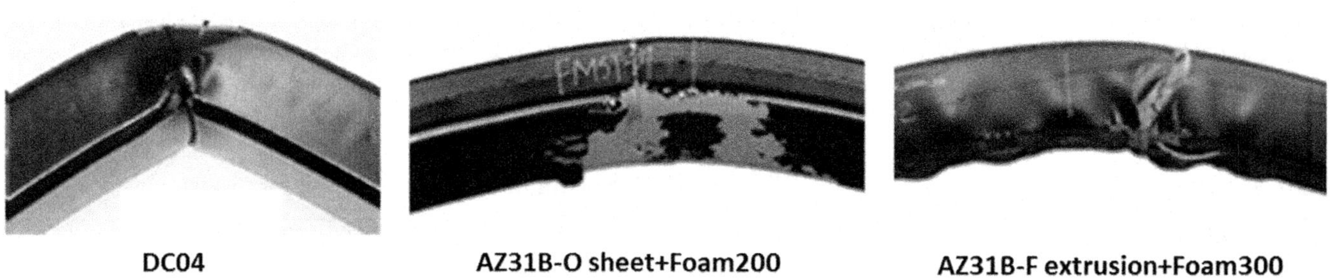


**Fig. 9** Comparison of the deformation of the different profiles at the onset of cracking

Figure 10 shows on the left a specimen after a quasi-static axial cutting test with a 4-blade cutting tool, performed on a hydraulically operated long-stroke Tinius Olsen compression testing machine. The extrusion was positioned such that the longitudinal axis was parallel to the motion of the crosshead and centered within the platen. The deformation behavior was noted to be stable, repeatable amongst additional tests, and also controlled the force response as a function of material thickness.

Figure 10 shows on the right the force/displacement response which exhibits a very constant force during deformation. It can be seen, that with the cutting principle, magnesium can be used to absorb energy in axial load cases without a tendency of fracture or failure. Together with the University of Waterloo, University of Windsor and DLR have developed a cutting absorbing device which can be applied for very precise energy absorption in crash test facilities. DLR further is working on axial crash absorbers using peeling of the outer skin of a metallic tube, which have been investigated successfully with Aluminium tubes. It is expected that Magnesium show similar properties like the Aluminium.

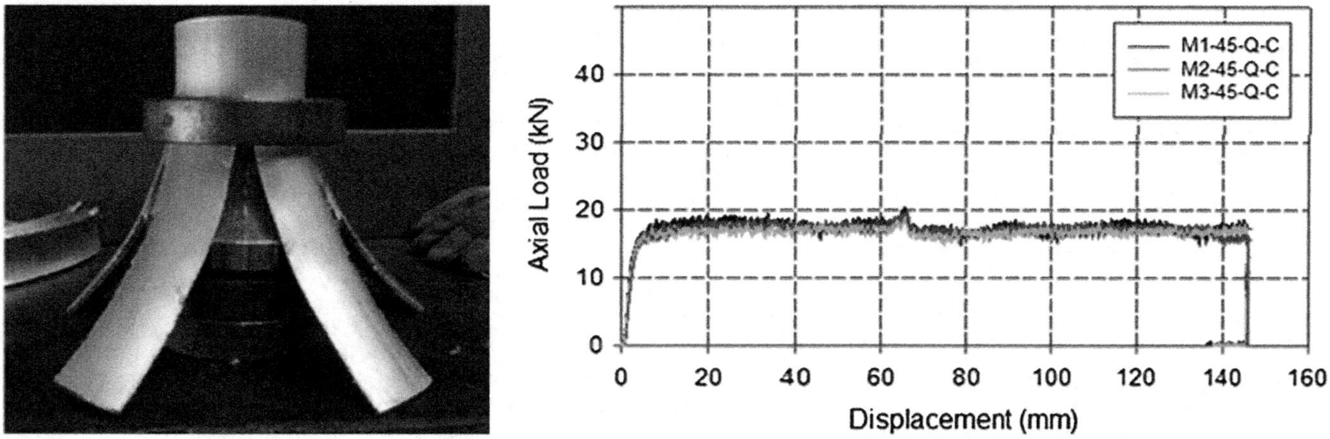


**Fig. 10** Quasi-static axial cutting of an AZ31B magnesium extrusion after test (left) and load/displacement response (right) [23]

## Summary

To apply magnesium bulk materials in lightweight structures, the special properties of these alloys, especially the tension/compression asymmetry and the strong work hardening under compression need to be considered. Beside applications in larger plane areas or as flat extrusions, e.g. in the vehicle floor area, especially applications with bending load cases, such as bumpers or door crash beams can benefit from the special properties of bulk magnesium. Tests with AZ31B-profiles show the potential of foam-filled profiles for that kind of applications. For axial load cases, a possible solution to apply magnesium is, to change the energy absorption mechanism from buckling to cutting/peeling. Results of tests, which have been performed at the University of Windsor, Ontario show the potential of this concept. The future objective of the partners in Canada and Germany is to investigate in detail the magnesium cutting and peeling tube potentials, because this will open up new application fields for magnesium profiles.

**Acknowledgements** The authors from the University of Windsor would like to thank the Natural Sciences and Engineering Research Council (NSERC) of Canada for financially supporting this work.

## References


1. H.E. Friedrich, and B. L.Mordike, Magnesium Technology - Metallurgy, Design.
2. M. Easton, M. Gibson, A. Beer, M. Barnett, C. Davies, Y. Durandet, S. Blacket, X. Chen, N. Birbilis, and T. Abbott, The Application of Magnesium Alloys to the Lightweighting of Automotive Structures, Sustainable Automotive Technologies, (2012) 17–23.
3. D.A. Wagner, S. Logan, K. Wang, T. Skszek, C.P. Salisbury, Test Results and FEA Predictions from Magnesium AM30 Extruded Beams in Bending and Axial Compression, in: N.R.N. S. R. Agnew, E. A. Nyberg, M. O. Pekguleryuz (Ed.) Magnesium Technology 2009, The Minerals, Metals and Materials Society, 2009, pp. 541–549.
4. D.A. Wagner, S.D. Logan, K. Wang, T. Skszek, Test Results and FEA Predictions from Magnesium AZ31 Sheet Beams in Bending and Axial Compression, in: S.R. Agnew, N.R. Neelameggham, E. A. Nyberg, W.H. Sillekens (Eds.) Magnesium Technology 2010, TMS (The Minerals, Metals & Materials Society), 2010, pp. 547–552.
5. G. Lu, T. Yu, Energy Absorption of Structures and Materials, Woodhead Publishing, 2003.
6. D. Steglich, X. Tian, J. Bohlen, S. Riekehr, N. Kashaev, K.U. Kainer, N. Huber, Experimental and numerical crushing analyses of thin-walled magnesium profiles, International Journal of Crashworthiness, 20 (2015) 177–190.
7. D. Ghaffari Tari, M.J. Worswick, U. Ali, and M.A. Gharghouri, Mechanical response of AZ31B magnesium alloy: Experimental characterization and material modeling considering proportional loading at room temperature, International Journal of Plasticity 55 (2014), pp. 247–267.
8. N.-T. Nguyen, O. Seo, C. Lee, M.-G. Lee, J.-h. Kim, and H. Kim, Mechanical Behavior of AZ31B Mg Alloy Sheets under Monotonic and Cyclic Loadings at Room and Moderately Elevated Temperatures, Materials 7 (2014), pp. 1271–1295.
9. D. Steglich, X. Tian, J. Bohlen, and T. Kuwabara, Mechanical Testing of Thin Sheet Magnesium Alloys in Biaxial Tension and Uniaxial Compression, Experimental Mechanics 54 (2014), pp. 1247–1258.
10. M.R. Barnett, Twinning and the ductility of magnesium alloys Part I: "Tension" twins, Materials Science and Engineering: A 464 (2007), pp. 1–7.
11. I. Ulacia, N.V. Dudamell, F. Gálvez, S. Yi, M.T. Pérez-Prado, and I. Hurtado, Mechanical behavior and microstructural evolution of a Mg AZ31 sheet at dynamic strain rates, Acta Materialia 58 (2010), pp. 2988–2998.
12. S. Xu, W.R. Tyson, R. Eagleson, R. Zavadil, Z. Liu, P.L. Mao, C. Y. Wang, S.I. Hill, and A.A. Luo, Dependence of flow strength and deformation mechanisms in common wrought and die cast magnesium alloys on orientation, strain rate and temperature, Journal of Magnesium and Alloys 1 (2013), pp. 275–282.
13. S. Kurukuri, M.J. Worswick, D. Ghaffari Tari, R.K. Mishra, and J. T. Carter, Rate sensitivity and tension-compression asymmetry in AZ31B magnesium alloy sheet, Philosophical Transactions of the Royal Society A: Mathematical, Physical and Engineering Sciences 372 (2014).

14. S. Kurukuri, M.J. Worswick, A. Bardelcik, R.K. Mishra, and J.T. Carter, Constitutive Behavior of Commercial Grade ZEK100 Magnesium Alloy Sheet over a Wide Range of Strain Rates, Metallurgical and Materials Transactions A 45 (2014), pp. 3321–3337.
15. S. Kurukuri, D.G. Tari, M.J. Worswick, R.K. Mishra, and J.T. Carter, Dynamic characterization of AZ31B and ZEK100 magnesium alloy sheets, International Conference of Magnesium Alloys and Applications, Vancouver, BC, Canada, (2012).
16. M. Easton, W. Qian Song, and T. Abbott, A comparison of the deformation of magnesium alloys with aluminium and steel in tension, bending and buckling, Materials & Design 27 (2006), pp. 935–946.
17. T. Hilditch, D. Atwell, M. Easton, and M. Barnett, Performance of wrought aluminium and magnesium alloy tubes in three-point bending, Materials & Design 30 (2009), pp. 2316–2322.
18. D. Steglich, X. Tian, J. Bohlen, S. Riekehr, N. Kashaev, K.U. Kainer, and N. Huber, Experimental and numerical crushing analyses of thin-walled magnesium profiles, International Journal of Crashworthiness 20 (2015), pp. 177–190.
19. Zhou, Ping und Beeh, Elmar und Kriescher, Michael und Friedrich, Horst E. und Kopp, Gundolf (2016) Dynamic bending behaviour of magnesium alloy rectangular thin-walled beams filled with polyurethane foam. International Journal of Crashworthiness. Taylor and Francis.
20. Zhou, Ping und Beeh, Elmar und Friedrich, Horst E. (2015) A novel testing method for uniaxial compression of thin sheet magnesium alloys. Experimental Mechanics. Springer. ISSN 0014–4851.
21. F. Berge, H.-P. Vogt, M. Ullmann, K. Neh, R. Kawalla (2017) Anforderungen an Magnesiumhalbzeuge und daraus gefertigte Komponenten. WerkstoffPlus Auto, 7. Tagung für neue Fahrzeug- und Werkstoffkonzepte, 15.-16. Februar 2017, Stuttgart.
22. Kriescher, Michael und Salameh, Walid und Beeh, Elmar und Roettger, Jan und Droste, Alexander und Otto, Manuel (2014) POSSIBILITIES FOR THE USE OF METAL-HYBRID-STRUCTURES FOR VEHICLE CRASH LOAD CASES. Euro Hybrid, Stade.
23. Shery, P., Altenhof, W., Smith, R., Beeh, E. et al., "Experimental Observations on the Mechanical Response of AZ31B Magnesium and AA6061-T6 Aluminum Extrusions Subjected to Compression and Cutting Modes of Deformation," SAE Technical Paper 2017-01-0377, 2017, doi: https://doi.org/10.4271/2017-01-0377.

# Mechanical Properties of Thermo-Mechanically Treated Extruded Mg–Zn-Based Alloys

Daria Drozdenko, Patrik Dobroň, Juraj Olejňák, Marius Hegedüs, Klaudia Horváth, and Jan Bohlen

**Abstract**

Mechanical properties of extruded Mg alloys are significantly influenced by the activation of extension twins during compression along the extrusion direction (ED) because of a strong texture with basal planes oriented parallel to ED. At the same time, the heat treatment is also supposed to tune mechanical properties via strengthening or softening mechanism. The influence of heat treatment on the mechanical behavior of Mg–Zn-based alloys with an addition of Ca and Nd in as-extruded state and after pre-compression (i.e. partly twinned microstructure) is discussed in the paper. Difference in distribution of precipitates for two materials after applying heat treatment at 200 °C for 16 h was observed. Isothermal ageing of pre-strained samples leads to strengthening in ZN10 alloy and softening in ZX10 alloy.

**Keywords**

Magnesium • Twinning • Annealing. precipitation

## Introduction

As one of the lightest material for engineering use, Mg alloys have found their application in automotive and aerospace industry. However, mechanical properties of wrought Mg alloys are influenced by several factors. Generally, the hexagonal close-packed (*hcp*) crystallographic lattice of Mg together with the texture of wrought material results in asymmetry and anisotropy of mechanical properties. Due to lack of sufficient independent slip systems to accommodate plastic deformation at room temperature (RT), twinning becomes a key important deformation mechanism [1]. The {10–12} <10–11> extension twinning provides strain along the *c*-axis and by reorientation of the lattice by 86° with respect to the original lattice leads to activation of dislocation slip in the reoriented lattice. It is known, that formation of extension twins strongly depends on grain orientation with respect to loading direction: extension {10–12} <10–11> twins are activated, when compression is applied perpendicular to *c*-axis of *hcp* lattice [2–4]. Thus, preferred grain orientation of the material has a significant influence on twinning activity. Extruded Mg alloys are characterized by texture with the *c*-axis of the *hcp* unit cells in the majority of grains oriented perpendicular to the extrusion direction (ED) [5]. Therefore, extension twinning activates during the in-plane compression along ED rather than during in-plain tension along ED [6]. Thus, due to the texture of the material and polar nature of twin formation, the twinning significantly influences the yield behavior resulting in compression-tension asymmetry. It has been reported [7, 8] that the Ca or rare-earth (RE) elements addition into Mg–Zn alloys can modify texture and make it weaker, with the resultant improvement of formability and ductility.

At the same time, pre-strain of textured material can also influence anisotropy of mechanical properties. For example, twins are activated during pre-compression along ED affects tension-compression asymmetry in yield strength (YS) [9]. Twin lamellae introduce (i) twin boundaries into the microstructure, which acts as obstacles for dislocation slip and can contribute to hardening of the material; (ii) fraction of the reoriented lattice, which as was mentioned above provides a new possibility for activation of dislocation slip.

It was reported that mechanical properties of Mg alloys are also influenced by solute segregation and precipitation

D. Drozdenko (✉) · P. Dobroň · J. Olejňák · M. Hegedüs · K. Horváth
Department of Physics of Materials, Charles University, Ke Karlovu 5, 12116 Prague 2, Czech Republic
e-mail: drozdenko@karlov.mff.cuni.cz

D. Drozdenko · K. Horváth
The Czech Academy of Sciences, Nuclear Physics Institute of the Czech Academy of Science, Řež 130, 250 68 Řež, Czech Republic

J. Bohlen
Magnesium Innovation Centre, Helmholtz-Zentrum Geesthacht, Max-Planck-Straße 1, 21502 Geesthacht, Germany


D. Orlov et al. (eds.), *Magnesium Technology 2018*, The Minerals, Metals & Materials Series, https://doi.org/10.1007/978-3-319-72332-7_40

[10–16]. For example, the effect of annealing time on yield stress asymmetry of MN11 alloy with respect to precipitation strengthening was studied in [16].

Combination of mechanical and heat treatment, i.e. thermo-mechanical treatment (TMT), was found to be promising for the strengthening of the material and reducing YS asymmetry. In [17] it was reported that ordered segregation of solute atoms (Zn, Gd) at twin boundaries provides strengthening these alloys via a pinning effect of twin boundaries. Similar annealing hardening phenomenon during compression tests was observed in pre-strained AZ31 [18]. In work of Zeng et. al. [19] the annealing strengthening phenomenon of pre-strained Mg–0.3Zn–0.1Ca sheet alloy was observed at 80–200 °C. It was found that different time (up to 24 h) of annealing leads to a difference in the magnitude of the strength increment. In case of increasing annealing temperature (from 200 to 250 °C) strengthening effect disappeared. The reasons for annealing strengthening in Mg–Zn–Ca were studied in detail by transmission electron microscopy [19]. It was concluded, that strengthening is caused by the pinning of gliding dislocations by GP zones and solute atoms have been segregated to the basal dislocations.

In our previous work [20] we analyzed the impact of TMT on solute segregations, precipitates along grain and twin boundaries and therefore on deformation behavior of extruded binary Mg–Zn alloy. As a continuation of previous work, present study is focused on using thermo-mechanical treatment (TMT), i.e. pre- compression (to introduce twins) followed by heat treatment, for ternary Mg–Zn-based alloys with an addition of Ca and Nd. Present paper provides preliminary characterization of effect of TMT on mechanical properties of material with pronounced texture.

Main aim of present work is to investigate the influence of pre-strain and heat treatment on mechanical properties, particularly yielding behavior, of Mg–Zn-based alloys with an addition of Ca and Nd.

## Experimental Procedure

The ZX10 (Mg 1wt% Zn 0.2 wt% Ca) and ZN10 (Mg 1wt% Zn 0.5wt% Nd) magnesium alloys were indirectly extruded at 400 and 480 °C, respectively. The extrusion speed was set to 2.3 and 0.5 mm/s for ZX10 and ZN10 alloys, respectively. The extrusion ratio for both materials was set to 1:25 to achieve a final diameter of extruded bar of 10 mm. Specimens with a length of 14 mm and a diameter of 9.5 mm were machined from the round bar parallel to ED. In order to investigate the influence of the thermo-mechanical treatment on mechanical properties of materials, samples were firstly pre-compressed up to a selected level of stress and then subjected to heat treatment (HT) at 200 °C for 16 h. Samples after solely pre-deformation and after pre-deformation with subsequent HT were again subjected to the compressive loading. All deformation tests were carried out at RT and a constant cross head speed of $10^{-3}$ $s^{-1}$ using a universal testing machine INSTRON 5882.

The texture of the alloys in an as-extruded state was determined by the X-ray PANalytical XPert diffractometer using CuKα radiation. Recalculation of full pole figures was carried out using the open-access computer code MTEX [21]. Microstructure analysis was performed by scanning electron microscope (SEM) Zeiss Auriga equipped with EDAX/TSL EBSD systems. The electron backscattered diffraction was used to reveal the distribution of precipitates in the microstructure of the investigated profiles in the as-extruded condition and after HT. To reveal microstructure and grain orientations EBSD patterns were collected with a step size of 0.3 μm and an acceleration voltage of 10 kV. The specimens for SEM observations were ground on SiC papers, polished by diamond pastes down to 0.25 μm particle size. Final polishing was done by Ar ion beam milling system Leica.

## Results and Discussion

Investigated ZN10 and ZX10 magnesium alloys are characterized by homogeneous microstructure with an average grain size of 13 ± 2 and 12 ± 2 μm, respectively (Fig. 1). Both materials exhibit texture with a distinct alignment of basal planes parallel to ED (Fig. 2). Such texture corresponds to above mentioned classical texture [5] for extruded bars and can be associated with the pronounced anisotropic behavior of Mg alloys.

Back scattered electron images of the samples in an as-extruded state are presented in Fig. 3. Small amount of precipitates are inhomogeneously distributed in the microstructure and segregation along the grain boundaries is observed in both ZN10 and ZX10 alloys.

Microstructures of the samples after isothermal ageing are presented in Fig. 4. Generally, higher number of precipitates is observed in both materials after HT. Both materials are characterized by homogeneous distribution of precipitates in the microstructure with significant segregation at the grain boundaries. Precipitates segregated along the grain boundaries are bigger in ZX10 alloy compared to ZN10 alloy, while precipitates inside the grains are finer in ZX10 alloy. Moreover, small amount of precipitates with bigger size (up to 1 μm) are observed in the microstructure of both alloys.

Parameters of different precipitates, such as shape, size play important role in strengthening and subsequent resulting mechanical properties. Effect of precipitate shape and size has been studied in detail by transmission electron microscopy in [16, 22]. For example, it was found that large

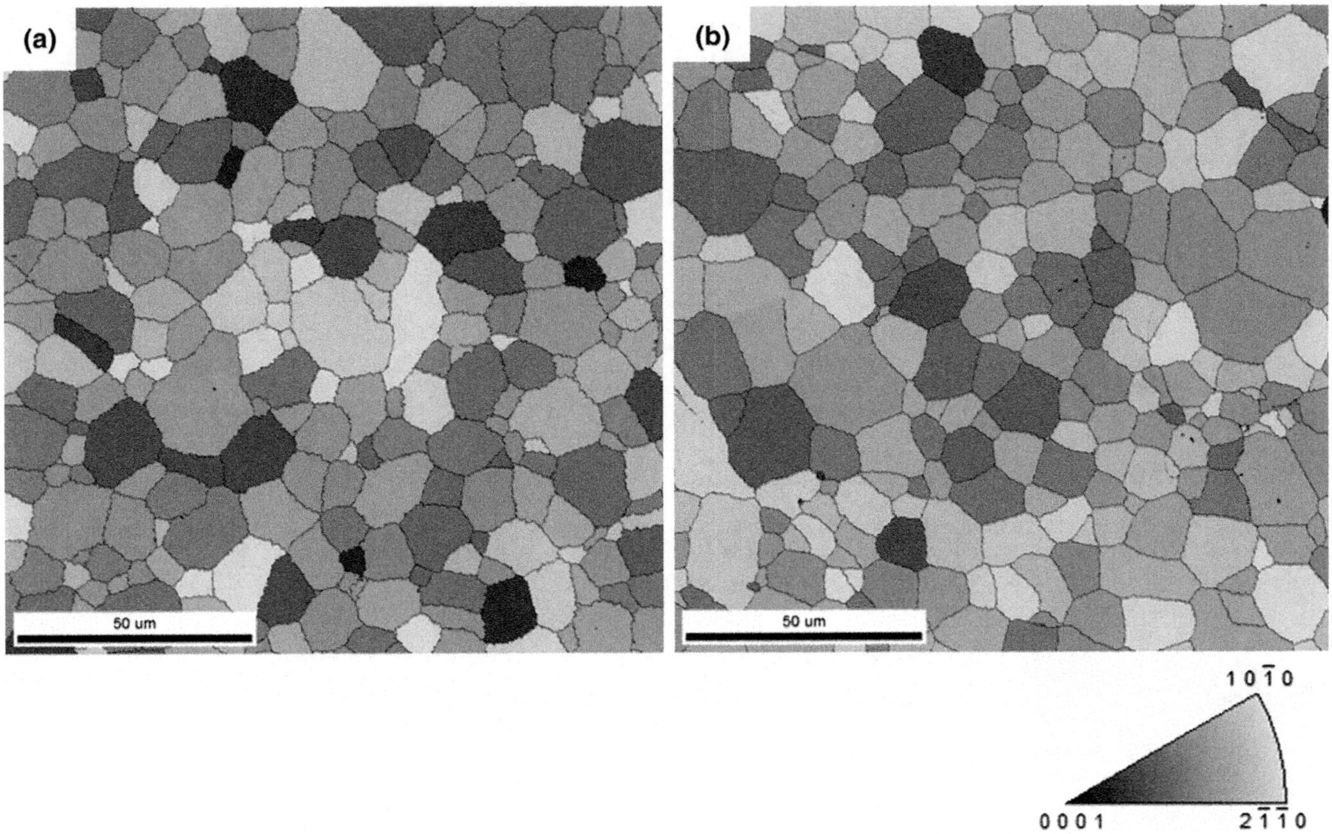


**Fig. 1** Orientation map of **a** ZN10 and **b** ZX10 magnesium alloys in an as-extruded state. ED is perpendicular to the image plane

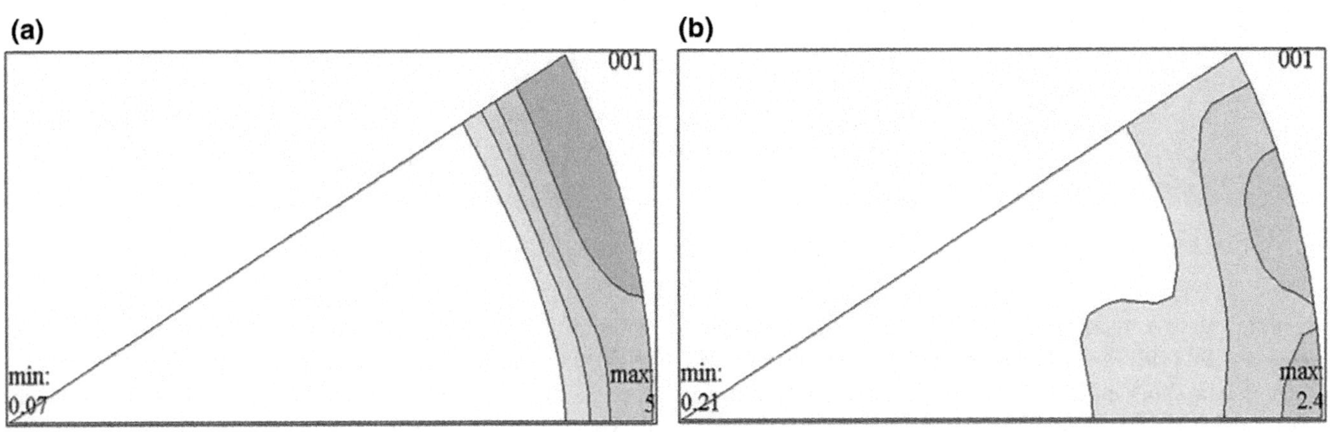


**Fig. 2** Inverse pole figure of **a** ZN10 and **b** ZX10 magnesium alloys in an as-extruded state. ED is perpendicular to the image plane

particles can significantly suppress the twin growth and new twins nucleate on the far side of the particle [23], while small particles can be inside the twins [24]. In present research, our attention was focused mainly on the mechanical properties of the investigated alloys after thermo-mechanical treatment rather than analysis of precipitates. Further detail analysis by means of transmission electron microscopy is needed.

Stress–strain curves are presented in Fig. 5 and values of YS are summarized in Table 1. Deformation curves for deformation of the samples in as-extruded state have a concave shape. The compression YS for the ZN10 and ZX10 alloys is 76 ± 2 and 112 ± 3 MPa, respectively. The sigmoidal shape of the both curves is associated with a massive twinning [25].

Deformation curves of the samples after HT at 200 °C for 16 h are comparable with those for samples in the as-extruded state with a slight difference in strain hardening for the ZX10 alloy. Values of YS for investigated alloys in

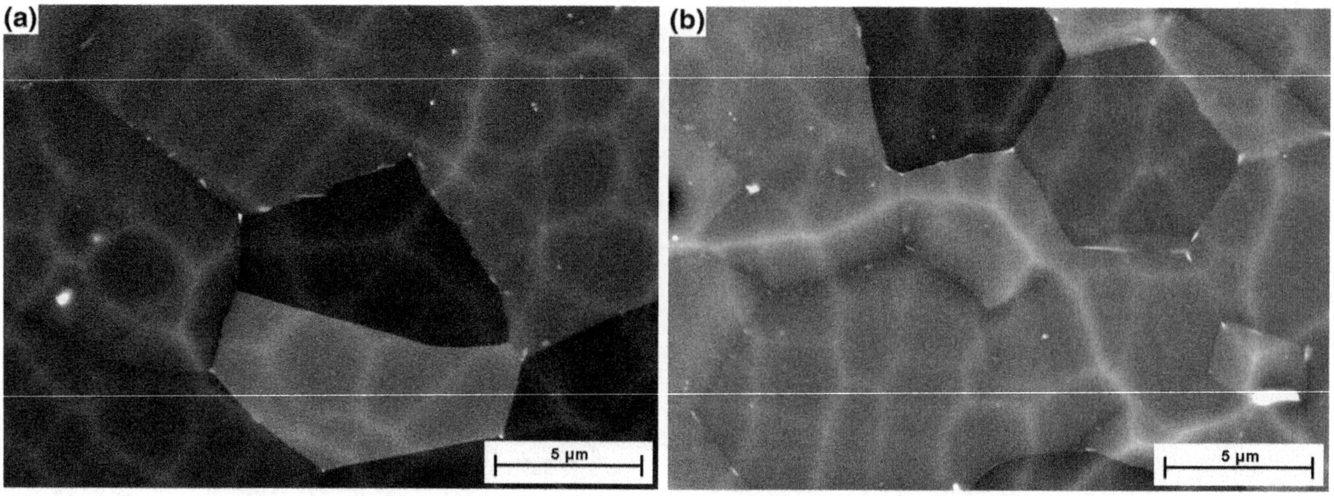


**Fig. 3** SEM back scattered electron images of the microstructure of **a** ZN10 and **b** ZX10 magnesium alloys in the as-extruded state. White "net-like" shadows are the result of ion milling and, therefore, correspond to the topography of the sample surface rather than to chemical composition of the material

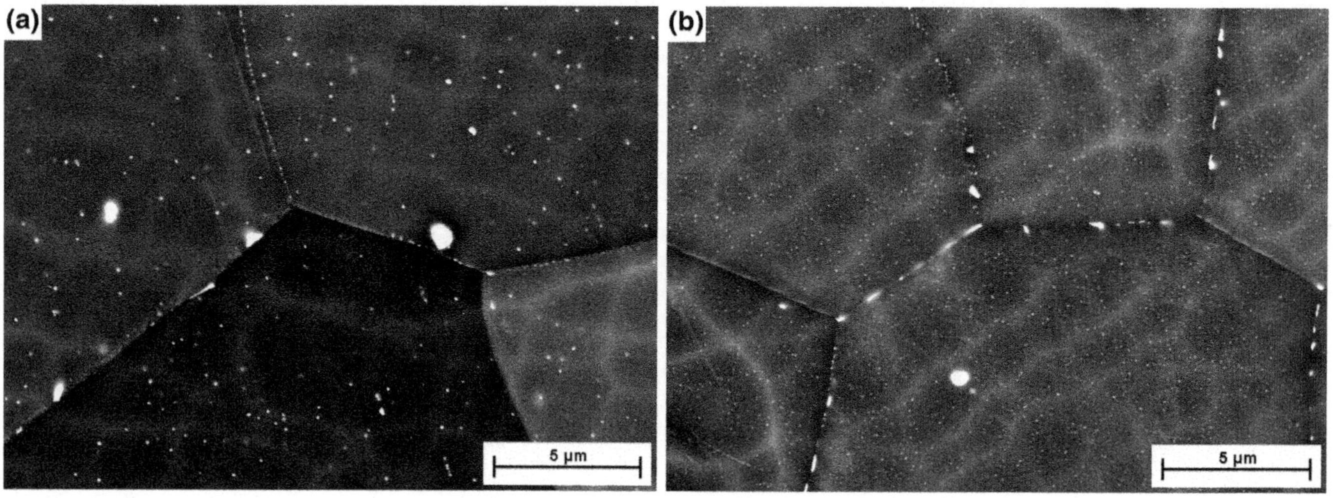


**Fig. 4** SEM back scattered electron images of the microstructure of **a** ZN10 and **b** ZX10 magnesium alloys after heat treatment at 200 °C for 16 h

the initial state and after HT were found to be the same (Table 1). In [19] it was reported that annealing of Mg–0.3Zn–0.1Ca sheet at 200 °C for 1 h prior the test did not lead to age hardening.

The main role of the pre-strain during thermo-mechanical treatment is introducing twin lamellae into microstructure. Thus, based on values of YS of samples in the as-extruded state, level of pre-compression along ED was chosen in a way to achieve partly twinned microstructure. Pre-compression was set at 105 and 130 MPa for the ZN10 and ZX10 magnesium alloys, respectively. This level of pre-compression corresponds approximately to 3% of strain. It was observed earlier [20, 26] that this level of pre-strain is enough to achieve partly twinned microstructure. Compression tests of samples after pre-compression and relaxation without intermediate HT showed that values of YS are slightly lower than the level of pre-compression (Table 1). It can be explained by stress relaxation effect. Stress–strain curves for both materials after pre-compression are also characterized by convex shape what is suspected to be result of continuation of twinning process.

In order to investigate the effect of HT on the deformation behavior of pre-strained materials samples after pre-compression were subjected to isothermal ageing at 200 °C for 16 h. As can be seen in Fig. 5 (blue lines), deformation behavior of the pre-compressed samples after the same HT is different for two investigated alloys. For the ZN10 magnesium alloy value of YS increases, while the ZX10 magnesium alloy significantly loses in the value of YS. Moreover, stress–strain curve of ZN10 alloy exhibits a

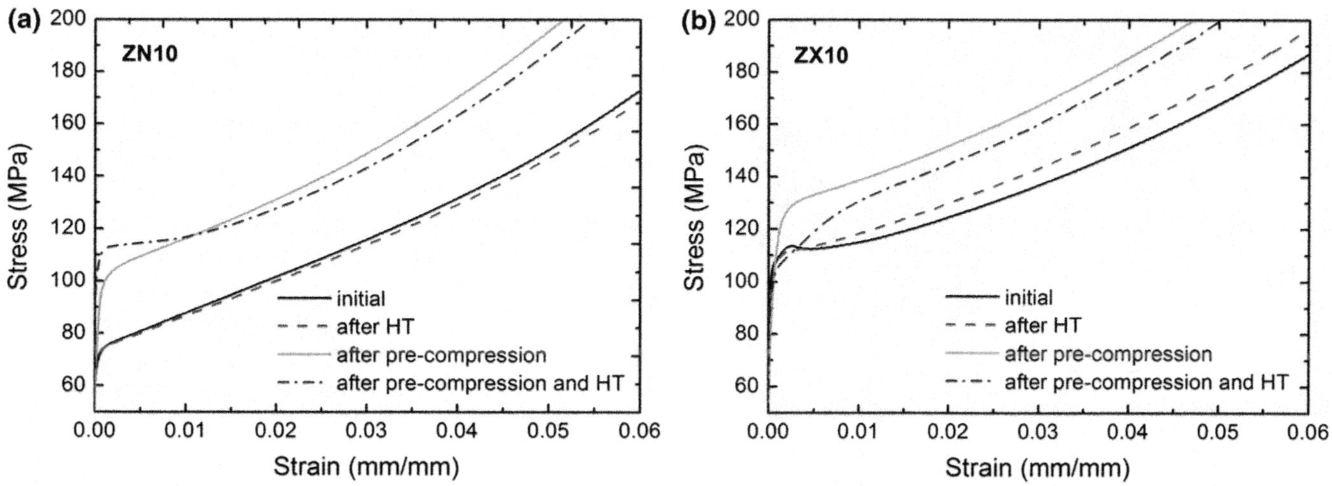


**Fig. 5** Stress versus strain curves for compression tests of **a** ZN10 and **b** ZX10 magnesium alloys in as-extruded state (black); after HT at 200 °C for 16 h (red); after pre-compression and relaxation (green); after pre-compression and subsequent HT at 200 °C for 16 h (blue). All curves have concave shape and to visualize the yielding part only beginning of deformation curves are presented in the figure

**Table 1** Compressive yield strength (YS) for ZN10 and ZX10 alloys in as-extruded state and after solely HT and thermo-mechanical treatment (TMT): pre-compression followed by HT at 200 °C for 16 h

| | Compressive yield strength (MPa) | | | | |
|---|---|---|---|---|---|
| | Initial | After HT | Level of pre-compression | After pre-compression | After TMT |
| ZN10 | 76 ± 2 | 76 ± 2 | 105 | 104 ± 2 | 113 ± 3 |
| ZX10 | 112 ± 3 | 112 ± 2 | 130 | 127 ± 2 | 109 ± 3 |

significant plateau after yielding. Therefore, it could be concluded that nucleation of new twins in ZN10 is more pronounced in comparison to ZX10 alloy. Further development of stress–strain curves for samples after solely pre-strain and with intermediate HT are comparable.

In general, the difference in the influence of Nd and Ca solutes on yielding behavior can be attributed to their different solute solution behavior in Mg, what affects the alteration of lattice parameters and different amount of the second phase. Significant difference in the distribution of the precipitates in the microstructure and along the grain boundaries can be seen in Fig. 4.

Similarly to fine segregation of precipitates along grain boundaries in case of HT at 200 °C for 16 h of ZN10 alloy in as-extruded state (Fig. 4a), HT of pre-compressed sample can lead to segregation of precipitates along twin boundaries. In our previous work [20] it was shown that isothermal ageing of binary Z1 alloys at 200 °C for 8 h leads to significant segregation of precipitates along twin boundaries. Therefore, mobility of induced during pre-compression twin boundaries is suppressed by pinning effect, and nucleation of new twins is requested for further plastic deformation. It can explain the increase in the value of YS and significant plateau of stress–strain curve in the pre-deformed ZN10 alloy.

In case of ZX10 alloy isothermal ageing of the pre-compressed sample results in softening of the material (Fig. 5b). In contrast, Zeng et al. in [19] observed remarkable strengthening rather than softening in pre-strained Mg–0.3Zn–0.1Ca alloy in shape of a sheet. By a set of experimental tests, it was shown that annealing strengthening is related to the plastic pre-strain and subsequent HT. The highest magnitude of the strength increment about 22 MPa, what makes 24% increase of the 0.2% proof strength of the initial test, was found. The annealing phenomenon was observed in the range of temperature 80–200 °C and depending on time magnitude of increment was changed. It is important to note that, the main deformation mechanism activated during pre-strain (tension) of sheet sample was established as basal dislocation slip, and twining played a minor role. Complementary HAADF-STEM analysis showed the formation of GP zones on the gliding dislocations. Based on this, the formation of the GP zones on the preferentially basal dislocations impedes the further gliding of these dislocations, what leads to some strengthening effects. In our investigated material, twinning is preferentially activated deformation mechanism due to mutual orientation of loading axis and crystallographic orientation of the majority of the grains. The microstructure of ZX10 alloy

**Fig. 6** Orientation map of the microstructure of ZX10 magnesium alloys after pre-compression up to 130 MPa. ED is perpendicular to the image plane

after pre-compression of up to 130 MPa contains many {10–12} <10–11> extension twins giving a twin area fraction of 23 ± 1% of the analyzed area, see Fig. 6.

Reorientation of the original lattice provides activation of basal slip in free-dislocation twinned fraction of the microstructure in the pre-strained material. Preferential activation of dislocation slip rather than nucleation of new twins can explain the absence of plateau of deformation curve of TMT ZX10 alloy.

Therefore, it could be concluded that in Mg–Zn–Ca alloy preferential deformation mechanism during pre-strain—dislocation glide or twinning—plays a significant role for the ability to strengthen the material. Annealing can lead to the strengthening via pinning of gliding dislocations by GP zones in the case when no twins are activated [19]. In contrast, pinned twins can provide a fraction of microstructure with a new orientation, which is favorable for dislocation glide, and therefore it could lead to the softening.

Moreover, in [27] it was observed that with isothermal ageing at 200 °C of Mg–1Zn–1Ca alloy, the average hardness increased gradually with increasing ageing time to a peak after 8 h and then subsequently decreased. Therefore, for further investigation shorter time of isothermal ageing will be applied for Mg–Zn-based alloys with addition of Ca. Detail microstructure analysis, including characterization of precipitates will significantly support the investigation.

## Conclusions

The effect of pre-compression and isothermal ageing of extruded Mg–Zn based alloys with Nd and Ca on mechanical properties was investigated. In the initial state microstructure of both materials is characterized by segregation at the grain boundaries. Isothermal ageing at 200 °C for 16 h leads to increasing amount of precipitates for two alloys and difference in the size of distributed precipitates for two alloys was observed. Heat treatment of as-extruded alloys only slightly affected mechanical properties. However, isothermal ageing of pre-compressed along extrusion direction samples resulted in the strengthening of ZN10 alloy and softening of ZX10 alloy. Based on comparison with literature, it was concluded that in Mg–Zn–Ca alloy preferential deformation mechanism during pre-strain, which is followed by heat treatment, plays important role in further deformation behavior of the material: either softening or strengthening will be observed.

**Acknowledgements** This work was supported by the Czech Science Foundation under the grant 17-21855S; the Grant Agency of the Charles University under grant Nr. 1262217; the Operational Programme Research, Development and Education, The Ministry of Education, Youth and Sports (OP RDE, MEYS) under the grant CZ.02.1.01/0.0/0.0/16_013/0001794.

## References

1. G. Proust, C.N. Tomé, A. Jain, S.R. Agnew, Int J Plasticity 25(5) (2009) 861–880.
2. J. Capek, K. Mathis, B. Clausen, J. Straska, P. Beran, P. Lukas, Mat Sci Eng a-Struct 602 (2014) 25–32.
3. Z.Z. Shi, Y.D. Zhang, F. Wagner, P.A. Juan, S. Berbenni, L. Capolungo, J.S. Lecomte, T. Richeton, Acta Mater 83 (2015) 17–28.
4. M.R. Barnett, Mat Sci Eng a-Struct 464(1–2) (2007) 1–7.
5. W.F. Hosford, R.M. Caddell, Metal Forming: Mechanics and Metallurgy (1993) 268.
6. J. Bohlen, S.B. Yi, J. Swiostek, D. Letzig, H.G. Brokmeier, K.U. Kainer, Scripta Mater 53(2) (2005) 259–264.
7. J. Bohlen, J. Wendt, M. Nienaber, K.U. Kainer, L. Stutz, D. Letzig, Mater Charact 101 (2015) 144–152.
8. S. Izumi, M. Yamasaki, Y. Kawamura, Corrosion Science 51(2) (2009) 395–402.
9. J. Bohlen, P. Dobron, L. Nascimento, K. Parfenenko, F. Chmelik, D. Letzig, Acta Phys Pol A 122(3) (2012) 444–449.
10. J. Bohlen, P. Dobron, J. Swiostek, D. Letzig, F. Chmelik, P. Lukac, K.U. Kainer, Materials Science and Engineering A 462 (1–2) (2007) 302–306.
11. E.W. Kelley, W.F. Hosford, Trans. AIME 242(1) (1968) 5–13.
12. H. Mirzadeh, Journal of Materials Research and Technology 5(1) (2016) 1–4.
13. S.M. Zhu, T.B. Abbott, M.A. Gibson, J.F. Nie, M.A. Easton, Materials Science and Engineering: A 656 (2016) 34–38.

14. L. Gao, R.S. Chen, E.H. Han, J Alloy Compd 472(1–2) (2009) 234–240.
15. L. Gao, R.S. Chen, E.H. Han, J Alloy Compd 481(1–2) (2009) 379–384.
16. M. Yamasaki, S. Izumi, Y. Kawamura, H. Habazaki, Appl Surf Sci 257(19) (2011) 8258–8267.
17. J.F. Nie, Y.M. Zhu, J.Z. Liu, X.Y. Fang, Science 340(6135) (2013) 957–960.
18. Y. Xin, X. Zhou, H. Chen, J.-F. Nie, H. Zhang, Y. Zhang, Q. Liu, Mat Sci Eng a-Struct 594 (2014) 287–291.
19. Z.R. Zeng, Y.M. Zhu, M.Z. Bian, S.W. Xu, C.H.J. Davies, N. Birbilis, J.F. Nie, Scripta Mater 107 (2015) 127–130.
20. D. Drozdenko, J. Bohlen, S. Yi, P. Dobroň, in: K.N. Solanki, D. Orlov, A. Singh, N.R. Neelameggham (Eds.), Magnesium Technology 2017, Springer International Publishing, Cham, 2017, pp. 619–624.
21. The Works of Geber, Translated into English by Richard Russel, A New Edition with Introduction by E.J. Holmyard, M.A., D. Litt., London, England: J.M. Dent and Sons Ltd. and New York, NY: E. P. Dutton and Company, 1928.
22. J.D. Robson, N. Stanford, M.R. Barnett, Metall Mater Trans A 44a (7) (2013) 2984–2995.
23. P. Minárik, E. Jablonská, R. Král, J. Lipov, T. Ruml, C. Blawert, B. Hadzima, F. Chmelík, Materials Science and Engineering: C 73 (2017) 736–742.
24. X. Zhang, Q. Wang, F. Chen, Y. Wu, Z. Wang, Q. Wang, Materials Letters 138 (2015) 212–215.
25. M.R. Barnett, M.D. Nave, A. Ghaderi, Acta Mater 60(4) (2012) 1433–1443.
26. D. Drozdenko, J. Bohlen, S. Yi, P. Minárik, F. Chmelík, P. Dobroň, Acta Mater 110 (2016) 103–113.
27. J.F. Nie, B.C. Muddle, Scripta Mater 37(10) (1997) 1475–1481.

# Influence of Low Temperature Forging on Microstructure and Low Cycle Fatigue Behavior of Cast AZ31B Mg Alloy

D. Toscano, S. K. Shaha, B. Behravesh, H. Jahed, B. Williams, and X. Su

**Abstract**

The effect of low temperature forging on the microstructure, quasi-static response and stress-controlled fatigue behavior of cast AZ31B Mg alloy was investigated. The forging process was conducted at a temperature of 275 °C and a forging rate of 20 mm/s. Fully reversed stress controlled cyclic tests were performed on cast and forged material under total stress amplitudes of 120–160 MPa. Neckless type bimodal grain structure, an indication of incomplete dynamic recrystallization was observed in the forged microstructure in addition to the development of a sharp basal texture. The obtained mechanical test results show that the forged material achieved significantly improved yield and tensile strengths along with longer fatigue life. The improvement in the quasi-static properties was attributed to the strengthening effect of partial grain refinement and activation of non-basal slip modes due to texture modification.

**Keywords**

Magnesium alloy • Forging • Microstructure
Texture • Fatigue

## Introduction

Human assisted climate change has been recognized as a major concern and vehicular emission are a major contributor to this climate change phenomenon. Therefore, the automotive industry has been investing heavily in lightweighting technologies in an attempt to lower emissions. Magnesium (Mg) alloys have been gaining interest as a substitute to more traditional steel and aluminum alloys due to their low density, excellent machinability and good recyclability [1, 2]. However, to date, the use of Mg alloys in mass market automotive applications has been limited to primarily cast components. To expand the potential applications of these alloys, the feasibility of producing complex load bearing Mg alloy components needs to be evaluated. Hot forging is a severe plastic deformation (SPD) method that may be used to fabricate such geometrically complex load bearing vehicle components. As is the case with other wrought processing techniques, hot forging is expected to alter the microstructures of the Mg alloys which in turn generally improves the mechanical response of the alloy [3, 4]. Such improvements have been observed during simple uniaxial upsets of several Mg alloys at high temperature [5–7]. However, all the studies [5–15] investigating the effect of forging on Mg alloys were performed at a relatively high temperature, specifically above 350 °C. These temperatures are typically chosen for Mg alloys due to their limited formability at temperatures below ~200 °C. However, a lower temperature forging operation may be desirable to obtain improved mechanical properties, dimensional control, low oxidation and energy consumption. Additionally, fatigue fracture is the primary failure mechanism in engineering components and therefore it is necessary to investigate the cyclic behavior of the forged components manufactured at the lower temperature. Thus, in this paper, the influence of low temperature close-die forging on the microstructure, quasi-static and cyclic behavior of cast AZ31B is examined. The forging temperature employed in this study was lower than those typically used for wrought

D. Toscano · S. K. Shaha (✉) · B. Behravesh · H. Jahed
Department of Mechanical and Mechatronics Engineering, University of Waterloo, 200 University Ave W, Waterloo, ON N2L 3G1, Canada
e-mail: skumarsh@uwaterloo.ca

B. Williams
CanmetMATERIALS, Natural Resources Canada, 183 Longwood Road South, Hamilton, ON L8P 0A1, Canada

X. Su
Ford Motor Company, 2101 Village Rd, Dearborn, MI 48124, USA


D. Orlov et al. (eds.), *Magnesium Technology 2018*, The Minerals, Metals & Materials Series, https://doi.org/10.1007/978-3-319-72332-7_41

processing of Mg alloys but was sufficiently high to allow for successful (crack free) component formation.

## Experimental Procedure

Cylindrical billets of 63.5 mm diameter and 65 mm length was extracted from 300 mm diameter cast AZ31B stock. These billets were heated to a temperature of 275 °C and placed in a hydraulic press equipped with isothermally heated dies. The cavities of both dies was symmetric about the parting line and included two flanges of differing depths —a tall flange with a relatively narrow width and a shorter, wider flange. Graphite lubricant was used during forging and the process was performed at a displacement rate of 20 mm/s. After forging, the components were air cooled to room temperature. A photograph of the forging and a schematic of the final forged shape with specimen locations is illustrated in Fig. 1.

Microstructure was captured using an Olympus BX51M metallurgical microscope whereas texture analysis was performed using a Bruker D8-Discover equipped with a VÅNTEC-500 area detector. The microstructure and texture analysis followed the procedures detailed in [16].

Tensile specimens were machined from the as-cast material as well as from the forgings (Fig. 1b). Details on specimen geometry and sample extraction from the cast condition can be found in [7]. These specimens were tested under both quasi-static and load-controlled cyclic loading using a MTS 810 uniaxial load frame with a load capacity of 25 kN. For quasi-static tests, a displacement rate of 1 mm/min was employed and strain was measured using Digital Image Correlation (DIC). Cyclic tests were performed at several stress amplitudes between 120 and 160 MPa in fully reversed (load ratio, $R = -1$) condition with the frequency of 30 Hz.

## Results and Discussion

### Microstructure and Texture

The as-cast AZ31B demonstrated equiaxed grain morphology with an average grain size of 278 μm ± 30. As seen in Fig. 2, the scanning electron microscopy revealed a dendritic morphology of $\alpha$-Mg and a eutectic phase containing $\beta$-$Mg_{17}Al_{12}$ particles, along with casting defects such as porosity. Texture analysis of as-cast AZ31B suggested no preferential c-axis orientation and a low texture intensity suggesting an overall lack of texture as is expected for as-cast product. A micrograph and (0002) pole figure of the as-cast condition are included in Fig. 2 and a more detailed examination may be found in [7].

Microstructural observations made after forging revealed a strong bimodal grain morphology with large deformed grains resembling those in the cast alloy surrounded by a neckless-shape refined structure, an indication of incomplete recrystallized grains as illustrated in Fig. 3. Qualitatively, the area fraction of recrystallized grains was higher in the web region (Fig. 3b) compared to in the two flanges (Fig. 3a, c). In the areas where dynamic recrystallization (DRX) was initiated, the average size of the new grains was 3.9 μm ± 2.1. The formation of inhomogeneous microstructures can be attributed to the occurrence of plastic deformation leading to DRX during forging. However due to the lower processing temperature and high deformation rate, there was insufficient time for the recrystallization process to reach completion [17].

Figure 4 depicts the (0002) and $(10\bar{1}0)$ pole figures of the forged components at different locations. Analysis of texture in the forged material indicated an evolution from the random texture as-cast material to a more complex textured state where the c-axis of the unit cells tended to be aligned with the localized deformation directions. For instance,

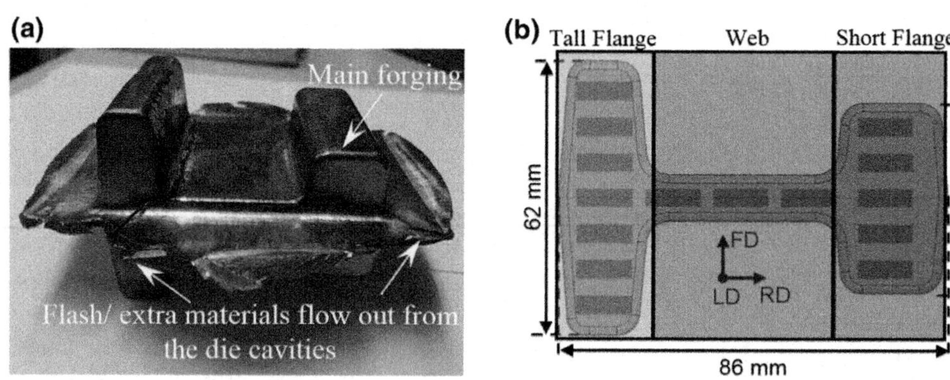


**Fig. 1** Photograph of the finished forging parts with flash (**a**) and a schematic of the final forged geometry looking along the Longitudinal Direction (LD) showing the specimen extraction locations (**b**)

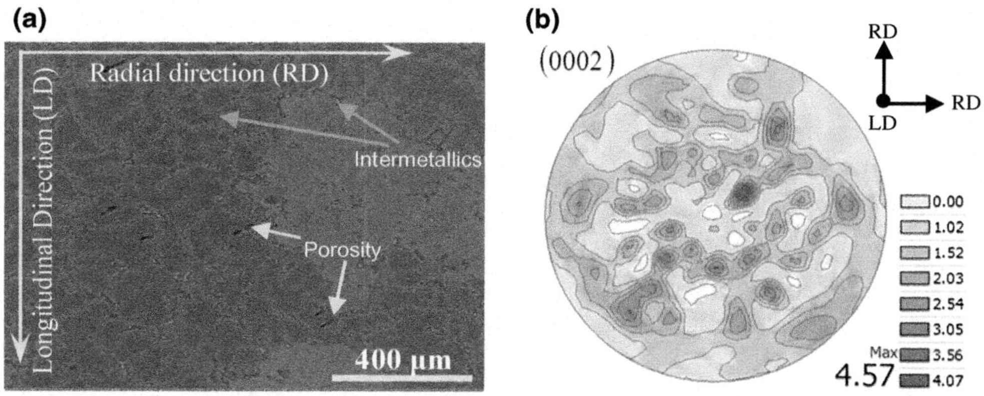


**Fig. 2** The initial microstructure (**a**) and (0002) pole figure (**b**) of cast AZ31B at the half radius (0.5R) of the 300 mm stock [7]

during forging the web region experiences near uniaxial loading along the FD which results in a sharp texture along the FD with a pole intensity of 13.88 MDR at that location. Similarly, texture measurements in the tall rib suggest that the majority of the c-axis are aligned towards the RD and pole intensity increased to ~11.16 Multiples of Random Distribution (MDR) compared to 4.57 MRD in the cast alloy. Such a modification in texture is due to the effect of Rotational Dynamic Recrystallization (RDRX) which results in a rotation in newly formed grains from a hard slip to stable slip directions [18–20].

It should be noted that the short flange pole figure above suggests some degree of out-of-plane c-axis rotation (i.e. towards LD). This is thought to be due to texture measurements being made approximately 10 mm into the depth of the forging where a LD loading component arising from die constraints may influence the final texture of the forging.

## Quasi-static Tension Properties

Figure 5 displayed isotropic tensile behavior of as-cast AZ31B, which is consistent with the random texture observed in Fig. 2. This material demonstrated a very low yield strength of approximately 56 MPa due to the presence of casting defects like porosity and activation of basal slip followed primarily by monotonic hardening behavior suggesting largely slip dominated plasticity throughout the test. However it should be noted that a mild sigmoidal behaviour (dotted ellipse in Fig. 5) was observed during tension suggesting that several grains were favorably oriented for the activation of extension twinning [21]. The quasi-static tensile curves for the cast and forged material along with results from open-die forged AZ31B [7] are presented in Fig. 5. It is seen that the material forged at lower temperature in a closed die obtained higher tensile properties compared to other testing conditions. The closed-die forged AZ31B showed a remarkable increase in both yield and ultimate strength compared to the as-cast material and open-die forged AZ31B published in [7]. Specifically, the average yield strength of the forged AZ31B in this study increased more than three-fold (from ~56 to ~178 MPa), whereas the tensile strength increased by about 26%. It is noteworthy that the tensile behaviors for the tall and short flanges are quite similar whereas that for the web show considerably higher strengths. This may be explained by the more severe strain accumulated in the web region as a consequence of the starting billet shape and die design resulting in a more refined microstructure (Fig. 3b) in the web section.

The primary tensile properties i.e. yield strength, ultimate strength and failure strain of the tested specimens are summarized in Fig. 6. It is interesting to note from Figs. 5 and 6 that the failure strain of the forged material is substantially lower than open-die forged AZ31B and the starting cast material. It is thought that the low failure elongation is a consequence of the low forging temperature and stronger texture in the closed-die forged alloy. More specifically, the measured texture in the current forging suggests a near perpendicular orientation between the c-axis and loading direction which inhibits basal slip which is freely activated in the cast material and instead requires activation of the more brittle prismatic and pyramidal slip modes or compression twinning [22–25]. Additionally, the considerably finer DRX grains in the current forged alloy would provide increased barriers to dislocation motion compared to the large-grained cast and less refined open-die forged (450 °C) conditions [7]. The increased impedance to dislocation motion would contribute to increased strength and reduced ductility of the low temperature forging [26].

## Fatigue Behavior

Stress amplitude versus cycles to failure for the cast and closed-die forged AZ31B are included in Fig. 7 along with results for rolled and extruded AZ31B published in literature for stress amplitudes ranging from 100 to 160 MPa.

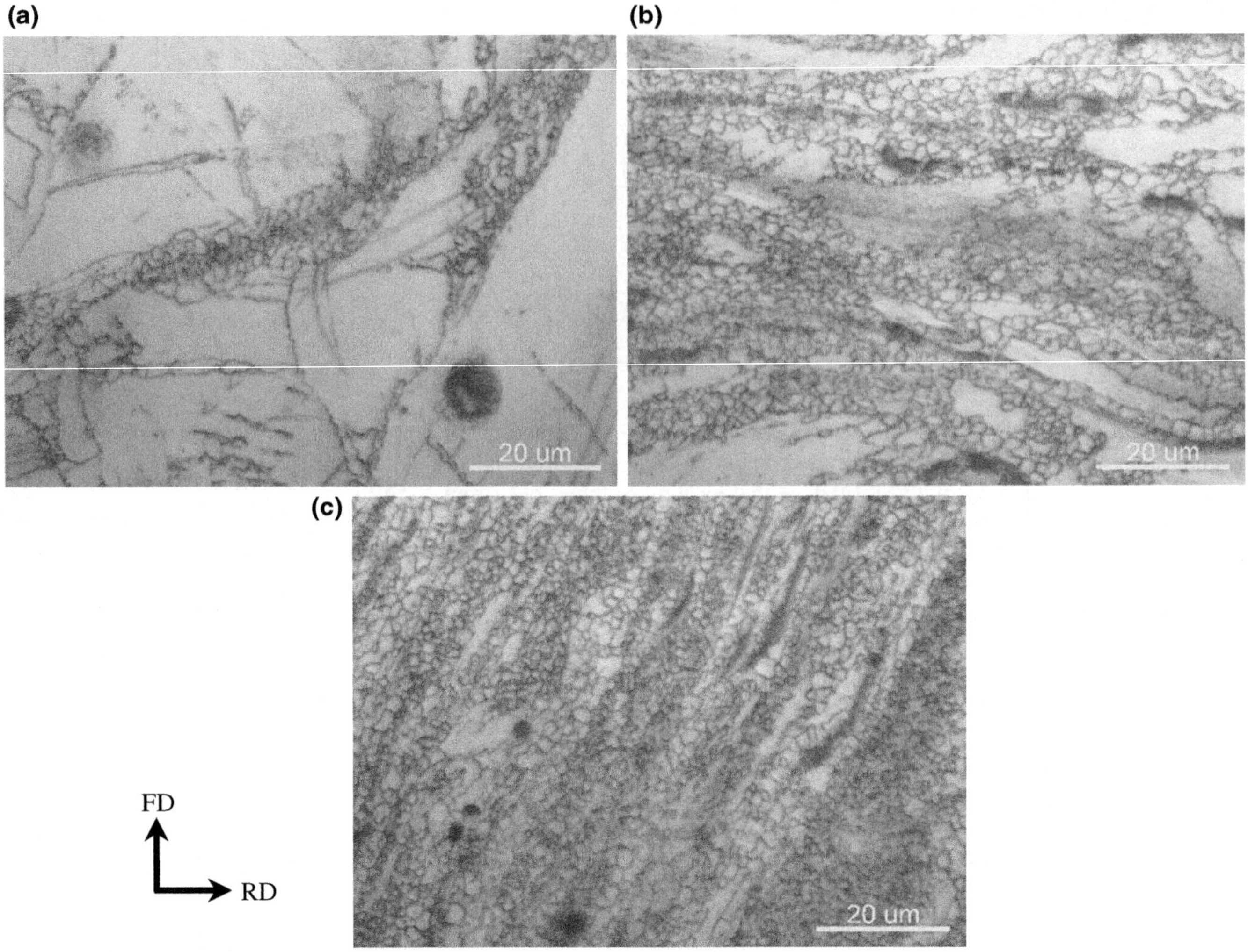


**Fig. 3** Optical micrographs of the forged AZ31B taken at the tall flange (**a**), web (**b**) and short flange (**c**). Note that the indicated directions apply to all images

At all the tested stress amplitudes, the as-cast material exhibited the lowest fatigue life whereas the closed-die forged AZ31B exhibited a substantially longer fatigue life (up to an order of magnitude). Note that the substantial scatter for the closed-die forged material is ascribed to microstructural variations in the forging (Fig. 3). In comparison, rolled and extruded AZ31 alloy investigated by Lv et al. [27] and Hasegawa et al. [28] respectively generally had lives between the cast and forged material considered in this study. The reason for the increase in the fatigue life of the alloy is thought to be two-fold: (i) due to the general strengthening of the material as shown in Fig. 6 delaying crack initiation and (ii) effect of grain refinement and texture modification on the crack growth rate. In regards to the strengthening effect, the considerably stronger forged material exhibits substantially reduced gross plastic deformation during cyclic loading compared to the cast alloy. As the process of fatigue crack initiation is related to the accumulation of damage, i.e., irreversible plastic deformation, the lower overall plastic strain induced in the forged material for a given stress amplitude increases the cycles required to initiate a fatigue crack [29]. Once a fatigue crack has generated, it propagates through the material with every additional cycle. The large uniform grains with low grain boundary density in the cast alloy provide relatively little resistance to this crack propagation. In contrast, the very fine grains observed in the DRX regions of the forging would serve to lower crack growth rate and increase overall life [30–32]. Further, some studies have suggested that surface roughness due to extension twinning along the crack path also delay crack growth due to roughness induced crack closure mechanism [33]. Such a mechanism may also aid in the overall improvement of the fatigue life observed in the textured closed-die forged material.

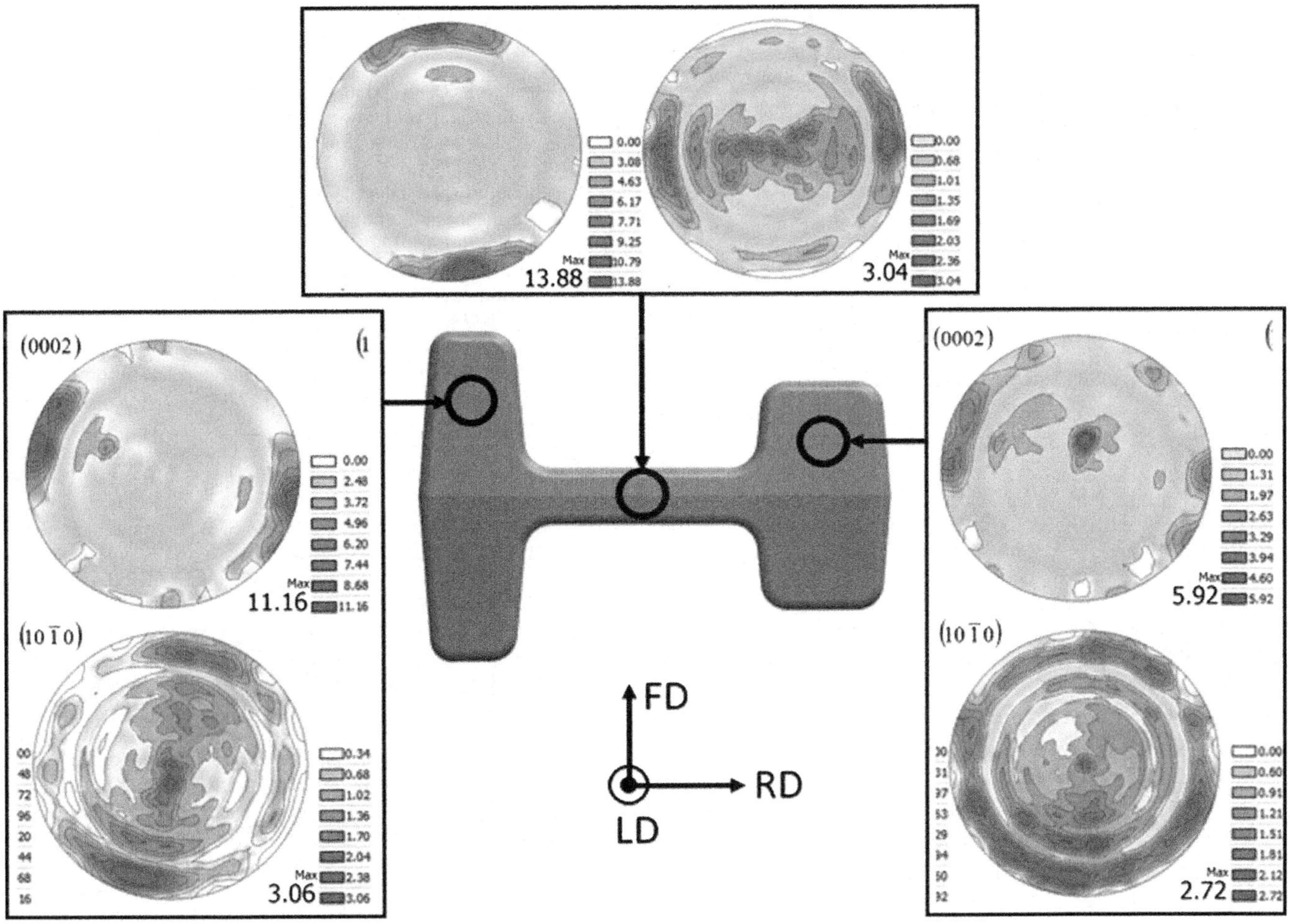


**Fig. 4** (0002) and $(10\bar{1}0)$ pole figures for forged AZ31B at different locations

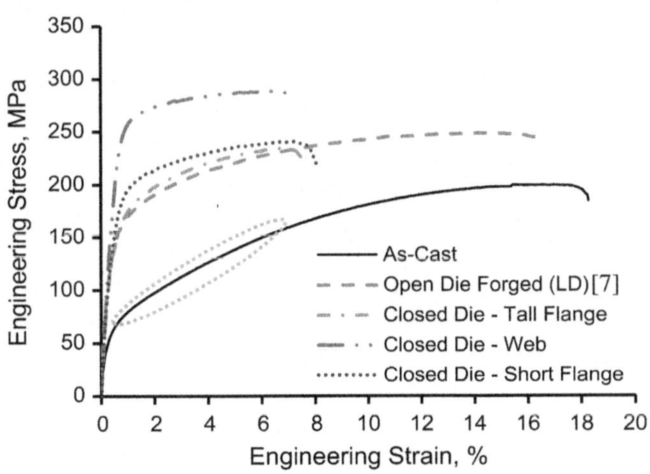


**Fig. 5** Typical engineering stress-strain tension curves for as-cast and forged AZ31B

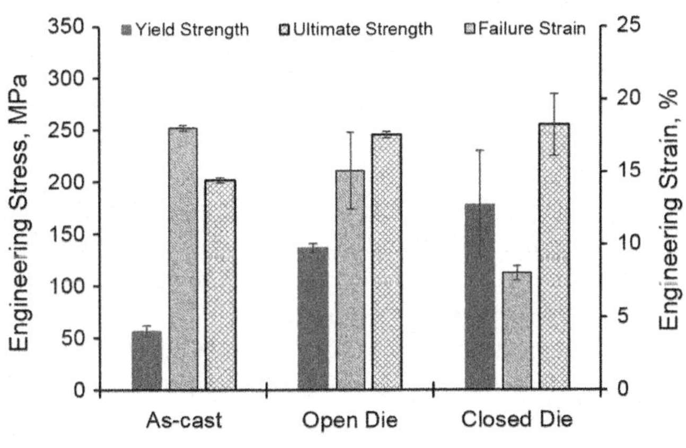


**Fig. 6** A comparison of tension properties between as-cast, open-die forged and closed-die forged AZ31B Mg-alloy

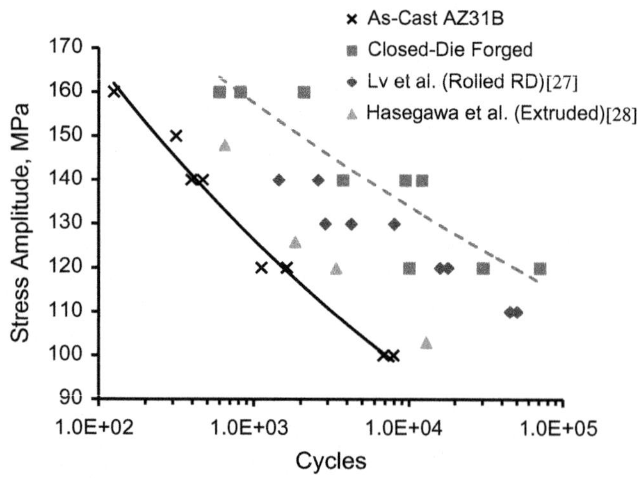


**Fig. 7** S-N results for cast and closed-die forged AZ31B for stress amplitudes between 100 and 160 MPa. Some results for extruded and rolled AZ31B from literature has also been included

## Conclusion

The effect of closed-die forging at low temperature on the microstructure, and mechanical properties of cast AZ31B was examined.

- Forging resulted in a modification of the cast microstructure to a bimodal grain morphology consisting of fine recrystallized grains surrounding deformed un-recrystallized parent grains.
- A sharp basal texture was developed by forging and c-axis alignment parallel to localized deformation direction was observed.
- Refined grain structure in addition to the inhibition of basal slip due to texture modification is thought to be responsible for the superior strength observed in the forged material.
- The considerable decrease in ductility compared to the cast material is thought to be a result of texture inhibited basal slip and increased impedance to dislocation motion due to the refined DRX grains.
- Improvement in the cyclic performance of the forged alloy was attributed to the delay in crack initiation due to the strengthening of the alloy and also due to retardation of crack growth rate due to grain refinement and twin boundaries.

**Acknowledgements** The financial support of the Natural Sciences and Engineering Research Council of Canada (NSERC) through the Automotive Partnership Canada (APC) program under APCPJ 459269–13 grant with contributions from Multimatic Technical Centre, Ford Motor Company, and Centerline Windsor are acknowledged. The authors would also like to thank J. McKinley and L. Blaga of CanmetMATERIALS for assistance with the forging trials.

## References

1. G. Huang, J. Li, T. Han, H. Zhang, and F. Pan, "Improving low-cycle fatigue properties of rolled AZ31 magnesium alloy by pre-compression deformation," *Mater. Des.*, vol. 58, pp. 439–444, 2014.
2. J. Albinmousa, H. Jahed, and S. Lambert, "Cyclic behaviour of wrought magnesium alloy under multiaxial load," *Int. J. Fatigue*, vol. 33, no. 8, pp. 1127–1139, 2011.
3. É. Martin and J. J. Jonas, "Evolution of microstructure and microtexture during the hot deformation of Mg-3% Al," *Acta Mater.*, vol. 58, no. 12, pp. 4253–4266, 2010.
4. G. Garcés, A. Müller, E. Oñorbe, P. Pérez, and P. Adeva, "Effect of hot forging on the microstructure and mechanical properties of Mg-Zn-Y alloy," *J. Mater. Process. Technol.*, vol. 206, no. 1–3, pp. 99–105, 2008.
5. A. Gryguc *et al.*, "Monotonic and cyclic behaviour of cast and cast-forged AZ80 Mg," *Int. J. Fatigue*, vol. 104, pp. 136–149, 2017.
6. S. M. H. Karparvarfard, S. K. Shaha, S. B. Behravesh, H. Jahed, and B. W. Williams, "Microstructure, texture and mechanical behavior characterization of hot forged cast ZK60 magnesium alloy," *J. Mater. Sci. Technol.*, 2016.
7. D. Toscano, S. K. Shaha, B. Behravesh, H. Jahed, and B. Williams, "Effect of Forging on Microstructure, Texture, and Uniaxial Properties of Cast AZ31B Alloy," *J. Mater. Eng. Perform.*, vol. 26, no. 7, pp. 3090–3103, 2017.
8. A. Gryguc, S. K. Shaha, H. Jahed, M. Wells, B. Williams, and J. McKinley, "Tensile and fatigue behaviour of as-forged AZ31B extrusion," *Frat. ed Integrita Strutt.*, vol. 10, no. 38, pp. 251–258, 2016.
9. A. Gryguc, S. K. Shaha, S. B. Behravesh, H. Jahed, M. Wells, and B. Williams, "Compression Behaviour of Semi-closed Die Forged AZ80 Extrusion," *Characterization of Minerals, Metals, and Materials. 2017*, pp. 361–369, 2017.
10. D. Toscano *et al.*, "Effect of Forging on Microstructure, Texture and Compression Behavior of Extruded AZ31B," *Proc. 3rd Pan Am. Mater. Congr.*, pp. 347–354, 2017.
11. S. M. H. Karparvarfard *et al.*, "Characterization of Semi-Closed Die-Forged ZK60 Mg Alloy Extrusion," *Magnes. Technol. 2017. Springer Int. Publ.*, pp. 329–334, 2017.
12. A. Hadadzadeh, S. K. Shaha, M. A. Wells, H. Jahed, and B. W. Williams, "Recrystallization behavior and texture evolution during hot deformation of extruded ZK60 magnesium alloy," *Mater. Sci. Technol. 2016*, pp. 281–288, 2016.
13. A. Hadadzadeh, S. K. Shaha, M. A. Wells, H. Jahed, and B. W. Williams, "Microstructure and Texture Evolution During Hot Deformation of Cast-Homogenized ZK60 Magnesium Alloy," *Magnes. Technol. 2017*, pp. 513–519, 2017.
14. A. Hadadzadeh, M. A. Wells, S. K. Shaha, H. Jahed, and B. W. Williams, "Role of compression direction on recrystallization behavior and texture evolution during hot deformation of extruded ZK60 magnesium alloy," *J. Alloys Compd.*, vol. 702, pp. 274–289, 2017.
15. D. Toscano, S. K. Shaha, B. Behravesh, H. Jahed, and B. Williams, "Effect of forging on the low cycle fatigue behavior of cast AZ31B Alloy," *Mater. Sci. Eng. A*, 2017.
16. A. A. Roostaei and H. Jahed, "Role of loading direction on cyclic behaviour characteristics of AM30 extrusion and its fatigue damage modelling," *Mater. Sci. Eng. A*, vol. 670, pp. 26–40, 2016.
17. G. Z. Quan, Y. Shi, Y. X. Wang, B. S. Kang, T. W. Ku, and W. J. Song, "Constitutive modeling for the dynamic recrystallization evolution of AZ80 magnesium alloy based on stress-strain data," *Mater. Sci. Eng. A*, vol. 528, no. 28, pp. 8051–8059, 2011.

18. I. Ulacia *et al.*, "Texture Evolution of AZ31 Magnesium Alloy Sheet at High Strain Rates," *4th Int. Conf. High Speed Form.*, pp. 189–197, 2010.
19. A. Jain and S. R. Agnew, "Modeling the temperature dependent effect of twinning on the behavior of magnesium alloy AZ31B sheet," *Mater. Sci. Eng. A*, vol. 462, no. 1–2, pp. 29–36, 2007.
20. X. yue Yang, Z. sheng Ji, H. Miura, and T. Sakai, "Dynamic recrystallization and texture development during hot deformation of magnesium alloy AZ31," *Trans. Nonferrous Met. Soc. China (English Ed.*, vol. 19, no. 1, pp. 55–60, 2009.
21. R. Abbaschian, L. Abbaschian, and R. E. Reed-Hill, *Physical Metallurgy Principles*, 4th ed. Boston: KWS-Kent Publishing Company, 1994.
22. Y. Wang and H. Choo, "Influence of texture on Hall-Petch relationships in an Mg alloy," *Acta Mater.*, vol. 81, pp. 83–97, 2014.
23. M. R. Barnett, "Twinning and the ductility of magnesium alloys. Part II. 'Contraction' twins," *Mater. Sci. Eng. A*, vol. 464, no. 1–2, pp. 8–16, 2007.
24. B. Q. Shi, R. S. Chen, and W. Ke, "Influence of grain size on the tensile ductility and deformation modes of rolled Mg-1.02 wt.% Zn alloy," *J. Magnes. Alloy.*, vol. 1, no. 3, pp. 210–216, 2013.
25. W. Guo, Q. Wang, B. Ye, and H. Zhou, "Microstructure and mechanical properties of AZ31 magnesium alloy processed by cyclic closed-die forging," *J. Alloys Compd.*, vol. 558, pp. 164–171, 2013.
26. C. M. Cepeda-Jimenez, J. M. Molina-Aldareguia, and M. T. Perez-Prado, "Effect of grain size on slip activity in pure magnesium polycrystals," *Acta Mater.*, vol. 84, pp. 443–456, 2015.
27. F. Lv *et al.*, "Fatigue properties of rolled magnesium alloy (AZ31) sheet: Influence of specimen orientation," *Int. J. Fatigue*, vol. 33, no. 5, pp. 672–682, 2011.
28. S. Hasegawa, Y. Tsuchida, H. Yano, and M. Matsui, "Evaluation of low cycle fatigue life in AZ31 magnesium alloy," *Int. J. Fatigue*, vol. 29, no. 9–11, pp. 1839–1845, 2007.
29. F. Yang, S. M. Yin, S. X. Li, and Z. F. Zhang, "Crack initiation mechanism of extruded AZ31 magnesium alloy in the very high cycle fatigue regime," *Mater. Sci. Eng. A*, vol. 491, no. 1–2, pp. 131–136, 2008.
30. S. Zheng, Q. Yu, and Y. Jiang, "An experimental study of fatigue crack propagation in extruded AZ31B magnesium alloy," *Int. J. Fatigue*, vol. 47, pp. 174–183, 2013.
31. M. Tsushida, K. Shikada, H. Kitahara, S. Ando, and H. Tonda, "Relationship between Fatigue Strength and Grain Size in AZ31 Magnesium Alloys," *Mater. Trans.*, vol. 49, no. 5, pp. 1157–1161, 2008.
32. M. Kamakura, K. Tokaji, H. Shibata, and N. Bekku, "Grain Refinement and Improvement of Fatigue Strength due to Controlled Extrusion in Magnesium Alloys," *J. Soc. Mater. Sci. Japan*, vol. 54, no. 3, pp. 245–250, 2005.
33. Y. J. Wu, R. Zhu, J. T. Wang, and W. Q. Ji, "Role of twinning and slip in cyclic deformation of extruded Mg-3%Al-1%Zn alloys," *Scr. Mater.*, vol. 63, no. 11, pp. 1077–1080, 2010.

# The Recrystallization and Grain Growth Behavior of Magnesium

Aeriel D. Murphy and John E. Allison

**Abstract**

Magnesium is an important structural material due to its attractive properties including having a low density and an excellent strength-to-weight ratio. Grain refinement is used as a method to increase the strength and ductility of Mg and therefore knowing the recrystallization and grain growth kinetics is important. The experimental results showed that inhomogeneous nucleation created a two-stage nucleation process during recrystallization and that alloying slowed down grain growth significantly in Mg. A complete article on this work can be seen in a future issue of Metallurgical and Materials Transactions A.

**Keywords**

Recrystallization • Grain-growth • Magnesium

## Extended Abstract

Magnesium alloys have gained increased attention over the past years due to their attractive properties including having a low density and an excellent strength-to-weight ratio [1–3]. Light-weighting of structural components has increased research efforts regarding Mg alloys [1]. Despite the advantages Mg alloys offer, they have poor ductility, room temperature deformation, a limited number of independent slip systems, display a strong deformation texture post processing, and highly directional anisotropy, therefore studying microstructural evolution (i.e. recrystallization and grain growth) is important to control grain size [4–8]. The microstructural evolution that occurs during annealing (grain refinement, grain growth) has a huge impact on the material's properties. Grain refinement is used as a method to increase the strength and ductility of wrought Mg alloys and strength/ductility decrease with increasing grain size (Hall-Petch Relationship). Therefore, knowing the recrystallization and grain-growth kinetics is necessary for their use in structural applications [6–9].

The initiation of twins has a significant influence on the static recrystallization behavior in Mg alloys. Due their limited number of independent slip systems, Mg and Mg alloys have a strong propensity to form twins during deformation [10]. Recent studies have shown that twin and twin boundaries serve as preferred nucleation sites during recrystallization of Mg alloys. During deformation, twins are barriers to dislocation leading to highly localized strains in their vicinity. One study in AZ31 by Su et al found that during recrystallization, complete recrystallization of twinned regions occurred before large, deformed grains began to recrystallize. They also reported that after long annealing times, deformed grains still remained in the material [11]. Guan et al found that twins were preferred nucleation sites during recrystallization of the rare-earth Mg alloy, WE43 [12].

Inhomogeneous nucleation of recrystallized grains due to the addition of twins has a strong influence on recrystallization behavior and differs from the Johnson-Mehl-Avrami-Kolmogorov (JMAK) relationship. The JMAK relationship assumes that the growth rate remains constant during recrystallization due to random distribution of nucleation sites and site-saturated nucleation and can be represented by a single Avrami exponent value of three [13]. Studies have reported that inhomogeneous nucleation leads to an Avrami exponent of three [11, 12]. Inhomogeneous nucleation due to twinning can create a two-stage recrystallization process where initially recrystallization occurs in regions with a high stored energy (twins and grain boundaries). Once those regions are fully-recrystallized, recrystallization continues in regions with a lower stored energy [11]. One study found that the two-stage recrystallization process could be described by two different Avrami exponents: $n_1$

A. D. Murphy (✉) · J. E. Allison
Department of Materials Science and Engineering, University of Michigan, Ann Arbor, MI, USA
e-mail: aerielm@umich.edu


D. Orlov et al. (eds.), *Magnesium Technology 2018*, The Minerals, Metals & Materials Series, https://doi.org/10.1007/978-3-319-72332-7_42

representing the first stage and $n_2$ for the second stage [11]. Previous studies in Mg have focused on dynamic recrystallization and there is limited information available on their static recrystallization behavior [14, 15].

In this current study, the static recrystallization, grain-size, and texture evolution were investigated using Electron Back-Scatter Diffraction (EBSD) in pure Mg and Mg-4%Al. Pure Mg samples of two grain sizes were used in this study: 45 and 350 μm. Static recrystallization, grain growth and texture evolution have been investigated on cylinders compressed uniaxially at room temperature and annealed at 150–400 °C for 1–1000 min. The level of dynamic recrystallization in the as-received condition was also characterized in these materials.

The experimental results showed that inhomogeneous nucleation created a two-stage nucleation process represented by two different Avrami exponents, $n_1$ and $n_2$. Stage I involved rapid nucleation of recrystallized grains at high stored energy twins and grain boundaries and during Stage II nucleation continued slowly at low stored energy grain interiors. The value of $n_1$ was significantly higher than $n_2$, indicating that recrystallization was faster during Stage 1 and slower during Stage 2.

It was also found that in pure Mg recrystallization was increased in the fine grain size material when compared to the coarse grain size sample, which suggests that grain boundaries play a dominant role during recrystallization of Mg. The addition of Al had little effect on the recrystallization kinetics, but slowed down grain growth significantly in Mg. A complete article on this work can be seen in a future issue of Metallurgical and Materials Transactions A.

**Acknowledgements** This work was supported by the U.S. Department of Energy, Office of Basic Energy Science, Division of Materials Science and Engineering under Award #DE-SC0008637 as part of the Center for PRedictive Integrated Structural Materials Science (PRISMS) at the University of Michigan and the National Science Foundation Graduate Research Fellowship Program.

## References

1. S.R. Agnew and J.F. Nie: Scripta Mater., 2010, vol. 63, pp. 671–673.
2. B.C. Suh, M.S. Shim, K.S. Shin, and N.J. Kim: Scripta Mater., 2014, vol. 84–85, pp. 1–6.
3. F.A Mirza and D.L. Chen: Fat. Fract. Eng. Mater. Struct., 2014, vol. 37, pp. 831–853.
4. S. Sandlobes, Z. Pei, M. Friak, L.F. Zhu, F. Wang, S. Zaefferer, D. Raabe, and J. Neugbauer: Acta Mater., 2014, vol. 70, pp. 92–104.
5. H.Y. Chao, H.F. Sun, W.Z. Chen, and E.D. Wang: Mater. Characterization, 2011, vol. 62, pp. 312–320.
6. D. Griffiths: Mater. Sci. Tech., 2015, vol. 31, pp. 10–24.
7. Z.R. Zeng, Y.M. Zhu, S.W. Xu, M.Z. Bian, C.H.J. Davies, N. Birbilis, and J.F. Nie: Acta Mater., 2016, vol. 105, pp. 479–494.
8. J. Hirsch and T. Al-Samman: Acta Mater., 2013, vol. 61, pp. 818–843.
9. A.G. Beer and M.R. Barnett: Scripta Mater., 2009, vol. 61, pp. 1097–1100.
10. C. Drouven, I. Basu, T. Al-Samman, and S. Korte-Kerzel: Mater. Sci. Eng. A, 2015, vol. 647, pp. 91–104.
11. J. Su, M. Sanjari, A.S.H. Kabir, J.J. Jonas, and S. Yue: Mater. Sci. Eng. A, 2016, vol. 662, pp. 412–425.
12. D. Guan, W.M. Rainforth, L. Ma, B. Wynne, and J. Gao: Acta Mater., 2017, vol. 126, pp. 132–144.
13. F.J. Humphreys and M. Hatherlys: Recrystallization and Related Annealing Phenomena, 2nd ed., Elsevier, Kidlington, Oxford, 2004.
14. T. Al-Samman, K.D. Molodov, D.A. Molodov, G. Gottstein, and S. Suwas: Acta Mater., 2012, vol. 60, pp. 537–545.
15. K.D. Molodov. T. Al-Samman, D.A. Molodov, and G. Gottstein: Acta Mater., 2014, vol. 76, pp. 314–330.

# Strengthening of a Biodegradable Mg–Zn–Ca Alloy ZX50 After Processing by HPT and Heat Treatment

A. Ojdanic, E. Schafler, J. Horky, D. Orlov, and M. Zehetbauer

**Abstract**

This study investigates a biodegradable Mg–5Zn–0.3Ca alloy (ZX50) during HPT-processing and long-term heat treatments, the latter with respect to the evolution of intermetallic precipitates and vacancy clusters. Both the precipitates as well as the vacancy clusters achieve strength increases as the Zn atoms may act as potential trapping sites not only for HPT-induced dislocations but also vacancies. So far, overall increases of strength of up to 200% have been reached while keeping the Young's modulus unchanged, thus representing an attractive improvement of mechanical properties for the actual alloy.

**Keywords**

Mg-alloys • Intermetallic particles • Vacancy agglomerates • Corrosion • Biodegradability • Severe plastic deformation

## Introduction

### Magnesium Alloys as Biodegradable Implants

Magnesium alloys, as the lightest structural materials, are becoming increasingly popular in large amount of applications, especially as biodegradable implants [1–3]. The main driving force to develop biodegradable implants is their degradation properties in the physiological environment. The clinical function of permanent implants can be achieved and, once complete, the devices may disappear completely when they are no longer useful. Mg–Zn-based alloys have been proposed as very suitable biodegradable materials for load-bearing applications due to their superior combination of strength and ductility. So far, the application of Mg products is limited due to difficulties in processing of alloys, their relatively low strength and ductility at ambient temperatures and extremely high reactivity leading to unacceptable levels of corrosion in many environments. Mechanical and corrosion properties are generally linked to alloy composition and microstructure. In general, improvements in mechanical properties can be achieved through alloying and processing, but for Mg the solubility of many alloying elements is limited [2, 3]. An effective method for improving the mechanical characteristics of Mg alloys is Severe Plastic Deformation (SPD) [4–8]. SPD processes cause grain refinement via dislocation generation, a redistribution of solutes, and also generates a large concentration of vacancies within the microstructure [9–11]. With additional ageing the generated vacancies form agglomerates which impede the dislocation movement and thus contribute to the macroscopic strength [12–15]. Besides, at special ageing conditions intermetallic precipitates form, as it was demonstrated by Orlov et al. in the ZK60 Mg alloy [16]. Mima and Tanaka [17] identified three important low-temperature ranges for Mg–Zn systems: (i) below 60 °C, the formation of stable Guinier-Preston (GP1) zones; (ii) 60–110 °C, the formation of stable rod-type and basal

A. Ojdanic (✉) · E. Schafler · M. Zehetbauer
Physics of Nanostructured Materials, Faculty of Physics, University of Vienna, 1090 Vienna, Austria
e-mail: andrea.ojdanic@univie.ac.at

J. Horky
Center for Health & Bioresources, Biomedical Systems, AIT Austrian Institute of Technology, 2700 Wr. Neustadt, Austria

D. Orlov
LTH, Lund University, P.O. Box 118 22100 Lund, Sweden

D. Orlov
Materials Research Laboratory, University of Nova Gorica, 5270 Ajdovscina, Slovenia


D. Orlov et al. (eds.), *Magnesium Technology 2018*, The Minerals, Metals & Materials Series, https://doi.org/10.1007/978-3-319-72332-7_43

platelet-type precipitates along with unstable GP1 zones followed by growth of the former at the expense of dissolution of the latter; and (iii) above 110 °C, the formation of stable rod-type and basal platelet-type precipitates, the most stable ones being the rod-type [17, 18].

## Motivation of the Study

This study aims at the optimization of mechanical performance of novel Mg alloys (Mg–Zn–Ca). Emphasis is given to a special ternary alloy consisting of Mg 5% Zn and 0.3% Ca. Ca has been shown to improve corrosion resistance when included in low quantities (0.3–0.8% being optimum). It also has a positive effect on grain refining and aids creep resistance. Zn has the ability not only to improve the strength of Mg alloys, but also to reduce the corrosion enhancing effects of all of the common impurities including Fe, Ni, and Cu. In addition, Zn and Ca are both essential trace elements in the human body and are known for positive influence on bone healing and cell reactions. In further steps the alloys have to be manipulated by severe plastic deformation in order to increase the strength. Still, the elastic modulus should be kept as low as possible, close to the one of natural bone (~20 GPa) in order to avoid stress shielding [19].

By the control of dislocation density and the **structure at the nanoscale**, the control of mechanical characteristics indicating the material performance for potential customers should be ensured. This study mainly focuses on special intermetallic precipitations [16] that form in Mg alloys at special conditions. They include basal platelets and rods that can be formed in a precipitation-free alloy at very long lasting heat treatments of few weeks at temperatures of approximately 50 and 80 °C. The application of deformation procedure should allow the control of solute atoms' concentration and distribution, precipitate particle sizes, shape, orientation and density, with particular attention paid to the formation of prismatic platelets [16, 20].

**At the microscale**, the structural optimization of grain sizes and their distribution as well as their boundaries ensure the control of mechanical characteristics. Spatial distribution of grains will be controlled through the formation of homogeneous ultra-fine grained structure.

## Experimental

### Materials and Processing

In the present study a specially cast magnesium alloy ZX50 having a chemical composition of Mg–5Zn–0.3Ca (wt%) was used. Discs with a diameter of 50 mm and a thickness of 12 mm were obtained from cast bars (50 mm in diameter and 120 mm in length) by a circular saw. The discs were then cut by spark erosion in proper pieces and then turned to rods with a diameter of 10 mm. These were then again cut by spark erosion in discs with a diameter of 10 mm and a thickness of 0.8 mm and subsequently cleaned and annealed at 450 °C for 24 h in argon atmosphere to obtain a nearly supersaturated solid-solution condition (SSSS) [21]. The samples were then furnace cooled. A nanostructured state was then formed by high-pressure torsion (HPT) at room temperature (Fig. 1). Discs of 10 mm in diameter and 0.8 mm in thickness were processed at a hydrostatic pressure of 4 GPa for 0.5 and 2 rotations. By applying a hydrostatic pressure of 4 GPa, torsional shear strains of $\gamma T = 1–94$ have been achieved, with $\gamma T = r\varphi/d$ where r is the radius and d the thickness of the sample, $\varphi$ is the amount of rotation of the HPT process in radiant. The equivalent strain $\varepsilon_{equ}$ can be evaluated using von Mises-criterion by $\varepsilon_{equ} = \gamma T/\sqrt{3}$. After HPT-processing, the samples were heat treated at temperatures between 50 and 460 °C by different annealing times, in an oil bath with a thermal stability of ±0.5 °C. After the heat treatment the samples were dropped into water at room temperature thus ensuring fast cooling and high accuracies of the periods of the heat treatments. Heat treatments at higher temperatures up to 460 °C were performed in a calorimeter Netzsch DSC 204. In these cases the cooling period was considerably longer.

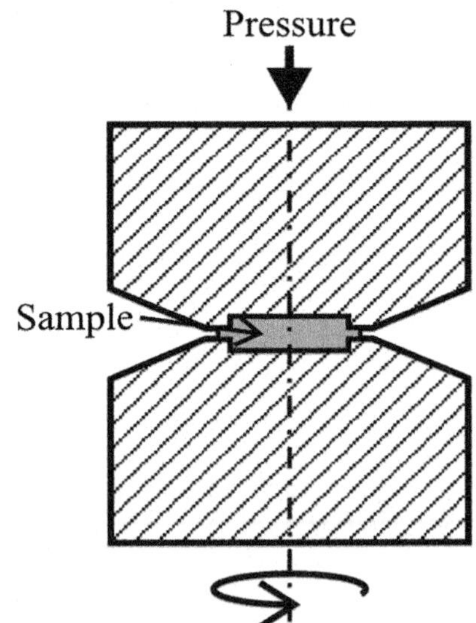


**Fig. 1** Illustration of high pressure torsion (HPT). The sample is located between two anvils, which undergo numerous rotations under enhanced hydrostatic pressure thus achieving large torsional strains

## Mechanical Properties

Vickers hardness test was done as a measure of strength. The surfaces of the sections were ground and mechanically polished to a mirror-like finish. Values of static hardness at ambient temperatures were derived from the area of indentation. An Anton Paar Microhardness Tester (MHT) equipped with a Vickers diamond indenter was used as part of a Light Microscope ZEISS Axioplan with a CCD camera. The maximum load applied was 0.5 N, with a gradient of 0.1 N/s and a constant loading time of 10 s. For the reported distributions of HV, an average over at least seven separate measurements taken. Young's modulus was measured using a Nanoindenter ASMEC Unat with QCSM module for depth dependent measurements with a maximum load of 0.2 N. An average value over at least 20 separate measurements was taken.

## Results

### Evolution of Vickers Hardness

The microhardness of as-cast Mg5Zn0.3Ca amounted to 65 HV. After annealing and furnace cooling, the microhardness decreased to 58 HV. The microhardness of HPT-processed Mg5Zn0.3Ca and an alloy with a similar composition Mg0.2Zn0.5Ca [20] is shown in Fig. 2 as a function of the equivalent strain during HPT-processing and compared to the as-extruded or IS condition. Different numbers of rotations and measurement radii were investigated. After HPT-processing the HV value increased up to 130 HV for Mg5Zn0.3Ca samples for an equivalent strain of ~5–10. For higher strains HV reached values of ~160 HV. The HV of Mg0.2Zn0.3Ca after annealing is 56 HV.

Heat treatments (Fig. 3) were performed to study the thermal stability of the alloys and microhardness dependence on the annealing temperature. It can be seen that heat treatments are capable of significantly increasing the hardness of HPT-processed Mg alloys. The hardness is increasing with increasing temperature up to a maximum and then decreasing below the initial value until a final saturation occurs at temperatures beyond approximately 350 °C. The visible fall of microhardness was observed after annealing at 140 °C. Further annealing at 160, 180, 200, 220, 260, 380 and 420 °C resulted in even lower values of microhardness of HPT-processed samples. The maximum peak temperature for the Mg5Zn0.3Ca samples is 100 °C. For Mg0.2Zn0.5Ca the peak temperature is shifted to 140 °C. Figure 4 shows the results for the Mg5Zn0.3Ca alloy after annealing (initial state, "IS"), HPT-processing by 0.5 and 2 rotations, and additional heat treatment for 1 h at the peak temperature. For the sample deformed at 2 rotations and heat treated, the HV increased by 250%. The homogenized samples were also heat treated for 1 h without processing by HPT, leading to a hardness of 75HV for Mg5Zn0.3Ca at the peak temperature of 100 °C, which results in a HV increase of 40% (Fig. 3). For the as-extruded and heat treated Mg0.2Zn0.5Ca samples, the hardness stays constant at 70 HV and drops rapidly at a temperature of 280 °C [13–15].

With the further heat treatments, the annealing time was raised to 24 h for temperatures of 75, 100, 125 and 150 °C (Fig. 5). Also in this case, the peak temperature for Mg5Zn0.3Ca appears at 100 °C. For non-processed, homogenized samples (IS), the peak is much smaller and slightly shifts to 75 °C.

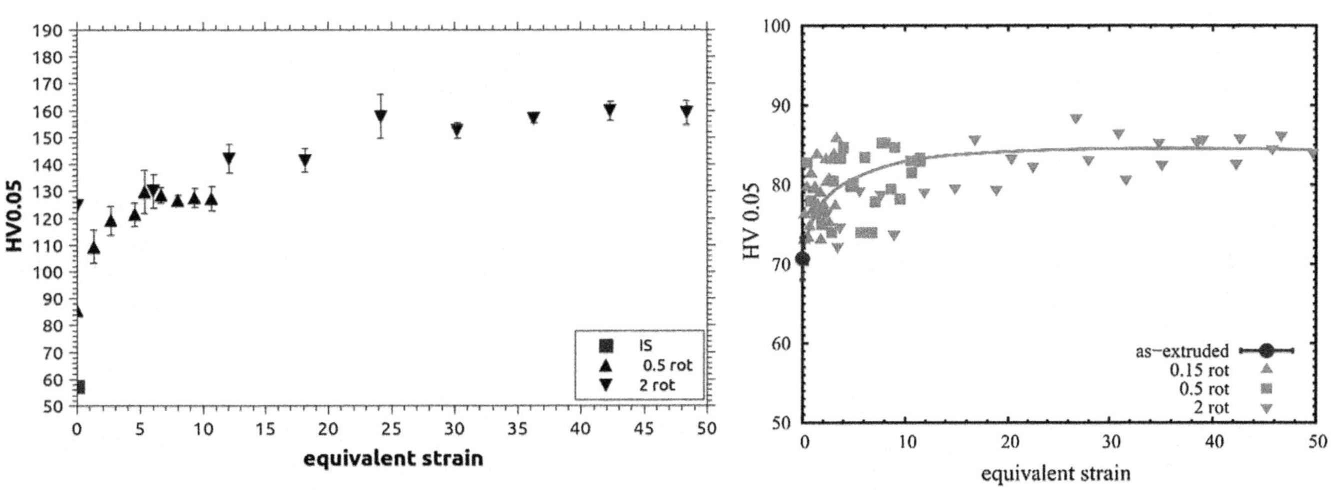


**Fig. 2** HV measurements after homogenizing (IS) and after HPT-processing by various rotation numbers. Left: HV results for Mg5Zn0.3Ca after homogenizing and HPT-processing by 0.5 and 2 rotations, right: HV results for Mg0.2Zn0.5Ca after extrusion, and HPT-processing by 0.15, 0.5 and 2 rotations [15]

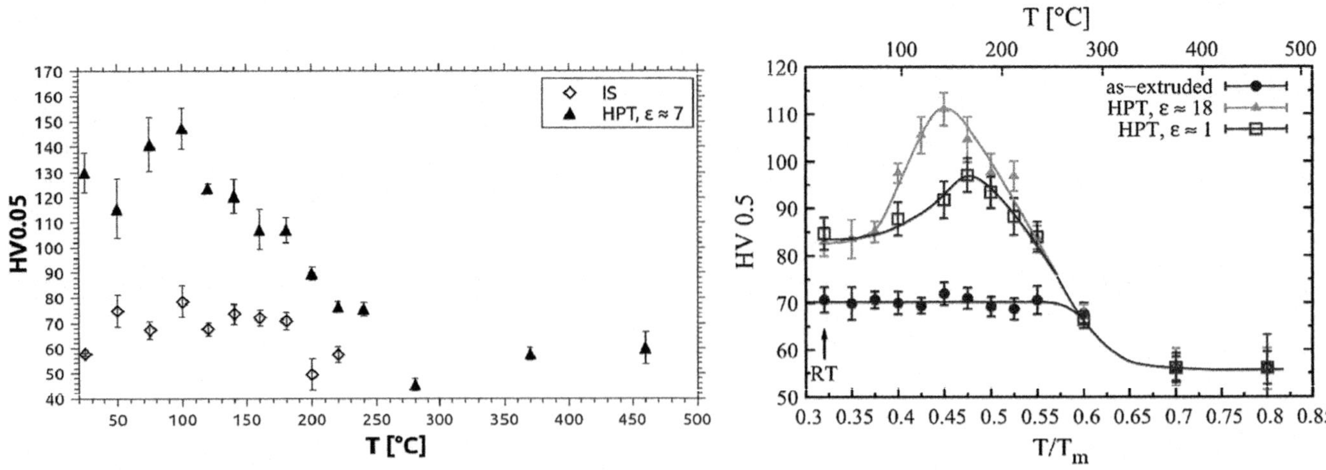


**Fig. 3** HV measurements for Mg alloys after HPT and additional heat treatment at various temperatures for 1 h. Left: HV results for Mg5Zn0.3Ca, homogenized (IS) and HPT-processed at 0.5 rotations and additionally heat treated. Right: HV results for Mg0.2Zn0.5Ca as-extruded and HPT-processed at 0.15 and 2 rotations and additionally heat treated (from [13])

**Fig. 4** HV data of Mg5Zn0.3Ca after homogenizing (IS), after HPT-processing at 4 GPa and 0.5 and 2 rotations, and after additional heat treatment at 100 °C for 1 h (HT)

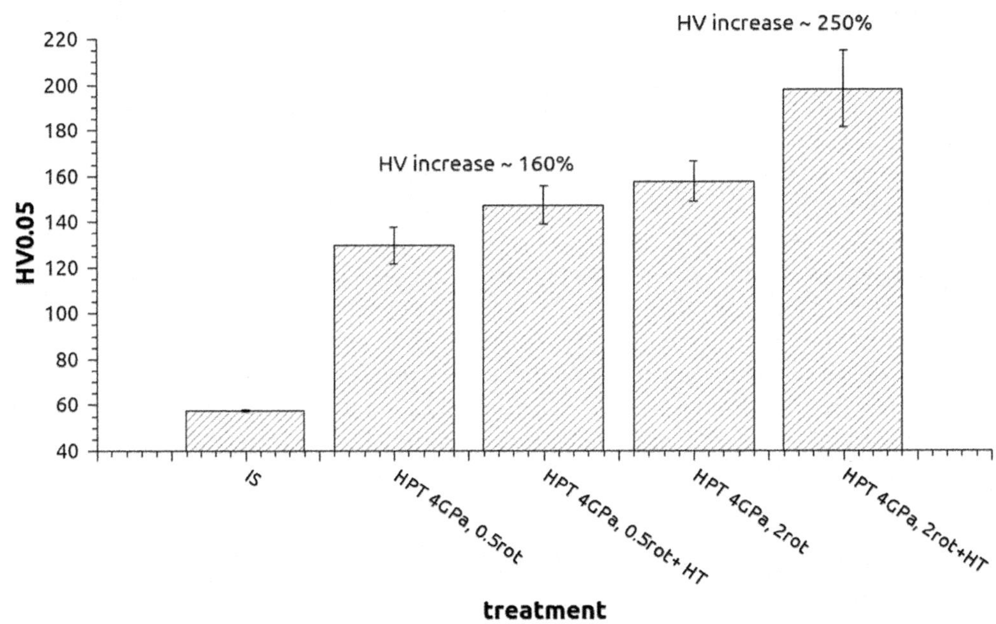


## Evolution of Young's Modulus

The Young's Modulus was measured by nanoindentation (Fig. 6). Along processing history including thermal treatment it almost stays constant, as there occurs only a slight increase from 37 to 47 GPa.

## Discussion

At first we discuss the evolution of hardness as a function of strain accumulated by HPT-processing. At the same time, we compare the results from ZX 50 alloy representing Mg5Zn0.3Ca with those of Mg0.2Zn0.5Ca which has been published elsewhere [13–15], Fig. 2. As expected, the hardness depends uniquely on the equivalent strain value applied irrespective of the actual sample radius chosen for the hardness measurement, although the scattering of data is relatively large compared to dependences measured in more isotropic pure metals like Cu and Al. Most remarkably, the saturation strength in case of the ZX50 alloy is almost twice that of the other alloy with much lower Zn content although the strength of homogenized initial state is comparable. It means that the HPT-induced hardening in the ZX50 alloy is tremendously higher, which must be attributed to the higher content of solved Zn atoms due to solid solution hardening

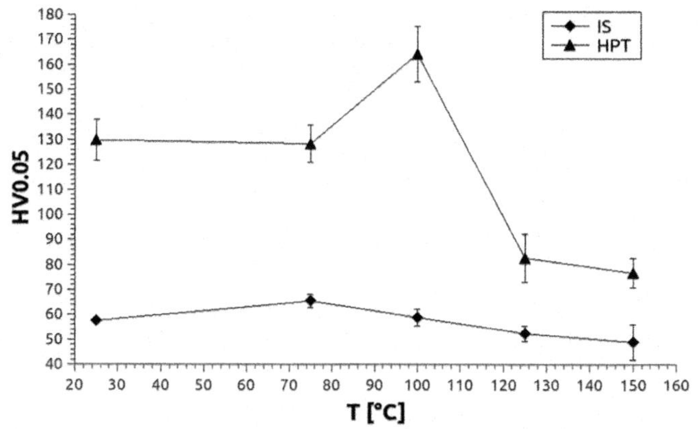


**Fig. 5** HV data of homogenized (IS, below) and Mg5Zn0.3Ca samples HPT-processed at 4 GPa and 0.5 rotations (above) which were also heat treated for 24 h at the temperatures indicated

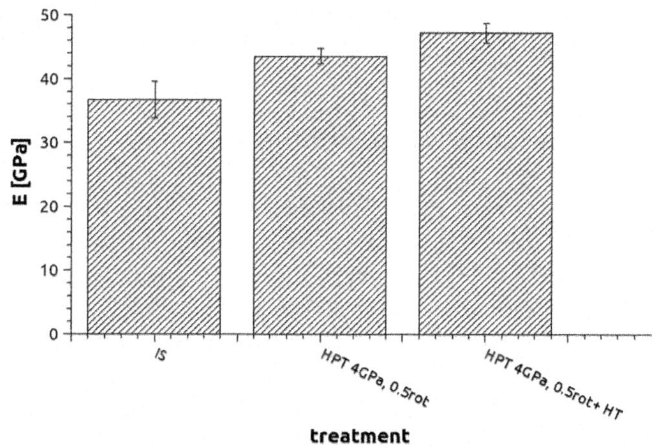


**Fig. 6** Measurement of Young's Modulus using a nanoindentation device for Mg5Zn0.3Ca along the following processing path: Initial state (IS), then also HPT-processed by 0.5 rotations and 4 GPa, and finally also heat treated at 100 °C

and increased strain induced evolution of dislocation density. This increased solid solution hardening obviously far outweighs the difference in Ca concentration which is smaller in the ZX50 alloy.

When additionally heat treating the HPT-processed samples for 1 h (Fig. 3) at the temperatures indicated, another increase of hardness occurs which—in case of Mg5Zn0.3Ca—reaches its maximum at 100 °C, while for Mg0.2Zn0.5Ca it does not show up before 140 °C. In both cases the hardness increase amounts to about 30% at least at the highest HPT-strains achieved (Figs. 3 and 4, 2 rotations, eps = 18). According to a structural investigation and estimations of Orowan precipitation hardening done in [14, 15], this maximum is to be attributed (1) to precipitates which have rod shape along the <c> axis and form at annealing temperatures around 100 °C during short-term heat treatments. However, a major part of hardening arises from vacancy agglomerates. Single vacancies are introduced during HPT-processing by interaction of HPT-induced dislocations and start to agglomerate as thermal treatment occurs [9, 10, 12]. The contribution of precipitation hardening may be concluded also from the hardness measurements done at heat-treated samples without the application of HPT (Fig. 3 left), which also showed a hardness maximum around 100 °C of Mg5Zn0.3Ca but not in case of the alloy Mg0.2Zn0.5Ca previously studied [13–15] confirming there the dominance of vacancy hardening in the HPT processed samples.

For the understanding of the hardness drops shown in Fig. 3, the discussion appears sensible only down to the hardness level of initial state (IS). Considering the HPT processed samples followed by heat treatment, vacancy agglomerates may coarsen and then anneal till about 150 °C in case of the ZX50 sample but till about 210 °C in case of Mg0.2Zn0.5Ca alloy again indicating a higher stability of SPD vacancy type defects there. The continued drop beyond those temperatures may be attributed to the annealing of HPT induced free- and/or grain boundary dislocations, at least to a temperature of 200 °C in case of ZX50. Again, this characteristic annealing temperature is higher in case of Mg0.2Zn0.5Ca, namely about 260 °C which indicates that the Ca atoms rather than the Zn atoms stabilize the HPT induced lattice defects. Any further drop of the hardness, especially that one seen in ZX50 alloy, may reflect the dissolution of rod precipitates as this is also observed in the samples without application of HPT (Fig. 3, left).

When increasing the annealing time for all temperatures from 1 to 24 h (Fig. 5), the hardness even increased from 145 to 163 HV, for the sample HPT-processed with 0.5 rotations. During the 24 h-heat treatment at 100 °C the rod-shape particles change into basal precipitates. However, as the hardness maximum in the non-processed sample of ZX50 after 24 h-treatment (Fig. 5) is lower than that after 1 h-treatment (Fig. 3, left), it can be concluded that the extra increase of hardness occurring during 24 h annealing time arises from continued vacancy agglomeration and not from basal precipitates. Obviously, 1 h heat treatment is too short to entirely form vacancy agglomerates out of single and/or double vacancies as being generated immediately after HPT-processing.

Concerning the Young's Modulus measurements (Fig. 6), there is only a slight increase observed during the processing history which may arise from two influences: (i) an increase because of the texture evolution of HPT processing, see the middle bar of Fig. 6, (ii) an increase because of formation of second phase precipitates. With respect to possible application for biodegradable implants, it should be noted that both increases seem to be rather low and are not expected to contribute to stress shielding [19] in significant way.

## Summary and Conclusions

The present study shows methods for the optimization of Mg–Zn–Ca alloys concerning its strength. The Mg alloy ZX50 (Mg5Zn0.3Ca) has been investigated with respect to its Vickers hardness (HV) and Young's modulus (E), and the results were compared to former investigations made at a similar alloy Mg0.2Zn0.5Ca. The combined application of Severe Plastic Deformation by High Pressure Torsion (HPT) and heat treatments at low temperatures provided the production of both special intermetallic precipitates as well as agglomerates and thus tremendous increases of the alloys' strength.

- For ZX50, total hardness increases up to 250% after homogenization (IS state), further processing by HPT and/or heat treatment could be reached.
- Hardening effects typical of solid solution hardening and precipitation hardening were found for both conditions (IS and HPT). Depending on the heat treatment times applied, both rod-shape and basal platelets form and contribute to the HV increase. For the HPT processed samples, however, the increase in HV is higher due to both HPT induced dislocations and especially HPT induced vacancy agglomerates.
- The content of Zn atoms has strong consequences for the number of HPT induced dislocations and vacancy agglomerates, and thus for the final strength of the Mg alloy in question. The content of Ca atoms, however, affects the corrosion rate and the thermal stability of the HPT induced defects, and thus the hardness reached.
- The Young's modulus varies slightly during the processing history due to the evolutions of texture and precipitations, but still remains small to cause stress shielding.

**Acknowledgements** The authors gratefully acknowledge the use of equipment within the Faculty Center for Nanostructure Research, the Machine Shop and Technical Services at the Faculty of Physics, University Vienna, Austria, as well as the Austrian Institute of Technology (AIT), Wiener Neustadt, Austria for the casting of the Mg bars. We also appreciate financial support from the Austrian Science Funds (FWF) within project I2815-N36 and Slovenian Research Agency (ARRS) through Research Project J2-7157.

## References

1. Polmear IJ, Light alloys. Metallurgy of the Light Metals. 3rd ed., London: Arnold; 1995.
2. Mordike BL, Ebert T. Mater Sci Eng A 2001;302:37.
3. Agnew SR, Nie JF. Scripta Mater 2010;63:671.
4. Alexander D, J Mater Eng Perform 2007;16:360.
5. Zehetbauer MJ, Zhu YTWiley VCH Weinheim 2009.
6. Valiev R, Nature Mater 2004;3:511.
7. Somekawa H, Singh A, Mukai T. Mater Trans 2007;48:1422.
8. Langdon TG, Acta Mater 2013;61:7035.
9. Setman D., Schafler E., Korznikova E., Zehetbauer M, J Mater. Sci. Eng. A 2008; 493, 116–122.
10. Schafler E, Steiner G, Korznikova E, Kerber M, Zehetbauer MJ, Mater. Sci. Eng. A 2005;410–411, 169–173.
11. Korznikova E, Schafler E, Steiner G, Zehetbauer MJ, Proc. 4th Int. Symp. Ultrafine Grained Mater. 2006; 97–102.
12. Zehetbauer M, Key Eng.Mater. 1994;97–98:287–306.
13. Valiev RZ, Estrin Y, Horita Z, Langdon TG, Zehetbauer MJ, Zhu YT, Mater. Res. Lett. 2016; 4:1–21.
14. Werbach K, M.Sc. Thesis 2016, University of Vienna, Austria.
15. Horky J, Ghaffar A, Werbach K, Mingler B., Pogatscher S., Schäublin R., Setman D., Uggowitzer P., Löffler JF., Zehetbauer MJ. 2017, submitted for publication.
16. Orlov D, Pelliccia D, Fang X, Bourgeois L, Kirby N, Nikulin AY, Ameyama K, Estrin Y, Acta Mater 2014;72:110.
17. Mima G, Tanaka Y, Trans JIM 1971;12:71.
18. Clark JB, Acta Metall 1965;13:1281.
19. Sumner DR, Turner TM, Igloria R, Urban RM, Galante JO, J. Biomech. 1998; 31:909–917.
20. Nie JF, Metall. Mater. Trans. A 2012;43,3891–3939.
21. Hofstetter J, Becker M, Martinelli E, Weinberg AM, Mingler B, Kilian H, Pogatscher S, Uggowitzer P J, Loeffler F, JOM 2016; 66:4:566–572.

# Strain Heterogeneity Structures in Wrought Magnesium AZ31 Under Reversed Loading

C. Can Aydıner

**Abstract**

Localized strain accommodation, a concern for any structural material with regard to formability and failure, is a particularly prominent issue for Magnesium alloys that possess the unipolar twinning mechanisms and the large differentials between activity ease of slip mechanisms. In this study, for both rolling and extrusion textures of a Magnesium AZ31 alloy, strain heterogeneity levels are characterized by in situ scanning microscopic image correlation. With a nominal 10 μm subset size and spanning over $10^5$ grains, grain-scale strain mapping is conducted over a [−2, 2]% reversed loading cycle. The rolled sample exhibits much sharper patterns that transcend to upper length scales with collaborative activity. The nature of these patterns change depending on the load sense as operational twin mechanisms switch. The extruded sample exhibits the more typical several-grain-long strain localization patterns that follow the grain boundaries. For each case, strain quantification of the localization structures are provided via histograms.

**Keywords**

Magnesium alloys • Twinning • Strain heterogeneity
Cyclic loading

## Introduction

For wrought light Magnesium alloys, profuse 10–12 tensile twinning leads to the formation of very sharp crystallographic textures during rolling and extrusion processes. Rolling texture is described by the c-axes of the crystallites aligned with the normal direction (ND) of the plate and extrusion texture entails c-axes axis-symmetrically distributed about the extrusion direction (ED). These sharp textures, once more combined with the unipolar activation of the tensile twin, lead to a mechanical behavior that has extreme loading path dependence at the component level. Both extruded and rolled materials undergo twin-dominated deformation for specific combinations of material axis and load path. In mechanical testing, this regime (epitomized with a 'twin plateau' in the stress-strain curve) is typically realized with compression normal to the c-axes of the abundant orientations—applied along ED for the extruded material and applied along rolling/transverse directions (RD/TD) for the rolled material.

In the twin-dominated deformation of the rolled and extruded materials, however, the nature of deformation show obvious differences in plastic anisotropy levels. The extruded sample is transversely isotropic about ED by virtue of its texture description while the rolled material shows severe plastic anisotropy that has been macroscopically described with the r-ratio [1, 2], namely, the ratio of transverse strains under uniaxial loading. For a mechanistic understanding, Kapan et al. [3] compare the evolution of spatial distribution of strain in the two cases using macroscopic digital image correlation (DIC). Magnesium alloys are prone to yield point elongation when they deform with tensile twinning and DIC lays out the nature of strain localization. This study shows the physical root of heavy plastic anisotropy in the rolled sample that is compressed along RD: macroscopic ±45° shear banding but in specific planes of shear (a maximum-shear-stress based argument would lead to shear activity of all ±45° planes axis-symmetrically distributed about the uniaxial load). More precisely, the shear banding planes are strictly aligned with the nominal activation planes of the 10–12 twin in the abundant orientation (this is a specific pair that bisects RD and ND), leading to extreme plastic anisotropy as the bands sequentially cover the gage volume with abrupt emergence/expansion. The supposed transverse isotropy of the extruded sample, on the other hand, already precludes macroscopic shear bands to occur at an angle on specific planes. In Kapan et al. [3], this is

C. C. Aydıner (✉)
Department of Mechanical Engineering, Boğaziçi University, 34342 Bebek Istanbul, Turkey
e-mail: can.aydiner@boun.edu.tr


D. Orlov et al. (eds.), *Magnesium Technology 2018*, The Minerals, Metals & Materials Series, https://doi.org/10.1007/978-3-319-72332-7_44

confirmed via a measurement of near-equal Poisson's ratios on both observation planes (equivalent to r-ratio = 1) for the extruded sample. The shear localization is also quantitatively much milder in the extruded sample on the twin plateau for macro-scale strain distributions. When the load is reversed to tension, both materials appear to reach a near zero strain background after detwinning. With further tensile straining after this point, strain is accommodated with no clear macroscopic strain localization in both rolled and extruded samples.

Strain localization, on the other hand, is a multi-scale phenomenon. Typically, the more one zooms into microstructural length scales, the more strain heterogeneity is captured [4]. At the intergranular length scale, any polycrystalline aggregate, even a low-plastic-anisotropy face-centered cubic metal [5], exhibits strain localization with some typical features of, e.g., bands following grain boundaries and stemming from triple points. Such crystallite-scale observations match the observation length scale of DIC with optical microscopy for moderate to large grained samples (roughly, for grain size >10 μm). (Fine intra-grain resolution requires electron microscopy [6].) The in situ scanning version of microscopic DIC has been used to investigate the character of the sharp twin plateau bands of the rolled sample by Aydıner and Telemez [7], resolving another set of ±45° intergranular bands inside the macroscopic bands. This study only considered the rolling texture with a monotonic loading path in compression.

The current study uses optical microscopic DIC to present deformation structures over representative grain neighborhoods in the reversed loading scheme of Kapan et al. [3], again comparatively considering extruded and rolled samples. Accordingly, the loading path first runs into the twin plateau with compression and then reverse loading is commenced at nominal −2%. This leads to detwinning first followed by tensile plasticity. Statistically significant selected-area regions are considered at each load sector (twinning, detwinning, tensile plasticity), presenting both the morphology and statistics (via strain histograms) of strain accommodation. The specifics of results for both textures are then comparatively interpreted.

## Materials and Methods

Rolled stock material for Magnesium alloy AZ31 is procured from Alfa Aesar, USA, in the form of a 9.5 mm thick hot-rolled plate in annealed condition. Its grain size has been found to be a nominal 12 μm using the lineal intercept method. The source of the extruded stock material for the same alloy is Xian Yuechen Magnesium, China, purchased in the form of a hot extruded billet with 2 cm diameter.

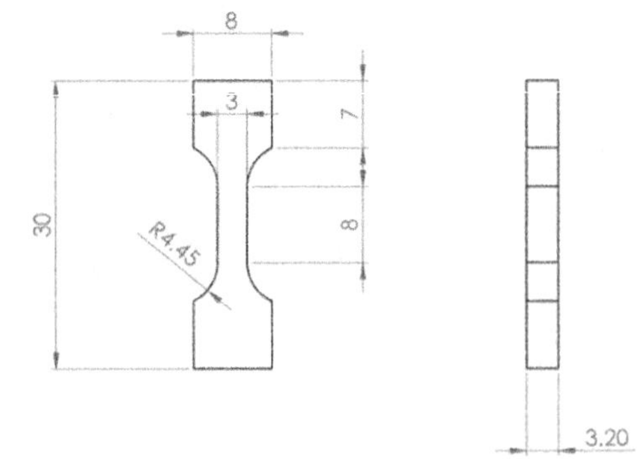


**Fig. 1** Dog-bone sample geometry utilized for reversed loading

Grain size is found to be 17 μm, similar order as the rolled material. For both materials, equiaxed grain morphology is observed. Tension-compression samples with the geometry shown in Fig. 1 are wire-electric-discharge machined from bulk materials such that load axes coincide with RD and ED of the rolled and extruded materials, respectively. For the rolled sample, imaging surface is chosen as the $TD^{\perp}$ plane. For sufficient buckling resistance about both transverse axes, the thickness and width of the sample is set to be a nominal 3 mm in the 8 mm long gage volume. The shown 0.2 mm excess in the thickness direction is margin that is eroded during metallography. Metallography allows viewing the grain structures in the DIC images in correlation to the calculated strain maps. Accordingly, observation surfaces of both specimens are grinded and polished progressively using 1200 P sandpaper down to 50 nm colloidal silica solution. This is followed by etching with a solution of 0.56 gr picric acid, 1.5 ml acetic acid, 1.5 ml distilled water and 12 ml ethanol of 99% purity.

The final stage of sample preparation is the application of a fine mist of black paint using an Iwata airbrush. Together with metallography artifacts and features, the paint speckles provide the speckle pattern for the microscopic image correlation. Details of the area-scanning microscopic DIC implementation are provided elsewhere [7]. Here, we cite the optical resolution (0.2 μm/pixel), subset size for the used first-order elements (61 × 61 pixels; 12 × 12 μm) and the corresponding strain uncertainty of 0.1%. The DIC grid spacing is 2 μm. A central difference scheme is utilized for finding strains over the displacement field; consequently the spacing between two grid points used for numerical differentiation is 4 μm. Small strain/rotation components will be presented in this treatise, however their equivalence with Lagrangian strains and rotation through spectral decomposition have been confirmed.

The setup utilizes computer-controlled X-Y-Z positioning elements that locate a microscopic loading stage. Hence, the positioners place the loaded sample in situ under the desired optical elements as directed by an in-house computer code. The details are provided in Fig. 2 with more specific element descriptions provided in the caption.

## Results

For the sample-scale stress-strain curves shown in Fig. 3, axial strains fields calculated from the images of the macro-DIC line (Fig. 2a, [3]) are used. For the microscopic DIC investigation, three points will be considered: point *a* for comparing deformation structures over the twin plateau; point *b* to compare effectiveness of detwinning at the microstructural level (i.e., how completely strain is reversed locally) and point *c* for comparing tensile strain accommodation structures.

**Fig. 2** Microscopic full-field deformation measurement setup: **a** AVT camera with a macroscopic lens (here, Edmund optics telecentric 0.5x), **b** AVT camera with Navitar microscopic lens **c** Newport X, Y, Z positioning stages with submicron accuracy **d**, **e** Kammrath and Weiss micro tensile tester and its data acquisition unit **f** strain-gage data collector **g** data and image acquisition computers that run in-house Python scripts **h** high-moment-inertia U-frame as a camera hanger. The physical components of the setup sits over a Newport RS 4000 vibration isolation table

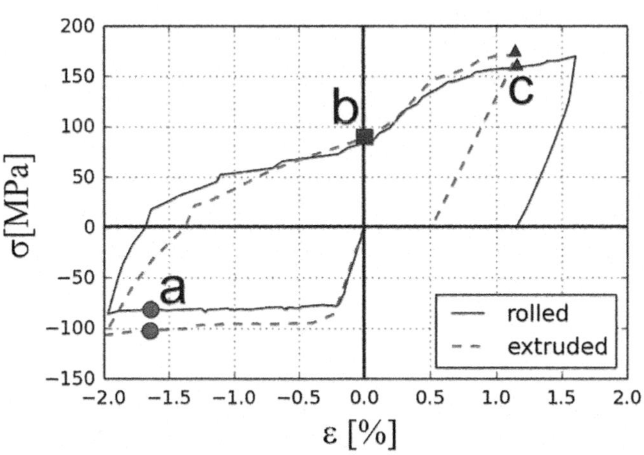


**Fig. 3** Macroscopic stress-strain curves of extruded and rolled samples subjected to compression-tension cycles. Indicated load points **a**, **b**, **c** will be the subject of comparative discussion with microscopic DIC data

Figures 4 and 5 show the selected-area microscopic DIC maps for axial and transverse strains ($\varepsilon_{yy}$, $\varepsilon_{xx}$) as well as in-plane rotation, $\omega_{xy}$. In-plane rotation is used to identify simple shear structures when/if they exist at the particular measurement length scale. Its use has many caveats discussed in detail elsewhere [7]. In a nutshell, rotation maps auto-distinguish +θ simple shear structures since +θ and −θ simple shear entails opposite sense rotations. To increase the clarity, red tones are assigned to positive values and blue tones to negative values, a scheme once again adopted from Refs. [7] and [3]. This means, under compression, a +θ simple shear band will appear red and a −θ band will appear blue; while under tension, appearance colors for both orientation bands will reverse (along with the sense of material point rotation) [8, 9]. In the maps of Figs. 4 and 5, strain/rotation contours are plotted with a degree of transparency to show the microstructure images plotted under them. The apparent incompleteness of the data is a combination of data absence among microscopic frames and DIC evaluation failures. Nonetheless, deformation structures can easily be identified and the statistics is representative, i.e., the histograms for ($\varepsilon_{yy}$, $\varepsilon_{xx}$) shown in the first column checks with those derived from larger areas of the sample (not shown due to dimensional constraints here). The areas considered (roughly 0.5 × 0.8 mm) contain over $10^3$ grains for both samples. To support the data in the histograms quantitatively, averages ($\bar{\varepsilon}_{yy}$, $\bar{\varepsilon}_{xx}$) and standard deviations $(\delta\varepsilon_{yy}, \delta\varepsilon_{xx})$ that belong to the selected area strain maps are summarized in Table 1.

### Tensile Twin Plateau (Load Point A)

At this stage, for the rolled sample, recall deformation progresses by shear banding with macroscopic ±45° structures [7]. This means, until the shear structures cover the volume, strain distribution is composed of heavily-sheared regions and neighboring dormant regions. On the chosen load point, most of the sample has undergone banded deformation and only a limited dormant region has been left. Including a small section of this dormant region in the area selection (Fig. 4a, top left corner) is purposeful. The data of this region inserts a sharp near-zero-strain peak in the histograms demonstrating the dual nature of this deformation. The mirrored appearance of the axial and transverse strain histograms about the zero strain axis (also see Table 1, for point *a* $\bar{\varepsilon}_{yy} \cong -\bar{\varepsilon}_{xx}$ and local Poisson's ratio around 1.) is consistent with strain accommodation by 2-D simple shear

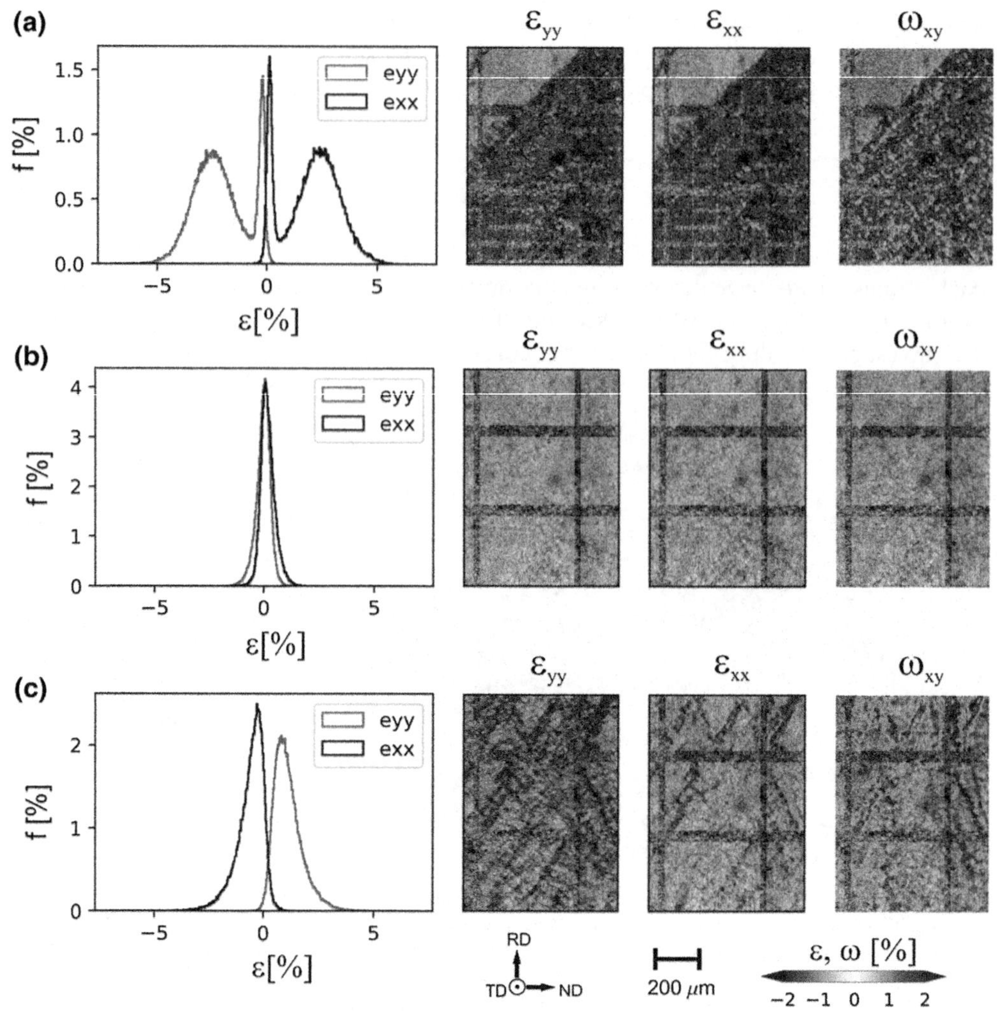


**Fig. 4** Selected area (nominal 0.5 mm × 0.8 mm) microscopic DIC results of the rolled sample at points **a**, **b**, **c** indicated on Fig. 3. (Color figure online)

structures, whose plane of shear matches the observation plane $TD^{\perp}$. Excluding the dormant peak, the histograms of the deformation band show a bell-shaped character and on its extreme edge reach ±5%, comparable to the tensile-twin transformation strain (±6.5% principals).

For the extruded sample (Fig. 5a), at the same macroscopic strain, deformation has covered the entire gage volume [3] and the average axial strain of Table 1 for the selected area (−1.67%) is equivalent to the sample-averaged strain (Fig. 3). There are no dormant regions; strain localization features are uniformly distributed over the maps. Rotation likely picks up a pattern of some small shear bands, since the bands appear to be consistent with the expected rotation senses. Here, one should carefully note that simple shear structures that are, e.g., into the observation plane will not be picked up by this $\omega_{xy}$ map since they would cause $\omega_{yz}$. The reason this is a serious concern for the extruded sample is clearly its transverse isotropy, with the c-axes of preferred orientations axis-symmetrically distributed about *y*. This means tensile twin and its networks operate in a 3-D sense [3]. Another aspect that is consistent with transverse isotropy is an average Poisson's ratio of approximately 0.5 (Table 1, 0.86/1.67 = 0.46). However, the histograms of both strain components are extremely broad and interesting in shape. The strain heterogeneity in $\varepsilon_{yy}$ extends to near −6.5%, further than that of the rolled sample. The $\varepsilon_{yy}$ distribution also has opposite sense content, reaching values of up to 2%. The opposite sense content, in a way, has to exist arithmetically to drop the average value to its mild 1.67%. On the other hand, it means immense strain heterogeneity with a strain span of over 8%. $\varepsilon_{xx}$ is even more notorious in terms of opposite sense content, taking values between roughly [−3, 5]%. The skewed shape of the histograms in Fig. 5a is also notable.

## Detwinning Content (Load Point B)

Figures 4b and 5b shows the closest available points to zero sample-averaged strain after load reversal. Kapan et al. [3] report a quite complete strain recovery in macroscopic distributions. The microscopic-resolution maps here also

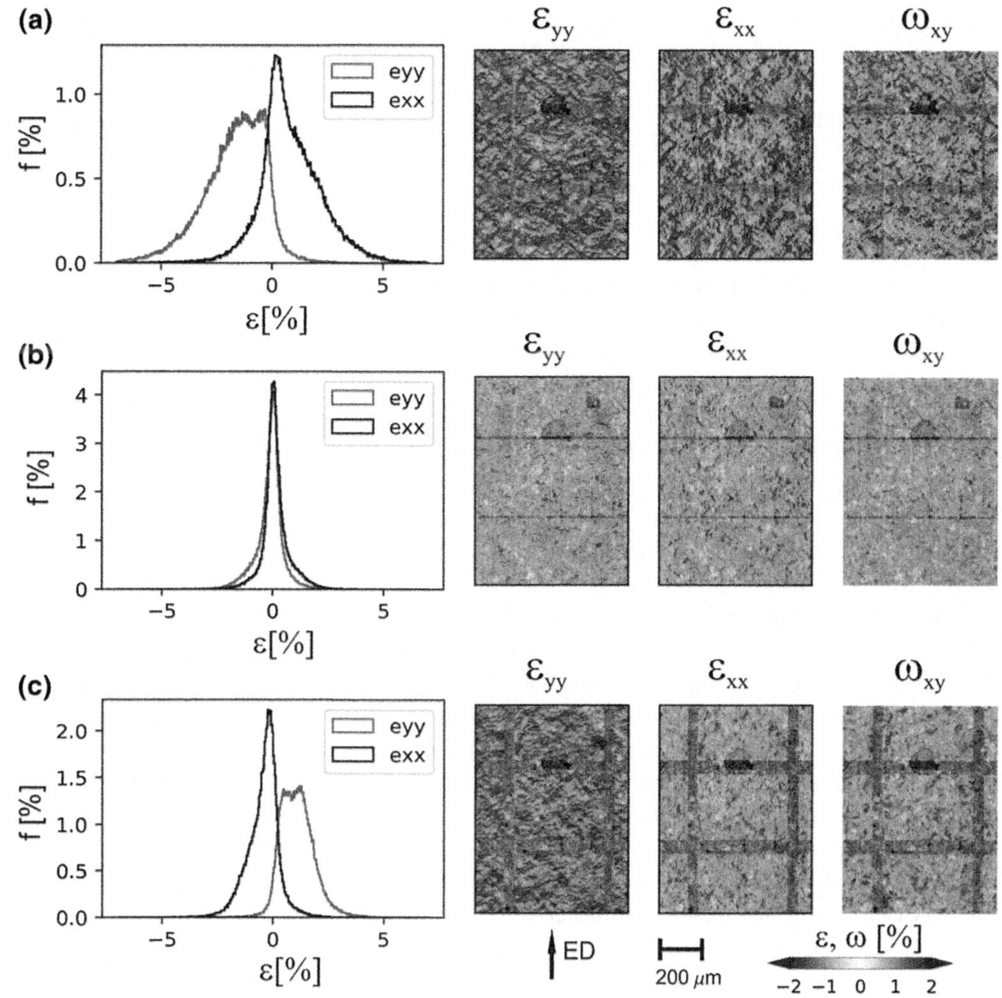


**Fig. 5** Selected area (nom. 0.5 mm × 0.8 mm) microscopic DIC results of the extruded sample at points **a**, **b**, **c** indicated on Fig. 3. (Color figure online)

**Table 1** Averages $\bar{\varepsilon}_{yy}$, $\bar{\varepsilon}_{xx}$ and standard deviations $\delta\varepsilon_{yy}$, $\delta\varepsilon_{xx}$ of axial and transverse strains for the selected areas (Figs. 4 and 5) of the rolled and extruded samples

| All values in % | Rolled | | Extruded | |
|---|---|---|---|---|
| Load point | $\bar{\varepsilon}_{yy} \mid \bar{\varepsilon}_{xx}$ | $\delta\varepsilon_{yy} \mid \delta\varepsilon_{xx}$ | $\bar{\varepsilon}_{yy} \mid \bar{\varepsilon}_{xx}$ | $\delta\varepsilon_{yy} \mid \delta\varepsilon_{xx}$ |
| a | −2.02 \| 1.99 | 1.23 \| 1.25 | −1.67 \| 0.86 | 1.42 \| 1.42 |
| b | −0.05 \| 0.23 | 0.37 \| 0.37 | −0.136 \| 0.135 | 0.67 \| 0.65 |
| c | 1.12 \| −0.56 | 0.67 \| 0.63 | 1.08 \| −0.43 | 0.79 \| 0.74 |

present a substantial strain reversal with few localization remnants. Not to be misguided by the remaining strain (strain is still not exactly zero; attesting to that, axial and transverse strains have opposite senses in both samples); let us focus the comparison on the standard deviation values of Table 1. The fact that point-*b* standard deviation is about two times higher in the extruded sample shows that it harbors more strain heterogeneity. This points to a less perfect pseudoelasticity (detwinning recovery) in a local sense for the extruded sample. This might be deemed consistent with the higher microscopic strain heterogeneity levels at the end of the twinning cycle. As Figs. 4b and 5b histograms indicate, the statistical difference among the two textures is more at the basin of the histogram, i.e., the difference stems from the values of remnant extreme strain points.

## Tensile Plasticity (Load Point C)

In terms of selected area Poisson's ratio, a plastic anisotropy indicator, rolled sample yields a value about 0.5 (Table 1, point c) in this sector; no more indicating a 2-D deformation. Nevertheless, there are high-angle shear structures [10, 11] that reflect in the rotation maps (Fig. 4c), which are striking morphological features that don't exist in, e.g., the maps of the extruded sample. Extruded sample's histograms are again more skewed in shape and their spread (see standard deviations, Table 1) is slightly higher. Also, once again, extruded sample's opposite sense strain content in the histograms is visibly more pronounced.

## Conclusion

Local strain structures of rolled and extruded Magnesium are discussed over a reversed loading cycle. One should carefully note the length scale of this type of measurement (12 μm subsets, 4 μm difference intervals) in interpreting the histograms. It is typical that histograms widen (more strain heterogeneity is captured by evading effective averaging) as the measurement length scale is reduced [4]. At this length scale, values that are comparable to the tensile-twin-transformation strain (6.5% principal) are recorded, particularly over the tensile twin plateau.

The most striking observation in this study was the fact that locally, extruded material harbors more strain heterogeneity than the rolled sample although the other is notorious for macroscopic strain localization with clear dormant and strained regions. Kapan et al. [3] noted mild macroscopic strain heterogeneity for the extruded material. This means that the more intense microscopic strain heterogeneity of the extruded sample is simply averaged out at the length scale of a macroscopic DIC measurement. The skewed nature of the strain histograms also deserve a more detailed study.

On tensile straining (point *c*) of the rolled sample, the apparent high-angle strain localization structures are under careful investigation, prepared in a publication [8]. For these, active ideas are (i) compressive/double twinning and (ii) bands of easy-slip grains that were formed during rolling. The two ideas are not necessarily mutually exclusive. The results clearly show the value of multi-scale investigations of strain distributions in developing a predictive understanding for these complex materials.

**Acknowledgements** This work was supported by the Scientific and Technological Research Council of Turkey, TÜBİTAK, Grant No: 114M215. The author acknowledges the experimental contributions of past and present members of Mechanics of Advanced Mechanics Laboratory, Bogazici University.

## References

1. S. R. Agnew and Ö. Duygulu, "Plastic anisotropy and the role of non-basal slip in magnesium alloy AZ31B," *Int. J. Plast.*, vol. 21, no. 6, pp. 1161–1193, Jun. 2005.
2. X. Y. Lou, M. Li, R. K. Boger, S. R. Agnew, and R. H. Wagoner, "Hardening evolution of AZ31B Mg sheet," *Int. J. Plast.*, vol. 23, no. 1, pp. 44–86, 2007.
3. E. Kapan, N. Shafaghi, S. Uçar, and C. C. Aydıner, "Texture-dependent character of strain heterogeneity in Magnesium AZ31 under reversed loading," *Mater. Sci. Eng. A*, vol. 684, no. August 2016, pp. 706–711, 2017.
4. C. Efstathiou, H. Sehitoglu, and J. Lambros, "Multiscale strain measurements of plastically deforming polycrystalline titanium: Role of deformation heterogeneities," *Int. J. Plast.*, vol. 26, no. 1, pp. 93–106, 2010.
5. M. Sachtleber, Z. Zhao, and D. Raabe, "Experimental investigation of plastic grain interaction," *Mater. Sci. Eng. A*, vol. 336, no. 1–2, pp. 81–87, 2002.
6. A. Orozco-Caballero, D. Lunt, J. D. Robson, and J. Quinta da Fonseca, "How magnesium accommodates local deformation incompatibility: a high-resolution digital image correlation study," *Acta Mater.*, vol. 133, pp. 367–379, 2017.
7. C. C. Aydiner and M. A. Telemez, "Multiscale deformation heterogeneity in twinning magnesium investigated with in situ image correlation," *Int. J. Plast.*, vol. 56, pp. 203–218, 2014.
8. B. Üçel, C. C. Aydıner, et al. "Mesoscale strain localization structures in Magnesium AZ31 tensioned normal to c-axes with multiscale image correlation", in preparation.
9. N. Shafaghi, E. Kapan, C. C. Aydıner, "Evolution of strain heterogeneity in reversed loaded Mg AZ31 with in-situ multiscale image correlation", in preparation.
10. M. R. Barnett, M. D. Nave, and C. J. Bettles, "Deformation microstructures and textures of some cold rolled Mg alloys," *Mater. Sci. Eng. A*, vol. 386, no. 1–2, pp. 205–211, 2004.
11. S. Sandlöbes, S. Zaefferer, I. Schestakow, S. Yi, and R. Gonzalez-Martinez, "On the role of non-basal deformation mechanisms for the ductility of Mg and Mg-Y alloys," *Acta Mater.*, vol. 59, no. 2, pp. 429–439, 2011.

# Hot Forging Behavior of Mg−8Al−4Ba−4Ca (ABaX844) Alloy and Validation of Processing Map

K. P. Rao, C. Dharmendra, Y. V. R. K. Prasad, H. Dieringa, and N. Hort

**Abstract**

Newly developed die-cast alloys based on Mg–Al−Ba–Ca (ABaX) system show promise for high temperature creep resistance. ABaX844 alloy is one of them and it has limited workability due to high alloy content. To identify the optimum processing conditions, processing map for this alloy was developed earlier, which exhibited two workability domains in the temperature and strain rate ranges: (1) 340–410 °C and 0.0003–0.005 $s^{-1}$, and (2) 425–500 °C and 0.0003–0.1 $s^{-1}$. Dynamic recrystallization (DRX) occurs in these domains. The map also exhibited extensive flow instability mainly at strain rates > 0.01 $s^{-1}$ up to a temperature of 400 °C and at strain rates >0.1 $s^{-1}$ beyond 400 °C. The aim of the present study is to validate the findings of processing map by performing forging tests in the temperature range 300–500 °C (at an interval of 40 °C) and forging speeds of 0.01, 0.1, 1 and 10 mm $s^{-1}$ to produce a rib-web (cup) shape component. Finite-element (FE) simulations were performed for obtaining the variations of strain and strain rate in the components during forging. The microstructures of forged specimens deformed under optimum process conditions derived from the processing map revealed the formation of dynamically recrystallized grains. The alloy specimens forged under the conditions of flow instability have fractured and/or exhibited flow localization. The results validated the predictions of the processing map and the load-stroke curves obtained by FE simulation correlated well with the experimental curves.

**Keywords**

Magnesium alloy • Hot forging • Dynamic recrystallization • Simulation

## Introduction

In view of their lightweight, magnesium alloys are being developed for aerospace, automobile, and biomedical applications. Mg–Al-Zn series of alloys (AZ31, AZ61, and AZ91) are developed initially [1, 2] but their creep resistance needed improvement. Rare-earth elements were added for this purpose [3], but they are expensive. In recent years, the addition of relatively cheaper alkaline-earth elements like Ca, Sr, and Ba [4–6] is found to be beneficial since these elements form stable intermetallic particles in the matrix, which cause dispersion strengthening. In the commercial MRI 230D alloy [5], which contains 7 wt% Al, 2.1 wt% Ca, and 0.3 wt% Sr, three different thermally stable intermetallic phases $Al_4Sr$, $Mg_2Ca$, and $(Mg,Al)_2Ca$ form in the matrix [7] and enhance the creep resistance. While Sr addition is not favored because of its detrimental effect on the corrosion resistance of several magnesium alloys [8], addition of Ba resulted in higher temperature strength and increased ignition temperature [6]. Two Ba + Ca containing creep resistant alloys Mg−4Al−2Ba−1Ca (ABaX421) and Mg−4Al−2Ba−2Ca (ABaX422) had better creep resistance than Mg−4Al−2RE (AE42) and Mg−3Sn−2Ca (TX32) alloys [9]. When Ba is added along with Ca, it forms a eutectic phase

K. P. Rao (✉) · C. Dharmendra
Department of Mechanical and Biomedical Engineering, City University of Hong Kong, Tat Chee Avenue, Kowloon, Hong Kong
e-mail: mekprao@cityu.edu.hk

C. Dharmendra
e-mail: chalasanidharmendra@gmail.com

Y. V. R. K. Prasad
Processingmaps.com (Formerly at City University of Hong Kong), Kowloon, Hong Kong
e-mail: prasad_yvrk@hotmail.com

H. Dieringa · N. Hort
Magnesium Innovation Centre, Helmholtz-Zentrum Geesthacht, Max-Planck-Str. 1, 21502 Geesthacht, Germany
e-mail: hazo.dieringa@hzg.de

N. Hort
e-mail: norbert.nort@hzg.de


D. Orlov et al. (eds.), *Magnesium Technology 2018*, The Minerals, Metals & Materials Series, https://doi.org/10.1007/978-3-319-72332-7_45

$Mg_{17}Ba_2$ and a ternary compound (Al, $Mg)_2Ca$ at the grain boundaries, preventing their sliding during creep deformation. By increasing the concentration of Al, Ba and Ca, the creep strength of ABaX alloys may be improved but their workability suffers. However, the processing map developed for ABaX844 revealed that the alloy possesses good hot workability [10]. The aim of the present investigation is to assess the hot workability of ABaX844 alloy by forging it to form a cup-shape component in accordance with the findings of the processing map, and validate its predictions on the formability and microstructural mechanisms. The methodology used for this purpose is briefly described below and is similar to that developed earlier to validate the processing maps on electrolytic copper [11] and rolled AZ31alloy [12–14]. It essentially consists of process simulation using finite element method (FEM) to estimate the effective stress and strain rates in forging and conduct actual forging experiments under controlled conditions using die and punch on MTS-servo-hydraulic machine.

## Methodology

### Processing Map Development

The processing map for a material is developed based on the variation of flow stress with temperature and strain rate in a wide range. The detailed principles and procedures have been described earlier [15, 16]. The efficiency of power dissipation occurring through microstructural changes during deformation is given by:

$$\eta = 2m/(m+1) \tag{1}$$

where, $m$ is the strain rate sensitivity of flow stress of the material. A three-dimensional plot of efficiency variation with temperature and strain rate gives a power dissipation map, which may be viewed as a contour map drawn with iso-efficiency contours.

Flow instability criterion is derived by exploring the extremum principles of irreversible thermodynamics as applied to continuum mechanics of large plastic flow [17], and is given by instability parameter and flow instability occurs when:

$$\xi(\dot{\varepsilon}) = \frac{\partial \ln[m/(m+1]}{\partial \ln \dot{\varepsilon}} + m \leq 0 \tag{2}$$

The variation of instability parameter with temperature and strain rate gives the instability map, which may be superimposed on the power dissipation map for obtaining a processing map. The map is conveniently viewed as a contour map, where iso-efficiency contours are drawn on a frame of temperature and logarithm of strain rate. Accurate experimental data of flow stress as a function of temperature and strain rate are required to generate the processing map, the procedure for which was explained earlier [15]. The processing map reveals domains, where different microstructural mechanisms like dynamic recrystallization (DRX) operate, and also the limiting conditions for regimes where flow instability occurs.

## FEM Simulation Model

The FEM model used in this study is the DEFORM program, which is a code developed based on a rigid viscoplastic model using the principle of analysis of large plastic incremental deformation (ALPID) by Kobayashi et al. [18]. The basis for the FEM simulation program has been presented by Oh [19]. Briefly, FEM uses extremum principles, which state that for a plastically deforming body of volume, $V$, under the traction, $F$, prescribed on a part of the surface, $SF$, and the velocity, $u$, prescribed on the surface, $SU$, and uses variational principles in minimizing the function:

$$\varphi = \int \bar{\sigma}\bar{\varepsilon}dV - \int \frac{\alpha}{2}(\dot{\varepsilon}_v)^2 dV - \int F.udS \tag{3}$$

where, $\dot{\varepsilon}$ is the effective strain rate, $\bar{\sigma}$ is the effective stress, $\bar{\varepsilon}$ is the effective strain, $u$ is the velocity vector, $\dot{\varepsilon}_v$ is the volumetric strain rate, and $\alpha$ is a large positive constant (penalty). The program accepts experimental constitutive equation data, the geometry of the die and punch, and the friction coefficient. The program also has an automatic remeshing option when the finite element mesh distorts and the nodes meet (Jacobian goes negative). The output of the simulation model consists of the local values of stress, strain, strain rate, and velocity vectors. In addition, the load-stroke curves may be obtained for each forging simulation.

## Experimental

The magnesium alloy Mg−8Al−4Ba−4Ca (ABaX844) billet was prepared by conventional casting method using elemental metals. While the molten AZ alloy was kept at 720 °C under $SF_6$-argon mix cover gas, Ca and Ba were added. The melt was held at the same temperature for another 5 min before pouring in a preheated permanent steel mold. A cast billet with 100 mm diameter obtained using the above procedure was diametrically sliced into disks. Slugs with 12.5 mm diameter and 14 mm height were machined for forging tests. One end of the slug was chamfered for specimen aligning and centering on the forging die. Isothermal forging experiments were conducted on cylindrical specimens of the geometry shown in Fig. 1a to produce a rib-web (cup) shape shown in Fig. 1b. The load train assembly is shown in Fig. 1c.

The experiments were conducted at different temperatures in the range of 300–500 °C and at speeds of 0.01, 0.1, 1.0, and 10 mm $s^{-1}$ using MTS 810 servo-hydraulic machine. These parameters were chosen such that both the workability domain and instability regimes are covered for validation. Details of the test set-up and procedure have been described in earlier publications [11–14]. Briefly, the procedure involved heating the die and punch to the testing temperature before loading the specimen on the die. Before the experiment, the specimen was lubricated with a grease paste containing graphite and heated for 15 min. However, during deformation, contact surface friction had no influence on the metallurgical structure development. As the tapered punch pierced the material, the latter flowed opposite the direction of punch movement, wrapping around the punch and producing a cup-shape at the end of the stroke, which was limited to about 11 mm. The load-stroke curve was recorded in each experiment. The forged specimens were quenched in water and prepared for microstructural examination by sectioning the deformed specimens at the center, parallel to the compression axis. The cut surface was mounted, polished, and etched with a acetic-nitric acid solution. An optical microscope (OLYMPUS/PMG3) was used to record the microstructure in the rib regions where strains were significant but gradients were smaller.

## Results and Discussion

### Initial Microstructure

The as-cast microstructure of ABaX844 alloy is shown in Fig. 2. The intermetallic phases consisting of Al, Ba and Ca have been identified to be $Mg_{21}Al_3Ba_2$ and $(Al,Mg)_2Ca$, all of them being present mainly at the grain boundaries in the form of lamelle and block form, respectively. The grain size of the as-cast alloy is in the range of 25–40 μm, and can be considered as fine-grained in the case of magnesium alloys.

## Processing Map and Its Interpretation

Details of the processing map obtained on ABaX844 alloy and interpretation of the domains, in terms of microstructural mechanisms, are given in an earlier publication [10]. For the purpose of ready reference, the processing map obtained at a strain of 0.5 is shown in Fig. 3. The numbers marked on the contours represent efficiency of power dissipation expressed in percent. The forging speeds estimated for the different strain rates are marked on the right side. The shaded areas represent the regimes of flow instability. The map exhibits two domains in the temperature and strain rate ranges as described below:

(1) 340–410 °C and 0.0003–0.005 $s^{-1}$ with a peak efficiency of about 32% occurring in the vicinity of 380 °C, and
(2) 425–500 °C and 0.0003–0.1 $s^{-1}$ with a peak efficiency of about 43% occurring at 500 °C/0.003 $s^{-1}$.

Based on the microstructural observations, the two domains are interpreted [10] to represent dynamic recrystallization (DRX), which replaces the as-cast microstructure with a wrought equiaxed grain structure. The grain size is finer in Domain 1 than in Domain 2. In the first domain, basal + prismatic slip causes the plastic flow and simultaneous recovery occurs by climb process. In the second domain that occurred at higher temperatures, second order

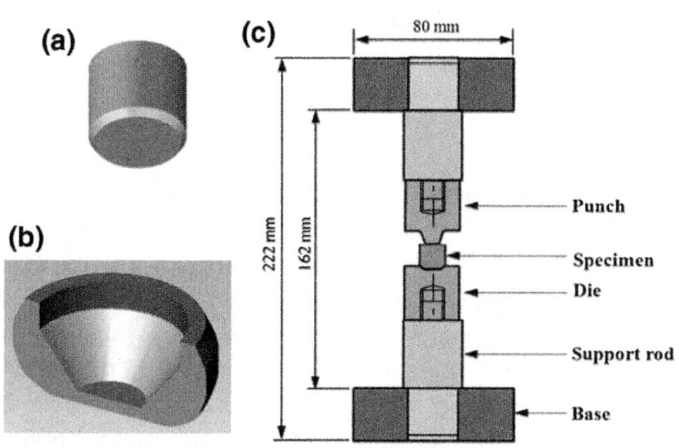


**Fig. 1** Sketches of (**a**) starting specimen, (**b**) final shape of the forging and (**c**) design of load train assembly for forging using servo-hydraulic machine [12]

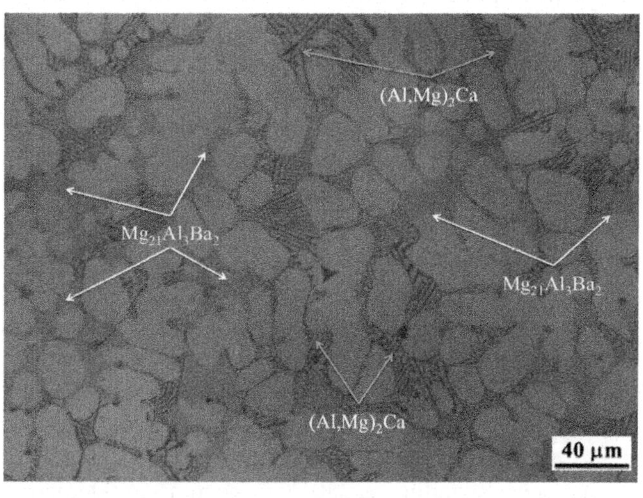


**Fig. 2** Microstructure of as-cast ABaX844 alloy

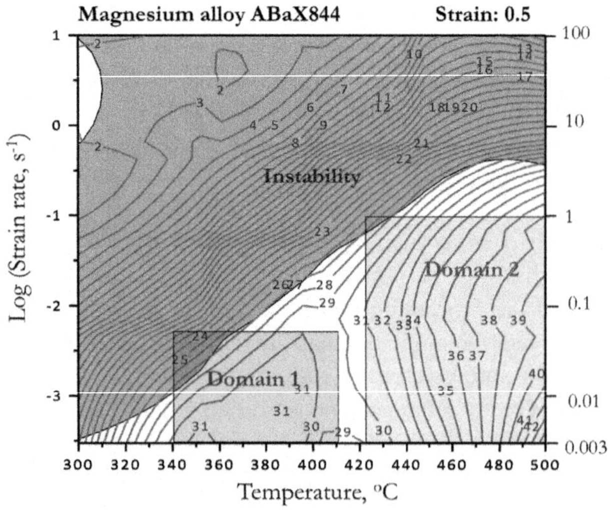


**Fig. 3** Processing map for ABaX844 alloy developed at a strain of 0.5. The numbers against the contours represent efficiency of power dissipation in percent

pyramidal slip occurs with simultaneous recovery by cross-slip process. Flow instability occurs mostly at lower temperatures and higher strain rates (shaded in Fig. 3), and its microstructural manifestation is flow localization.

## Process Simulation

The process simulation based on FEM has become an effective tool in flow investigation during material forming, as demonstrated by Sivaprasad et al. [20] for optimizing the extrusion of a stainless steel and by Ou et al. [21] for the forging of aerofoils using a nickel alloy. Such simulations provide information on the flow pattern during die filling as well as on the local values of state-of-stress, strain, and strain rate. These values may be used in predicting the microstructural developments. This model also predicted the forging loads so that the required forging equipment capacity may be determined [18]. A finite element simulation of the forging process was conducted under isothermal conditions (the work piece and the die remained under the same temperature) using the software DEFORM 2D axisymmetric version equipped with a pre-processor to input material data and object definition. A post-processor was also used, which provides information on deformed geometry, state-of-stress, velocity vectors, strain, and strain rate.

Process simulations were conducted at the temperature range of 300–500 °C at speeds of 0.01–10 mm $s^{-1}$ until the stroke reached 11 mm in 0.1 mm increments. As an example, the effective strain distributions in the forged component at the end of the stroke are shown in Fig. 4a and b for forgings corresponding to 380 °C/0.01 mm $s^{-1}$ (Domain 1) and 500 °C/0.01 mm $s^{-1}$ (near peak conditions in Domain 2), respectively. From the simulations, the minimum and maximum effective strains range between 0.1 and 4. The average strain rates corresponding to the forging speeds of 0.01, 0.1, 1.0, and 10 mm $s^{-1}$ are 0.001, 0.01, 0.1, and 1.0 $s^{-1}$, respectively.

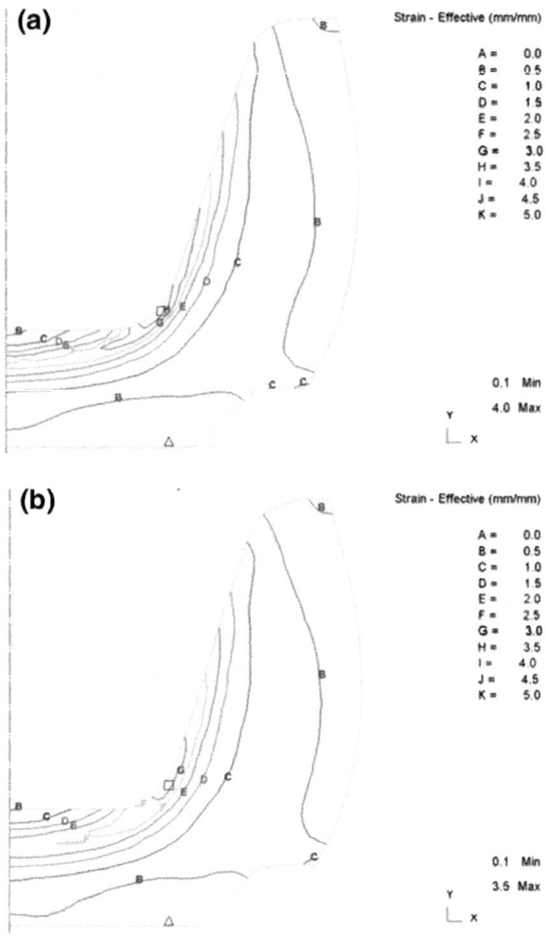


**Fig. 4** Strain contours obtained in finite element simulation of forging of ABaX844 at the end of stroke (11 mm) obtained at temperature and speed of **a** 380 °C/0.01 mm $s^{-1}$ (Domain 1) and **b** and 500 °C/0.01 mm $s^{-1}$ (Domain 2)

## Load-Stroke Curves

The curves representing semi-close die forging involve three stages as follows: (1) increase in the load until plastic flow initialization, (2) material flow until the cup formation is complete, and (3) direct material compression in the bottom of the cup, resulting in a steep increase in the load with stroke. The load-stroke curves recorded during the forging experiments at 380 °C (Domain 1), 500 °C (Domain 2), at different forging speeds are shown in Fig. 5a and b, respectively. In both the DRX domains conditions (i.e. at

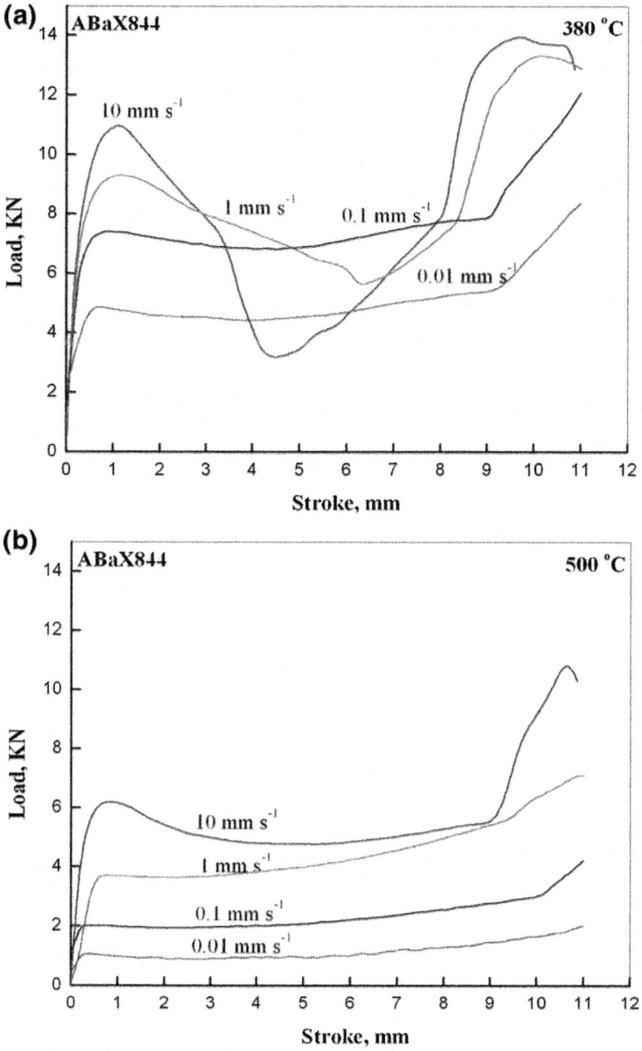


**Fig. 5** Load-stroke curves obtained for forging of ABaX844 alloy at different speeds and at temperatures of **a** 380 °C and **b** 500 °C

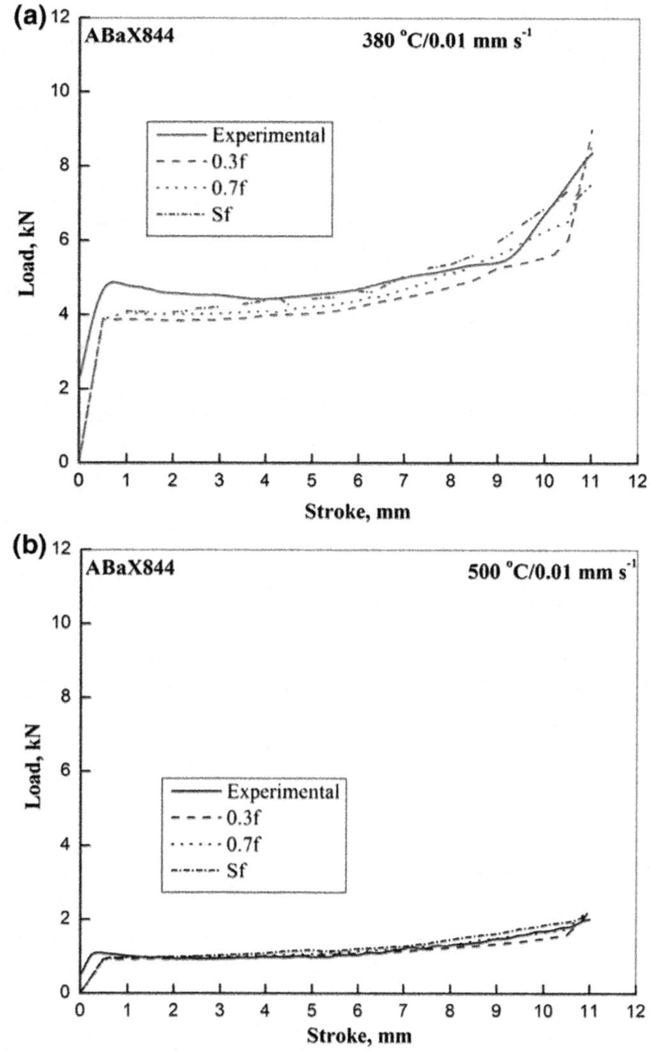


**Fig. 6** Comparison of load-stroke curves obtained on ABaX844 alloy from the finite element simulation and forging experiments at a temperature and speed of **a** 380 °C/0.01 mm $s^{-1}$ (Domain 1) and **b** 500 °C/0.01 mm $s^{-1}$ (Domain 2)

lower strain rates), the load-stroke curves are of steady state type.

Load-stroke curves obtained for forging simulations 380 °C/0.01 mm $s^{-1}$ (Domain 1) and 500 °C/0.01 mm $s^{-1}$ (Domain 2) are shown in Fig. 6a and b, respectively and are compared with those obtained from forging experiments. A reasonable agreement exists between the simulated and experimental curves validating the simulation model. Simulation with a higher friction factor marginally changes the loads, indicating that the contribution of friction is insignificant.

## Shapes of Forged Specimens

The bottom views of ABaX844 forged specimens at different temperature and forged speeds are shown in Fig. 7. Clearly, the specimens forged under conditions corresponding to the two DRX domains: Domain 1, namely, 380 °C and a speed 0.01 mm $s^{-1}$ and Domain 2, namely, 460 and 500 °C and speeds of 0.01, 0.1, and 1.0 mm $s^{-1}$, possess the expected regular shapes for this component under good workability conditions. The specimens forged at 300 °C, and 500 °C and the highest speed were fractured because of the flow instability.

## Microstructural Correlation

The microstructures recorded on the specimens forged at temperatures and speeds of 380 °C/0.01 mm $s^{-1}$ (Domain 1) and 500 °C/0.1 mm $s^{-1}$ (Domain 2) are shown in Fig. 8a and b, respectively.

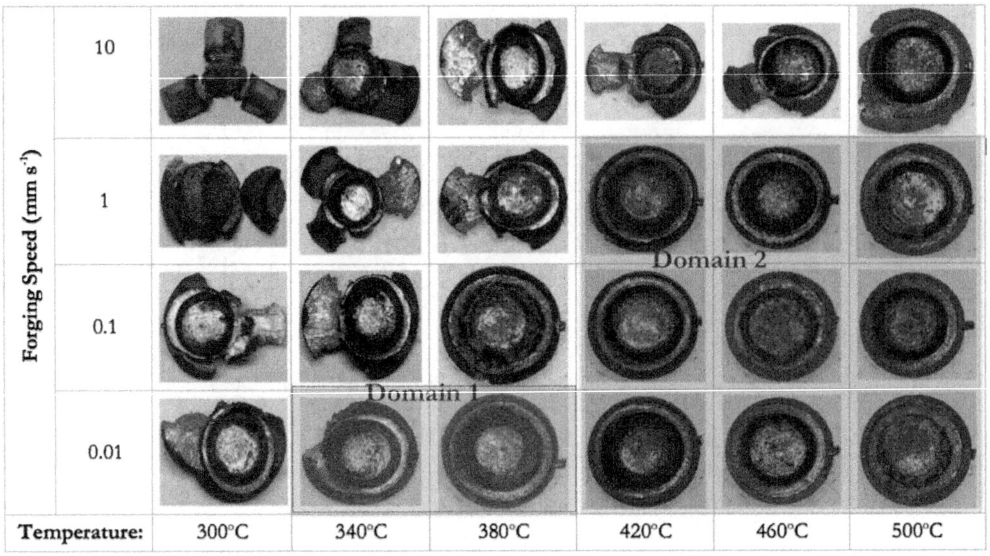


**Fig. 7** Geometry of the ABaX844 specimens forged at different temperatures and speeds —Bottom view

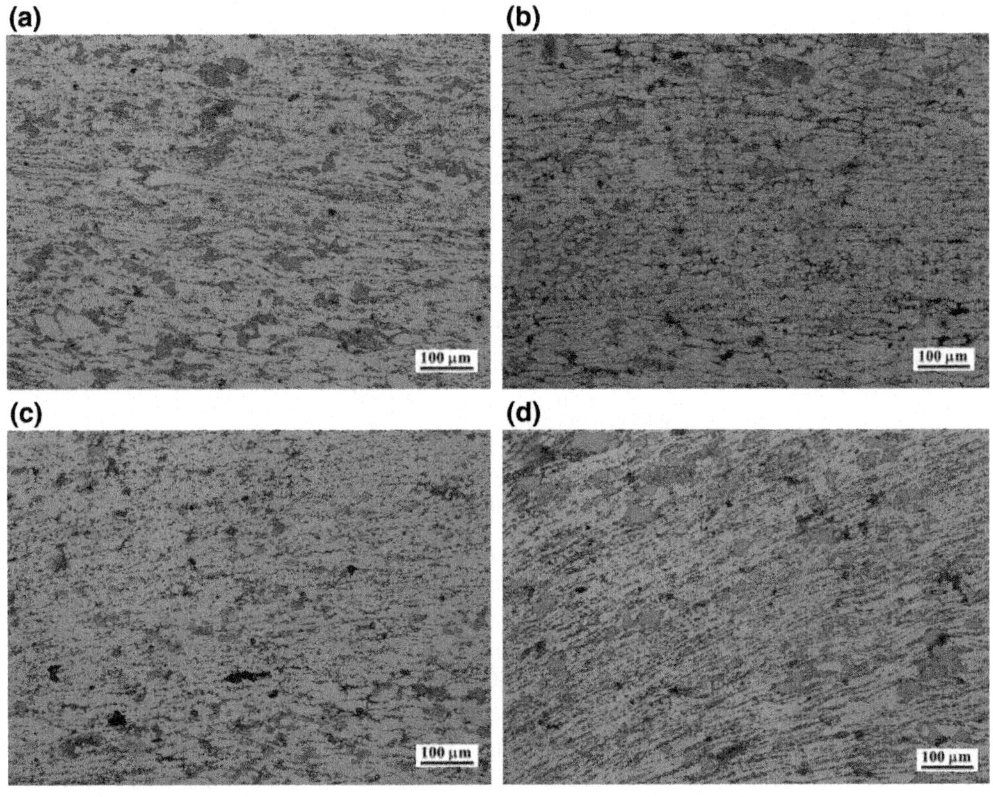


**Fig. 8** Microstructures of ABaX844 forging done at **a** 380 °C/0.01 mm $s^{-1}$ (Domain 1), **b** 500 °C/0.1 mm $s^{-1}$ (Domain 2), **c** 500 °C/0.01 mm $s^{-1}$ (at a lower speed in Domain 2), and **d** 460 °C/0.1 mm $s^{-1}$ (at a lower temperature in Domain 2)

These microstructures were recorded in the bottom region of the cup and those examined at the radius and the inside wall regions are not any different, signifying microstructural homogeneity. However, the microstructure recorded in the outside rib region did not undergo significant change when compared with the starting as-cast microstructure (Fig. 2). This is because the local strain value in this region is too low to cause dynamic recrystallization. The microstructures clearly showed the occurrence of DRX and the average grain size obtained in Domain 1 (18 μm) is smaller than that in Domain 2 (22 μm). The microstructures obtained on specimens forged in Domain 2 at lower strain rates (500 °C/0.01 mm $s^{-1}$) and at lower temperatures (460 °C/0.1 mm $s^{-1}$) are shown in Fig. 8c and d, respectively. These confirm that dynamic recrystallization occurs in entire region of Domain 2.

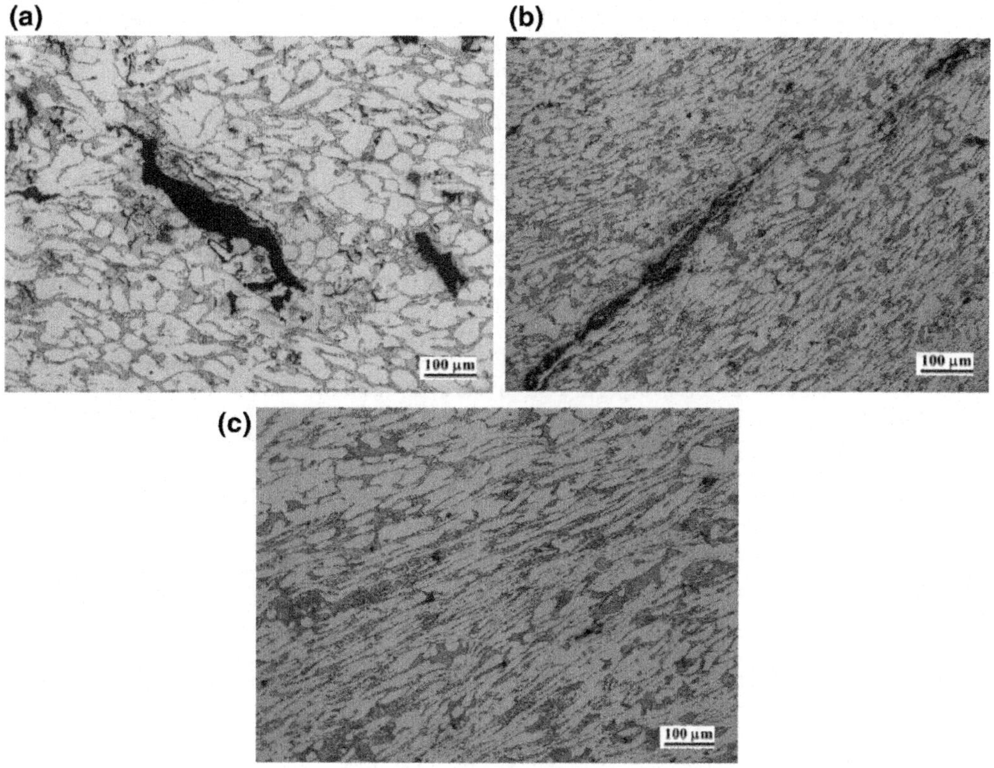


**Fig. 9** Microstructures of ABaX844 forging done at the conditions of: **a** 380 °C/10 mm $s^{-1}$ (instability regime—flow localization and cracking), **b** 420 °C/1 mm $s^{-1}$ (instability regime—flow localization and cracking), and **c** 380 °C/0.1 mm $s^{-1}$ (instability regime—localized flow). The compression axis is vertical

The microstructures obtained at various locations of the forging conducted at 380 °C and 10 mm $s^{-1}$, which falls within the instability regime of the processing map (Fig. 3), is shown in Fig. 9a. The flow localization is clearly observed in the bottom, radius and inside-wall regions and is associated with cracking. The microstructures obtained at similar locations on the forging conducted at 420 °C and 1 mm $s^{-1}$ and 380 °C and 0.1 mm $s^{-1}$ are shown in Fig. 9b and c. The flow localization is manifested as an intense shear band in the former case and is as shear localization in the latter. Thus, the microstructural observations completely validate the workability regions where DRX occurs as well as the flow instability regions of the processing map.

## Conclusions

Cup-shaped hot forging of Mg alloy ABaX844 has been modeled by FEM simulation under different conditions predicted by the processing map. The results have been experimentally validated in the temperature range of 300–500 °C and at a speed range of 0.01–10 mm $s^{-1}$. The following conclusions have been drawn:

(1) Processing map exhibits two domains within the following temperature and strain rate ranges: (1) 340–410 °C and 0.0003–0.005 $s^{-1}$, and (2) 425–500 °C and 0.0003–0.1 $s^{-1}$, both of which show the occurrence of DRX.

(2) Finite element simulation of a rib-web (cup-shaped) forging in ABaX844 alloy conducted in the DRX domains accurately predicts load-stroke curves that correlate well with experimental data.

(3) The microstructures obtained in the components forged under conditions in Domain 1 and Domain 2 of the processing map exhibit DRX, and validate the processing map predictions.

(4) The load-stroke curves obtained for forgings done under DRX conditions are of steady state type.

(5) Forging in the regimes of flow instability of the processing map results in flow localization in the microstructure and disintegration of specimens before the final shape is forged.

**Acknowledgements** The work presented in this paper has been supported by a strategic research grant from City University of Hong Kong (Project Ref. No. 7004454).

## References

1. Pekguleryuz M, Celikin M (2010) Creep resistance in magnesium alloys, International Materials Reviews, 55:197–217.
2. Ham Q, Kad B.K, Viswanathan S (2004) Design perspectives for creep-resistant magnesium die-casting alloys, Philosophical Magazine, 84:3843–3860.
3. Dieringa H, Hort N, Kainer K.U (2009) Barium as alloying element for a creep resistant magnesium alloy. In: *Magnesium: 8th*

*International Conference on Magnesium Alloys and Their Applications, (Ed.) K.U. Kainer*, Weimar, Germany, 26–29 Oct, pp. 62–67.

4. Abu Leil T, Hort N, Dieringa H, Blawert C, Huang Y, Kainer K.U, Rao K.P, (2006) Development and characterization of a series of Mg–Sn–Ca alloys, *Magnesium Technology in Global Age, (Eds.) M.O. Pekguleryuz and L.W.F. Mackenzie,* Canadian Institute of Mining, Metallurgy and Petrolelum, Montreal, Canada, pp. 739–749.
5. Bronfin B, Moscovitch N, Trostenetsky V, Gerzberg G, Nagar N, Yehuda R, (2008) High performance HPDC alloys as replacements for A380 aluminum alloy, In: *Magnesium Technology 2008, (Eds.) M.O. Pekguleryuz, et al.,* TMS, Warrendale, PA, pp. 411–415.
6. Wang Z, Geng H, Li J, and Teng X (2007) Influence of Y and Ba on microstructure and mechanical properties of AZ91 magnesium alloys, *Key Engineering Materials,* 353–358:1593–1596.
7. Jing B, Yangshan S, Shan X, Feng X, Tianbai Z (2006) Microstructure and tensile creep behavior of Mg-4Al based magnesium alloys with alkaline-earth elements Sr and Ca additions, *Materials Science and Engineering* A, 419:181–188.
8. Abu Leil T, Rao K.P, Hort N, Kainer K.U (2006) Corrosion behavior and microstructure of a broad range of Mg–Sn–X alloys, In: *Magnesium Technology 2006, (Eds.) A.A. Luo et al.* TMS, Warrendale, PA, pp. 281–286.
9. Dieringa H, Zander D, Gibson M.A (2013) Creep behavior under compressive stresses of calcium and barium containing Mg–Al based die casting alloys, *Materials Science Forum*, 765:69–73.
10. Rao K.P, Lam S.W, Hort N, Dieringa H (2016) High temperature deformation of a newly developed Magnesium alloy containing Al, Ba and Ca, *Proceedings of the 7th TSME International Conference on Mechanical Engineering*, 13–16 December 2016, Paper no. AMM0023, 6 pages.
11. Rao K.P, Prasad Y.V.R.K (2011) Materials modeling and finite element simulation of isothermal forging of electrolytic copper, *Materials and Design*, 32:1851–1858.
12. Rao K.P, Prasad Y.V.R.K, Suresh K (2012) Anisotropy of flow during isothermal forging of rolled AZ31B magnesium alloy plate in three orthogonal directions: Correlation with processing maps, *Materials Science and Engineering A*, 558:30–38.
13. Rao K.P, Prasad Y.V.R.K, Suresh K (2011) Anisotropy of flow during forging of rolled AZ31B plate in transverse direction: Validation of Materials Models, *Materials Science Forum*, 690:57–60.
14. Rao K.P, Prasad Y.V.R.K, Suresh K (2011) Materials modeling and simulation of isothermal forging of rolled AZ31 magnesium alloy: Anisotropy of flow, *Materials and Design*, 32:2545–2553.
15. Prasad Y.V.R.K, Rao K.P, Sasidhara S (2015) *Hot working guide: A compendium of processing maps*, 2nd Edition, ASM International, Materials Park, OH.
16. Prasad Y.V.R.K, Seshacharyulu T (1998) Modeling of hot deformation for microstructural control, *International Material Reviews*, 44:243–258.
17. Ziegler Z (1965) Some extremum principles in irreversible thermodynamics with applications to continuum mechanics. In: Progress in Solid Mechanics, Volume 4, (Eds.) I.N. Sneddon, R. Hill, John Wiley, New York; pp. 91–193.
18. Kobayashi S, Oh S-I, Altan T, (1989) Metal forming and the finite-element method, Oxford and New York: Oxford University Press, pp. 1–41.
19. Oh S-I, (1982) Finite element analysis of metal forming processes with arbitrarily shaped dies, *International Journal of Mechanical Sciences*, 24:479–493.
20. Sivaprasad P.V, Venugopal S, Davies C.H.J, Prasad Y.V.R.K, (2004) Identification of optimum processing parameters for hot extrusion using finite element simulation and processing maps, Modell. Sim. Mater. Sci. Eng., 12:285–291.
21. Ou H, Lan J, Armstrong C.G, Price M.A, Walloe S.J, Ward M.J, (2006) Reduction in postforging errors for aerofoil forging using finite element simulation and optimization, Modell. Sim. Mater. Sci. Eng., 14:179–193.

# Effect of Ca on Oxidation Resistance of Magnesium Alloys (AZ91)

Shima Paridari, Hassan Saghafian Larijani, and Ghasem Eisaabadi. B

**Abstract**

Since magnesium has very low density, it is always attractive in aerospace and transportation industries. However, due to high reactivity with environment, melting, alloying and casting of magnesium is always faced with problems about melt quality. So these alloys have low resistance to oxidation and combustion. Recent studies show that addition of alkali earth metals reduces the oxidation tendency of molten magnesium alloys. This study aimed to investigate the effect of calcium on oxidation resistance of magnesium alloys melt through microstructural studies, XRD and melt oxidation test. Results show that addition of calcium forms intermetallic compounds such as $AL_2Ca$. Formation of these type of compounds in grain boundaries with surface oxide film of CaO, prevents the melt oxidation and minimizes the oxidation during atmosphere exposure. The most important advantage of these alloys is minimizing the use of shielding gases like $SF_6$ which is considered a greenhouse gas that causes many environmental problems.

**Keywords**

Magnesium alloys • Oxidation • Calcium $AL_2Ca$ phase

S. Paridari (✉) · H. S. Larijani
Iran University of Science & Technology, Tehran, Iran
e-mail: shimaparidari@yahoo.com

H. S. Larijani
e-mail: saghafian@iust.ac.ir

G. Eisaabadi. B
Arak University, Arak, Iran
e-mail: g-eisaabadi@araku.ac.ir


D. Orlov et al. (eds.), *Magnesium Technology 2018*, The Minerals, Metals & Materials Series, https://doi.org/10.1007/978-3-319-72332-7_46

## Introduction

Magnesium alloys having density of about 1.8 g/cm$^3$, high specific strength and proper castability is considered the lightest engineering material for transportation, aerospace and electronics industries [1–3]. However, since magnesium has a high tendency to react with oxygen, utilization of magnesium alloys faces with serious limitation.

In recent decades, liquid and solid state oxidation of magnesium alloys have been subject of many studies. In temperatures less than 400 °C, oxidation occurs slowly by forming a protective and weak oxide layer of MgO, in temperatures higher than 450 °C, oxidation occurs more quickly and in temperatures higher than 500 °C, in some areas, sparks are locally propagated before their expansion on the whole surface [4, 5]. To prevent the oxidation and combustion of magnesium alloys, melt must be protected during melting and casting. One of the common industrial protection methods is using sulfur from the early stages of melting until the final stages of casting. In this method, using sulfur and creating an $SO_2$ atmosphere reduces oxygen content of the melting environment considerably. Using sulfur is a cost-effective method, but accelerates the corrosion of equipment, and on the other hand, it is very dangerous and carcinogenic in terms of safety and health. Another method for protection of molten magnesium is using $SF_6$ gas. In recent decades, application of $SF_6$ gas has been reported to be successful [6]. Combined gas including fluoride protects the melt and creates a protective layer on the melt surface. When combination of dry air and $SF_6$ is used, the surface of magnesium oxide and magnesium fluorine is formed. The more magnesium fluoride particles connected to magnesium oxide are, the more Pilling-Bedworth ratio will be and, as a result, the layer does not decompose and is used as a melt protector to prevent oxidation. The expensive $SF_6$ gas is considered to be a greenhouse gas which has potential of warming-up of planet atmosphere 24,000 times more than $CO_2$ [7]. Recently studies show that using alkaline earth elements can delay the oxidation of magnesium alloys in both molten and solid states. Elements such as calcium forms a protective, dense and compact film on the surface that prevents oxygen from entering into the bulk of the alloy. Oxide film formed on the surface includes MgO and CaO that increases Pilling-Bedworth ratio (which is $\alpha = 0.64$ in the absence of calcium, as the calcium increases, it increases to more than 1) [8–12].

In this paper, the effect of calcium on oxidation resistance of magnesium alloys has been investigated by performing oxidation tests, microstructural analysis and GI-XRD.

## Materials and Methods

AZ91 alloy (Mg-9 wt%Al-1 wt%Zn) was prepared by melting pure Mg, Al and Zn in an electric resistance furnace and crucible made from carbon steel. The alloy was kept in constant temperature of 700 °C. After complete melting, Ca was added to the melt in amounts of 0.5, 1.5 and 2%wt. The melt was kept at 700 °C for 30 min in order to complete the reaction between the alloy and added Ca. After completion of the reaction between added Ca and molten alloy and careful casting of the melt, samples (in diameter of 50 mm and height of 30 mm) were poured in a steel mold (pre-heated up to 200 °C) for each melt. These samples were used to evaluate the oxidation resistance of the alloy in liquid state. Later, these samples were heated up to 700 °C and were kept in this temperature for 25 min. Due to oxidation of magnesium alloy during melting time and subsequent variation of the temperature, a K-type thermocouple was use for temperature measurement.

Microstructural studies were carried out on the $10 \times 10 \times 10$ mm$^3$ cubes that were machined from each oxidation resistance test samples. These samples were machined in a manner they could represent the oxidation surface of the corresponding oxidation sample (Fig. 1). These cubes were carefully grinded and polished and etched with Acetic-Picral. The microstructure of the samples was studied by SEM (TESCAN) and GI-XRD (Phillips PW3830) at 1° intervals.

## Result and Discussion

Figure 2 shows the surface of casting samples without protection. As you can see, the surface of AZ91 sample contains burned oxides and black and opaque surfaces. While in the samples containing calcium, surface of the samples is free of any burns or oxides.

Figure 3 shows the surface of samples AZ91, AZ91-0.5% Ca, AZ91-1.5% Ca and AZ91-2% Ca after an oxidation test at 700 °C and duration of 20 min. After a short time, the surface of AZ91 sample began to oxidize until complete burning of the sample. (a) In the sample containing 0.5 wt% of calcium, oxidation resistance is low, and this amount of calcium is not responsive to oxidation over a long period of time and after 2–3 min, the melt surface begins to oxidize and spark (b).

By addition of calcium up to 2%, oxidation resistance of the samples was increasing, in a way that in the sample containing 2% calcium, burned oxides are not visible one the sample surface. This is due to the presence of calcium in form of CaO in surface oxide film. At temperatures above 500 °C, the weak surface oxide layer of MgO in AZ91 alloy cannot act as a melt protector, and oxide film breaks down and oxygen penetrates easily and accelerates the oxidation process. And, as shown in Fig. 3a, the surface contains burned oxides and agglomerates. However, by addition of calcium, the pilling-bedworth ratio is increased to more than 1 and the melt protection action is performed by a CaO + MgO surface oxide film.

Figure 4 shows the oxidation diagram at 700 °C and duration of 20 min. As you can see, AZ91 alloy oxidizes quickly when it melts at normal atmosphere and the sharp gradient seen in AZ91 diagram reflects this. If different amounts of calcium are added to the alloy, oxidation resistance increases and gradient increase in the diagrams of alloys containing calcium is much slower than that of pure sample or even containing little amount of beryllium.

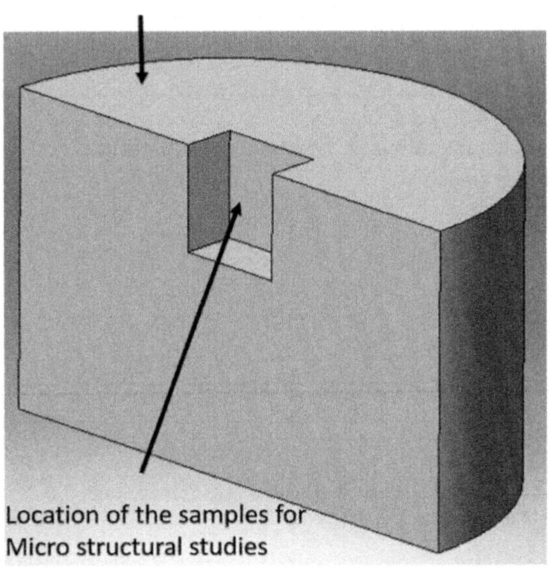


**Fig. 1** Surface of the sample exposed to atmosphere during the test

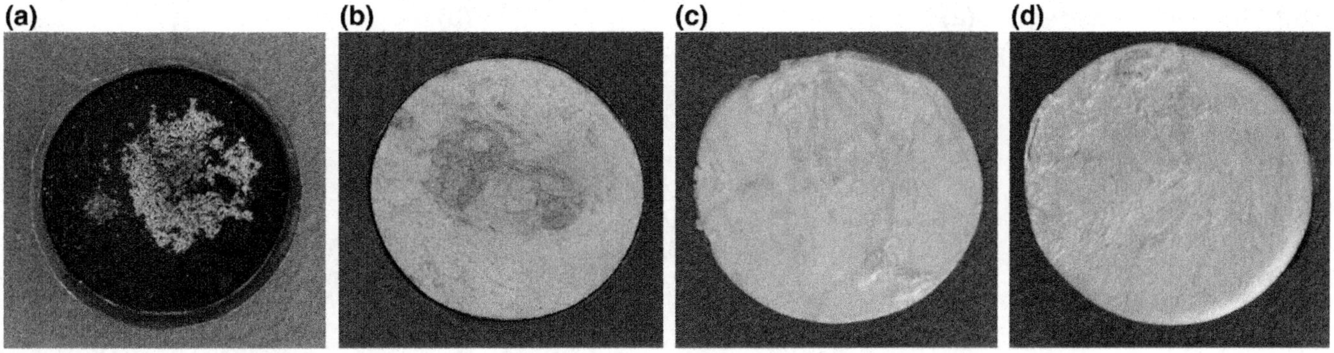


**Fig. 2** Surface of the casted samples (**a**) AZ91 (**b**) AZ91+0.5%Ca (**c**) AZ91+1.5%Ca (**d**) AZ91+2%Ca without protective atmosphere

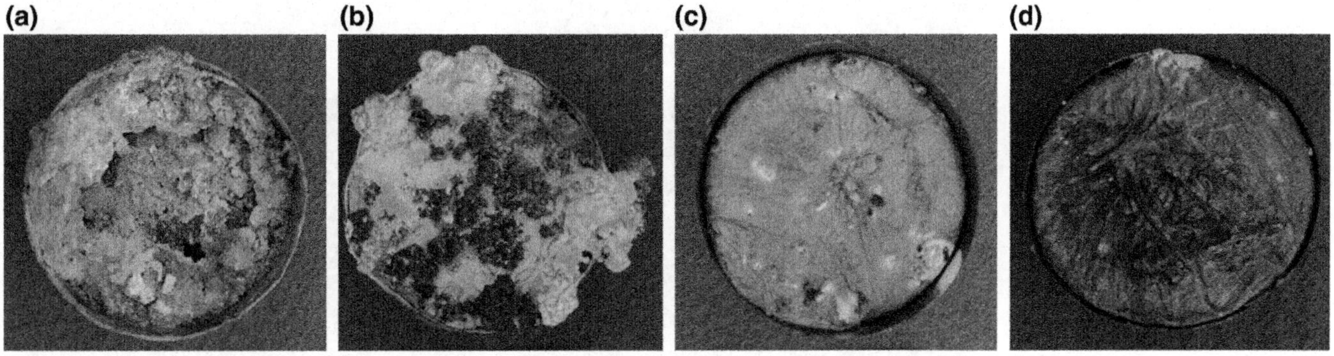


**Fig. 3** Exterior surface of the samples (**a**) AZ91 (**b**) AZ91+0.5%Ca (**c**) AZ91+1.5%Ca (**d**) AZ91+2%Ca kept at 700 °C and duration of 20 min

Figure 5 shows SEM images of AZ91, AZ91-0.5%Ca, AZ91-1.5%Ca and AZ91-2%Ca samples. In grain boundaries of AZ91 alloy (a) the $Mg_{17}Al_{12}$ phase forms which is a crisp and brittle intermetallic compound. Also, this phase has a low melting temperature, which is why it shows a low resistance to temperature rise. By addition of calcium, strong intermetallic compound of $Al_2Ca$ or $Mg_2Ca$ with high melting temperature forms in surface boundaries or internal grain boundaries which prevent the growth and formation of $Mg_{17}Al_{12}$ phase. The presence of $Al_2Ca$ or $Mg_2Ca$ phases leads to increased oxidation resistance. These phases in Fig. 5d, with bright colors are clearly distinguished from the background Mg-$\alpha$ phase (marked with C) and the other phases.

Figure 6 shows the results of X-ray test at a low angle of 1 (GI-XRD). From this image, it can be concluded that at the surface of AZ91 allow, the melt containing calcium, surface oxides of MgO and CaO forms. The formation of these phases together leads to formation of a dense, oxidation-resistant protective film.

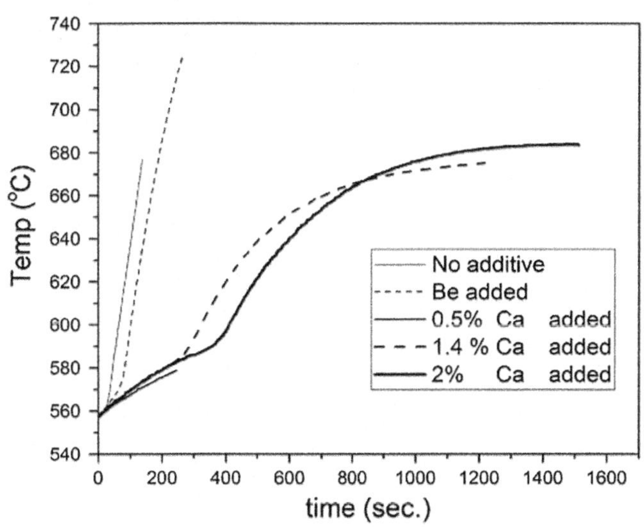


**Fig. 4** Oxidation diagram in molten magnesium alloys

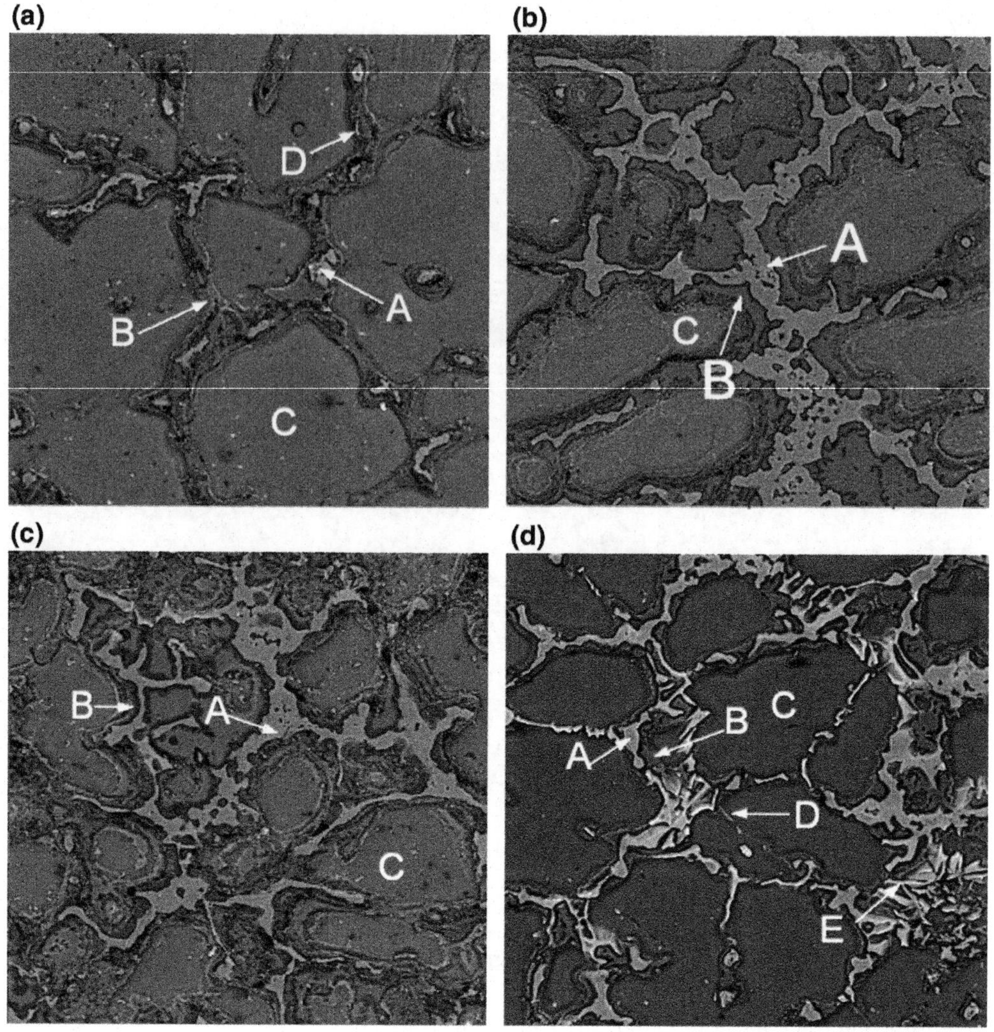


**Fig. 5** SEM images of (**a**) AZ91 (**b**) AZ91+0.5%Ca (**c**) AZ91 +1.5%Ca (**d**) AZ91+2%Ca

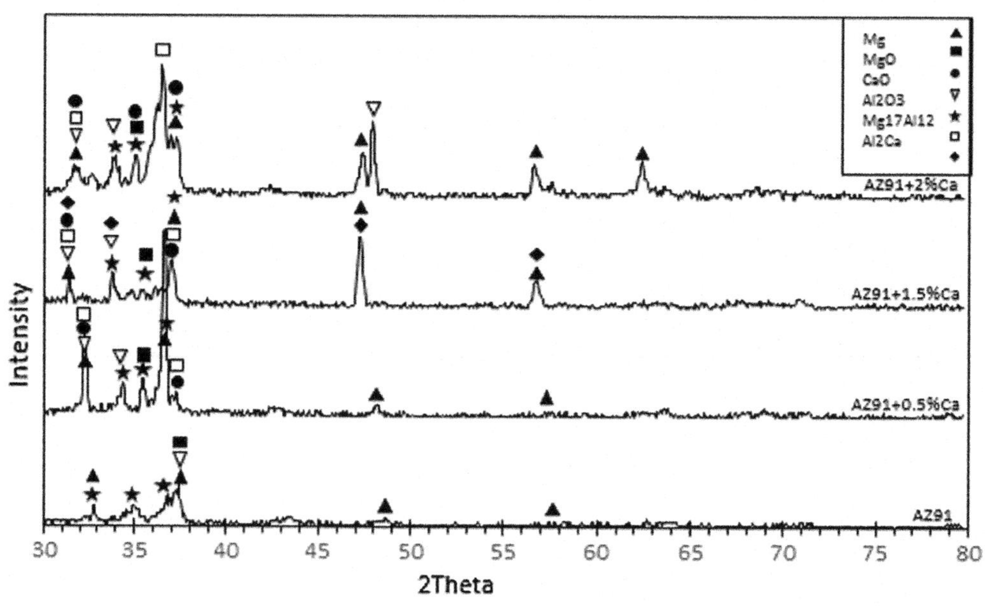


**Fig. 6** Results of X-ray diffraction pattern of the samples

## Conclusion

The purpose of this paper is to improve the oxidation resistance of magnesium alloys in the presence of calcium. By adding calcium with an optimum amount of 2%, magnesium melt can be stored for up to 40 min at 700 °C without the need for a protective atmosphere. While the magnesium alloys oxidize rapidly in less than few minutes below their melting point in the normal state and without the presence of protective gas, and then they burn. The action of protecting the melt by calcium is accompanied by the formation of a dense and resistant CaO-oxide film in the surface, which prevents the penetration of oxygen into the melt and, as a result, oxidation action is postponed. The most important advantage of these alloys is not applying protective gases like $SF_6$ which is considered a greenhouse gas and causes many environmental problems.

## References


1. Materials and Manufacturing Engineering, vol. 24, pp. 99–102.
2. B.C., H.N., 2004, "Automotive Applications of Magnesium and Its Alloys". Trans. Indian Inst. Met., vol. 57, pp. 397–408.
3. Shuang-Shou Li, Bin Tang and Da-Ben Zeng, 2006, "Effects and mechanism of Ca on refinement of AZ91D alloy". Journal of Alloys and Compounds, pp. 317–321.
4. Dong-Bok Lee, Lee-Seok Hong and Young-Jig Kim, 2008, "Effect of Ca and CaO on the High Temperature Oxidation of AZ91D Mg Alloys". Materials Transactions, Vol. 49, No. 5, pp. 1084–1088.
5. S. kim, 2014, "Magnesium based alloy for high temperature and manufacturing of them". U.S patent.
6. Ha Seong-Ho, LEE Jin-Kyu and JUNG Seong-Boo, 2006, "Behavior of CaO and Calcium in pure Magnesium". Rare Metals, vol 25, pp. 150.
7. M. Holtzer and. A. Bobrowski, 2008, "Magnesium melt protection by covering gas". Archives of Foundry Engineering, Vol 8, pp. 131–136.
8. T. W. Lee, 2015, "Microstructural evaluation of oxide layers in CaO-added Mg alloys". Journal of Alloys and Compounds, pp. 5–10.
9. B. H. Choi, B. S. You and I. M. Park, 2006, "Characterization of Protective Oxide Layers Formed on Molten AZ91 Alloy Containing Ca and Be". Metals and Materials International, vol. 12. pp. 63–67.
10. J. Hwan, 2013, "Effect of CaO Addition on Microstructure and Damping Capacity of AM50 Magnesium Alloy". Materials Transactions.
11. S. K. Kim, 2011, "Design and development of high performance Mg alloys". Design, Processing and Properties, no. 978-953-307-520-4.
12. Y. Chen, 2013, "Effects of the addition of Ca and Sb on the microstructure and mechanical properties of AZ91 magnesium". Materials Science & Engineering, pp. 262–267.

# Evolution of Microstructure and Mechanical Properties During Casting and Rolling of the ZEK100 Sheet

A. Javaid and F. Czerwinski


## Abstract

The use of magnesium sheet in automotive light-weighting initiatives is low due to its poor formability at room temperature. In this investigation the cast ZEK100 (Mg-1.2Zn-0.35Zr-0.17Nd, in wt%) alloy was subjected to rolling at temperatures from 350 to 450 °C and for each rolling temperature the evolution of microstructure and tensile properties were assessed. To assess strengthening during rolling, properties were also measured after annealing following each rolling. The results show that increasing the rolling temperature caused reduction of tensile strength, yield stress but increasing elongation. A correlation was established between mechanical properties and grain size influenced by the rolling temperature. The results are discussed in terms of the role the rare earth elements play in controlling the formability of magnesium alloys.



## Keywords

Magnesium alloys • Solidification • Microstructure
Rare earths • Sheet rolling


## Introduction

Advanced magnesium sheet technology offers great opportunities for automobile designers to significantly reduce the vehicle weight. However, the use of magnesium sheet in automotive applications is very limited due to its low formability at room temperature caused by poor plastic flow properties. This is because magnesium has a hexagonal close packed (HCP) structure and active slip systems at low temperatures are mainly limited to those involving basal planes [1]. As a result, the critical resolved shear stress (CRSS) for basal plane slip in magnesium single crystal is about two orders of magnitude lower than that for non-basal plane slip involving prismatic or pyramidal planes near room temperature [1]. Thus, the distribution of basal planes (0001) in magnesium plays an important role in determining formability at low temperatures. As the temperature increases, the CRSS of the non-basal slip systems decreases and, therefore, at about 220–250 °C and above, there is a significant increase in formability [1]. Sheet forming at elevated temperatures is one possibility that is being contemplated, and there is considerable research being done in this general area, including experiments concerned with the viability of superplastic forming [1].

Magnesium-rare earth alloys are increasingly investigated due to promising results of weakening the basal texture and improving the formability of magnesium. Of the elements that are known to weaken the basal texture of magnesium, all have large atomic radii that give them this different behaviour [2]. A wide range of rare earth elements are predicted to segregate strongly to grain boundaries due to the large atomic size misfit with magnesium [3]. It was proposed [2] that a strong interaction of solutes with dislocations and grain boundaries is responsible for the significant impact rare earth additions have on the extruded grain size and texture at very low alloying levels. Previous experimental work [3] has identified segregation of rare earth elements to grain boundaries and dislocations as being potentially important in producing this change in behavior. Automotive light-weighting applications of magnesium depend much on the cost of alloys, which is largely determined by the cost of the alloying elements. One of the main limitations in using rare earth elements is their high cost. The remaining elements with large atomic radii outside the rare earth element group that are likely to be effective texture modifiers are Ca, Sr and Ba with Ca at the center of interest because of low density,




A. Javaid (✉) · F. Czerwinski
CanmetMATERIALS, Natural Resources Canada, 183 Longwood Rd South, Hamilton, ON L8P 0A5, Canada
e-mail: Amjad.javaid@Canada.ca





D. Orlov et al. (eds.), *Magnesium Technology 2018*, The Minerals, Metals & Materials Series, https://doi.org/10.1007/978-3-319-72332-7_47

relatively low cost and abundant supply. In addition, Ca also weakens the strong basal texture and refines the grain size of Mg–Zn–Ca alloys in the as-cast condition [4]. It was found [5] that simultaneous alloying with Ca and Ce enhances precipitation hardening, refines grain size and improves texture and ductility.

The most promising magnesium sheet alloys contain small amount of rare earth elements. Recently, considerable research has been conducted towards understanding the texture development in wrought magnesium alloys because weakened textures are known to improve the room temperature formability. The conventional wrought magnesium alloys currently used for sheet manufacturing, such as AZ31, have relatively poor formability at room temperature due to their strong basal texture. In particular, research by NSERC Magnesium Network (MagNET) has shown that the commercial magnesium alloy ZEK100 (Mg-1.2Zn-0.35Zr-0.17Nd, in wt%) containing a small amount of rare earth element neodymium exhibits a very different texture compared to the strong basal texture of conventional alloys such as the AZ31 grade. The ZEK100 chemistry translates to improvements in elevated temperature formability [6, 7]. Using this alloy, automobile parts were successfully formed at elevated temperature at industrially acceptable forming rates and temperatures. In contrast, an annealed AZ31 alloy could not be formed through conventional rolling within technically viable experimental settings [7].

This paper describes activities at CanmetMATERIALS aimed at manufacturing magnesium sheet. The experiments were focused on assessing the role of temperature on rolling behaviour of the ZEK100 alloy containing additions of rare earths.

## Experimental Procedure

### Alloy Melting and Plate Casting

The material of this investigation was commercial magnesium alloy ZEK100 (Mg-1.2Zn-0.35Zr-0.17Nd, in wt%). The 60 kg melt was prepared in a resistance furnace using a mild steel crucible under a protection of $CO_2$ + 0.5% $SF_6$ gas mixture. No grain refiner was added to the melted alloy that already contained 0.35% Zr. After verifying the chemical composition and removal of any dross/oxide films, 25 mm thick plate castings were poured at 730 °C in a water-cooled copper mold with characteristics shown in Fig. 1.

### Sheet Rolling and Its Heat Treatment

The hot rolling of the ZEK100 plate to the 1.65 mm thick sheet was done at three different temperatures using a pilot-scale rolling mill (Fig. 2). The aim was to optimize the hot rolling schedule for the ZEK100 alloy in order to produce sheet with a fine and homogeneous microstructure. The cast plates were heated to 400 °C under argon and kept at that temperature for 10 h before rolling. Three different hot rolling temperatures of 350, 400 and 450 °C were examined at one rolling speed of 30 rpm. The rolls were preheated to 100 °C. Initial dimensions of the as-cast ZEK100 plate section used for rolling were 25 × 25 × 160 mm. The 25 mm thick strips were rolled to 1.65 mm sheet using a single stand reversing mill. Before hot rolling, the sliced plates were soaked for 1 h at the hot rolling temperature to

(a) Water-cooled copper mold (b) Plate casting (25.4 x 165.1x 393.7 mm)

**Fig. 1** Experimental setup of water-cooled copper mold (**a**) used for casting magnesium plates (**b**) being precursors for sheet rolling

*Stanat Rolling Mill Specifications:*

| | |
|---|---|
| Maximum Roller Temperature: | 200°C |
| Roller Diameter: | 156.3 mm (top roll) / 156.9 mm (bottom roll) |
| Roller Length: | 203.2 mm |
| Motor Rating: | 15 hp with a maximum speed of 1.775 rpm |
| Maximum Load: | 50-ton |
| Maximum Torque: | 19,063 N-m (168.720 lbf-in or 14,060 lbf-ft) |

**Fig. 2** Experimental set-up of laboratory rolling mill used for manufacturing of magnesium sheet

ensure the proper through heating. The entry temperature was kept constant, by reheating the plate between passes. Typically, a thickness reduction of 10–15% per pass was obtained.

Usually, commercial ZEK100 sheet is produced in H24 temper to obtain optimum combination of strength and ductility, which are intermediate between those of the fully strain hardened and the fully annealed conditions. For severe forming, sheet in the annealed, O temper, condition is preferred. Since most of the forming operations on the wrought magnesium alloys are done at elevated temperature, the need for fully annealed condition is less than that for other metals. After hot rolling, a flash annealing at 450 °C for 5 min was conducted with an objective of texture weakening and refining microstructure.

## Property and Microstructure Characterization

The standard tensile tests were carried out on as-cast samples machined from 20 mm diameter cast bars made in pre-heated permanent mold. To test the rolled sheet, the sub-sized tensile samples were extracted from the as-rolled and annealed state. Each value of strength and elongation is the average of three specimens tested. Tensile tests were performed at room temperature and a strain rate of $5 \times 10^{-3}\ s^{-1}$.

To assess microstructure, samples were subjected to conventional metallographic preparation from grinding to polishing and then chemically etched using 10% HF or acetic glycol to reveal the dendritic structure or grain size, respectively. The optical microscope, scanning electron microscope (SEM) along with EDX analyzer were used for microstructural and micro-chemical characterizations of phases present in the ZEK100 alloy.

## Results and Discussion

### Alloy Phase Composition

The FACTsage software allows predicting the alloy phase composition formed under equilibrium and non-equilibrium (Scheil) cooling conditions. Figure 3 shows the calculated phase diagrams with predicted equilibrium phase compositions of the ZEK100 alloy. The red line corresponds to the alloy composition. Under equilibrium solidification conditions the FACTsage predicted for the ZEK100 alloy the phases of α-Mg, α-Zr, $Mg_{12}Zn_{13}$ and $Nd_xMg_y$. For the particular chemistry of the ZEK100 alloy examined here, the calculated phase composition consists of 98.62% α-Mg, 0.25% α-Zr, 0.90% $Mg_{12}Zn_{13}$ and 0.23% $Nd_xMg_y$. Changing conditions to non-equilibrium (Scheil) solidification led to changes in both the amount and nature of phases,

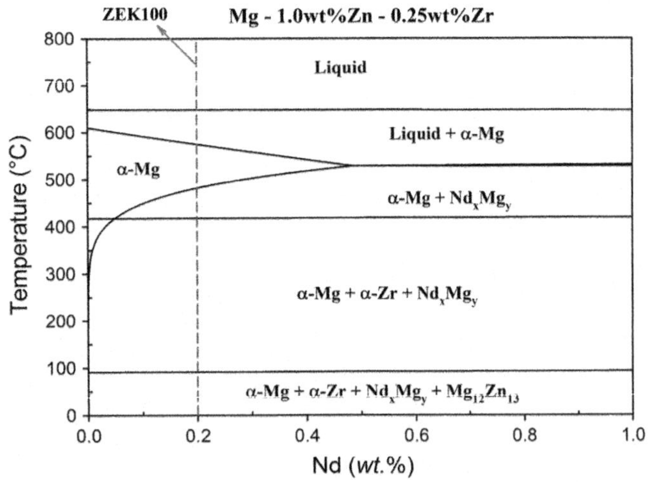


**Fig. 3** Phase diagram showing the effect of Nd on phase formation during equilibrium solidification of the ZEK100 (Mg-1.2Zn-0.17Nd-0.35Zr) alloy

specifically the precipitates. The calculated composition of the ZEK100 alloy for non-equilibrium (Scheil) solidification consists of 98.78% α-Mg+0.69% $Mg_{12}Zn_{13}$+0.42% NdMgZn+0.11% $Mg_{51}Zn_{20}$.

In general, rare earth elements have very low solubility in magnesium and usually form there intermetallic compounds. Among the rare earth elements, neodymium has the highest solid solubility in magnesium and for this reason shows best response to age hardening [8, 9]. Some authors [9] have attributed the presence of $Mg_3Nd$ hard phase to better strength of neodymium containing magnesium alloys as compared to those containing cerium and lanthanum. Others reported [8] that the microstructure of Mg–Nd alloys consists of dendritic α-Mg and divorced eutectic $Mg_{12}Nd$. The improvement in strength with increase in neodymium content was attributed to both solid solution hardening and precipitation hardening [8]. As a major alloying element in combination with Zr and/or rare earth elements, Zn is used in the ZEK100 alloy. It has very low solid solubility in magnesium and forms many intermetallic compounds causing strengthening through precipitation hardening. Zn improves the strength, ductility and castability of magnesium alloys [10, 11]. Zirconium has a powerful grain-refining effect on magnesium alloys and is usually added to alloys containing zinc, rare earths, or a combination of these elements, where it serves as a grain refiner up to its limit of solid solubility [10, 11]. Good formability is achieved by fine grain size and a uniform dispersion of secondary phases.

## Microstructural Characteristics

To show the microstructure evolution, metallographic samples were sectioned from the 75 mm thick ZEK100 ingot used for melting, 25 mm thick plate castings and 1.65 mm thick rolled sheet. As depicted in Fig. 4 at the same magnification, processing affected essentially the alloy grain size. The grain size of the 75 mm thick ingot (Fig. 4a) experienced multiply reduction after casting into the 25 mm thick plate used as a precursor for sheet rolling (Fig. 4b). The grain size reduction in as-cast state is explained through an increased solidification rate observed in the 25 mm plate and a presence of the grain refiner. In case of the ZEK100 alloy, the presence of zirconium grain refiner is responsible for preventing grain coarsening. The sheet rolling process let to further substantial reduction in the alloy grain size (Fig. 4c). In this case, grain refinement was achieved due to thermo-mechanical processing that led to an increase in the overall mechanical properties of the magnesium sheet.

The general microstructure of the ZEK100 alloy in as-cast state consisted of the α-Mg dendritic matrix and intermetallic precipitates, forming a network between inter-dendritic regions. Some individual precipitates or their clusters were also present inside dendritic cells. In the ZEK100 alloy, precipitates form a chain-like morphology of near-globular individual particles in the inter-dendritic regions. The precipitates, located at triple junctions of dendritic cells, were generally larger, frequently with irregular shapes.

The alloy grain size was influenced by the rolling temperature with increasing temperature in the range from 350 to 450 °C causing grain coarsening and the highest growth seen for the temperature range of 400–450 °C (Fig. 5, left side). To assess the thermal stability of the as rolled microstructure the alloy was subjected to post rolling annealing at 450 °C, corresponding to the highest rolling temperature. As shown in right side of Fig. 5, for rolling temperatures of 350 and 400 °C, annealing introduced microstructural changes. For rolling temperature of 450 °C no evident modifications were observed at magnifications of optical microscope.

## Alloy Mechanical Properties

Table 1 shows the tensile properties of as-cast ZEK100 alloy measured at room temperature. The samples for tensile testing were machined from 20 mm diameter cast bars made in a pre-heated permanent mold. There is some variation in strength and elongation of as-cast alloy with tensile strength scatter between 192 and 162 MPa. A larger scatter was seen for elongation values, which for quite high average of 13% oscillated from 9 to 16%.

The rolling process led to an increase in both the alloy strength and elongation as compared to the as-cast-state. Each value of strength and elongation in Table 2 is the average of three specimens tested. As shown in Table 2, the

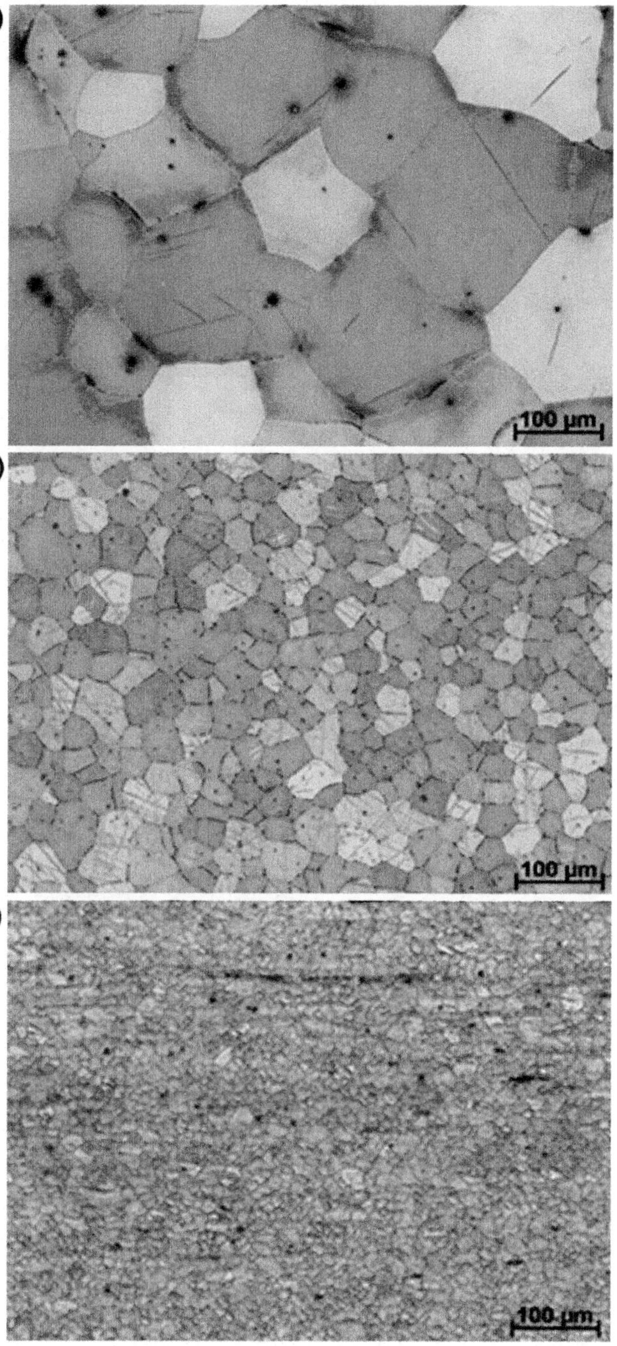


**Fig. 4** Microstructure of the ZEK100 alloy in the form of 75 mm thick ZEK100 ingot used for melting (**a**), 25 mm thick plate castings used for rolling (**b**) and 1.65 mm thick rolled sheet (**c**). All images are shown at the same magnification

as-rolled tensile strength increased to the range of 228–257 MPa and elongations from 17 to 21%. The largest increase was achieved for yield stress which values reached from 185 to 237 MPa. As anticipated, the alloy tensile properties after rolling were highly influenced by the rolling temperature and the largest gains were recorded for lower rolling temperatures (Fig. 6). According to the literature, in addition to weakening the basal texture and improving formability, the advantages of rare earth containing magnesium alloys are increasing strength, creep resistance, corrosion and flammability resistance [10, 11]. The presence of rare earth elements in magnesium alloys also improves castability, grain refining and age hardening [10, 11]. Neodymium is a beneficial alloying addition in the ZEK100 magnesium alloy as it reduces intensity of the basal texture while increasing strength and ductility [8]. Similarly, magnesium when micro-alloyed with small amount of rare-earth element cerium weakens the basal texture of Mg–Zn alloys resulting in improved room temperature formability [12, 13]. Furthermore, the formation of highly stable eutectic phases by the addition of rare earth elements act as effective strengtheners in magnesium alloys [9].

## Microstructure-Property Correlation

To examine the microstructure influence on properties as a parameter describing microstructure, grain size was selected. It is seen from presented earlier micrographs that the higher rolling temperature also increased the grain size what is shown quantitatively in Fig. 7. After conversion from the rolling temperature, the full correlation of alloy grain size with tensile properties is shown in Fig. 8 where a reduction in grain size is accompanied in an increase of both the tensile strength and yield strength. At the same time, a reduction in alloy grain size is accompanied by a reduction in elongation. This observation contradicts trends recorded for other materials. As a possible explanation, it was reported [14] that with increasing extrusion temperature, the yield strength gradually decreased with an increase in DRXed grain size according to the Hall-Petch relation, while the elongation increased due to the decreased unDRX fraction, suppressing crack initiation at twins in coarse unDRXed grains, and weakened basal texture.

According to [14, 15], reported for the Mg-1.58Zn-0.52Gd and Mg-2Zn-0.5Ce, Mg-1Zn-1Mn-0.5Ce, Mg-2Zn-1.5Gd alloys, the grain size of the extruded alloys increased with increasing extrusion temperatures due to grain growth during hot deformation and which can be attributing to the dynamic recrystallization. The occurrence of recrystallization during the thermo-mechanical processing of magnesium alloys is a fundamental process that influences directly the microstructure and mechanical properties of semi-finished products. Thermomechanical processing can lead to considerable grain refinement by recrystallization. The grain refinement, achieved as a result of thermo-mechanical processing, leads to an increase in strength and improves ductility. In fact, the elongation of pure magnesium increases exponentially as its grain size decreases [16]. In many materials, this has been attributed to the reduction of microscopic stress concentration,

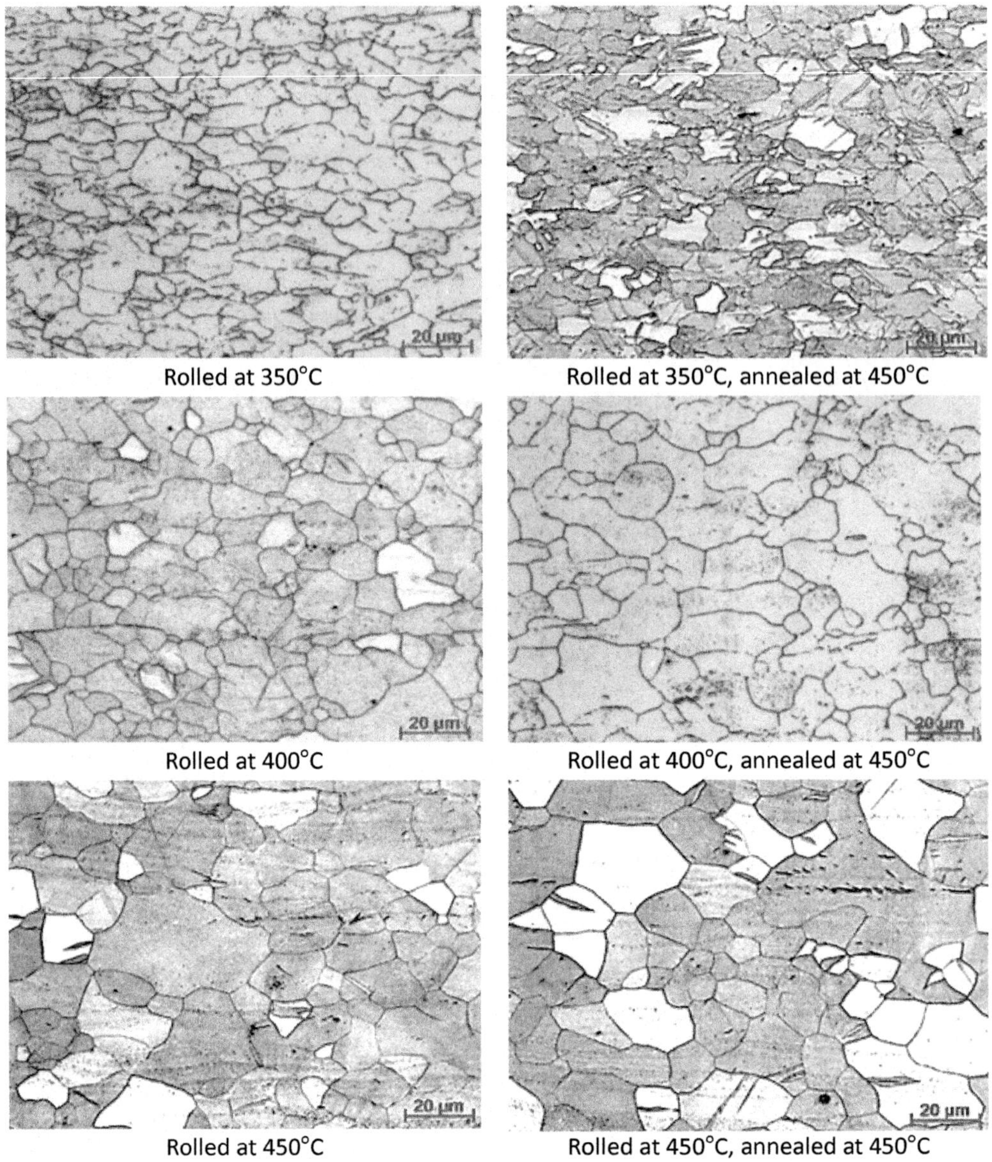


**Fig. 5** Microstructure of the rolled and annealed sheets of the ZEK100 alloy showing an influence of rolling temperature on alloy grain size. All photomicrographs are displayed at the same magnification

**Table 1** Room temperature tensile properties of as-cast ZEK100 alloy

| Sample | Condition | UTS (MPa) | YS (MPa) | Elongation (%) |
|---|---|---|---|---|
| 1 | As-cast | 192 | 59 | 16 |
| 2 | As-cast | 165 | 54 | 13 |
| 3 | As-cast | 162 | 74 | 9 |
| 4 | As-cast | 184 | 63 | 15 |
| Average | As-cast | 176 | 63 | 13 |

**Table 2** Room temperature tensile properties of rolled and annealed ZEK100 sheet

| Experiment | Condition | UTS (MPa) | YS (MPa) | Elongation (%) |
|---|---|---|---|---|
| 1 | Rolled at 350 °C | 257 | 237 | 17 |
| 2 | Rolled at 350 °C, annealed at 450 °C | 254 | 211 | 15 |
| 3 | Rolled at 400 °C | 231 | 169 | 23 |
| 4 | Rolled at 400 °C, annealed at 450 °C | 226 | 154 | 19 |
| 5 | Rolled at 450 °C | 228 | 185 | 21 |
| 6 | Rolled at 450 °C, annealed at 450 °C | 207 | 116 | 22 |

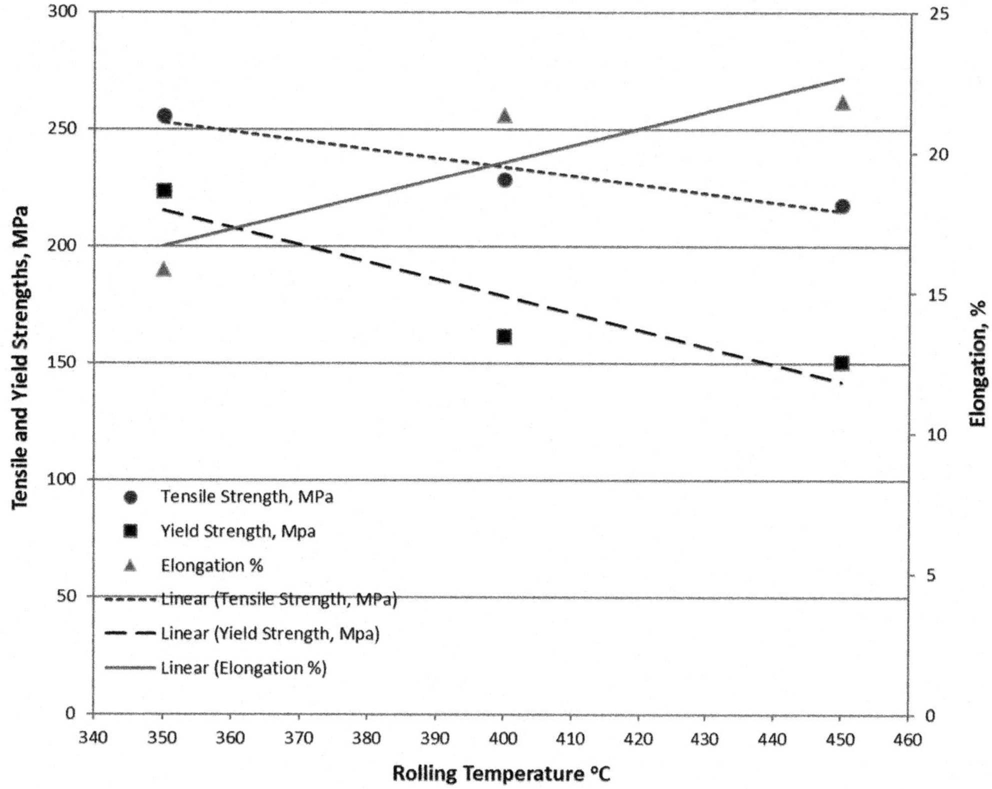


**Fig. 6** Effect of rolling temperature on tensile strength, yield stress and elongation of the ZEK100 alloy

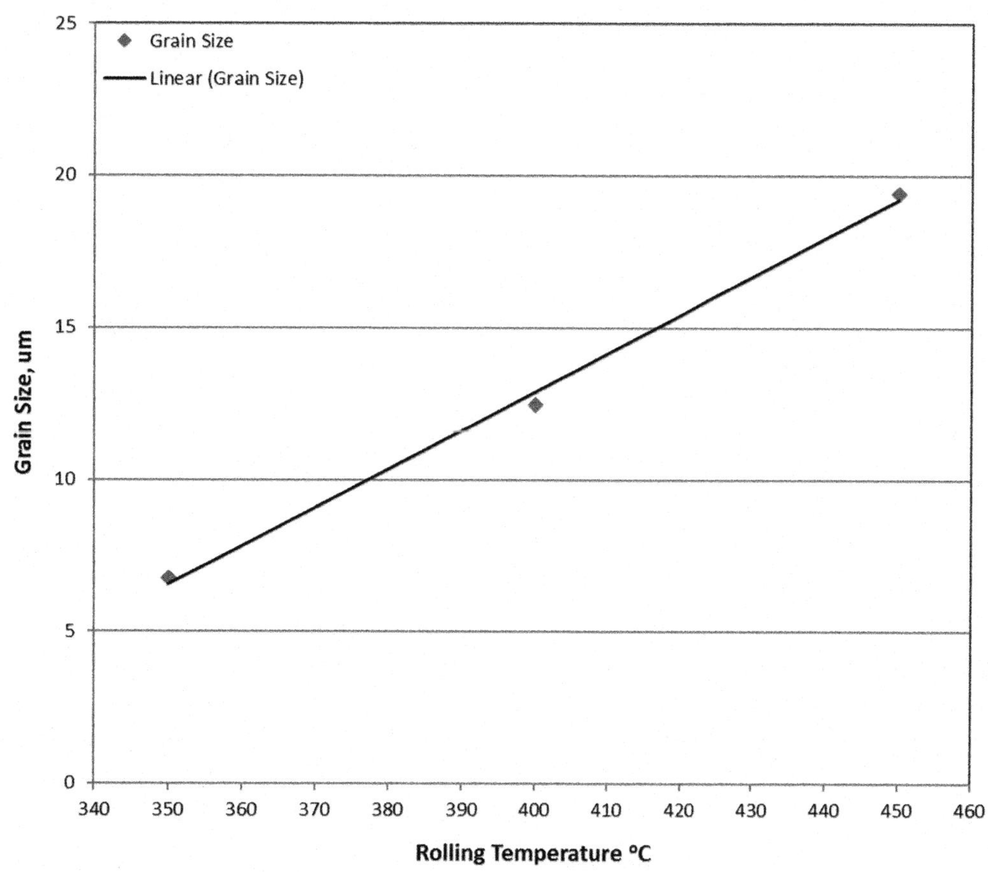


**Fig. 7** Effect of rolling temperature on grain size of the ZEK100 alloy

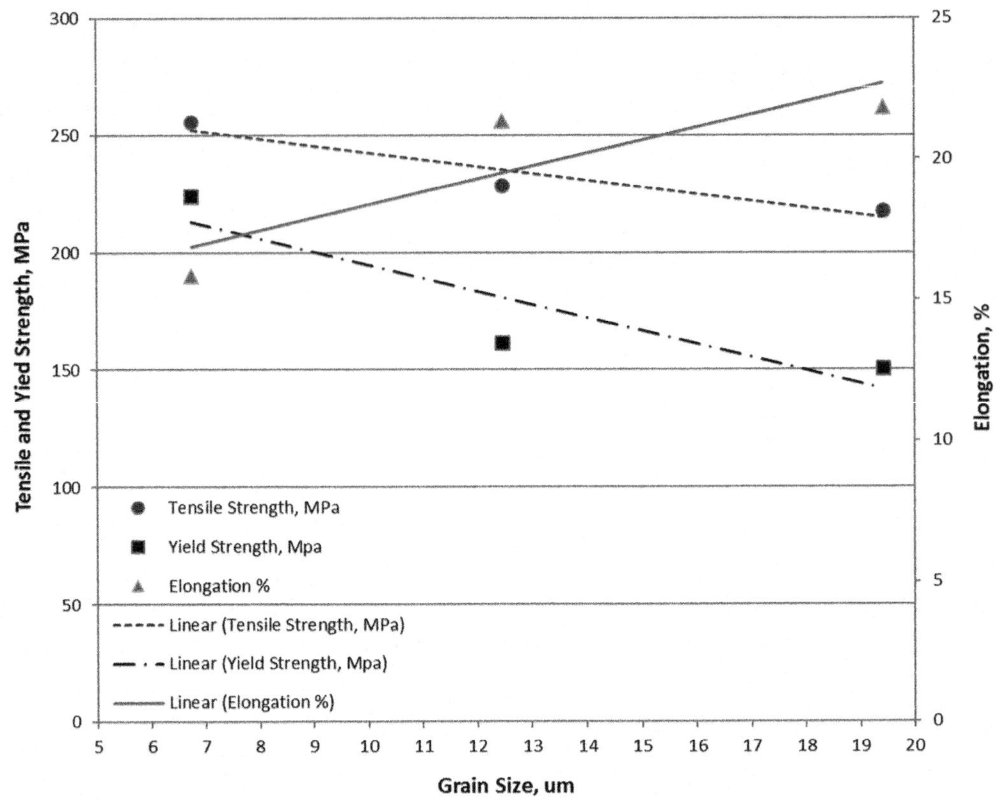


**Fig. 8** A correlation between grain size and tensile strength, yield stress and elongation of the ZEK100 rolled sheet

which is responsible for micro-crack nucleation [17]. In other alloy systems, grain refinement can be brought about by recrystallization due to hot working and solid phase transformation [1]. Unfortunately, magnesium alloys do not undergo any phase transformation during processing, thus recrystallization during deformation (dynamic recrystallization, or DRX) and during the period of deformation interruption (static recrystallization, or SRX) at elevated temperatures is essential for refining grain structure of magnesium wrought alloys [1]. The literature observations reported above coincide with our results recorded for the ZEK100 alloy.

The alloy grain size was influenced by the rolling temperature with increasing temperature in the range from 350 to 450 °C causing grain coarsening with the highest growth seen for the temperature range of 400–450 °C.

The correlation of alloy grain size after rolling with tensile properties was revealed where a reduction in grain size was accompanied in an increase of both the tensile strength and yield stress. At the same time, a reduction in alloy grain size was accompanied by a reduction in elongation. The latter observation contradicts trends recorded for other materials and the possible explanation of this finding was provided.

## Conclusions

The ZEK100 alloy was hot rolled at temperatures from 350 to 450 °C and the effect of rolling temperature on the microstructure and mechanical properties was investigated.

The phases present, determined by metallography and EDX, were verified by calculations using the FACTsage software. For the specific chemistry of the ZEK100 alloy, the calculated phase composition consisted of 98.62% α-Mg, 0.25% α-Zr, 0.90% $Mg_{12}Zn_{13}$ and 0.23% $Nd_xMg_y$.

## References


1. F. Czerwinski: Magnesium Alloys - Design, Processing and Properties, Magnesium Sheet; Challenges and Opportunities, InTech, January 14, 2011, p. 297.
2. N. Stanford, "Micro-alloying Mg with Y, Ce, Gd and La for texture modification-A comparative study" Materials Science Engineering A 527 (2010) 2669–2677.
3. Joseph D. Robson," Effect of Rare-Earth Additions on the Texture of Wrought Magnesium Alloys: The Role of Grain Boundary Segregation", Metallurgical and Materials Transactions A, Volume 45A, July 2014, pp. 3205–3212.

4. A. Zilberov et al," High Temperature Behaviour of a New Mg-Ca-Zn Alloy", Proceedings of the 6th International Conference on Magnesium Alloys and Their Applications, 2006, pp. 134–140.
5. B. Langelier, P. Donnadieu, and S. Esmaeili, "Characterization of Precipitation in Mg-Zn-Ce-(Ca) Alloys", Mg2012: 9th International Conference on Magnesium Alloys and their Applications, July 2012, pp. 485–491.
6. M. Boba, M. J. Worswick, and R. Mishra, "Formability of AZ31B and ZEK100 Magnesium Alloy Sheets," Proceedings of 9th International conference on magnesium alloys and their applications, Vancouver, Canada, 2012.
7. N. Panahi, M. Worswick, and T. Skszek, "Determination of warm forming limit diagram for ZEK100," in NUMISHEET 2014: The 9th International Conference and Workshop on Numerical Simulation of 3D Sheet Metal Forming Processes: Part A Benchmark Problems and Results and Part B General Papers, 2013, pp. 366–369.
8. Yan, J.; Sun, Y.; Xue, F.; Xue, S.; Tao, W. Microstructure and mechanical properties in cast magnesium–neodymium binary alloys. Mater. Sci. Eng. A 2008, 476, 366–371.
9. Chia, T.L.; Easton, M.A.; Zhu, S.M.; Gibson, M.A.; Birbilis, N.; Nie, J.F. The effect of alloy composition on the microstructure and tensile properties of binary Mg-rare earth alloys. Intermetallics 2009, 17, 481–490.
10. Magnesium and Magnesium Alloys, ASM Specialty Handbook, ASM International, September 1999.
11. Magnesium Technology: Metallurgy, Design Data, Applications, edited by Horst E. Friedrich, Barry L. Mordike, 1999.
12. A. A. Luo, R. K. Mishra, and A.K. Sachdev," High-Ductility Magnesium-Zinc-Cerium Extrusion Alloys", Scripta Materialia, 64 (2011), pp. 410–413.
13. R. K. Mishra et al," Influence of Cerium on Texture and Ductility of Magnesium Extrusions, Scripta Materialia, 59 (2008), pp. 562–565.
14. M.G. Jiang et al," Microstructure and Mechanical Properties of an Extruded Mg-1.58Zn-0.52Gd Alloy, Magnesium Technology 2017, p 299.
15. Sung Hyuk Park, Bong Sun You, Raja K. Mishra, Anil K. Sachdev," Effects of extrusion parameters on the microstructure and mechanical properties of Mg–Zn–(Mn)–Ce/Gd alloys" Materials Science & Engineering, A598 (2014) 396–406.
16. E.F. Emley, Principles of Magnesium Technology, Pergamon Press, Oxford, 1966.
17. G.E. Dieter, Mechanical Metallurgy, McGraw-Hill Book Company, New York, 1976.

# Part II

# Magnesium Alloy Development: An LMD Symposium in Honor of Karl Kainer

# Recent Developments in the Application of the Interdependence Model of Grain Formation and Refinement

D. H. StJohn, X. Hu, M. Sun, L. Peng, and H. Dieringa


## Abstract

The Interdependence model will be briefly reviewed and then applied to two different casting situations. One is the solidification of Mg–Al–Sm alloys to determine the optimum composition for achieving a fine as-cast grain size. Because the size range of the nucleant particles can be measured, the key factors describing the potency of the particle can be calculated providing a more complete description of the grain formation mechanisms operating for this alloy. This approach should be relevant for other Mg–Al–RE alloys. The other casting situation is where the melt of an AM60-AlN nanoparticle composite was treated ultrasonically producing a fine grain size on solidification. The limitations to grain size reduction by nanoparticles are discussed in terms of the Interdependence and Free Growth models.



## Keywords

Interdependence model • Nucleation • Magnesium alloys
Solidification • Ultrasonic treatment


## Introduction and the Interdependence Model

To obtain optimum mechanical performance a uniform fine equiaxed grain size is usually required. Equiaxed solidification and refinement of the as-cast grain size can be accomplished by: the addition of solute elements with a high value of the growth restriction factor $Q$*; inoculation by potent nucleant particles with low nucleation undercooling $\Delta T_n$ (i.e. high nucleation potency); a fine distribution of these particles within the melt; and controlling solidification parameters. (* $Q = mC_0(k - 1)$ where $m$ is the slope of the liquidus temperature in a binary phase diagram at a given alloy composition $C_0$, and $k$ is the partition coefficient of the solute element between the solid and liquid phases.)

An example where the above factors play a role is provided by the application of UltraSonic Treatment (UST) of a range of Mg–Al alloys [1]. The results of this study are presented in Fig. 1 where grain size decreases as the value of $Q$ increases and the intensity of UST is increased. In this example, it is not known precisely which compounds act as the nucleants but Fig. 1 shows that as the UST intensity increases the number of nucleant particles that can be activated (related to the y-axis intercept) also increase. Figure 1 highlights the significant effect of alloy composition, nucleant particles and casting environment on the grain size achieved. The grain size data allows a linear relationship between grain size and $1/Q$ to be plotted and this simple linear form assists the determination of the mechanisms affecting the final as-cast grain size. The slope of the $1/Q$ plots indicates the potency of the nucleant particles where a lower slope means a more potent particle. To conclude that a change in slope means a change in potency assumes that the casting conditions remain constant because other factors such as the temperature gradient can also change the slope. In order to understand why these factors affect grain size we need to understand the role of constitutional supercooling (CS).

As a grain grows, solute is rejected and builds up in the liquid at the solid-liquid interface generating a concentration


D. H. StJohn (✉)
School of Mechanical and Mining Engineering, Centre for Advanced Materials Processing and Manufacturing (AMPAM), The University of Queensland, St Lucia, QLD, Australia
e-mail: d.stjohn@uq.edu.au

X. Hu · L. Peng
Shanghai Jiaotong University, Shanghai, China

M. Sun
University of Shanghai for Science and Technology, Shanghai, China

H. Dieringa
Institute of Materials Research, Helmholtz-Zentrum Geesthacht, Max-Planck-Str. 1, 21502 Geesthacht, Germany



D. Orlov et al. (eds.), *Magnesium Technology 2018*, The Minerals, Metals & Materials Series, https://doi.org/10.1007/978-3-319-72332-7_48

gradient in front of the interface as shown in Fig. 2a. In Fig. 2b, this concentration gradient is converted to a gradient of the equilibrium temperature $T_E$. The CS zone exists where the actual temperature gradient of $T_A$ is lower than the equilibrium temperature $T_E$. The temperature difference between $T_E$ and $T_A$ is $\Delta T_{CS}$. The size of the $\Delta T_{CS}$ zone and the peak value of $\Delta T_{CS}$ govern the extent of equiaxed grain nucleation [2]. When the gradient of $T_A$ is relatively steep as shown schematically in Fig. 2b, the casting conditions favour directional solidification of columnar grains. In contrast, Fig. 2c represents the ideal situation for the production of a fully equiaxed structure. Also required are potent nucleant particles to trigger the nucleation of grains. The important properties of the nucleant particles, whether naturally present in the alloy melt or deliberately added as inoculants, are their nucleation potency, defined by $\Delta T_n$, and their distribution and number density, which define the spacing $x_{Sd}$ between nucleated grains.

Figure 2 highlights a challenge to overcome which was revealed during the development of the Interdependence Theory [3]. This is the formation of a Nucleation-Free Zone (NFZ) around each successfully nucleated grain where nucleation is unlikely to occur. The length of NFZ, $x_{nfz}$, includes the length of the diffusion field as marked on the schematic of Fig. 2c plus the amount of growth of the grain needed to generate sufficient $\Delta T_{CS}$. The size of $x_{nfz}$ depends on a number of factors as shown by the Interdependence equation, Eq. 1.

The Interdependence equation (Eq. 1) [3] shows the relationship between constitutional supercooling and particle

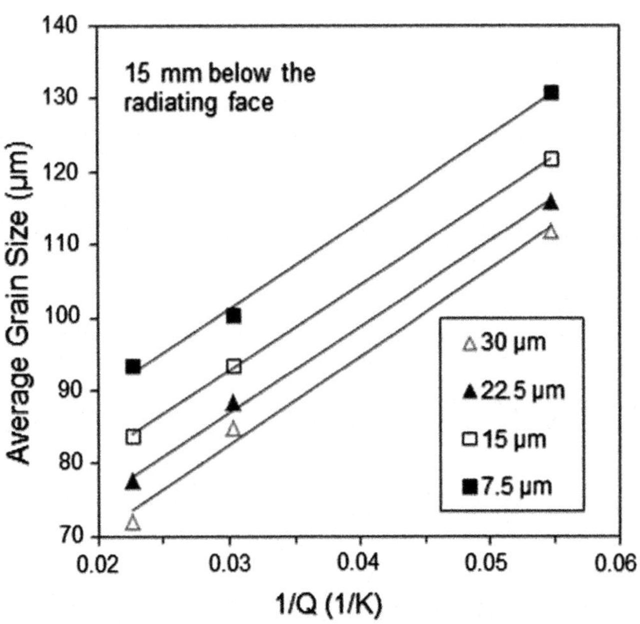


**Fig. 1** Grain size versus Mg–Al alloy Q values, subjected to UT amplitudes between 7 and 30 μm [1]

characteristics in defining the distance between nucleation events and, therefore, is a predictor of the relative grain size, $d_{gs}$.

$$d_{gs} = \frac{D \cdot z\Delta T_{n-min}}{vQ} + \frac{4.6D}{v} \cdot \left( \frac{C_l^* - C_0}{C_l^*(1-k)} \right) + x_{Sd} \quad (1)$$

where $D$ is the solute diffusion coefficient, $z\Delta T_{n-min}$ is the incremental amount of undercooling needed to trigger nucleation on the most potent particle of potency $\Delta T_{n-min}$, $v$ the growth velocity of the interface, $C_0$ the alloy composition, $C_l^*$ the liquid composition at the interface, and $k$ the partition coefficient. The term $z$ is related to the temperature gradient of $T_A$.

The three terms of Eq. 1 define the distance between nucleation events where

$$d_{gs} = x_{CS} + x'_{dl} + x_{Sd} \quad (2)$$

- $x_{CS}$ is the amount of growth of a previous grain to generate $\Delta T_{CS} = \Delta T_n$,
- $x'_{dl}$ is the length of the diffusion field to where $\Delta T_n$ is achieved, and
- $x_{Sd}$ is the average distance to the next most potent particle that successfully nucleates a grain.

Figure 3 illustrates the relationship between these distances and shows the effect of alloy composition as represented by $Q$, on the grain size. Therefore, a key strategy for generating a fine equiaxed grain size is to reduce both the size of the nucleation free zone $x_{nfz}$ and the spacing between the most potent particles $x_{Sd}$.

Figure 4 illustrates the effect of the particle potency $\Delta T_n$ on $x_{nfz}$ and on the achievable grain size. As the nucleation temperature decreases (i.e. $\Delta T_n$ increases) the size of $x_{nfz}$ increases as more growth is needed for $\Delta T_{CS}$ to equal or exceed $\Delta T_n$.

The Interdependence model is so named because the grain size is dictated by the interaction between CS and the particle characteristics $\Delta T_n$ and $x_{Sd}$ in Eq. 1. Since there is a range of particle sizes as shown by the example presented in Fig. 5a, the average distance between particles will differ for particles of different values of $\Delta T_n$. According to the Free Growth model $\Delta T_n$ is related to the size of the particles by the following equation

$$\Delta T_n = 4\sigma/(\Delta S_v.d) \quad (3)$$

where $\sigma$ is the solid-liquid interfacial energy and $\Delta S_v$ the entropy of fusion. $S_d$ is calculated for each value of $\Delta T_n$ from the distribution in Fig. 5a where $S_d$ = 100 μm/$N_d$ and $N_d$ is the number of particles within 100 μm. Bringing these calculations together generates Fig. 5b. The addition of further master alloys shifts the curves to the left reducing $S_d$ and therefore the grain size. This approach was applied to the

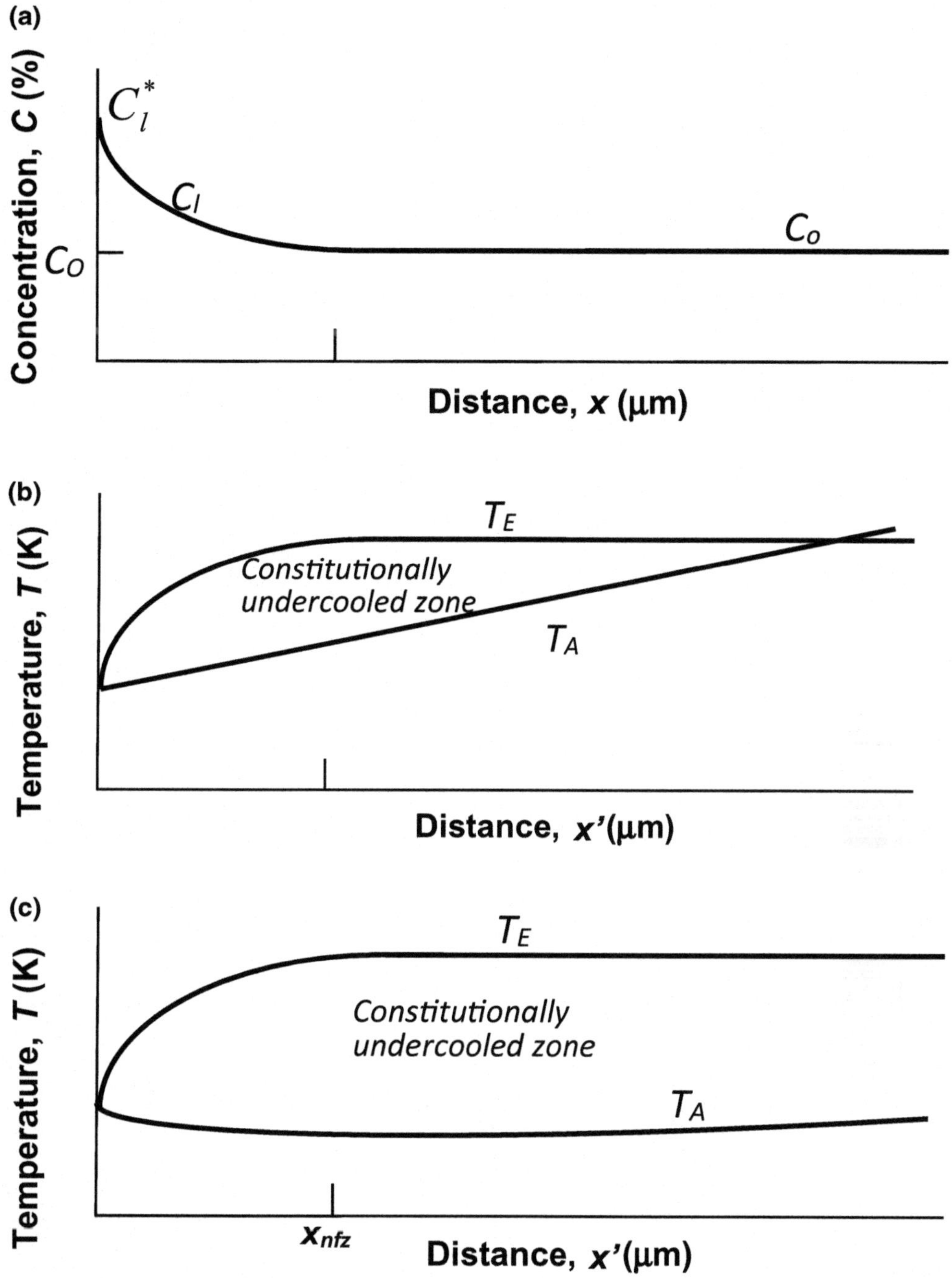


**Fig. 2** Schematics of (**a**) the solute concentration in front of a growing solid-liquid interface; (**b**) the concentration gradient converted to the equilibrium liquidus temperature $T_E$ where the constitutionally supercooled zone is the difference between $T_E$ and the actual temperature $T_A$ and the gradient of $T_A$, $G$, is typical of directionally solidified alloy; and (**c**) represents the case of equiaxed solidification in a low temperature gradient. $x_{nfz}$ denotes the end of NFZ (from [4])

addition of Zr refiner to magnesium to compare the refinement efficiency of three Zr master alloys [5].

Figure 6 brings together the development of CS and the $\Delta T_n - S_d$ curve highlighting that nucleation occurs at the intersection of the gradient of $T_A$ and the $\Delta T_n - S_d$ curve at time $t_2$.

A recent review [7] discussed the benefits and limits of constitutional supercooling's effect on grain nucleation. The main limitation is the formation of NFZ. However, this is balanced by the facilitation of the formation of an equiaxed zone of fine grain size by the generation of $\Delta T_{CS}$ and the protection of newly nucleated grains from remelting while these grains are subjected to convection and transport throughout the melt during casting. The following two examples show how application of the Interdependence model provides an explanation of the mechanisms occurring during solidification that lead to the refinement of equiaxed grains. These examples were chosen because the mechanisms that make the largest contribution to the as-cast grain size are different. In the first example the number of nucleant

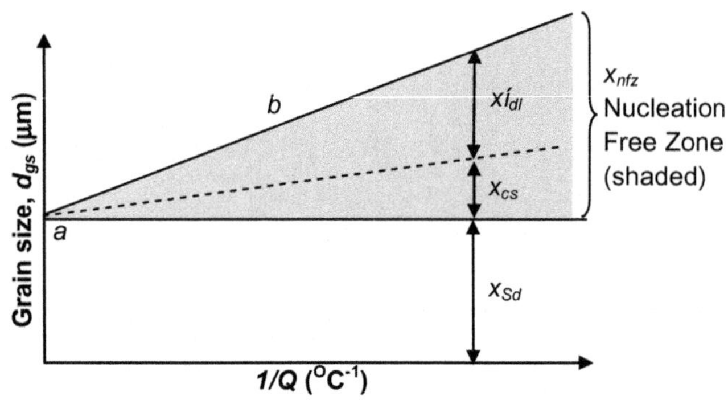


**Fig. 3** A representation showing the relationship between composition as defined by $Q$ and the components in Eq. 2 that contribute to the grain size. $x_{Sd}$ is constant if the number density of nucleant particles does not change with composition as shown in this figure (from [3])

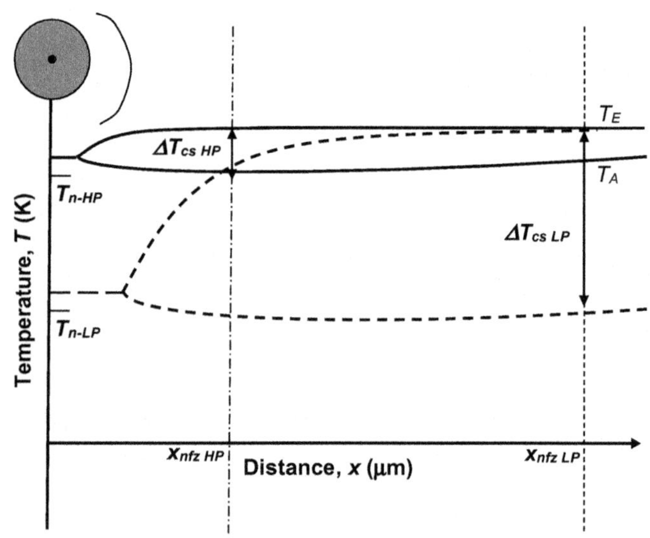


**Fig. 4** Schematic of the formation of the CS zone where a small $\Delta T_{CS\text{-}HP}$ forms for nucleant particles of high potency $T_{n\text{-}HP}$ and $\Delta T_{CS\text{-}LP}$ for low potency $T_{n\text{-}LP}$ particles. The high potency particles lead to a much smaller NFZ of $x_{nfz\text{-}HP}$ than the low potency particles at $x_{nfz\text{-}LP}$

particles plays an important role and in the second example the alloy chemistry and potency of the particles have a significant impact on the as-cast grain size.

## Grain Size Variation of an Mg–Al–Sm Alloy

In a recent publications [8, 9] it was shown (Fig. 7) that a ternary addition of Sm (Samarium) to the Mg-3Al alloy firstly coarsened the grain size and then refined the grain size as the Sm composition was increased from 0 to 2.1 wt%. A thermodynamic analysis [10] of the Mg–Al–Sm system calculated that the pro-eutectic $Al_2Sm$ phase begins to form above 1.3 wt%Sm. This factor plus observation of $Al_2Sm$ particles in the center of grains in both 1.4 and 2.1 wt% Sm samples and a crystallographic misfit of 0.45% with magnesium [8] clearly indicates that $Al_2Sm$ is a good nucleant for magnesium. (The cause of coarsening from 0 to 0.7 wt% Sm was found to be due to the transformation of the native Al–Fe–C–O particles to lower potency Al–Fe–Sm–C–O particles [8].)

Because the $Al_2Sm$ particles are clearly visible in the microstructure (Fig. 7b), the size and their number were readily measured (Fig. 8a). Figure 8b converts the sizes in Fig. 8a into values of $\Delta T_n$ by Eq. 2. Figure 8c converts the number distribution in Fig. 8a to an average distance $S_d$ versus particle size and Fig. 8d generates a plot of $S_d$ versus $\Delta T_n$ which is used in Fig. 9.

Figure 9 was constructed based on the schematic in Fig. 6 and the data in Fig. 8d. This was the first time that there has been sufficient data to make a quantitative figure of the representation in Fig. 6. Figure 8d indicates the value of $\Delta T_n$ is less than 0.2 K which is reasonable as the crystallographic misfit is very low. The low value of $\Delta T_n$ plus the small value of $z$ due to a relatively low temperature gradient in the melt means that $x_{nfz}$ will be small. Thus, the grain size above 1.3 wt%Sm is controlled by the size of $x_{Sd}$. Regarding $x_{Sd}$, the average spacing between the largest most potent particles is 120 μm. However, the as-cast grain size is 69 μm. To achieve this value all of the particles with a potency above 0.2 K must be activated which gives an $x_{Sd}$ of 60 μm. Therefore, $x_{nfz}$ must be small at about 10 μm as indicated above. Thus, in this example the major contributor to grain size is the number of particles able to successfully nucleate a grain.

From Fig. 9, Eq. 1 can be simplified to be a predictive equation for compositions above 1.3 wt% Sm to

$$d_{gs} = 10\left(x_{nfz}\right) + (1000\text{–}1175 \times (C_o - 1.3))\,(x_{Sd} \text{ where } 1.3 < \text{Co} < 2.2 \text{ wt.\%Sm}).$$

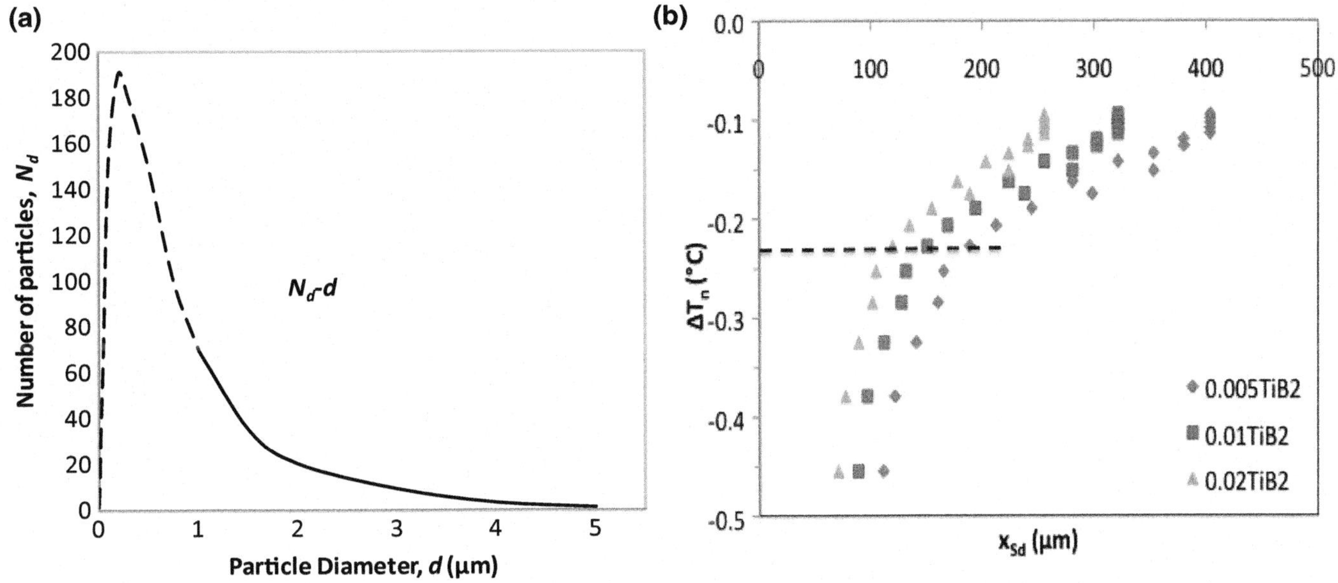


**Fig. 5** **a** The size distribution of particles based on $TiB_2$ size data from an Al5Ti1B master alloy [6]. **b** Using Eq. 3, the particle size and its distribution in (**a**) are transformed to a plot of nucleation undercooling $\Delta T_n$ versus the average distance between particles $x_{Sd}$ of that value of $\Delta T_n$

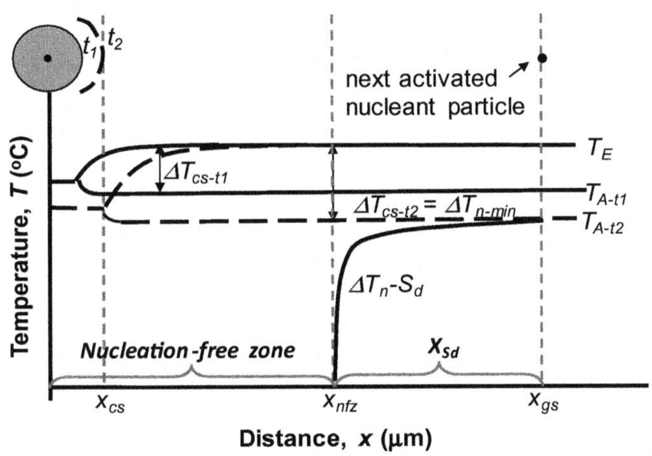


**Fig. 6** Schematic showing the relationship between the development of $\Delta T_{CS}$ from time $t_1$ to $t_2$ and the $\Delta T_n - S_d$ curve of the distribution of particles over a range of $\Delta T_n$ values. When the $\Delta T_n - S_d$ curve intersects $T_A - t_2$ temperature gradient a nucleation event will occur

## UST Assisted Grain Refinement of an AM60 Alloy Containing AlN Nanoparticles

Ultrasonic treatment can refine the grain size by treating the molten alloy or by application during the nucleation stage of solidification or by combining both methods [11]. In this example the former approach is taken where UST is applied above the liquidus temperature to ensure well-wetted nanoparticles and their uniform distribution throughout the melt [12]. One weight percent of AlN powder was added to the AM60 melt, stirred, UST applied for 5 min, and then the melt was poured into a 100 mm diameter cylinder which was lowered into a water bath for directional solidification from the bottom of the cylinder. In one gram of AlN powder there are about $1.15 \times 10^{15}$ particles with an average size of 80 nm (Fig. 10). The largest particles are 162 nm in size and represent 0.003% of the total number of particles.

Figure 11 shows the effect of nanoparticles on grain size where UST of AM60 has a grain size of 1277 μm while UST of AM60 with AlN particles has a grain size of 85 μm. By converting the number of particles per gram to the number of particles per volume of the nanocomposite, the average spacing between the largest particles is 2.1 μm. Thus, the grain size would be 2.1 μm if we assume all of the largest particles are the most potent and nucleate a grain. However, the measured grain size is 85 μm. This means that only a very small fraction (approximately 0.002%) of the largest 0.003% of particles successfully nucleate a grain.

Considering Eq. 1, the very high number density of nucleant particles implies $x_{Sd}$ would be small, probably <5 microns. Thus, changes in grain size would be largely affected by changes to the size of the nucleation-free zone which was also found to be the case for Mg–Al alloys without particle additions [3]. A key factor in determining

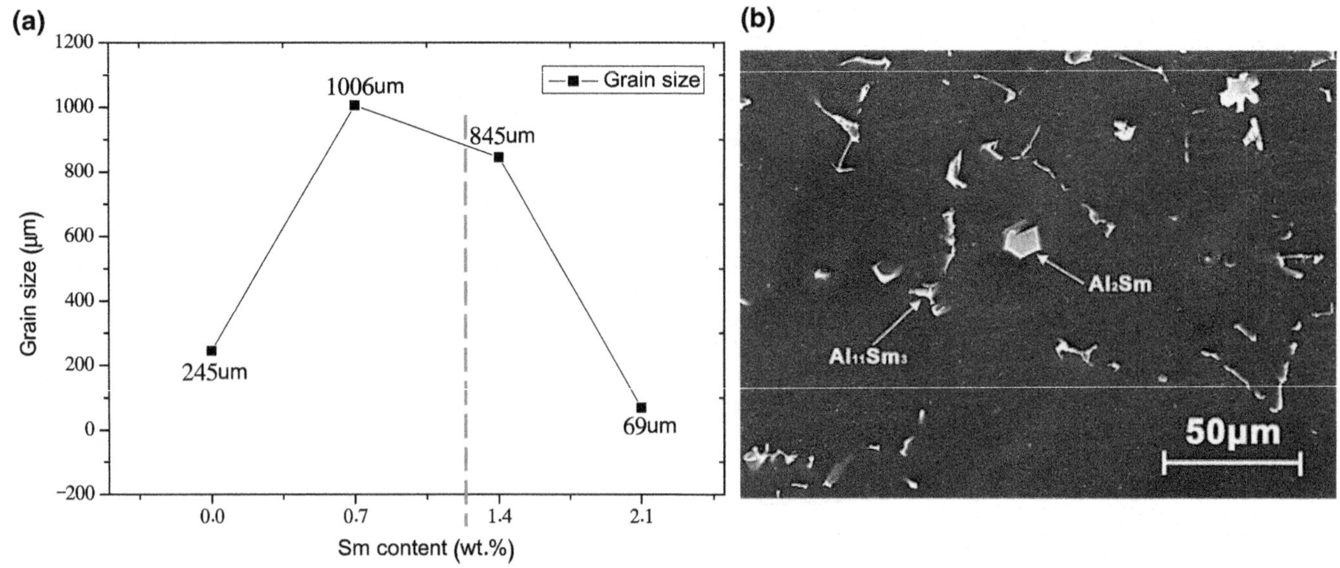


**Fig. 7** (**a**) Grain size versus Sm content, (**b**) SEM image of the Mg-3Al-2.1Sm alloy [8]

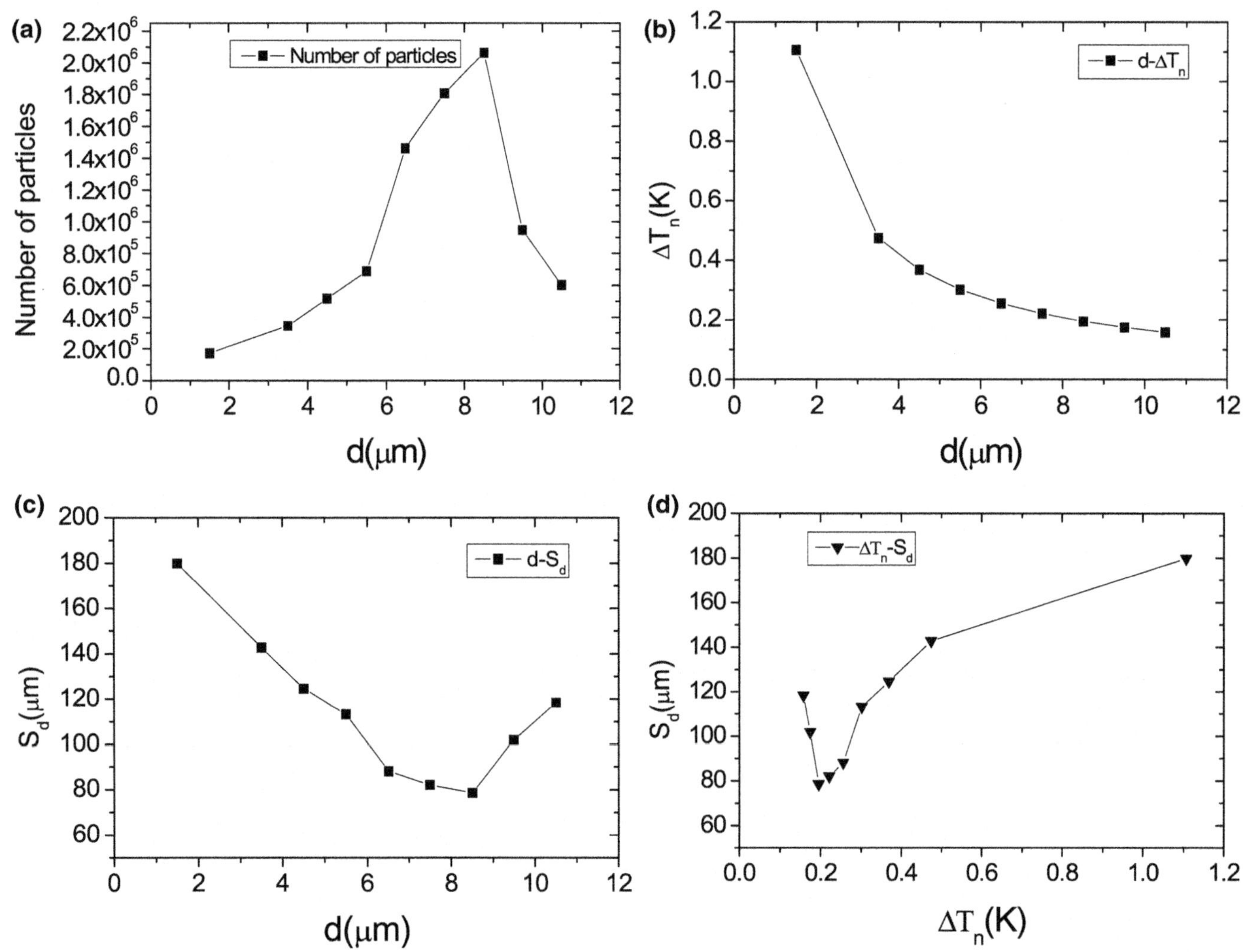


**Fig. 8** Characteristics of the $Al_2Sm$ particles in the Mg-3Al-2.1Sm alloy. **a** The distribution $N_d$ of the number of potent particles for each value of diameter *d*. **b** The relationship between nucleation undercooling $\Delta T_n$ and *d*. **c** Relationship between the average particle spacing *Sd* and particle size *d*. **d** $S_d$ versus $\Delta T_n$. [8]

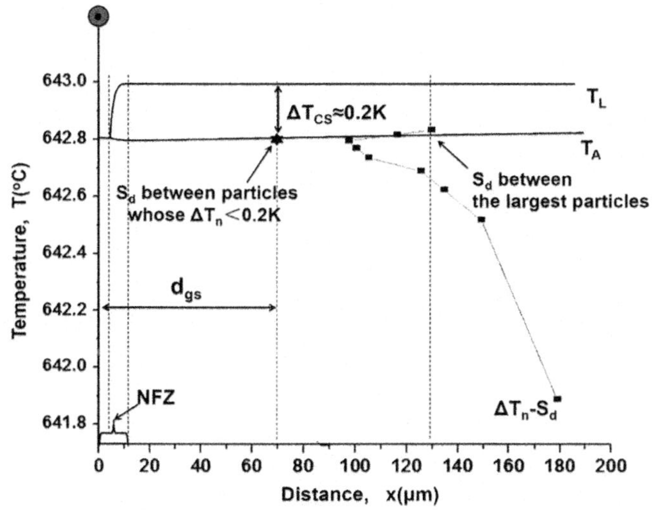


**Fig. 9** A schematic that illustrates the interrelationship between the development of $\Delta T_{CS}$ between the equilibrium liquidus temperature $T_E$ and the actual temperature of the melt $T_A$, and the distribution of particles for the range of particle sizes converted to their nucleation undercooling ($\Delta T_n - S_d$) that together establish the grain size of the Mg-3Al-2.1Sm alloy [8]

the size of NFZ is the nucleation undercooling of the particles. $\Delta T_n$ was measured by DSC to be 14 K. The high value of $\Delta T_n$ would be expected according Eq. 2 for very small particle sizes. Based on calculations of $x_{nfz}$ used previously [3] $x_{nfz}$ is predicted to be ~600 μm which is much larger than 85 μm. The main parameters in Eq. 1 that could reduce the size of $x_{nfz}$ are $D$ and $v$ as $Q$ has not changed. V may be faster due to the casting rate of 3 mm/s. It is also possible that high levels of nanoparticles reduce the effective diffusion coefficient. An indication of a decrease in $D$ can be caused by a decrease in viscosity [13]. This is supported by spiral fluidity measurements where the spiral length decreased by about 14% from 96.2 to 83.5 cm when AlN particles are added. In order to decrease $x_{nfz}$ to less than 90 μm $D$ needs to be reduced from $5 \times 10^{-9}$ to $7 \times 10^{-10}$ $m^2$/sec. This hypothesis regarding a relationship between viscosity and diffusion coefficient needs verification by further research.

Understanding the formation of NFZ is critical to improving the performance of nanoparticles as nucleants because the dominant effect of NFZ will impact on the

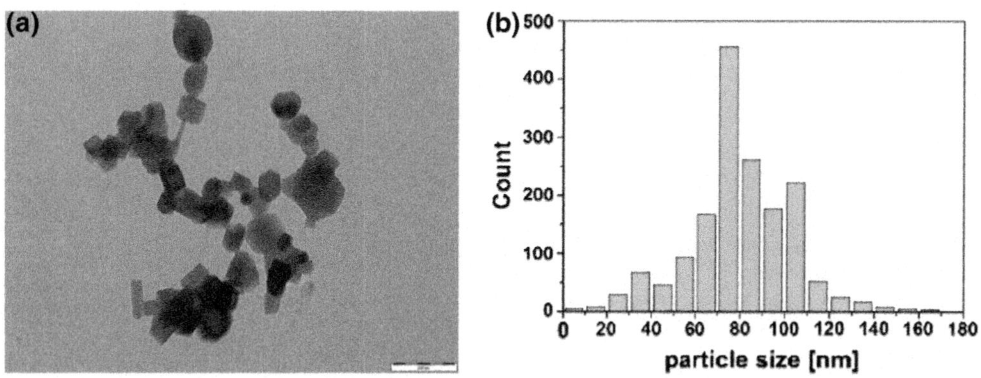


**Fig. 10** (**a**) Typical AlN nanoparticles and (**b**) particle size distribution of the AlN nanoparticles [12]

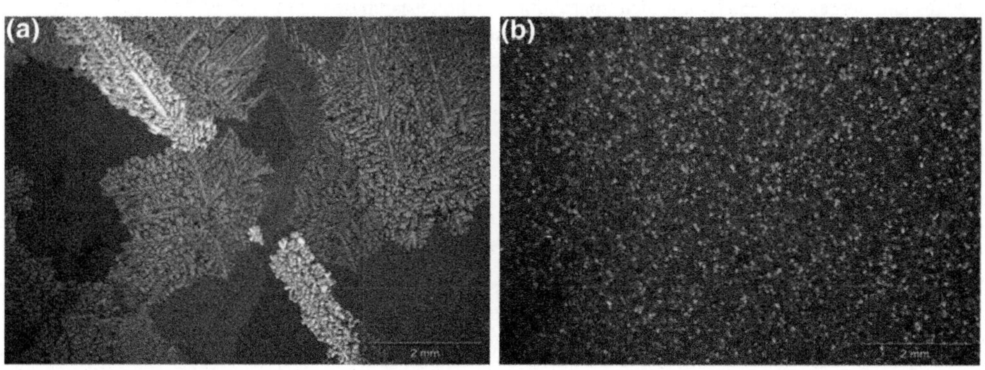


**Fig. 11** Microstructure and grain size of (**a**) AM60, 1277 ± 300 μm and (**b**) AM60 + AlN, 85 ± 6.2 μm [12]

mechanical properties through the Hall-Petch relationship. In a broader study of the AM60-AlN system, a Hall-Petch relationship with grain size was found despite the very large number density of nanoparticles [12].

## Concluding Remarks

The two examples described above show that the Interdependence model is a framework for determining whether $x_{nfz}$ or $x_{Sd}$ make a greater contribution to grain size, and therefore provides a focus for identifying methods to decrease the grain size. In the Mg–Al–Sm example $x_{Sd}$ has the predominant effect on grain size while for the AM60-AlN nanocomposite example the size of $x_{nfz}$ controls the grain size. In the case of the nanocomposite the very small particle size results in a large NFZ preventing very small grain sizes being formed. This effect may be applicable to other nanocomposite alloy systems.

The Interdependence model can be used to:

- understand the sometimes complex interaction between factors affecting nucleation and grain refinement;
- highlight the importance of NFZ;
- understand why NFZ prevents the formation of very fine grain sizes in AlN nanocomposites (assuming particle pushing does not occur); and
- analyse the effect of a range of solidification conditions on grain size.

**Acknowledgements** DStJ acknowledges the support of Australian Research Council grants DP 140100702 and DP160100560. XH acknowledges the support of the National Natural Science Foundation of China (No. 51201103), SJTU Special Funds for Science and Technology Innovation (No. 13X100030018), and the Program for Outstanding Academic Leader of Shanghai (14XD1425000). The authors thank Penghuai Fu, Mingxing Zhang, Lydia Katsarou, Ricardo Buzolin, Gabor Szakacz, Manfred Horstmann, Martin Wolff, Chamini Mendis and Sergey Vorozhtsov for their contribution to the research on the two examples. Stuart McDonald and Arvind Prasad are thanked for calculating the composition of the $Al_2Sm$ phase boundary and providing advice on the relationship between viscosity and the diffusion coefficient, respectively. HD and DStJ wish to acknowledge financial support from the European Commission (ExoMet Project, 7th Framework Programme, contract FP7-NMP3-LA-2012-280421).

## References

1. A. Ramirez A, Das A, StJohn DH (2010) The effect of solute on ultrasonic grain refinement of magnesium alloys. Journal of Crystal Growth 312: 2267–2272.
2. Prasad A, Yuan L, Lee P, Easton M, StJohn D (2016) The effect of the melt thermal gradient on the size of the constitutionally supercooled zone, in: IOP Conference Series: Materials Science and Engineering, IOP Publishing, 012001.
3. StJohn DH, Qian M, Easton MA, Cao P (2011) The Interdependence Theory: the relationship between grain formation and nucleant selection. Acta Mater, 59:4907–4921.
4. Sun M, Easton MA, StJohn DH, Wu G, Abbott TB, Ding W (2013) Grain Refinement of Magnesium Alloys by Mg-Zr Master Alloys: the Role of Alloy Chemistry and Zr Particle Number Density, Advanced Engineering Materials 15(5): 373–378.
5. StJohn DH, Prasad A, Easton MA, Qian M (2015) The contribution of constitutional supercooling to nucleation and grain formation. Metall. Mater. Trans. A 46:4868–4885.
6. Hu X, Fu P, StJohn D, Peng L, Sun M, Zhang M (2016) On grain coarsening and refining of the Mg-3Al alloy by Sm. Journal of Alloys and Compounds 663: 387–394.
7. Ming Sun, XiaoyuHu, LimingPengn, PenghuaiFu, YinghongPeng (2014) Effects of Sm on the grain refinement, microstructures and mechanical properties of AZ31 magnesium alloy. Mater. Sci. &Eng. A 620:89–96.
8. Jin L, Kevorkov D, Medraj M, Chartrand P (2013) Al–Mg–RE (RE = La, Ce, Pr, Nd, Sm) systems: Thermodynamic evaluations and optimizations coupled with key experiments and Miedema's model estimations. J. Chem. Thermodynamics 58: 166–195.
9. Wang G, Dargusch MS, Eskin DG, StJohn DH (2017) Identifying the Stages during Ultrasonic Processing that Reduce the Grain Size of Aluminum with Added Al3Ti1B Master Alloy. Advanced Engineering Materials DOI:https://doi.org/10.1002/adem.201700264.
10. Dieringa H, Katsarou L, Buzolin R, Szakacz G, Horstmann M, Wolff M, Mendis C, Vorozhtsov S, StJohn D (2017) Ultrasound assisted casting of an AM60 based metal matrix nanocomposite, its properties and recyclability. Submitted for publication August 2017.
11. D.R. Poirier (2014) Density, viscosity, and diffusion coefficients in hypoeutectic Al-Si liquid alloys: an assessment of available data. Met. Mater. Trans. B 45B:1345–1354.
12. Polmear I, StJohn D, Nie JF, Qian M (2017) Light Alloys: Metallurgy of the Light Metals, 5th ed., Butterworth-Heinemann.
13. Greer AL, Bunn AM, Tronche A, Evans PV, Bristow DJ (2000) Modelling of inoculation of metallic melts: application to grain refinement of aluminium by Al–Ti–B. Acta Mater. 48:2823–2835.

# Thermodynamics of Phase Formation in Mg–Al–C Alloys Applied to Grain Refinement

G. Deffrennes, B. Gardiola, M. Lomello, J. Andrieux, O. Dezellus, and R. Schmid-Fetzer

**Abstract**

Grain refinement of Mg–Al based alloys is challenging because it is known that Zr, which is extremely effective in many Al-free alloys, cannot be used. The addition of carbon through various routes by using carbon-containing sources is considered as an option. The grain refinement mechanisms are still under debate. The present work is focused on the ternary base system Mg–Al–C, including the potential nucleants $Al_4C_3$ and $Al_2MgC_2$, presently without consideration of $Al_2CO$. The ternary carbide $Al_2MgC_2$ was synthesized and characterized using sealed Ta crucibles. The decomposition of the carbide was measured at 1290 °C by Differential Thermal Analysis under a pressure of 8 bar. Practical difficulties, including high vapor pressure of Mg and high affinity of Mg with oxygen, as well as rapid hydrolysis of the $Al_2MgC_2$ carbide have been overcome.

**Keywords**

Al–C–Mg • $Al_2mgc_2$ • Mg–Al alloys • Grain refinement

G. Deffrennes (✉) · B. Gardiola · J. Andrieux · O. Dezellus
Laboratoire des Multimatériaux et Interfaces, Université Claude Bernard Lyon 1, 22 avenue Gaston Berger, 69100 Villeurbanne, France
e-mail: guillaume.deffrennes@univ-lyon1.fr

B. Gardiola
e-mail: bruno.gardiola@univ-lyon1.fr

J. Andrieux
e-mail: jerome.andrieux@univ-lyon1.fr

O. Dezellus
e-mail: olivier.dezellus@univ-lyon1.fr

M. Lomello
SYMME, Université Savoie Mont Blanc, Maison de la mécatronique 7 chemin de bellevue, 74944 Annecy le Vieux, France
e-mail: marc.lomello@univ-smb.fr

R. Schmid-Fetzer
Institute of Metallurgy, Clausthal University of Technology, Robert-Koch-Str. 42, 38678 Clausthal-Zellerfeld, Germany
e-mail: schmid-fetzer@tu-clausthal.de


D. Orlov et al. (eds.), *Magnesium Technology 2018*, The Minerals, Metals & Materials Series, https://doi.org/10.1007/978-3-319-72332-7_49

## Introduction

A promising route to improve the mechanical properties of Mg and Mg alloys and to sustain the development of competitive magnesium based materials is the grain refinement of Mg alloys [1]. In fact, the Hall-Petch strengthening coefficient of magnesium is roughly four times greater than the one of aluminum [2]. Indeed, a refined microstructure enhances most of the mechanical properties of magnesium alloys [3–6], including mechanical properties at higher temperature [7] as well as corrosion resistance [8].

Zirconium is widely accepted as the most successful and effective grain refiner currently known for magnesium [1]. However it is not a viable solution for Mg–Al alloys due to the formation of intermetallic phases with aluminum. Therefore, carbon addition methods have become the major industrial grain refinement technique for Mg–Al alloys [9]. The mechanism behind this grain refinement has raised many conflicting hypothesis over the last 15 years as the phases involved in the process such as $Al_4C_3$ and $Al_2MgC_2$ are micron-sized and may react strongly with water to form oxides [10] prior to characterization.

The most widely accepted hypothesis in the literature is that $Al_4C_3$ forms in the melt and leads to the heterogeneous nucleation of Mg grain during cooling [4, 5, 7, 11–15]. Indeed, $Al_4C_3$ would be a suitable nucleant for α-Mg [13]. The facts supporting this hypothesis is the systematic observation of micrometric Al–Mg–C–O particles in the final microstructure [4–6, 11–14, 16, 17], notably in the center of the grains [4, 6, 13, 14, 16].

Several conflicting explanation of the refining phenomena are usually considered. Duplex nucleation theory was proposed considering $Al_4C_3$ as a nucleant for the $Al_8Mn_5$ phase in a first stage that would lead to the microstructure refinement in a second stage [7, 18]. Because of the presence of oxygen in the characterized particles, $Al_2CO$ was considered as a potential nucleant [19]. However the formation of the oxide phase in the melt is not favorable from a thermodynamic point of view [13, 14, 20]. Jin et al. proposed that a segregation phenomenon of carbon would be the reason behind the refining by affecting the constitutional undercooling and restricting the grain growth during solidification [21]. This hypothesis was debated between Qian et al. [19] and Jin et al. [20].

Recently, strong experimental evidences [22] suggest that the formation of $Al_2MgC_2$ particles in the melt would lead to the heterogeneous nucleation of Mg grain in a later stage. In addition, first-principles calculations conducted by Wang et al. emphasized that $Al_2MgC_2$ is a suitable crystal nucleus for α-Mg [23]. From a thermodynamic point of view the work of Viala et al. [10] also support this hypothesis as the authors found that at 727 °C alloys containing from 0.6 to 19 wt%Al were in a two-phase equilibrium with $Al_2MgC_2$. Indeed, the carbon inoculation process temperatures found in the literature for Mg–Al alloys vary from 700 to 790 °C and the alloy composition range from 3 to 9 wt%Al. It is to note that equilibrium between Mg–Al liquid and $Al_4C_3$ is only reached above 19 wt%Al at 727 °C [10].

Despite the interest of grain refinement of Mg–Al alloys by carbon inoculation the Al–C–Mg system is not satisfactorily assessed as the carbide phase $Al_2MgC_2$ is presently not described in any thermodynamic database for Mg alloys [24, 25]. Currently, the only information available related to $Al_2MgC_2$ are an isothermal section at 727 °C [10] as well as the structure of the allotropic form of $Al_2MgC_2$ stable above 727 °C determined by Rietveld refinement from X-ray powder diffraction data [26, 27]. This lack of experimental data is a common issue for any magnesium-based system above 1000 °C. It is a direct consequence of the high vapor pressure of Mg coupled with its high reactivity making experimental work delicate. In addition, $Al_2MgC_2$ strongly reacts with water, making extraction and characterization even more difficult.

An extensive knowledge of phase equilibria between carbon and Mg–Al alloys as well as the determination of the structural, thermal and thermodynamic properties of $Al_2MgC_2$ is compulsory to make a reliable thermodynamic assessment of the Al–C–Mg system and, in fine, to sustain the development of grain refined Mg–Al alloys. In the present study $Al_2MgC_2$ was synthesized and the thermal decomposition of the carbide was characterized by DTA.

## Experimental Procedure

### Materials

Samples used in the study were prepared from commercial powders of magnesium (purity > 99.8 wt%, grain size 150 < d < 850 μm, Alfa Aesar), aluminum (purity > 99.8 wt%, grain size 44 < d < 420 μm, Alfa Aesar) and graphite (synthetic, d < 20 μm, Sigma Aldrich).

### Sample Preparation

The powders were ball-milled during 20 min in a tungsten carbide mortar and cold-pressed under 250 MPa. All the preparation steps were performed under protective Ar atmosphere. Due to the high vapor pressure of Mg, reaching almost 6 bar at 1350 °C [28], as well as the high reactivity of Mg regarding oxygen, the syntheses were carried out using sealed Ta crucibles (h = 8 mm, d = 7 mm, thickness = 0.5 mm, purity > 99.95wt%, Concept Metal) arc welded under 0.6 bar of Ar. Thermal shields made of Ta were placed over the sample in order to protect magnesium from the radiations during the welding of the crucible lid that would otherwise lead to evaporation of Mg.

### Synthesis and Thermal Treatments

For the thermal treatments made at 1000(±4) °C the Ta crucibles were sealed inside silica vessels under 0.2 bar of Ar in order to avoid oxidation of the Ta crucibles during the thermal treatment and to offer a second protection in case the crucibles would fail under the Mg vapor pressure. A conventional horizontal tube furnace with Eurotherm controller was used as heat source. Owing to the small volume of the crucibles the vapor pressure of Mg could be reached without altering the initial sample composition, as less than 0.1 wt% of Mg was needed to supply the vapor phase at 1000 °C. At the end of the isothermal heat-treatments the silica vessels were broken in water as soon as being driven out of the furnace so that the samples were efficiently quenched.

### SEM-EDS

Quenched samples were cut using a diamond saw without adding any lube, and polished using a water-free polishing procedure to prevent the hydrolysis of $Al_2MgC_2$. SEM observations were performed directly after polishing on a Zeiss SEM under a high voltage of 10 kV and at a working distance of 8.5 mm. Mass fraction of the three elements were obtained from electronic standards.

### Differential Thermal Analysis

Thermal analysis was performed from room temperature to 1350 °C in a Setaram TAG92 using a B-type Pt–Rh DSC sensor. The sensor was slightly deformed to match the Ta crucible geometry, and an yttrium oxide paste was used to prevent any reaction at the crucible/DSC-sensor interface. As reference a cylinder made of tungsten with a thermal mass (mCp) as close as possible to that of the sample was used. The DSC was calibrated in temperature using gold standard.

## Results and Discussion

### Synthesis of $Al_2MgC_2$

The synthesis of the $Al_2MgC_2$ carbide was carried at 1000 °C during 240 h from the composition 70Mg–19Al–11C wt%. After the synthesis stoichiometric $Al_2MgC_2$ crystals were obtained inside a 98.9Mg-1.1Al at.% matrix as well as with unreacted graphite located at the carbide—matrix interface as shown in Fig. 1. Smaller crystals were hexagonal-shaped whereas largest ones were rectangular-shaped. As witnessed by Viala et al. [10] $Al_2MgC_2$ hydrolysis was observed and was found to be significant after half an hour in air. Huang et al. [22] proposed the following hydrolysis reaction:

$$2Al_2MgC_2 + 8H_2O = 2MgAl_2O_4 + 4CH_{4(g)} \quad (1)$$

Phases were characterized by EDS and results are displayed in Table 1. $Al_2MgC_2$ composition was determined from a set of crystals selected so that oxygen lies below the detection limit. No solubility of carbon could be measured in the matrix within the detection range.

The fact that the matrix was very poor in aluminum suggests that thermodynamic equilibrium have been reached. Plus the composition of 1.1 wt%Al for the liquid involved in the three-phase equilibria with graphite and $Al_2MgC_2$ at 1000 °C is coherent with the value of 0.6 wt% Al obtained at 727 °C by Viala et al. [10].

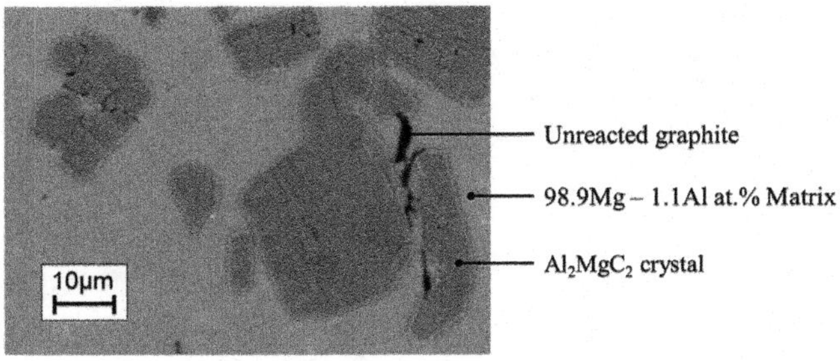


**Fig. 1** SEM observation of a 70Mg–19Al–11C wt.% sample heat-treated 240 h at 1000 °C. The microstructure is characterized by $Al_2MgC_2$ crystals, unreacted graphite and a 98.9Mg-1.1Al at.% matrix

### Thermal Decomposition of $Al_2MgC_2$

Differential Thermal Analysis was performed on samples (70Mg–19Al–11C wt% powder compact equilibrated at 1000 °C during 240 h, Sect. 3.1). The as-quenched samples were made of $Al_2MgC_2$, unreacted graphite and a 98.9Mg–1.1Al at.% matrix (Fig. 1). Results are displayed in Fig. 2. Melting of the matrix was observed at 615.8 °C during heating and the solidification started at 640.12 °C during cooling. Those temperatures are coherent with the matrix composition found before (Table 1) and after (Table 2) DTA as they correspond to the fusion and solidification of a 98Mg–2Al at.% alloy [28]. In addition, a sharp exothermic signal was detected at 1290 °C.

The exothermic signal found at 1290 °C was attributed to the decomposition of $Al_2MgC_2$, and the reaction of decomposition is proposed as follows:

$$Al_2MgC_2 = Al_4C_3 + C + L \quad (2)$$

Equation (2) denotes a four-phase equilibrium, a ternary peritectic on cooling, as opposed to the stoichiometric reaction equation in Eq. (1). It is interesting to note that after the DTA the $Al_2MgC_2$ crystals morphology changed significantly compared to the synthesis carried out at 1000 °C. Indeed $Al_2MgC_2$ crystals were one order of magnitude bigger after the thermal analysis and platelet-shaped instead of being hexagonal-shaped or rectangular shaped. This new morphology was very similar in both size and shape to the $Al_4C_3$ crystals that would be obtained by the authors when performing synthesis with Al rich liquids. As a matter of fact, in the center of the biggest $Al_2MgC_2$ crystals the $Al_4C_3$ carbide could be found after the thermal analysis as shown in Fig. 3 as a record of the decomposition reaction. $Al_4C_3$ crystals contained 5.0 at.% of Mg present in solid solution similarly to the findings of Viala et al. at 727 °C [10].

One should note that the decomposition of $Al_2MgC_2$ was measured under a pressure of 8 bar, half of it being due to the vapor pressure of Mg and the other half to the Ar pressure initially filled during arc-welding of the crucible increasing with the temperature as an ideal gas.

**Table 1** Composition in at.% of the phases formed in 70Mg–19Al–11C wt.% samples heat-treated 240 h at 1000 °C

| Phases | at.% Mg | at.% Al | at.% C |
|---|---|---|---|
| Matrix | 98.9 | 1.1 | – |
| $Al_2MgC_2$ | 20.9 | 40.3 | 38.8 |

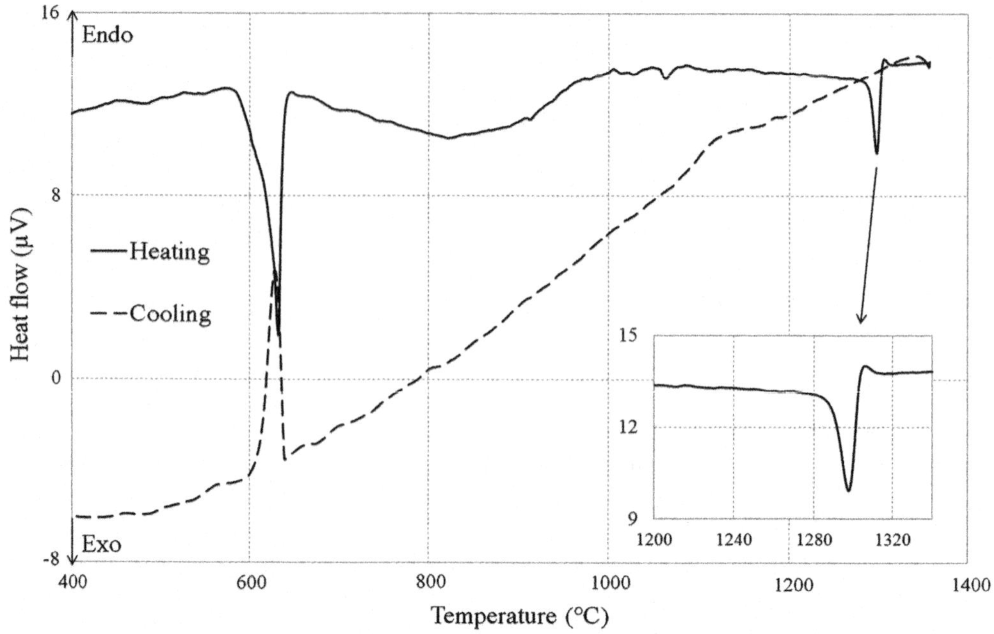


**Fig. 2** Heat flow versus temperature for thermal analysis performed on a sample of composition 70Mg–19Al–11C wt.% equilibrated at 1000 °C during 240 h

**Table 2** Composition in at.% of the phases formed in a 70Mg–19Al–11C wt.% sample heat-treated 240 h at 1000 °C after DTA was performed

| Phases | at.% Mg | at.% Al | at.% C | at.% O |
|---|---|---|---|---|
| Matrix (a) | 98.4 | 1.6 | – | – |
| $Al_2MgC_2$ (b) | 20.3 | 39.3 | 40.4 | – |
| $Al_4C_3$ (c) | 5.0 | 52.7 | 42.3 | – |
| Al–C–Mg–O (d) | 8.2 | 19.1 | 5.0 | 67.7 |

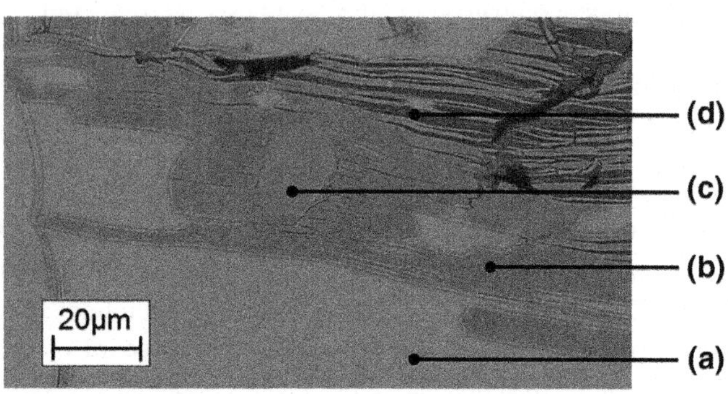


**Fig. 3** SEM observation of a 70Mg–19Al–11C wt.% sample equilibrated 240 h at 1000 °C after DTA. The microstructure is characterized by (**a**) 98.4Mg-1.6Al at.% Matrix, (**b**) $Al_2MgC_2$ crystals, (**c**) $Al_4C_3$ crystals in the center of the biggest $Al_2MgC_2$ crystals and (**d**) $Al_2MgC_2$ crystals undergoing oxidation. EDS results are displayed in Table 2

## Conclusion

The $Al_2MgC_2$ phase has been synthesized in sealed Ta crucibles capable of withstanding up to 15 bar of internal pressure. The ternary carbide decomposition temperature in (Mg, Al) liquid was measured at 1290 °C by DTA. The developed experimental procedure is promising as it allowed working with magnesium at temperatures up to 1350 °C where the pressure inside the crucible reaches 10 bar. Further experiments including the determination of phase equilibria at 1000 and 1290 °C are being carried out by the authors to provide a complete thermodynamic description of the Al–C–Mg system. Indeed, the thermodynamic modeling of the $Al_2MgC_2$ carbide is of interest to support grain refining of Mg-Al alloys by carbon inoculation under optimized processing conditions by quantitative simulations.

**Acknowledgements** The authors wish to thank the GDR CNRS n° 3584 (TherMatHT) and the CALPHAD communities where fruitful discussions led to collaborations on this project. The assistance of members of the "Centre Technologique des Microstructures, Université Lyon 1" (CTµ, http://microscopies.univ-lyon1.fr) for SEM characterizations is also gratefully acknowledge.

## References

1. Ali Y, Qiu D, Jiang B, Pan F, Zhang MX (2015) Current Research Progress in Grain Refinement of Cast Magnesium Alloys: A Review Article. J. Alloys. Compd. 619:639–51
2. King JF (2007) Magnesium: commodity or exotic? Mater. Sci. Technol. 23(1):1–14

3. Caceres CH, Mann GE, Griffiths JR (2011) Grain Size Hardening in Mg and Mg-Zn Solid Solutions. Metall. Mater. Trans. A 42 (7):1950–1959
4. Du J, Yang J, Kuwabara M, Li W, Peng J (2008) Effects of Carbon and/or Alkaline Earth Elements on Grain Refinement and Tensile Strength of AZ31 Alloy. Mater. Trans. 49(10):2303–2309
5. Suresh M, Srinivasan A, Ravi KR, Pillai UTS, Pai BC (2011) Microstructural Refinement and Tensile Properties Enhancement of Mg-3Al Alloy Using Charcoal Additions. Mater. Sci. Eng., A 528(6):2502–2508
6. Suresh M, Srinivasan A, Ravi KR, Pillai UTS, Pai BC (2013) Mechanism for Grain Refinement and Mechanical Properties of AZ91 Mg Alloy by Carbon Inoculation. Procedia Eng. 55:93–97
7. Nimityongskul S, Jones M, Choi H, Lakes R, Kou S, Li X (2010) Grain Refining Mechanisms in Mg–Al Alloys with Al4C3 Microparticles. Mater. Sci. Eng., A 527(7–8):2104–2111
8. Ralston KD, Birbilis N (2010) Effect of Grain Size on Corrosion: A Review. Corrosion 66(7):075005-075005-13
9. Lee YC, Dahle AK, StJohn DH (2000) The role of solute in grain refinement of magnesium. Metall. Mater. Trans. A 31(11):2895–2906
10. Viala JC, Claveyrolas G, Bosselet F, Bouix J (2000) The chemical behaviour of carbon fibres in magnesium base Mg-Al alloys. J. Mater. Sci. 35:1813–1825
11. Chen TJ, Ma Y, Lv WB, Li YD, Hao Y (2010) Grain refinement of AM60B magnesium alloy by SiC particles. J. Mater. Sci. 45 (24):6732–6738
12. Chen T, Wang R, Huang H, Ma Y, Hao Y (2012) Grain refining technique of AM60B magnesium alloy by MgCO3. Trans. Nonferrous Met. Soc. China 22(7):1533–1539
13. Lu L, Dahle AK, StJohn DH (2005) Grain refinement efficiency and mechanism of aluminium carbide in Mg–Al alloys. Scripta Mater. 53(5):517–522
14. Lu L, Dahle AK, StJohn DH (2006) Heterogeneous nucleation of Mg–Al alloys. Scripta Mater. 54(12):2197–2201
15. Wang L, Kim YM, Lee J, You SB (2010) Effect of hafnium carbide on the grain refinement of Mg-3wt.% Al alloy. J. Alloys Compd. 500(1):L12–L15
16. Du J, Wang M, Zhou M, Li W (2014) Evolutions of Grain Size and Nucleating Particles in Carbon-Inoculated Mg–3% Al Alloy. Alloys Compd. 592:313–318
17. Yano E, Tamura Y, Motegi T, Sato E (2003) Effect of pure carbon powder on grain refining of cast magnesium alloy AZ91. Mater. Trans. 41(1):107–110
18. Kim YM, Yim CD, You BS (2007) Grain refining mechanism in Mg–Al base alloys with carbon addition. Scripta Mater. 57 (8):691–694
19. Qian M, Cao P (2005) Discussions on grain refinement of magnesium alloys by carbon inoculation. Scripta Mater. 52(5):415–419
20. Jin Q, Eom JP, Lim SG, Park WW, You BS (2005) Reply to comments on "Grain refining mechanism of a carbon addition method in a Mg–Al magnesium alloy". Scripta Mater. 52(5):421–423
21. Jin Q, Eom JP, Lim SG, Park WW, You BS (2003) Grain refining mechanism of a carbon addition method in a Mg–Al magnesium alloy. Scripta Mater. 49(11):1129–1132
22. Huang Y, Kainer KU, Hort N (2011) Mechanism of Grain Refinement of Mg–Al Alloys by SiC Inoculation. Scripta Mater. 64(8):793–96
23. Wang HL, Tang JJ, Zhao YJ, Du J (2015) First-Principles Study of Mg/Al2MgC2 Heterogeneous Nucleation Interfaces. Appl. Surf. Sci. 355:1091–1097
24. Schmid-Fetzer R, Gröbner J (2012) Thermodynamic Database for Mg Alloys—Progress in Multicomponent Modeling. Metals 2 (3):377–398
25. Bale CW, Bélisle E, Chartrand P, Decterov SA, Eriksson G, Hack K, Jung IH, Kang YB, Melançon J, Pelton AD, Robelin C, Petersen S (2009) FactSage thermochemical software and databases—recent developments. Calphad 33:295–311
26. Bosselet F, Mentzen BF, Viala JC, Etoh MA, Bouix J (1998) Synthesis and Structure of T2-Al2MgC2. Eur. J. Solid State Inorg. Chem. 35:91–99
27. Kubus M, Meyer HJ (2013) Convenient Preparation of Ternary Carbidoaluminates – The Example MgAl2C2. Eur. J. Inorg. Chem. 2013(31)5450–5453
28. Ansara I, Dinsdale AT, Rand MH (1998) COST 507 – Thermochemical database for light metal alloys. ed. European Commission, rue de la Loi 200 (SOME 1/44) B-1049 2

# Development of Magnesium-Rare Earth Die-Casting Alloys

Mark Easton, Mark A. Gibson, Suming Zhu, Trevor Abbott, Jian-Feng Nie, Colleen J. Bettles, and Gary Savage

**Abstract**

An overview of the development of a high-performance Mg–RE based alloy, HP2+, is presented, which has a good combination of die-castability and mechanical properties at ambient and elevated temperatures. The original alloy, HP2, was a die-casing version of the sand-cast alloy SC1 developed for powertrain applications. However, HP2 tended to crack substantially, leading to unusable castings due to its high Nd content. It was found that the solidification path of Mg–RE alloys can be engineered to reduce the propensity to hot tearing by changing the mixture of RE elements towards La-rich, which leads to an increase in the amount of eutectic and a reduction of the solidification range. Precipitate-forming RE elements, such as Nd or Y, were optimized for HP2+ to meet the requirement for high temperature creep resistance. Whilst some challenges remain with the commercial application of HP2+, the learnings from the alloy design process can be applied to other alloy development programs.

**Keywords**

Mg–RE alloys • Creep resistance • Hot tearing
High-pressure die-casting

M. Easton (✉) · M. A. Gibson · S. Zhu · T. Abbott
School of Engineering, RMIT University, Melbourne 3001, Australia
e-mail: mark.easton@rmit.edu.au

M. A. Gibson · G. Savage
CSIRO Manufacturing, Clayton 3168, Australia

M. A. Gibson · J.-F. Nie · C. J. Bettles
Department of Materials Science and Engineering, Monash University, Melbourne 3800, Australia

T. Abbott
Magontec Ltd, Potts Point, Sydney 2011, Australia

## Introduction

Alloy design is a very complex challenge, as obtaining a successful alloy is really a process of multi-variant optimization. A successful alloy must be easily processable to shape, have an excellent balance of properties, be inexpensive and minimize impact to the environment over the life cycle [1–3]. Magnesium alloys have the inherent advantage of being the lowest density structural metal, although this is only an advantage if the property profile is good. Hence there has been a substantial effort to improve the properties of magnesium alloys over many years to be able to realise promised weight savings [2, 4, 5].

Most high-pressure die-cast alloys are based on the Mg–Al alloy system. Mg–Al alloys are the most commonly used alloys with AZ91 being the alloy most commonly used and AM50/60 used in applications that require greater energy absorption [6]. These alloys have inherently limited creep resistance and whilst some improvement can be made by the introduction of other alloying elements such as Si, rare earth (RE) elements, Ca and Sr, there appears to be a limit on the amount of improvement possible.

Magnesium–RE alloys have been a candidate alloy system for high end mechanical and elevated temperature properties. There are a number of systems that are age-hardenable (e.g. Nd, Y and Gd) [7], and it has been commonly known that adding rare earths to alloys improves the creep resistance, whether that be to Mg–Al alloys, Mg–Zn alloys or as a primary alloying element itself [8].

The process which Mg-alloys most easily lend themselves to is casting, particularly high-pressure die-casting [9, 10]. The inherent advantages include shorter production times due to lower latent and specific heat, reduced die soldering and excellent fluidity. There have been a number of successful Mg–RE based alloys developed for sand-casting and permanent-mould casting, including WE54/43 [11] and AM-SC1, but it has proven very challenging to make these alloys castable in high-pressure die-casting.


D. Orlov et al. (eds.), *Magnesium Technology 2018*, The Minerals, Metals & Materials Series, https://doi.org/10.1007/978-3-319-72332-7_50

A Mg–RE rich high-pressure die-cast alloy was proposed in about 2000 [12, 13] but interest was lost rapidly due to its poor castability. The CAST Co-operative Research Centre developed the alloy AM-SC1 which was a successfully trialled sand-cast/permanent-mould cast alloy [14–16] but when an alloy of similar composition (primarily without Zr) named HP2 [17] was high-pressure die-cast it was found to hot tear and could not reliably produce sound castings. The difficulties associated with this led to a long-term alloy development program to design a high-performance high-pressure die-cast alloy without it being too expensive.

## Alloy Development

This section describes the approach taken to develop a high-pressure die-castable Mg–RE alloy. The main consideration is how to get the RE mix correct, but minor elements such as Zn and Al are also critically important to the performance of the alloy.

### Rare Earths

Understanding the roles of individual RE elements was the key factor in being able to improve the performance of Mg-RE alloys. Until recently, it has been most common for rare earth additions to be made as a Ce-rich misch metal, often containing approximately 60%Ce, 30% La, and up to 10%Nd with some Pr and other rare earths also usually present in small amounts. Typically, the nomenclature for RE additions has been 'E' and to a large extent it was assumed that all rare earth elements behaved similarly. The main factor that was important to understand was how the different RE elements behaved when added to Mg in castability [18, 19], and in various properties such as strength [20], creep [21] and corrosion resistance [22, 23].

Studying binary alloy systems, allowed us to understand the behaviour of each particular RE and provided the building blocks to understand multi-RE systems. Whilst the mechanical properties of Mg–RE alloys did not vary too much between the different REs [20], other properties did vary considerably. The different behaviour that could be easily explained by their different phase diagrams (Fig. 1). Mg–Nd alloys have a much higher propensity to hot tearing than Mg–Ce and even more so compared to Mg–La alloys, which showed little hot tearing even at around additions of 1–2% where hot tearing tends to peak [19]. This was because of the larger solidification range for Mg–Nd alloys. Mg–Nd alloys however, had by far the best creep properties, which appears to be due to the greater solubility of Nd in Mg leading to increased levels of precipitation during creep [21]. Yield strength increased and ductility decreased with the addition of RE elements for all alloys and is related to the type of intermetallic and its volume fraction. Corrosion rate tended to also increase with the amount of intermetallic although there also appeared to be an effect of the changes in intermetallic morphology [22, 23].

Whilst some work has been performed to develop multi-component phase diagrams for these alloys [24–26], what was observed experimentally was that multi-component Mg–RE alloys were found to act as pseudo-binary eutectic alloys. All alloys containing La, Ce and Nd, except for binary Mg–Nd alloys [20, 27] consisted of primary α-Mg and a eutectic of α-Mg and a (La, Ce, Nd) $Mg_{12}$ intermetallic phase, where the different REs almost were interchangeably soluble in the intermetallic phase (Fig. 2). This meant that the RE mix can be used to tailor the solidification range, eutectic composition and the intermetallic morphology. For example, increasing the La content in the RE mix reduces the solidification range and increases the amount of eutectic but reduced the amount of solute in solid solution. This leads to a reduction in hot tearing and a decrease in creep resistance.

Since the RE elements were found to predominantly form the $REMg_{12}$ phase, a reasonably consistent relationship of strength and ductility was found with the total RE (TRE) content (Fig. 3). The slightly lower amount of $REMg_{12}$ phase in the Nd-rich alloys due to the greater amount of Nd in solid solution did not significantly affect the relationship. From data from over 30 alloys with different TRE contents and RE ratios it was observed that at TRE contents above 4 wt%, the alloy became too brittle to be useful. Hence it was clear that the TRE content needed to be in the range 3–4 wt%. Lower RE contents, depending on the mix, could lead to an increased hot tearing susceptibility.

The creep response and the hot tearing susceptibility of the alloys were found to be very dependent on the RE type as would be expected from the results of the binary alloys

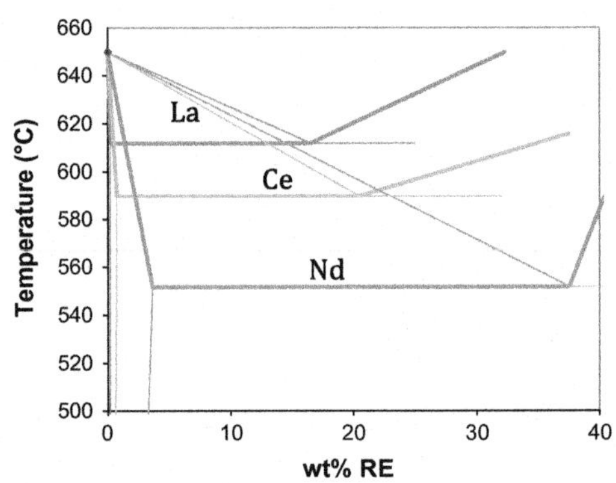


**Fig. 1** Mg–La, Mg–Ce, and Mg–Nd phase diagrams [28] overlaid on one another and assuming straight lines between invariant points to demonstrate the differences between the phase diagrams [19]

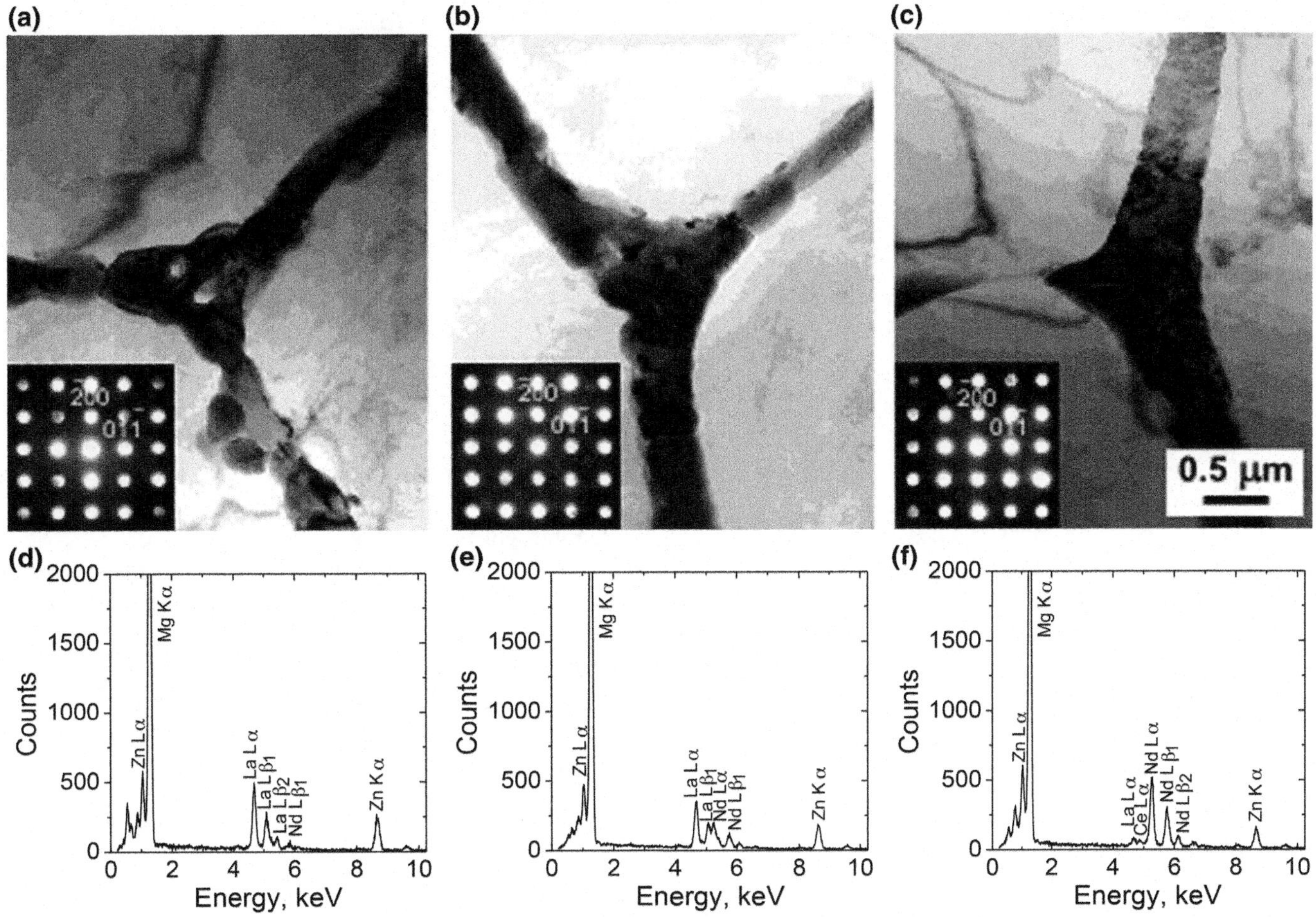


**Fig. 2** TEM bright field images, micro-beam electron diffraction patterns (B ≈ [011]) and EDX spectra from the interdendritic intermetallic phase in as-cast alloys with a TRE content comprising 7.3% Nd (**a** and **d**), 38.6% Nd (**b** and **e**) and 87% Nd (**c** and **f**). Images modified from [29]

[19, 21], where in an alloy containing approximately 4 wt% RE, alloys that are Nd-rich have high creep resistance and alloys that are La rich had high resistance to hot tearing (Fig. 4). It is an interesting observation that the initial creep response is highly influenced by the RE addition, with the primary creep level (defined here as the strain after 1 h) is an order of magnitude larger in the alloy with the lowest Nd content than with the highest Nd content. This is despite the alloy with 7.3% of its RE addition being Nd having the highest proof strength at the test temperature. Two of the conventional critical factors for describing creep behaviour are the creep strain at 100 h and the steady-state creep rate. There is clearly a dramatic improvement in creep response as the Nd content is increased and there is a transition in the behaviour at about the alloy containing 38.6% Nd in its RE mix for both of these indicators. It can be inferred from these results that the benefits of the Nd on the creep properties are virtually fully realised once Nd becomes the dominant RE element in the composition. There appears to be little to gain by increasing the Nd proportion significantly above 50% Nd in the RE mixture for optimum creep performance. However, below about 40% Nd the creep behaviour becomes extremely sensitive to composition and the creep performance deteriorates exponentially with decreasing Nd content in the alloy composition. It is clear that there are combinations that had relatively low hot tear susceptibility and very good creep resistance (27–38%Nd).

A castability die was developed to provide a comprehensive assessment of castability for magnesium alloys [1, 30]. The changes from HP2 to HP2+ were the increase in the amount of La present, but also a reduction in the Ce and Nd content along with a small addition of Y, which also assists with the creep performance [31] but can also affect hot tearing behaviour [32]. It is clear that there is a substantial improvement in castability (Fig. 5) with the modification in alloy composition [33].

Given the understanding of the role of RE elements the following alloy design considerations were developed:

- The total RE content needs to be less than 4.0 wt%. If it is higher than this the elongation to fracture drops below the 2% which can make the alloy too brittle.

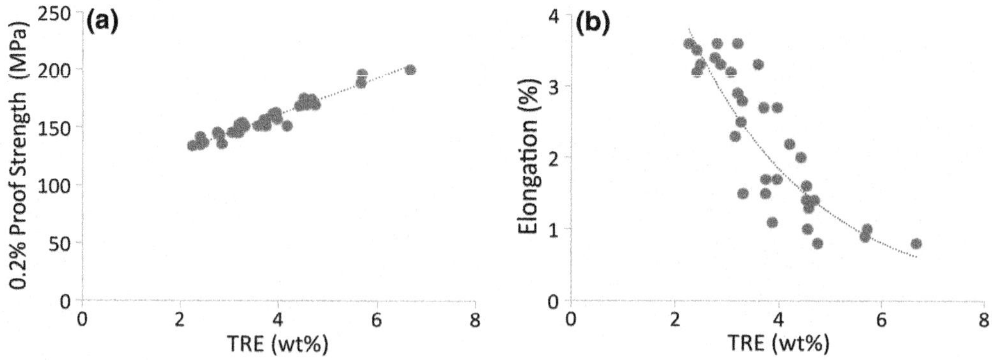


**Fig. 3** **a** 0.2% Proof strength and **b** Elongation to fracture plotted against the total rare earth content (La, Ce and Nd) for over 30 alloys that also contain 0.5–0.7 Zn and up to 0.06 Al

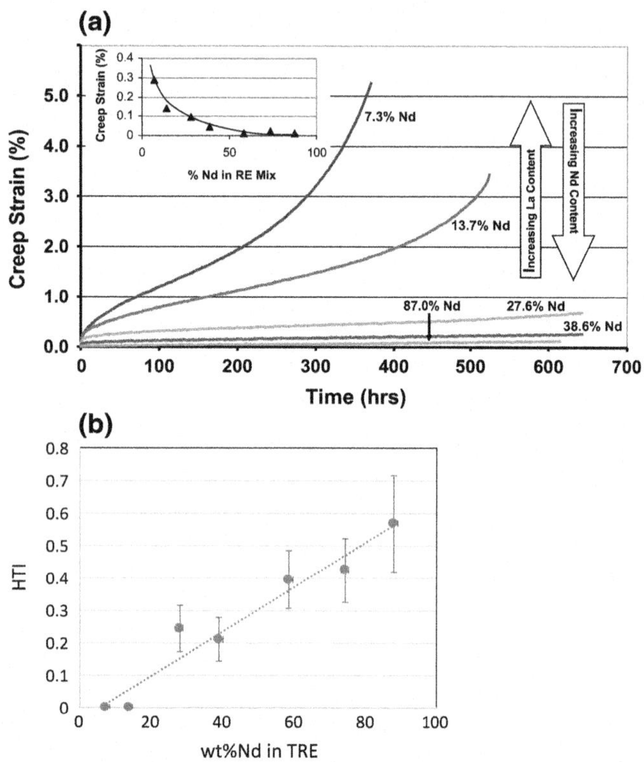


**Fig. 4** **a** Creep curves at 177 °C and 90 MPa. The inset shows the creep strain (%) after 1 h as a function of the percentage of Nd in the rare earth mixture. Modified from [29]. **b** Hot tearing index measurement out of a total of 4. To obtain measurement details see [19]. An alloy needs to have a value of less than 0.5 and preferably less than 0.2 to be castable. The original HP2 alloy had an HTI of 0.8–1.0. The alloys contain approximately 0.2 Ce so the Mg–RE alloys with different ratios of La to Nd in the TRE mix (Note: The percentage of Nd content in the alloy is indicated with the associated creep curve). The alloys also contained 0.5–0.6% Zn

- The total RE content needs to be above 2.5–3.0 wt%. This is particularly important for hot tearing, as unless the alloy contains almost entirely La then some hot tearing will be observed.
- La is preferred to be the major RE mixture. This is because Mg–La alloys are much less susceptible to hot tearing.
- RE elements with some solubility do need to be added as these are key to obtaining the excellent creep properties. Nd, Y and Gd are all possible alloying additions. See Gavras et al. [31, 32] for considerations.

## Zn Additions

Zn additions have also been found to be critical to the performance of Mg–RE alloys. However, again there is a 'trade-off' between hot tearing and creep performance. It has been found that Zn additions improve creep by reducing the minimum creep rate by an order of magnitude and to maintain very low creep-strain for prolonged duration of testing (Fig. 6). With the addition of Zn to Mg–Nd–La alloy a very high number density of $\gamma''$ precipitated on both basal and prismatic planes compared to that observed in the base alloy [34, 35]. Such higher number density is expected to enhance the load bearing capacity of the Zn-containing alloy. Examination of dislocation substructures (of Mg–Nd–La–Zn) at minimum creep rate indicated that both intragranular precipitates and interdendritic phases are contributing the creep resistance by regarding dislocation motion. In addition it has been shown that the presence of

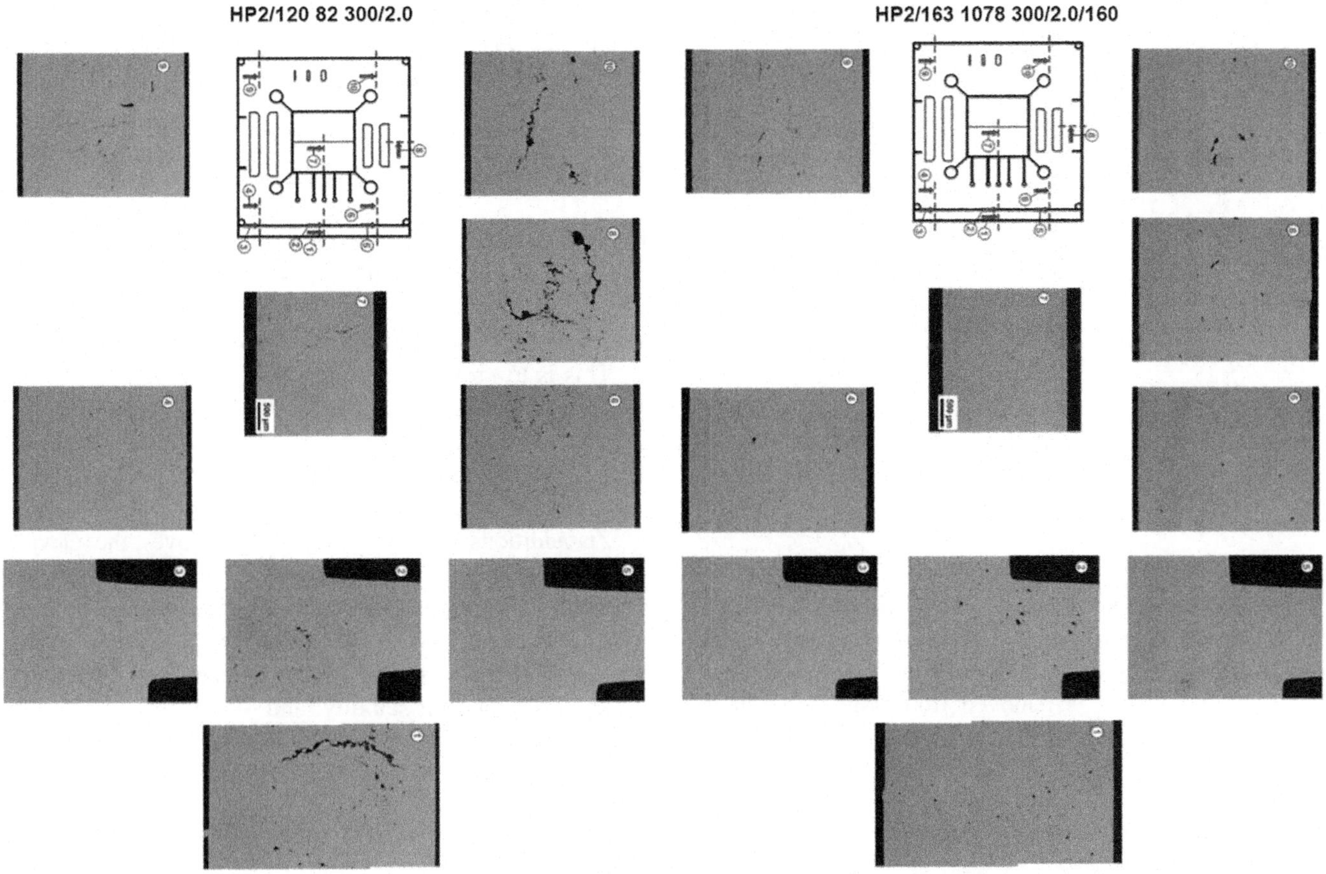


**Fig. 5** Typical castings in a castability die of the original Ce-rich alloy HP2 (Mg-1.62Nd-0.66Ce-0.37La-0.5Zn) and the modified HP2+ alloy (Mg-0.72Nd-0.34Ce-2.81La-0.4Zn-0.29Y) showing a dramatic reduction in the amount of solidification defects

both Zn and RE elements strengthen the interplanar bonding by imparting covalent bond character within the Mg-matrix and also at the γ″/Mg interfaces [36]. As a result, a distribution of such local covalently bonded regions inside the Mg-matrix increases the out-of-plane vacancy migration energy barrier by an order of magnitude compared to that without the presence of Zn and this makes dislocation climb more difficult. Therefore, the alteration of electronic structure, together with the influence on the precipitate morphology and distribution, highlights Zn as an effective alloying element for developing creep resistant Mg–RE alloys, including HP2+.

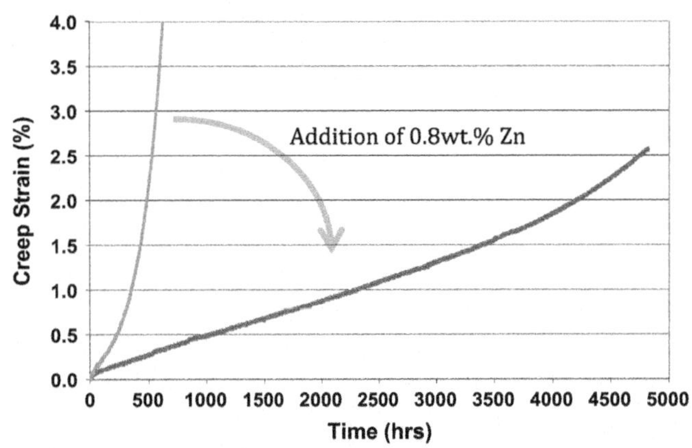


**Fig. 6** Creep curves obtained under the same test conditions for Mg-3.5 wt%Nd-2.49 wt%La (red curve) and Mg-3.3 wt%Nd-2.43 wt% La 0.77 wt%Zn (blue curve). Data from [34]

## Al Additions

One of the more surprising discoveries of the alloy development program was the importance of the amount of Al on creep. It is known that Al additions decrease the creep performance of Mg-based alloys. However, in these alloys it was found that a small amount of Al between 0.05–0.20 wt % affected the creep response dramatically (Fig. 7).

The addition of Al to HP2-type compositions appears to result in the formation of a second intermetallic phase within the interdendritic regions and in association with the Mg-RE compound observed in low Al content alloys. This Al-containing intermetallic phase has a more blocky

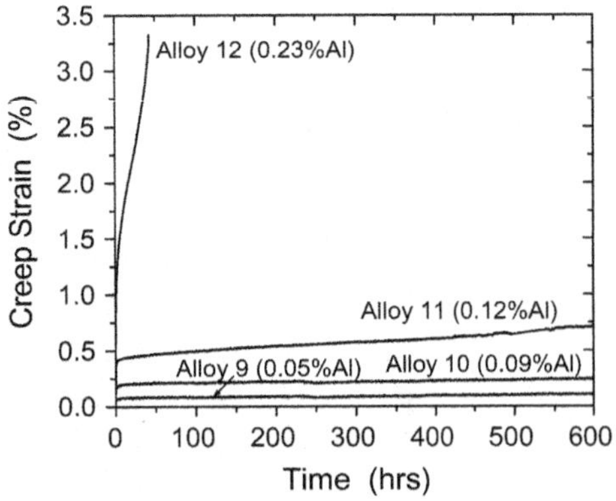


**Fig. 7** Creep curves at 177 °C/90 MPa showing the influence of Al in a base alloy composition of Mg-1.43 wt%Nd-0.55 wt%Ce-0.33 wt% La-0.59 wt%Zn

morphology than the $REMg_{12}$ phase (Fig. 8). It appears that the Al-containing intermetallic is also richer in Nd than the Mg–RE phase. As such, it is therefore likely that the Nd that is available to precipitate during creep in the alloy containing 0.05 wt% Al, Fig. 8, is removed from solid solution and tied-up in the second compound that forms on solidification in the 0.23 wt% Al alloy and that this results in a dramatic deterioration in the creep response.

## Alloy Composition and Properties

From these considerations an alloy called HP2+ was defined as containing the following elements:

- 2.0–2.8 wt% La/Ce with La being the major element. This is to provide a castable base to the alloy;
- Up to 1.0 wt%Nd and 0.2 wt%Y can be added to the alloy to improve alloy performance, particularly at elevated temperatures but this needs to be 'traded off' against an increase susceptibility to hot tearing;
- Zn additions at up to 0.5 wt% improves the elevated temperature mechanical properties but again Zn additions increase hot tearing susceptibility;
- Al less than 0.2 wt% to improve the creep resistance;
- Y also improves melt stability, Be can also be added at a few ppm for melt stability; and

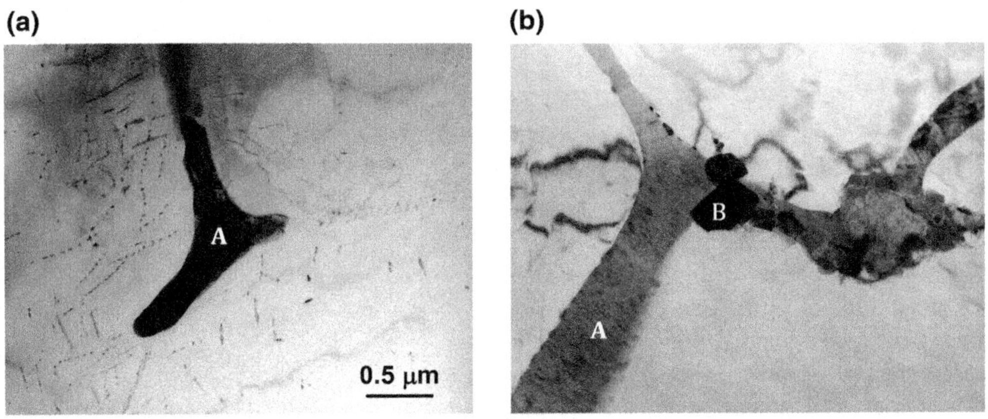


**Fig. 8** Bright-field micrographs from crept specimens (at 177 °C, 90 MPa) from **a** an Al content of 0.05% showing a triple point, with arrays of fine precipitates decorating dislocations in close proximity to grain boundaries and **b** an Al content of 0.23%, showing an absence of precipitates and two types of interdendritic phases, the Mg–RE-based continuous intermetallic (labelled 'A') and a small blocky phase that is Al–RE (Nd-rich) (labelled 'B')

**Table 1** A comparison of the main properties of HP2+ with other die-cast magnesium alloys and aluminium alloy A380

| | Room Temp | | 177 °C | | Creep Strength[a] @177 °C (MPa) | Material Cost (Relative to AZ91)[b] |
|---|---|---|---|---|---|---|
| Alloy | 0.2% YS (MPa) | Elong (%) | 0.2% YS (MPa) | Elong (%) | | |
| HP2+ | 174.6 | 3.7 | 125.9 | 10.6 | 102 | 1.30 |
| AE44 | 129.6 | 13.2 | 96.1 | 34.2 | 82 | 1.30 |
| AXJ530 | 187.7 | 4.2 | 146.1 | 13.4 | 92 | 1.02 |
| A380 | 165.6 | 4.4 | 165.5 | 6.5 | 72 | ~1.0 |

[a]Stress to produce 0.1% creep strain after 100 h
[b]Based on metal prices on July 15, 2014

- Small additions of Zr and/or Mn can be used to reduce the detrimental effects of transition metals, especially Fe, on corrosion.

HP2+ has been shown to have premium high temperature performance compared with other commercial and semi-commercial alloys (Table 1) [2]. The addition of Nd and Y also put it on the premium end of the cost curve with approximately 40% increase in the cost of the elemental additions over AZ91 [37]. It should be noted that the due to the use of RE elements and the tight control on some of the elemental additions, there may be some additional costs in the manufacture of the alloy. At room temperature, its ductility is a little lower than many of the other alloys, however, the reason for this is the continuous intermetallic phase, which is the dominant phase in the eutectic, and eutectic is required to reduce the hot tearing susceptibility of the alloy. It is likely that this intermetallic scaffold improves the strength especially at high temperatures [38], which is also a feature of die-cast Al alloys used at elevated temperatures.

## Conclusions

The path to develop a new alloy is a very complicated one. Often the optimization of one property leads to a reduction in another. In the development of a high-pressure die-cast alloy HP2+, important learnings included:

- The choice of the RE element mixture has a significant effect on the hot tearing susceptibility during high-pressure die-casting. To achieve suitable castability it is necessary to have sufficient La in the RE mixture to allow for the formation of a volume fraction of eutectic phase and to reduce the freezing range, both are required to prevent hot tearing.
- The choice of the RE element mixture has a significant effect on the creep behaviour of HPDC Mg–RE alloys. To achieve good elevated temperature properties it is necessary to have Nd (or Y) as the dominant RE addition. It is likely that the improved creep resistance, defined by the primary creep response and the steady state creep rate, is related to the presence of neodymium in solid solution.
- The selection/control of appropriate minor alloying elements can have dramatic effects on the alloy property balance, particularly the creep response. In particular Zn and Al are of great importance.

**Acknowledgements** The authors would particularly like to acknowledge the late Prof. Gordon Dunlop, who over the expanse of his career as the CEO of the CAST CRC and at Advanced Magnesium Technologies as an industry partner, started and always supported this alloy development journey. There were many people that contributed to this work over the years, but two of the most important were Andy Yob and Maya Gershenzon, who made tens of thousands of castings on the Toshiba high-pressure die-casting machine at CSIRO.

## References


1. Easton, MA, SM Zhu, TB Abbott, M Dargusch, MT Murray, G Savage, N Hort and MA Gibson (2016). Evaluation of Magnesium Die-casting Alloys for Elevated Temperature Applications: Castability. Adv. Engng. Mater. 18(6): 953–962.
2. Zhu, SM, MA Easton, TB Abbott, JF Nie, M Dargusch, N Hort and MA Gibson (2015). Evaluation of Magnesium Die-casting Alloys for Elevated Temperature Applications: Microstructure, Tensile Properties and Creep Resistance. Metall. Mater. Trans. A 46(8): 3543–3554.
3. Cáceres, CH (2007). Economical and Environmental Factors in Light Alloys Automotive Applications Metall. Mater. Trans. A 38 (7): 1649–1662.
4. Luo, A (2004). Recent Mg alloy development for elevated temperature applications. Int. Mater. Rev. 49(1): 13–30.
5. Pekgüleryüz, MÖ and M Celikin (2010). Creep resistance in magnesium alloys. Int. Mater. Rev. 55(4): 197–217.
6. Aune, TK, H Westengen and T Ruden (1993). Mechanical Properties of Energy Absorbing Magnesium Alloys: SAE paper no. 930418.
7. Apps, PJ, H Karimzadeh, JF King and GW Lorimer (2003). Precipitation reactions in magnesium-rare earth alloys containing yttrium, gadolinium or dysprosium. Scripta Mater. 48: 1023–1028.
8. Nie, JF (2012). Precipitation and Hardening in Magnesium Alloys. Metall. Mater. Trans. A 43(11): 3891–3939.
9. King, JF (2007). Magnesium: commodity or exotic? Mater. Sci. Technol. 23(1): 1–14.
10. Abbott, TB (2015). Magnesium: Industrial and Research Developments Over the Last 15 Years. Corrosion 71(2): 120–127.
11. Ahmed, M, GW Lorimer, P Lyon and R Pilkington (1992). The Effect of Heat Treatment and Composition on the Microstructure and Properties of Cast Mg-Y-RE Alloys. Magnesium Alloys and Their Applications: 301–308.
12. Moreno, IP, TK Nandy, JW Jones, JE Allison and TM Pollock (2001). Microstructural Characterisation of a Die-Cast Magnesium-Rare Earth Alloy. Scripta Mater. 45: 1423–1429.
13. Moreno, IP, TK Nandy, DS Jones, JE Allison and TM Pollock (2003). Microstructural stability and creep of rare-earth containing magnesium alloys. Scripta Mater. 48: 1029–1034.
14. Gibson, MA, CJ Bettles, SM Zhu, MA Easton and JF Nie (2009 of Conference). Microstructure and mechanical properties of a Mg-rare earth based alloy AM-SC1. Magnesium Technology 2009, The Minerals, Metals and Materials Society, Warrendale PA: 243–248.
15. Kunst, M, A Fisherworring-Bunk, MA Gibson and GL Dunlop (2008 of Conference). Mechanical property evaluation of permanent-mould cast AM-SC1 Mg-alloy. SAE 2008 world congress, Detroit, USA: SAE technical paper 2008010375.
16. Bettles, CJ, MA Gibson and SM Zhu (2009). Microstructure and Mechanical Behaviour of an Elevated Temperature Mg-Rare Earth Based Alloy. Mater. Sci. Engng. A 505(1–2): 6–12.
17. Gibson, MA, CJ Bettles, MT Murray and GL Dunlop (2006). AM-HP2: A new magnesium high pressure die-casting alloy for automotive powertrain applications. Magnesium Technology 2006, The Minerals, Metals and Materials Society, Warrendale, PA: 327–332.

18. Easton, MA, S Gavras, MA Gibson, S Zhu, JF Nie and T Abbott (2016 of Conference). Hot tearing in Magnesium - Rare Earth Alloys. Magnesium Technology 2016, Nashville, TN, USA, TMS (The Minerals, Metals & Materials Society): 123–128.
19. Easton, MA, MA Gibson, S Zhu and T Abbott (2014). An a priori hot tearing indicator applied to die-cast magnesium-rare earth alloys. Metall. Mater. Trans. A 45A(8): 3586–3595.
20. Chia, TL, MA Easton, SM Zhu, MA Gibson, N Birbilis and JF Nie (2009). The effect of alloy composition on the microstructure and tensile properties of binary Mg-rare earth alloys. Intermetallics 17: 481–490.
21. Zhu, SM, MA Gibson, MA Easton and JF Nie (2010). The relationship between microstructure and creep resistance in die-cast magnesium-rare earth alloys. Scripta Mater. 63(7): 698–703.
22. Birbilis, N, MK Cavanaugh, AD Sudholz, S Zhu, MA Easton and MA Gibson (2011). A combined neural network and mechanistic approach for the prediction of corrosion rate and yield strength of magnesium-rare earth alloys. Corr. Sci. 53(1): 168–176.
23. Birbilis, N, MA Easton, AD Sudholz, SM Zhu and MA Gibson (2009). On the corrosion of binary magnesium-rare earth alloys. Corr. Sci. 51(3): 683–689.
24. Gröbner, J, A Kozlov, R Schmid-Fetzer, MA Easton, S Zhu, MA Gibson and JF Nie (2011). Thermodynamic analysis of as-cast and heat-treated microstructures of Mg–Ce–Nd alloys. Acta Mater. 59 (2): 613–622.
25. Gröbner, J, M Hampl, R Schmid-Fetzer, MA Easton, S Zhu, MA Gibson and JF Nie (2012). Phase analysis of Mg-La-Nd and Mg-La-Ce alloys. Intermetallics 28: 92–101.
26. Schmid-Fetzer, R, J Gröbner, A Kozlov, M Hampl, MA Easton, S Zhu, MA Gibson and JF Nie (2013 of Conference). Thermodynamics of phase formation in Mg-La-Ce-Nd alloys. Magnesium Technology 2013, San Antonio, Texas, TMS (The Minerals, Metals & Materials Society): 243–248.
27. Easton, MA, MA Gibson, D Qiu, SM Zhu, J Gröbner, R Schmid-Fetzer, JF Nie and MX Zhang (2012). The role of crystallography and thermodynamics on phase selection in binary magnesium–rare earth (Ce or Nd) alloys. Acta Mater. 60: 4420–4430.
28. Rokhlin, LL (2003). Magnesium alloys containing rare earth metals. New York, USA, Taylor & Francis.
29. Zhu, SM, MA Gibson, JF Nie, MA Easton and CJ Bettles (2009). On the Creep Resistance of HPDC Mg-RE based alloys. Mater. Sci. Forum 618–619: 453–458.
30. Strobel, K, MA Easton, V Tyagi, MT Murray, MA Gibson, G Savage and T Abbott (2010). Evaluation of the castability of high pressure die cast magnesium based alloys. Int. J. Cast Met. Res. 23 (2): 81–91.
31. Gavras, S, S Zhu, JF Nie and MA Gibson (2016). On the microstructural factors affecting creep resistance of die-cast Mg–La-rare earth(Nd, Y. or Gd) alloys. Mater. Sci. Engng. A 675: 65–75.
32. Gavras, S, MA Easton, MA Gibson, S Zhu and JF Nie (2014). Microstructure and property evaluation of high-pressure die-cast Mg–La–rare earth (Nd, Y or Gd) alloys. J. Alloy. Compd. 597: 21–29.
33. Gibson, MA, MA Easton, V Tyagi, MT Murray and GL Dunlop (2008 of Conference). Further improvements in HPDC Mg alloys for powertrain applications. Magnesium Technology 2008, New Orleans, LA, The Metals, Minerals and Materials Society: 227–232.
34. Choudhuri, D, D Jaeger, MA Gibson and R Banerjee (2014). Role of Zn in enhancing the creep resistance of Mg–RE alloys. Scripta Mater. 86: 32–35.
35. Choudhuri, D, D Jaeger, S Srivilliputhur, MA Gibson and R Banerjee (2015 of Conference). Creep response of a Zn containing Mg-Nd-La alloy. Magnesium Technology 2015, John Wiley & Sons Inc., NJ, USA, 2015: 35–39.
36. Choudhuri, D, S Srivilliputhur, MA Gibson and R Banerjee (2017 of Conference). Bond environments in a creep resistant Mg-RE-Zn alloy. Magnesium Technology 2017, Springer International Publishing, Cham, Switzerland: 471–475.
37. Easton, MA, S Zhu, MA Gibson, T Abbott, HQ Ang, XB Chen, N Birbilis and G Savage (2017 of Conference). Performance Evaluation of High-Pressure Die-Cast Magnesium Alloys. Magnesium Technology 2017, San Diego, CA, The Minerals, Metals and Materials Society: 123–129.
38. Zhang, B, S Gavras, AV Nagasekhar, C Caceres and MA Easton (2014). The Strength of the Spatially Interconnected Eutectic Network in HPDC Mg-La, Mg-Nd, and Mg-La-Nd Alloys. Metall. Mater. Trans. A 45(10): 4386–4397.

# Creep Resistant Mg–Mn Based Alloys for Automotive Powertrain Applications

Mert Celikin and Mihriban Pekguleryuz

**Abstract**

Quaternary Mg–Sr–Mn–Ce alloys were developed for automotive powertrain applications. The design strategy was based on previous studies of the authors studies on the creep behaviour of subsystems (Mg–Mn, Mg–Ce–Mn, Mg–Sr–Mn) and the role of Mn in the dynamic precipitation of fine nano-scale dispersoids. In the present work, the creep resistance and the microstructural evolution of the selected quaternary Mg–Sr–Mn–Ce compositions were investigated. The final creep strain in the quaternary alloys was seen to be four times lower than the ternary Mg–Sr–Mn alloys creep tested at 200 °C under 50 MPa stress. The creep strengthening was attributed mainly to the dynamic co-precipitation of $Mg_{12}Ce$ and α-Mn phases.

**Keywords**

Magnesium alloy • Rare-Earth • Creep • Transmission electron microscopy • Dynamic precipitation

## Introduction

Environmental concerns have dictated the reduction in weight to be the key factor in alloy development strategies for automotive applications. The most significant weight reduction can be achieved by replacing steel (7.75–8.05 g/cm$^3$) and aluminium (2.70 g/cm$^3$) powertrain components (engine block and transmission case) with the lightest structural metal, magnesium (Mg) with a density of 1.74 g/cm$^3$. Today, the use of Mg alloys in powertrain components is limited due to the low creep resistance of the most common Mg alloys (AZ91, AM50) [1–3]. It is known that silicon, rare-earths (RE) and alkaline earths improve the creep resistance of Mg–Al alloys. The important examples in the past are the use of Mg–Al–Si alloys by VW in the 70s, and more recently the use of AE (Mg–Al–RE) alloy in the Corvette engine cradle and the AJ (Mg–Al–Sr) alloy by all BMW models between 2004 and 2014 in the Mg engine block with aluminium insert. AE and AJ alloys may not offer adequate performance for many monolithic engine blocks exposed to 40–50 MPa at temperatures of 175–200 °C. Extensive studies in the last two decades focused on RE alloying additions to Mg for improving the creep resistance at the expense of production cost [4–6]. Hence, the development of cost-efficient Mg alloys to be used by the industry necessitates limiting the use of RE additions at low concentrations through the combined use of alloying additions.

Mn-bearing Mg–Sr–Mn alloys previously studied by the authors revealed that the dynamic precipitation of α-Mn from the supersaturated α-Mg matrix is the main factor for the creep resistance of the alloys up to 175 °C [7, 8]. Dislocation pinning by dynamically formed α-Mn precipitates is the governing factor in creep resistance, but an increase in the amount of $Mg_{17}Sr_2$ dendritic phase also improves the creep performance.

This study aims at understanding the effects of minor cerium (Ce) additions on the creep resistance of ternary Mg–Sr–Mn alloys. The effect of Ce on Mg has been studied in the past. The main intermetallic phase, $Mg_{12}Ce$, in binary Mg–Ce alloys exists both as coarse interdendritic eutectics and as intradendritic solid state precipitates [5, 6, 9, 10]. Studies on Mg–Ce alloys show that the dynamically formed intradendritic $Mg_{12}Ce$ precipitates have a more dominant effect than the coarse interdendritic $Mg_{12}Ce$ eutectic on creep resistance [6, 11]. The removal of interdendritic phase via solution treatment in Mg–Ce alloys is reported not to have any effect on the creep properties at 200 °C [4], however, substantial improvement in creep strength was obtained with the refinement of precipitate size via Mn additions to the binary Mg–Ce system [4–6, 12]. The authors have

M. Celikin (✉) · M. Pekguleryuz
Mining and Materials Engineering, McGill University, 3610 University St., Wong Bldg., Rm 2140, Montreal, QC H3A 2B2, Canada
e-mail: mert.celikin@mail.mcgill.ca

M. Pekguleryuz
e-mail: mihriban.pekguleryuz@mcgill.ca


D. Orlov et al. (eds.), *Magnesium Technology 2018*, The Minerals, Metals & Materials Series, https://doi.org/10.1007/978-3-319-72332-7_51

previously reported that dynamically co-precipitated α-Mn and $Mg_{12}Ce$ phases mutually affect the nucleation and growth of the precipitates leading to both a decrease in precipitate size and an increase in the number of precipitates in ternary Mg–Ce–Mn alloys [6].

The present work investigates the dynamic precipitates and their role in creep resistance in two quaternary compositions, Mg–5Sr–2Mn–0.5Ce and Mg–3Sr–2Mn–0.5Ce (% wt.), and compares to the dynamic precipitation observed in Mg–(3–5) Sr–2Mn ternary alloys through creep tests and microstructural characterization.

## Experimental Procedure

Mg–Sr–Mn–Ce alloys were prepared in a Norax induction furnace using pure Sr (99.99%wt.) and pure Ce (99.7%wt.) supplied by HEFA, pure Mg (99.9%wt.) and Mg–Mn (5%wt.) master alloy (Applied Magnesium—formerly Timminco). Consecutive additions of Mg–Mn master alloy, pure Sr and pure Ce was done under a protective atmosphere of $CO_2$ + 0.5%SF6 at 700 °C. The melt was cast from a temperature of 735–740 °C into a boron-nitride coated steel-mold (preheated to 400 °C prior to casting) to produce cylindrical castings 200 mm in length and 30 mm diameter. Cast bars were machined into tensile creep specimens according to the ASTM E8-04 standard (gauge length: 50 mm and gauge diameter: 12.7 mm) [13]. The chemical composition of the cast bars was determined via ICP-OES analysis (Table 1). Compositions in this work are presented in wt% unless otherwise stated. Tensile creep tests were performed under 40–50 MPa stress at a temperature of 175–200 °C on as-cast alloys. Two extensometers (B1-error (ε) $\lesssim$ 0.0001 mm/mm) were attached for displacement measurements during each creep test. Thermocouples with an error of ±0.1 °C were connected to the creep specimens for temperature control. Scanning electron microscopy (SEM) analysis are conducted (Hitachi SU3500 SEM at 20 kV) on samples that are ground and polished up to 1200 grid and 1 μm diamond paste respectively. Samples for (scanning) transmission electron microscopy (TEM) investigation were first cut as plates of 1.0 mm thickness, and then ground down to 1200 grid to obtain 0.2 mm thick plates. Discs, 3 mm in diameter, were punched and then Ion-beam Polishing (PIPS) was conducted using GATAN 691 PIPS at 4 keV (Gun tilt ±4°). While TEM analysis was conducted using Philips CM200 microscope at 200 kV, STEM analysis for high resolution EDS-mapping was conducted using Hitachi SU-8500 SEM. For XRD analysis, Bruker D-8 Discovery X-ray Diffractometer was used with Cu–Kα source at 40 kV and 40 mA (step size: 0.005, step time: 300 s/frame, 2θ range: 3.8–103°).

## Results and Discussion

### Microstructural Evolution

As-cast Mg–Sr–Mn–Ce alloys show typical dendritic solidification structures. SEM/EDS analysis indicated that Ce is present in dissolved state both within the matrix and in interdendritic phases, whereas Mn is mainly concentrated in intradendritic regions (Fig. 1a). The concentration of Mn solely in intradendritic regions is due to its peritectic nature in Mg. It was reported that the as-cast microstructures of ternary Mg–Sr–Mn alloys with 3–5 Sr and 2 Mn (wt%) consist of α-Mg matrix (supersaturated in Mn), primary α-Mn particles and $Mg_{17}Sr_2$ intermetallic phase in interdendritic regions [7, 8]. XRD and EDS analyses (Fig. 1b) confirm that these three phases are also present in the Ce containing quaternary alloys. No precipitates were found in intradendritic regions of the quaternary alloys in the as-cast condition.

Upon annealing at 250 °C for 200 h, no significant coarsening was observed in the interdendritic $Mg_{17}Sr_2$ phase while primary α-Mn phases increased in size (and amount) consistent with the appearance of α-Mn peak in XRD spectra (Fig. 1b). The most notable change in the microstructure upon annealing is the precipitation from supersaturated α-Mg matrix in intradendritic regions.

Apart from morphology and size variation, precipitates can be distinguished with their difference in effective atomic number. Two types of precipitates with different effective atomic numbers were observed (Fig. 2) in annealed structures: (i) Precipitates with higher effective atomic number (darker) are present both as polygonal, (100 nm) and rod-like (5–20 nm in width and 100–200 nm in length); (ii) precipitates with lower effective atomic number (less dark) formed both as thin plates, (~1 μm in length and ~5 nm in width) and as polygones of irregular morphology, (~100 nm in size). High resolution EDS maps showed a distinct elemental distribution for the precipitates (Fig. 2b);

**Table 1** Alloy compositions, wt% (ICP-OES analysis)

| Alloy | Designation | Mn | Ce | Sr |
|---|---|---|---|---|
| Mg–3Sr–2Mn | JM32 | 2.09 | 0.06 | 3.27 |
| Mg–3Sr–2Mn–0.5Ce | JME321 | 1.90 | 0.43 | 3.50 |
| Mg–5Sr–2Mn | JM52 | 1.91 | 0.06 | 5.56 |
| Mg–5Sr–2Mn–0.5Ce | JME521 | 2.01 | 0.59 | 6.16 |

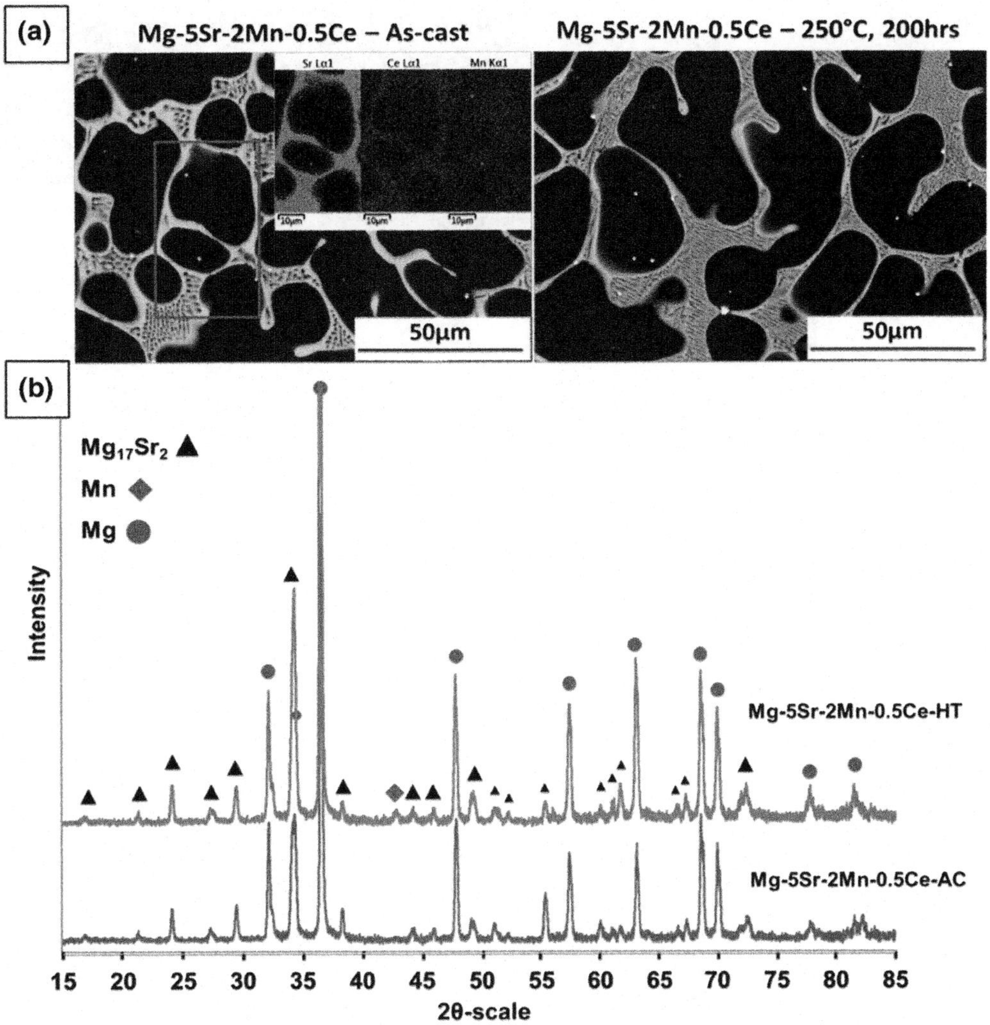


**Fig. 1** **a** SEM images (with EDS maps) and **b** XRD patterns of Mg–5Sr–2Mn–0.5Ce alloy in as-cast (AC) and heat treated (HT-250 °C, 200 h) conditions

Ce is concentrated in lower atomic number regions (Ce-rich precipitates), whereas darker precipitates mainly consist of Mn (Mn-rich precipitates). Sr is also present in trace levels (Fig. 3b) in the intra-dendritic regions mainly in precipitates consistent with its low solubility in Mg (0.11 wt% at 585 °C [14]).

*Mn-rich precipitates*: The precipitates are most likely α-Mn (Sr), in line with the binary Mn–Sr phase diagram reporting the solubility limit of Sr in α-Mn as ~0.25 wt% at 700 °C [15]. The morphology of rod-like precipitates (Fig. 2a) is similar to the α-Mn precipitates observed in Mg–Mn [16] binary and Mg–Sr–Mn ternary alloys previously studied [7, 8]. On the other hand, larger polygonal Mn-precipitates closely associated with Ce-rich precipitates (Fig. 2b) seen in this present work agree with α-Mn precipitate morphology found in Mg–Ce–Mn alloys [6]. Hence, it can be seen that the presence of Ce-rich precipitates alters the size and morphology of α-Mn precipitates dynamically formed from the supersaturated Mg matrix.

*Ce-rich precipitates*: In the Mg–Ce and Mg–Ce–Mn systems, at Ce concentrations (up to ~32 wt%Ce in the Mg–Ce binary and ~25 wt%Ce in the Mg–Ce–Mn ternary systems) the stable phase is β-$Mg_{12}Ce$. Since no phase forms between Ce and Mn [14], Ce-rich precipitates with thin-plate morphology are likely $Mg_{12}Ce$ phase. Long plate-like morphology was reported for $Mg_{12}Ce$ phase in numerous studies [6, 12, 17]. Moreover, the equilibrium β precipitates with irregular (leaf-like) morphology as shown in Fig. 2.b, were also found in earlier studies in association with $\beta_1$ ($Mg_3Ce$) phases [18].

## Creep Strengthening

The results of creep tests conducted at temperature and stress ranges of 175–200 °C and 40–50 MPa are shown in Fig. 3. The addition of a low level of Ce to the ternary Mg–5Sr–2Mn results in four times lower strain value in

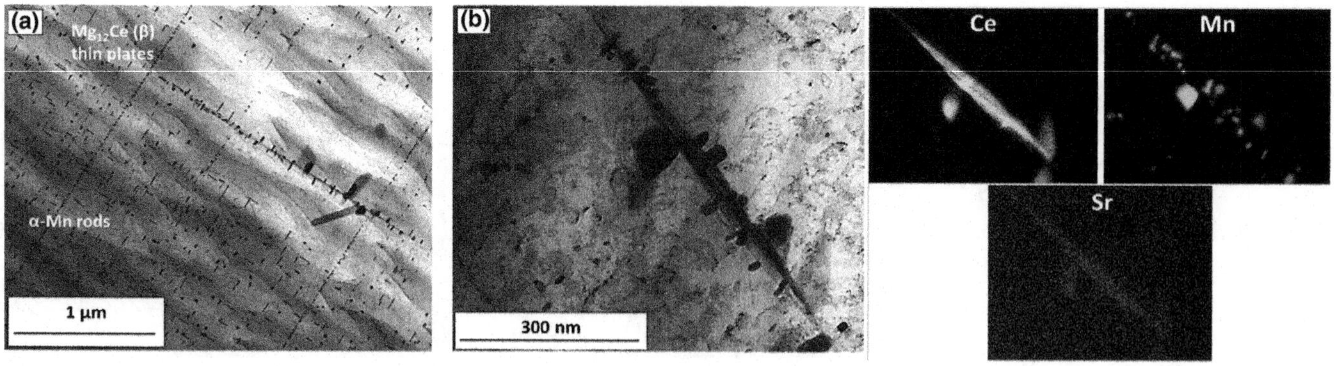


**Fig. 2** **a** STEM images of intra-dendritic precipitates with changing size and morphology formed upon annealing at 250 °C for 200 h in Mg–5Sr–2Mn–0.5Ce. EDS-maps of precipitates **b** indicating presence of two separate precipitates with concentration of Mn or Ce with trace Sr presence

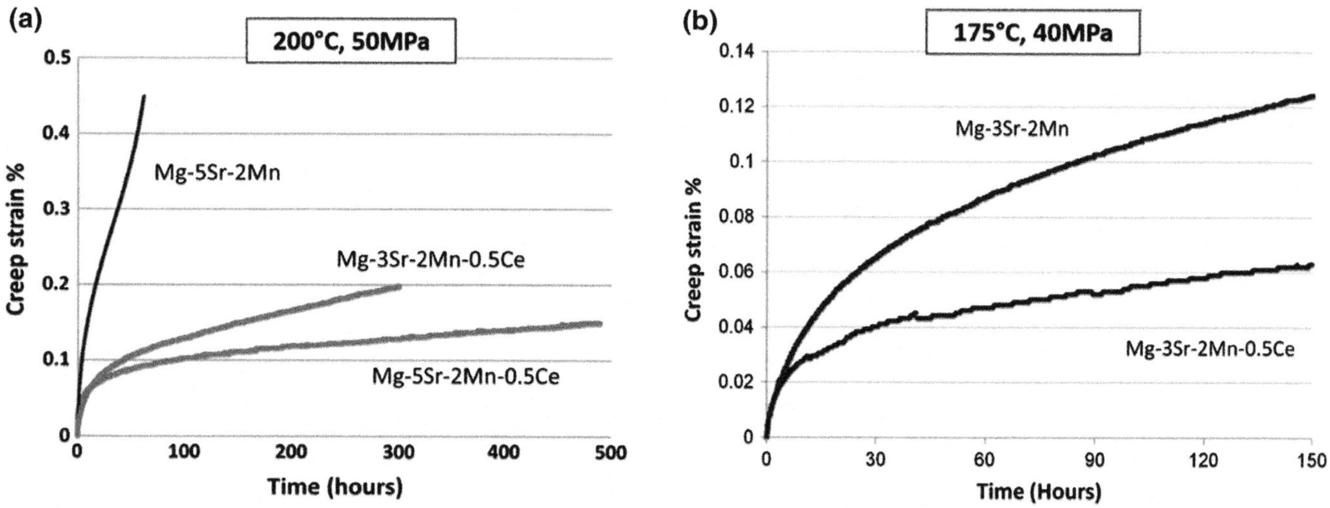


**Fig. 3** The creep results for two different conditions; **a** 200 °C, 50 MPa and **b** 175 °C, 40 MPa indicating the effectivity of dynamic $Mg_{12}Ce$/α-Mn co-precipitation in quaternary Mg–Sr–Mn–Ce alloys

Mg–5Sr–2Mn–0.5Ce (Fig. 3a). With the increase in Sr level from 3 to 5%, an improvement in creep resistance (200 °C, 50 MPa, 300–500 h) was also observed. However, it should be noted that Ce concentration in Mg–5Sr–2Mn–0.5Ce was ~0.15% higher compared to the Mg–3Sr–2Mn–0.5Ce alloy (Table 1), hence, the increase in creep resistance can also be attributed to the additional Ce. The minimum creep rate for Mg–5Sr–2Mn–0.5Ce (JME521) at 200 °C under 50 MPa was 2.78 * $10^{-10}$/s; this represents higher creep performance than commercial Mg alloys tested under similar conditions [2].

TEM/Selected area electron diffraction (SAED) analysis showed that $Mg_{12}Ce$ thin-plates which were dynamically co-precipitated with α-Mn (Sr) phase in the quaternary alloys had a preferred orientation. Diffraction patterns obtained at $B = [01\bar{1}1]_{Mg}$ indicate that they lie on the $\{2\bar{1}\bar{1}0\}_{Mg}$ planes (Fig. 4a). This is consistent with the orientation of the $Mg_{12}Ce$ precipitates (observed in Mg–Ce binary [12] and Mg–Ce–Mn ternary [6] alloys) where plate-like precipitates lie along $[0001]_{Mg}$. Moreover, coherency strain fields perpendicular to the surface of $Mg_{12}Ce$ thin-plates $(g = 2\bar{1}\bar{1}0)$ were also observed. TEM analysis of creep tested JME521 alloy also shows that dislocations are pinned via both the $Mg_{12}Ce$ and α-Mn (Sr) precipitates (Fig. 4c). Hence, the dynamic co-precipitation of $Mg_{12}Ce$ along with α-Mn, as well as α-Mn precipitation itself act as obstacles against dislocation motion during creep

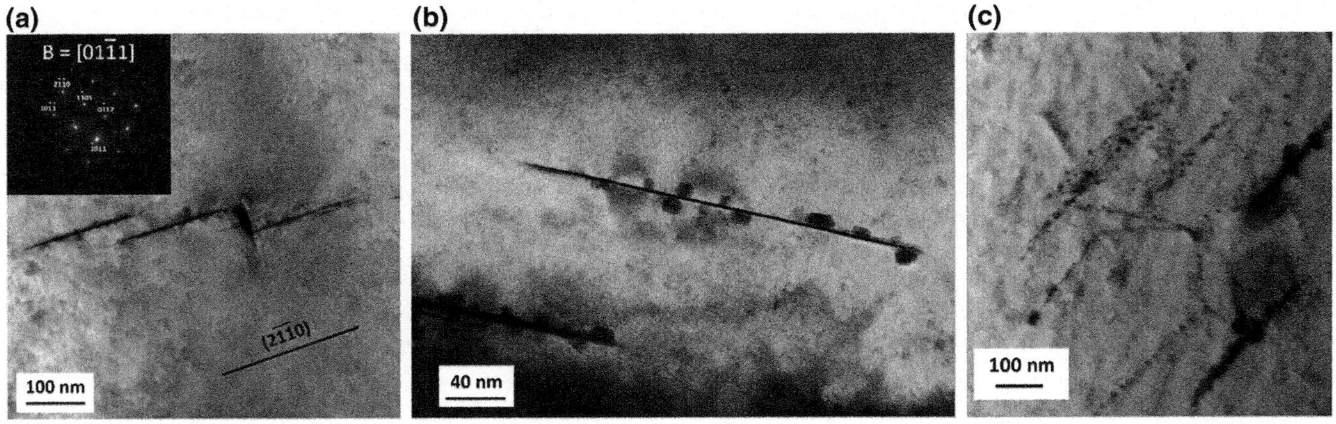


**Fig. 4** **a** TEM Bright-field images of dynamic co-precipitates of $Mg_{12}Ce$ and α-Mn observed in JME521 after creep testing at 200 °C, 50 MPa, 300 h with corresponding SAED at $[01\bar{1}1]_{Mg}$, **b** Elastic coherency strain fields $(g = 2\bar{1}\bar{1}0)$ perpendicular to the surface of $Mg_{12}Ce$ thin plates and **c** Dislocation pinning via dynamic co-precipitates

deformation. This study shows that dynamic precipitation is one of the key factors that can positively alter the creep resistance of Mg alloys.

## Conclusions

Mg–Sr–Mn–Ce alloys exhibit lower creep strain than the ternary Mg–Sr–Mn alloys tested at 200 °C, 50 MPa. The microstructure of the quaternary alloy is characterized by interdendritic $Mg_{17}Sr_2$, primary α-Mn phase and Mg matrix in the as-cast state. Upon annealing, co-precipitation of α-Mn and $Mg_{12}Ce$ was observed in intradendritic regions. Dynamically formed $Mg_{12}Ce$ thin-plates were oriented preferentially on the $\{2\bar{1}\bar{1}0\}_{Mg}$ planes found to be effective in dislocation pinning. Hence, dynamic co-precipitation of $Mg_{12}Ce$ and α-Mn was found to be responsible for creep strengthening.

This study has been supported through a Discovery Grant from the Natural Sciences and Engineering Research Council (NSERC) of Canada. The authors thank Pierre Vermette of McGill University for assistance in casting experiments.

## References

1. Celikin, M., F. Zarandi, D. Sediako, and M.O. Pekguleryuz, (2009) Compressive creep behaviour of cast magnesium under stresses above the yield strength and the resultant texture evolution, Canadian Metallurgical Quarterly, **48**(4): p. 419–432.
2. Luo, A.A., (2004) Recent magnesium alloy development for elevated temperature applications, International Materials Reviews, **49**(1): p. 13–30.
3. Pekguleryuz, M. and M. Celikin, (2010) Creep resistance in magnesium alloys, International Materials Reviews, **55**(4): p. 197–217.
4. Raynor, G.V., (1959) *Physical Metallurgy of Magnesium and Its Alloys*, London: Pergamon Press.
5. Rokhlin, L.L., (2003) *Magnesium Alloys Containing Rare Earth Metals Structure and Properties*: Taylor and Francis.
6. Celikin, M., A.A. Kaya, R. Gauvin, and M. Pekguleryuz, (2012) Effects of manganese on the microstructure and dynamic precipitation in creep-resistant cast Mg-Ce-Mn alloys, Scripta Materialia, **66**(10): p. 737–740.
7. Celikin, M., A.A. Kaya, and M. Pekguleryuz, (2012) Microstructural investigation and the creep behavior of Mg-Sr-Mn alloys, Materials Science and Engineering A, **550**: p. 39–50.
8. Celikin, M. and M. Pekguleryuz, (2012) The role of α-Mn precipitation on the creep mechanisms of Mg-Sr-Mn, Materials Science and Engineering A, **556**: p. 911–920.
9. Easton, M.A., M.A. Gibson, D. Qiu, S.M. Zhu, J. Gröbner, R. Schmid-Fetzer, J.F. Nie, and M.X. Zhang, (2012) The role of crystallography and thermodynamics on phase selection in binary magnesium-rare earth (Ce or Nd) alloys, Acta Materialia, **60**(11): p. 4420–4430.
10. Chia, T.L., M.A. Easton, S.M. Zhu, M.A. Gibson, N. Birbilis, and J.F. Nie, (2009) The effect of alloy composition on the microstructure and tensile properties of binary Mg-rare earth alloys, Intermetallics, **17**(7): p. 481–490.
11. Zhu, S.M., M.A. Gibson, M.A. Easton, and J.F. Nie, (2010) The relationship between microstructure and creep resistance in die-cast magnesium-rare earth alloys, Scripta Materialia, **63**(7): p. 698–703.
12. Weiss, D., A.A. Kaya, E. Aghion, and D. Eliezer, (2002) Microstructure and creep properties of a cast Mg-1.7%wt rare earth-0.3%wt Mn alloy, Journal of Materials Science, **37**(24): p. 5371–5379.
13. ASTM Standard E8-04, "Standard Test Methods for Tension Testing of Metallic Materials", ASTM International, West Conshohocken, PA, 2004, www.astm.org.

14. *ASM Metals Handbook Alloy Phase Diagrams.* 1992, ASM International.
15. Bale, C.W., E. Bélisle, P. Chartrand, S.A. Decterov, G. Eriksson, K. Hack, I.H. Jung, Y.B. Kang, J. Melançon, A.D. Pelton, C. Robelin, and S. Petersen, (2009) FactSage thermochemical software and databases — recent developments, Calphad, **33**(2): p. 295–311.
16. Celikin, M., A.A. Kaya, and M. Pekguleryuz, (2012) Effect of manganese on the creep behavior of magnesium and the role of α-Mn precipitation during creep, Materials Science and Engineering A, **534**: p. 129–141.
17. Wei, L.Y., G.L. Dunlop, and H. Westengen, (1996) Age hardening and precipitation in a cast magnesium - Rare-earth alloy, Journal of Materials Science, **31**(2): p. 387–397.
18. Saito, K. and H. Kaneki, (2013) TEM study of real precipitation behavior of an Mg-0.5at%Ce age-hardened alloy, Journal of Alloys and Compounds, **574**: p. 283–289.

# Solutions for Next Generation Automotive Lightweight Concepts Based on Material Selection and Functional Integration

Horst E. Friedrich, Elmar Beeh, and Carmen S. Roider

**Abstract**

Various aspects of global trends like energy saving or safety, economic and production targets or requirements due to the automation give high and challenging targets for the development of future vehicles. With respect to these requirements weight saving is also a must during an early phase of the concept and structural development process. Within the Next Generation Car project the DLR is looking for solutions for future integrated functions like health monitoring and the combination of passive and active safety functions. New material systems and novel combinations of design methods are also developed. The integrated approach of combining development and optimization methods with tools for the design is the key to the development of concepts for holistic lightweight solutions and new vehicle concepts utilizing highly automated and autonomous driving abilities. Challenges and potentials for next generation lightweight design are shown and solutions and results from the DLR project Next Generation Car are presented.

**Keywords**

Lightweight concepts • Automotive structures
Integrated functions • Magnesium alloy

H. E. Friedrich (✉) · E. Beeh · C. S. Roider
German Aerospace Center (DLR)—Institute of Vehicle Concepts, Pfaffenwaldring 38-40, 70569 Stuttgart, Germany
e-mail: horst.friedrich@dlr.de

E. Beeh
e-mail: elmar.beeh@dlr.de

C. S. Roider
e-mail: carmen.roider@dlr.de

## Introduction

Even if the topics of electrification and atomization of vehicles are stronger in the current focus of public discourse, lightweight technologies are an enabler for improving fuel economy, enhancing the vehicle performance, extending the range of future electric or fuel cell vehicles and for complying with legal regulations. Caused by cost sensitive large-production volumes in the automotive industry, there is an ongoing competition between lightweight materials and designs which is not only decided by performance but also by costs. New requirements, e.g. through the implementation of autonomous vehicles, new manufacturing routes, new material developments and new concepts offer chances for all available lightweight technologies. The DLR-Institute of Vehicle Concepts as a neutral institution has decided to do research on a large number of different technological lightweight routes to compare them and to show possible application areas. Therefor not only mechanical performance and costs but also the life cycle assessment (LCA) of different materials and production options are used as differencing criteria.

## Lightweight Design Strategy

For finding capable solutions to a lighter vehicle structure, DLR is using a holistic lightweight strategy (see Fig. 1). Starting with a detailed analysis of requirements for the application, this first step allows to identify areas, where e.g. new regulations offer new concept options or reduce their number. In the second phase, concept ideas are developed. Within this step the final solution is strongly influenced. A concept has to use all possible improvements, like modified design space or functional integration, to make the following steps of material selection and the shape optimization as effective as possible. The decision on the final concept has to be made in a loop by taking into account the


D. Orlov et al. (eds.), *Magnesium Technology 2018*, The Minerals, Metals & Materials Series, https://doi.org/10.1007/978-3-319-72332-7_52

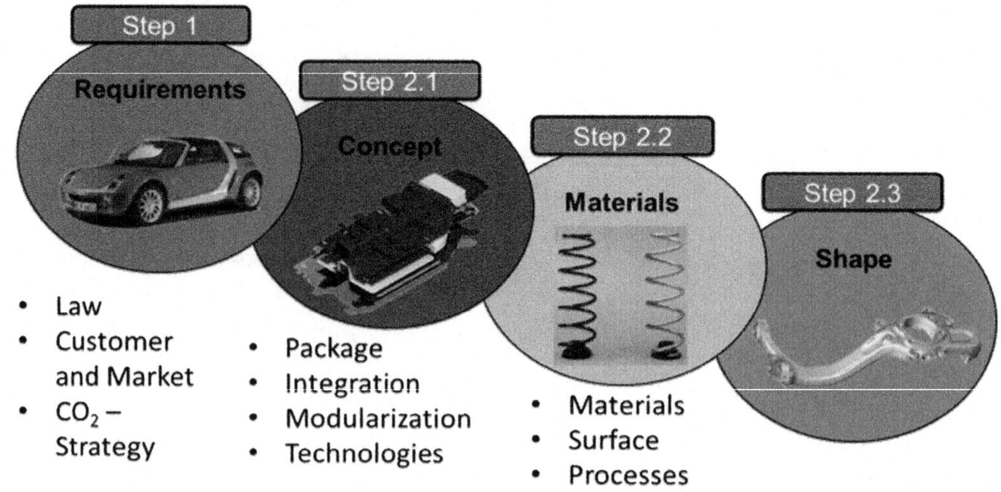


**Fig. 1** Steps of a holistic lightweight strategy, including material selection

material and shape since e.g. manufacturing options for different materials have also an influence on the concept.

To select the right material for the right application, a deep understanding of material potentials and properties is necessary. Figure 2 shows greenhouse gas emissions (GHG) of Steel, Aluminum, Magnesium, Glass- and Carbon-fiber reinforced plastics and the influence of production routes, e.g. for magnesium casting. It can be seen, that improving the energy efficiency of manufacturing routes and the availability of a recycling route are important factors to make a lightweight material solution beneficial regarding greenhouse gas emissions.

## Next Generation Car (NGC)

At the German Aerospace Center in the research project NGC, which is part of the program topic of terrestrial vehicles, three different vehicle concepts are being developed according to the technical requirements and capabilities of 2030: the Urban Modular Vehicle (UMV), the Safe Light Regional Vehicle (SLRV) and the Interurban Vehicle (IUV). The aim of NGC is to consolidate the technologies, methods and tools for the various vehicle concepts researched at the DLR in the field of transport, to generate synergies and to demonstrate research results. The NGC project is divided

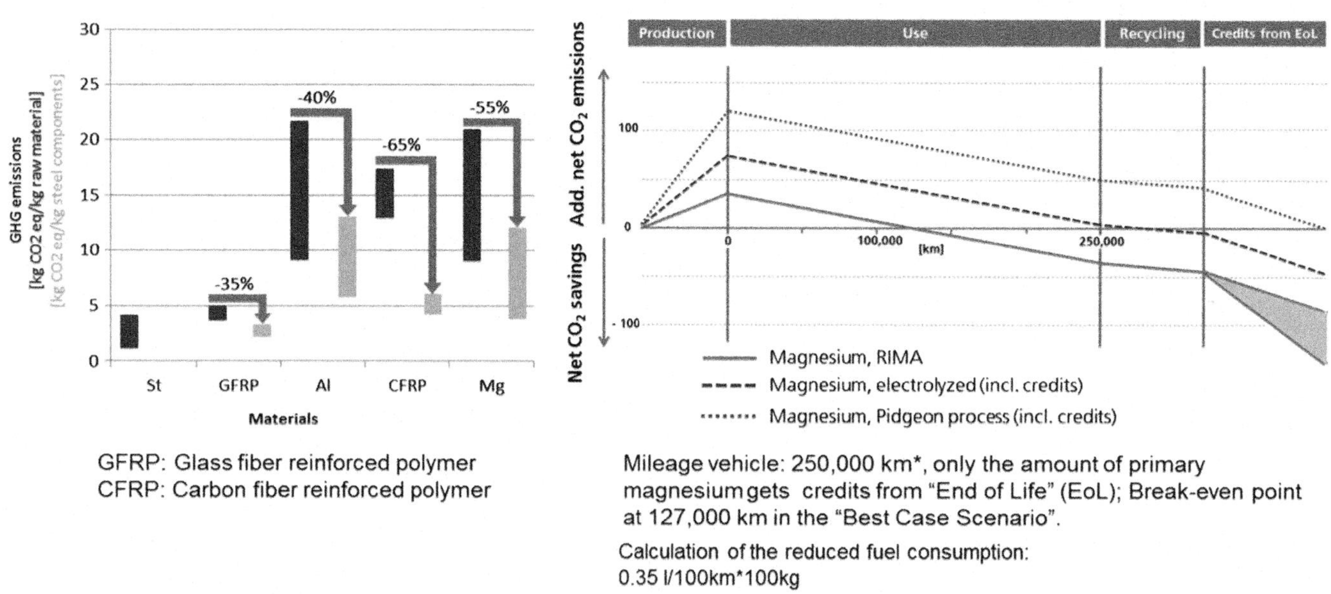


**Fig. 2** Greenhouse gas emissions of lightweight materials (left) and a magnesium gearbox (right) [3]

into the following areas: Vehicle Concepts, Vehicle Structure, Energy Management, Drive Train, Chassis and Vehicle Intelligence [1]. Figure 3 shows the different objectives being addressed with the three vehicle concepts, leading to different requirements and different technological lightweight solutions being suggested for the respective vehicle body structure.

## Safe Light Regional Vehicle (SLRV)—Safe and Light by Sandwich Design

The NGC-SLRV is a very light regional car designed for the vehicle class L7e. These kinds of small vehicles do not have to fulfill special crash requirements by the regulations. DLR has designed a vehicle body which nevertheless offers the same crash safety level of a medium size passenger car. Therefor novel technological approaches have been developed. The main idea is to build the vehicle body in a sandwich intensive design and additionally utilize the significant improvement of crash energy absorption by stabilizing hollow profiles by a foam core. For the front and rear structure flat sandwich structures with thin metallic skins and a foam core are used to build a very light and stiff structure. Through precise investigation into the crash behavior of these sandwich structures, the front or rear structure assembly can be used as a large energy absorbing structure in the case of an accident.

To achieve a high safety level in case of a side impact, like e.g. the pole impact, a foam filled ring structure protects the passengers. Within the current SLRV concept the material of the ring structure is a HSD-steel which offers both high strength and ductility. The specific energy absorption within the bending load case can be increased from 140 to 500 J/kg by filling a ring structure, made from a soft steel DC04, with a polyurethane foam with a density of 400 kg/m$^3$. By using HSD-steel the specific energy absorption increases to 714 J/kg [2].

DLR has conducted further investigations with alternative profile materials. Magnesium AZ31B is usually a very brittle material, which therefore performs weak, when it comes to larger deformations, like in the bending load case. Furthermore, magnesium bulk materials show a significant tension/compression asymmetry in their material properties, which leads to challenges in designing components out of this material. With the concept of foam filling these magnesium profiles the bending behavior can be significantly improved in such a way that it was possible to further increase the specific energy absorption to a value of 772 J/kg. Figure 4 shows the comparison between bending tests of a foam-filled steel profile and two different foam-filled magnesium profiles.

## Urban Modular Vehicle (UMV)—Side Crash Structure

The NGC-UMV represents a modular vehicle platform, where a modular body-in-white for electrified vehicles allows designing derivations from people to cargo vehicle and from assisted to fully automated vehicle. Hence the vehicle structure has to be safe, light and modular. To meet

| | NGC SLRV | NGC UMV | NGC IUV |
|---|---|---|---|
| Customer benefit | NGC SLRV: safe, economic, efficiency | NGC UMV[1]: mobility-as-a-service, adaptable, intermodal | NGC IUV: comfortable, range, flexible |
| Autonomous driving | up to SAE-Level 3 | up to SAE-Level 5 | up to SAE-Level 4 |
| Emission | local zero, $H_2$ FC | local zero, BEV | local zero, $H_2$ FC / PHEV |
| Energy percentage for traction | - | +20% | +10% |
| Mass | 450 kg / < 90 kg BIW | -25% / 680 kg | -30% / < 250 kg BIW |
| Range | 400 km | urban / modular battery | up to 1000 km |
| Safety | equivalent to M1 | active & passive | active & passive |
| Life cycle costs | -25% comp. to SoA | Reduction by modular platform | -25% comp. to SoA |

**Fig. 3** Objectives of the NGC vehicle concepts

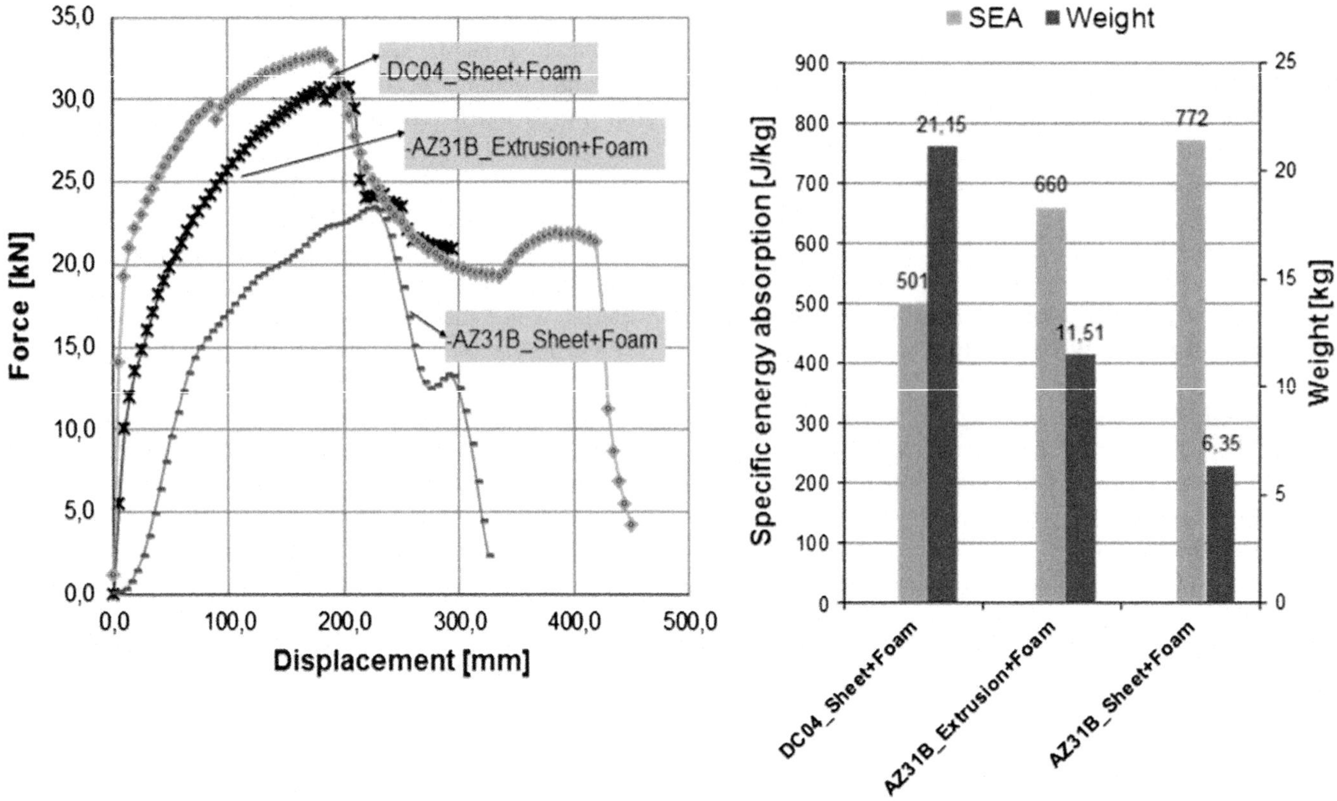


**Fig. 4** Comparison of foam-filled profiles made out of DC04 and AZ31B

these requirements, a new approach for a holistic method for urban vehicle concepts with electric powertrain has been developed.

To address the challenging side crash requirements, especially when carrying a large and heavy battery in the floor area, a novel side impact protection system has been developed and evaluated in a component crash test. Figure 5 shows the floor concept with an energy absorbing zone made of an inner longitudinal rail and the rocker with metallic sandwich structures in between.

The crash test shows a high energy absorption capability and a very good correlation between test and simulation. The design fulfills the requirements of scalability for being used in a modular platform.

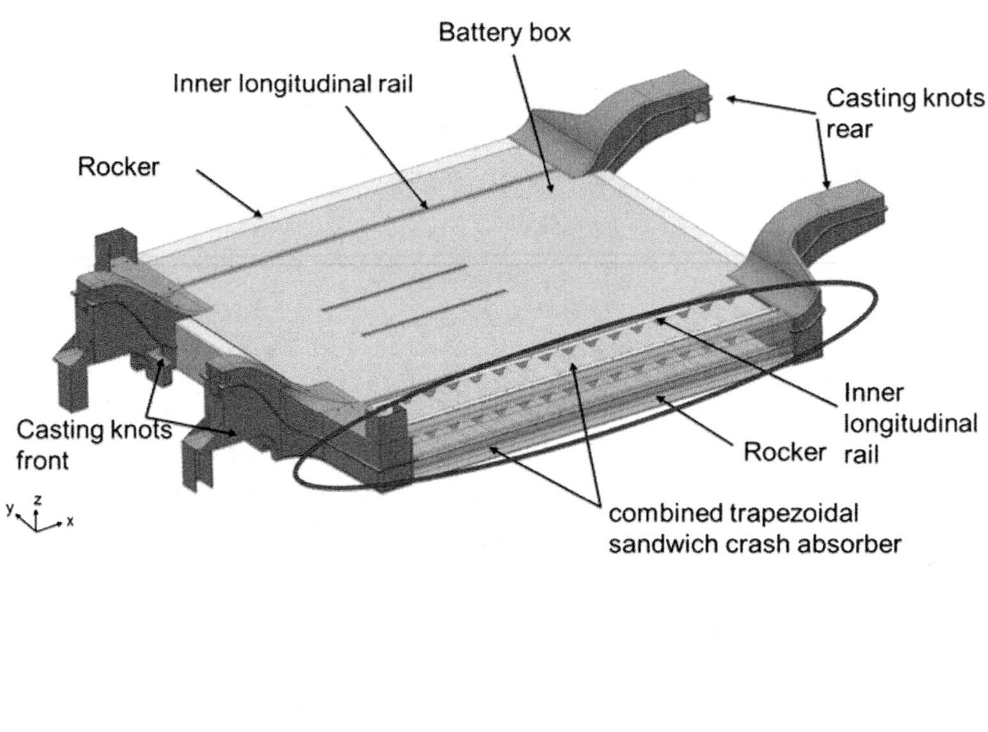


**Fig. 5** Floor concept of the NGC-UMV with energy absorbing area (red) [3]

## Inter Urban Vehicle (IUV)—Lightweight Design with Functional Integration

With the NGC-IUV, DLR is addressing a large and comfortable vehicle for travelling longer distances. Therefor this concept is equipped with a fuel cell drive train which allows the range of up to 1000 km (see Fig. 3). The vehicle body is designed as a carbon fiber reinforced plastic (CFRP) structure with high content of functional integration, like e.g. the wiring. The CFRP allows to lower the weight of the vehicle body to less than 250 kg. This corresponds to a weight reduction of more than 30% compared to current vehicles of this size. Furthermore the CFRP leads to a high stiffness of the structure and good crash safety.

## Cross-Section Topics for Different Applications

Besides the three NGC vehicle concepts the DLR-Institute of Vehicle Concepts is working on cross-section technologies, which can be applied in road and rail vehicles. Examples are the topics of digitalization, additive manufacturing and joining technologies. In the field of multi-material joining, the research is focusing on so called cold joining techniques. In areas, where different materials need to be joined, the widely used welding technologies cannot be applied. Therefor our research focusses on adhesive bonding and on mechanical joints, like self-pierced rivets (SPR) or flow-drill screws (FDS). Figure 6 shows examples of different joining technologies investigated to be used e.g. in the three NGC-concepts.

## Summary

Future vehicles will still have to address a wide range of use cases and user requirements. Therefor different technological solutions for different segments are needed. Within the 'Next Generation Car'-Project, DLR has been developing technological solutions based on different materials, designs and manufacturing technologies. A technology portfolio is provided for car manufacturer, from which a solution can be selected e.g. for crashworthiness or weight saving or material requirements. The Institute of Vehicle Concepts has specialized competence in utilizing complex material properties, like magnesium bulk materials, sandwich structures or other new material systems. With cross-section technologies DLR offers solutions, which can be applied in a wide range of vehicles.

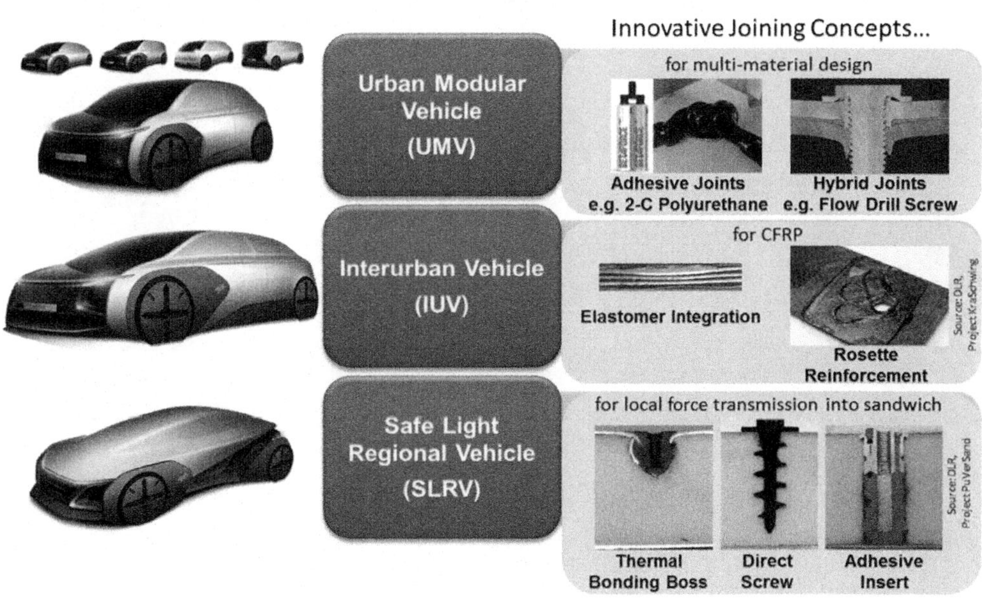


**Fig. 6** Research on different joining technologies for the three NGC-concepts

## References


1. Münster, Marco and Schäffer, Michael and Kopp, Gerhard and Kopp, Gundolf and Friedrich, Horst E. (2016): *New approach for a comprehensive method for urban vehicle concepts with electric powertrain and their necessary vehicle structures.* In: Transport Research Arena. TRA 2016, 18.–21. April 2016, Warschau, Polen.
2. Kriescher, Michael and Brückmann, Simon and Salameh, Walid and Otto, Manuel and Roettger, Jan and Oswald, Katja (2015): *Entwicklung einer leichten, funktionsintegrierten Karosserie in Metall-Sandwich-Bauweise.* Symposium Lightweight Solutions, 03.–04.11.2015, Hannover.
3. Friedrich, Horst and Kopp Gundolf and Beeh, Elmar (2017): *Challenges and potentials for next generation automotive light-weight concepts.* EUROMAT Light Weight Metals Symposium, Thessaloniki, Greece.

# Magnesium Pistons in Engines: Fiction or Fact?

N. Hort, H. Dieringa, and Karl Ulrich Kainer

**Abstract**

Magnesium alloys are already widely used in numerous applications in transportation and consumer products. Ways have been found to improve corrosion and creep resistance, formability in general, and processing routes have been optimized. But would Mg alloys also be suitable for use in an environment where friction, corrosion, thermal fatigue and creep resistance at elevated temperatures are issues? Due to lightweighting benefits, pistons would be an ideal application for Mg based materials. It is much more efficient to accelerate and to decelerate a lightweight material compared to a heavier one. Al alloy pistons are already fairly well established. But Mg could provide further benefits compared with Al due to its specific strength and mass. We will report the state of the art in Mg pistons, with our own and others approaches to improve properties and the challenges that Mg pistons have to face.

**Keywords**

Magnesium alloy • Piston • Corrosion • Creep

## Introduction

The question in the title is if Mg pistons are fiction or fact. Both can actually be answered with yes. The fact is that there are Mg pistons in service. But their application is limited to racing. In today's real life applications Mg pistons in internal combustion engines are still fiction regardless of its history. And this story leads back to the early days of Mg alloy development and its use in several applications [1–4].

After the synthesis of metallic Mg by Sir Humphrey Davy in 1808 it took a long time to establish production processes, to produce Mg in larger amounts (in the range of tons) and to find proper applications. Due to its affinity to oxygen and the bright flame of burning Mg, it was proposed as a possible flashlight during the early days of photography. This offered completely new possibilities and Mg was used in this area until the 1980s when Mg flashlights were replaced by electronic flashes.

However, flashlights are not a lightweight application where Mg is utilizing one of its advantages, its good specific strength. Also if there are other challenges like poor corrosion behavior especially in contact with other metals (and an electrolyte), its relatively poor creep performance and poor deformation behavior, Mg still has advantages.

The benefits of Mg as a lightweight material was already shown in the early 20th century [3]. Chemische Fabrik Griesheim used their patented Elektron alloy (Mg–Al) for an aircraft engine (1. Internationale Luftfahrtausstellung 1909, Frankfurt, Germany). However, after World War I, in Germany Mg got an increased attraction and alloy and process development for several industrial applications started successfully. The reason why Germany especially focused on Mg alloys and process development is not completely clear but resources, experiences and available processing opportunities were developed and magnesium was widely applied.

1921 seemed to be the year of Mg pistons. There are reports that DOW produced Mg pistons [4]. But more importantly, the German Ministry of Transportation (Reichsverkehrsministerium) issued a competition to develop a lightweight piston. Two Mg pistons and two Al pistons were awarded. The first two places went to:

1. Chemische Fabrik Griesheim (later bought by W. Mahle), Mg piston
2. K. Schmidt (became later Kolbenschmidt), Al piston.

In the years after the competition, the Chemische Fabrik Griesheim sold around half a million of Mg pistons mostly for internal combustion engines for cars.

N. Hort (✉) · H. Dieringa · K. U. Kainer
Magnesium Innovation Centre (MagIC), Helmholtz-Zentrum Geesthacht, Max-Planck-Strasse 1, 21502 Geesthacht, Germany
e-mail: hort_tms@hzg.de


D. Orlov et al. (eds.), *Magnesium Technology 2018*, The Minerals, Metals & Materials Series, https://doi.org/10.1007/978-3-319-72332-7_53

It seemed that Mg pistons would be a superior or even equal alternative to Al pistons. Racing motor cycles had internal combustion engines with Mg piston. Mg pistons also were applied in different engines for automobiles at this time. Even in engines for airplanes Mg piston were widely used (Fig. 1). At this time, the regular use of Mg pistons in common internal combustion engines was a fact and companies used the achieved world records etc. for advertisements. But things changed and now Al pistons are most common in high performance internal combustion engines. What are the reasons that Mg pistons are now rarely used when they were widely used in the 1920s to the 1940s?

On reason could be that molten Mg cannot be handled as easy as molten Al. Molten Al always has a protective surface layer in the solid and in the molten state (Pilling-Bedworth ratio > 1). For molten Mg the Pilling-Bedworth ratio is in the range of 0.8. This means that the surface layer of MgO always cracks and allows further oxidation. Additionally the MgO film does not stick to either molten or solid Mg. The result is an ongoing oxidation that could lead to the ignition of molten Mg. Therefore, melt protection is necessary and increases the efforts during melt handling compared to Al alloys [5].

**Fig. 1** Advertisement for Elektron piston from Mahle for airplane engines, Germany, 1940

The obvious sensitivity of Mg and its alloys regarding oxygen is also one obstacle to its use in piston applications. Internal combustion engines combust fuel in a reaction that repeats several times per second. This also could lead to improved oxidation of a Mg piston in service. This especially important for the piston head. However, as the fuel is even more prone to oxidation (the reason why fuel is used in internal combustion engines) normally oxygen should be consumed by the combustion and is in general not available for the reaction with Mg.

The piston head is frequently heated up by the combustion, followed by rapid cooling immediately afterwards. This continued expansion and contraction is also known as thermal fatigue. The more heat resistant the piston material is, the better the performance of a piston in service. To overcome this problem metal matrix composites were already considered for piston applications.

In general, it also has to be stated that properties of standard Mg alloys at elevated temperatures often cannot compete with Al alloys. Creep resistance is here of importance. To improve creep resistance several methods are possible. Alloying is the first method. A combination of alloying elements that form intermetallic phases on grain boundaries and within grains have been state of the art for decades. WE series alloys work like this. According to the ASTM designation system [6] for Mg alloys, W represents Yttrium. E stands for rare earth elements and in this case mostly for Ce and Nd or a mixture of both (mischmetal).

Y forms intermetallics that are dissolved and re-precipitated during heat treatments. Mg–Y intermetallics can therefore be regarded as a useful measure to improve creep resistance. They form in the grains and can act as obstacles for dislocation movement. In WE additionally Zr is added to achieve a homogeneous grain size. Ce/Nd form intermetallic particles mainly on grain boundaries during solidification. Due to their low solid solubility these particles remain stable even at elevated temperatures and can pin grain boundaries, preventing grain boundary sliding.

Figure 2 shows tomographic image of a WE43 cross section. Here an enrichment of Y towards the grain boundary can be observed as well as the fact that Nd is mainly observed at grain boundaries. Moreover, the Zr particles are in the center of the grains.

Nevertheless, due to the use of rare earth elements (RE) it is relatively expensive. Moreover, WE series alloys have been designed as sand cast alloys and are not really suitable for high pressure die casting (HPDC). Due to the nature of the casting processes production of pistons from WE by sand casting does not fit industry requirements regarding mass

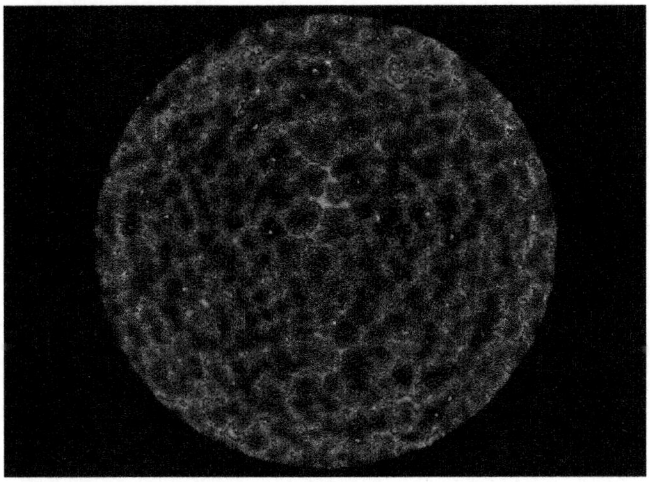

**Fig. 2** Microtomography of a WE43 alloy: blue is Y, red is Nd, yellow is Zr, the diameter of the structure is 1 mm [7]

production, dimensional accuracy, and cost efficiency. HPDC fits these requirements but the WE alloy system (like other creep resistant Mg alloy systems) have a relatively short freezing interval. This leads to difficulties regarding castability and mold filling and prevents its use in HPDC.

The mechanisms of alloying to improve creep performance can be easily transferred to other alloys to improve creep resistance. For instance, Ca or Sr can be added to Al containing Mg alloys, reducing the amount of $Mg_{17}Al_{12}$ by forming Al containing, producing high temperature stable intermetallics within grains and on grain boundaries. However, REs are expensive and any intermetallic forming during solidification has an impact on castability.

Using reinforcements is another possibility to improve creep resistance. Options available include long or short fibers, whiskers, micron or nano-sized particles or a mixture of fibers/whiskers and particles. Long fibers have been used in a project funded by the German Ministry of Research and Education (Long-fiber reinforced magnesium alloys, 03M0361, 1994–1997) [8]. Preforms made from long carbon fibers have been manufactured and were infiltrated by squeeze casting. It could be shown that this process is suitable to produce pistons. The pistons showed acceptable creep resistance and thermal fatigue was also not an issue.

However, the reinforced pistons suffered from other problems. The carbon fibers ended in the surface of the piston head and the piston shaft. Therefore, they could react with the environment. In the head, the fibers burned off during combustion. This resulted in a loss of properties and affected the function of the piston. Moreover, corrosion occurred at the body of the piston.

Another approach was undertaken in the project "Matalend—Materials and Process Engineering of new creep resistant Mg alloys for Thixomolding", also funded by the German Ministry of Research and Education (03N3085, 2000–2004) [9]. Thixomolding (TM) was the chosen production process and carbon short fibers reinforcements with a length of 250 μm as well as micron sized Si particles (100 mesh) have been used in combination with AZ91 and creep resistant MRI alloys. It could be shown that addition of Ca to Mg–Al alloys in combination with thixomolding is suitable to improve creep resistance by two orders of magnitude compared to materials produced by HPDC (Table 1) [10]. Moreover, corrosion resistance is also improved [11].

The carbon fibers align parallel to injection direction during thixomolding. Unfortunately, the process broke the short carbon fibers into smaller pieces (Fig. 3a) due to the screw movement in the thixomolding device. Therefore, the fibers lost their effectiveness and did not contribute to an improvement in strength at elevated temperatures.

The addition of Si particles also could be regarded as successful. They partially reacted with Mg from the matrix alloy. An interface consisting of $Mg_2Si$ formed and further reactions were avoided. Figure 3b shows Si particles in a matrix of MRI 230D. The Si particles appear in grey color, while the $Mg_2Si$ at the interface between matrix and particles appears in a blueish color. However, the Si particle were also broken during thixomolding and had final sizes in the range of 20–50 μm and therefore the efficiency of reinforcement was not as high as expected.

The project "Nano-MMC—Ultra-lightweight materials based on spray-formed Al alloys with high volume fraction of $Mg_2Si$ and nano-particle reinforcement", again funded by German Ministry of Education and Research (03X3007, 2005–09), applied spray forming in combination with forging to produce pistons [12]. To further improve properties, micron and nano sized particles were introduced. Even if in this project Al was the matrix, the new processing chain could be of interest for Mg as well. Again, feasibility was shown. But in this project the consortium failed to

**Table 1** Secondary creep rates ($s^{-1}$) of AZ91-D and MRI153 [10]

| Appl. stress (MPa) | AZ91-D | | MRI153 | |
|---|---|---|---|---|
| | HPDC | TM | HPDC | TM |
| 50 | $6.3 \times 10^{-8}$ | $4.3 \times 10^{-8}$ | $2.7 \times 10^{-9}$ | $4.2 \times 10^{-9}$ |
| 60 | $6.6 \times 10^{-8}$ | $6.2 \times 10^{-8}$ | $4.01 \times 10^{-8}$ | $1.3 \times 10^{-8}$ |
| 70 | $1.2 \times 10^{-6}$ | $8.9 \times 10^{-7}$ | $9.80 \times 10^{-8}$ | $5.0 \times 10^{-8}$ |
| 85 | $1.2 \times 10^{-6}$ | $1.2 \times 10^{-6}$ | $4.96 \times 10^{-7}$ | $3.3 \times 10^{-7}$ |

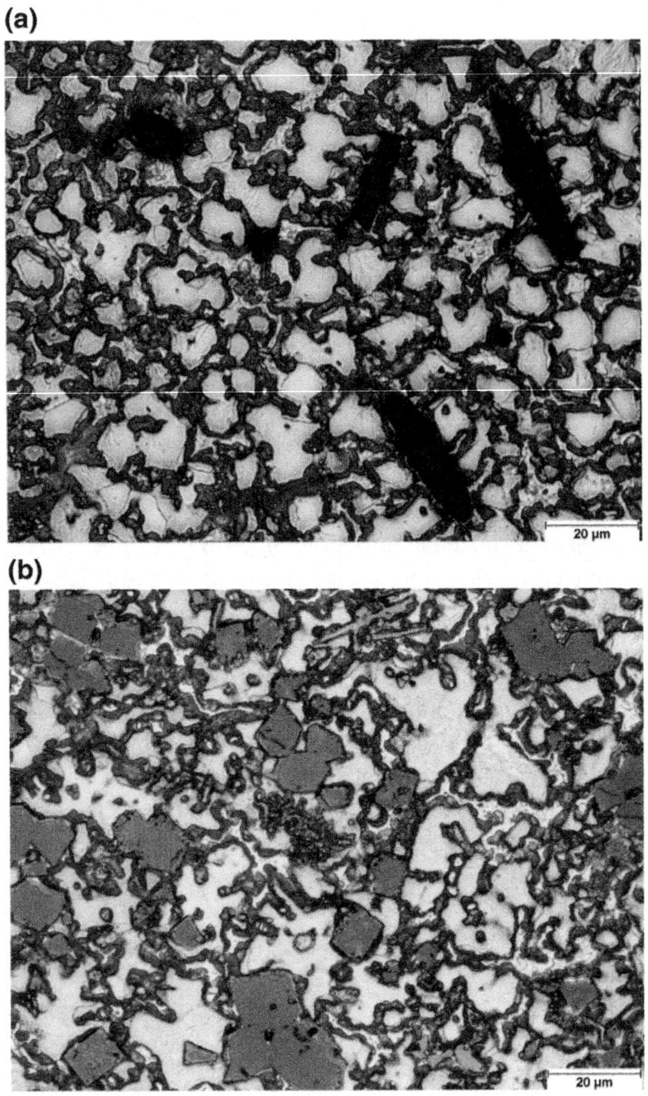


**Fig. 3** MRI 230 D with (**a**) carbon fibers and (**b**) Si particles

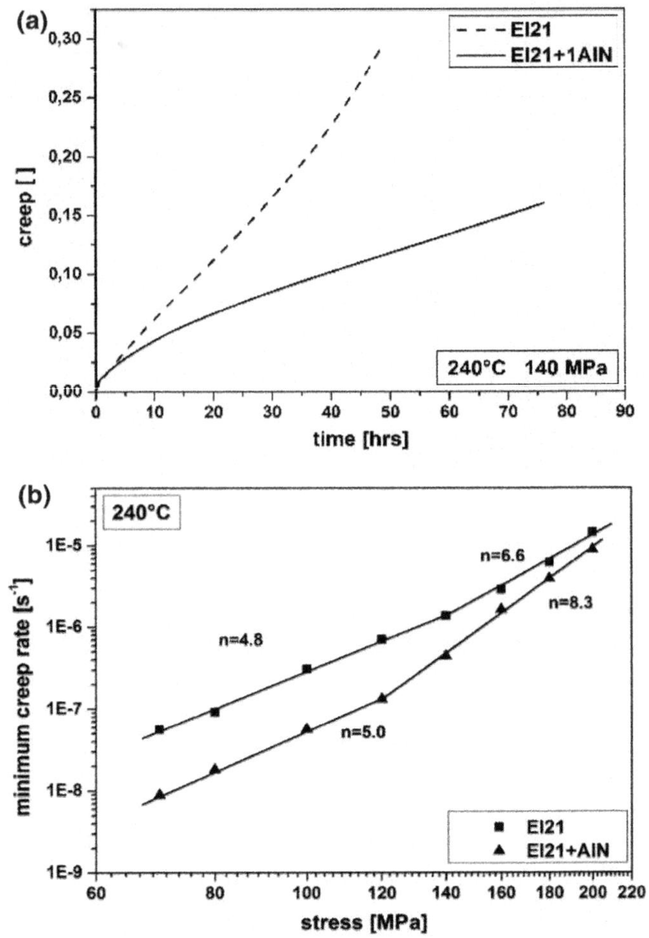


**Fig. 4** (**a**) Creep curves of Elektron21 and Elektron21+1 wt% AlN (**b**) and double logarithmic plot of minimum creep rate versus applied stress from tests performed at 240 °C [14]

incorporate the nano-particles into the matrix during spray forming. Around 80% of the nano-particles were lost.

The problem of incorporating nano-particles into a melt was solved within the EU project "Exomet—Physical processing of molten light alloys under the influence of external fields" (EU Grant Agreement 280421) [13]. This project dealt with the improvement of properties of Al and Mg alloys. External fields (electromagnetic, ultrasound) were successfully used to incorporate nano-particles in the lightweight metal matrix. The nano-particles could be homogeneously distributed. Their use in combination with the already creep resistant alloy also improved creep resistance further in the case of Elektron 21 (Fig. 4). Additionally the application of ultrasound did not only distribute the nano-particles homogeneously in the melt. It also led to a substantial grain refinement (Fig. 5).

Besides improved properties at elevated temperatures, wear has to considered. Obviously the body of the piston has to avoid direct contact with the cylinder liner. Therefore piston rings were established, to mitigate the effects of wear. However, these rings are also in contact with the piston itself and a can cause wear.

Last but not least there is the issue of corrosion. During combustion water is created. And Mg easily reacts with water and corrodes. While this is not a problem for the piston head, corrosive attacks can occur on cold parts of the piston e.g. the piston body.

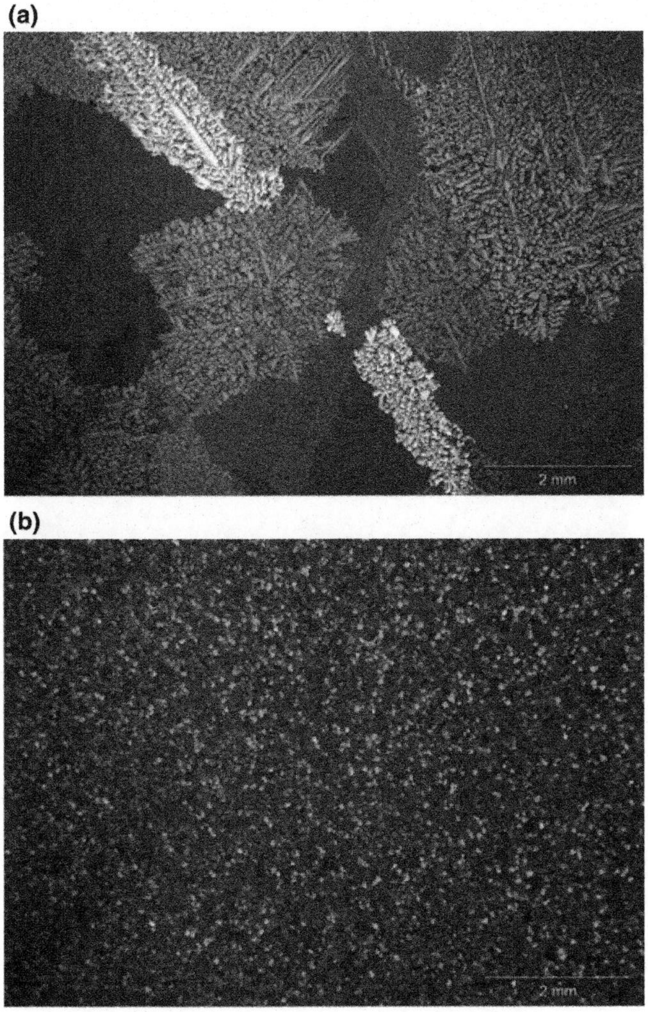


**Fig. 5** Microstructure of (**a**) AM60, and (**b**) AM60+AlN [15]

## Conclusions

What needs to be done to successfully introduce Mg pistons to modern internal combustion engines?

- Use a creep resistant alloy
- Prevent corrosion
- Select appropriate materials that withstand thermal fatigue
- Install a suitable production process.

A number of creep resistant alloys exist and can be processed without too many difficulties. Feedstock material can also be produced. Nano-particles can be effectively incorporated into the melt for further improvement of properties. And like Al pistons, Mg pistons should be forged to obtain an improved microstructure with an optimal property profile. Where necessary a coating should be applied to locally improve wear resistance and corrosion behaviour.

Even if e-mobility will be important in the future, internal combustion engines will not completely disappear in the near future. And due to this fact, Mg pistons can definitely contribute to a more efficient use of fuel. It is time for it!

## References

1. E. F. Emley, Principles of Magnesium Technology, Pergamon Press, Oxford, UK, 1966
2. H. E. Friedrich, B. L. Mordike, Magnesium Technology, Springer, Berlin, Germany, 2006
3. A. Beck, Magnesium und seine Legierungen, Springer, Berlin, 1939
4. A. A. Luo, Magnesium casting technology for structural applications, Journal of Magnesium Alloys 1 (2013) 2–22
5. N. Hort et al., Protecting molten Magnesium and its alloys, La Metallurgia Italiana 6 (2016) 105–108
6. ASTM B93 / B93 M – 15, Standard Specification for Magnesium Alloys in Ingot Form for Sand Castings, Permanent Mold Castings, and Die Castings
7. N. Hort et al., Casting: Theory and Practice, European Cells and Materials (2017), in press
8. Final report, https://www.tib.eu/en/search/id/TIBKAT%3A24933464X/Herstellung-und-Charakterisierung-von-langfaserverst%C3%A4rkten/?tx_tibsearch_search%5Bsearchspace%5D=tibub
9. Final report, https://www.tib.eu/de/suchen/id/TIBKAT%3A507819497/MATALEND-Schlussbericht-des-Verbundprojektes-Werkstoff/
10. H. Frank et al., Creep resistant magnesium alloys for power train applications, Proc. MCWASP Conference on Modeling of Casting, Welding and Advanced Solidification Processes XI, (2006) 669–676
11. R. Buzolin et al., Corrosion and Creep Resistance of Thixomolded® Magnesium Alloys, TMS 2017, Magnesium Technology 2017, Eds. K. N. Solanki et. al., (2017) 381–389
12. Final report, https://doi.org/10.2314/GBV:654037914
13. http://cordis.europa.eu/project/rcn/103659_en.html
14. L. Katsarou et al., Microstructure, mechanical properties and creep of magnesium alloy Elektron21 reinforced with AlN nanoparticles by ultrasound-assisted stirring; Materials Science & Engineering A 659 (2016) 84–92
15. H. Dieringa et al., Ultrasound assisted casting of an AM60 based metal matrix nanocomposite, its properties and recyclability; submitted to Metals

# Development of Magnesium Sheets

Dietmar Letzig, Jan Bohlen, Gerrit Kurz, Jose Victoria-Hernandez, Roland Hoppe, and Sangbong Yi

**Abstract**

Innovative semi-finished products require tailoring mechanical properties and improving formability. Due to science-based alloy design as well as optimization of processing parameters (process-property relationship), it is now possible to control microstructures, phase distributions and texture development of magnesium sheets. A combination of new technologies and innovative alloys could help to alleviate the strong textures formed in semi-finished Mg products. The presentation will show Magnesium sheet development via twin roll casting technology and the how the subsequent warm rolling process influences the properties of the final magnesium sheet metal. Further optimisation of process parameters of twin roll casting and of the rolling process in combination with alloy design lead to a microstructure showing promising mechanical properties like high formability und high strength.

**Keywords**

Magnesium-sheets • Twin roll casting • Rare earth element free Mg-sheets

## Introduction

At present, magnesium alloys used in the automobile industry are mainly processed by die casting. This technology allows components with a complex geometry to be manufactured. However, the mechanical properties of die cast materials often do not meet essential requirements with regard to endurance, strength, ductility, etc. A promising alternative for thin, large area parts, such as automotive body components is to utilize sheet material. Sheet metal formed parts are characterized by superior mechanical properties and high quality surfaces without pores in comparison to die cast components. Substitution of conventional sheet materials such as steel or aluminum by magnesium sheets could lead to significant weight savings. However, it will be necessary to produce sheet material with competitive properties in an economic production process.

Twin-roll casting is an economic production process for the generation of fine-grained feedstock materials that can subsequently be warm rolled to thin sheets. This production process for thin strips combines solidification and rolling into one single production step. Thus, it saves a high number of rolling and annealing passes in comparison to the conventional rolling process. The twin roll casting technology is already well established for example in the aluminum industry. However, applications to magnesium alloys are in their infancy at present [1]. Worldwide, there are a small number of industrial or laboratory scale twin roll casters installed at universities, companies and research facilities. Initial results from these activities on conventional wrought and cast alloys have shown promising sheet properties [2–6].

Conventional Mg-based sheet alloys such as AZ31, with strong basal type textures, have been investigated in many studies in order to evolve potential for the limited sheet formability by restricted activities of slip systems. To improve the formability of Mg sheet, it is required to develop weaker textures, especially a less distinct alignment of basal planes parallel to the sheet plane. Texture weakening has been revealed in Mg alloy containing yttrium (Y), cerium (Ce), or neodymium (Nd), i.e. rare earth (RE) elements, during rolling. It is reported that their addition in binary Mg–RE or ternary Mg–Zn–RE alloys contribute to higher accommodation of deformation by active basal <a> slip [7–10]. Furthermore, several studies have shown that texture weakening contributes to the activation of various deformation mechanisms, e.g. the formation of different twin types, shear bands, and the activation of <c + a> slip, which also leads to an improvement of the formability [7, 11, 12].

D. Letzig (✉) · J. Bohlen · G. Kurz · J. Victoria-Hernandez
R. Hoppe · S. Yi
Magnesium Innovation Centre, Helmholtz-Zentrum Geesthacht, Max-Planck Str. 1, 21502 Geesthacht, Germany
e-mail: dietmar.letzig@hzg.de


D. Orlov et al. (eds.), *Magnesium Technology 2018*, The Minerals, Metals & Materials Series, https://doi.org/10.1007/978-3-319-72332-7_54

The aim of the national funded project "Mobil mit Magnesium (M3)" which ended in 2010 was to qualify technically and economically competitive light-weight solutions with magnesium sheets, using advanced materials, semi-finished products and design concepts, in order to meet the requirements of the automotive industry with respect to stress-tolerant design, availability of the semi-finished product as well as component quality. Furthermore, the focus was on the industrial feasibility of production and assembly strategies as well as their process stability and environmental compatibility. All this in the face of the customer's benefit as well as the ability of magnesium sheets to compete in the competition to material alternatives such as fiber-reinforced plastics, magnesium die-casting or aluminum sheet. As part of the project, a roof element Porsche was implemented and tested as a demonstrator component (Fig. 1) [13].

Magnesium sheet materials with the addition of rare earths show a randomized texture and fine-grained microstructure [2–4]. So the addition of rare earths as alloying elements improves the ductility of these alloys.

## Experimental Procedures

The development of wrought magnesium alloys and their introduction into industrial, structural applications are the main goal of the activities at the Magnesium Innovation Centre MagIC of the Helmholtz-Centre Geesthacht (HZG). The current focus of the research work is on alloy design and the development of processing technologies for semi-finished magnesium products. In the particular case of sheet materials, it has been recognized that the feedstock for the warm-rolling process needs to be in the form of thin bands as they are produced via twin roll casting, if thin magnesium sheets are to become competitive industrial products. For this purpose HZG has installed the twin roll caster shown in Fig. 2.

The twin roll casting line consists of a furnace line (Striko-Westofen) and a twin roll caster (Novelis, Jumbo 3CM). The twin roll caster is designed to produce thin strips of magnesium with a maximum width of 650 mm and a thickness in the range of 4–12 mm at a maximum rolling speed of 6 m/min. The furnace line is configured to allow

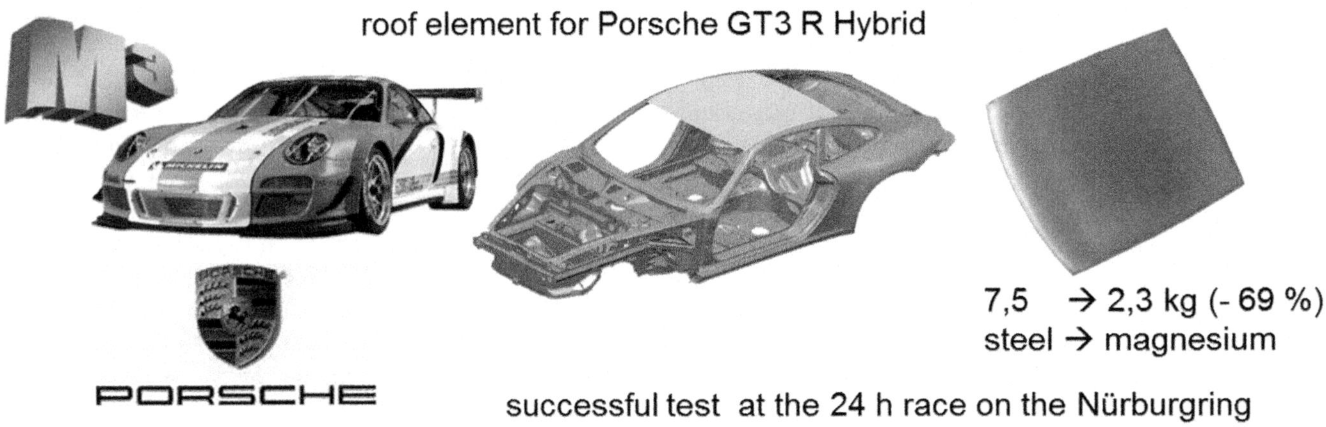


**Fig. 1** Realized roof element (result of M3—Mobil mit Magnesium) [13]

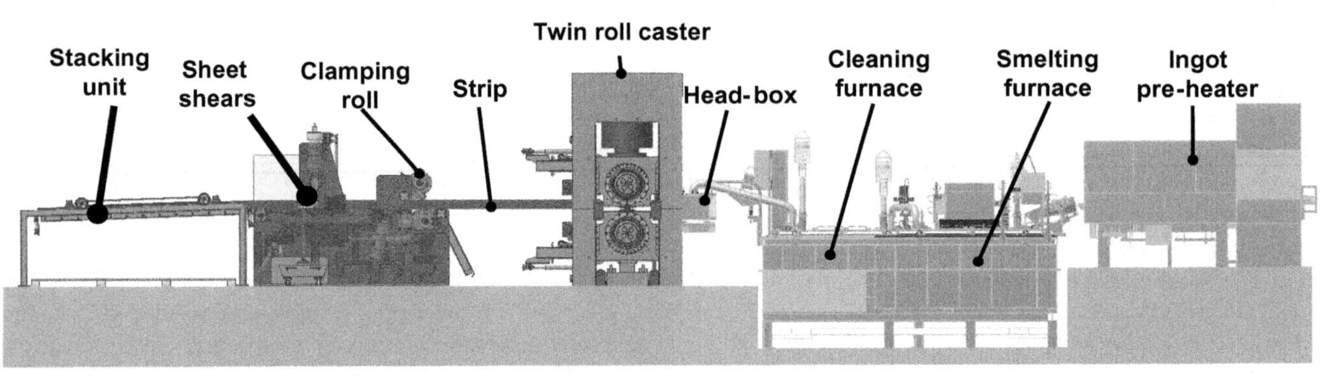


**Fig. 2** Twin roll casting equipment at HZG

manual loading of raw materials in ingots and the possibility to vary the alloy composition, i.e. cast strips can be made from different alloys. The furnace line includes an ingot pre-heater to remove humidity from the ingots and accelerate fusion in the melting furnace. After pre-heating, the feedstock is transferred to the melting furnace. A transfer tube conveys the molten metal to the cleaning furnace. From the cleaning furnace the melt is transferred to the head-box which is connected to the tip. The liquid metal flows by gravity through the tip into the gap of the rolls of the twin roll caster (Fig. 3).

The metal exiting the tip solidifies on the rolls into a strip that is further deformed by the rolls. Important features of the strip such as microstructure and texture are influenced by the position of the solidification front and therefore the resulting degree of deformation as a result of the rolling pass. The solidification starts when the magnesium melt enters the rolling gap and is in contact with the surface of the cooled rolls, the melt temperature drops to the solidus temperature at the strip surface (Fig. 3).

Ongoing along rolls surface, more heat flows from the strip in the rolls as it fully solidifies. Then the heat is extracted to the solid strip. If the contact time of the melt at the rolls increases or the melt temperature decreases the solidification starts earlier and the solidification front is located near the exit of the tip. Its position influences the degree of deformation in the strip. If the solidification front is located closer to the exit of the tip, the degree of deformation increases because the strip is fully solidified before entering the rolls (Fig. 4a). If the solidification front moves towards the kissing point of the rolls due to changes in the processing parameters, the strip is not completely solidified and the degree of solid-state deformation decreases (Fig. 4b).

All twin roll cast experiments were performed with the above mentioned equipment. The trials were carried out at melt temperatures of 650 and 710 °C. The rolling speed was at 650 °C 1.8 m/min and at 710 °C 3.7 m/min. The setback and the rolling gap were constant. In these initial trials, the commercial magnesium alloy AZ31 was used and strips cast with a width of 350 mm. After twin roll casting and rolling, the microstructure of the material was analyzed using optical microscopy.

Standard metallographic sample preparation techniques were employed and an etchant based on picric acid was used to reveal grains and grain boundaries [14]. Texture measurements in the strips and sheets were performed on the sheet mid-planes using a Panalytical X-ray diffractometer setup and CuK$\alpha$ radiation. 6 pole figures were measured up to a tilt of 70° which allows recalculation of full pole figures using an open source software routine MTEX [8]. The (0001) and (10–10) pole figures are used in this work to present the texture of the strips and sheets at midplane.

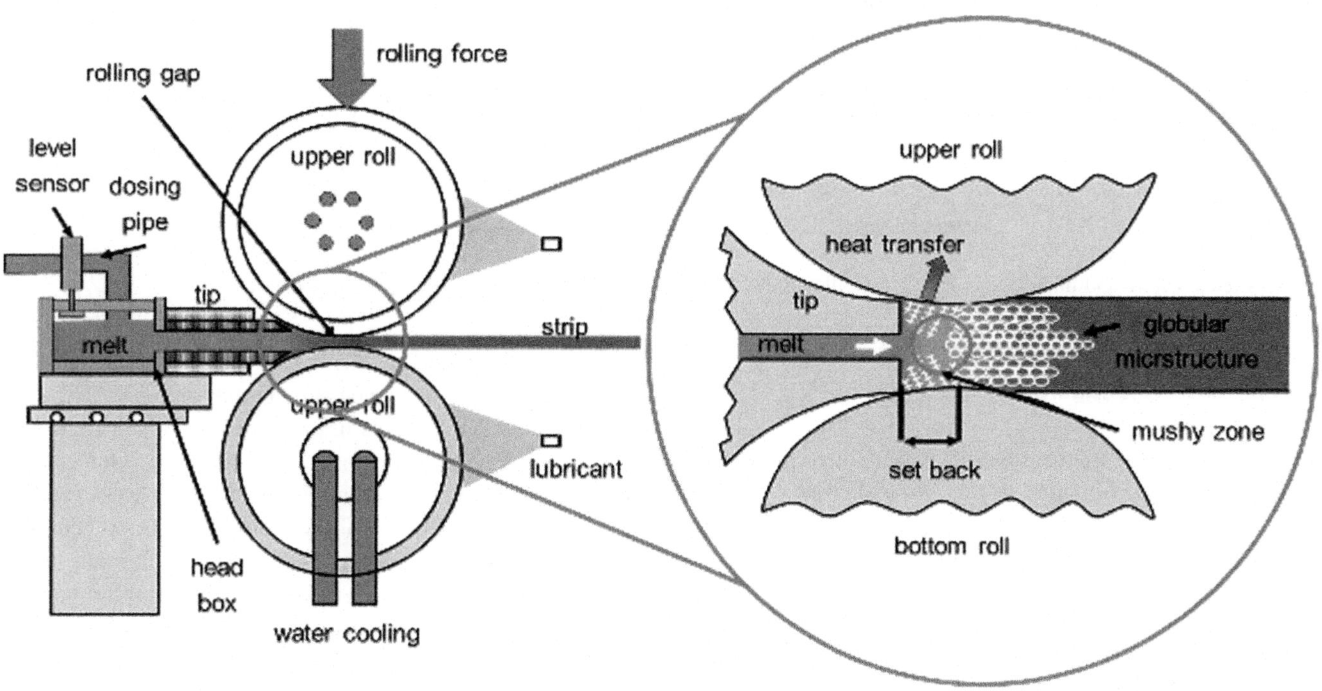


**Fig. 3** Principle and solidification of the metal during twin roll casting

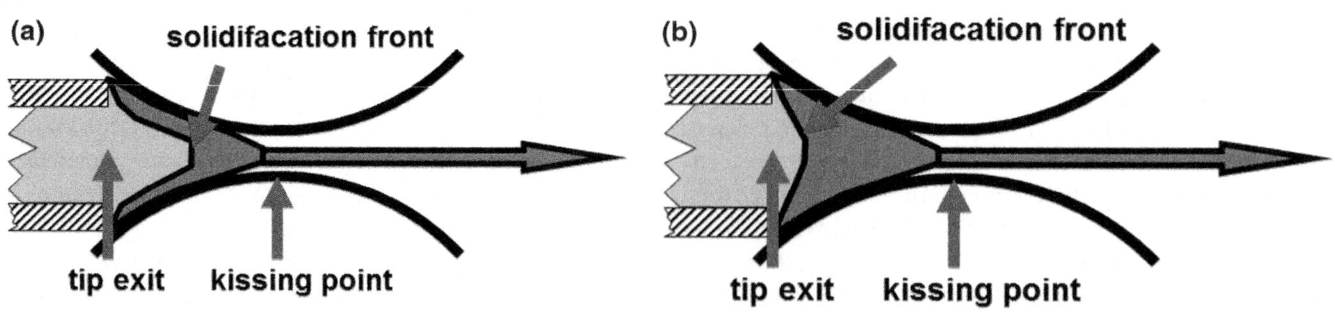


**Fig. 4** Position of the solidification front

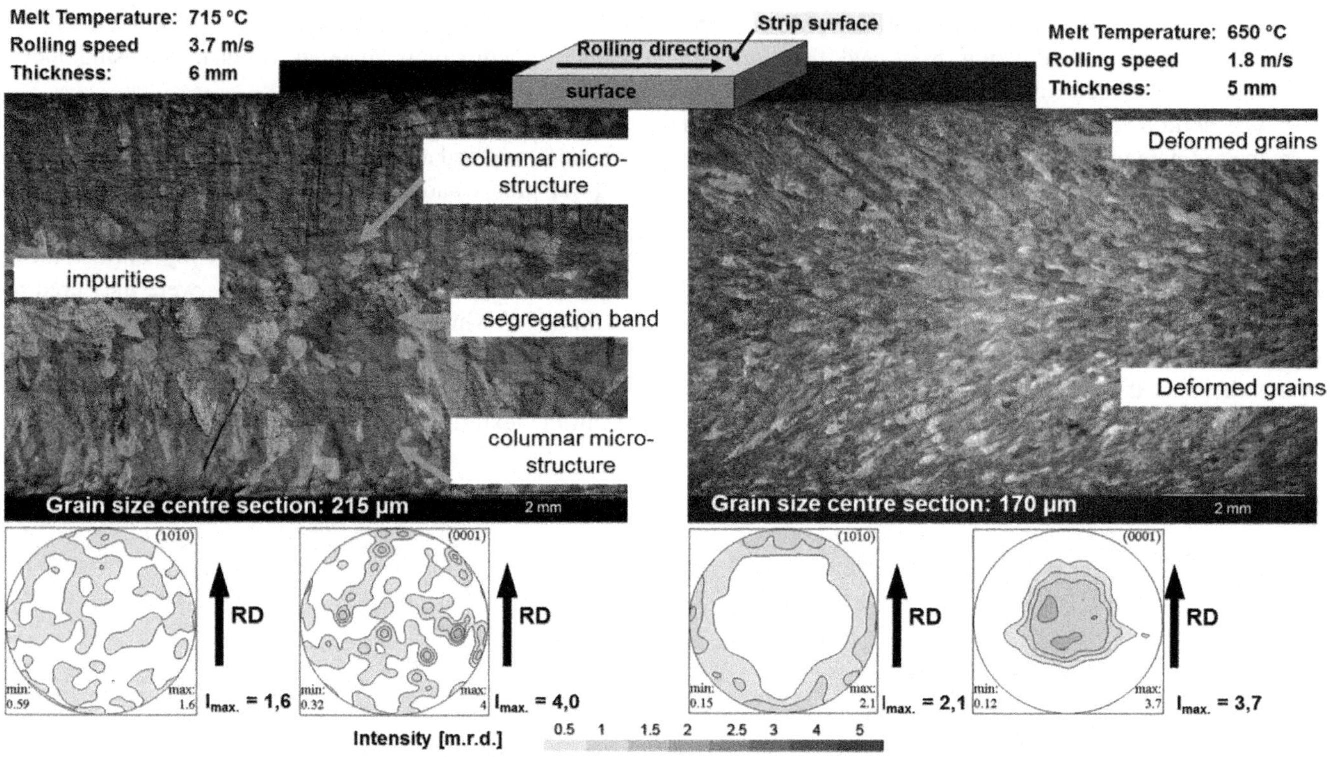


**Fig. 5** Microstructure and mid-plane texture of the alloy AZ31 after twin roll casting different process parameters [19]

## Results and Discussion

Figure 5 depicts the microstructures and textures of samples taken from strips obtained from two different twin roll casting trials [9]. All figures show the longitudinal sections of the strips. The microstructure on the left in Fig. 5 exhibits a columnar structure growing from the upper and bottom surfaces of the strip towards the centre and a region containing equiaxed grains at the centre. A remarkable feature is the strong segregation band containing a large amount of impurities at the centre-line of the strip. Such segregation is a common effect in this type of processing. The formation of this segregation band can only be avoided by optimisation of the process parameters: casting speed, roll gap and melt temperature which influence the thermal gradient. The large number of dendritic grains in the upper and lower regions of the strip is an indicator for a low degree of deformation during the rolling operation. Figure 5 also shows recalculated (0001) and (10–10) pole figures of this strips. In case of this high temperature casting at high speed basically a random texture is revealed at midplane. The random texture corresponds to no deformation at all, i.e. a globular cast microstructure.

The microstructure and texture of a twin roll cast strip with a melt temperature of 650 °C and a thickness of 5 mm are shown on the right side of Fig. 4. This microstructure differs significantly from the microstructures of the strip shown on the left side. It consists of smaller equiaxed grains, which are elongated in the rolling direction in the near-surface regions of the strip. This microstructure is typical for deformed material.

Both, the (0001) and (10–10) pole figures presented on the right in Fig. 5 are characterized by a weak alignment of basal planes parallel to the strip plane. Such a weak alignment of basal planes corresponds with a low degree of deformation [9]. In general, these results demonstrate that the strip cast at 650 °C and low speed experienced more deformation than the strip cast at 715 °C and high speed which corresponds well with the consideration in Fig. 4.

The ongoing aim of the development is therefore to achieve the properties of the rare earth element containing alloys or even surpass them and at the same time to avoid the use of these elements, or at least greatly reduce it and thus create the basis for an economic production of high strength and ductile magnesium materials and components.

Alloying addition of Ca into Mg-based alloys is known as an effective method to improve the high temperature properties and the ignition-proof behaviour [15], which has attracted attention for casting technology and its application.

Recently it has been shown that the Ca addition leads to a texture weakening and grain refinement in Mg alloys after wrought processes, such as rolling [16, 17], or extrusion [18].

Kim et al. [16] showed that the Ca addition in Mg alloys with different Zn contents (Mg–1Zn and Mg–6Zn alloys) modify the sheet textures produced via twin roller casting. The as-rolled sheets show the textures having a splitting of basal poles toward the sheet rolling direction (RD) in the sheet having lower Zn contents, while the high Zn containing alloy sheet shows a basal pole split toward the sheet transverse direction (TD). The recrystallization annealing of the sheets leads to development of very weak texture with the basal pole split in TD in the both sheets. These Ca containing sheets have not only higher yield strength but also an excellent stretch formability comparable to that of Al 5052 alloy. The authors suggested that the texture modification in the ZX11 and ZX61 sheets are related to the twin activities, such as double twinning and extension twinning. Furthermore, a texture randomization effect in deformation bands consisting of twins as well as a accentuation of grains with TD split orientation as a result of recovery and recrystallization has been suggested. The effect of Ca addition on the texture and microstructure of extruded Mg–Mn alloys was studied by Stanford [18]

The global aim of an actual project at the HZG is also the development of calcium containing, rare earth element-free Mg sheet materials with advantageous properties, which are currently reachable only by the use of rare earth element-containing alloys by using innovative production and processing methods like twin roll casting. This work is funded by the German Federal Ministry of Education and Research and entitled "Substitution of rare earth elements in high strength and ductile Magnesium sheet material—SubSEEMag".

Various approaches are being used in this, which include the complete process chain of the alloy development through production of semi-finished products and components to be integrated into the final product. The results of this project show that in Al–Zn based alloys the addition of around 1w% of Ca leads to significant advantages in formability and strength of the sheets. Figure 6 displays the microstructure, texture and mechanical properties of the in this project developed alloy ZAX210. This new alloy ZAX210 forms a very fine grained and homogeneous microstructure without local deformed areas after rolling. Also can be determined that the texture of ZAX210 is similar to Zn-based RE containing alloys like ZE10. Because of this properties the new alloy ZAX210 depict a high formability and a good strength.

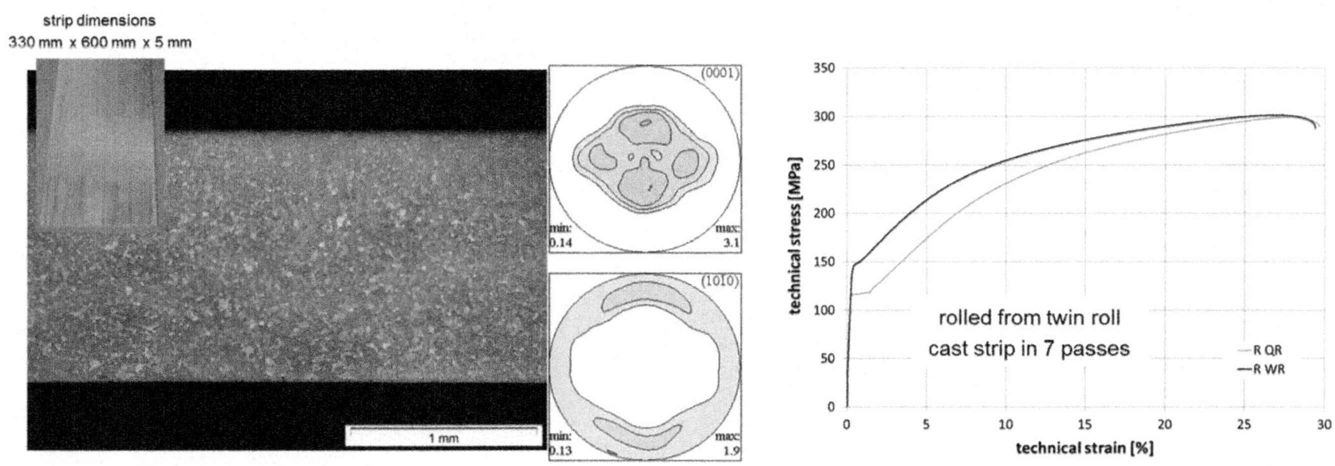


**Fig. 6** Microstructure, texture and mechanical properties of ZAX210 sheets produced via twin roll casting

## Conclusion

The problems in using magnesium as sheet material could be relieved by suitable alloy compositions in conjunction with the appropriate manufacturing processes. In comparison to the AZ31 sheets, the ZE10 sheets show enhanced formability at all rolling conditions. So the alloy composition and the complexity of the rolling process both offer chances to tune the product by modifying the process parameters (e.g. temperature or degree of deformation). Using Calcium containing alloys helps saving rare earth elements, and choosing twin roll casting as a high efficient production method helps saving energy during the production of sheet material for improving the sustainability in vehicle production. The prototype components produced and successfully tested in project the national funded project "Mobil mit Magnesium" demonstrate that a process chain from the production of magnesium components to the use of these components is technically and economically competitive and qualified for automotive industry. Due to the further development of twin roll casting technology in the last years, magnesium sheet metal components have become more and more attractive economically and technically, so that the first applications for magnesium sheet metal components have been introduced to the market.

**Acknowledgements** This research was performed as a joint research project in the research program "r$^3$-Innovative Technologien für Ressourceneffizienz-Strategische Metalle und Mineralien" within the framework "Forschung für nachhaltige Entwicklung". The project „Substitution of rare earth elements in high strength and ductile Magnesium sheet material - SubSEEMag" was financially supported by the German Federal Ministry of Education and Research.

## References

1. F. Basson, D. Letzig, Aluminium International Today, November/December 2010, 19–21.
2. D.H. St John, J. Advanced Materials Research, 29/30, 2007, 3–8.
3. S.S. Parks, G.T. Bae, J.G. Lee, D.H. Kang, K.S. Shin, N.J. Kim, Materials Science Forum, 539/543, 2007, 119–126.
4. R Kawalla, M Oswald, C. Schmidt, M. Ullmann, H.P. Vogt, N.D. Cuong, Magnesium Technology 2008, New Orleans, Louisiana, TMS, 2008, 177–182.
5. H. Watari, T. Haga, R. Paisarn, N. Koga, K. Davey, Key Engineering Materials, 345/346, 2007, 165–168.
6. M. Aljarrah, E. Essadiqi, D.H. Kang, I.H. Jung, Light Metals Technology V, 5th International Conference on Light Metals Technology, 19–22 July 2011, Lueneburg, Germany, 331–334.
7. K. Hantzsche et al. (2010) Effect of rare earth additions on microstructure and texture development of magnesium alloy sheets. Scripta Materialia 63:725–730.
8. J. Bohlen et al. (2015) The Influence of the Combination of Alloying Elements on the Microstructure and Texture Development During Rolling and Annealing of Magnesium. Paper presented at the 10th International Conference, Mg 2015, Jeju, 11–16 October 2015.
9. J. Bohlen et al. (2007) The texture and anisotropy of magnesium–zinc–rare earth alloy sheets. Acta Materialia 55:2101–2112.
10. Y. Chino et al. (2011) Effects of Ca on Tensile Properties and Stretch Formability at Room Temperature in Mg–Zn and Mg–Al Alloys. Materials Transactions 52(7):1477–1482.
11. D.W. Kim et al. (2013) Texture Evolution in Mg-Zn-Ca Alloy Sheets. Metallurgical and Materials Transactions A 44A:2950–2961.
12. S.B. Yi et al. (2010) Mechanical anisotropy and deep drawing behaviour of AZ31 and ZE10 magnesium alloy sheets. Acta Materialia 58:592–605.
13. Final report Mobil mit Magnesium: [Forschungsvorhaben]; Laufzeit: 01.09.2006–30.09.2010, Doi https://doi.org/10.2314/GBV:717510913.
14. V. Kree, J. Bohlen, D. Letzig, K.U. Kainer, Practical Metallography 5, 2004, 233–246.
15. You B-S, Park W-W, Chung I-S, (2000) Effect of calcium additions on the oxidation behavior in magnesium alloys. Scripta Mater Vol. 42. pp 1089–1094.
16. Kim D-W, Suh B-C, Shim M-S, Bae J H, Kim D H, Kim N J (2013) Texture evolution in Mg-Zn-Ca alloy sheets. Metall. Mater. Trans. A. Vol. 44. pp 2950–2961.
17. Suh B-C, Shim M-S, Shin K S, Kim N J (2014) Current issues in magnesium sheet alloys: Where do we go from here?. Scripta Mater. Vol. 84–85. pp 1–6.
18. Stanford N (2010) The effect of calcium on the texture, microstructure and mechanical properties of extruded Mg–Mn–Ca alloys. Mater. Sci. Eng. A, Vol. 528. pp 314–322.
19. G. Kurz, J. Bohlen, D. Letzig, K.U. Kainer, Materials Science Forum Vol. 765 (2013), 205–209. Doi:https://doi.org/10.4028/www.scientific.net/MSF.765.205.

# Development of Heat-Treatable High-Strength Mg–Zn–Ca–Zr Sheet Alloy with Excellent Room Temperature Formability

M. Z. Bian, T. T. Sasaki, B. C. Suh, T. Nakata, S. Kamado, and K. Hono


**Abstract**

Lightweight magnesium (Mg) alloys have attracted considerable attention for potential applications in the automotive industries. However, the low strength or poor formability at room temperature (RT) hinders the wider applications of wrought Mg alloy sheets. The newly developed a heat-treatable magnesium sheet alloy, Mg–0.6Zn–0.3Ca–0.1Zr (at. %), shows a large Index Erichsen value of 8.0 mm at RT in a solution treated condition. The excellent RT formability can be ascribed to a weak basal texture. Subsequent artificial aging at 170 °C for 4 h (T6) increases the 0.2% proof strength from 165 to 213 MPa. This improvement in strength by the T6 treatment is associated with a dense distribution of Guinier–Preston zones lying on the basal planes of the Mg matrix. Our finding overcomes the trade-off relationship of the room temperature formability and proof strength in Mg alloy sheets and this can be used as "bake hardenable" magnesium sheet alloy with excellent RT formability.



**Keywords**

Magnesium alloy • Formability • Strength Guinier–preston zone



M. Z. Bian (✉) · T. T. Sasaki · B. C. Suh · K. Hono
National Institute for Materials Science, 1-2-1 Sengen, Tsukuba, 305-0047, Japan
e-mail: BIAN.Ming-Zhe@nims.go.jp

T. Nakata · S. Kamado
Nagaoka University of Technology, 1603-1, Kamitomioka, Nagaoka 940-2188, Japan

K. Hono
Graduate School of Pure and Applied Science, University of Tsukuba, 1-1 Tenno-Dai, Tsukuba, 305-0001, Japan




D. Orlov et al. (eds.), *Magnesium Technology 2018*, The Minerals, Metals & Materials Series, https://doi.org/10.1007/978-3-319-72332-7_55


## Introduction

Increasing demands of weight reduction in automobiles is expected to lead to the gradual application of lightweight magnesium (Mg) alloys in commercial vehicles [1, 2]. However, the vast majority of their applications are confined to cast alloys, and the applications of wrought alloys, particularly sheet alloys are very limited due to the poor formability or low strength at room temperature (RT) [3, 4]. Recently, we have demonstrated that the development of heat-treatable Mg sheet alloy is a promising approach to overcome the current issues of Mg alloy sheets [5]. Specifically, the solution treated (T4) Mg–1.1Al–0.3Ca–0.2Mn–0.3Zn (at. %) alloy sheet shows a large Index Erichsen (I.E.) value of 7.7 mm and its yield strength is significantly increased from 144 to 204 MPa after a T6 treatment. Therefore, it can be easily formed in a T4 condition and thereafter strengthened by the artificial aging. Another notable feature of this alloy is the rapid age-hardening response. It can achieve the peak hardness at 200 °C only for an hour because the main strengthening phase is Al and Ca enriched monoatomic layer Guinier–Preston (G.P.) zones.

Mg–Zn–Ca (ZX) based alloys have been reported to form the Zn and Ca enriched monoatomic layer G.P. zones [6, 7]. Moreover, dilute ZX alloy sheets were demonstrated to show excellent RT formability due to the weak basal texture [8, 9]. Although there is a high possibility to develop heat-treatable high-strength ZX sheet alloys with excellent RT formability, little attention has been paid to develop heat-treatable sheet alloys so far [10]. This motivated us to develop heat-treatable sheet alloys from the ZX system. A small amount of Zr was added to a ZX alloy as Zr is a commonly used grain refiner of Mg alloys [11, 12].

## Experimental Procedure

Mg–0.6Zn–0.3Ca–0.1Zr in at. % or Mg–1.6Zn–0.5Ca–0.4Zr in wt% (ZXK210) alloy ingot was prepared from high-purity Mg, ZK60 (Mg–5.5Zn–0.5Zr in wt%), Mg–30 wt%Ca and Mg–34 wt% Zr master alloys by an induction melting furnace in an Argon atmosphere. 10 mm thick plates were machined from the cast ingot and subsequently thermomechanically treated. They were first homogenized at 300 °C for 4 h and then slowly heated to 450 °C with a heating rate of 7.5 °C/h, and maintained at 450 °C for 6 h, followed by water quenching. The homogenized plates were primarily rolled to ~5 mm thick plates at 300 °C by 4 passes, with ~15% thickness reduction per pass. The 5 mm thick plates were further fine-rolled to ~1 mm thick sheets at 100 °C by 6 passes, with ~23% thickness reduction per pass. In the case of the fine rolling, the sheets were reheated at 450 °C for 5 min prior to subsequent rolling. Tensile specimens having the gage length of ~12.5 mm and width of ~5 mm were machined from the as-rolled sheets along the rolling direction (RD) and then solution treated at 430 °C for 1 h (T4). Some of the T4 treated tensile samples were subjected to a T6 (aged in an oil bath at 170 °C for 4 h) treatment.

Tensile tests were conducted at RT using a screw driven Instron 5567 tensile testing machine at an initial strain rate of $1.0 \times 10^{-3}\ s^{-1}$. To evaluate the stretch formability of the T4 treated sheet, Erichsen cupping tests were carried out on rectangular specimens using a hemispherical punch with a diameter of 20 mm at room temperature (RT). Punch speed and blank-holder force were ~6 mm/min and ~10 kN, respectively. At least three samples were tested to ensure repeatability of tensile and Erichsen cupping test results. The age hardening response was measured by the Vickers hardness tests under a load of 300 g. Microstructures and textures were characterized using FEI Helios Nanolab 650 equipped with Bruker electron backscatter diffraction (EBSD) detector and FEI Titan G2 80–200 transmission electron microscopes (TEM) operating at 200 kV.

## Results and Discussion

Figure 1a shows the age-hardening response of the ZXK210 alloy sheet during isothermal aging at 170 °C. The T4 treated sheet has a hardness value of 54.2 ± 0.6 HV, which exhibits a rapid age-hardening to a peak hardness of 65.8 ± 1.0 HV only for 4 h. Figure 1b shows the tensile stress-strain curves obtained from the samples in T4 and T6 conditions stretched along the RD. The T4 heat treated sample exhibits a moderate 0.2% proof strength (hereafter proof strength) of 165 MPa with a large elongation of 30%. Subsequent artificial aging at 170 °C for 4 h (T6) increases the 0.2% proof strength from 165 to 213 MPa while maintaining good elongation of 30%. The tensile properties of the T4 and T6 treated alloy sheets are provided in Table 1. The T4 treated sample after the Erichsen cupping test is also shown in Fig. 1b. It can be seen that this sample exhibits a large Index Erichsen (I.E.) value of 8.0 mm.

To understand the reasons for the excellent formability of the ZXK210 alloy sheet, the evolution of microstructure and texture during stretch forming was investigated. Figure 2a shows the EBSD inverse pole figure (IPF) map and texture of the T4 treated sample. The T4 treated sample has an average grain size of about 8.7 μm with a weak basal texture intensity of 3.1 mrd. Figures 2b and c show the IPF maps and textures of the samples stretch formed to the dome heights of 3 and 8.0 mm (fracture dome height), respectively. As compared to the IPF map and the (0002) pole

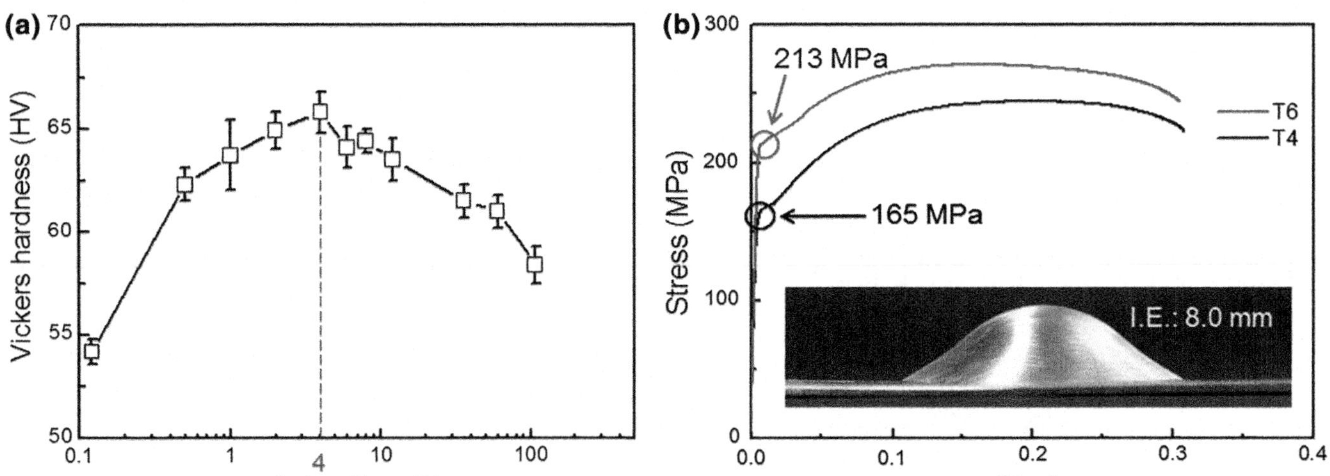


**Fig. 1** **a** Age hardening response of ZXK210 alloy sheet at 170 °C, and **b** tensile curves of T4 and T6 treated sheets stretched along the RD

**Table 1** Room temperature tensile properties of T4 and T6 treated ZXK210 alloy sheet

| Condition | 0.2% proof strength (MPa) | UTS (MPa) | Elongation (%) | I.E. (mm) |
|---|---|---|---|---|
| T4 | 165 ± 2 | 245 ± 1 | 30 ± 1 | 8.0 ± 0.2 |
| T6 | 213 ± 2 | 270 ± 2 | 30 ± 0 | – |

figure of the un-stretched sample, there are essentially no significant changes in grain orientation with the progress of stretch forming. The only noticeable difference is that the angular distribution of basal poles gets gradually narrower toward ND. Additionally, the basal texture intensity of the top half region is found to be stronger than that of the bottom half region, but virtually no difference in the distribution of basal poles. Figure 2d shows the variation of the misorientation angle distribution with progression of stretch forming. It can be seen that the fraction of misorientation angles lower than 10° increases substantially with the progression of stretch forming. In contrast, the fraction of misorientation angles near 86° gradually decreases. In general, the formation of misorientation angles near 86° is intimately related to $\{10\bar{1}2\}$ tensile twins to a large extent during deformation [13, 14]. Therefore, it is concluded that the $\{10\bar{1}2\}$ tensile twin does not play an important role in accommodating the plastic strain during stretch forming. The misorientation angles near 30° developed after the T4 treatment almost remain unchanged with the progress of the stretch forming. The formation of misorientation angle of 30° is reported to be intimately related to the preferential growth of <$11\bar{2}0$> recrystallized grains [15, 16].

The microstructure of the T6 treated sample was examined by TEM to understand the reasons for the improved proof strength. Figure 3 shows the high-angle annular dark-field scanning transmission electron microscopy (HAADF-STEM) image taken with the incident beam along the $[2\bar{1}\bar{1}0]_\alpha$ zone axis of the T6 treated sample. The microstructure is found to contain a dense distribution of nanoscale plate-like precipitates on the basal plane of the Mg matrix. Closer inspection from the atomic-scale HAADF-STEM image reveals that these nanoscale plate-like precipitates have a monolayer structure (marked with an arrow). According to Oh-Ishi et al. the precipitates having such features are monoatomic layer G.P. zones

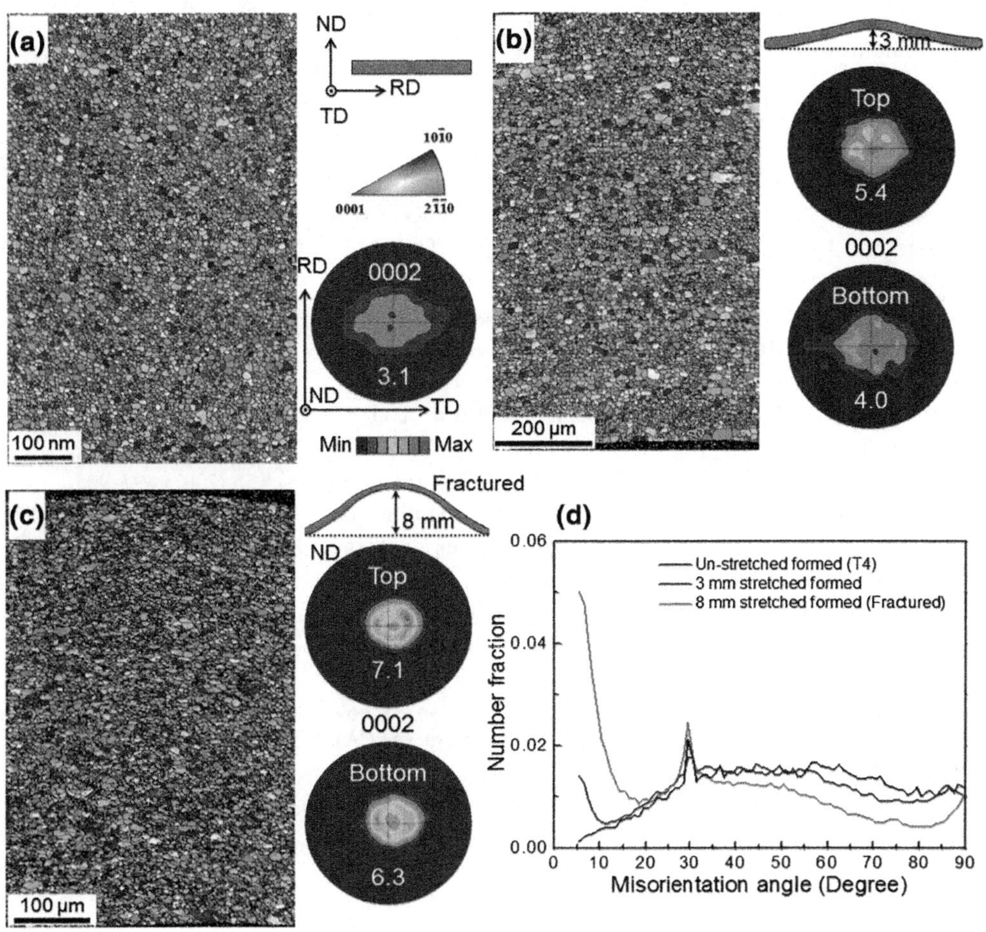


**Fig. 2** The EBSD inverse pole figure (IPF) maps and textures showing microstructures and textures of **a** un-stretched (T4 treated), **b** 3 mm stretch deformed, **c** 8 mm stretch defomred (fractured) ZXK210 samples, and **d** evolution of the distribution for misorientation angles with progression of stretch forming

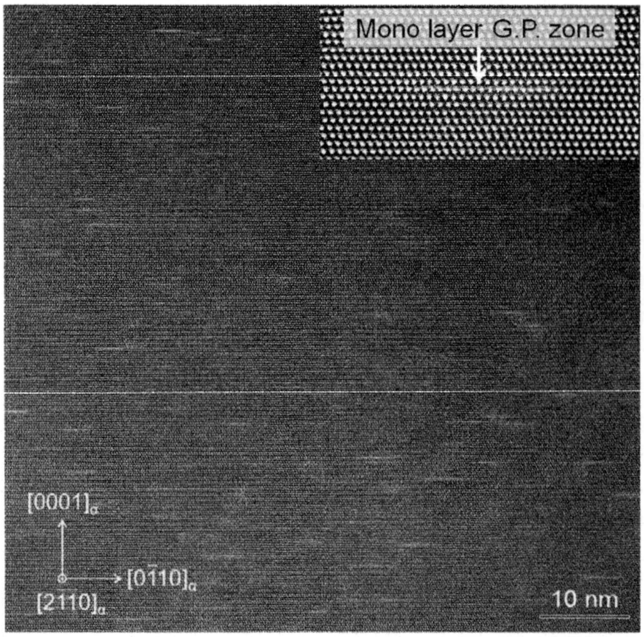


**Fig. 3** A HAADF-STEM image showing the presence of monoatomic layer G.P. zones in the T6-treated ZXK210 alloy sheet. The HAADF-STEM image was taken from the $[2\bar{1}\bar{1}0]_\alpha$ direction

enriched with Ca and Zn atoms [17]. Thus, the strength increase by the T6 treatment can be ascribed to the dense distribution of these monoatomic layer G.P. zones (Fig. 3).

## Summary

A heat-treatable Mg–Zn–Ca–Zr sheet alloy with good room temperature formability has been developed. The Mg–0.6Zn–0.3Ca–0.1Zr (at. %) exhibits a large Index Erichsen value of 8 mm in a T4 condition, which is associated with the weak basal texture. Moreover, the ZXK210 sheet alloy has a considerable age hardening response and its 0.2% proof strength is significantly enhanced from 165 to 213 MPa by artificial aging at 170 °C for only 4 h. TEM observations reveal that the monoatomic layer G.P. zones on the basal planes are responsible for the age-hardening and strength improvement. This study demonstrates that the dilute ZX system is a promising candidate for bake hard-enable high-strength magnesium sheet alloys with excellent room temperature formability.

**Acknowledgements** This work was supported by Advanced Low Carbon Technology Research and Development Program (ALCA), 12102886, JSPS KAKENHI Grant Number JP26709055 and the Structural Materials for Innovations of the Cross Ministrial Strategic Innovation Program (SIP) of Japan Science and Technology (JST).

## References

1. W.J. Joost, P.E. Krajewski, Towards magnesium alloys for high-volume automotive applications, Scr. Mater. 128 (2017) 107–112.
2. K. Hono, C.L. Mendis, T.T. Sasaki, K. Oh-ishi, Towards the development of heat-treatable high-strength wrought Mg alloys, Scr. Mater. 63 (2010) 710–715.
3. A.A. Luo, Magnesium casting technology for structural applications, J. Magnes. Alloy. 1 (2013) 2–22.
4. B.C. Suh, M.S. Shim, K.S. Shin, N.J. Kim, Current issues in magnesium sheet alloys: Where do we go from here?, Scr. Mater. 84–85 (2014) 1–6.
5. M.Z. Bian, T.T. Sasaki, B.C. Suh, T. Nakata, S. Kamado, K. Hono, A heat-treatable Mg–Al–Ca–Mn–Zn sheet alloy with good room temperature formability, Scr. Mater. 138 (2017) 151–155.
6. J.F. Nie, B.C. Muddle, Precipitation hardening of Mg-Ca(-Zn) alloys, Scr. Mater. 37 (1997) 1475–1481.
7. J.C. Oh, T. Ohkubo, T. Mukai, K. Hono, TEM and 3DAP characterization of an age-hardened Mg–Ca–Zn alloy, Scr. Mater. 53 (2005) 675–679.
8. Y. Chino, X. Huang, K. Suzuki, M. Mabuchi, Enhancement of stretch formability at room temperature by addition of Ca in Mg–Zn alloy, Mater. Trans. 51 (2010) 818–821.
9. Y. Chino, K. Sassa, X.S. Huang, K. Suzuki, M. Mabuchi, Effects of zinc concentration on the stretch formability at room temperature of the rolled Mg–Zn–Ca alloys, J. Japan Inst. Met. Mater. 75 (2011) 35–41.
10. M.Z. Bian, Z.R. Zeng, S.W. Xu, S.M. Zhu, Y.M. Zhu, C.H. J. Davies, N. Birbilis, J.F. Nie, Improving formability of Mg–Ca–Zr sheet alloy by microalloying of Zn, Adv. Eng. Mater. 18 (2016) 1763–1769.
11. M. Qian, D.H. StJohn, M.T. Frost, Characteristic zirconium-rich coring structures in Mg–Zr alloys, Scr. Mater. 46 (2002) 649–654.
12. T. Bhattacharjee, T. Nakata, T.T. Sasaki, S. Kamado, K. Hono, Effect of microalloyed Zr on the extruded microstructure of Mg–6.2Zn-based alloys, Scr. Mater. 90 (2014) 37–40.
13. J.W. Christian, S. Mahajan, Deformation twinning, Prog. Mater. Sci. 39 (1995) 1–157.
14. Y.N. Wang, J.C. Huang, The role of twinning and untwinning in yielding behavior in hot-extruded Mg–Al–Zn alloy, Acta Mater. 55 (2007) 897–905.
15. G. Gottstein, T. Alsamman, Texture development in pure Mg and Mg alloy AZ31, Mater. Sci. Forum. 195–197 (2005) 623–632.
16. M.A. Steiner, J.J. Bhattacharyya, S.R. Agnew, The origin and enhancement of {0001} ⟨11–20⟩ texture during heat treatment of rolled AZ31B magnesium alloys, Acta Mater. 95 (2015) 443–455.
17. K. Oh-ishi, R. Watanabe, C.L. Mendis, K. Hono, Age-hardening response of Mg-0.3 at.%Ca alloys with different Zn contents, Mater. Sci. Eng. A. 526 (2009) 177–184.

# Interaction Between Propagating Twins and Non-shearable Precipitates in Magnesium Alloys

Matthew R. Barnett and Huan Wang

**Abstract**

The impact of non-shearable particles on the stress required to propagate a twin is examined using numerical and analytical techniques. Of particular note is the nature of this effect in magnesium alloys where twinning can count for a significant fraction of the plastic strain. Dislocation Dynamics simulations are employed to calculate the Orowan bypass stress. It is seen that the stress for bypass can exceed that for a single twinning dislocation but under some circumstances it may be similar. The conditions that give rise to different regimes of behaviour are mentioned and the significance for magnesium alloy design is described.

**Keywords**

Twinning • Magnesium • Dislocations

## Body

It is possible to consider a propagating twin as an expanding super dislocation loop [1]. The approximation is employed here to make some points about the propagation of {10–12} twins through an array of non-shearable particles.

In last year's (2017) Magnesium Technology proceedings [2], we estimated the stress required to bow a super dislocation approximating the twinning front between non-shearable obstacles with spacing $\lambda$. We represented the twin by a super dislocation with Burgers vector $nb$. The 'height' of each dislocation is $h$, where the twinning shear is $s = b/h$, and so the effective twin thickness, $t$, is $nb/s$. We set $n = 100$ based on calculations of the number of dislocations needed in the twin front for the Peach-Koheler force to meet the opposing back force arising from the twin surface energy. For an obstacle spacing of 100 nm, the by-pass stress was estimated to be 270 MPa. It was pointed out that this value is significantly higher than observed hardening levels, which are 30–40 MPa for particle spacings in the range 80–150 nm [3, 4]. The discrepancy was explained in terms of relaxation effects.

Here, we show that there are other important factors to be considered. These include the need to properly account for (i) the effective size of the core of the super dislocation and (ii) the number of dislocations that partake in the bypass event.

Previously, we employed Equations 6.51 in [5] for the elastic line energy of a dislocation loop. To obtain a solution more relevant to the bowing of a dislocation between obstacles, we employ here instead the expression suggested by Bacon et al. [6], which allows the by-pass stress for a super screw dislocation to be given as:

$$\tau = \frac{Gnb}{2\pi(1-\nu)\lambda}\ln\left(2\frac{\left(\lambda^{-1}+d^{-1}\right)^{-1}}{r}\right) \quad (1)$$

where $d$ is the diameter of the (spherical) obstacles and $r$ is the inner cut-off radius.

To assess this solution, Dislocation Dynamics simulations were performed using the DDlab code developed by Cai et al. [7]. We examine a pile-up of dislocations against a line of obstacles as shown in Fig. 1. Screw dislocations are considered, the simulation is isotropic and the core energy (within radius $b$) is set to zero. The obstacles are pinned single dislocation loops. A back stress is applied to the leading dislocation to mimic the force from the twin surface energy. A twin surface energy of 0.12 J/m$^2$ is employed [8]. There are two twin surfaces so that 'back' force is 0.24 N/m. For $|b| = 0.3$ nm, this translates to a back stress of 800 MPa. We insert this directly into the simulations. Equilibrium configurations were calculated using $|b| = 0.3$ nm, $G = 17{,}000$ MPa, $\nu = 0.33$ and $n = 3$ and 16 for $\lambda = 140$ nm and $d = 10$ nm. The Burgers vector of the {10–12}

M. R. Barnett (✉) · H. Wang
Institute for Frontier Materials, Deakin University, Geelong, VIC 3217, Australia
e-mail: barnettm@deakin.edu.au


D. Orlov et al. (eds.), *Magnesium Technology 2018*, The Minerals, Metals & Materials Series, https://doi.org/10.1007/978-3-319-72332-7_56

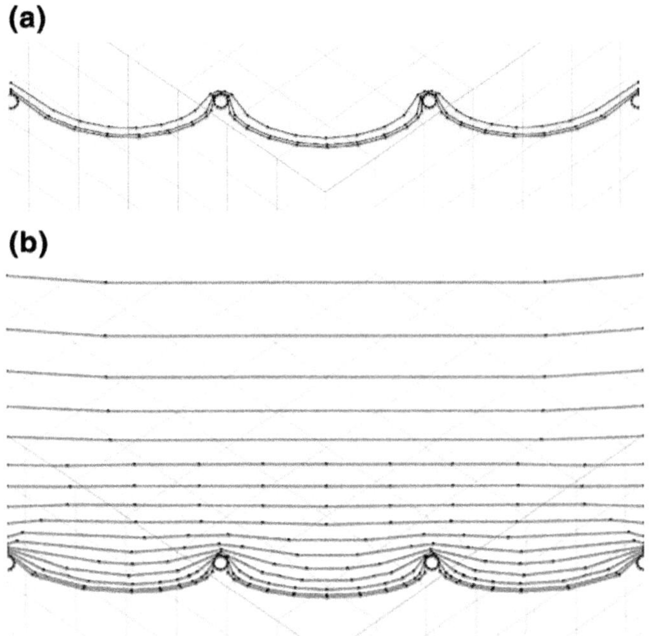

**Fig. 1** Equilibrium positions of dislocations a prior to by-pass for **a** n = 3 and **b** n = 16

twinning dislocations in magnesium is 0.049 nm so each of the present simulated dislocations represents approximately 6 twinning dislocations. The corresponding by-pass shear stresses are found to be 75 and 37 MPa for these two cases, respectively.

Equation 1 gives a breakthrough stress of 79 MPa for $n = 3$, using the parameters given above, if the inner cut-off radius is set to $nb$. This value agrees reasonably well with the simulated value (75 MPa). We thus propose that a single super dislocation model is a reasonable approximation for small $n$ and that $nb$ is an appropriate choice for the effective core radius of a super dislocation in the limit of zero core energies for the individual dislocations that make up the super dislocation.

If we set $n = 16$ (which is equivalent to 98 twinning dislocations) we see that Eq. 1 predicts a by-pass stress of 189 MPa. This considerably over-estimates the simulated value (37 MPa), in keeping with the opening point of the paper; i.e. that the super dislocation model provides an over-prediction of the by-pass stresses for $n \sim 100$ twinning dislocations.

Interestingly, the stress increment required for bypass in the simulation for 98 twinning dislocations is within the range of stress increments seen in experiment (i.e. 30–40 MPa for $\lambda = 80$–150 nm). This suggests that the relaxation stresses invoked previously to explain the low hardening increments seen in experiment may not be as important as we thought. Such relaxation is mediated by slip events and this was not something incorporated in the present simulations.

Inspection of Fig. 1 reveals that not all of the dislocations present in the twin will participate in the initial by-pass event. Dislocations in the rear of the pile-up serve to push the leading dislocations between the obstacles. The problem is probably better approximated by two super dislocations rather than one. The leading super dislocation contains the dislocations participating in the by-pass event and the trailing dislocations provide the additional stress needed for the by-pass.

According to Eq. 1, the stress required for a single dislocation to bow through the gap between obstacles 140 nm apart ($|b| = 0.3$ nm and $d = 10$ nm) is 36 MPa. Thus it is seen that the pile-up model of a twin predicts that the twin may in principal propagate through an array of obstacles at stresses similar to that required for a single twinning dislocation to do so. Indeed, the more twinning dislocations present during propagation, the lower the by-pass stress is likely to be.

This leads to an interesting observation: hardening mechanisms that force twins to propagate at higher thicknesses—and thus with higher numbers of co-operatively moving twinning dislocations—will find it easier to propagate through an array of non-shearable obstacles. To increase the effectiveness of an array of non-shearable particles in providing strengthening against twin propagation, one should seek to encourage twins to propagate as thinner rather than thicker lamellae. How to achieve this in a practical sense is not entirely clear but the higher the stress the twin propagates at, the thinner the twin is likely to be during the propagation step. Multiple hardening mechanisms are thus to be favoured.

However, we need to recall that the present model is in two dimensions. It therefore only holds for thin large diameter twins. When the twin is thicker, as pointed out by Gharghouri et al. [9], the interaction with precipitates changes. This is an aspect of our ongoing investigation.

In conclusion, the super dislocation model of twinning sheds interesting light on the propagation of a twin through non-shearable obstacles. Particularly it is found that:

- The bowing between obstacles of a twin tip comprising a few dislocations can be approximated by a single super-dislocation model.
- The effective core employed for the super dislocation model can be approximated by $nb$.
- For higher numbers of dislocations in the twin, propagation occurs at lower levels of stress because the dislocations at the twin tip are forced along by those further behind.

- Simulations reveal stress increments of similar order to those seen in experiment.
- The model predicts that encouraging twins to propagate as thin lamellae will favour hardening.

**Acknowledgements** The authors would like to acknowledge valuable discussions on the topic with Laurent Capolungo and Marc Fivel. The Australian Research Council Discovery Grant Scheme and the Deakin University DUPRS scheme are also gratefully acknowledged.

## References

1. Friedel, J., *Dislocations*. International series of Monographs on solid State Physics, ed. R. Smoluchowski and N. Kurti. 1964, Oxford: Pergamon.
2. Barnett, M.R. *Twinning Super Dislocations to Help Understand Strength*. in *Magnesium Technology 2017*. 2017. TMS.
3. Jain, J., et al., *The role of back stress caused by precipitates on {101–2} twinning in a Mg-6Zn alloy*. Materials Science and Engineering A, 2015. **647**: pp. 66–73.
4. Stanford, N. and M.R. Barnett, *Effect of particles on the formation of deformation twins in a magnesium-based alloy*. Materials Science and Engineering A, 2009. **516**(1–2): pp. 226–234.
5. Hirth, J.P. and J. Lothe, *Theory of dislocations*. 1992, Florida: Krieger.
6. Bacon, D.J., U.F. Kocks, and R.O. Scattergood, *The effect of dislocation self-interaction on the orowan stress*. Philosophical Magazine, 1973. **28**(6): pp. 1241–1263.
7. Cai, W. 2013; Available from: http://micro.stanford.edu/~caiwei/Forum/2005-12-05-DDLab/.
8. Wang, Y., et al., *First-principles calculations of twin-boundary and stacking-fault energies in magnesium*. Scripta Materialia, 2010. **62**(9): pp. 646–649.
9. Gharghouri, M.A., G.C. Weatherly, and J.D. Embury, *The Interaction of Twins and Precipitates in a Mg-7.7at.%Al Alloy*. Philosophical Magazine A, 1998. **78**(5): pp. 1137–1149.

# Effects of Severe Plastic Deformation on Mechanical Properties and Corrosion Behavior of Magnesium Alloys

Ahmad Bahmani and Kwang Seon Shin

**Abstract**

Magnesium alloys were produced with extrusion, screw rolling (SR) and multi-directional forging (MDF) processes. Various temperatures were used for SR and MDF. The tensile test was carried out to measure the mechanical properties and immersion test in 3.5 wt% NaCl saturated with $Mg(OH)_2$ was run to measure the corrosion rate. It was found that the grain size increased gradually with increasing the MDF or SR temperature. The increase in the MDF or SR temperature resulted in decrease in yield strength due to increase in grain size. However, the corrosion rate decreased after MDF and SR. Furthermore, the increase in MDF or SR temperature resulted in more reduction in the corrosion rate due to reduction of galvanic cells by dissolution of the intermetallic phases into the matrix.

**Keywords**

Magnesium alloy • Multi-directional forging
Screw rolling • Corrosion rate • Yield strength

## Introduction

Mg alloys are attractive structural materials for light-weight vehicles and electronic devices due to their high strength-to-weight ratio. Fuel efficiency can be improved by the application of Mg alloys to automobiles through weight reduction. Mg alloys are also identified as promising materials for biodegradable implants in the human body. However, the poor corrosion resistance of existing Mg alloys limits their applications. Severe plastic deformation (SPD) is one of the techniques to refine the microstructure of Mg alloys through ultra-fine grains (UFG) by applying force in different directions. Several SPD techniques such as equal channel angular pressing/extrusion (ECAP/ECAE) [1–3], accumulative roll bonding (ARB) [2], accumulative back extrusion (ABE) [4], three roll planetary milling or screw rolling (SR) [5] and multi directional forging (MDF) [6] have been studied. The majority of the researchers studied the effects of the process parameters such as the strain rate, number of passes and processing temperature to improve the mechanical properties of alloys, but studies on the corrosion behavior have been limited.

Multi-directional forging is reported to effectively control the properties of alloys by controlling the microstructure parameters. The highest yield strength in AZ alloys was achieved by MDF of AZ80 combined with aging treatment [6].

Despite several studies on improving the mechanical properties of Mg alloys by MDF and a few studies on enhancing the mechanical properties by SR, the effects of MDF and SR on corrosion behavior have not yet been extensively explored. Therefore, in this work, MDF and SR processes at different temperatures have been utilized to manufacture alloys with enhanced mechanical properties and improved corrosion resistance. It was found that SPD can not only increase the mechanical properties but also significantly improve corrosion resistance. Detailed studies on the microstructure and corrosion behavior were carried out.

## Methods

The ZAXM4211 magnesium alloy was prepared in an electric furnace. The cast alloy was homogenized at 340 °C for 12 h and then quenched in water. Cylindrical samples were machined from the homogenized billets for extrusion. Indirect extrusion was carried out at an extrusion ratio of 10:1 and a ram speed of 5 mm/s. The final cross-sectional area was 22 × 22 mm for MDF and the diameter was

A. Bahmani · K. S. Shin (✉)
Magnesium Technology Innovation Center, School of Materials Science and Engineering, Seoul National University, Seoul, Korea
e-mail: ksshin@snu.ac.kr


D. Orlov et al. (eds.), *Magnesium Technology 2018*, The Minerals, Metals & Materials Series, https://doi.org/10.1007/978-3-319-72332-7_57

25 mm for SR. In the case of MDF, rectangular cubes were cut from the extruded alloy and subjected to MDF. The alloy was placed in the furnace during MDF to maintain desired temperatures of 180, 220, 260 and 300 °C and the samples were denoted as MDF180, MDF220, MDF260 and MDF300, respectively. In the case of SR, the samples were preheated at a desired temperature and then inserted into the screw rolling machine to reduce the diameter in several passes. The samples were denoted as SR180, SR220, SR260, SR300 and SR340 according to the preheating temperature.

## Results

The relationship between yield strength and grain size for MDF and SR samples processed at different temperatures has been presented in Fig. 1. It can be seen that the increase in the process temperature resulted in decreasing yield strength values which is attributed to the grain size increase at high temperature processing. Grain boundaries and intermetallic compounds impede dislocation motion. Since the samples processed at low temperatures have smaller grain sizes, they showed higher tensile yield strength.

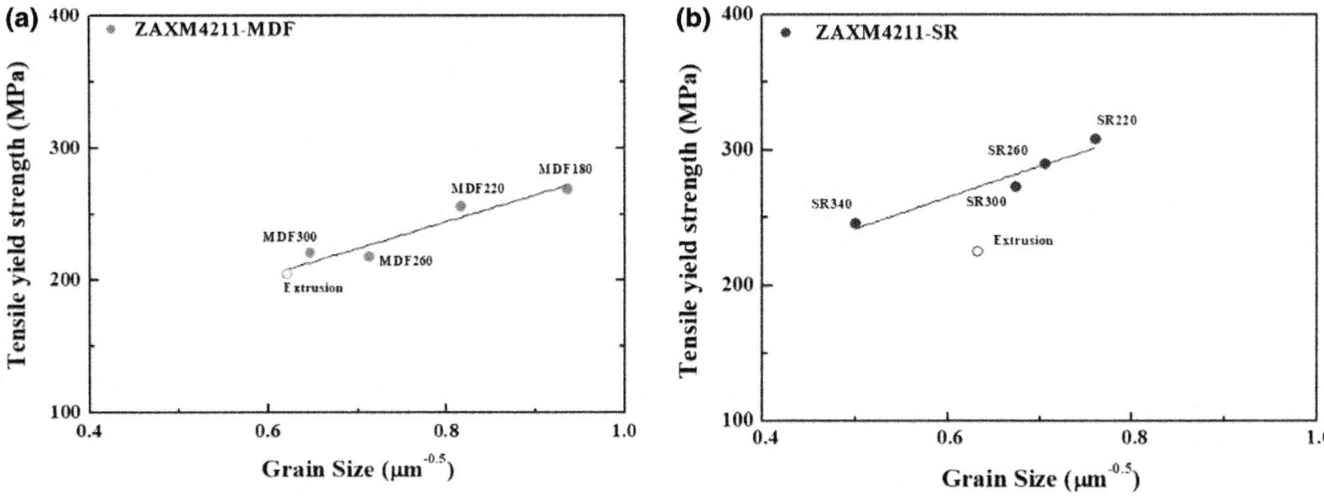


**Fig. 1** The relationship between yield strength and grain size: **a** MDF and **b** SR

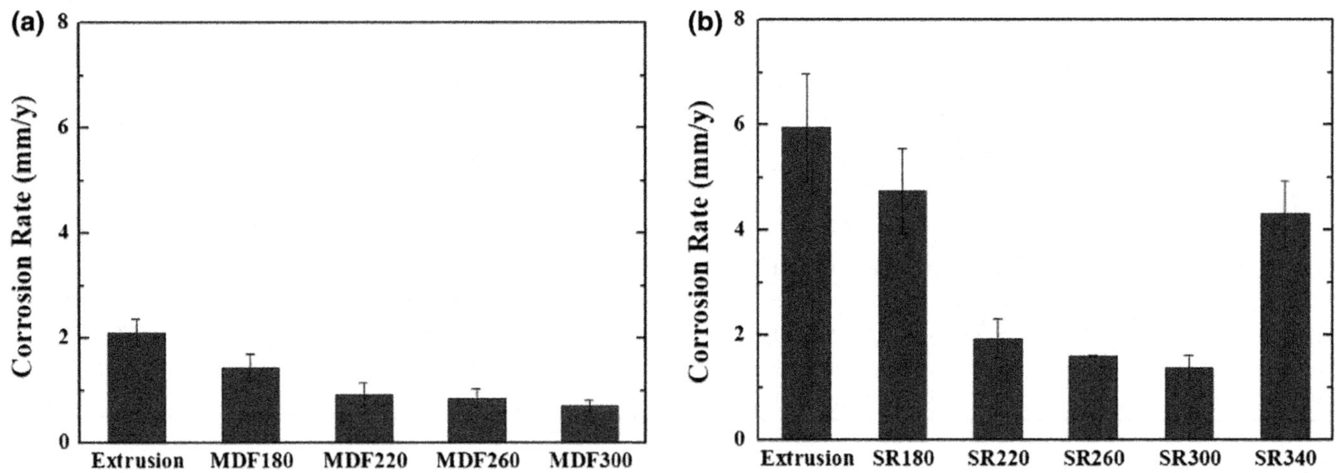


**Fig. 2** The relationship between corrosion rate and processing temperature: **a** MDF and **b** SR

A Hall-Petch equation can be written using the obtained experimental data as:

$$\sigma_{y(MDF)} = 201.5 + 83.3\,GS^{-0.5} \quad (1)$$

$$\sigma_{y(SR)} = 227.8 + 128.7\,GS^{-0.5} \quad (2)$$

where $\sigma_y$ is the yield strength and GS is the grain size.

The immersion test in 3.5 wt% NaCl solution saturated with $Mg(OH)_2$ was carried out for seven days, and the corrosion rate was obtained using the hydrogen evolution technique. Figure 2 shows the relationship between the corrosion rate obtained by the hydrogen evolution technique and the processing conditions. The corrosion rates of MDF and SR samples are significantly less than that of the extruded alloy. Among the SPD samples, those produced at 300 °C show the lowest corrosion rate.

## Conclusions

The mechanical properties and corrosion behavior of the ZAXM4211 alloy produced by extrusion, MDF and SR were studied. The following conclusions were made:

1. The grain size increased gradually by increasing the MDF or SR temperature and the intermetallic phase fraction decreased due to their dissolution into the matrix.
2. It can be seen that the increase in the MDF or SR temperature results in decreasing yield strength which is attributed to the grain size increase at high temperature processing.
3. The corrosion rate decreased after MDF and SR. Furthermore, the increase in MDF temperature resulted in more reduction in the corrosion rate. This behavior is attributed to the reduction of galvanic cells by dissolution of the intermetallic phases into the matrix.

**Acknowledgements** This work was financially supported by the Ministry of Trade, Industry and Energy (10052196) and the Basic Science Research Program through the National Research Foundation of Korea (NRF) grant funded by the MEST (No. 2015R1A2A1A01006795), through the Research Institute of Advanced Materials and Magnesium Technology Innovation Center.

## References

1. D. Song, A.B. Ma, J.H. Jiang, P.H. Lin, D.H. Yang, J.F. Fan, Corrosion behaviour of bulk ultra-fine grained AZ91D magnesium alloy fabricated by equal-channel angular pressing, Corrosion Science 53(1) (2011) 362–373.
2. A. Nikfahm, Effect of grain size changes on corrosion behavior of copper produced by accumulative rol bonding process, Mater Res 16 (2013) 1379–1386.
3. G.B. Hamu, D. Eliezer, L. Wagner, The relation between severe plastic deformation microstructure and corrosion behavior of AZ31 magnesium alloy, Journal of Alloys and Compounds 468(1–2) (2009) 222–229.
4. S.M. Fatemi-Varzaneh, A. Zarei-Hanzaki, Processing of AZ31 magnesium alloy by a new noble severe plastic deformation method, Materials Science and Engineering A 528(3) (2011) 1334–1339.
5. M. Diez, H.-E. Kim, V. Serebryany, S. Dobatkin, Y. Estrin, Improving the mechanical properties of pure magnesium by three-roll planetary milling, Materials Science and Engineering A 612 (2014) 287–292.
6. H. Miura, W. Nakamura, M. Kobayashi, Room-temperature Multi-directional Forging of AZ80 Mg Alloy to Induce Ultrafine Grained Structure and Specific Mechanical Properties, Procedia Engineering 81 (2014) 534–539.

# Alloy Design for the Development of Heat Treatable High Strength Mg Sheet Alloy with Excellent Room Temperature Formability

B.-C. Suh, M.-Z. Bian, T. Nakata, T. T. Sasaki, S. Kamado, and K. Hono

**Abstract**

To widen the application of the magnesium sheet alloys, good room temperature (RT) formability and satisfactory strength need to be achieved. The development of heat treatable alloy can be an effective approach to simultaneously achieve the good RT formability and satisfactory strength. In this work, we have investigated the effects of Zn content on the microstructure, stretch formability and mechanical properties in the Mg–xZn–0.36Zr–0.32Ca (x = 3, 4, 5 wt%) sheet alloys. The as-rolled samples showed strong basal textures regardless of the Zn content. However, the decrease in the Zn content resulted in the significant texture weakening in the solution treated samples, and this leads to the improvement of the stretch formability up to 7.3 mm in Index Erichsen value in the Mg–3.1Zn–0.36Zr–0.32Ca alloy. Subsequent artificial aging at 160 °C for 16 h slightly increased the tensile yield strength of Mg–4Zn–0.36Zr–0.32Ca alloy sheet from 176 to 194 MPa. This work has demonstrated that the Mg–Zn system is promising to achieve excellent RT formability and high strength, however, the slow age hardening kinetics needs to be improved to make it industrially viable.

**Keywords**

Mg–Zn–Zr–Ca alloy • Formability • Strength
Zn content

B.-C. Suh · M.-Z. Bian · T. T. Sasaki (✉) · K. Hono
National Institute for Materials Science, 1-2-1
Sengen, Tsukuba, 305-0047, Japan
e-mail: sasaki.taisuke@nims.go.jp

T. Nakata · S. Kamado
Nagaoka University of Technology, 1603-1, Kamitomioka,
Nagaoka 940-2188, Japan

K. Hono
Graduate School of Pure and Applied Science, University of Tsukuba, 1-1 Tenno-Dai, Tsukuba, 305-0001, Japan


D. Orlov et al. (eds.), *Magnesium Technology 2018*, The Minerals, Metals & Materials Series, https://doi.org/10.1007/978-3-319-72332-7_58

## Introduction

To meet increasing demands for the development of lightweight magnesium (Mg) sheet alloys applicable as structural components of transportation vehicles such as trains and automobiles [1, 2], Mg sheet alloys with excellent RT formability have been developed recently by the optimization of the thermomechanical processing conditions and trace addition of alloying elements [3–8]. However, many of the Mg sheet alloys suffer from achieving excellent RT formability and satisfactory strength simultaneously [3–10]. This is because the excellent RT formability and yield strength are inversely correlated [2]. Development of heat treatable sheet alloys may be a promising approach to resolve this issue since excellent RT formability is anticipated after solution treatment and the strength can be increased by subsequent artificial aging.

Mg–Zn–Zr alloy microalloyed with Ca is one of the feasible candidates as a heat treatable Mg alloy as demonstrated by a Mg–6.2Zn–0.5Zr–0.2Ca alloy (wt%) [11]. This particular alloy exhibits high yield strength of 286 MPa after a the artificial aging; however, its Index Erichsen value, an indicator of the stretch formability, is only about 5.2 mm due to a strong basal texture. This means that there is still a room for the improvement in the stretch formability by weakening the strong basal texture. According to previous studies, reducing this Zn content is a plausible way to weaken the sheet texture of the Mg–Zn–Zr–Ca alloys [3, 7, 12–16]. This motivated us to develop heat-treatable and RT formable Mg–Zn–Zr–Ca sheet alloys with lower Zn contents. Another issue of Mg sheet alloys is the high processing cost. In order to achieve the good stretch formability and suppress the crack formation during the hot-rolling process of existing Mg sheet alloys, high temperature reheating prior to subsequent rolling is necessary. Unfortunately, this results in a low production efficiency and ultimately high price of sheet products [6].

To overcome this engineering issue, we continuously rolled the Mg–xZn–0.36Zr–0.32Ca (x = 3, 4, 5 wt%) sheet alloys at 200 °C without high temperature reheating and investigated their microstructures and mechanical properties.

## Experimental Procedure

Alloy ingots with 4.2 kg in weight were prepared by an induction melting using steel crucible under an Ar atmosphere and casting into an iron molds. A high-purity Mg, Mg–5.5 wt%Zn–0.5 wt%Zr (ZK60) and Mg-30 wt%Ca master alloys. The pure Mg and ZK60 ingots were heated to 800 °C. Then, the Mg–Ca master alloy was alloyed sample was added into the melt and kept for 0.5 h followed by casting into iron molds. Table 1 summarizes the chemical compositions of the prepared ingots measured by inductively coupled plasma atomic (ICP) emission spectroscopy.

The cast ingots were cut into plates with 10 mm in thickness and the plates were subjected to two-step homogenization to avoid the partial melting in the early stage of the homogenization; the sample was homogenized at 350 °C for 24 h and 450 °C for 24 h in an Ar atmosphere followed by air cooling. During the homogenization, the temperature increased from 350 to 450 °C at a heating rate of 5 °C/h. The homogenized samples were then hot-rolled at 200 °C to obtain final sheet samples with 1 mm in thickness. The samples were solution treated in an electric furnace at 450 °C for 2 h. Some solution treated ZKX400 tensile samples were aged in an oil bath to achieve the peak hardness condition (T6).

Microstructure characterization was performed by optical microscope and X-ray diffraction (XRD). Mechanical properties were evaluated by tensile tests and Erichsen cupping tests at RT. Grain sizes of the samples were measured by liner interception method. Tensile tests were conducted using specimens with a gauge length, width, and thickness of 12.5, 5 and 1 mm at a strain rate of $1 \times 10^{-3}$ $s^{-1}$. The tensile loading axis was parallel to the rolling direction (RD). Erichsen cupping test were done using the square shape specimen with over 70 × 70 $mm^2$. The punch diameter and speed were 20 mm and 0.1 mm/s, respectively.

## Results and Discussion

Figure 1 shows engineering stress-strain curves of the solution treated ZKX300, ZKX400 and ZKX500 alloys, and Table 2 summarizes the mechanical properties obtained from the tensile tests and Index Erichsen values for each composition. ZKX300 alloy shows yield strength, $\sigma_{ys}$, ultimate tensile strength, $\sigma_{UTS}$, and elongation to fracture, ε, of 148 ± 3.0 MPa, 262 ± 2.2 MPa and 26 ± 4.3%, respectively. The increase in the Zn content to 4 wt% resulted in the significant increase in the $\sigma_{ys}$ and $\sigma_{UTS}$ to 173 ± 4.5 and 243 ± 1.0 MPa at the expense of elongation. The further addition of Zn to 5 wt% slightly decreased the $\sigma_{ys}$ and $\sigma_{UTS}$ to 133 ± 2.5 and 243 ± 0.95 MPa. The Index Erichsen value decreases from 7.3 ± 0.79 mm to 5.2 ± 1.1 mm with increasing the Zn content from 3 to 5 wt%.

Figure 2 shows the optical micrographs and (0002) pole figures of the as-rolled and solution treated ZKX300, ZKX400 and ZKX500 alloys. The (0002) pole figures obtained from the as-rolled samples show that these samples have a typical strong basal textures with similar maximum intensity in the range of 10.8–13.4 m.r.d. The optical micrographs obtained from these samples show that these samples mainly consist of the deformed grains. The solution treatment leads to the formation of the fully recrystallized microstructure, and the microstructure consists of fully static recrystallized grains, and the average grain sizes are 10.1, 7.7 and 13.4 μm for the ZKX300, ZKX400 and ZKX500 alloys, respectively. These samples have a bimodal grain structure, and this feature is more significant in the ZKX500 alloy consisting of coarse grains with >50 μm and fine grains with <5 μm in diameter. On the other hand, significant texture weakening can be found in all alloys by the solution treatment. Additionally, the texture intensity of (0002) pole figures is significantly decreased to 3.0, 4.8 and 7.3 m.r.d. for the ZKX300, ZKX400 and ZKX500 alloys, which means that the maximum intensity of the basal poles decreases with decreasing the Zn content. Another interesting finding from the present study is that the basal poles tilt towards the transverse direction (TD) of the rolled sheet is more significant with decreasing the Zn content.

Since the ZKX400 alloy showed relatively high strength and good stretch formability in the solution treated condition, we measured the variation in the Vickers hardness

**Table 1** Chemical compositions of the ingots used in this work

| | | | | (wt.%) |
|---|---|---|---|---|
| Sample | Zn | Zr | Ca | Mg |
| ZKX300 | 3.1 | 0.36 | 0.32 | Bal. |
| ZKX400 | 3.9 | 0.36 | 0.32 | Bal. |
| ZKX500 | 5.2 | 0.36 | 0.32 | Bal. |

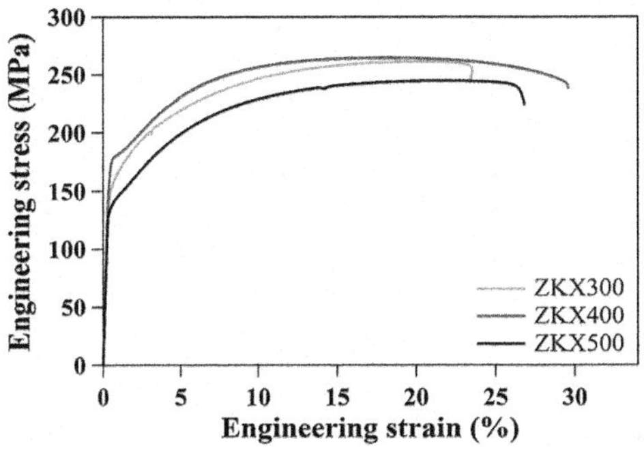


**Fig. 1** Stress-strain curves of the solution treated ZKX300, ZKX400 and ZKX500 alloys stretched along the RD

during the artificial aging at 160 °C, and evaluated the mechanical properties. As shown in the variation in Vickers hardness as a function of aging time, Fig. 3a, the sample shows slight age hardening. The Vickers hardness is 59.0 ± 1.3 VHN in the solution treated condition and increase slightly up to 61.7 ± 1.1 VHN by aging at 160 °C for 16 h. Figure 3b shows the engineering stress-strain curves of the solution treated and peak aged samples, and Table 3 summarizes the tensile properties. The T6 treatment resulted in the slight increase in the strength. Specifically, the $\sigma_{ys}$ and $\sigma_{UTS}$ increased from 173 ± 4.5 to 192 ± 1.7 MPa and 243 ± 1.0 to 267 ± 2.0 MPa at the expense of the elongation.

In this work, we have investigated the effect of Zn content on the microstructure and mechanical properties in Mg–xZn–0.36Zr–0.32Ca (x = 3, 4, 5 wt%) alloys, and demonstrated that the decrease in the Zn content is an effective approach to improve the stretch formability. As shown in Fig. 2, the solution treatment resulted in the texture weakening, and the texture weakening effect is more significant with decreasing the Zn content. The maximum texture intensity decreases from 10.8 to 13.4 m.r.d. with decreasing the Zn content from 5 to 3 wt%, and the decrease in the Zn content also resulted in the alignment of the (0002) poles towards the TD. The improvement of the stretch formability due to the decrease in the Zn content can be explained by this weakened texture feature. The artificial aging could increase the strength of the ZKX400 alloy, Fig. 3 and Table 3; $\sigma_{ys}$, and ultimate tensile strength, $\sigma_{UTS}$, increased from 176 to 194 MPa and 265 to 271 MPa due to the dispersion of the $\beta_1$ ′ phase [17].

Figure 4 summarizes the relationship between yield strengths and Index Erichsen values in various Mg sheet alloys [2]. The yield strengths tend to decrease with increasing the Index Erichsen values. The ZKX400 alloy show slightly better stretch formability compared to the sheet alloys with similar strengths in the solution treated condition. The artificial aging can give rise the strength and leads to the good balance of the strength and stretch formability compared to the existing sheet alloys. In order to achieve good room

**Table 2** Summary of mechanical properties and index Erichsen values of solution treated Mg–0.36Zr–0.32Ca alloys with different Zn content

| Sample | Yield strength, ($\sigma_{ys}$, MPa) | Ultimate tensile strength, ($\sigma_{UTS}$, MPa) | Elongation, (%) | Index Erichsen value, (mm) |
|---|---|---|---|---|
| ZKX300 | 148 ± 3.0 | 262 ± 2.2 | 26 ± 4.3 | 7.3 ± 0.79 |
| ZKX400 | 173 ± 4.5 | 243 ± 1.0 | 24 ± 1.5 | 7.0 ± 0.44 |
| ZKX500 | 133 ± 2.5 | 243 ± 0.95 | 28 ± 3.7 | 5.2 ± 1.1 |

**Fig. 2** Optical micrographs and (0002) pole figures of as-rolled and solution treated ZKX300, ZKX400 and ZKX500 alloy sheets

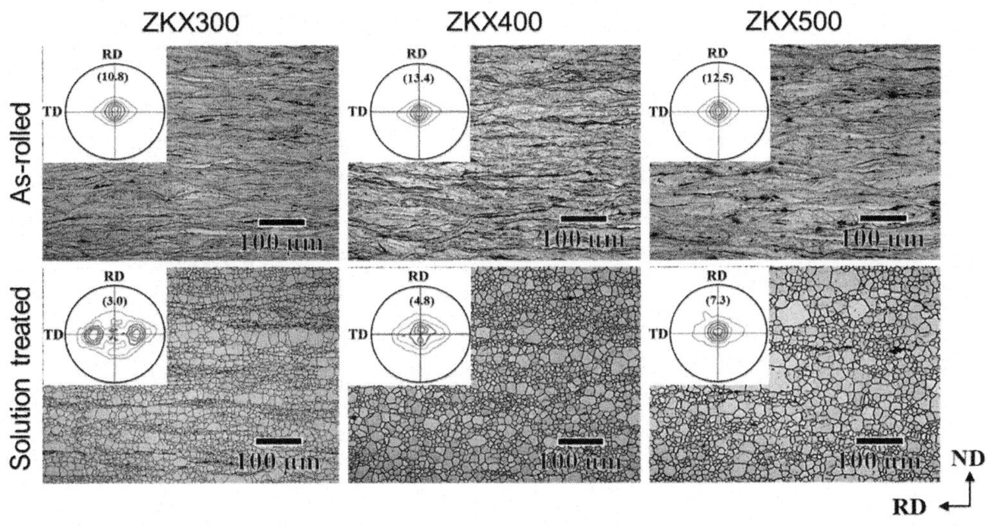

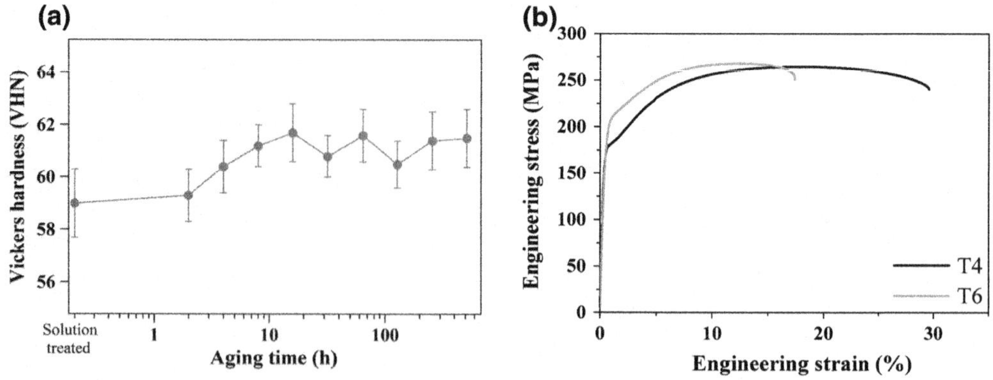


**Fig. 3** **a** Age hardening curve of ZKX400 alloy at 160 °C, and **b** engineering stress-strain curves of the solution treated and peak aged ZKX400 alloy stretched along the RD

**Table 3** Summary of mechanical properties of the solution treated and peak aged ZKX400 alloy

| Sample | Yield strength, ($\sigma_{ys}$, MPa) | Ultimate tensile strength, ($\sigma_{UTS}$, MPa) | Elongation, (%) |
|---|---|---|---|
| Solution treated | 173 ± 4.5 | 243 ± 1.0 | 24.0 ± 1.5 |
| Peak aged | 192 ± 1.7 | 267 ± 2.0 | 18.2 ± 2.0 |

temperature stretch formability with the Index Erichsen values over 8 mm in the Mg–Al based alloys, high temperature rolling or specimen reheating between the rolling was needed. This causes the increase in the processing cost and may hinder the application. Unlike these samples, the good RT formability could be obtained in the ZKX400 alloy that was continuously rolled sample at relatively low temperature of 200 °C, which makes more attractive as structural applications. To make the heat treatable Mg–Zn based alloy truly industrially viable heat treatable alloy, future work will shed light on to decrease the aging time.

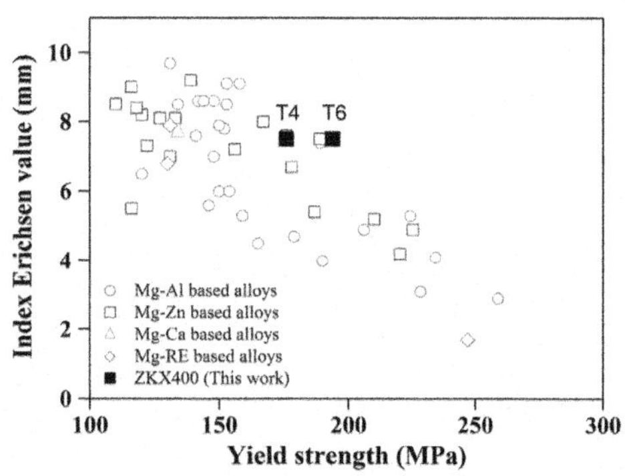


**Fig. 4** Relationship between yield strengths and Index Erichsen values in various Mg sheet alloys [2]

## Summary

We have developed a ZKX400 alloy, which shows relatively good stretch formability with Index Erichsen value of 7.0 ± 0.44 mm in the solution treated condition and higher yield strength of 192 ± 1.7 MPa in the peak aged condition compared to the existing samples. The excellent room temperature formability is attributed to the weakened textured feature, and the dispersion of the precipitates is believed to give rise to the strength. As demonstrated in this work, the precipitation hardenable alloy is promising to achieve excellent room temperature formability and high strength, however, the slow kinetics needs to be improved to make it industrially viable.

**Acknowledgements** This work was supported by Advanced Low Carbon Technology Research and Development Program (ALCA), 12102886, JSPS KAKENHI Grant Number JP26709055 and the Structural Materials for Innovations of the Cross Ministrial Strategic Innovation Program (SIP) of Japan Science and Technology (JST).

## References

1. W.J. Joost, P.E. Krajewski, Towards magnesium alloys for high-volume automotive applications, Scr. Mater. 128 (2017) 107–112.

2. B.-C. Suh, M.-S. Shim, K.S. Shin, N.J. Kim, Current issues in magnesium sheet alloys: Where do we go from here?, Scr. Mater. 84 (2014) 1–6.
3. Y. Chino, K. Sassa, M. Mabuchi, Texture and stretch formability of a rolled Mg–Zn alloy containing dilute content of Y, Mater. Sci. Eng. A. 513–514 (2009) 394–400.
4. Y. Chino, M. Mabuchi, Enhanced stretch formability of Mg–Al–Zn alloy sheets rolled at high temperature (723 K), Scr. Mater. 60 (2009) 447–450.
5. X. Huang, K. Suzuki, Y. Chino, Influences of initial texture on microstructure and stretch formability of Mg–3Al–1Zn alloy sheet obtained by a combination of high temperature and subsequent warm rolling, Scr. Mater. 63 (2010) 395–398.
6. X. Huang, K. Suzuki, Y. Chino, M. Mabuchi, Improvement of stretch formability of Mg–3Al–1Zn alloy sheet by high temperature rolling at finishing pass, J. Alloys Compd. 509 (2011) 7579–7584.
7. T. Bhattacharjee, B.-C. Suh, T.T. Sasaki, T. Ohkubo, N.J. Kim, K. Hono, High strength and formable Mg–6.2Zn–0.5Zr–0.2Ca alloy sheet processed by twin roll casting, Mater. Sci. Eng. A. 609 (2014) 154–160.
8. M.Z. Bian, Z.R. Zeng, S.W. Xu, S.M. Zhu, Y.M. Zhu, C.H. J. Davies, N. Birbilis, J.F. Nie, Improving Formability of Mg–Ca–Zr Sheet Alloy by Microalloying of Zn, Adv. Eng. Mater. 18 (2016) 1763–1769.
9. Y. Chino, H. Iwasaki, M. Mabuchi, Stretch formability of AZ31 Mg alloy sheets at different testing temperatures, Mater. Sci. Eng. A. 466 (2007) 90–95.
10. D. Wu, R.S. Chen, E.H. Han, Excellent room-temperature ductility and formability of rolled Mg–Gd–Zn alloy sheets, J. Alloys Compd. 509 (2011) 2856–2863.
11. T. Bhattacharjee, B.C. Suh, T.T. Sasaki, T. Ohkubo, N.J. Kim, K. Hono, High strength and formable Mg-6.2Zn-0.5Zr-0.2Ca alloy sheet processed by twin roll casting, Mater. Sci. Eng. A. 609 (2014) 154–160.
12. Y. Chino, X. Huang, K. Suzuki, M. Mabuchi, Enhancement of Stretch Formability at Room Temperature by Addition of Ca in Mg–Zn Alloy, Mater. Trans. 51 (2010) 818–821.
13. Y. Chino, X. Huang, K. Suzuki, K. Sassa, M. Mabuchi, Influence of Zn concentration on stretch formability at room temperature of Mg–Zn–Ce alloy, Mater. Sci. Eng. A. 528 (2010) 566–572.
14. Y. Chino, T. Ueda, Y. Otomatsu, K. Sassa, X. Huang, K. Suzuki, M. Mabuchi, Effects of Ca on Tensile Properties and Stretch Formability at Room Temperature in Mg–Zn and Mg–Al Alloys, Mater. Trans. 52 (2011) 1477–1482.
15. D.H. Kang, D.W. Kim, S. Kim, G.T. Bae, K.H. Kim, N.J. Kim, Relationship between stretch formability and work-hardening capacity of twin-roll cast Mg alloys at room temperature, Scr. Mater. 61 (2009) 768–771.
16. S.J. Park, H.C. Jung, K.S. Shin, Deformation behaviors of twin roll cast Mg–Zn–X–Ca alloys for enhanced room-temperature formability, Mater. Sci. Eng. A. 679 (2017) 329–339.
17. C.J. Bettles, M.A. Gibson, K. Venkatesan, Enhanced age-hardening behaviour in Mg-4 wt.% Zn micro-alloyed with Ca, Scr. Mater. 51 (2004) 193–197.

# Co-precipitation on the Basal and Prismatic Planes in Mg–Gd–Ag–Zr Alloy Subjected to Over-Ageing

Jiehua Li, Fredrik S. Hage, Ali Gholinia, Pan Xie, Yu Zhang, Yujuan Wu, Liming Peng, Sarah J. Haigh, Quentin M. Ramasse, and Peter Schumacher

**Abstract**

Precipitation hardening is one of the dominant strengthening mechanisms for Mg alloys, especially for Mg alloys containing rare earth elements. However, precipitation hardening of conventional Mg alloys (such as AZ91D, ZK60 or even WE54/43) is relatively weaker than in high performance Al alloys [such as Al–Cu–Mg based alloys (2024) and Al–Zn–Mg–Cu based alloys (7075, 7050)]. Further avenues for improving precipitation hardening of Mg alloy are therefore being actively explored. In this paper, co-precipitation on the basal and prismatic planes of the α-Mg matrix in a Mg–2.4Gd–0.4Ag–0.1Zr alloy after over-ageing at 200 °C for 2048 h was investigated using high-angle annular dark field (HAADF) scanning transmission electron microscopy (STEM) imaging and electron energy loss spectroscopy (EELS). Ag was observed within precipitates on the basal plane, while Gd was observed within precipitates on both the basal and prismatic planes. Furthermore, Ag, Gd-rich clusters were also observed in the vicinity of these precipitates, which were proposed to form during solution treatment and to coarsen during subsequent ageing treatment.

**Keywords**

Mg alloy • Precipitation • Precipitation hardening HAADF-STEM • EELS

## Introduction

Precipitation hardening is one of the dominant strengthening mechanisms for Mg alloys, especially for Mg alloys containing rare earth elements (RE) [1]. It is generally accepted that precipitation hardening is mainly dependent on the number density, size, distribution, shape (morphology) and volume fraction of precipitates. In order to improve the performance of Mg alloys, it is desirable to form precipitates with a high number density, a small size, a uniform network (i.e. perpendicular to each other) distribution, a plate-shaped morphology and a high volume fraction. However, precipitation hardening of conventional Mg alloys (such as AZ91D, ZK60 or even WE54/43) is relatively weaker than in high performance Al alloys [such as Al–Cu–Mg based alloys (2024) and Al–Zn–Mg–Cu based alloys (7075, 7050)]. Further avenues for improving precipitation hardening of Mg alloy are therefore being actively explored.

In order to further improve precipitation hardening of Mg alloys, it is desirable to not only increase the number density and thereby the volume fraction of precipitates but also to decrease the size and the tendency for coarsening of the precipitates. To this effect, special Mg binary alloy systems (e.g. the Mg–Gd binary alloy system) are used. Gd has a significant difference of solute solubility in Mg at different temperatures and thereby the Mg–Gd binary alloy system has a strong potential to achieve precipitation hardening. Indeed, the Mg–Gd binary alloy system has been reported to have a remarkable precipitation hardening due to the formation of β-type precipitates (i.e. β″ $Mg_3Gd$, β′ $Mg_7Gd$, $\beta_1$ $Mg_3Gd$ and β $Mg_5Gd$) on the prismatic plane [1]. Furthermore, some alloying elements (i.e. Zn [2, 3], Ag [4, 5], Zn and Ag together [6]) are also often added into the Mg–Gd

J. Li (✉) · P. Xie · P. Schumacher
Institute of Casting Research, Montanuniversität Leoben, 8700 Leoben, Austria
e-mail: jiehua.li@unileoben.ac.at

J. Li · Y. Zhang · Y. Wu · L. Peng
National Engineering Research Center of Light Alloy Net Forming, State Key Laboratory of Metal Matrix Composites, Shanghai Jiaotong University, 200240 Shanghai, People's Republic of China

F. S. Hage · Q. M. Ramasse
SuperSTEM Laboratory, SciTech Daresbury Campus, Keckwick Lane, Daresbury, WA4 4AD, UK

A. Gholinia · S. J. Haigh
School of Materials, The University of Manchester, Manchester, M13 9PL, UK

P. Schumacher
Austrian Foundry Research Institute, Leoben, Austria


D. Orlov et al. (eds.), *Magnesium Technology 2018*, The Minerals, Metals & Materials Series, https://doi.org/10.1007/978-3-319-72332-7_59

alloy. It has thus been reported that the addition of Zn or Ag into Mg–Gd alloy causes the formation of precipitates on the basal plane and thereby enhances the precipitation hardening [2–5]. This is also the case for combined addition of Zn and Ag into the Mg–Gd alloy [6]. It should be noted here that the formation of precipitates on the basal plane is believed to be beneficial to for improve improving the alloy performance. It is generally accepted, however, that precipitation in Mg alloys such as the aforementioned Mg–Gd binary alloy very often occurs on the prismatic plane, which can significantly hamper the basal slip [i.e. <11–20> (0002)]. Such precipitates on the prismatic plane have no significant strengthening effect on other slip systems (i.e. <11–2> {11–22}), which can be activated at elevated temperatures. It is, therefore, of great necessity to form precipitates on the basal plane which can significantly hamper the slip at elevated temperatures. In previous research work [5], the role of Ag on the formation of precipitates on the basal plane in Mg–Gd–Ag–Zr alloy during the early stage of ageing (up to 2 h) has been investigated using atom probe tomography (APT). Ag-rich clusters were detected immediately after solution treatment and water quenching. The presence of Ag-rich clusters has also been proposed to be involved in the formation of a high density of basal precipitates during the early stage of ageing at 200 °C. Furthermore, APT also confirmed that these basal precipitates were enriched with Ag and Gd. However, it should be pointed out that this confirmation is not straightforward due to the fact that it is challenging to determine unambiguously the orientation relationship between precipitates and the α-Mg matrix using APT, despite the technique's very high sensitivity for chemistry analysis. More importantly, to date, there lacks a detailed investigation on the distribution of Ag and Gd within precipitates in Mg–Gd–Ag–Zr alloy after over-ageing, which would be of great importance to elucidate the role of Ag and Gd during the precipitation and thereby to improve the thermal stability of precipitates. Transmission electron microscopy (TEM) is particularly well-suited to the determination of orientation relationships between precipitates and a host matrix. Here, we therefore apply TEM and associated spectroscopy techniques to elucidate the trace Ag distribution after over-ageing.

In this paper, the co-precipitation on the basal and prismatic planes in Mg–2.4Gd–0.4Ag–0.1Zr alloy after over-ageing at 200 °C for 2048 h was investigated using high-angle annular dark field (HAADF) scanning transmission electron microscopy (STEM) imaging and electron energy loss spectroscopy (EELS), with a special focus on the distribution of Ag and Gd within the precipitates on the basal and prismatic planes.

## Experimental

The Mg–2.4Gd–0.4Ag–0.1Zr alloy (at.%) was prepared from pure Mg (99.9%), Ag (99.9%), Mg–30Gd (wt%) and Mg–30Zr (wt%) in an electric resistance furnace under the protection of a mixed gas ($CO_2$ and $SF_6$) and then cast into a die mould. The casting temperature was about 760 °C. Solution treatment was performed at 500 °C for 6 h, followed by quenching into cold water. Ageing treatment was performed at 200 °C for up to 2048 h. In this paper, the over-aged sample at 200 °C for 2048 h was chosen for detailed TEM investigation as an extreme case of over-aged alloy.

Thin foil samples for TEM were prepared using a Focused Ion Beam (FIB) instrument. Prior to FIB thinning, electron backscattering diffraction (EBSD) was performed on the bulk sample to select a region with a suitable orientation. HAADF STEM imaging and EELS were performed using a Nion UltraSTEM100 aberration corrected dedicated STEM. The microscope was operated at an acceleration voltage of 100 kV and an electron probe convergence semi-angle of 31 mrad, which resulted in an estimated minimum electron probe size of 0.1 nm. The cold field emission gun of the microscope has a native energy spread of 0.35 eV. The HAADF detector collection semi-angle was 83–185 mrad and the spectrometer collection semi-angle was 36 mrad. EELS elemental maps were then created by integrating the EELS signal of each edge above their nominal edge onset energy: Gd $M_{4,5}$ (1185 eV), Ag $M_{4,5}$ (367 eV) and Mg K (1305 eV) over an energy window of suitable width after subtraction of the preceding decaying background using a power law model. All EELS edges were identified following reference [7]. EEL spectra were de-noised using Principal Component Analysis (PCA) as implemented in the MSA plugin [8] for Gatan's Digital Micrograph software suite. The intensities of the EELS maps were displayed on a false colour scale, so that within each map, a low intensity (black) corresponds to a lower relative concentration, while increased contrast (bright) corresponds to an increase in (relative) elemental concentration.

## Results and Discussion

Figure 1 shows HAADF STEM images (Fig. 1a) and EELS maps of Mg (Fig. 1b) and Gd (Fig. 1c) of a region containing a grain boundary (as marked with GB) and a $Mg_5Gd$ phase (in the lower-right section of Fig. 1a) in Mg–2.4Gd–0.4Ag–0.1Zr alloy aged at 200 °C for 2048 h. Contrast in HAADF STEM images scales to a good approximation with the atomic number Z of the observed material as $\sim Z^{1.7}$ so that atomic columns enriching with Gd and Ag atoms within the α-Mg matrix are clearly distinguishable with a much brighter intensity (Fig. 1a), this is in particular the case of the numerous bright features observed in the alloy, which are likely to be Gd-rich or Gd, Ag-rich precipitates. As an aside, we note that the presence of the $Mg_5Gd$ phase (Fig. 1a) is due to the formation of intermetallic $Mg_5Gd$ phase during solidification, which cannot be completely dissolved into the Mg matrix during solution treatment (500 °C for 6 h). The presence of a precipitation free zone along the grain boundary (Fig. 1a) can be attributed to the solute diffusion during solution treatment (500 °C for 6 h) and ageing treatment (200 °C for 2048 h). These two structural features, while interesting in their own right, are not a focus of the present investigation. Instead, the formation of the precipitates within the α-Mg matrix will be investigated in detail here. These precipitates appear to be perpendicular to each other (Fig. 1a) and with a high number density, as can be clearly seen in Fig. 2.

Figure 2 shows HAADF STEM images (Fig. 2a, b, c) and EELS map for Gd (Fig. 2d) of the precipitates on the prismatic plane in Mg–2.4Gd–0.4Ag–0.1Zr alloy aged at 200 °C for 2048 h, clearly demonstrating that the precipitates on the prismatic plane are enriched with Gd (Fig. 2d). No significant presence of Ag was observed within the precipitates on the prismatic plane, indicating that Ag may not be involved into the formation of the precipitates on the prismatic plane. In contrast, a significant presence of Ag was observed within the precipitates on the basal plane (Fig. 3), indicating that Ag was involved into the formation of the precipitates on the basal plane.

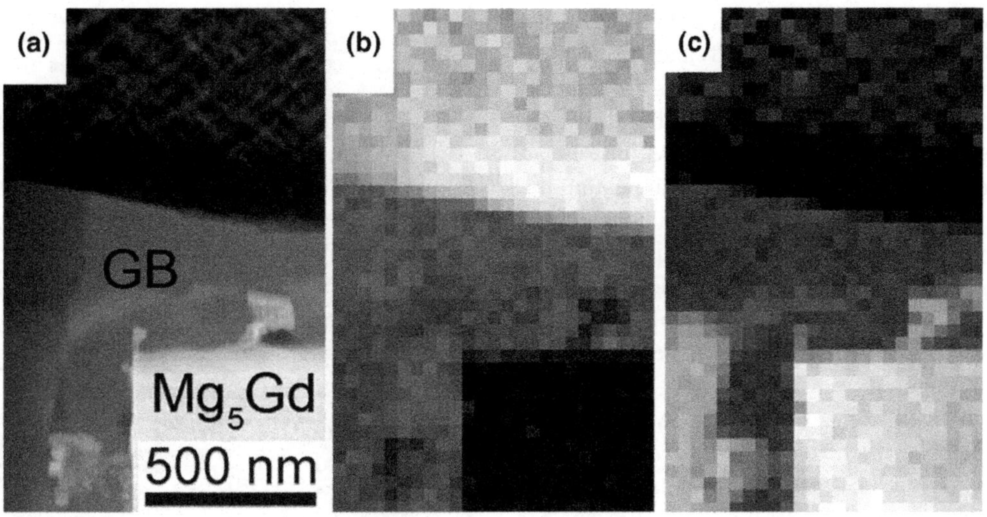


**Fig. 1** HAADF STEM image (**a**) and EELS maps of Mg (**b**) and Gd (**c**) showing the presence of a $Mg_5Gd$ phase along the grain boundary in Mg–2.4Gd–0.4Ag–0.1Zr alloy aged at 200 °C for 2048 h

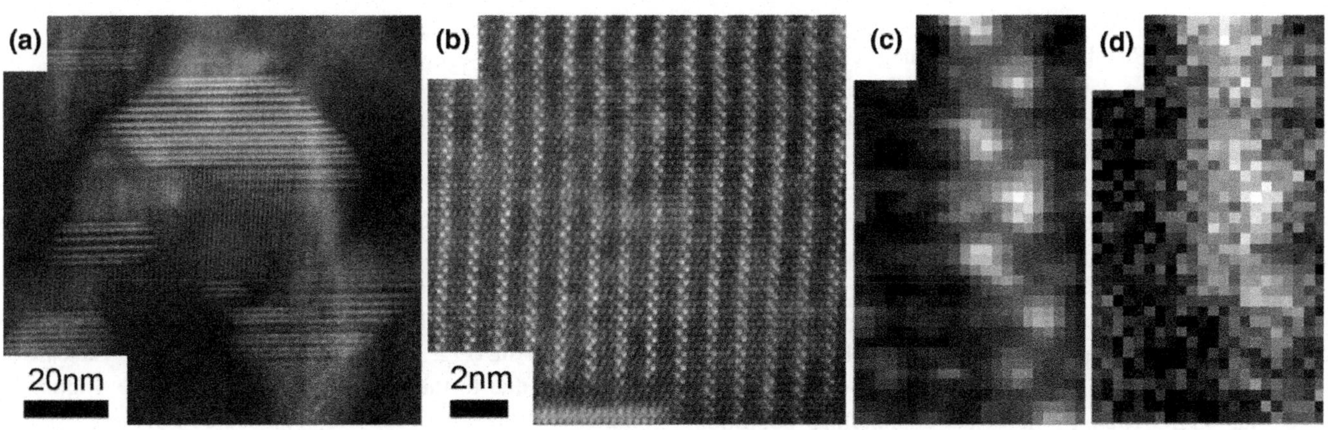


**Fig. 2** HAADF STEM images (**a**, **b**, **c**) and EELS map of Gd (**d**) of the precipitates on the prismatic plane in Mg–2.4Gd–0.4Ag–0.1Zr alloy aged at 200 °C for 2048 h. The precipitates on the prismatic plane are enriched with Gd (**d**) (B//<21–10> $_{Mg}$)

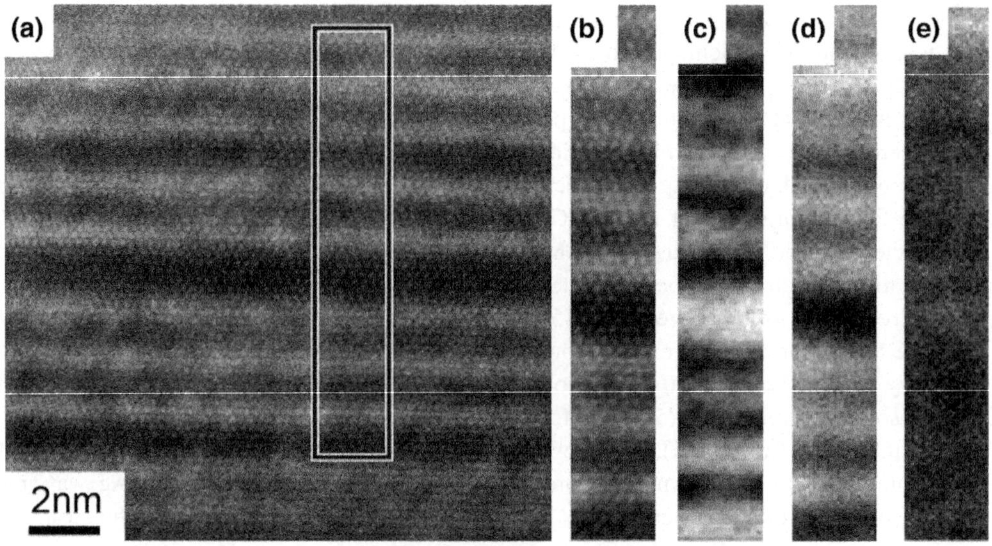


**Fig. 3** HAADF STEM images (**a**, **b**) and EELS maps of Mg (**c**), Gd (**d**) and Ag (**e**) of the precipitates on the basal plane in Mg–2.4Gd–0.4Ag–0.1Zr alloy aged at 200 °C for 2048 h. The precipitates on the basal plane are enriched with Gd (**d**) and Ag (**e**). (B//<21–10> $_{Mg}$)

Figure 3 shows HAADF STEM images (Fig. 3a, b) and EELS maps for Mg (Fig. 3c), Gd (Fig. 3d) and Ag (Fig. 3e) of the area indicated on Fig. 3a, which crosses a number of precipitates observed on the basal plane of the alloy. The EELS maps reveal that the basal plane precipitates are enriched with Gd (Fig. 3d) and Ag (Fig. 3e), indicating that both Gd and Ag may be involved in the formation of the precipitates on the basal plane, which is also fully consistent with the APT result reported in Ref. [5]. We note, however, that the EELS maps did not reveal any observable difference in Gd and Ag between the three atomic planes of the precipitates. It was therefore not possible to conclusively associate the contrast differences in the HAADF image in Fig. 3 with a difference in chemistry. Furthermore, the Ag distribution in Fig. 3e does not necessarily coincide with the location of each precipitate. The reason for this is unclear. Thus further work is clearly needed to unambiguously determine the degree to which Ag contributes to basal plane precipitate formation.

Figure 4 shows a high resolution HAADF image of a number of basal plane precipitates. Similar to the observations reported in Ref. [5], the basal plane precipitates shown in Fig. 4 appear to consist of three atomic layers, with the two outermost layers exhibiting a contrast differing significantly from that of the middle layer, suggesting they may have different structures. Specifically, the two outmost atomic columns of the precipitates appear elongated along the direction perpendicular to the precipitate axis, while there is a significant "smearing" of the middle layer contrast (making it appear more like a continuous line). The crystal structure of these basal precipitates is visually similar with that of the $\gamma''$ phase in Mg–Gd–Zn(−Zr) alloys [2]. Here we note that slight changes in beam channelling conditions might also affect the observed HAADF intensities in Fig. 4, at least in part. Thus the present structural analysis should be regarded as tentative.

**Fig. 4** HAADF STEM image of basal plane precipitates

Furthermore, Ag-rich clusters were observed in the vicinity of precipitates, as illustrated in Fig. 5, which shows an HAADF STEM image (Fig. 5a) and EELS maps of Mg (Fig. 5b), Gd (Fig. 5c) and Ag (Fig. 5d) over a large field of view of the sample. For clarity, a representative spectrum is also shown in (Fig. 5e), where the presence of Ga can be ignored as due to the FIB preparation procedure. Figure 5d highlights in the Ag map the presence of a large cluster of Ag-rich material (lower part of the map). Such large Ag-rich

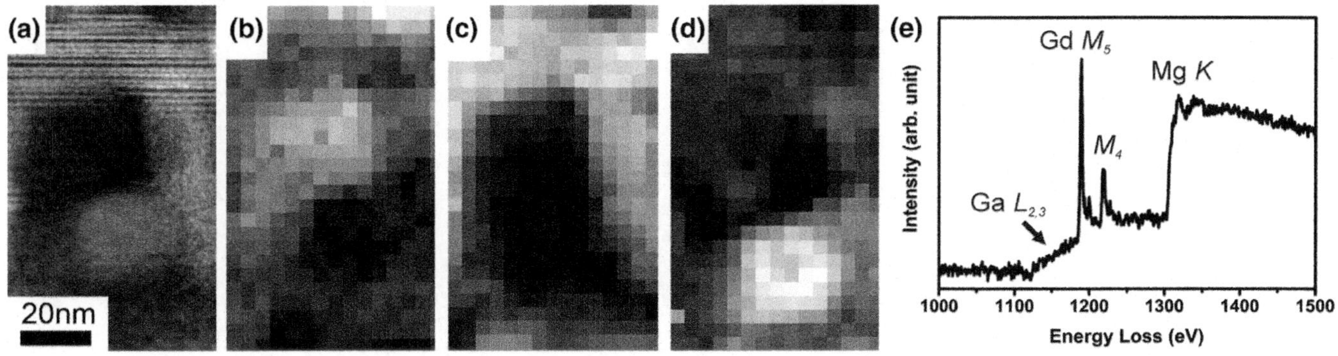


**Fig. 5** HAADF STEM image (**a**) and EELS maps of Mg (**b**), Gd (**c**) and Ag (**d**) of the precipitates on the basal plane in Mg–2.4Gd–0.4Ag–0.1Zr alloy aged at 200 °C for 2048 h. For clarity, the spectrum is also shown in (**e**). (B//<21–10> $_{Mg}$)

clusters may form during solution treatment and coarsen during subsequent ageing treatment.

## Conclusion

We have carried out an atomic-scale experimental investigation on co-precipitation on the basal and prismatic planes in a Mg–Gd–Ag–Zr alloy aged at 200 °C for 2048 h. Two types of precipitates on the basal and prismatic planes were observed. Ag is proposed to be involved in the formation of the precipitates on the basal plane, while Gd is responsible for the formation of the precipitates not only on the prismatic plane but also on the basal plane.

**Acknowledgements** The author (J.H. Li) acknowledges the casting sample preparation from Shanghai Jiaotong University and the financial support from the Major International (Regional) Joint Research Project (No. 51420105005) from China and the projects (EPU 04/2017, CN 09/2016, mmc-kf15-11). The authors (Y.J. Wu and L.M. Peng) acknowledges the financial support from the National Key Research and Development Program of China (No. 2016YFB0301000) and National Natural Science Foundation of China (No. 51671128). The SuperSTEM Laboratory is the U.K. National Facility for Advanced Transmission Electron Microscopy, supported by the Engineering and Physical Sciences Research Council (EPSRC). SJH and AG thank EPSRC for funding under grant EP/M010619/1.

## References

1. Nie JF (2012) Precipitation and hardening in Magnesium alloys. Metall. Mater. Trans. A. 43: 3891–3939.
2. Nie JF, Oh-ishi K, Gao X, Hono K (2008) Solute segregation and precipitation in a creep-resistant Mg–Gd–Zn alloy. Acta Mater. 56: 6061–6076.
3. Li JH, Hage FS, Zhong XL, Wu YJ, Peng LM, Haigh SJ, Ramasse QM, Schumacher P (2017) Elemental distribution within the long-period stacking ordered structure in a Mg–Gd–Zn–Mn alloy. Mater. Charact. 129: 247–251.
4. Yamada K, Hoshikawa H, Maki S, Ozaki T, Kuroki Y, Kamado S, Kojima Y (2009) Enhanced age-hardening and formation of plate precipitates in Mg–Gd–Ag alloys. Scripta Mater. 61: 636–639.
5. Zhang Y, Alam T, Gwalani B, Rong W, Banerjee R, Peng LM, Nie JF, Birbilis N (2016) On the role of Ag in enhanced age hardening kinetics of Mg–Gd–Ag–Zr alloys. Philos. Mag. Lett. 96: 212–219.
6. Gao X, Nie JF (2008) Enhanced precipitation-hardening in Mg–Gd containing Ag and Zn alloys. Scripta Mater. 59: 619–622.
7. Ahn CC, Krivanek OL (1983) EELS Atlas, Gatan Inc., Warrendale.
8. Watanabe M, Kanno M, Ackland D, Kiely C, Williams D (2007) Microsc. Microanal. 13 (S2): 1264.

# Evolution of the Dislocation Structure During Compression in a Mg–Zn–Y Alloy with Long Period Stacking Ordered Structure

Kristian Máthis, Moustafa El-Tahawy, Gerardo Garcés, and Jenő Gubicza

**Abstract**

Evolution of the dislocation structure in $Mg_{97}Y_7Zn_5$ (at. %) alloy having long period stacking ordered (LPSO) structure was studied during compression tests. Two materials, an as-cast and an extruded one were deformed up to the applied strain of ~25%. The evolution of the crystallite size, the dislocation density and the population of the particular slip systems were determined by the evaluation of the X-ray diffraction peak profiles. A very low dislocation density with the order of magnitude $10^{12}$–$10^{13}$ $m^{-2}$ was detected in the compressed specimens. This dislocation density did not increase considerably with increasing strain. At the same time, a significant decrease of the crystallite size occurred during compression. These observations can be explained by the arrangement of dislocations into low energy dipolar configurations, such as kink walls, which do not contribute to the dislocation density measurable by X-ray diffraction peak profile analysis, however they yield a fragmentation of the crystallites.

**Keywords**

LPSO structure • Diffraction line profile analysis
Dislocation density • Non-basal slip

K. Máthis (✉)
Department of Physics of Materials, Faculty of Mathematics and Physics, Charles University, Ke Karlovu 5, 121 16 Prague, Czech Republic
e-mail: mathis@met.mff.cuni.cz

M. El-Tahawy · J. Gubicza
Department of Materials Physics, Faculty of Science, Eötvös Loránd University, Pázmány P. Sétány 1/A, 1117 Budapest, Hungary

M. El-Tahawy
Department of Physics, Faculty of Science, Tanta University, 31527 Tanta, Egypt

G. Garcés
Department of Physical Metallurgy, CENIM-CSIC, Avenida Gregorio Del Amo 8, 28040 Madrid, Spain


D. Orlov et al. (eds.), *Magnesium Technology 2018*, The Minerals, Metals & Materials Series, https://doi.org/10.1007/978-3-319-72332-7_60

## Introduction

Magnesium alloys with long-period stacking ordered (LPSO) structure belong to the new class of lightweight alloys, offering excellent strength-to-weight ratio. The high mechanical performance is ensured by the LPSO phase, which acts as short-fiber reinforcement, similarly to the composite materials [1]. The final strength is influenced particularly by the manufacturing parameters (e.g., extrusion ratio, rate and temperature) and the volume content and crystallographic structure of the LPSO phase. Many polytypes of LPSO phase was reported, including 10H, 14H, 18R and 24R [2], where the H and R refer to the hexagonal or rhombohedral symmetry, respectively, and the numbers indicate the periodicity of the structure. In general, the LPSO structures consist of combination of hexagonal closed packed Mg and Y or Zn enriched layers. The latter have a local face-centered cubic stacking sequence on the close packed planes [3]. There is a general agreement that one of the main deformation mechanisms in the LSPO phase is the so-called kinking, when wedge-like bands forms in the structure as a consequence of correlated movement of basal <*a*>-type dislocation dipoles. However, molecular dynamics simulations of Matsumoto et al. indicates that several further mechanisms should be considered, including twinning and non-basal slip [4]. The available experimental observations are usually performed by scanning-(SEM) or transmission electron microscopy (TEM) [5]. The results partially confirm the theoretical calculations, but the investigation at higher strain levels is rather difficult owing to the high dislocation density. The X-ray diffraction line profile analysis (XLPA) has been found as a reliable method for the investigation of the dislocation structure [6]. During the data processing, the experimental profiles are evaluated for the dislocation density, crystallite size or dislocation population in the particular slip systems [7, 8]. A doubtless advantage of this approach is the significantly larger investigated sample volume than that for TEM. In addition, XLPA is a non-destructive testing method. Thus, the sample preparation does not alter the dislocation structure formed during deformation.

In the present work, we studied the evolution of the dislocation structure in $Mg_{97}Y_7Zn_5$ (at. %) alloy, where the volume content of LPSO phase is very high (about 85%). Compression tests on both the as-cast and extruded samples were performed. The influence of the initial microstructure on the dislocation structure is discussed in detail.

## Experimental

The $Mg_{97}Y_7Zn_5$ (at.%) alloy used for this study was prepared by melting of high purity elements Mg, Y and Zn and a Mg-22%Y (wt%) master alloy in a graphite crucible coated with boron nitride under a protective Ar atmosphere. A part of the ingot was further extruded at 350 °C with an extrusion ratio of 18:1. Rectangular samples with the cross section of 5 mm × 5 mm and a height of 8 mm were cut parallel and perpendicular to the longitudinal axis of the casted and extruded specimens for compression tests. The compression experiments were performed at room temperature and an initial strain rate of $10^{-3}$ $s^{-1}$ for the engineering strains of ~5 and ~25% using an MTS 810 universal mechanical testing machine. The list of the samples is shown in Table 1. It is noted that for the extruded samples the samples failed before the desired maximum strain of 25% was achieved. Therefore, for the extruded samples compressed parallel and perpendicular to the extrusion direction the maximum strains were ~21 and ~13, respectively. The dislocation structures in the LPSO phase for the compressed specimens were studied by XLPA. The surface of the samples was mechanically polished and etched before the X-ray experiments. The diffraction peaks were measured on the surfaces lying parallel to the compression direction by a high-resolution rotating anode diffractometer (type: RA-MultiMax9, manufacturer: Rigaku) using $CuK\alpha_1$ radiation with a wavelength of $\lambda = 0.15406$ nm. The line profiles were evaluated for the crystallite size, the dislocation density and the fractions of different slip systems using the modified Williamson-Hall method.

## Results

Figure 1 shows a part of the X-ray diffraction pattern obtained on sample Extr-PA-5%. Both Mg and LPSO phases were detected. The structure of the major LPSO phase was identified as 18R. The fractions of the LPSO and Mg phases were 85 and 15% respectively. The lattice constants obtained from the peak positions are listed in Table 2. The peaks of the LPSO phase are broad which may be caused either by the small crystallite size or the significant density of dislocations formed during extrusion and compression. These two effects can be separated by the modified Williamson-Hall method. In this procedure, the integral breadth or the Full Width at Half Maximum (FWHM) of the peaks are plotted as a function of the product of square of the magnitude of the diffraction vector ($g$) and the dislocation contrast factor ($\overline{C}_{hkl}$) in accordance with the following equation [9].

$$FWHM = \frac{0.9}{d} + \Omega^* g^2\overline{C}_{hkl} + O\left(g^4\overline{C}^2_{hkl}\right), \quad (1)$$

where $d$ is the volume-weighted mean column length in the crystallites (it can be regarded as an apparent crystallite size), $\Omega^*$ depends on both the density and arrangement of dislocations and $O$ stands for higher order terms in $g^2\overline{C}_{hkl}$. The magnitude of the diffraction vector can be obtained as $g = 2\sin\theta/\lambda$ where $\theta$ is the half of the Bragg angle of the diffraction peak. The dislocation contrast factor, $\overline{C}_{hkl}$, can be expressed as:

$$\overline{C}_{hkl} = \overline{C}_{hk0}\left(1 + q_1 z + q_2 z^2\right), \quad (2)$$

where $q_1$ and $q_2$ are two parameters depending on the anisotropic elastic constants of the crystal and the type of dislocation slip system. $z = (2/3)(l/ga)^2$, where $a$ is the

**Table 1** The samples and their notations studied in this work

| Sample name | Description |
|---|---|
| As-cast-PA-5% | As cast material, compressed parallel to the cast direction for the strain of 5% |
| As-cast-PA-25% | As cast material, compressed parallel to the cast direction for the strain of 25% |
| As-cast-PE-5% | As cast material, compressed perpendicular to the cast direction for the strain of 5% |
| As-cast-PE-25% | As cast material, compressed perpendicular to the cast direction for the strain of 25% |
| Extr-PA-5% | Extruded material, compressed parallel to the extrusion direction for the strain of 5% |
| Extr-PA-21% | Extruded material, compressed parallel to the extrusion direction for the strain of 21% |
| Extr-PE-5% | Extruded material, compressed perpendicular to the extrusion direction for the strain of 5% |
| Extr-PE-13% | Extruded material, compressed perpendicular to the extrusion direction for the strain of 13% |

*PA*—parallel, *PE*—perpendicular direction with respect to the cast and extrusion axes

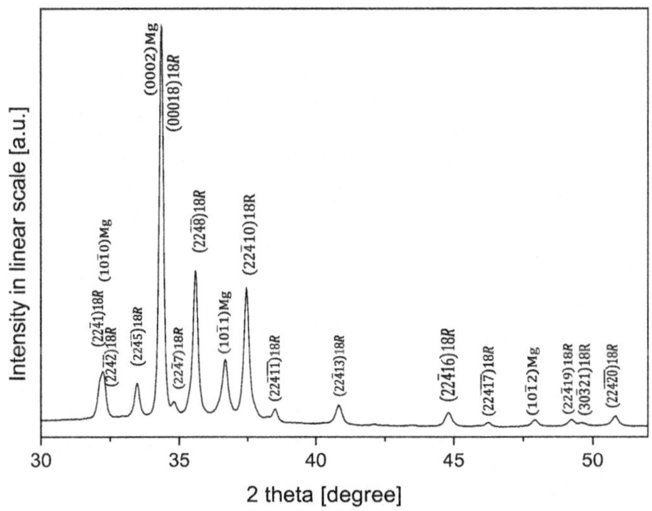


**Fig. 1** A part of the X-ray diffractogram obtained on sample Extr-PA-5%. The indices of reflections for the LPSO and Mg phases are given above the peaks

lattice constant in the basal plane. $\overline{C}_{hk0}$ is the average dislocation contrast factor of the *hk0* type reflections. The values of $\overline{C}_{hk0}$, $q_1$ and $q_2$ are listed for the eleven most common slip systems of the 18R LPSO phase in Table 3. We note that this is the first study reporting these contrast factor values which were calculated by the ANIZC software [6] using the single crystalline elastic constants listed in Table 2.

Figure 2 shows the modified Williamson-Hall plot for sample Extr-PA-5%. Only those LPSO peaks were used in this evaluation procedure which do not overlap strongly with Mg peaks or other LPSO reflections. It is noted that the FWHM values were corrected for the instrumental broadening. The instrumental correction of the peak breadth was performed in accordance with the method described in [9] using the profile widths measured on $LaB_6$ standard material. The values of $q_1$ and $q_2$ in Eq. (2) are fitted in the modified Williamson-Hall plot in order to get the most smooth arrangement of the datum points along a straight line. From the comparison of the experimental values of $q_1$ and $q_2$ with the theoretical values listed for the different slip systems in Table 3, the fractions of <a>, <c> and <c + a> -type dislocations were determined using the method described in details in [8]. It was found that for all compressed samples studied in this work the majority of dislocations is of <a>-type with the fractions between 70 and 96%. The rest of dislocations is of <c + a> -type with the fraction of 4–30%. The amount of <c> dislocations is negligible. The intercept and the slope of the straight line fitted to the data points in the modified Williamson-Hall plot can be used for the estimation of the apparent crystallite size and the dislocation density, respectively [see Eq. (1)]. The slope of the straight line in the modified Williamson-Hall plot, $\Omega^*$, is proportional to $\rho \times b^2$, where $\rho$ is the dislocation density and $b$ is the Burgers vector [9]. For all the samples studied in this work, the slope was $0.0003 \pm 0.0001$ nm, except for the as-cast specimens compressed for the strain of 5%, as for the latter samples the crystallite size and the dislocation density were larger and smaller, respectively, than the detection limits of the present X-ray line profile analysis. Therefore, the modified Williamson-Hall could not be applied for these samples. The slope of the modified Williamson-Hall plots determined for the other LPSO samples ($0.0003 \pm 0.0001$ nm) was similar to that obtained for other Mg alloys with the dislocation density of $\sim 10^{14}$ $m^{-2}$ [11]. However, for the LPSO phase the Burgers vector is larger with a factor of 3–7 than for the common Mg alloys (such as for AX41 or AZ31), due to the much larger lattice constants of the LPSO material. It is noted that the average Burgers vectors for the different specimens were calculated from the fractions of <a>, <c> and <c + a> -type dislocations and their values were found to be 1–2 nm. Since the slope of the straight line in the modified Williamson-Hall plot is proportional to $\rho$ $b^2$, a similar slope in the modified Williamson-Hall plot for the LPSO phase corresponds to a one-two orders of magnitude lower dislocation density than that in the mentioned Mg alloys. Therefore, the estimated dislocation density in the present LPSO specimens is $10^{12}$–$10^{13}$ $m^{-2}$. It is noted that the order of magnitude of the dislocation density was the same in the extruded samples for both low and high strains as well as for the as-cast specimens compressed up to the strain of $\sim$25%.

Although, in the extruded samples the dislocation density measurable by X-ray line profile analysis did not change considerably with increasing strain, there was a decrease in the crystallize size (see Table 4) which indicates dislocation activity during compression. The dislocation structures with strongly shielded strain fields, such as small dipoles, cannot be detected by X-ray line profile analysis. Former studies [12] suggested the formation of kink walls during deformation of hexagonal materials. If the spacing between the dislocations with opposite signs in the kink walls is small, the strain field of these dislocations are strongly shielded, therefore they are practically invisible by X-ray line profile analysis. However, these kink walls result in a fragmentation

**Table 2** The lattice constants (a and c) of the LPSO phase and its elastic constants taken from Ref. [10]

| a (nm) | c (nm) | $C_{11}$ (GPa) | $C_{12}$ (GPa) | $C_{13}$ (GPa) | $C_{33}$ (GPa) | $C_{44}$ (GPa) |
|---|---|---|---|---|---|---|
| 1.118 | 4.694 | 68.4 | 25.8 | 19.6 | 75.2 | 21.6 |

**Table 3** The dislocation contrast factors calculated for the eleven dislocation slip system in the LPSO phase

| Notation | Burgers vector type | Burgers vector | Slip plane | $\bar{C}_{hk0}$ | $q_1$ | $q_2$ |
|---|---|---|---|---|---|---|
| PrE | <a> | $1/3\ \langle\bar{1}2\bar{1}0\rangle$ | $\{\bar{1}0\bar{1}0\}$ | 0.334 | −0.1702 | 0.0072 |
| PrE | <a> | $1/3\ \langle\bar{1}2\bar{1}0\rangle$ | $\{10\bar{1}1\}$ | 0.327 | −0.1630 | 0.0066 |
| BE | <a> | $1/3\ \langle\bar{1}2\bar{1}0\rangle$ | {0001} | 0.2016 | −0.0016 | −0.0041 |
| S1 | <a> | $1/3\ \langle\bar{1}2\bar{1}0\rangle$ | n.a. | 0.1259 | 0.1654 | −0.0213 |
| Pr2E | <c> | $\langle 0001\rangle$ | $\{1\bar{1}00\}$ | 0.03622 | 0.8318 | 0.0245 |
| S3 | <c> | $\langle 0001\rangle$ | n.a. | $4 \times 10^{-6}$ | −4450 | 29552 |
| Pr3E | <c + a> | $1/3\ \langle\bar{2}113\rangle$ | $\{01\bar{1}0\}$ | 0.04081 | 0.0259 | 1.3281 |
| Py3E | <c + a> | $1/3\ \langle\bar{2}113\rangle$ | $\{1\bar{2}11\}$ | 0.04533 | 0.0320 | 1.0262 |
| Py4E | <c + a> | $1/3\ \langle\bar{2}113\rangle$ | $\{10\bar{1}0\}$ | 0.0485 | 0.6871 | −0.0072 |
| Py2E | <c + a> | $1/3\ \langle\bar{2}113\rangle$ | $\{2\bar{1}\bar{1}2\}$ | 0.0451 | 0.6326 | 0.0175 |
| S2 | <c + a> | $1/3\ \langle\bar{2}113\rangle$ | n.a. | 0.0249 | 2.6611 | −0.2191 |

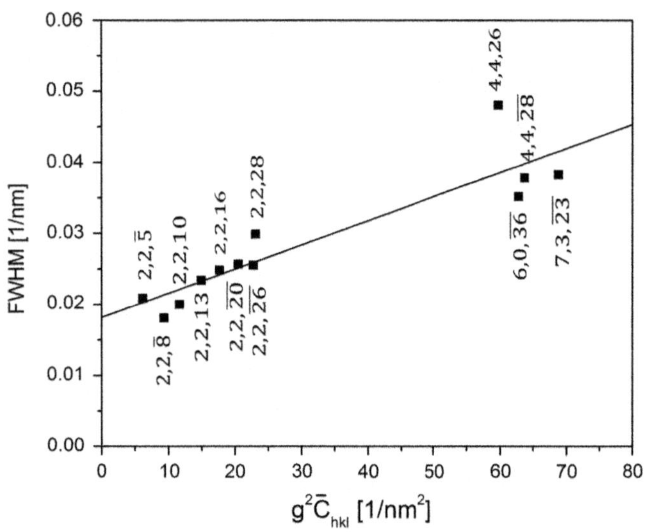


**Fig. 2** The modified Williamson-Hall plot for sample Extr-PA-5%

of grains into smaller coherently scattering domains, i.e., the dislocations in the kink walls can be observed indirectly by the reduction of the crystallite size measurable by X-ray peak profile analysis. Therefore, according to our experiments it seems that there was a dislocation multiplication during compression of the LPSO samples, however a part of dislocations were arranged into low energy kink walls. This process yielded an unchanged low value of the dislocation density ($10^{12}$–$10^{13}$ m$^{-2}$) obtained by X-ray peak profile analysis and a reduction of the apparent crystallite size. The low value of the dislocation density in the compressed LPSO samples can be attributed to the large Burgers vector, which yields a high shear strain during gliding even if the dislocation density is small. It should also be noted that this is the first study in the literature, which reports the type and density of dislocations in plastically deformed LPSO phase.

## Summary

In this work, the evolution of the dislocation structure in as-cast and extruded magnesium alloys with 85% LPSO volume content was studied. The following conclusions can be drawn:

- The dislocation density was in the order of $10^{12}$–$10^{13}$ m$^{-2}$ after the compression of the extruded samples up to the strain of ~5%. The dislocation density did not change considerably with increasing the strain to ~25%. Similar dislocation density value was observed for the

**Table 4** The crystallite size of the LPSO phase for the different samples

| Sample name | Crystallite size (nm) |
|---|---|
| As-cast-PA-5% | n.a. |
| As-cast-PA-25% | 42 ± 4 |
| As-cast-PE-5% | n.a. |
| As-cast-PE-25% | 39 ± 4 |
| Extr-PA-5% | 56 ± 6 |
| Extr-PA-21% | 37 ± 4 |
| Extr-PE-5% | 49 ± 5 |
| Extr-PE-13% | 33 ± 3 |

as-cast LPSO materials compressed to the strain of $\sim 25\%$, i.e., the dislocation density was not sensitive to the initial state of material.

- Although, the dislocation density measurable by X-ray line profile analysis did not change considerably with increasing strain, the crystallite size decreased which can be explained by the arrangement of dislocations into kink walls. The majority of dislocations was found to be of <a>-type with the fractions between 70 and 96% for the different samples. The rest of dislocations is of <c + a> -type with the fractions of 4–30%.

**Acknowledgements** The authors are grateful for the support of the Czech Grant Agency under grant Nr. 16-12075S.This work was supported by the Hungarian Scientific Research Fund, OTKA, Grant no. K-109021.

## References

1. Matsuda, M, Ii, S, Kawamura, Y, Ikuhara, Y and Nishida, M, (2005) Variation of long-period stacking order structures in rapidly solidified Mg97Zn1Y2 alloy. Mater. Sci. Eng. A 393: 269–274.
2. Nie, JF, Zhu, YM and Morton, AJ, (2014) On the Structure, Transformation and Deformation of Long-Period Stacking Ordered Phases in Mg-Y-Zn Alloys. Metallurgical and Materials Transactions A 45a: 3338–3348.
3. Zhu, YM, Weyland, M, Morton, AJ, Oh-Ishi, K, Hono, K and Nie, JF, (2009) The building block of long-period structures in Mg–RE–Zn alloys. Scripta Mater. 60: 980–983.
4. Matsumoto, R and Uranagase, M, (2015) Deformation Analysis of the Long-Period Stacking-Ordered Phase by Using Molecular Dynamics Simulations: Kink Deformation under Compression and Kink Boundary Migration under Tensile Strain. Mater. Trans. 56: 957–962.
5. Yamasaki, M, Hagihara, K, Inoue, S, Hadorn, JP and Kawamura, Y, (2013) Crystallographic classification of kink bands in an extruded Mg–Zn–Y alloy using intragranular misorientation axis analysis. Acta Mater. 61: 2065–2076.
6. Borbély, A, Dragomir-Cernatescu, J, Ribárik, G and Ungár, T, (2003) Computer program ANIZC for the calculation of diffraction contrast factors of dislocations in elastically anisotropic cubic, hexagonal and trigonal crystals. J. Appl. Cryst. 36: 160–162.
7. Máthis, K, Gubicza, J and Nam, N, (2005) Microstructure and mechanical behavior of AZ91 Mg alloy processed by equal channel angular pressing. J. Alloy Comp. 394: 194–199.
8. Máthis, K, Nyilas, K, Axt, A, Dragomir-Cernatescu, I, Ungár, T and Lukáč, P, (2004) The evolution of non-basal dislocations as a function of deformation temperature in pure magnesium determined by X-ray diffraction. Acta Mater. 52: 2889–2894.
9. Gubicza, J, (2014) X-ray line profile analysis in materials science. IGI Global, Hershey.
10. Tane, M, Kimizuka, H, Hagihara, K, Suzuki, S, Mayama, T, Sekino, T and Nagai, Y, (2015) Effects of stacking sequence and short-range ordering of solute atoms on elastic properties of Mg–Zn–Y alloys with long-period stacking ordered structures. Acta Mater. 96: 170–188.
11. Krajňák, T, Minárik, P, Strásk á, J, Gubicza, J, Máthis, K and Janeček, M, (2017) Influence of equal channel angular pressing temperature on texture, microstructure and mechanical properties of extruded AX41 magnesium. J. Alloys Comp. 705: 273–282.
12. Hess, JB and Barrett, CS, (1949) Structure and Nature of Kink Bands in Zinc. T. Am. I. Min. Met. Eng. 185: 599–606.

# Intermetallic Phase Characteristics in the Mg–Nd–Zn System

Domonkos Tolnai, Samuel A. Hill, Serge Gavras, Tungky Subroto, Ricardo Buzolin, and Norbert Hort

**Abstract**

Neodymium, a Rare Earth with low solid solubility in Mg is an ideal alloying element to improve the yield strength and creep resistance cost effectively. The addition of Zn achieves a further improvement; however, its influence on the intermetallic phases in the Mg–Nd–Zn ternary system is not yet fully understood. A Mg-5Nd alloy modified with 3, 5 and 7 wt% of Zn was investigated with in situ synchrotron radiation diffraction during cooling from the molten state to 200 °C in order to investigate the phase-formation and -transformation characteristics of the alloys. The synchrotron diffraction results have been complemented with TEM investigations on the as-solidified samples. The results suggest that Zn has a strong effect on the microstructure by stabilizing the $Mg_3Nd$ phase and accelerating the precipitation formation. The experimental results do not fully comply with the theoretical calculations, indicating the necessity of improving the thermodynamic databank for this alloy system.

**Keywords**

Mg–Nd–Zn alloys • Intermetallic phases • In situ synchrotron diffraction • Solidification

D. Tolnai (✉) · S. A. Hill · S. Gavras · T. Subroto · R. Buzolin · N. Hort
Magnesium Innovation Centre, Helmholtz-Zentrum Geesthacht, Max-Planck Str. 1, Geesthacht, D21502, Germany
e-mail: domonkos.tolnai@hzg.de

S. A. Hill
Department of Materials Science & Metallurgy, University of Cambridge, 27 Charles Babbage Rd, Cambridge, CB30FS, UK

R. Buzolin
Institue of Materials Science, Joining and Forming, Graz University of Technology, Kopernikusgasse 24/I, 8010 Graz, Austria


D. Orlov et al. (eds.), *Magnesium Technology 2018*, The Minerals, Metals & Materials Series, https://doi.org/10.1007/978-3-319-72332-7_61

## Introduction

The constant strive on energy efficiency has increased the importance of weight saving through light-weight design in the automotive and aerospace industries [1]. The high specific strength and stiffness of Mg and its alloys [2] complemented by their generally good castability would ensure a prominent place among the materials used for transportation. Despite their advantages, the wide usage is hindered by their poor absolute properties, insufficient formability at ambient temperatures [3] and weak corrosion resistance. The latter, however can be exploited in the field of degradable implant development [4].

There are two approaches to improve the poor absolute strength of Mg; generating secondary phases with reinforcing characteristics and/or thermo-mechanical processing to obtain ultrafine structures. An effective solution to enhance the mechanical property profile of Mg is the alloying with the combination of Zn and Rare Earth (RE) metals [5]. The addition of Zn to the Mg–RE systems accelerates the age hardening response and increases the peak strength due to the uniformly distributed basal precipitates [6,7]. The addition of RE improves the castability further and increases the elevated temperature strength [8]. Furthermore, it weakens the anisotropy [9, 10] that leads to improved ductility of Mg alloys.

These alloys are produced mainly in the form of castings therefore their property profile is determined during the solidification [11, 12] and the following thermo-mechanical processing [13]. Thus, controlling the sequence of formation and evolution of the meta-stable and stable phases during solidification is a prerequisite to the design of the microstructure.

Performing in situ diffraction experiments while cooling the molten sample to the fully solidified state, allows identifying the secondary phases, to determine their solidification sequence [14, 15] and to use these results for experimental validation of the existing thermodynamic databases on the investigated system.

## Experimental Methods

The alloys were prepared by permanent mould indirect chill casting [16]. Pure Mg was melted in an electric resistance furnace under protective atmosphere of 2 vol.% $SF_6$ and Ar. The Zn and Nd alloying elements were added as pure materials. After mixing, the melt was held at 720 °C for 10 min then was poured into a steel mould preheated to 660 °C. After 5 min isothermal holding the mould was quenched into water at a rate of 10 mm $s^{-1}$ until the top of the melt was in line with the cooling water level. The composition of the ingots was measured by Spark Optical Emission Spectroscopy and X-ray Fluorescence Spectroscopy.

The metallographic preparation of the samples was done by grinding using SiC paper and polishing with OPS solution. For the optical microscopy (OM) investigation the samples were etched with acetic-picral solution. The OM analyses were performed using a Leica DMI 500 light optical microscope. The average grain size was calculated based on measuring 100 grains using the line intercept method.

Scanning Electron Microscopy (SEM) was done with a Tescan Vega 3 SEM. A Phillips CM200 Transmission Electron Microscope (TEM) with accelerating voltage of 200 kV was used to obtain bright field (BF) micrographs and microbeam selected area diffraction (SAED) patterns of the intermetallic phases. TEM foils were prepared via electropolishing using a 1.5% perchloric acid solution in ethanol.

In situ solidification synchrotron diffraction experiments were performed at the P07 (HEMS) beamline of PETRA III at DESY (Deutsches Elektronen-Synchrotron). A monochromatic beam with a cross-section of 1.1 mm 1.1 mm was used, and a beam energy of 100 keV ($\lambda$ = 0.0124 nm) was used. The acquisition time for each image was 0.5 s. The samples were encapsulated in stainless steel crucibles under Ar atmosphere. The experiments were performed in the chamber of a modified DIL805 A/D dilatometer (Bähr-Thermoanalyse GmbH, Hüllhorst, Germany). The diffraction patterns were acquired with a PerkinElmer 1622 Flatpanel detector with pixel size of (200 μm)$^2$ The distance between sample and detector was 1603 mm (calibrated with a $LaB_6$ standard powder sample). The sample was heated to 800 °C with a then held at this temperature for 5 min before cooling to room temperature with a cooling rate of 50 K/min. The recorded diffraction patterns were analyzed using Fit2D.

Solution treatments were performed at 460 °C for 24 h in an electric resistance furnace and were immediately followed by room temperature water quenching. Samples were solution treated. Ageing treatments were performed at 250 °C for 168 h and were immediately followed by room temperature water quenching.

For the hardness measurements samples were ground using SiC paper up to P2500 before testing. The vickers hardness (HV5/30) was measured on an EMCO TEST M1C 010 hardness machine using 5 kg with a loading time of 30 s. 10 indentations were taken per sample.

Phase diagrams were produced with PANDAT$^{TM}$ 2016.1 software using the PanMg 2017 database. The Nd content was set to 4.3 wt%. A diagram was constructed with no suppressed phases, and one with the phases $Mg_{41}Nd_5$ and $Mg_{12}Nd$ suppressed.

## Results and Discussion

The actual alloy compositions measured with spark analyser and XRF are listed in Table 1.

### As-Cast Microstructure

The as-cast microstructures of the investigated alloys are shown in Fig. 1a–c. In all the alloys, there is a semi-continuous network of intermetallic particles at the grain boundaries. The grain sizes are measured on optical metallographic images as 0.56 ± 0.03 mm for Mg5Nd3Zn, 0.36 ± 0.02 mm for Mg5Nd5Zn and 0.2 ± 0.02 mm for Mg5Nd7Zn, respectively.

To analyse the influence of Zn additions on the volume fraction and morphology of the intermetallic phases, SEM investigations were used (Table 2).

In all the alloys two types of intermetallic phases are present, a eutectic lamellar and a continuous intermetallic morphology. The latter has an increasing volume fraction as the Zn content is increased to 5 wt% and further to 7 wt% as shown in Fig. 1.

**Table 1** Chemical composition of the alloys

| Composition | Nd wt%(XRF) | Zn wt%(spark) |
|---|---|---|
| Mg5Nd3Zn | 4.35 | 3.20 |
| Mg5Nd5Zn | 4.20 | 5.20 |
| Mg5Nd7Zn | 4.34 | 8.00 |

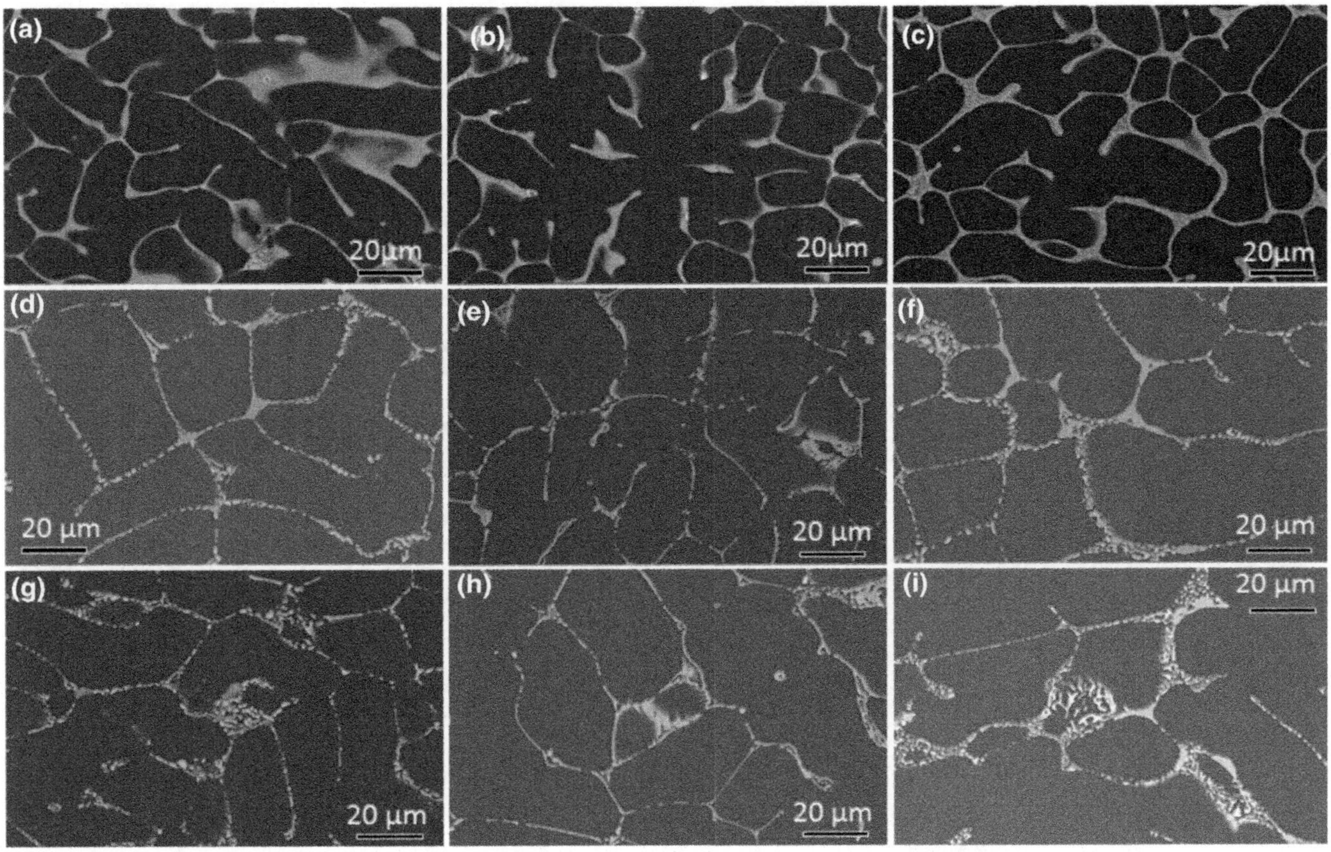


**Fig. 1** SEM micrographs of Mg5Nd3Zn (**a**, **d**, **g**) Mg5Nd5Zn (**b**, **e**, **h**) and Mg5Nd7Zn (**c**, **f**, **i**) in as-cast (**a–c**), solution heat treated (**d–f**), and aged (**g–i**) conditions

**Table 2** Average volume fraction of the intermetallic phase in the investigated alloys

| Composition (wt%) | Average volume fraction (%) |
|---|---|
| Mg5Nd3Zn | 13.9 (± 1.7) |
| Mg5Nd5Zn | 13.4 (± 0.4) |
| Mg5Nd7Zn | 15.2 (± 0.7) |

## Phase Identification

In order to characterize the intermetallic phases, synchrotron radiation diffraction was used. The line profiles of the samples in as-cast condition are shown in Fig. 2.

The results show that in the case of Mg5Nd3Zn only one intermetallic, the $Mg_3(Nd, Zn)$ quasi binary phase is present. Although on the SEM images two distinct phases are observed, it is possible that the continuous morphology has such a low volume fraction that is below the detection limit of the synchrotron diffraction. In the case of the higher Zn containing alloys the $Mg_{50}Nd_8Zn_{42}$ phase is clearly detectable. To support the results of the synchrotron radiation diffraction investigations, TEM phase identification was performed on the intermetallic particles. Using micro-beam electron diffraction, it was possible to obtain selected area electron diffraction patterns from the intermetallic phases. Following zone axis identification, it was confirmed that the intermetallic lamellar phase was quasi-binary face-centred cubic (FCC) phase, $Mg_3(Zn, Nd)$ while the continuous intermetallic phase was a c-centered orthorhombic phase $Mg_{50}Nd_8Zn_{42}$. Figure 3 shows a region of the sample from Mg5Nd5Zn, where these two different intermetallics lie adjacent to each other.

## Thermal Analysis

Differential Thermal Analysis (DTA) was used to obtain the liquidus and solidus temperatures of the alloys. The samples were subjected to three subsequent heating-cooling cycles with a rate of 10 K/min. The results show that with the increasing Zn content the formation temperatures of the phases decrease and the freezing range increases as shown in Table 3.

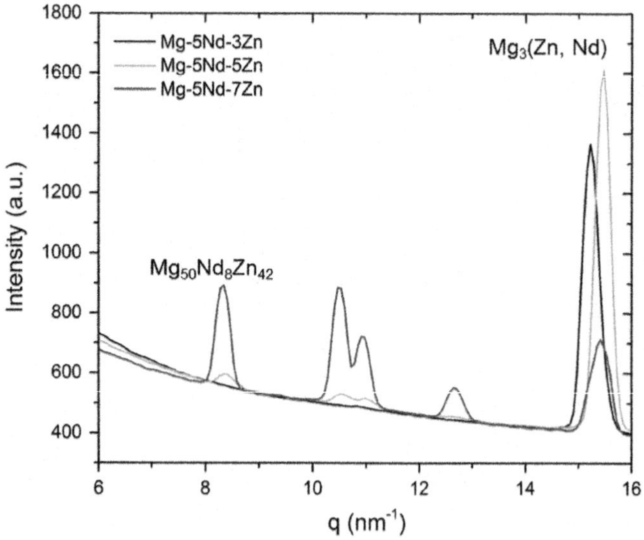


**Fig. 2** Section of the line profiles of the investigated Mg–Nd–Zn alloys

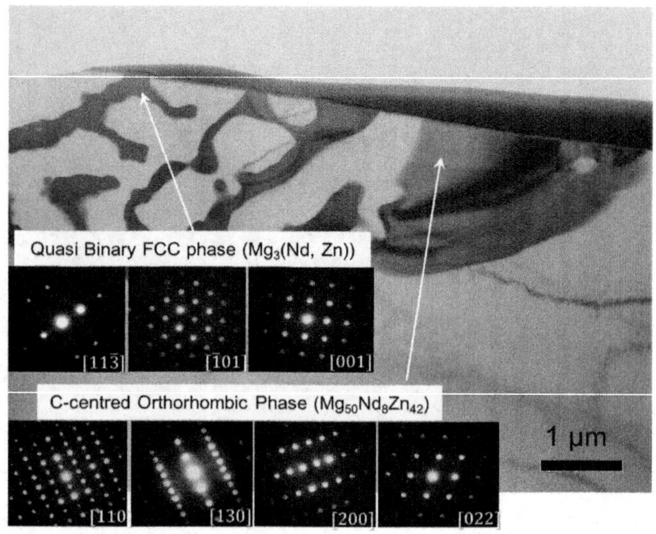


**Fig. 3** Mg5Nd5Zn bright field micrograph of showing 2 adjacent intermetallic phases the lamellar FCC and the continuous intermetallic with c-centred orthorhombic phase with corresponding zones axes showed respectively

**Table 3** Liquidus and solidus temperatures measured with DTA and phase formation temperatures measured by in situ synchrotron diffraction at 50 K/min cooling rate

| | Mg5Nd3Zn | Mg5Nd5Zn | Mg5Nd7Zn |
|---|---|---|---|
| Liquidus (°C) | 628 | 623 | 616 |
| Solidus (°C) | 510 | 495 | 487 |
| α-Mg (°C) | 633 | 633 | 623 |
| $Mg_3$(Nd, Zn) (°C) | 511 | 508 | 488 |
| $Mg_{50}Nd_8Zn_{42}$ (°C) | N/A | 462 | 469 |

**Table 4** Changes in Vickers hardness through heat treatment

| | As-cast | Solution treated | Aged |
|---|---|---|---|
| Mg5Nd3Zn | $57.4 \pm 4.5$ | $42.9 \pm 3$ | $40.3 \pm 2$ |
| Mg5Nd5Zn | $50.3 \pm 3$ | $42.6 \pm 3.4$ | $43 \pm 3.4$ |
| Mg5Nd7Zn | $65.1 \pm 3.9$ | $48.6 \pm 3.7$ | $51.7 \pm 2.9$ |

In situ synchrotron diffraction experiments were performed during melting and subsequent cooling of the samples in order to investigate the solidification sequence of the alloys and to determine whether phase transformations occur in liquid or solid state. The results indicate that first α-Mg dendrites form followed by the $Mg_3$(Nd, Zn) quasi binary phase. The solidification ends with the formation of $Mg_{50}Nd_8Zn_{42}$ phase detected only in the higher Zn containing alloys, Mg5Nd5Zn and −7Zn as shown in Fig. 4.

During the post mortem metallographic investigation of the as-solidified samples from the in situ solidification experiments, precipitations were found in the matrix as shown in Fig. 5. From the initial investigation of these precipitates, it is likely that they are $Mg_3$RE ($Mg_3$Nd in this investigation). However, further analysis needs to be completed before the phase of the precipitates can be decisively stated. The presence of precipitates may be explained by the reduction in solubility within the hcp matrix upon cooling. The final solid hcp Mg will likely form with the maximum solid solubility of Nd and Zn. This coring effect would explain the tendency of precipitates to form close to the intermetallics on the grain boundaries, as these are regions of greater solute enrichment.

## Heat Treatments

The evolution of the microstructure through solution heat treatment and ageing is shown in Fig. 1d, e, f, g, h, i, respectively. The volume fraction of intermetallics decreases during solution treatment as they dissolved into solid solution in the α-Mg matrix. SEM micrographs show both intermetallic phases present in Mg5Nd3Zn (Fig. 1d) and in Mg5Nd7Zn (Fig. 1f). The Mg5Nd5Zn alloy is likely to

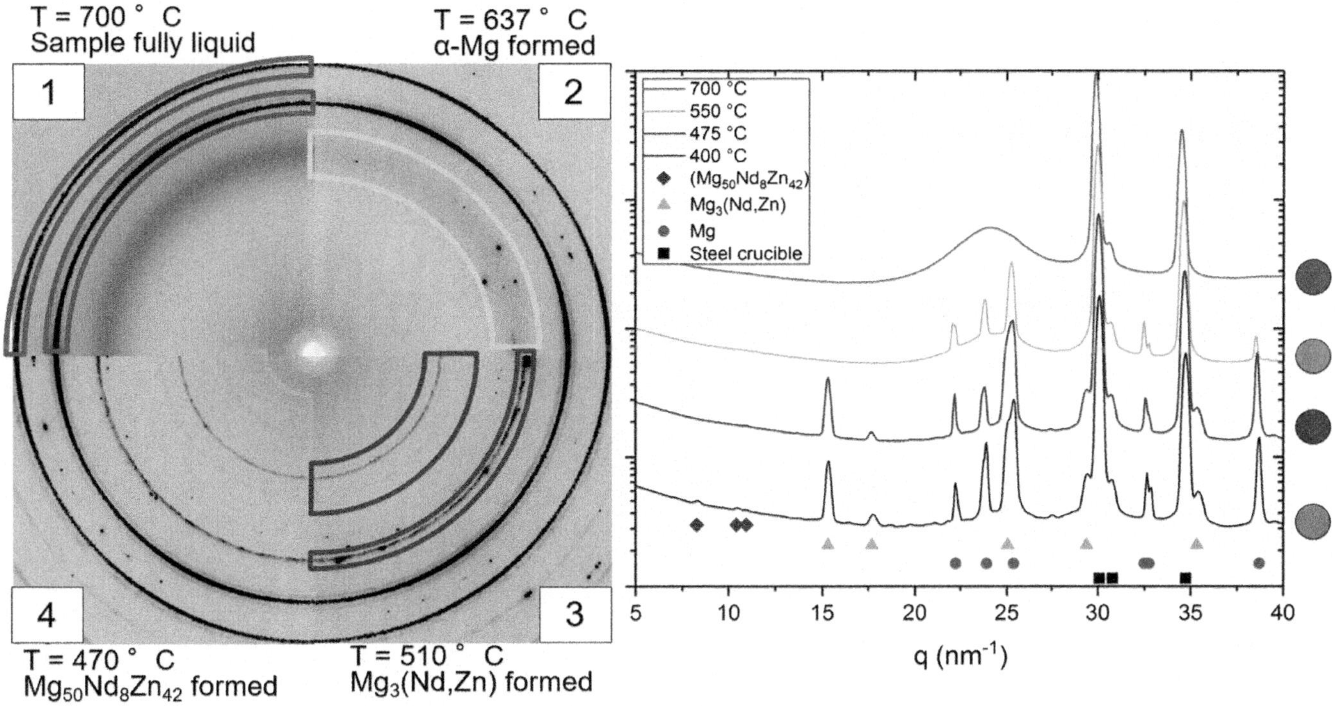


**Fig. 4** Solidification sequence of Mg5Nd5Zn, obtained by in situ synchrotron radiation diffraction

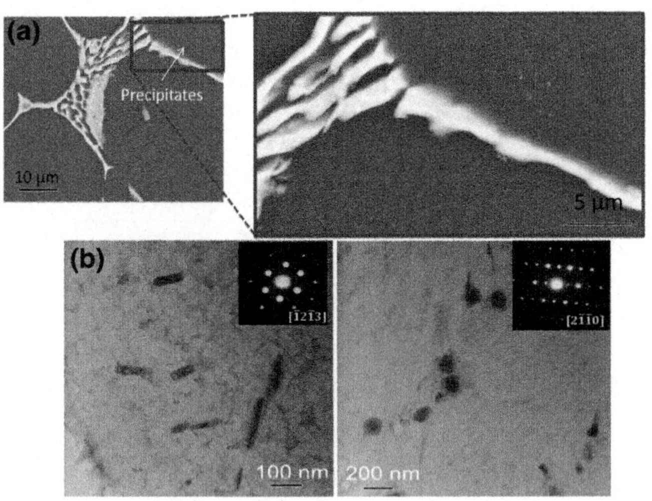


**Fig. 5** **a** SEM BSE micrograph overview of Mg5Nd3Zn following melting and solidification at 100 K/min indicating the location of precipitates **b** TEM BF $Mg_3RE$ precipitates from two different zone axes present following melting and solidification at 100 K/min in Mg5Nd5Zn

contain both intermetallic phases, but only the quasi-binary phase is evident in Fig. 1e.

Micrographs show the continuous phase to form a different shape after the solution heat treatment. The quasi-binary phase appears to form between successive dendrite arms, whereas the continuous phase is likely to be located between successive dendrites, or in larger voids. 2D micrographs show this as the quasi-binary phase linking the continuous regions. Precipitates were observed in all microstructures after ageing. They appear to be more evenly distributed. This is as expected, as the heat treatments allow diffusion within the matrix, leading to a more even solute distribution. They are seen to form arrays, likely along defects such as a dislocation. With increasing Zn content the precipitates appear to become finer.

The change in Vickers hardness of the alloys throughout the heat treatments is shown in Table 4.

In case of all three alloys a decrease in hardness can be observed after solution heat treatment. Although intermetallic particles precipitate, there are no significant differences between the solution treated and the aged samples. The T6 heat treatment is seen to decrease the hardness of all alloys, indicating that such a high alloying content is already detrimental to the age hardening of the system [17]. In this case the hardness is not strongly dependant on the precipitation strengthening, instead relies on the volume fraction of the intermetallic phases. These hard particles can effectively strengthen the matrix, and as during the heat treatment their volume fraction decreases their overall strengthening capability shows the same trend.

## Thermodynamic Modeling

A quasi-binary section of the Mg–Nd–Zn phase diagram at 4.3 wt% Nd content calculated with Pandat is shown in Fig. 6.

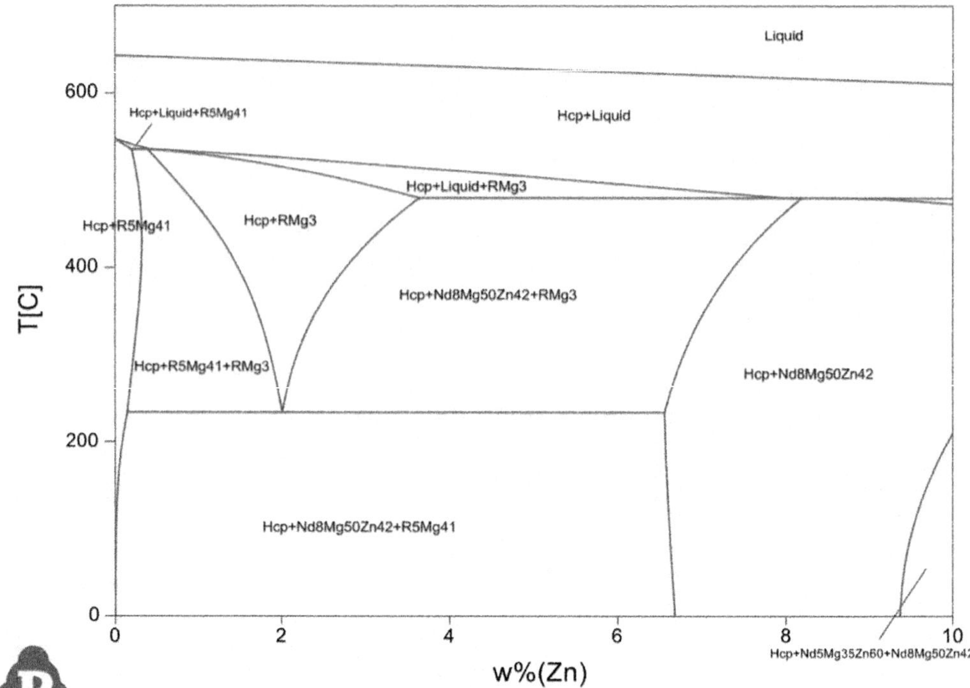


**Fig. 6** Quasi-binary section of the Mg–Nd–Zn phase diagram at 4.3 wt% Nd

It is evident that the equilibrium phase diagram cannot describe the reality in this case. However, the continuous intermetallic phase ($Mg_{50}Nd_8Zn_{42}$) is predicted in the case of all alloys, the investigations did not reveal the presence of $Mg_{41}Nd_5$, which is reported to be the stable phase in Mg–Nd binary system. When the simulation is performed with the $Mg_{41}Nd_5$ phase suppressed, $Mg_{12}Nd$ is predicted instead, which is in good correlation with previous findings under realistic solidification conditions [15]. Suppressing this phase as well, leads to the correct description of the phases as shown in Fig. 7.

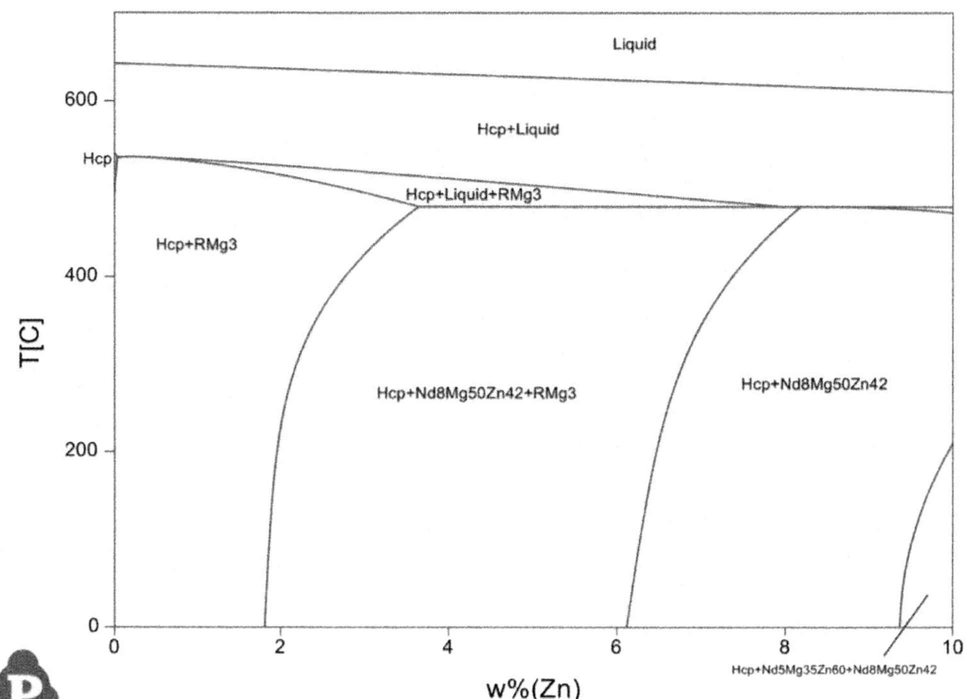


**Fig. 7** Quasi-binary section of the Mg–Nd–Zn phase diagram at 4.3 wt% Nd excluding the $Mg_{41}Nd_5$ and $Mg_{12}Nd$ phases

However, the latter thermodynamic calculation yields good result in the case of the lower Zn content alloys, the description of Mg4NdZn8 (Mg5Nd7Zn) is still incorrect, since it does not predict the $Mg_3$(Zn, Nd) phase. This suggests, the necessity of improving the corresponding thermodynamic database.

## Conclusions

From the phase identification and thermal-analyses of the alloys from the Mg–Nd–Zn system the following conclusions can be drawn:

- The increasing Zn content decreasing the grain size of the alloys.
- There are two different intermetallic phases present in the alloys, the face centered cubic $Mg_3$(Nd, Zn) and the c centered orthorhombic $Mg_{50}Nd_8Zn_{42}$ phase. The latter has a negligible volume fraction in the lowest Zn content Mg5Nd3Zn alloy.
- Precipitation have been found in the as-solidified samples, close to the grain boundaries and the intermetallic phases. This can be a result of solute enrichment of these regions during solidification.
- The solution and ageing heat-treatments lead to the worsening of the hardness, indicating that precipitation strengthening has no decisive effect on the overall strength of the material.
- The addition of Zn to binary Mg-Nd alloys stabilizes the $Mg_3$(Nd, Zn) phase.
- The thermodynamic databases not correlate with the experimental results, however in the case of Mg5Nd3Zn and -5Zn the calculations give acceptable results when the $Mg_{41}Nd_5$ and the $Mg_{12}Nd$ phases are suppressed.

**Acknowledgements** The authors acknowledge the Deutsches Elektronen-Synchrotron for the provision of facilities within the framework of the proposal I-20150471 and the Deutsche Forschungsgemeinschaft for the funding in the framework of the proposals TO817/4-1 and ME4487/1-1.

## References

1. M. Pekguleryuz, K. Kainer, A. Kaya, Fundamentals of magnesium alloy metallurgy, first ed. Woodhead, Philadelphia, 2013.
2. M.M. Avedesian, H. Baker, Magnesium and magnesium alloys, ASM Speciality Handbook, ASM International, United States of America, 1999.
3. M.H. Yoo, J.R. Morris, K.M. Ho, S.R. Agnew, Metall. Mater. Trans. A 41 (2002) 813–822.
4. H. Hermawan, D. Dubé, D. Mantovani, “Developments in metallic biodegradable stents” Acta Biomaterialia 6 (2010) 1693–1697.
5. T. Itoi et al.: Microstructure and mechanical properties of Mg-Zn-Y alloy sheet prepared by hot-rolling, Mater Sci Eng A, 560 (2013) 216–223.
6. J. F. Nie et al.: Enhanced age hardening response and creep resistance of Mg–Gd alloys containing Zn: Scr. Mater., 53, (2005) 1049–1053.
7. T. T. Sasaki et al.: Enhanced age hardening response by the addition of Zn in Mg–Sn alloys, Scripta. Mater., 55, (2006). 251–254.
8. Q. Li, et al.: Effect of Nd and Y addition on microstructure and mechanical properties of as-cast Mg–Zn–Zr alloy, J. Alloys Compd., 427, (2007) 115–123.
9. N. Stanford et al.: Effect of microalloying with rare-earth elements on the texture of extruded magnesium-based alloys Scripta. Mater., 59, (2008) 772–775.
10. B. Langelier et al.: Improving microstructure and ductility in the Mg–Zn alloy system by combinational Ce–Ca microalloyingMat. Sci. Eng. A, 620, (2015) 76–84.
11. D. Tolnai et al., “In situ synchrotron tomographic investigation of the solidification of an AlMg4.7Si8 alloy” Acta Materialia 60 (2012) 2568–2577.
12. S. Terzi et al., “In situ study of nucleation and growth of the irregular $\alpha$-Al/$\beta$-$Al_5FeSi$ eutectic by 3-D synchrotron X-ray microtomography” Acta Materialia 58 (2010) 5370–5380.
13. D. Tolnai et al., “Sub-micrometre holotomographic characterisation of the effects of solution heat treatment on an AlMg7.3Si3.5 alloy” Materials Science and Engineering A 550 (2012) 214–221.
14. O. Shuleshova et al., “In situ observations of solidification in Ti–Al–Ta alloys by synchrotron radiation” Intermetallics 19 (2011) 688–692.
15. D. Tolnai et al., “In situ synchrotron diffraction of the solidification of Mg4Y3Nd” Materials Letters 102–103 (2013) 62-64.
16. F.R. Elsayed et al., Magnesium permanent mold Castings optimization Materials Science Forum 690 (2011) 65–68.
17. J.F. Nie, “Precipitation and hardening in magnesium alloys” Metallurgical and Materials Transactions A 43(11) (2012) 3891–3939.

# Biodegradable Mg–Y and Mg–Li Alloys with the Addition of Ca and Zn for Medical Applications

Sonia Boczkal, Michał Karaś, Anna Maria Osyczka, and Marzena Lech-Grega

**Abstract**

The study describes structure characteristics as well as mechanical properties, corrosion behavior and biocompatibility of Mg–Y and Mg–Li alloys with 0.2% Ca, 1% Ca and 1% Zn cast into permanent molds. The structure showed a significant effect of the amount of alloying additives on the morphology of phases occurring mainly at the grain boundaries. Alloy composition also affects the possibility of grain structure refinement. The mechanical properties obtained the highest level in alloys with the addition of 1% Zn and 0.2% Ca. The results of biodegradation tests carried out in Ringer's solution showed that the lowest dissolution rate had the alloys with low Ca content. High Ca content raised the content of the $Mg_2Ca$ phase in the alloy, which acted as a catalyst of the grain boundary corrosion. The examination of Mg alloys carried out under sterile conditions at 37 °C after 144 h of the test in α-MEM solution has proved a high biocompatibility of these alloys when containing additions of Li, Ca and Zn.

**Keywords**

Mg–Y alloys • Mg–Li alloys • Ca and Zn additions Structures • Properties • Immersion tests

## Introduction

The currently used in medicine metallic implants as well as those developed by scientists must exhibit a number of necessary properties, including mechanical strength, resistance to cyclic fatigue and biocompatibility. Biomaterials such as steel, titanium or Co–Cr alloys have stiffness and durability far exceeding the natural properties of human bones. Additionally, these materials introduced into the body generate the risk of reduced immunological resistance as a result of corrosion process and the release of toxic metal ions [1]. Magnesium alloy implants are an interesting alternative to non-resorbable metallic biomaterials as regards both physical and chemical properties. Magnesium and its alloys are lightweight materials with mechanical properties and modulus of elasticity similar to the natural bone, and what is even more important is their full biodegradability in living organisms. Magnesium is also an essential component of human metabolism and the fourth most abundant cation in the human body [2]. In spite of this, the development of biomaterials for implants fully resorbable by the human body encounters many problems that require solutions in the field of both materials engineering and biomedicine. One of the issues to be solved is the effect of various elements included in the composition of magnesium alloys on the properties of these alloys, the corrosion rate in physiological fluids in particular, and on the in vitro and in vivo biocompatibility [3]. Elements such as Ca, Zn, Li and small amounts of RE are generally considered to be the most biocompatible additives introduced into magnesium alloys. The composition and the amount of alloying additives contribute to the change of mechanical properties and corrosion behavior. Calcium is the most important mineral in human bones. The addition of over 1% Ca to magnesium alloys improves the corrosion resistance and to some extent also the mechanical properties [4, 5]. Calcium in magnesium alloys forms phases such as $CaMg_2$ [4–6]. The addition of Ca and Y to magnesium alloys reduces the evolution of hydrogen and thus increases the corrosion resistance [1]. The addition of yttrium is particularly beneficial for the mechanical properties. Combined with magnesium, yttrium forms $Mg_{24}Y_5$ and $Mg_2Y$ phases. It has to be remembered, however, that increasing yttrium content in the alloy, and thus increasing the content of the $Mg_{24}Y_5$ phase, will

S. Boczkal (✉) · M. Karaś · M. Lech-Grega
Light Metals Division, Institute of Non-Ferrous Metals in Gliwice, Skawina, Poland
e-mail: sboczka@imn.skawina.pl

A. M. Osyczka
Faculty of Biology and Earth Sciences, Jagiellonian University, Cracow, Poland


D. Orlov et al. (eds.), *Magnesium Technology 2018*, The Minerals, Metals & Materials Series, https://doi.org/10.1007/978-3-319-72332-7_62

accelerate the micro-galvanic corrosion [7]. Like yttrium, zinc also improves mechanical properties, especially during heat treatment, and reduces the corrosion rate in alloy compared to pure Mg. Introduced to Mg alloys, it forms the $MgZn_2$ phase [7, 8].

Mg–Li alloys are characterized by satisfactory plastic properties and moldability; they also offer high strength and stiffness. Added to magnesium alloys, lithium reduces the c/a ratio of lattice parameters, resulting in easier slip of non-basal systems [9]. Lithium content in magnesium alloys raised to 8.5% contributes to higher biodegradation rate. The high corrosion rate in Mg–Li alloys due to the high lithium content can be attributed to an increase in the proportional content of β phase with regular lattice [10]. In addition to the mechanical properties and corrosion rate, another important feature of magnesium alloys is their biocompatibility determined in immersion test. This test is carried out in simulated body fluids (SBF), physiological buffered saline (PBS) and α-MEM solution. Monitoring of the pH level and time-based observations allow considering this material as fully biocompatible. Pure magnesium dissolves in a solution with the pH value of 7.4–7.6 [11]. Similar in vitro studies of the MgY2Zn1Ca0.25 alloy performed in α-MEM solution showed after the lapse of 100 h the initial pH value of 7.5 and the final value of 8.5 [12].

Controlling magnesium properties through the best choice of alloy components should give an optimum chemical composition of the alloy with strictly defined properties comparable to natural bone tissue. In order to design a biocompatible material, the structure and mechanical properties of castings made from the Mg–Y and Mg–Li alloys with different content of Ca and Zn were characterized. The impact of chemical composition on the degradation rate in Ringer's solution and biocompatibility in α-MEM solution was determined by the time-related pH measurements. Sample structure and surface changes after the immersion test were studied.

## Methodology

Studies were made on eight as-cast Mg–Y and Mg–Li alloys with the addition of Ca and Zn (Table 1). Mg–Y alloys were prepared on the basis of ZK60 alloy and Mg–Li alloys on technically pure components. Casting into permanent steel molds of 40 mm diameter and 160 mm length was carried out in a protective atmosphere of argon in a magnesium melting furnace with a capacity of 10 kg. The alloys were cast at a temperature above 700 °C to the molds heated to about 280–300 °C. Castings were sampled for structural, properties, corrosion and biocompatibility studies.

Structure examinations and grain size measurements were made on as-cast metallographic sections using an Olympus GX70 optical microscope and Inspect F50 scanning electron microscope with EDS and EBSD. Average grain size studies were performed by incisal techniques in three images from an optical microscope. Mechanical properties (yield strength, ultimate tensile strength and elongation) were tested on an Instron 5582 100 kN machine at room temperature. The tests were performed on three samples with two deformation rates $\dot{\varepsilon} = 2,8 \cdot 10^{-3} s^{-1}$ and $\dot{\varepsilon} = 2,5 \cdot 10^{-4} s^{-1}$. The corrosion behavior was observed in Ringer's solution with an osmolarity of 309 mOsmole/l and pH of 5.0–7.5 at a temperature of 37 °C using samples with a total area of approximately 280 mm$^2$. The test time was 24, 48, 72 and 96 h. The corrosion products were removed in accordance with [13] using $CrO_3 + AgNO_3 + Ba(NO_3)_2$ solution. Based on the results of weight measurements, the weight loss was calculated. Biocompatibility was determined under sterile conditions on disc-shaped samples of 2 mm thickness and 8 mm diameter at a temperature of 37 °C in α-MEM solution with the addition of 10% FBS and 1% penicillin/streptomycin. After 24 h, 72 h and 144 h, the pH of the solution was measured, corrosion changes were recorded, and the solution was replaced.

## Results

Structure was studied on castings made from the Mg–Y and Mg–Li alloys with varying Ca and Zn content, revealing the grain structure (Table 2) in the examined materials. In the Mg–Y alloy containing 0.2% Ca, very large grains of the size of even a few millimeters were grouped around the sample axis (so called "bamboo structure") (Fig. 1a). The smallest mean grain size was observed in the Mg-Li alloys. The values ranged from 100 to 200 μm (Fig. 1b).

SEM analysis of the structure showed that in the examined alloys the intermetallic phases occurred mainly within the grain boundaries (Fig. 2). Increasing the Ca and Zn

**Table 1** Comparison of the chemical composition of castings [wt%]

| Alloy | Y | Ca | Zn | Alloy | Li | Ca | Zn |
|---|---|---|---|---|---|---|---|
| MgY1Ca0.2 | 0.98 | 0.20 | 0.01 | MgLi1Ca0.2 | 0.57 | 0.22 | 0.00 |
| MgY1Zn1Ca0.2 | 0.80 | 0.16 | 0.88 | MgLi1Zn1Ca0.2 | 0.68 | 0.23 | 0.81 |
| MgY1Ca1 | 0.80 | 1.10 | 0.01 | MgLi1Ca1 | 0.55 | 0.90 | 0.00 |
| MgY1Zn1Ca1 | 1.05 | 1.25 | 1.10 | MgLi1Zn1Ca1 | 0.68 | 0.92 | 0.82 |

**Table 2** Average grain size ($d$) and standard deviation ($\sigma$)

| Alloy | $\bar{d}$(mm) | $\sigma$ (mm) | Alloy | $\bar{d}$(mm) | $\sigma$ (mm) |
|---|---|---|---|---|---|
| MgY1Ca0.2 | 2.07 | 2.03 | MgLi1Ca0.2 | 0.22 | 0.09 |
| MgY1Ca0.2Zn1 | 1.97 | 1.88 | MgLi1Ca0.2Zn1 | 0.17 | 0.07 |
| MgY1Ca1 | 1.20 | 0.50 | MgLi1Ca1 | 0.16 | 0.07 |
| MgY1Ca1Zn1 | 0.90 | 0.34 | MgLi1Ca1Zn1 | 0.11 | 0.04 |

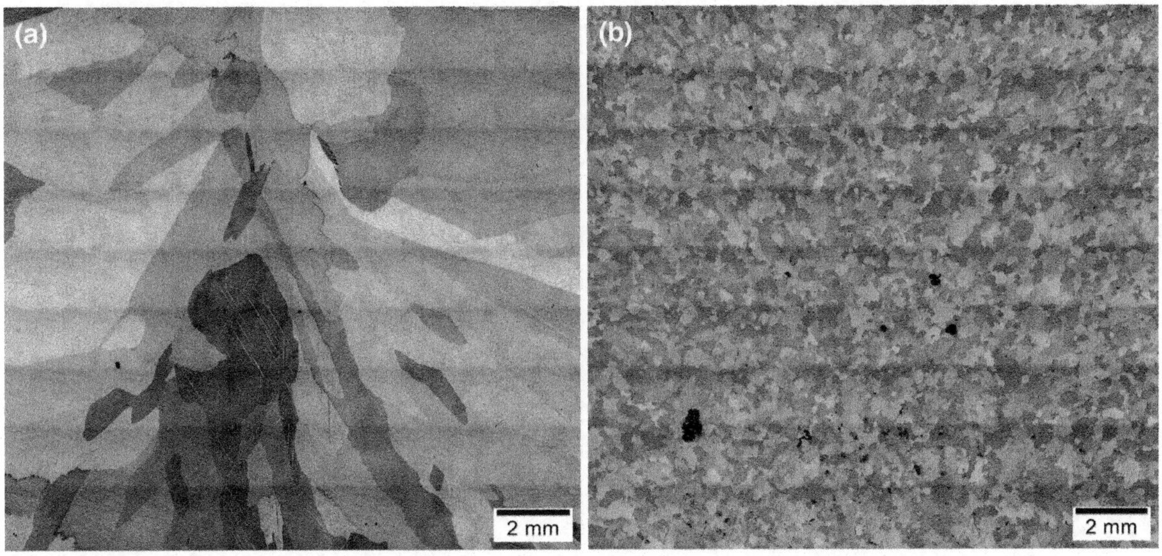


**Fig. 1** Grains in alloys: **a** MgY1Ca0.2 and **b** MgLi1Ca0.2

content increased the volume fraction of these phases. Higher content of Ca and Zn in Mg-Li alloys was observed to lead to the formation of eutectic $Mg_2Ca$ and α-Mg phases. The addition of Zn changed the shape and morphology of phases within the grains. Chemical analysis and phase analysis using the EBSD crystallographic orientation analyzer showed the occurrence of $Mg_{24}Y_5$, $Mg_3Y_2Zn_3$ and $YZn_5$ phases, closely related to the alloy chemical composition. The phases were either isolated or "anchored" to the $Mg_2Ca$ phase.

The results of mechanical tests (Table 3) have indicated that zinc added to the alloy causes a considerable increase of the tensile strength but only slight increase of the yield strength. The highest yield strength was obtained in MgY1Ca1Zn1 and MgLi1Ca1Zn1 alloys, while MgY1-Ca0.2Zn1 and MgLi1Ca0.2Zn1 alloys showed the highest tensile strength and elongation.

Studies of the weight loss changes in magnesium alloys after the exposure to Ringer's solution (Fig. 3) lead to the conclusion that Ca content increased to 1% increases the dissolution rate. The MgY1Ca1 alloy dissolved completely after 24 h (Fig. 3a). The second fastest dissolving magnesium alloy was MgY1Ca1Zn1, which after 24 h had lost nearly 20% of its initial weight. It can therefore be concluded that 1% zinc addition to these alloys reduces the dissolution rate of the material. Compared to Mg–Y alloys, Mg-Li alloys dissolved much more slowly, since the loss of weight in time was for these materials the lowest (Fig. 3b). Large dispersion of the weight loss results in MgLi1Ca1Zn1 alloy was caused by the difficult cleaning of the sample from the reaction products after the immersion test.

As a result of biocompatibility studies carried out by immersion test in α-MEM solution with the addition of 10% FBS and 1% penicillin/streptomycin (Fig. 4), it was found that MgYCa0.2Zn1, MgLiCa0.2Zn1 and MgLi1Ca1Zn1 alloys exhibited the lowest pH of the solution after 144 h and no traces of infection. The pH level of these materials increased at the start of the test but remained unchanged after the 24th hour of the test. In other samples, foam has appeared and the value of pH rose to 9.4. The samples of MgYCa1 and MgY1Ca1Zn1 alloys were completely dissolved.

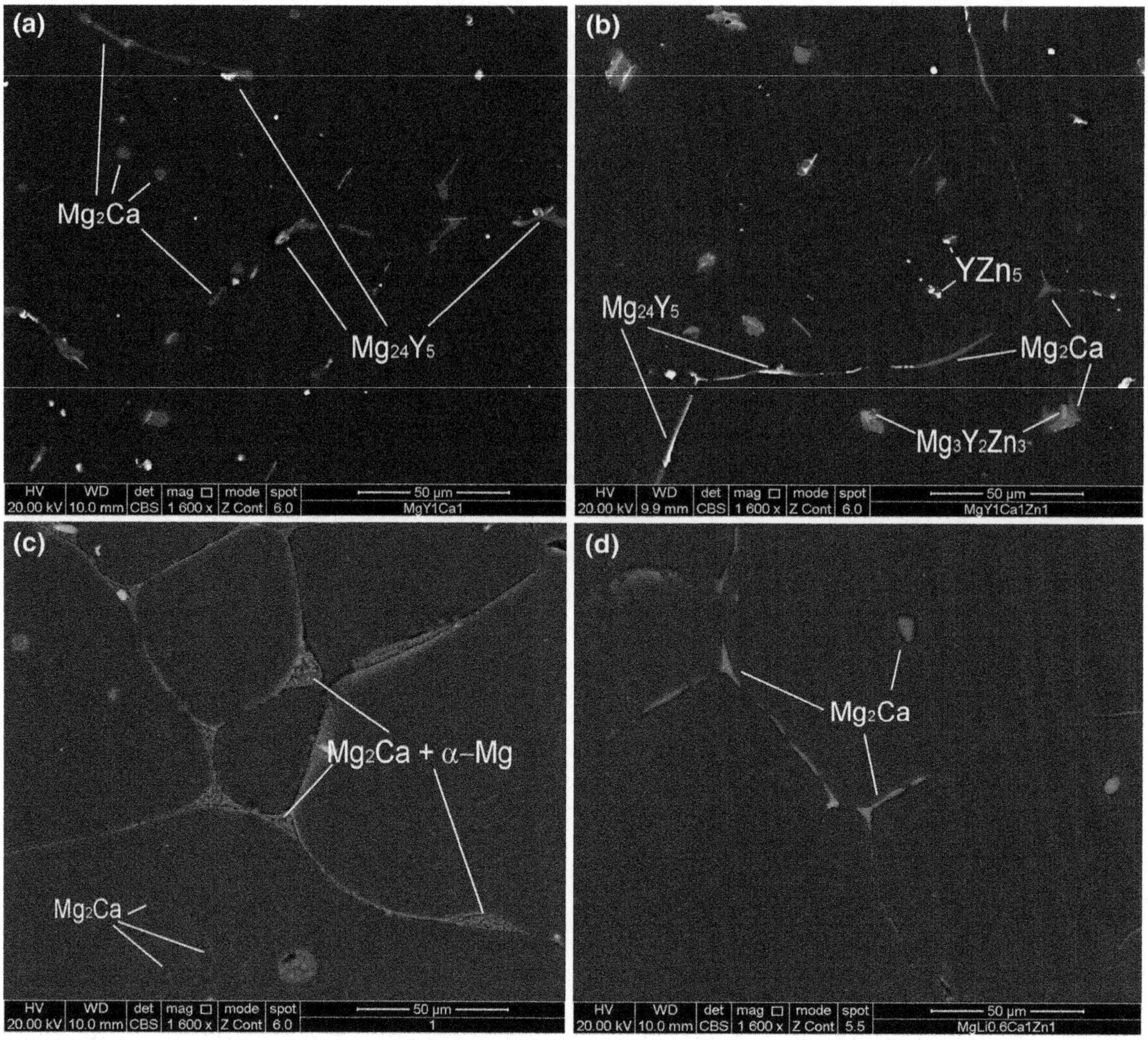


**Fig. 2** SEM structure of magnesium alloys: **a** MgY1Ca1, **b** MgY1Ca1Zn1, **c** MgLi1Ca1 and **d** MgLi1Ca1Zn1

**Table 3** Mechanical properties obtained in static tensile test

| Alloy | YS (MPa) | UTS (MPa) | EL (%) | Alloy | YS (MPa) | UTS (MPa) | EL (%) |
|---|---|---|---|---|---|---|---|
| MgY1Ca0.2 | 34 ± 2.8 | 78 ± 8.4 | 6.0 ± 0.6 | MgLi1Ca0.2 | 45 ± 6.1 | 109 ± 21.7 | 5.2 ± 2.4 |
| MgY1Ca0.2Zn1 | 58 ± 10.6 | 135 ± 30.4 | 6.3 ± 1.3 | MgLi1Ca0.2Zn1 | 68 ± 2.0 | 181 ± 17.7 | 15.4 ± 1.3 |
| MgY1Ca1 | 59 ± 2.1 | 82 ± 28.9 | 1.6 ± 1.3 | MgLi1Ca1 | 61 ± 0.7 | 96 ± 4.2 | 1.9 ± 0.4 |
| MgY1Ca1Zn1 | 89 ± 4.9 | 120 ± 2.1 | 1.4 ± 0.2 | MgLi1Ca1Zn1 | 72 ± 4.0 | 129 ± 7.0 | 3.9 ± 0.9 |

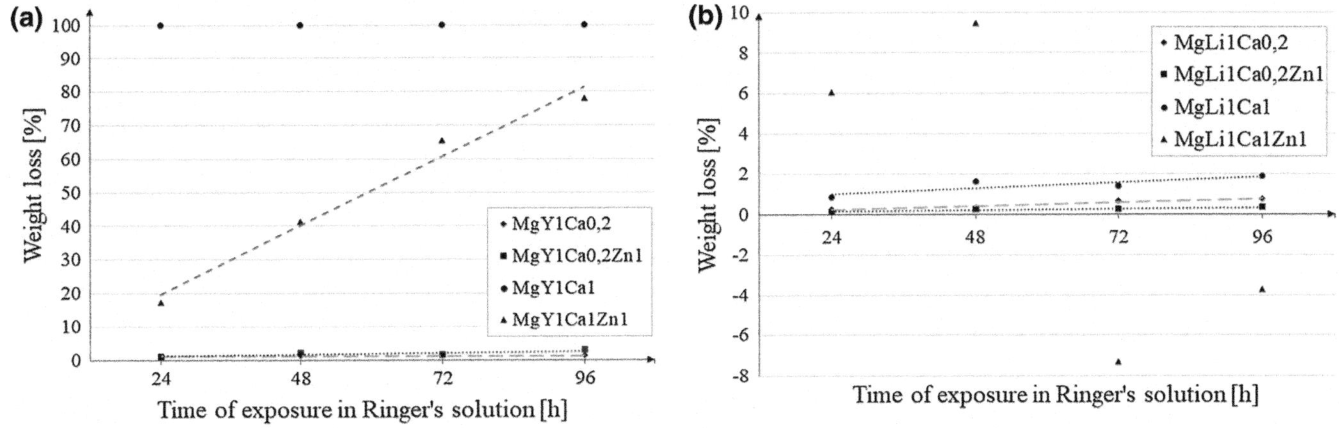


**Fig. 3** Weight loss during immersion test in Ringer's solution for alloys: **a** Mg–Y and **b** Mg-Li

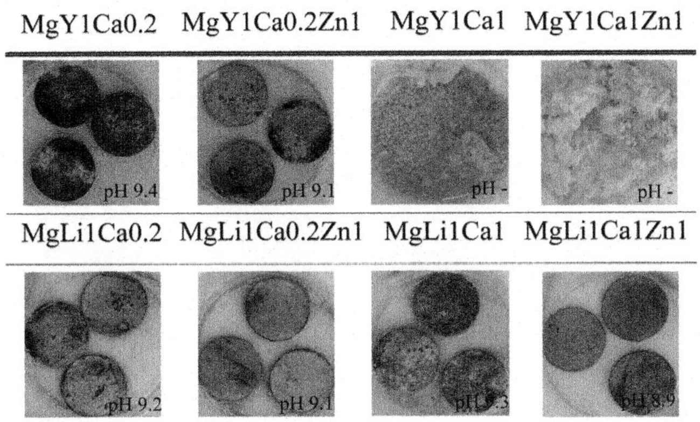


**Fig. 4** Surface of magnesium alloy samples after the 144 hour-lasting immersion test in α-MEM solution with additives

## Conclusions

Tests were carried out on eight as-cast Mg alloys, including four Mg–Y alloys and four Mg-Li alloys with varying Ca and Zn content. As a result of structure examinations, the smallest grains were found in Mg–Li alloys. In the MgLi1Ca1Zn1 alloy, the mean particle size was 110 μm. The grain size in magnesium alloys had no serious impact on the mechanical properties, as these depended mainly on the alloy chemical composition. In both tested alloy series, the tensile strength and ductility have reached the highest values when the addition of Ca was 0.2% and of Zn—1%. Hence follows the conclusion that higher Ca content in alloys leads to lower tensile strength and higher yield strength. The immersion tests carried out in Ringer's solution and in α-MEM solution with the addition of 10% FBS and 1% penicillin/streptomycin have shown that Mg-Li alloys dissolve much more slowly than Mg–Y alloys and maintain a relatively low pH of the solution after 144 h. Additionally, it was found that lower corrosion rate was achieved in alloys containing 0.2% Ca and not in those with 1% Ca. As a result of structure examinations, mechanical tests, and studies of the dissolution rate and biocompatibility, three alloys, i.e. MgYCa0.2Zn1, MgLiCa0.2Zn1 and MgLi1Ca1Zn1, were identified as biomaterials with the greatest potential for use in bone tissue implantology.

**Acknowledgments** The authors acknowledge gratefully the financial support from the project no. 3787/E-138/2015 financed by Ministry of Science and Higher Education.

## References

1. Chou DT, Hong D, Saha, Ferrero J, Lee B, Tan Z, Dong Z, Kumta PN (2013) In vitro and in vivo corrosion, cytocompatibility and mechanical properties of biodegradable Mg–Y–Ca–Zr alloys as implant materials: Acta Biomater 9 (10):8518–8533.
2. Gu XN, Zheng YF (2010): A review on magnesium alloys as biodegradable materials: Front Mater Sci China, 4(2):111–115.
3. Gu XN, Li SS, Li XM, Fan YB (2014) Magnesium based degradable biomaterials: A review: Front Mater Sci, 8(3):200–218.
4. Wan Y, Xiong G, Luo H, He F, Huang Y, Zhou X (2008) Preparation and characterization of a new biomedical magnesium–calcium alloy: Mater and Design 29:2034–2037.
5. Z Li, Gu X, Lou S, Zheng Y (2008) The development of binary Mg-Ca alloys for use as biodegradable materials within bone: Biomat 29:1329–1344.
6. Boczkal S, Lech-Grega M, Żelechowski J (2015) The effect of additions of Zn and Zr on the microstructure and texture of cast and wrought Mg alloys with different content of Ca: Sol St Phen 231:73–79.
7. Zeng R, Qi W, Zhang F, Cui H, Zheng Y (2014) In vitro corrosion of Mg–121Li–112Ca–1Y alloy: Prog Nat Sci-Mater 24:492–499.

8. Rosalie J M, Somekawa H, Singh A, Mukai T (2013) Increasing volume fraction of precipitates and strength of a Mg–Zn–Y alloy by pre-ageing deformation: Magnesium Technology 323–328.
9. Bao L, Le Q, Zhang Z, Cui J, Li Q (2013) Effect of homogenization treatment on microstructure evolution and the distributions of RE and Zr elements in various Mg–Li–RE–Zr alloys: J Magnes and Alloys, 1(2):139–144.
10. Zheng Y (2015): Magnesium Alloys as Degradable Biomaterials: October 9, CRC.
11. Peng Q, Huang Y, Zhou L, Hort N, Kainer KU (2010) Preparation and properties of high purity Mg–Y biomaterials: Biomaterials 31 (3):398–403.
12. Hänzi A C, Gerber I, Schinhammer M, Löffler J F, Uggowitzer P J (2010) On the in vitro and in vivo degradation performance and biological response of new biodegradable Mg–Y–Zn alloys: Act Biomat 6:1824–1833.
13. EN ISO 8407:2014-06.

# Degradable Magnesium Implants—Assessment of the Current Situation

R. Willumeit-Römer, N. Ahmad Agha, and B. Luthringer

**Abstract**

Mg and its alloys degrade under physiological conditions. The great challenge here is to tailor the degradation in a manner that is suitable for a biological environment. Fast or uncontrolled corrosion is associated with strong hydrogen and ion release and severe pH changes, which can lead to a fast loss of mechanical stability and undesirable biological reactions. Since these processes are highly complex in a living system and sufficient data describing the degradation in vivo is missing, it is very difficult to produce knowledge based new alloys. Still, the endeavour is successful: one CE certified Mg-alloy compression screw (Magnezix, Syntellix AG, Germany) and a Mg-based drug-eluting stent (Magmaris, Biotronik AG, Germany) are on the market. In addition, in China and Korea patient trials (hip surgery and hand fracture) are reported. This paper gives a brief outline of the current status of Mg-implants and which obstacles still have to be mastered. As an example for the special nature of Mg and its interaction with cells, a comparison is made between the influence of osteoblasts (bone forming cells) and fibroblasts (the most abundant cells in connective tissue) on the degradation layer underneath the cells.

**Keywords**

Magnesium (Mg) • Degradation • Osteoblasts
Fibroblast • Degradation layer

R. Willumeit-Römer (✉) · N. Ahmad Agha · B. Luthringer
Helmholtz-Zentrum Geesthacht, Institut Für Materialforschung, Division Metallische Biomaterialien, Geesthacht D, Germany
e-mail: regine.willumeit@hzg.de

N. Ahmad Agha
e-mail: nzha.agha586@gmail.com

B. Luthringer
e-mail: berengere.luthringer@hzg.de


D. Orlov et al. (eds.), *Magnesium Technology 2018*, The Minerals, Metals & Materials Series, https://doi.org/10.1007/978-3-319-72332-7_63

## Introduction

For almost 100 years, Mg has been tried to be used as biomaterial [1]. As a metal having mechanical properties similar to the ones of bone, Mg can be used as load bearing implant. Furthermore, Mg has the specialty to degrade under physiological conditions, making it a very good candidate for applications were no permanent support is needed. Still, almost as long it is known that using Mg as biodegradable implant can be quite problematic. Fast or uncontrolled corrosion is associated with strong hydrogen and ion release and severe pH changes, which can lead to a fast loss of mechanical stability and undesirable biological reactions. Despite several attempts to bring Mg into the clinics for orthopaedic, trauma or internal medicine: when other materials like steel, titanium and even polymers entered the field, Mg implants lost the interest of researchers.

A renascence of Mg as biomaterial was observed when the technology advanced in other fields such as e.g. automotive industry and studies showed that from the patient's point of view, degradable implants would clearly be preferred [2]. In addition, aging populations, active adults exhibiting risky sporting activities, a predicted rise in osteoporosis-related fractures and special and better care for juvenile patients will sustain a need for degrading load bearing implants for especially orthopaedic intervention. In the case of cardiovascular interventions, where Mg also can play an important role, the total worldwide volume is forecast to be over 18.73 million in the year 2022 [3].

To date, there are several Mg-based products on the market. The first company, which launched in 2013 a compression screw for Halux valgus indications, was Syntellix GmbH, Germany. Their product portfolio encompasses now screws and pins of different sizes. With the Magmaris stent, the company Biotronik, USA/Cortronik GmbH, Germany, placed in early 2016 a second class of products on the market. Asia as well is very active in implant development. In China the approval process was started recently for a Mg screw with a purity of 99.99 wt% Mg. The clinical trials with these screws treating osteonecrosis of the femoral head (ONFH) showed very good results, however, the application is not a load-bearing one [4]. In parallel several first-in-man studies were performed in Korea [5]. For this study, a Mg-5 wt%Ca-1 wt%Zn alloy was shaped into screws for the surgical fixation of distal radius fracture. In this case, load is applied to the implant.

Still, even though the reported human results are impressive and show the enormous potential of Mg as a new class of implant material, first papers also indicate that in some rare cases the expected outcome of the procedure does not meet the expectations [6]. The reason for this different behaviour changing from patient to patient and from application to application is completely unclear and reveals the still strong need for fundamental understanding since these processes are highly complex in a living system and sufficient data describing the degradation in vivo is missing [7].

To establish biodegradable implants as a standard therapy in medicine is inherently connected to detailed knowledge about the interaction of degradation and its products with the specific biological environment and about the dependence of degradation on materials characteristics like alloy composition, impurities and microstructure. Even more, in the case of mechanically loaded Mg-implants the effect of stress on degradation and on cell reactions has to be elaborated. Complete understanding of all these effects and interactions leads to the possibility of future computational modelling and by this designing the ideal material and geometry for a specific implant at a specific location in the body in advance. It is obvious that to reach this aim a holistic approach of physical, chemical, biological and medical ground is necessary.

## Obstacle 1—The Material

Worldwide, numerous alloy compositions have been tried to obtain an optimal material in terms of mechanical properties, degradation behaviour, and biocompatibility [8, 9]. Furthermore, probably even more processes were studied. By combining different cast technologies with varying thermomechanical treatments and post processing steps (e.g., extrusion, rolling, equal channel angular pressing (ECAP), and equal channel angular rolling (ECAR)), including powder metallurgy approaches, a plethora of materials was produced. The focus in all attempts was on obtaining homogeneous and well-defined microstructures, which should lead to application-specific mechanical properties, very good formability and machinability, and homogenous degradation coupled with excellent biocompatibility of the released ions. That finally materials can be obtained to be suitable not only for animal experiments but also human applications was mentioned already above. The general trend goes to alloys which contain Ca and Zn because these elements are found in the body and seem not to be critical upon degradation [10]. Still, other alloying elements should not be neglected as the composition of the Syntellix material shows: They use an alloying system based on the composition MgYREZr (Mg > 90%). The implants are made by a powder metallurgical route, leading to a homogeneous microstructure with a grain sizes of less than 10 μm [11].

## Obstacle 2—The Surface

Control of the surface properties by surface design is another very powerful tool to either generate application-specific degradation rates or introduce functionality such as drug release [12, 13]. Current approaches include degradable natural or synthetic polymers [14, 15], inorganic coatings [16, 17], functionalisation via plasma-electrolytic or micro-arc oxidation (PEO, MAO) [18], and superhydrophobic coating [19], to mention only a few of them. The functionalisation usually leads to the suppression of the initially high degradation rates and a general improvement of cell adhesion. For the superhydrophobic coating, even the stimulation of the mineralisation activity of osteoblasts was detected. Combining these coatings with biocompatible corrosion inhibitors or drug release systems offers a great potential for tissue engineering.

However, in most cases it is of interest to work directly with the "bare" materials. Then the question comes up how to condition the surface in a proper way that impurities but also disturbances in the microstructure (which might accelerate the initial degradation rate) are removed. Several surface conditioning procedures as part of the production process are being considered. This should diminish surface alterations due to machining, which mutates the original material through possible potential contamination with foreign material and deforms the surface layer, e.g., by twinning [20]. Therefore extensive studies were performed on the effect of post-treatments such as cleaning with different solvents and applying etching agents were performed in a systematic and detailed manner [21, 22].

## Obstacle 3—Degradation and Cytotoxicity Tests

Mg degradation in a biological corrosive environment (in vitro and in vivo) is a very complex process [23]. Consequently, the predictability of in vivo performance based on in vitro observation is still very limited [7]. Despite the fact that ISO or ASTM guidelines exist and consider the specialties imposed by Mg degradation, many different approaches are chosen for in vitro degradation tests. The degradation solutions used range from simple salt solution, simulated body fluid, and cell culture media towards blood. Other conditions, such as temperature (room temperature vs 37 °C), flow, $CO_2$ or $O_2$ concentrations ranging from ambient conditions to more physiological conditions, influence the degradation behaviour and thus increase again the variety of methods and experimental conditions which can be selected. Different techniques exist to measure the degradation rate: hydrogen release, weight loss, and electrochemical approaches which make a comparison between materials very difficult.

If we move forward to the assessment of cytotoxicity and biocompatibility—which is always connected to the degradation of the material—most groups use single cell culture models to study the effect of the material on cells. These simplified cell culture tests mainly use cell lines (e.g. MG63, Saos2, U2OS and others) which are well documented cells. They are easy to handle and show relatively fast growth rates. The reproducibility of the results is higher than for patient derived cells. On the other hand they might react more sensitive to the material [24] and, since they are usually cancer-derived immortal cells, they show not the same gene expression profile as healthy patient derived cells [25].

Therefore, for the understanding which metabolic changes might be introduced by the degrading material primary (= patient derived) cells are necessary. These cells show a high patient specific and unsystematic variability in their response to the material and have to be extracted from tissue, which is delivered freshly from the operation theatre. Most studies for orthopaedic applications have been performed with osteoblasts, the cells that form new bone. But also stem cells of different origin serve as a good model to study the interaction of primary cells which the material. Still, to fully understand the complex interplay between materials, cells and tissue, co-cultures models of at least two different cells types are needed because the communication between cells can severely influence the acceptance of the implant material. One of the pioneers in this field is Prof. James Kirkpatrick who detected the expression of tissue factors during the co-culture of endothelial cells and monocytes [26, 27]. For the understanding of the influence of a degrading Mg-based implant, it is quite important to study the cross talk between cells. It was shown that while in mono-culture osteoclasts are very sensitive to Mg concentrations which exceed 6 mM, they can survive 25 mM when incubated together with osteoblasts [28].

The implementation of 3D scaffolds for more reliable cell data is nowadays out of question [29, 30]. However, in most cases these approaches are not easy to follow when Mg-based implant materials are studied. 3D cell culture on a metal implant is always challenging because the access to the cells when ingrown into the material is rather limited. Therefore, if these promising technologies are followed, only extracts of the material can so far be studied.

## Obstacle 4–The in Vivo Assessment

In vivo studies are essential to understand the behaviour of the implant in the highly complex environment of a living organism. Many different materials were tested worldwide, and—as long as the degradation rate was in tolerable range—new bone formation observed [31]. However, the underlying mechanisms are not easy to understand [32, 33]. Furthermore, the choice of the animal model will determine a specific outcome and the implant geometry as well as the implantation site have to be considered [34, 35]. In addition the temporal and spatial resolution of data obtained throughout the degradation of the material during the life time of the animals (not the mention in the human case) is very limited. By multimodal imaging and blood samples a certain idea about the behaviour of the material can be obtained. However, due to the high radiation dosage high resolution μCT can not be applied on a daily (or even hourly) basis. MRI or MR spectroscopy which can give insight into some physiological changes in the tissue around the Mg implant still needs improvement to de-convolute artefacts from the images [36]. Other e.g. ultrasound [37], or optical coherence tomography [38] have a too low spatial resolution to study the processes in the vicinity of the implant surface. Only when the access is given to explants the full power of analytical techniques can be applied [39]. But then, the data is valid only for a limited number of time points. Therefore, a massive improvement of in vivo analytical techniques—not only for the pre-clinical studies but especially also for the clinical application—is needed.

## Scope of This Work

Despite the well documented fact that the degradation of Mg sensitively depends on environmental conditions [40], the influence of cells on the degradation is usually neglected. The study of whether and how cells can actively influence the degradation of the material by changing their metabolism is of interest. Therefore, here the influence of human primary osteoblast and fibroblast on the material degradation was analysed via the examination of the elemental composition of the degradation layer. The osteoblast cell model was chosen for its active role in bone remodelling and its ability to express extracellular matrix as well as mineralisation capacity. Fibroblast, on the contrary, was selected as control as they are part of connective tissue and should not be able to modify the degradation layer.

## Materials and Methods

Pure magnesium (Mg; 99.95%; Magnesium Elektron, Manchester, UK) was prepared by permanent mould gravity casting (Helmholtz Zentrum Geesthacht, Geesthacht, Germany). The ingots were then T4 treated, extruded into rods of 1.2 cm diameter (Strangpreßzentrum Berlin, Berlin, Germany, for extrusion parameter see [41]), and finally machined to obtain a diameter of 1 cm. Discs of 1.5 mm thickness were then cut and gamma sterilised with a total dosage of 29 kGy (BBF Sterilisationsservice GmbH, Kernen- Rommelshausen, Germany).

Dulbecco's Modified Eagle's Medium (DMEM, Life technologies, Darmstadt, Germany) with 10% Foetal Bovine Serum (FBS; PAA laboratories, Linz, Austria) and 1% Penicillin/Streptomycin was used as a cell culture and immersion media. The degradation rate of the Mg samples without cells was determined after 14 days of immersion to be 0.327 ± 0.002 mm/year [41]. It slightly decreased when incubated for 14 days with osteoblasts (0.325 ± 0.004 mm/year) while it was significantly reduced when fibroblasts were grown for two weeks on the surface (0.044 ± 0.021 mm/year).

To study the direct influence of cells on the degradation, two types of cells were selected here, osteoblast (mediates mineralisation) and fibroblast (no mineralisation). Human primary osteoblasts (OB) were obtained from patients undergoing hip arthroplasty (Hospital Eilbek, Hamburg, Germany; written consent documented) with the approval of the local ethics committee, but without passing patient data (Ethikkommission der Ärztekammer Hamburg, Germany). Cells of at least 5 donors (up to the third passage) were used. Primary human osteoblasts were isolated according to the protocol of Gallagher [42]. L929 mouse fibroblasts cell line (Sigma-Aldrich, Taufkirchen, Germany) was used as model for connective tissue.

Samples to be cultured with cells or without cells were treated similarly. Mg samples were pre-incubated for 24 h under cell culture condition (37 °C, 20% $O_2$, 5% $CO_2$, 95% rH) to avoid the initial high degradation rate and to allow the formation of a degradation layer suitable for further cell culture experiment. Afterwards, 100 000 human primary osteoblasts or L929 were seeded on the surface and allowed to adhere for 30 min. Then, medium was added to continue the immersion. Cell culture was pursuit for up to 14 days with medium change every 3–4 days.

To analyse the active effect of cells on the degradation layer, the samples were first critically dried (i.e., after a glutaraldehyde fixation step, an alcoholic dehydration row was performed, then samples were critical point dried in 2-propanol (Sigma-Aldrich Chemie GmbH, Munich, Germany) to preserve cell morphology by a Leica EM CPD300 (Leica Mikrosysteme Vertrieb GmbH, Wetzlar, Germany). The degradation interface was then investigated using a Scanning Electron Microscope (Auriga microscope, Zeiss, Oberkochen, Germany) equipped with Energy Dispersive X-Ray Spectroscopy and a Focused Ion Beam (SEM/EDX/FIB). Samples were then critically point dry (see [43] for more detailed procedure) for further analysis. Cross sections were performed on samples with cells (directly beneath the cells) and on sample without cells (negative control samples). Five EDX line scans were performed to define the vertical element (carbon (C), nitrogen (N), oxygen (O), sodium (Na), Mg, phosphorus (P), sulphur (S) and calcium (Ca)) distribution of the degradation layer on the cross-sections.

## Results

The EDX line measurements of the degradation interface are shown in Fig. 1. The degradation layer measured in the presence of osteoblast is thicker than the one observed for fibroblast (22.23 ± 0.38 and 9.59 ± 0.42 µm, respectively). The thickness of the degradation layer without cell was about 7 µm, thickness which is also closer to the one of fibroblast than to the osteoblasts one. Furthermore, it can be appreciated on the SEM pictures that the osteoblasts exhibited a rather flat surface and are well attached to the material surface. Contrary, the fibroblasts are appearing more rounded and it can be also observed that the cell layer has the tendency to detach during the preparation of the specimen. At the beginning of the line-scan (i.e., close to the A region), slight increases in S, N, and C can be noticed, especially for osteoblasts, which can coincide with the physical presence of cells. Except for Mg and Ca, an enrichment of measured elements can be seen for osteoblasts—especially for phosphorous deeper in the degradation layer (i.e., closer to B).

## Discussion

The degradation layer thickness on Mg-based materials indicated that especially osteoblasts significantly influenced the degradation interface without altering the degradation rate while fibroblasts can form an effective protective layer without changing the thickness or composition of the degradation layer. The element quantification of sections showed relatively higher S, P, N, O and C concentrations in the degradation layer for sample with osteoblasts as compared to specimen which were incubated without cells or with fibroblasts. As it was shown in [41] this can mean that osteoblasts have the potential to actively modify the degradation layer with respect to the formation of Ca-P- rich

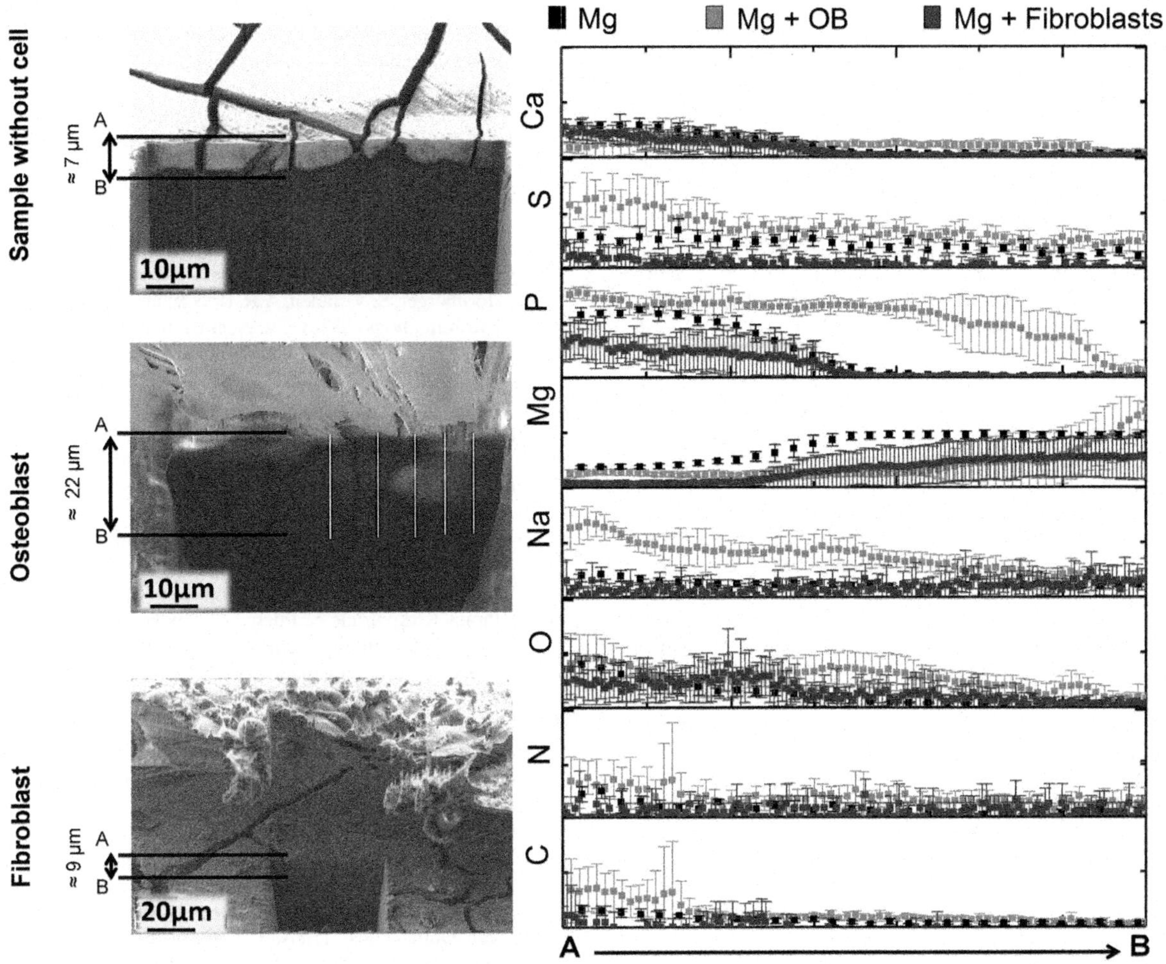


**Fig. 1** Comparison of the corrosion layer underneath of osteoblasts (OB) and fibroblasts. SEM images of the processed cross sections by FIB of samples without cell, with human primary osteoblast, and with mouse fibroblasts L929. The chart present the EDX line scan measurements on the processed cross sections. [AB] values represent the cross section thickness, which were further scaled (abscissa) for the chart representation. EDX line scans start at point **a** and end at point **b**. SEM pictures of the osteoblast & without cell degradation layers were previously published here [41]

phases as they can be found in bone. In addition, the degradation layer thickness as well as the composition was comparable for Mg incubated without cell and for Mg specimen which were incubated with fibroblast. This is not unexpected as fibroblasts will not actively influence the degradation layer to form anorganic "bone-like" matrix as osteoblasts do. However, when properly attached to the surface they passively form "protective" cell layer to reduce the degradation.

## Conclusions

It was shown that bone forming osteoblasts grown on Mg-based materials changed the degradation the composition of the forming corrosion layer next to and underneath the cell [41] while fibroblast do not metabolise the material. This is the first direct proof that bone forming cells can actively modify the implant degradation and synthesize Ca-P—like inorganic compounds.

**Acknowledgements** This research was carried out with funding from the People Program (Marie Curie Actions) of the European Union's Seventh Framework Program FP7/2007-2013/under REA grant agreement n° 289163.We wish to thank Dr. Daniel Laipple for the support with the FIB sample preparation and analysis.

## References

1. F. Witte, The history of biodegradable magnesium implants: a review, Acta Biomater 6(5) (2010) 1680–92.
2. R. Mittal, J. Morley, H. Dinopoulos, E.G. Drakoulakis, E. Vermani, P.V. Giannoudis, Use of bio-resorbable implants for stabilisation of distal radius fractures: the United Kingdom patients' perspective, Injury 36(2) 333–338.

3. pdriscoll, Cardiovascular procedure volume growth (interventional and surgical). https://blog.mediligencecom/2017/02/13/cardiovascular-procedure-volume-growth-interventional-and-surgical/. 2017.
4. D. Zhao, S. Huang, F. Lu, B. Wang, L. Yang, L. Qin, K. Yang, Y. Li, W. Li, W. Wang, S. Tian, X. Zhang, W. Gao, Z. Wang, Y. Zhang, X. Xie, J. Wang, J. Li, Vascularized bone grafting fixed by biodegradable magnesium screw for treating osteonecrosis of the femoral head, Biomaterials 81 (2016) 84–92.
5. J.-W. Lee, H.-S. Han, K.-J. Han, J. Park, H. Jeon, M.-R. Ok, H.-K. Seok, J.-P. Ahn, K.E. Lee, D.-H. Lee, S.-J. Yang, S.-Y. Cho, P.-R. Cha, H. Kwon, T.-H. Nam, J.H.L. Han, H.-J. Rho, K.-S. Lee, Y.-C. Kim, D. Mantovani, Long-term clinical study and multiscale analysis of in vivo biodegradation mechanism of Mg alloy, Proceedings of the National Academy of Sciences 113(3) (2016) 716–721.
6. C. Plaass, S. Ettinger, L. Sonnow, S. Koenneker, Y. Noll, A. Weizbauer, J. Reifenrath, L. Claassen, K. Daniilidis, C. Stukenborg-Colsman, H. Windhagen, Early results using a biodegradable magnesium screw for modified chevron osteotomies, Journal of Orthopaedic Research 34(12) (2016) 2207–2214.
7. A.H. Martinez Sanchez, B.J. Luthringer, F. Feyerabend, R. Willumeit, Mg and Mg alloys: how comparable are in vitro and in vivo corrosion rates? A review, Acta Biomater 13 (2015) 16–31.
8. Y. Chen, Z. Xu, C. Smith, J. Sankar, Recent advances on the development of magnesium alloys for biodegradable implants, Acta Biomaterialia 10(11) (2014) 4561–4573.
9. D. Zhao, F. Witte, F. Lu, J. Wang, J. Li, L. Qin, Current status on clinical applications of magnesium-based orthopaedic implants: A review from clinical translational perspective, Biomaterials 112 (2017) 287–302.
10. F. Amerstorfer, S.F. Fischerauer, L. Fischer, J. Eichler, J. Draxler, A. Zitek, M. Meischel, E. Martinelli, T. Kraus, S. Hann, S.E. Stanzl-Tschegg, P.J. Uggowitzer, J.F. Löffler, A.M. Weinberg, T. Prohaska, Long-term in vivo degradation behavior and near-implant distribution of resorbed elements for magnesium alloys WZ21 and ZX50, Acta Biomaterialia 42 (2016) 440–450.
11. Syntellix, MAGNEZIX®—a milestone in material research. http://www.syntellix.de/en/products/technology.html. 2017.
12. M.S. Uddin, H. Colin, M. Peter, Surface treatments for controlling corrosion rate of biodegradable Mg and Mg-based alloy implants, Science and Technology of Advanced Materials 16(5) (2015) 053501.
13. P. Tian, X. Liu, Surface modification of biodegradable magnesium and its alloys for biomedical applications, Regenerative Biomaterials 2(2) (2015) 135–151.
14. S. Agarwal, J. Curtin, B. Duffy, S. Jaiswal, Biodegradable magnesium alloys for orthopaedic applications: A review on corrosion, biocompatibility and surface modifications, Materials Science and Engineering: C 68 (2016) 948–963.
15. S. Heise, S. Virtanen, A.R. Boccaccini, Tackling Mg alloy corrosion by natural polymer coatings-A review, J Biomed Mater Res A 104(10) (2016) 2628–41.
16. P. Tian, X. Liu, Surface modification of biodegradable magnesium and its alloys for biomedical applications, Regen Biomater 2(2) (2015) 135–51.
17. X. Li, X. Liu, S. Wu, K.W.K. Yeung, Y. Zheng, P.K. Chu, Design of magnesium alloys with controllable degradation for biomedical implants: From bulk to surface, Acta Biomaterialia 45 (2016) 2–30.
18. T.S.N. Sankara Narayanan, I.S. Park, M.H. Lee, Strategies to improve the corrosion resistance of microarc oxidation (MAO) coated magnesium alloys for degradable implants: Prospects and challenges, Progress in Materials Science 60 (2014) 1–71.
19. Y. Zhang, F. Feyerabend, S. Tang, J. Hu, X. Lu, C. Blawert, T. Lin, A study of degradation resistance and cytocompatibility of super-hydrophobic coating on magnesium, Materials Science and Engineering: C 78 (2017) 405–412.
20. B. Denkena, A. Lucas, Biocompatible Magnesium Alloys as Absorbable Implant Materials—Adjusted Surface and Subsurface Properties by Machining Processes, CIRP Annals—Manufacturing Technology 56(1) (2007) 113–116.
21. T.S.N.S. Narayanan, I.S. Park, M.H. Lee, eds. *Surface Modification of Magnesium and its Alloys for Biomedical Applications*. 2015, Woodhead Publishing Series in Biomaterials.
22. H. Hornberger, S. Virtanen, A.R. Boccaccini, Biomedical coatings on magnesium alloys—a review, Acta Biomater 8(7) (2012) 2442–55.
23. J. Reifenrath, A.K. Marten, N. Angrisani, R. Eifler, A. Weizbauer, In vitro and in vivo corrosion of the novel magnesium alloy Mg–La–Nd–Zr: influence of the measurement technique and in vivo implant location, Biomed Mater 10(4) (2015) 045021.
24. J. Fischer, D. Pröfrock, N. Hort, R. Willumeit, F. Feyerabend, Improved cytotoxicity testing of magnesium materials, Materials Science and Engineering: B 176(11) (2011) 830–834.
25. A. Burmester, B. Luthringer, R. Willumeit, F. Feyerabend, Comparison of the reaction of bone-derived cells to enhanced MgCl2-salt concentrations, Biomatter 4(1) (2014) e967616.
26. C. James Kirkpatrick, S. Fuchs, M. Iris Hermanns, K. Peters, R.E. Unger, Cell culture models of higher complexity in tissue engineering and regenerative medicine, Biomaterials 28(34) (2007) 5193–5198.
27. J.C. Lewis, N.L. Jones, M.I. Hermanns, O. Röhrig, C.L. Klein, C. J. Kirkpatrick, Tissue Factor Expression during Coculture of Endothelial Cells and Monocytes, Experimental and Molecular Pathology 62(3) (1995) 207–218.
28. L. Wu, F. Feyerabend, A.F. Schilling, R. Willumeit-Romer, B. J. Luthringer, Effects of extracellular magnesium extract on the proliferation and differentiation of human osteoblasts and osteoclasts in coculture, Acta Biomater 27 (2015) 294–304.
29. V. van Duinen, S.J. Trietsch, J. Joore, P. Vulto, T. Hankemeier, Microfluidic 3D cell culture: from tools to tissue models, Current Opinion in Biotechnology 35 (2015) 118–126.
30. R.E. Unger, S. Halstenberg, A. Sartoris, C.J. Kirkpatrick, Human endothelial and osteoblast co-cultures on 3D biomaterials, Methods Mol Biol 695 (2011) 229–41.
31. A. Chaya, S. Yoshizawa, K. Verdelis, N. Myers, B.J. Costello, D.-T. Chou, S. Pal, S. Maiti, P.N. Kumta, C. Sfeir, In vivo study of magnesium plate and screw degradation and bone fracture healing, Acta Biomaterialia 18 (2015) 262–269.
32. S. Ding, J. Zhang, Y. Tian, B. Huang, Y. Yuan, C. Liu, Magnesium modification up-regulates the bioactivity of bone morphogenetic protein-2 upon calcium phosphate cement via enhanced BMP receptor recognition and Smad signaling pathway, Colloids and Surfaces B: Biointerfaces 145 (2016) 140–151.
33. Y. Zhang, J. Xu, Y.C. Ruan, M.K. Yu, M. O'Laughlin, H. Wise, D. Chen, L. Tian, D. Shi, J. Wang, S. Chen, J.Q. Feng, D.H.K. Chow, X. Xie, L. Zheng, L. Huang, S. Huang, K. Leung, N. Lu, L. Zhao, H. Li, D. Zhao, X. Guo, K. Chan, F. Witte, H.C. Chan, Y. Zheng, L. Qin, Implant-derived magnesium induces local neuronal production of CGRP to improve bone-fracture healing in rats, Nat Med 22(10) (2016) 1160–1169.
34. A.I. Pearce, R.G. Richards, S. Milz, E. Schneider, S.G. Pearce, Animal models for implant biomaterial research in bone: a review, Eur Cell Mater 13 (2007) 1–10.
35. M.J. Byrom, P.G. Bannon, G.H. White, M.K.C. Ng, Animal models for the assessment of novel vascular conduits, Journal of Vascular Surgery 52(1) (2010) 176–195.

36. L. Sonnow, S. Könneker, P.M. Vogt, F. Wacker, C. von Falck, Biodegradable magnesium Herbert screw—image quality and artifacts with radiography, CT and MRI, BMC Medical Imaging 17(1) (2017) 16.
37. R. Waksman, R. Erbel, C. Di Mario, J. Bartunek, B. de Bruyne, F. R. Eberli, P. Erne, M. Haude, M. Horrigan, C. Ilsley, D. Böse, H. Bonnier, J. Koolen, T.F. Lüscher, N.J. Weissman, Early- and Long-Term Intravascular Ultrasound and Angiographic Findings After Bioabsorbable Magnesium Stent Implantation in Human Coronary Arteries, JACC: Cardiovascular Interventions 2(4) (2009) 312–320.
38. M. Haude, H. Ince, A. Abizaid, R. Toelg, P.A. Lemos, C. von Birgelen, E.H. Christiansen, W. Wijns, F.-J. Neumann, C. Kaiser, E. Eeckhout, S.T. Lim, J. Escaned, H.M. Garcia-Garcia, R. Waksman, Safety and performance of the second-generation drug-eluting absorbable metal scaffold in patients with de-novo coronary artery lesions (BIOSOLVE-II): 6 month results of a prospective, multicentre, non-randomised, first-in-man trial, The Lancet 387(10013) 31–39.
39. S. Galli, J.U. Hammel, J. Herzen, I. Damm, R. Jimbo, F. Beckmann, A. Wennerberg, R. Willumeit-Römer, Evaluation of the degradation behavior of resorbable metal implants for in vivo osteosynthesis by synchrotron radiation based x-ray tomography and histology, Developments in X-Ray Tomography X SPIE, 2016.
40. N.T. Kirkland, J. Lespagnol, N. Birbilis, M.P. Staiger, A survey of bio-corrosion rates of magnesium alloys, Corrosion Science 52(2) (2010) 287–291.
41. N. Ahmad Agha, R. Willumeit-Romer, D. Laipple, B. Luthringer, F. Feyerabend, The Degradation Interface of Magnesium Based Alloys in Direct Contact with Human Primary Osteoblast Cells, PloS one 11(6) (2016) e0157874.
42. J.A. Gallagher, Human osteoblast culture, Methods Mol Med 80 (2003) 3–18.
43. B.J.C. Luthringer, F. Ali, H. Akaichi, F. Feyerabend, T. Ebel, R. Willumeit, Production, characterisation, and cytocompatibility of porous titanium-based particulate scaffolds, J Mater Sci: Mater Med 24 (2013) 2337–2358.

# Study on Mg–Si–Sr Ternary Alloys for Biomedical Applications

Omer Van der Biest, Andrea Gil-Santos, Norbert Hort, Rainer Schmid-Fetzer, and Nele Moelans

**Abstract**

The Mg rich corner of the ternary Mg–Si–Sr alloy system is experimentally and thermodynamically investigated in this work. Thermodynamic simulation of the ternary phase diagram required modelling of the ternary phases that occur in these alloys, namely the ternary intermetallics MgSiSr and $MgSi_2Sr$ for which only scarce information is available in literature. The Mg-rich side of the Mg–Si–Sr phase diagram is constructed based on descriptions of the binary phase diagrams Mg–Si, Mg–Ca and Ca–Si from literature and assuming complete solubility (i.e. a line compound) between the ternary phase MgSiSr and the binary phase $Sr_2Si$. It is also assumed that $MgSi_2Sr$ is a stoichiometric compound. A good agreement is found between the constructed phase diagram and the experimental microstructural results for a range of cast and heat treated alloys.

**Keywords**

Magnesium alloys • Intermetallics • Thermodynamic modelling • Phase diagrams • Microstructure Scanning electron microscopy • SEM

O. Van der Biest (✉) · A. Gil-Santos · N. Moelans
Department of Materials Engineering, KU Leuven, Kasteelpark Arenberg, 44, 3001 Louvain, Belgium
e-mail: omer.vanderbiest@mtm.kuleuven.be

N. Hort
Magnesium Innovation Centre (MagIC), Helmholtz-Zentrum Geesthacht, Max-Planck-Strasse 1, 21502 Geesthacht, Germany

R. Schmid-Fetzer
Technical University of Clausthal, Institute of Metallurgy, Robert-Koch-Str. 42, 38678 Clausthal-Zellerfeld, Germany

## Introduction

The potential of Mg rich alloys as resorbable materials for biomedical implants has been widely investigated in the last decade [1–3]. Alloying magnesium appears to be necessary in order to improve its mechanical properties, degradation behavior or processing aspects. In our approach to select suitable alloying elements we have chosen only those elements which are already in the human body in order to minimize the occurrence of short term or long term toxic effects. Also we have considered favorable influences the alloying elements may have on the functioning of the implant in particular for orthopedic and related applications. For the present work we selected silicon and strontium as alloying elements. Si is known to contribute to the growth and development of bone and connective tissue [4] and Sr addition enhances new bone formation and increases bone quality around the implant [5, 6]. From the biological point of view, it is favorable that the implant material induces new bone formation, showing a good biocompatibility between the implant and the tissue [7, 8]. This biocompatibility has been observed for Sr or Si in previous in vivo studies [4, 9–11].

When developing a new alloy system like is the case here for Mg–Si–Sr, the knowledge of the phase diagram as function of temperature is a prerequisite to select suitable compositions. Unfortunately, the published information on the Mg–Si–Sr ternary system is scarce and it does not include any thermodynamic data. The available information collected from literature refers to the existence of some ternary compounds, related sometimes with respect to their physical properties [12]. Five different compounds have been recognized within the Mg–Si–Sr system, they are: MgSiSr [13], $MgSiSr_2$, $Mg_2Si_{10}Sr_{11}$ [14], $Mg_2Si_{20}Sr_{13}$ [15] and $Mg_{16.6}Si_{13}Sr_{6.3}$ [16]. However, only the crystal structure of these ternary compounds is known. Further information on the ternary Mg–Si–Sr system is absent.


D. Orlov et al. (eds.), *Magnesium Technology 2018*, The Minerals, Metals & Materials Series, https://doi.org/10.1007/978-3-319-72332-7_64

In this work, the phase equilibria of the two ternary phases MgSiSr and $MgSi_2Sr$ located in the Mg rich corner region are experimentally identified. Then a thermodynamic description of the 2 ternary phases MgSiSr and $MgSi_2Sr$ is proposed and used to generate a ternary database describing the Mg-rich side of the Mg–Si–Sr system. The calculated phase equilibria obtained with this database are compared with the experimental observations.

## Thermodynamic Calculations

In the works of Eisenman et al. and Currao et al. [13, 14], it is found that the ternary compounds MgSiSr and $MgSi_2Sr$ have both an orthorhombic crystal structure with Pnma-62 space group. Considering the chemical similarities between Ca and Sr we can assume that the Mg–Si–Sr phase diagram has similarities with the Mg–Si–Ca. Also Si and Sn belong to the same group IV in Mendeleev's table, and so we can expect similarities between the Mg–Si–Sr and Mg–Sn–Sr phase diagrams. Also the structure of the ternary MgSiSr phase is similar to the crystal structure of the MgSiCa compound reported in the Mg–Si–Ca system [17] and to the MgSnSr compound reported in the Mg–Sn–Sr system [18].

Since the MgSiSr and the $SiSr_2$ phase have the same crystal structure as well, it is assumed in this work that there is complete solubility between both compounds i.e. they form a line compound in the phase diagram. This assumption was also based on the observation that MgCaSi and $SiCa_2$ exhibit the same crystal structure and show full solubility in the Mg–Ca–Si and the same for MgSrSn and $SnSr_2$ in the Mg–Sr–Sn system. For the other ternary compound, $MgSi_2Sr$, a stoichiometric description was selected similar to that used for $MgSi_2Ca$ in reference [17]. The thermodynamic parameters chosen for the two ternary phases are given in Table 1.

With these descriptions for the ternary phases and using the most recently published thermodynamic descriptions of the three binaries Mg–Si, Mg–Sr and Si–Sr, a thermodynamic database for Mg–Si–Sr was composed and the Mg-rich side of the ternary diagram at 300 °C was calculated (Fig. 1). The binary descriptions with their thermodynamic parameters and models used in this work are taken from the assessments published by respectively, Schick et al., Zhong et al. and Garay et al. [19–21]. The Gibbs energies of pure Mg, Si and Sr are taken from the Scientific Group Thermodata Europe (SGTE) pure element database [22]. The programs Thermo-Calc 4.0 and Pandat were used to generate the database and for the phase diagram calculation.

Figure 1 shows the isothermal section calculated for 300 °C. Three regions can be discerned with different phase constitution. They are defined by the at.%Si:at.%Sr ratio. A combination of Mg, M17 and M121 is expected for the Sr rich alloys (at.%Si:at.%Sr <1). For the case of the Si rich alloys (at.%Si:at.%Sr >2), Mg, M2 and M111 are predicted. For the alloys with a at.%Si:at.%Sr ratio closer to 1.5 (1< at.%Si:at.%Sr <2), the expected phases are the two ternary compounds M111 and M121 together with the Mg matrix.

In order to facilitate the analysis the intermetallic names were shortened to a few characters, as listed in Table 2.

## Materials and Methods

The Mg–Si–Sr alloys were processed by permanent mold gravity casting at Magnesium Innovation Centre MagIC (Geesthacht, Germany). Cylinders of 1.8 cm diameter and 20 cm length were cast. Raw materials consisted of Mg ingots (99.9%), Si granulates (<6 mm, 99.9% purity from Chempur) and Sr chips (3–12 mm, 99.9% purity from Alfa Aesar packaged in high purity mineral oil). An electrical resistance furnace under a protective atmosphere of Ar and 2 vol.% $SF_6$ was used to melt the alloys. Initially, solid Mg was molten by heating it up to 700 °C. When molten, the pre-heated Si granulates at 400 °C were added and the mixture was stirred with a steel paddle every 5 min to improve the homogeneity. The mixture was maintained at 700 °C during 20 min in order to favor the solution of Si. Then preheated Sr chips at 400 °C were added after which an exothermic reaction occurred. The furnace is then switched off and the temperature of the melt increased to the range of 750–850 °C. The mixture was stirred and oxide products formed on top of the melt were removed. When the temperature of the melt was cooled down to 700 °C the molten alloy was poured into a steel die preheated at 400 °C during 45 min. Hexagonal BN was used as a mold release agent. After two minutes, the die was opened and the alloys were cooled down to room temperature (RT) by air cooling.

Because by casting the alloys do not reach their equilibrium phase content, cast alloys were heat treated under

**Table 1** Thermodynamic parameters used in this work, together with the descriptions of the binaries of Schick et al., Zhong et al. and Garay et al. [19–21], to describe the Mg-rich side of the Mg–Si–Sr ternary system

| Phase | Parameters | Reference |
|---|---|---|
| $Sr_2Si$ (Mg, Sr)(Si)(Sr) | $G^{0,Sr_2S_i}_{Mg,Sr:Si:Sr} = 2G^{0,fcc}_{Sr} + G^{0,diamond}_{Si} - 122329.65 + 5.48T$ | Garay et al. [21] |
| MgSiSr (Mg, Sr)(Si)(Sr) | $G^{0,MgSrSi}_{Mg,Sr:Si:Sr} = G^{0,hcp}_{Mg} + G^{0,fcc}_{Sr} + G^{0,diamond}_{Si} - 180000 + 50T$ | This work |
| $MgSi_2Sr$ (Mg)(Si)$_2$(Sr) | $G^{0,MgSi_2Sr}_{Mg:Sr:Si} = G^{0,hcp}_{Mg} + G^{0,fcc}_{Sr} + 2G^{0,diamond}_{Si} - 233827$ | This work |

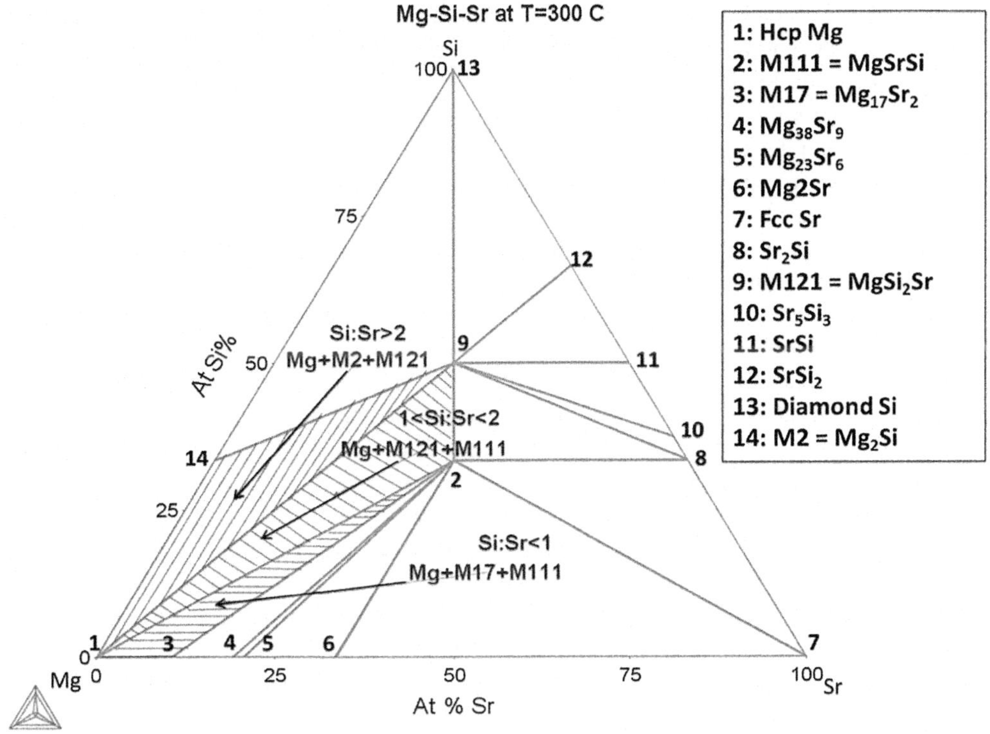


**Fig. 1** Calculated isothermal section at 300 °C from the Mg–Si–Sr system including the identification of the regions in the Mg rich corner, using the database developed in this work

**Table 2** Correspondence between intermetallic compound names and their identification used in this work

| Intermetallic compound | Name |
|---|---|
| $Mg_{17}Sr_2$ | M17 |
| $Mg_2Si$ | M2 |
| MgSiSr | M111 |
| $MgSi_2Sr$ | M121 |

argon atmosphere, followed by quenching in water at room temperature. A first heat treatment (HT1) was done during 48 h at 500 °C and then a second one (HT2) during longer time (240 h) and higher temperature (550 °C) was also applied to fully equilibrate the samples. Previous works involving Mg-rich intermetallic stability have been consulted to decide on the length of the anneal to attain equilibrium. For instance Mg–Sn alloys have been described as fully homogenized after 96 h at 320 °C [23]; complete dissolution of $Mg_2Si$ has been observed after 45 min at 525 °C [24] and good agreement with equilibrium calculations have been found for Mg–Ca–Zn alloys after a total of 147 h at different temperatures [25]. The longest HT of 240 h was found in the work of Nie and Muddle [26] were the thermal stability of compounds in Mg–Ca alloys was studied after heat treatments from 8 up to 240 h.

Disk shaped specimens for microstructural characterization were cut with 2 mm height and 18 mm diameter from the center of the 20 cm long cast specimens. These disk samples were then ground with silicon carbide emery papers by wet grinding from 800 down to 2500 grit size and later polished with water-free colloidal silica suspension (0.2 µm particle size). The etching agent used was picric acid based, containing 10 mL of acetic acid, 4.2 g of picric acid, 20 mL $H_2O$ and 50 mL ethanol and the etching time was between 5 and 10 s. In the cases where grains were not revealed after etching and under polarized light, the use of electron back scattered diffraction (EBSD) was needed in order to differentiate the grains. The average grain size was measured by Image-Pro Plus 6.0 image software (Media Cybernetics Inc., Rockville, USA) from optical microscope images of the etched surfaces. The linear intercept method according to ASTM: E112-13 (2013) was used.

The microstructure was analyzed using an optical microscope and a Philips XL30 FEI scanning electron microscope (SEM) equipped with an EDAX TSL energy dispersive X-ray spectroscopy (EDS) detector. Backscatter electron (BSE) mode was used for the phase identification. Due to the small size of the intermetallic phases an excess of Mg coming from the bulk is measured when identifying the compound's stoichiometry. The inflated Mg values are caused by the interaction volume reached during EDS analysis. For this reason the ratio at.%Si:at.%Sr was considered as more reliable than the Mg percentage values. For the very small intermetallic particles (smaller than 1 μm) field emission electron probe micro-analyzer (EPMA), FEG EPMA JXA-5830F was used for detailed chemical analysis. The size of the intermetallic phases was measured by image analysis of the micrographs using Image-J software and the average values are presented in this paper. Phase identification was further assisted by XRD analysis which was carried out on a XRD Seifert 3003-TT. Cu-Kα radiation was used at 40 kV and 40 mA, from 20° to 80° with a 0.02 step size and a measurement time of 2 s per step. The phase identification was performed by comparison with the JCPDS cards included in the Pearson database. The used cards for Mg, $Mg_2Si$, $Mg_{17}Sr_2$, MgSiSr and $MgSi_2Sr$ are identified with the numbers 00-035-0821, 00-035-0773, 03-065-3649, 01-089-1920 and 01-087-0897. Due to the low amount of compounds compared to the Mg matrix some of the peaks are covered by the Mg ones which are always the dominating peaks. Nevertheless, characteristic peaks for the different intermetallic phases could be identified in the XRD patterns.

Image analysis was carried out on SEM micrographs to measure the volume fraction of each phase using Image-J software (version 1.47v, Wayne Rasband). The different phases were identified based on their diverse shapes and atomic number contrast provided by the back scatter electron (BSE) mode. Several micrographs were used for every sample to estimate the average volume fraction of each phase.

## Results

### As Cast Condition

The compositions of the cast alloys (A to L) are given in Table 3 together with the phase identification in the as-cast condition. From the experimental data (SEM-EDS, EPMA and XRD) four sets of co-existing phases have been found within the studied composition range: Mg+M17+M111, Mg +M17+M111+M121, Mg+M121 and Mg+M17+M121. The identification of the four different intermetallic phases based on SEM-EDS analysis is presented in Fig. 2. Compositions obtained from SEM-EDX and EPMA are summarized in Table 4. The microstructure and the related XRD pattern of one representative sample from each set are discussed further. The rest of the samples in each set show similar structures as those presented. As an exception, samples C and F belonging to the same set are presented because they show different phases after equilibration.

The interaction volume measured during EDS analysis has been estimated by using electron flight simulator software (version 3.1E http://www.2spi.com) based on Monte Carlo simulations as a function of the energy applied and the selected elements. Based on these simulations the reliability of the phase composition was reduced in cases where the size of the intermetallic compounds was small. This happens for cases where the MgSiSr and $MgSi_2Sr$ particle size is smaller than 1 μm, which can be deduced from the calculations presented in Fig. 3. In the simulations a particle thickness of 1 μm is represented by the white band; the Mg bulk underneath is represented in grey. The electron trajectory is shown dark and represents the real volume from

**Table 3** List of the studied alloys with their phase identification in the as-cast condition. Bold phases indicate the primary phase as expected from the calculated phase diagram

| Name | Composition at.% | | Phase identification as-cast | Si/Sr ratio |
|---|---|---|---|---|
| | Si | Sr | | |
| A | 0.68 | 4.80 | Mg+M17+**M111** | 0.14 |
| B | 0.63 | 3.09 | Mg+M17+M111+**M121** | 0.20 |
| C | 1.20 | 3.75 | | 0.32 |
| D | 1.41 | 2.65 | | 0.53 |
| E | 2.50 | 1.62 | | 1.54 |
| F | 0.33 | 0.14 | | 2.36[a] |
| G | 0.12 | 0.02 | **Mg**+M121 | 6.00[a] |
| H | 0.41 | 0.07 | Mg+M2+**M121** | 5.86[a] |
| I | 5.31 | 0.96 | | 5.53 |
| J | 4.60 | 0.82 | | 5.61 |
| K | 3.40 | 0.75 | | 4.53 |
| L | 3.17 | 0.90 | | 3.52 |

[a]In alloys F, G and H the concentration of Sr is below the reliable detection limit. The amount of Sr is expected to be higher than the measured value. The presented values are based on extrapolated points from the standard ICP curves since interpolation was not possible for amounts smaller than 0.5 at.% Sr

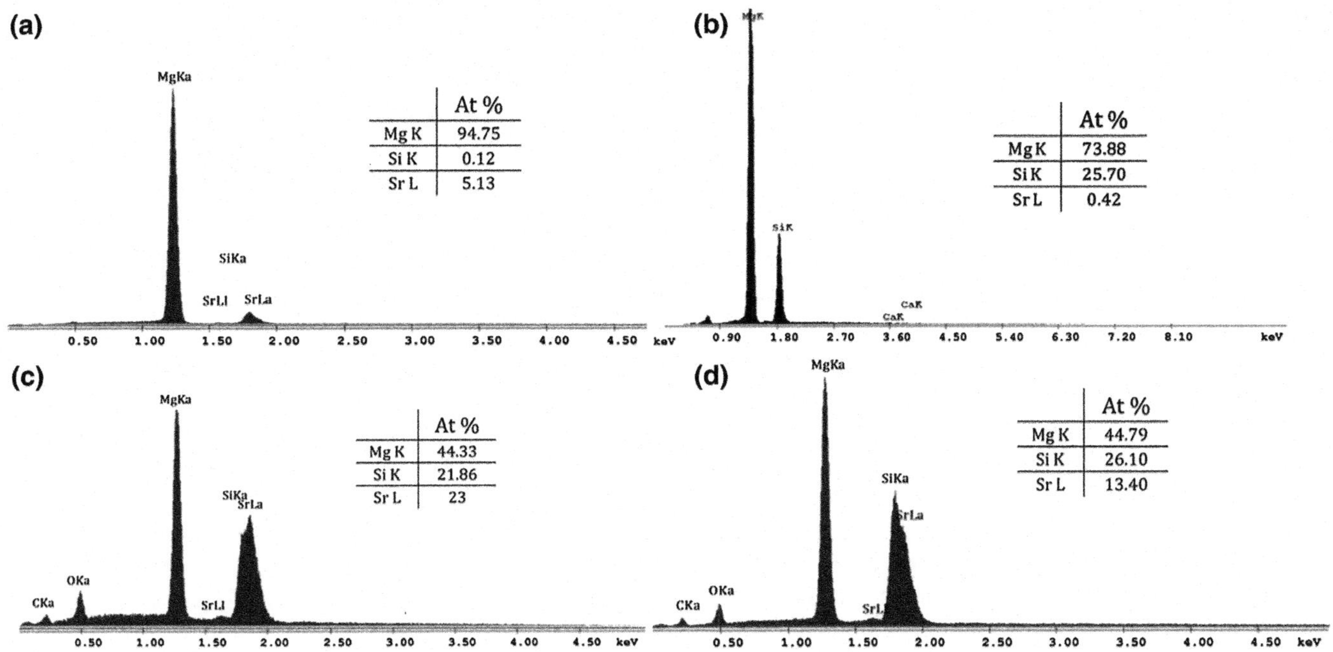


**Fig. 2** EDX spectra of the intermetallic phases in SEM at 10 keV **a** $Mg_{17}Sr_2$ **b** $Mg_2Si$ **c** MgSiSr and **d** $MgSi_2Sr$

**Table 4** Representative identification of the intermetallic phases in the different conditions based on the element quantification SEM-EDS analysis. One sample from each different phase combination is presented

| Alloy | Phase | Condition | Elements (at.%) | | | Ratio (at.%) Si:Sr | Alloy | Phase | Condition | Elements (at.%) | | | Ratio (at.%) Si:Sr |
|---|---|---|---|---|---|---|---|---|---|---|---|---|---|
| | | | Mg | Si | Sr | | | | | Mg | Si | Sr | |
| A | M111 | As Cast | 54.7 | 21.13 | 23.02 | 0.92 | F | M111 | As Cast | 74.76 | 13.82 | 14.38 | 0.96 |
| | | HT1 | 40.92 | 33.13 | 25.83 | 1.28 | | | HT1 | 46.85 | 29.95 | 23.23 | 1.29 |
| | | HT2 | 46.68 | 26.07 | 25.23 | 1.03 | | | HT2 | 36.80* | 32.59* | 29.1* | 1.12* |
| | M17 | As Cast | 94.13 | 0.01 | 5.87 | – | | M121 | As Cast | 59.68 | 26.54 | 13.77 | 1.93 |
| | | HT1 | 91.32 | 0.70 | 7.98 | – | | | HT1 | 56.33 | 13.01 | 12.62 | 1.03 |
| | | HT2 | 90.05 | 0.56 | 9.39 | – | | | HT2 | 68.90 | 21.46 | 9.64 | 2.23 |
| C | M111 | As Cast | 51.56 | 23.17 | 25.27 | 0.92 | | M17 | As Cast | 88.97 | 0.15 | 9.92 | – |
| | | HT1 | 47.75 | 28.23 | 24.02 | 1.18 | G | M121 | As Cast | 87.38 | 8.24 | 4.38 | 1.88 |
| | | HT2 | 36.65 | 32.60 | 27.63 | 1.18 | | | HT1 | 52.63* | 34.5* | 12.79* | 2.70* |
| | M121 | As Cast | 52.81 | 31.47 | 14.99 | 2.10 | | | HT2 | 93.38 | 4.71 | 1.90 | 2.48 |
| | | HT1 | 55.50 | 27.95 | 16.61 | 1.68 | K | M121 | As Cast | 57.68 | 30.25 | 12.07 | 2.51 |
| | M17 | As Cast | 91.56 | 0.01 | 8.44 | – | | | HT1 | 55.89 | 31.04 | 13.05 | 2.38 |
| | | HT1 | 91.51 | 0.40 | 8.09 | – | | | HT2 | 53.79 | 32.42 | 13.79 | 2.35 |
| | | HT2 | 91.57 | 0.67 | 7.76 | – | | M2 | As Cast | 67.08 | 30.79 | 2.13 | – |
| Mg matrix | | As Cast | 99.75 | 0.15 | 0.09 | – | | | HT1 | 80.36 | 18.84 | 0.80 | – |
| | | HT1 | 99.69 | 0.17 | 0.13 | – | | | HT2 | 69.83 | 29.73 | 0.44 | – |

*These values were obtained based on EPMA-WDS measurements

which the element analysis is coming. Due to the fact that the interaction volume is twice the thickness in case of $MgSi_2Sr$ particles (2.00 μm vs. 1.00 μm from Fig. 3b) and even bigger (2.50 μm vs. 1.00 μm from Fig. 3a) in case of MgSiSr an inflated Mg value is indeed expected for the measurements. In these cases the phase size may be much smaller than the interaction volume, making the results of the element analysis coming partially from the bulk instead of from the selected particle. Then, more accurate extra measurements with EPMA-WDS were done for detailed

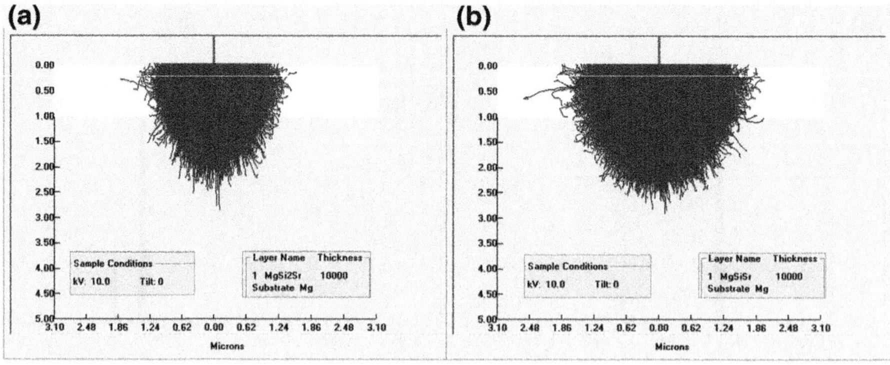

**Fig. 3** Interaction volume simulations of 1 μm thickness particles of **a** $MgSi_2Sr$ (M121) and **b** MgSiSr (M111) in Mg matrix under 10 keV accelerating voltage

chemical analysis of the particles. The results are also presented in Table 4 indicated with a star. The Si and Sr content have been also measured within the Mg matrix and average values are presented in as-cast and after HT1 conditions.

Figure 4 shows phase identification in alloy A. Both the SEM data and the XRD pattern (not shown) clearly show the presence of the two intermetallic phases M17 and M111. The binary eutectic with M17 as continuous phase and the dark grey Mg as distributed phase, is dominating the microstructure in the as-cast condition. The ternary M111 appears in a needle like shape with a brighter white tone. Size and thickness of the M111 particles vary from 5 to 20 μm lengths and from 0.2 to 1 μm thickness. M111 appears surrounded by the eutectic indicating its earlier formation.

In alloy C, the ternary phases M111 and M121 and the binary phase M17 were identified from both SEM micrographs (Fig. 5) and XRD. The binary eutectic (M17 + Mg) is present on the whole surface with an aligned disposition and pure Mg is visible as dark islands in between. The ternary M111 appears with the same morphology as it does in alloy A (Fig. 5a), in a white needle like shape. The size of the M111 particles ranges from 5 to 20 μm lengths with about 2 μm thickness. M121 particles have a darker grey tone and they show angular edges. M121 presents in general a coarser hexagonal polygonal shape with sizes from 10 to 50 μm in both length and thickness and usually angled holes in the central part. Also the M121 particles are enclosed by the eutectic indicating its earlier formation during solidification for this alloy.

For alloy F, the same phases as in alloy C are present but in a much lower amount and with a smaller size, as can be observed in the micrographs in Fig. 6a, b. In this case M121 is the primary phase as it took the Si and Sr from the liquid when growing, leaving regions rich in Mg behind which surround the polygonal ternary particles. M111 particles have a smaller size than in previous alloys due to the smaller amount of alloying elements in alloy F compared with A and C.

The microstructure of alloy G shows that Mg is the first phase that solidified with a dendritic morphology and that an intermetallic phase crystallized in the interdendritic spaces. This phase has been identified as the ternary M121 phase (Fig. 7). It is present as small dot shaped particles with a bright white color. Possibly these particles are the product of a eutectic reaction at the end of the solidification. In this case the identification of the intermetallic phase in the XRD pattern (Fig. 8b) is not evident since the amount of the M121 is minor compared with the pure Mg. The characteristic peaks are overlapping with the Mg ones or vanish in the

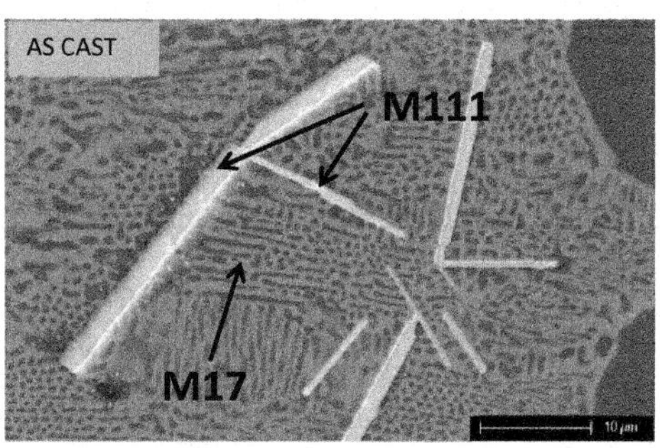


**Fig. 4** Identification of the intermetallic phases M17 and M111 in alloy A (Mg 0.68Si 4.80Sr)

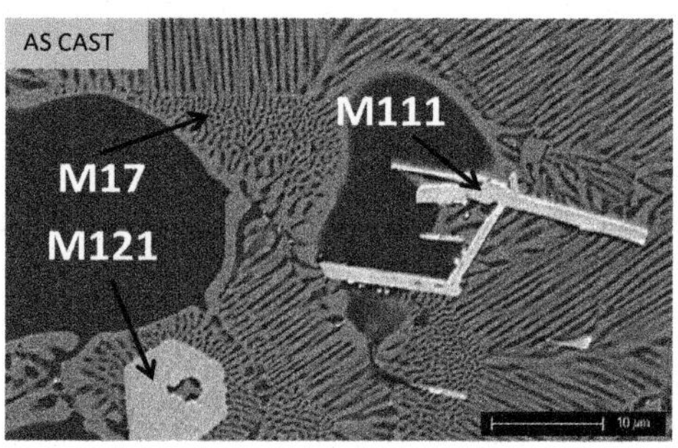


**Fig. 5** Identification of the secondary phases M17, M111 and M121 in alloy C (Mg 1.41Si 2.65Sr) in a BSE SEM micrograph

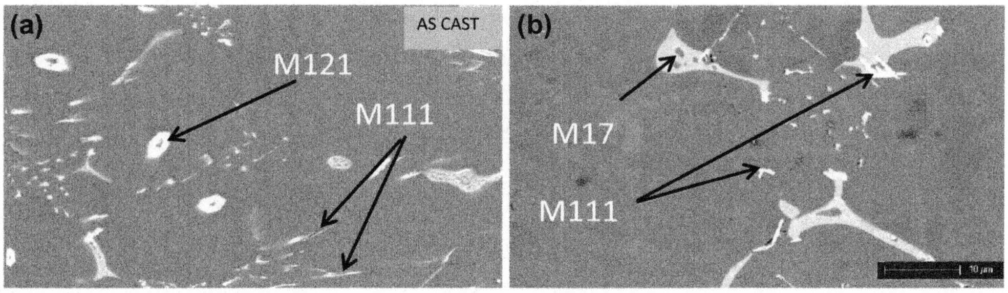


**Fig. 6** Identification of the intermetallic phases in alloy F (Mg 0.33Si 0.14Sr) **a** BSE SEM micrograph with M111 and M121, **b** BSE SEM micrograph with M17 and M111

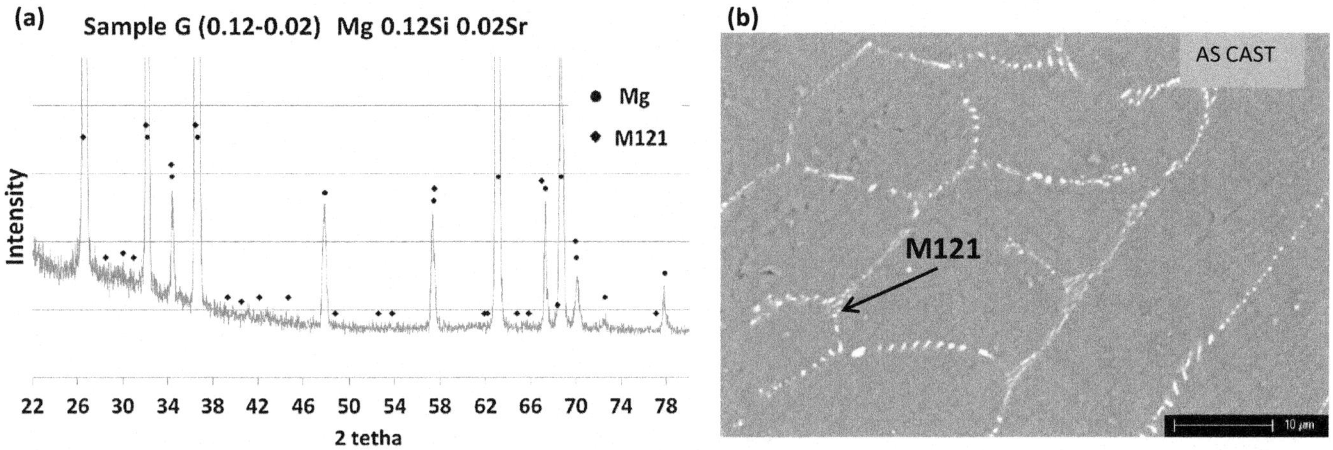


**Fig. 7** Identification of the ternary phase M121 in sample G (Mg 0.12Si 0.02Sr) from **a** its XRD pattern and **b** BSE SEM micrograph

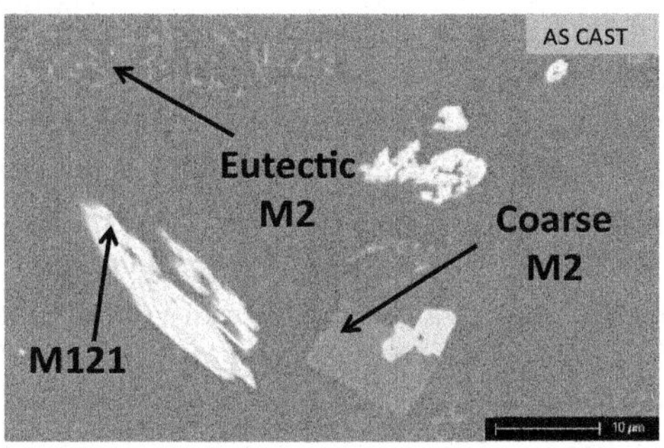


**Fig. 8** Identification of the secondary phases M2 and M121 in alloy K (Mg 3.40Si 0.75Sr) in a BSE SEM micrograph

background although one characteristic peak of the phase is prominent.

In alloy K the binary phase formed is M2 having a light grey color in the SEM micrograph shown in Fig. 8. Two typical morphologies of this phase that have been previously described, among others by Hu et al. [27], are observed. Those morphologies are the coarse polygonal particles and the rod-like shaped eutectic ones. The latter is smaller in size and more homogeneously distributed in the microstructure. M121 also exists in this alloy appearing bright white in the micrographs due to the Sr content. The shape of this ternary varies from close to polygonal to irregular. Both intermetallics have been identified in the XRD pattern as well.

## After Heat Treatment

A first heat treatment (HT1, 48 h at 500 °C) was carried out in an attempt to equilibrate the phase composition. In order to make sure that an equilibrium phase content was reached alloys were also heat treated at 550 °C for 240 h (HT2). As argued in Sect. 2 these temperatures and times are similar to those used to equilibrate other magnesium based alloys systems. The results are summarized in Table 5.

The identification of the phases with SEM-EDS analysis and from the XRD patterns for the alloys after HT1 and HT2 is discussed and compared with the as-cast condition in the following.

In alloy A, the same phases M17 and M111 as in the as-cast condition are identified from the micrographs (Figs. 9b, c) and XRD peaks (Fig. 9a). The binary eutectic has been drastically reduced and presents a more globular form instead of the lamellar one present immediately after

**Table 5** List of the studied alloys with their phase identification in the equilibrated condition

| Name | Composition at.% | | Phase identification in equilibrium conditions | Si/Sr ratio |
|---|---|---|---|---|
| | Si | Sr | | |
| A | 0.68 | 4.80 | Mg+M17+M111 | 0.14 |
| B | 0.63 | 3.09 | | 0.20 |
| C | 1.20 | 3.75 | | 0.32 |
| D | 1.41 | 2.65 | | 0.53 |
| E | 2.50 | 1.62 | Mg+M111+M121 | 1.54 |
| F | 0.33 | 0.14 | | 2.36[a] |
| G | 0.12 | 0.02 | Mg+M121 | 6.00[a] |
| H | 0.41 | 0.07 | Mg+M2+M121 | 5.86[a] |
| I | 5.31 | 0.96 | | 5.53 |
| J | 4.60 | 0.82 | | 5.61 |
| K | 3.40 | 0.75 | | 4.53 |
| L | 3.17 | 0.90 | | 3.52 |

[a]In alloys F, G and H the concentration of Sr is below the reliable detection limit. The amount of Sr is expected to be higher than the measured value. The presented values are based on extrapolated points from the standard ICP curves since interpolation was not possible for amounts smaller than 0.5 at.% Sr

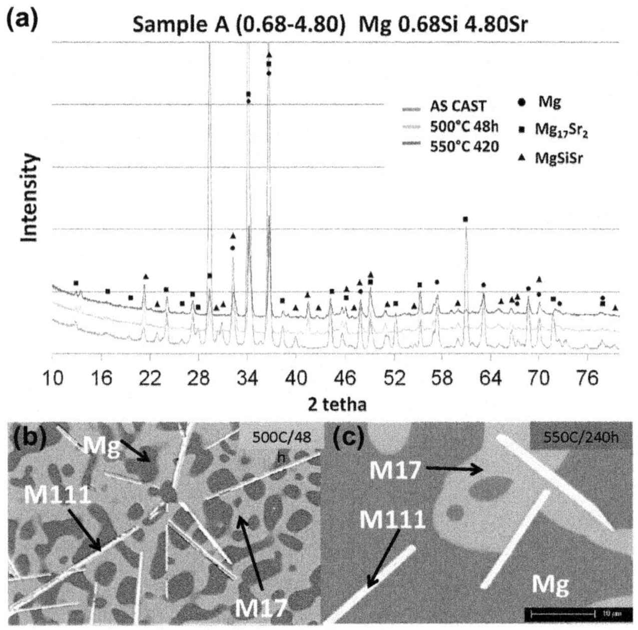


**Fig. 9** **a** XRD pattern of alloy A with the identified phases in as-cast condition and after the two HT. **b** BSE SEM micrograph with labelled phases of the alloy A after HT1 and (**c**) after HT2

casting before the equilibration (Fig. 4). It shows how the system strives for equilibrium not only with respect to phase content but also with respect to the interfacial energies. The intensity of the M17 peaks is gradually reduced when increasing temperature and duration of HT, indicating a reduction in the amount of this phase from the as-cast to the equilibrium condition. The amount of pure Mg has increased and the amount of ternary M111 appears to be only slightly reduced. The M111 particles are more homogeneous in size, with an average of 15 μm length and 1 μm thickness. Finally, the ternary M111 phase together with the M17 are the stable ones for this alloy composition.

In alloy C, three intermetallic phases together with the Mg matrix were identified in the as-cast condition (Fig. 10). When HT1 is applied, the M121 phase seems to lose the well-defined polygonal shape and starts dissolving (Fig. 10a). In fact after HT2, the M121 does no longer exist (Fig. 10b). The amount of binary M17 is reduced but the phase is still present after the HT2. Similar for the M111 particles, their size (max 10 μm length and 1 μm thickness) and number are reduced from to the as-cast condition but they are still present after HT2. From these observations and assuming the alloy has reached its equilibrium state after 240 h annealing at 550 °C, we can conclude that, in equilibrium state the alloy contains apart from the Mg phase, the M17 and M111 phases.

Alloy F presents a similar situation to alloy C, where from the four phases observed in the as-cast condition (Fig. 6) only three remain after HT1 and HT2. The binary M17 phase has dissolved and disappeared after the HT1 and HT2 (Fig. 11a, b). The two ternary phases and pure Mg form the equilibrium phase combination.

In alloy G the amount of ternary M121 phase is reduced with increasing time and temperature of the heat treatment as can be observed when comparing the Fig. 12a after HT1 and Fig. 12b after HT2 with Fig. 7b of the as-cast alloy. Particle size and shape are similar as in the as-cast condition. Also from the XRD pattern the intermetallic particles are still identified with the Mg matrix after equilibration.

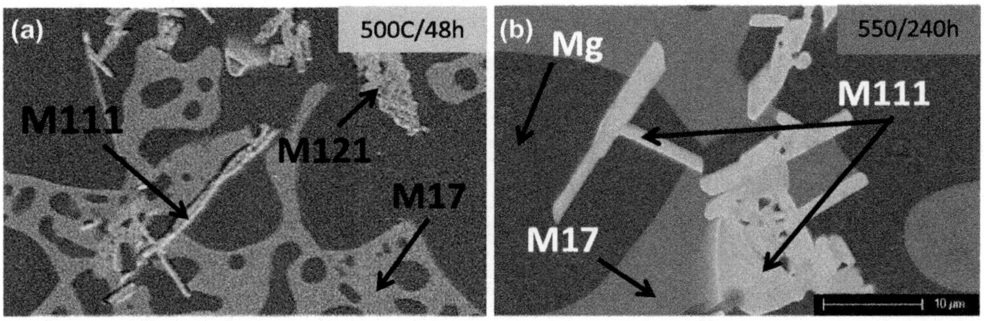


**Fig. 10** BSE SEM micrograph with labelled phases of the alloy C **a** after HT1 and **b** after HT2

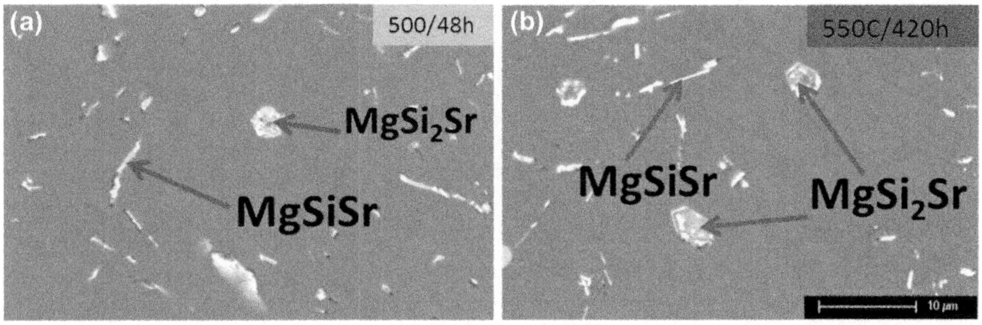


**Fig. 11** BSE SEM micrograph with labelled phases of alloy F **a** after HT1 and **b** after HT2. XRD pattern of alloy F showing the phase identification after the two HT

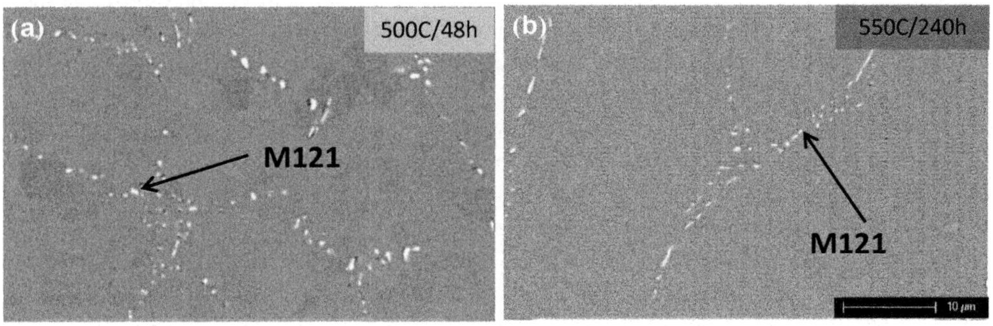


**Fig. 12** BSE SEM micrograph with labelled phases of the sample G **a** after HT1 and **b** after HT2

In alloy K, the same phases are present before and after the heat treatments. The microstructures can be compared in Figs. 8 and 13. The number and size of M2 phase particles, appearing in light grey in Fig. 13a, b, is reduced, but the phase remains present in the equilibrium microstructure. When analyzing the XRD patterns (Fig. 13c), the reduction in peak intensity is linked with the M2 phase, which has partially dissolved after HT. The ternary M121 did not undergo significant modifications in size or shape from the as-cast condition after equilibration, but some gradation in color is observed inside the particles (Fig. 13a, b). A brighter region can be distinguished in the central part of the M121 particle while the external part is darker. The external darker area is richer in Si showing a difference in composition from the center to the edges. When comparing the microstructure with the as-cast condition (Fig. 8) brighter and darker areas are also observed there. Hence some segregation within this phase took place either during solidification when the crystals formed from the liquid or afterwards during cooling. The heat treatments have further accentuated this segregation.

## Discussion

When comparing Tables 3 and 5, a difference in the phases present is observed between the equilibrium and as-cast alloys. This difference indicates that the as-cast alloys are not always in equilibrium. This can be concluded as well from the coexistence of four phases observed in the as-cast alloys B to F. More than three different phases in a ternary system indicates a non-equilibrium structure as it violates Gibbs phase rule.

After the heat treatments a change in microstructure has been observed for some of the studied alloys (B–F). The equilibrated microstructures show that the M111 and M17

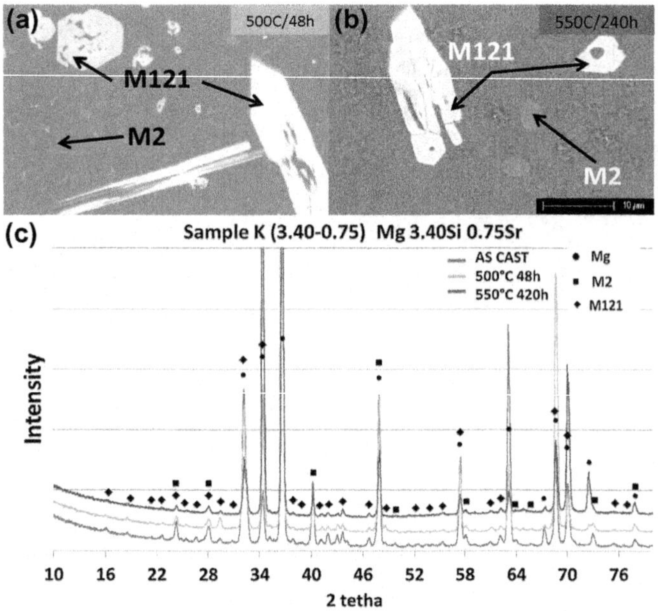


**Fig. 13** BSE SEM micrograph with labelled phases of the alloy K **a** after HT1 and **b** after HT2. **c** XRD pattern of alloy K showing the phase identification in the as-cast condition and after the two HT

phases present in the as-cast condition were sometimes unstable, and therefore dissolved during equilibration. In alloys B, C and D the disappearing phase was the ternary M111 while in alloys E and F it was the binary M17. Finally, in all equilibrated alloys a maximum of 3 phases coexists together, in agreement with Gibbs phase rule applied to a system with 3 elements. For all samples, the phase combination present at equilibrium is presented in Table 5.

The sets of coexisting phases in equilibrium can be related with the experimental ratio at.%Si:at.%Sr and the results are presented in Table 5. We can distinguish four different regions, namely Mg+M17+M111 when the ratio at.%Si:at.%Sr is below 1, Mg+M111+M121 when the ratio is between 1 and 2.5, Mg+M2+M121 when the ratio is between 3 and 6 and Mg+M121 for the highest Si content.

The experimentally obtained phase regions after equilibrium (from Table 5) can be compared with the predicted ones in the ternary diagram at 550 °C (Fig. 14) in order to validate the generated database. This phase diagram was calculated using the thermodynamic description established in this work. Figure 14 shows a zoom of the Mg-rich corner of the ternary diagram up to 6 at.% in both Sr and Si at 550 °C and it shows the compositions of the studied alloys as well.

From Fig. 14 the relation established between the combination of coexisting equilibrium phases and the ratio at.% Si:at.%Sr obtained from the theoretical phase diagram calculation agrees well with the experimentally observed phases in equilibrium. Both the calculations shown in Fig. 14

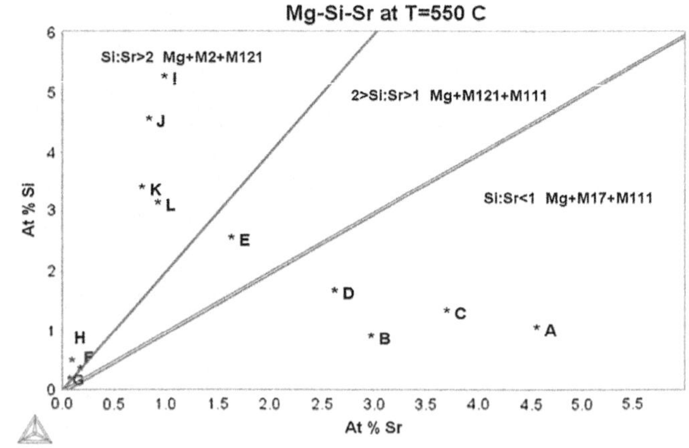


**Fig. 14** Distribution of the studied alloys (A–L) in the Mg rich corner at 550 °C of the ternary phase diagram within the different phase regions. The phase boundaries were calculated using the database developed in this work

and the heat treatment are done at 550 °C. It has to be taken into account that the at.%Si:at.%Sr ratio in alloys F, G and H were somewhat increased with respect to their EDS analyses due to the low Sr amount detected in the ICP measurements.

It can be read from the diagram that the region with Mg+M17+M111 is present when the ratio at.%Si:at.%Sr is below 1; the region with Mg+M111+M121 is present when the at.%Si:at.%Sr ratio is between 1 and 2; the one with Mg+M121 when the ratio is exactly 2 and the region with Mg+M2+M121 when the ratio is higher than 2. This good agreement between the theoretically predicted and the experimentally obtained phase regions indicates the soundness of the suggested phase diagram and the underlying assumptions. This can be considered as a good starting point for the full assessment of the ternary Mg–Si–Sr system.

## Conclusions

The presence of the two ternary compounds MgSiSr and $MgSi_2Sr$ in the Mg–Si–Sr system has been identified by microstructure characterization. These two ternary compounds together with the binaries $Mg_{17}Sr_2$ and $Mg_2Si$ make the four phases that cover the possible existing intermetallic combinations in the Mg rich corner of the Mg–Si–Sr alloy system.

The stability of the four intermetallic phases MgSiSr, $MgSi_2Sr$, $Mg_{17}Sr_2$ and $Mg_2Si$ has been investigated based on their evolution with temperature. For this purpose sample's microstructures have been analyzed including phase identification and morphology study in as-cast condition and at 500 and 550 °C after the sample's exposure to heat treatments at these temperatures. A comparison between as-cast condition and equilibrium condition was possible after the heat treatments.

For the $MgSi_2Sr$ ternary compound it is found to be present in most of the as-cast specimens. It is only not present when the alloys atomic Si/Sr ratio is lower than 0.14. Nevertheless, it turned out to be a metastable phase for compositions with an atomic Si/Sr ratio below 1; the $MgSi_2Sr$ ternary phase disappears during annealing for these compositions. In the rest of the cases, where the alloy's atomic Si/Sr ratio is higher than 1, the $MgSi_2Sr$ is an equilibrium phase and after some reduction in size and volume fraction it remains stable at both 500 and 550 °C.

A similar behavior than for the $MgSi_2Sr$ is observed for the binary $Mg_{17}Sr_2$ phase. It shows a tendency to disappear or decrease its amount after equilibration compared to the as-cast condition. The $Mg_{17}Sr_2$ phase is present in all the as-cast alloys with atomic Si/Sr ratio lower than 2.5 but it dissolves during heat treatment except for the alloy compositions with atomic Si/Sr ratio lower than 1. For these compositions where the eutectic $Mg_{17}Sr_2$ is not disappearing but stays in equilibrium it shows a change from lamellar structure to a globular one indicating how the system strives for equilibrium not only with respect to phase content but also with respect to the interfacial energies.

The ternary MgSiSr and the binary $Mg_2Si$ which are present in the as-cast specimens with an atomic Si/Sr ratio below 2 and above 2 respectively, they both remain in the equilibrium microstructures. They do not present modifications with heat treatments, showing their stability not only at 300 °C but also at high temperatures of 500 and 550 °C.

The expected combination of phases in equilibrium can be predetermined based on the at.%Si:at.%Sr ratio. A good agreement was found when comparing the experimental phases in the equilibrated alloys with the calculated isothermal section of the phase diagram at 550 °C (considered as equilibrium). It can be concluded that the hypothesis that both ternary systems Mg–Ca–Si and Mg–Sn–Sr could be used as a guide for the thermodynamic description of the two ternary compounds in the Mg–Si–Sr system is realistic. It is assumed here that the $MgSi_2Sr$ can be described as a stoichiometric compound. Also it is assumed that there is be a complete solubility between MgSiSr and $SiSr_2$. These hypotheses could be investigated in more detail.

The biodegradation performance of alloys within the Mg–Sr–Si system has recently been reported [28].

**Acknowledgements** The research was supported by the PEOPLE program (Marie Sklodowska-Curie Action) of the EU FP7 Program FP7/ 2007-2013/ under REA grant agreement N289163.

## References

1. Witte, F., A. Eliezer, and S. Cohen. *The history, challenges and the future of biodegradable metal implants*. in *Advanced Materials Research*. 2010. Trans Tech Publ.
2. Kirkland, N., et al., *A survey of bio-corrosion rates of magnesium alloys*. Corrosion Science, 2010. **52**(2): p. 287–291.
3. Xin, Y., T. Hu, and P. Chu, *In vitro studies of biomedical magnesium alloys in a simulated physiological environment: a review*. Acta biomaterialia, 2011. **7**(4): p. 1452–1459.
4. Zhang, E., et al., *Microstructure, mechanical properties and bio-corrosion properties of Mg–Si(–Ca, Zn) alloy for biomedical application*. Acta biomaterialia, 2010. **6**(5): p. 1756–1762.
5. Tie, D., et al., *An in vivo study on the metabolism and osteogenic activity of bioabsorbable Mg–1Sr alloy*. Acta biomaterialia, 2016. **29**: p. 455–467.
6. Suganthi, R.V., et al., *Fibrous growth of strontium substituted hydroxyapatite and its drug release*. Materials Science and Engineering: C, 2011. **31**(3): p. 593–599.
7. P.A. Revell, E.D., X.S. Zhang, P. Evans, C.R. Howlett, *The Effect of Magnesium Ions on Bone Bonding to Hydroxyapatite Coating on Titanium Alloy Implants*. Key Eng. Mater., 2004. **254**(2): p. 447–450.
8. Zreiqat, H., et al., *Mechanisms of magnesium-stimulated adhesion of osteoblastic cells to commonly used orthopedic implants*. Journal of Biomedical Materials Research, 2002. **62**(2): p. 175–184.
9. Tie, D., et al., *Antibacterial biodegradable Mg-Ag alloys*. European cells & materials, 2012. **25**: p. 284–98; discussion 298.
10. Bornapour, M., et al., *Biocompatibility and biodegradability of Mg–Sr alloys: The formation of Sr-substituted hydroxyapatite*. Acta biomaterialia, 2013. **9**(2): p. 5319–5330.
11. Dahl, S.G., et al., *Incorporation and distribution of strontium in bone*. Bone, 2001. **28**(4): p. 446–453.
12. Bennett, J.W., et al., *Orthorhombic A B C Semiconductors as Antiferroelectrics*. Physical Review Letters, 2013. **110**(1): p. 017603.
13. B. Eisenmann, H.S., A. Weiss, *Z. Anorg. Allg. Chem.* (1972). **391**: p. 241–264.
14. Currao, A., Curda, J., Nesper, R., Z. Anorg. Allg. Chem., 1996. **622** p. 85–94.
15. Nesper, R., Currao, A., Angew. Chem., Int. Ed., 37, 1998. **841**.
16. Nesper, R., Wengert, S., Zuercher, F., Currao, A., Chem. Eur. J., 1999. **5**: p. 3382.
17. Gröbner, J., I. Chumak, and R. Schmid-Fetzer, *Experimental study of ternary Ca–Mg–Si phase equilibria and thermodynamic assessment of Ca–Si and Ca–Mg–Si systems*. Intermetallics, 2003. **11**(10): p. 1065–1074.
18. Zhou, B.-C., S.-L. Shang, and Z.-K. Liu, *First-principles calculations and thermodynamic modeling of the Sn–Sr and Mg–Sn–Sr systems*. Calphad, 2014. **46**: p. 237–248.
19. Schick, M., et al., *Combined ab initio, experimental, and CALPHAD approach for an improved thermodynamic evaluation of the Mg–Si system*. Calphad, 2012. **37**: p. 77–86.
20. Zhong, Y., et al., *Thermodynamics modeling of the Mg–Sr and Ca–Mg–Sr systems*. Journal of Alloys and Compounds, 2006. **421** (1–2): p. 172–178.
21. Garay, A., et al., *Thermodynamic modeling of the Si–Sr system*. Calphad, 2009. **33**(3): p. 550–556.
22. Dinsdale, A.T., *SGTE data for pure elements*. Calphad, 1991. **15** (4): p. 317–425.

23. Avraham, S., A. Katsman, and M. Bamberger, *Optimization of composition and heat treatment design of Mg–Sn–Zn alloys via the CALPHAD method.* Journal of Materials Science, 2011. **46**(21): p. 6941–6951.
24. Chaudhury, S. and D. Apelian, *Fluidized bed heat treatment of cast Al-Si-Cu-Mg alloys.* Metallurgical and Materials Transactions A, 2006. **37**(7): p. 2295–2311.
25. Levi, G., et al., *Solidification, solution treatment and age hardening of a Mg–1.6 wt.% Ca–3.2 wt.% Zn alloy.* Acta Materialia, 2006. **54**(2): p. 523–530.
26. Nie, J.F. and B.C. Muddle, *Precipitation hardening of Mg-Ca(-Zn) alloys.* Scripta Materialia, 1997. **37**(10): p. 1475–1481.
27. Hu, J.-l., et al., *Modification of Mg 2 Si in Mg-Si alloys with neodymium.* Transactions of Nonferrous Metals Society of China, 2013. **23**(11): p. 3161–3166.
28. Gil-Santos A., Marco I., Moelans N., Hort N., Van der Biest O., Microstructure and degradation performance of biodegradable Mg-Si-Sr implant alloys, Materials Science and Engineering C71, 25–34 (2017).

# Solidification Analysis of Grain Refined AZ91D Magnesium Alloy via Neutron Diffraction

T. Davis, L. Bichler, D. Sediako, and L. Balogh

**Abstract**

Neutron diffraction (ND) remains an important tool for in situ analysis of material behavior during various stages of its lifespan, including fabrication, service, damage accumulation and failure. ND was also developed to study solidification of alloys, where minor phases at low solid fractions can be successfully detected, thus providing valuable information difficult to obtain using traditional characterization methods. In the present work, in situ ND solidification experiments were carried out with AZ91D Mg alloy. To this alloy, novel grain refiners were added and the phase evolution prior to the alloy's liquidus temperature, as well throughout the freezing range was recorded. The role of the refining elements and their interaction with the solidifying alloy was investigated.

**Keywords**

AZ91D • Magnesium alloys • Neutron diffraction
Grain refinement

## Introduction

Improving the as-cast properties of aluminum containing magnesium alloys via grain refinement remains a challenge, due to the undesirable formation of intermetallic compounds which poison the grain refining potency of various additives [1, 2]. Many studies on grain refinement mechanisms in magnesium alloys typically focused on evaluating the microstructure and properties of as-cast solidified samples [1–4]. Recently, present authors have developed a novel grain refiner which successfully reduced the grain size of AZ91D alloy by ≈75%; however, the exact mechanism responsible for this significant grain size reduction remains unclear [5].

Neutron diffraction (ND) is an important method for in situ analysis of crystalline materials [6, 7]. Diverse characteristics of Aluminum (Al) and Magnesium (Mg) alloys were successfully studied using ND and are reported in the literature [6, 8–11]. Interestingly, ND enabled the study of primary phase evolution and fraction of solid during alloy solidification. However, the technique has not been applied to study the grain refinement of Mg alloys.

In the present work, ND analysis of grain refined AZ91D Mg alloy during solidification was carried out. Specifically, in situ tracking of diffraction peaks associated with the primary and minority phases was used to quantify the effect of grain refinement on the freezing range of the alloy. The fraction of solid curve was generated using ND data, clearly showing the impact of grain refinement on the solidification kinetics of the alloy.

## Experimental Procedure

### Sample Preparation

The composition of the grain refiner used in this research is proprietary. The grain refiner was fabricated using a powder metallurgy process (spark plasma sintering) and consisted of a micron-scale Al powder matrix with uniformly dispersed ceramic particles. Prior to sintering, the powder blend was homogenized in a planetary ball mill, and subsequently sintered at approx. 400 °C, to form a "master alloy" pellet with a density of ≈80%. The effect of this grain refiner on the hot tearing susceptibility of the AZ91D alloy has been reported by the authors earlier [5].

Alloy specimens used for ND analysis were cast to near net shape dimensions to fit a graphite mold with a cavity of

T. Davis · L. Bichler (✉) · D. Sediako
School of Engineering, University of British Columbia—Okanagan, 3333 University Way, Kelowna, Canada
e-mail: lukas.bichler@ubc.ca

L. Balogh
Canadian Nuclear Laboratories, Canadian Neutron Beam Centre, 286 Plant Road Chalk River, Chalk River, Canada


D. Orlov et al. (eds.), *Magnesium Technology 2018*, The Minerals, Metals & Materials Series, https://doi.org/10.1007/978-3-319-72332-7_65

12 mm diameter and 50 mm in length. 30 g of the alloy (i.e., virgin AZ91D or grain refined AZ91D) was heated to 740 °C and poured into the graphite mold preheated to 400 °C. For the refined alloy, 1.0 wt% of grain refiner was mixed into the AZ91D alloy for 30 s and then held for 2 min before pouring into the mold. After the mold cooled to room temperature, the specimens were machined to final dimensions (10.6 mm diameter by 40 mm length).

## Neutron Diffraction (ND) Experiments

ND experiments were performed using the C2 powder diffractometer at the Canadian Neutron Beam Centre (CNBC) in Chalk River, Ontario, Canada. A neutron beam with a wavelength of 2.37 Å was used for all experiments. The angle of scattering (2θ) ranged from 30° to 110°, which captured all major diffraction peaks for the AZ91D alloy.

During ND experiments, the alloy specimen was loaded in an oscillating furnace under a positive pressure argon atmosphere. A complete description of the in situ solidification mold set-up was published earlier [8, 9].

The details pertinent to the solidification path imposed during the in situ solidification experiments are provided in Table 1. Each temperature was held for one hour to ensure statistically valid neutron counts. Before and after each ND experiment, a scan was also completed at room temperature to ensure that the sample did not chemically change during the experimental run.

## Data Analysis

After each in situ ND experiment, the obtained diffraction pattern was compiled and all peaks were indexed using the Inorganic Crystal Structure Database (ICSD, FIZ-2009).

The evolution of the fraction of solid (FS) during the cooling of the sample was calculated based on the change in the integrated intensity (i.e. area) as a function of temperature of the {10–10}, {0002} and {10–11} reflections corresponding to the hexagonal close packed (hcp) α-Mg phase. As the solid formed from the liquid phase the intensity of the peaks increased, corresponding to an increasing fraction of crystalline solid phase. The area of each α-Mg reflection was measured as a function of temperature and normalized relative to the area of the corresponding peak measured on the fully solidified sample. A correction factor based on the Debye-Waller (DW) effect, which describes the decrease of peak intensities as a function of temperature due to the thermal vibrations of atoms in the crystal structure, was applied in an iterative fashion to account for the phenomenon [6, 7, 12]. These calculations were completed for each of the {10–10}, {0002} and {10–11} α-Mg diffraction peaks and the average of their normalized and DW-corrected areas was used to calculate the overall FS values as a function of temperature. A thorough analysis and explanation of this technique for quantification of the solid fraction curve has been discussed and proven in literature by multiple authors conducting similar neutron diffraction experiments on Mg and Al alloys [8–11, 13].

# Results

## Diffraction Patterns

During solidification, the alloy's peaks associated with the matrix and the second phases of the AZ91D Mg alloy were measured and recorded. Figure 1 shows the evolution of the ND pattern of the (a) unrefined and (b) refined alloy during cooling of the samples from 630 to 40 °C; the {10–10}, {0002}, and {10–11} peaks correspond to the α-Mg main phase of the alloys. The 3D graph shows the evolution of the peak intensities which corresponds to the change in the FS as discussed in the data analysis section. The shift of the peak positions towards lower angles at higher temperatures indicates the thermal expansion of the α-Mg crystal lattice. Also, the evolution of the $Mg_{17}Al_{12}$ eutectic phase was observed, which is typically difficult to track given its low volume fraction [9].

Prior to the onset of solidification (at ≈630 °C), no peaks were detected, confirming that the sample was completely liquid. The ND results also confirmed the theoretical liquidus of AZ91D reported as ≈595 °C [2, 9]. As Fig. 1 shows, at 593 °C the {10–11} reflection, which is the strongest diffraction peak for α-Mg in a ND pattern, appears as a low intensity signal above the background indicating the presence of a small fraction of crystalline Mg. As cooling continued below 593 °C the weaker {10–10} and {0002} reflections also became detectable in addition to the {10–11} peak, and their intensities increased representing a growth in the volume fraction of the solid phase. Once ≈440–430 °C

**Table 1** Experimental procedure

| Sample | Temperature (°C) | Time to complete full experiment (hrs) |
|---|---|---|
| Unrefined/refined alloy | 630, 593, 585, 575, 565, 555, 545, 535, 525, 515, 495, 485, 475, 465, 455, 450, 445, 440, 435, 430, 425, 420, 405, 300 | 24 |

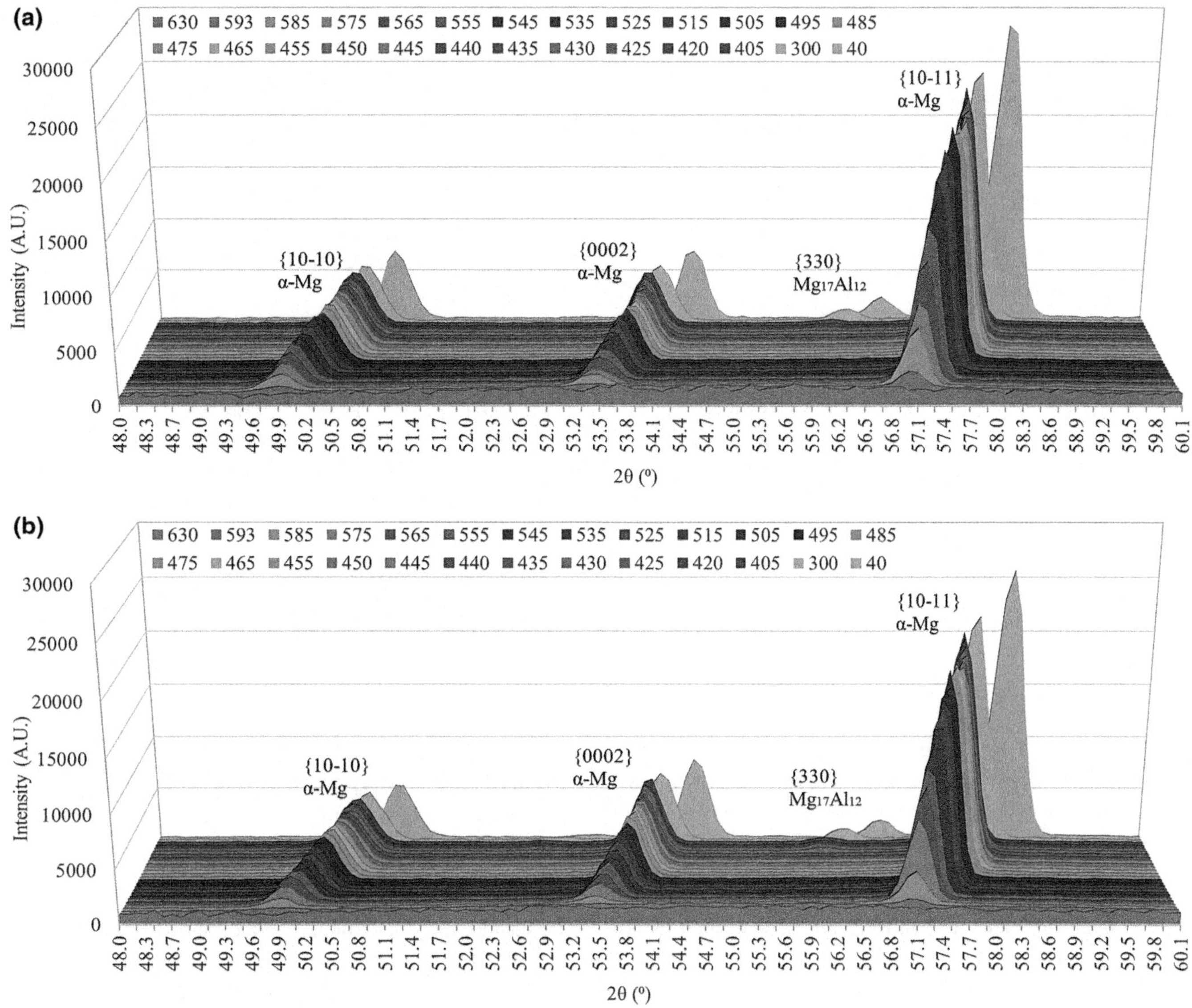


**Fig. 1** Diffraction pattern for {10–10}, {0002}, {10–11} α-Mg planes and {330} eutectic plane at each temperature, **a** unrefined, **b** refined

was reached, the increase of the peak intensities for each {hkil} plane was seen to remain constant as a function of temperature, indicating that the samples have fully solidified.

The liquidus was the same for both refined and unrefined alloys. At the initial stages of the cooling process, the refined alloy had a lower relative solid fraction suggesting a slower rate of solid formation at the beginning of solidification. At the end of solidification in the unrefined alloy, the solidus was between 440 and 435 °C, while the solidus in the refined alloy was at ≈430 °C.

The $Mg_{17}Al_{12}$ eutectic for both the unrefined and refined alloy was also captured using ND through the {330} reflection corresponding to $Mg_{17}Al_{12}$. The refined alloy showed the eutectic forming at ≈430 °C, while in the unrefined alloy the eutectic formed significantly later at 405 °C. The final eutectic peak intensity for both alloys was approximately equal, suggesting the same volume fraction evolved at the end of solidification. This could suggest that more effective liquid eutectic feeding occurred via a finer interdendritic structure in the grain refined alloy.

## Fraction of Solid

The fraction of solid (FS) curve for the α-Mg is shown in Fig. 2. The estimate of uncertainty based on the 95% confidence interval is ≈0.01 for both unrefined and refined FS values.

The major difference for the alloys was the slope at which the FS curve reached a value of 1.0. In the unrefined alloy, the curve slope was steeper and reached a high solid fraction (>0.9) approximately 40 °C earlier than in the refined alloy.

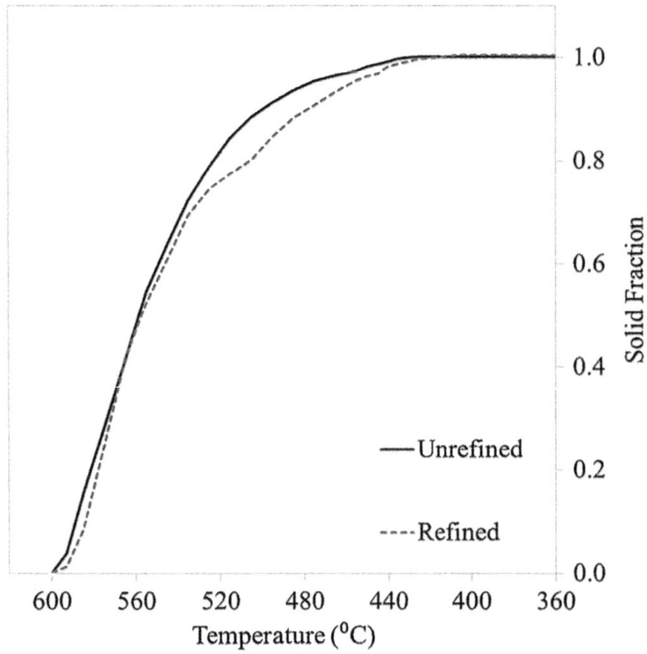


**Fig. 2** Average fraction of solid curve for unrefined and refined alloy

Also, the fraction of solid for both alloys followed the same path until ≈575 °C, where the curves began to diverge. This temperature could be related to the completion of nucleation and growth of the primary phase. After this point, the refined alloy was consistently at a lower solid fraction until the end of solidification, suggesting that a higher volume fraction of liquid was available at the end of solidification. Such liquid would improve interdendritic feeding which could provide an improved resistance to solidification defects such as hot tearing and porosity [5], [14].

## Conclusions

The following conclusions may be drawn based on the results of this research:

1. In situ solidification analysis using ND was successfully carried out to study grain refinement in AZ91D Mg alloy.
2. For both the unrefined and refined AZ91D Mg alloy, the evolution of the α-Mg reflections was tracked as a function of temperature to determine the volume fraction of the solidified phase. Data on the eutectic phase evolution was also collected, which is typically difficult given the weak diffraction signals due to its low volume fraction.
3. The fraction of solid curves for both unrefined and refined AZ91D Mg alloy were successfully developed. It was observed that the refinement had changed the solidification temperature evolution by reducing the rate at which a high solid fraction (i.e., >0.9) was reached.

**Acknowledgements** The authors would like to acknowledge the financial support of NSERC and the NSERC-Discovery grant. They would also like recognize the support from the Canadian Neutron Beam Centre in completing the neutron diffraction experiments at the Canadian Nuclear Laboratories, Chalk River, Ontario.

## References

1. D. H. StJohn, M. A. Easton, M. Qian, and J. A. Taylor, "Grain Refinement of Magnesium Alloys: A Review of Recent Research, Theoretical Developments, and Their Application," *Metall. Mater. Trans. A*, vol. 44, no. 7, pp. 2935–2949, 2013.
2. S. K. Saha, "A Study of Grain Refinement of AZ91E and Mg-9 Wt.% Al Alloys Using Zinc Oxide," Ryerson University, 2015.
3. T. J. Chen, X. D. Jiang, Y. Ma, Y. D. Li, and Y. Hao, "Grain refinement of AZ91D magnesium alloy by SiC," *J. Alloys Compd.*, vol. 496, no. 1–2, pp. 218–225, 2010.
4. E. Yano, Y. Tamura, T. Motegi, and E. Sato, "Effect of carbon powder on grain refinement of an AZ91E magnesium alloy," *Mater. Trans.*, vol. 44, no. 1, pp. 107–110, 2003.
5. T. Davis, L. Bichler, F. D'Elia, and N. Hort, "The Effect of Grain Refinement on Hot Tearing in AZ91D Magnesium Alloy," in *Magnesium Technology 2017*, 2017, pp. 653–660.
6. A. Elsayed, D. Sediako, and C. Ravindran, "Solidification Analysis of A Magnesium-Zinc Alloy Using in-Situ Neutron Diffraction," in *Shape Casting: 6th International Symposium*, 2016, pp. 167–174.
7. B. D. Cullity and S. R. Stock, *Elements of X-ray Diffraction*, Third. Upper Saddle River, NJ: Prentice Hall, 2001.
8. F. D'Elia, C. Ravindran, D. Sediako, and R. Donaberger, "Solidification analysis of Al–5 wt-%Cu alloy using in situ neutron diffraction," *Can. Metall. Q.*, vol. 54, no. 1, pp. 9–15, 2015.
9. A. Elsayed, D. Sediako, and C. Ravindran, "Investigation of solidification behaviour of Mg–6Al and Mg–9Al alloys using in situ neutron diffraction," *Can. Metall. Q.*, vol. 54, no. 1, pp. 16–23, 2015.
10. W. Kasprzak, D. Sediako, M. Walker, M. Sahoo, and I. Swainson, "Solidification analysis of an Al-19 Pct Si alloy using in-situ neutron diffraction," *Metall. Mater. Trans. A Phys. Metall. Mater. Sci.*, vol. 42, no. 7, pp. 1854–1862, 2011.
11. D. G. Sediako and W. Kasprzak, "In Situ Study of Microstructure Evolution in Solidification of Hypereutectic Al–Si Alloys with Application of Thermal Analysis and Neutron Diffraction," *Metall. Mater. Trans. A*, vol. 46, no. 9, pp. 4160–4173, 2015.
12. A. Lombardi, A. Elsayed, D. Sediako, and C. Ravindran, "Analysis of the solidification characteristics of a 319 type Al alloy using in-situ neutron diffraction," *J. Alloys Compd.*, vol. 695, pp. 2628–2636, 2017.
13. A. Elsayed, "Inclusion Removal and Grain Refinement of Magnesium Alloy Castings," Ryerson University, 2015.
14. J. Song, F. Pan, B. Jiang, A. Atrens, M. X. Zhang, and Y. Lu, "A review on hot tearing of magnesium alloys," *J. Magnes. Alloy.*, vol. 4, no. 3, pp. 151–172, 2016.

# Microstructure Evolution and Mechanical Properties of Thin Strip Twin Roll Cast (TRC) Mg Sheet

X. Yang, C. L. Mendis, J. B. Patel, and Z. Fan

**Abstract**

TRC Mg alloys are hot rolled to further reduce the thickness of the TRC strip which results in a sheet material with a relatively strong basal texture which produces anisotropic tensile properties based on the test direction. The low force TRC concept developed at BCAST provides a pathway to produce thin strip that does not require further hot rolling to achieve the final desired thickness. The as TRC and homogenised AZ31 show a refined grain and very little centre line. The as TRC AZ31 has a tensile yield strength of 220 MPa along the casting direction (CD) and 190 MPa perpendicular to CD showing a very small yield anisotropy.

**Keywords**

Twin roll casting • AZ31 • High shear melt conditioning

## Introduction

The use of Mg alloys in automotive and personal electronic applications could be increased by the production of Mg sheet or strip in a more cost-effective manner that reduces or even eliminates the need for hot or cold rolling which is an inherent problem with hexagonal materials [1, 2]. Twin roll casting (TRC) [3, 4] conventionally combines casting and hot rolling in a single step to produce a strip at or close to the final thickness. The solidification front of TRC strip has a complex dependence on alloy composition, feeding temperature of the melt, position of the tip of the tundish, speed of the rolls and the roll force on the solidifying strip [3]. As the melt leaves the tip region, solidification begins at the surface of the rolls and continues until the kissing point where the two strands of the solidified metal meet and the solidification of the strip is complete. This causes the solute rich liquid to segregate to the centre of the strip leading to the formation of central line segregation. This macro-segregation along the centre line decreases the ductility of as-cast strip, extends the homogenization time and increases the production cost [5, 6]. Therefore, majority of research into twin roll casting consider dilute alloys which minimise the centre line segregation and only relatively small fraction of research on TRC Mg alloys concentrates on the development of solute rich Mg alloys that can be strengthened further through precipitation hardening after TRC.

## High Shear Melt Conditioned TRC (HSMC-TRC)

The high shear melt conditioning (HSMC) developed within BCAST when applied to twin roll casting reduced the macro-segregation observed along the centre line of the TRC strip by changing the solidification front which cause the grain structure to change from columnar grains observed in conventional TRC to more equi-axed grain structure. In conventional TRC process columnar grains form near the roll surface and grow towards the centre of the strip. During growth, the columnar grains reject solute atoms towards the centre of the strip decreasing the solidification temperature of the centre of the strip creating a deep sump [7]. This promotes the formation of brittle intermetallic particles along the centre line. In the case of HSMC-TRC process, enhanced heterogeneous nucleation results in the advance of equi-axed grains and a uniform solidification front from the roll surface to the centre of the strip and the solute rejection into the liquid to create centre line segregation is insignificant. Figure 1a and b show the conventional TRC strip without melt conditioning for AZ31 and AZ91 respectively. Both TRC strips contain significant amount of centre line segregation while AZ91 shows increased segregation. The HSMC-TRC strip did not show similar segregation at the centre line and retained a finer equi-axed grain structure

X. Yang · C. L. Mendis (✉) · J. B. Patel · Z. Fan
BCAST, Brunel University London, Kingston Lane, Uxbridge, UB8 3PH, UK
e-mail: chamini.mendis@brunel.ac.uk


D. Orlov et al. (eds.), *Magnesium Technology 2018*, The Minerals, Metals & Materials Series, https://doi.org/10.1007/978-3-319-72332-7_66

from the surface of the strip to the centre, Fig. 1b and d respectively for AZ31 and AZ91. The HSMC-TRC strip shows significantly larger elongation to failure compared with the conventional TRC strip, even though the strips prepared through conventional TRC and MC-TRC both have similar yield strengths [8]. Both conventional TRC and HSMC-TRC strips were also tested at elevated temperatures and HSMC-TRC strip shows significant improvement in elongation to failure compared with the conventional TRC strip [8]. Unlike the conventional TRC strip, where elongation to failure decreased with the increased test temperature MC-TRC strip showed continued increase in the elongation to failure.

## Low Force Thin Strip TRC

The conventional TRC process produce a minimum strip thickness of approximately 3–10 mm, which needs to be subsequently hot rolled to produced Mg strip that may be used in personal electronics or automotive applications. Basal texture develops in majority of Mg alloys during the hot rolling reductions, reducing the formability of the TRC cast strip. Thus, removing the relatively randomised texture achieved during twin roll casting. The BCAST twin roll casting (BCAST-TRC) process is controlled by the solidification process and the low force on the rolls reduces the deformation of the solidified strip and retains the random texture distribution of the as-cast microstructure. The BCAST-TRC provides just enough rolling force for strip thickness and surface quality control. The thin strip, low force TRC process in conjunction with melt conditioning provides a novel pathway for Mg sheet production [10]. The in-house twin roll caster has a small roll diameter and low separation force, and is capable of casting thin strip (<2 mm thickness). The setback between the tundish tip and kissing point of twin rolls is reduced by the small roll diameter as well as the reduced tip size. With smaller setback, the strip thickness is closer to the final gap due to reduced path from liquid metal to solid strip.

The as TRC strip produced via the BCAST-TRC show a refined microstructure and reduced segregation as compared with conventional TRC, Fig. 2a while combining the BCAST-TRC with high shear melt conditioning provides further refinement of the microstructure, Fig. 2b. The reduction in the strip thickness changes the solidification front associated with the TRC reducing centreline segregation as the amount of heat to be extracted from the cross section of the strip is reduced. This is illustrated by the relatively 'smaller central region containing coarse grains, The introduction of HSMC to the reduced thickness strip further modifies the solidification front to produce a strip consisting entirely of finer equi-axed grains. The thin strip prepared through the BCAST-TRC has a higher yield and

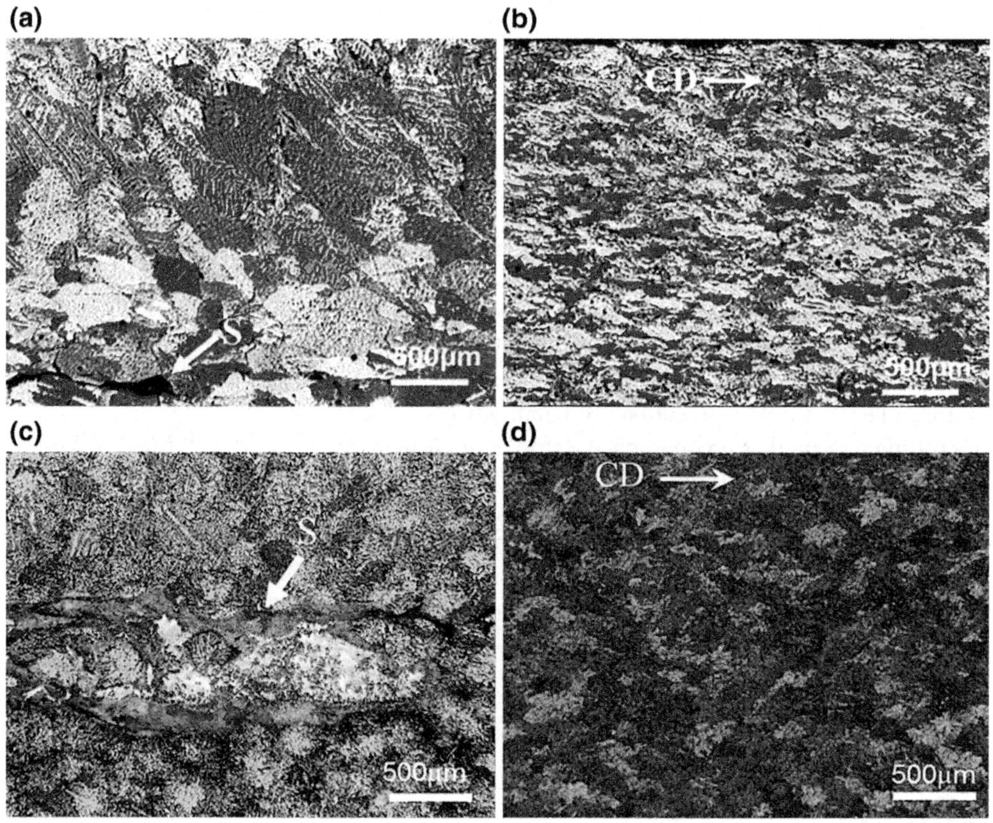


**Fig. 1** The microstructures of the as-cast TRC strips of (**a**, **b**) AZ31 and (**c**, **d**) AZ91 alloys prepared through (**a**, **c**) conventional TRC and (**b**, **d**) HSMC-TRC processes. Adopted from [9]

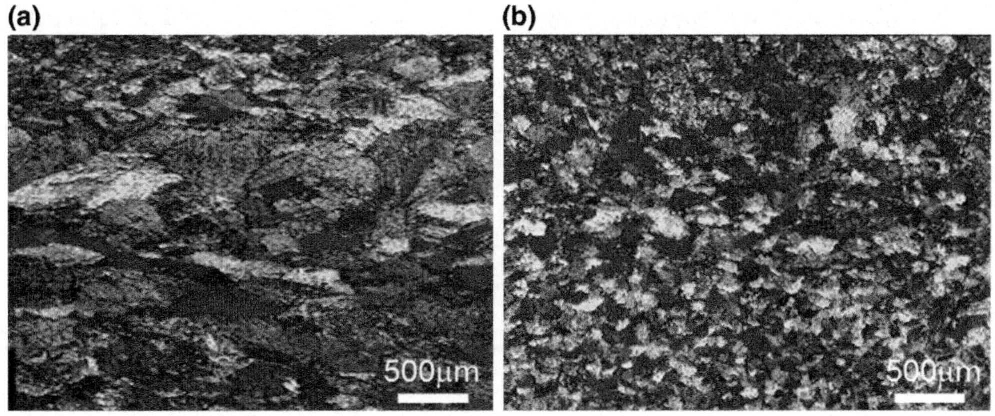


**Fig. 2** The microstructures of the thin strip AZ31 produced via BCAST-TRC process **a** without HSMC and **b** with HSMC. Adopted from [12]

tensile strengths along and perpendicular to the casting directions as compared with 5 mm TRC strip prepared under similar conditions, Fig. 3. The yield strength of the AZ31 strip produced via BCAST-TRC has a minimum yield strength of 180 MPa perpendicular to the casting direction and a maximum yield strength of 240 MPa parallel to the casting direction. The 5 mm TRC strip however has a maximum yield strength of 200 MPa parallel to the casting direction with the yield strength perpendicular to the casting direction reaching approximately 155 MPa. A yield strength of 240 MPa is generally reported for the TRC and hot rolled AZ31 strip using conventional TRC processes [11] which is achieved through the BCAST-TRC in the as-cast condition.

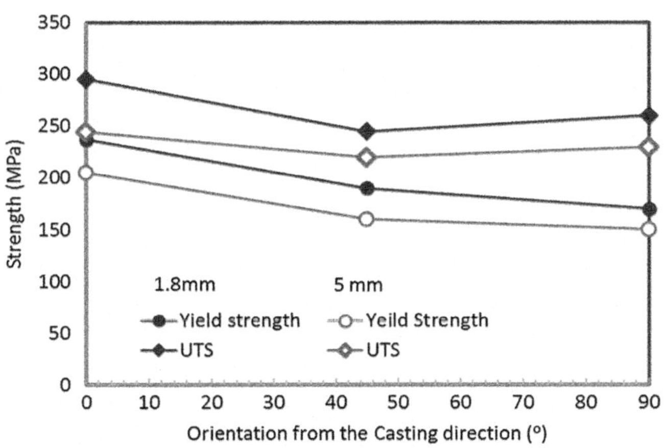


**Fig. 3** The yield and ultimate tensile strengths of as-TRC AZ31 alloy with a strip thickness of 5 and 1.8 mm based on the tensile test direction

## Conclusions

The HSMC-TRC process with the application of melt conditioning through intensive melt shearing results in a uniform fine grain structure with reduced/eliminated centreline segregation. The thin strip produced using a BCAST low force twin roll caster has the potential to enable the development of high strength Mg alloy strip without a need for subsequent hot rolling with large thickness reductions. Thin gauge strip in combination with melt conditioning provides novel way to produce high quality TRC Mg strip providing an economical pathway to producing Mg sheet material.

## References

1. E. F. Emley; (1966) 'Principles of Magnesium Technology', Pergamon Press; London.
2. https://www1.eere.energy.gov/vehiclesandfuels/pdfs/alm_05/2j_herling.pdf (accessed 14th April 2017).
3. M. Ferry; (2006) 'Direct strip casting of metals and alloys', Woodhead publishing, London.
4. N. J. Kim, (2014) Critical Assessment 6: Magnesium sheet alloys: viable alternatives to steels? Mater. Sci. Tech., 30: 1925.
5. P. Bradbury, (1994) A mathematical model for the twin roll casting process D. Phil. Thesis, University of Oxford.
6. D. Liang, C. B. Cowley, (2004) The twin-roll strip casting of magnesium JOM, 56: 26.
7. S. Das et al. (2015) Effect of melt conditioning on heat treatment and mechanical properties of AZ31 alloy strips produced by twin roll casting Mater. Sci. Eng., A620: 223.
8. S. Das et al., (2013) Melt Conditioned Twin Roll Casting (MC-TRC) of Thin Mg-Alloy Strips for Direct Stamping of Mg Components Mater. Sci. Forum, 765: 170.

9. I. Bayandorian, (2010) Magnesium alloy strip produced by a Melt-conditioned twin roll casting process D. Phil. Thesis, Brunel University London.
10. Y. Huang et al. (2012) Microstructure control during twin roll casting of an AZ31 magnesium alloy, IOP Conference Series Mater. Sci. Eng., 27: 012065.
11. G. Kruz et al (2015) Rolling twin roll cast magnesium strips with varied temperature and degree of deformation Materials Today: Proceedings 2S: S39.
12. X. Yang (2016) Particle dispersion in Aluminium and Magnesium Alloys, D. Phil Thesis, Brunel University London.

# Part III

# Environmental Challenges and Opportunities for the Magnesium Industry: Recycling and Sustainability Joint Session

# Repaired Algorithm for Nonlinear to Predict the Displacement of Copper Ion in the Absorption System of Treated Steal Slag

Shujing Zhu and Ying Qin

**Abstract**

The nonlinear properties of copper ion removal through the adsorption on treated steel slag is a function of pH and the election potential. Based on phase space reconstruction theory and the powerful nonlinear mapping ability of support vector machines, the information offered by the time series datum sets can be fully exploited and the trend of displacement evolution of adsorption system can be precisely predicted by making real-time predicting. The experimental results suggest that the methods based on phase space reconstruction and $\nu$-SVR algorithm are very accurate, and the study can help to build the displacement forecast system to analyze the removal rate of copper ion in the adsorption system.

**Keywords**

Copper ion • Phase space reconstruction • Algorithm for nonlinear • Predict

## Introduction

The removal rate of metal ions in the adsorption system varies with the change of metal ion and pH values. It is the most relevant reaction among the survival state, the mutual influence and evolution of the adsorption system. It is also the most stable control key parameters of the adsorption system [1]. The basic method of numerical calculation and inversion analysis are used to analyze metal ions removal rate in the solution with the change of pH values. It is a basic method for the numerical calculation and inversion analysis of the change rules of the composition of the adsorption system and its stability. In the further mapping, it will be the most important affect the adsorption parameters and process parameters to optimize the process. The change of the potential of the metal ion in the adsorption system can be regarded as a phase change sequence with the change of PH values of the solution. It is important to explore the whole process of heavy metal ion adsorption experiment by using this information to explore the change rule of heavy metal ion migration, so as to evaluate and forecast the change trend of metal ion adsorption behavior and removal rate in solution [2–4]. The adsorption process of metal ions in the solution system is a nonlinear dissipative power system which was controlled by many factors such as temperature and oscillation conditions, and is influenced by various factors such as feeding and human activities. Under the combined effect of these factors, the evolution of adsorption system can be regarded as a complex process with chaotic characteristics. In the phase space, its trajectory will return to the strange attractor orbit. Therefore, it is feasible to predict the adsorption rate of metal ions in the solution system by the phase space reconstruction technique within a certain pH range [5–10]. In recent years, the support vector machine (SVM) based on the SVM has been successfully applied in the field of environmental engineering technology.

Based on the theory of phase space reconstruction of chaotic dynamical system delay coordinates, this paper combines the non-linear mapping ability of $\nu$-SVR with the non-linear mapping ability of chaotic dynamical systems based on the inherent deterministic and non-linear regression rules of the chaotic variation sequence of the metal ion adsorption rate The prediction model of copper ion adsorption removal in copper ion solution with modified steel slag was investigated by means of the SVM. The results show that the experimental results are highly in agreement with the predicted results. The prediction results of the model predictors show that the support vector machine prediction model is accurate and reliable in the application of copper ion adsorption removal in solution system.

S. Zhu (✉)
School of Resource and Environment, Hubei University, Wuhan, 430070, Hubei, China
e-mail: shujingz@mtu.edu

Y. Qin
School of Science, Wuhan University of Technology, Wuhan, 430070, Hubei, China


D. Orlov et al. (eds.), *Magnesium Technology 2018*, The Minerals, Metals & Materials Series, https://doi.org/10.1007/978-3-319-72332-7_67

## Method and Procedure

### Material and Equipment

The materials used in this equipment are as follows, Ordinary mechanical strength of steel slag, Take Wuhan Iron and Steel mixed steel slag, dried and Hubei produced sodium bentonite by 1:4 mass ratio of mixed ball milling 4 h after drying finished products; Pickling mechanical modified steel slag; take Wuhan Iron and Steel mixed steel slag with 1/10 hydrochloric acid dip for 24 h, drying and Hubei produced sodium bentonite by 1/4 mass ratio of mixed ball for 4 h, dried finished products.

Adsorbent liquid: accurately weighed chemical pure anhydrous $CuSO_4$ 2.5120 g dissolved in deionized water prepared into 1000 ml.

Specific test operation in strict accordance with the (water and wastewater testing methods-the fourth edition) and the national standard GB13195-91 implementation.

#### Equipment

Atomic Absorption Spectrometer (TAS-900), Raytheon PHS-300 PH Meter; One-tenth Electronic Balance

#### Method

The initial concentration of solution was 1000 mg/L in the adsorption process, the temperature was 24.8 °C, the horizontal rotation oscillation speed was 200 r/min, and the change period of the recording potential was 3.50–5.50 PH value was recorded every 0.05 reading. In order to facilitate the record to determine the ordinary mechanical force to modify the steel slag for the delivery of a material, pickling mechanical modification of steel slag for the feeding of two.

## Prediction Model of Adsorption and Removal Rate of Copper Ion

### Phase Space Reconstruction

Figure 1 shows the relationship between the measured PH value and the removal rate. According to the analysis of the correlation data in Fig. 1, it is known that the PH value and the removal rate have a non-linear chaotic change certainty. It can be used to treat copper ions in the copper ion solution. The support vector machine prediction model of the adsorption removal rate and the PH value of the solution is used to reconstruct the phase variables of the experimental variables. Based on the chaotic dynamic system phase space delay coordinate reconstruction theory and the non-linear mapping capability of $\nu$-SVR, The adsorption removal rate of copper ions was modeled.

The purpose of phase space reconstruction is to restore the high correlation chaotic attractor in multidimensional phase space. Chaotic attractor is one of the important characteristics of chaos, with the high regularity of chaotic system, so chaotic system will fall into a certain trajectory at the end point, and this kind of proprietary trajectory is called chaotic attractor. And the complex characteristics of the phase sequence is the chaotic attractor after similar stretching and folding and other complex evolution of the results after the change. The evolution of any system component is controlled and influenced by the other components associated with it. Thus, the information of these related components is implied in the evolution of any component associated with the component. Thus, we can extract and restore the basic laws of the system based on a number of sets of phase-varying sequence data, which is a specific trajectory in the high-dimensional space. Since the relevant motivating factors of the chaotic system are interacting and interacting, the data points that have been generated on the taking phase must also be associated with each other. Packard et al. Used to reconstruct the phase space with the delay coordinates of a variable in the original system. Takens also proved that a suitable embedding dimension could be found in which the regular trajectory (attractor) was restored out, this is the dimension of the delay coordinates $m \geq 2d + 1$ ($d$ is the dimension of the power system).

In this test model prediction process, with a single variable travel potential measured PH value sequence $\{x(t_i), i = 1, 2, \cdots, N\}$, Set the PH value of this sequence to change the interval $\Delta t$. The phase space can be expressed as:

$$X_i(t) = (x(t_i), x(t_i + \tau), \cdots, x(t_i + (m-1)\tau)), i = 1, 2 \cdots, n \quad (1)$$

Here $X_i(t)$ is the definite $m$ phase in the phase space; $m$ is the embedded dimension; $\tau$ is the delay variable; $n$ is the number of points, and satisfies the condition; $n = N - (m-1)\tau$. The set of trajectories of the evolution of

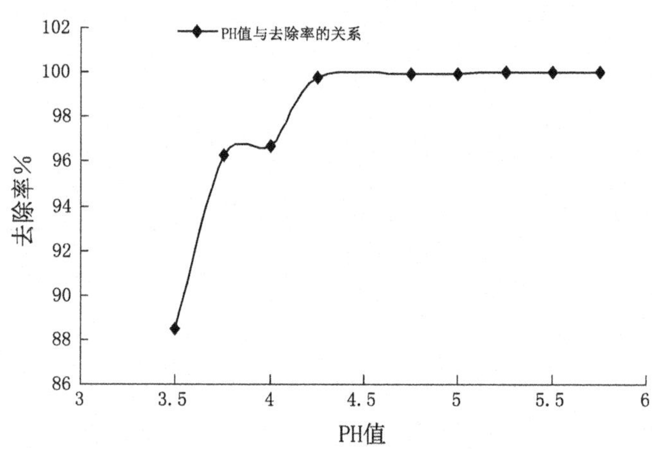


**Fig. 1** Relation between PH and ratio of remove

the system in the phase space $\{X_i(t), i = 1, 2, \cdots, N\}$ is constructed. According to Takens et al. set the embedding theorem, as long as the position and numerical $m$, $\tau$ selection are appropriate, reconstructing the "trajectory" in the phase space of the embedded space is the "dynamic equivalent" of the original system in the sense of topology. In general, the correlation dimension $d$ is calculated according to Grassberger-Procaccia algorithm, and then the $m$ embedded dimension value is determined $m \geq 2d + 1$ by the change of the autocorrelation function of the PH value change sequence to the initial value $1 - \frac{1}{e}$ or the mutual information component reaches the minimum value domain for the first time Set the value to determine the PH value of the delay $\tau = k \cdot \Delta t$ [7–10].

## Nonlinear Regression Theory of Support Vector Machines

The original idea of support vector machine SVM is to use the nonlinear transformation defined by the inner product function to map the input space into a high dimension feature space, and then use the function analysis to obtain the optimal linear classifier in the high dimension feature space The boundary between the classification plane and the nearest point (support vector) is the largest, and then the SVM problem is transformed into another quadratic programming problem, and the solution is solved.

Through the autonomous programming improved regression support vector machine $v$-SVR algorithm can be described as follows, Set the training sample as $\{x_i, y_i\}_{i=1}^{l}$, set the input variable $x_i \in R^n$, set $y_i \in R$ to the corresponding output value, set $l$ the number of training samples, the regression problem at this time is to find a input space from the input space to the mapping domain set $f : R^n \rightarrow R$ to make $f(x) = y$ satisfy. The purpose of the $v$-SVR is to find the regression function equation:

$$y - f(x) = (w \cdot x) + b \tag{2}$$

Here, $w$, $x \in R^n$; $b \in R$

The optimization problem at this time is;

$$\text{Min}\, \tau\left(w, \xi^{(*)}, \varepsilon\right) = \frac{1}{2}\|w\|^2 + C \cdot \left(v\varepsilon + \frac{1}{l}\sum_{i=1}^{l}\left(\xi_i + \xi_i^*\right)\right) \tag{3}$$

The constraints are:

$$((\omega \cdot x_i) + b) - y_i \leq \varepsilon + \xi_i \quad y_i - ((\omega \cdot x_i) + b) \leq \varepsilon + \xi_i^* \tag{4}$$

$$\xi_i^{(*)} \geq 0,\ \varepsilon \geq 0,$$

In the formula (3), $C > 0$, is set to the weight parameter, $\frac{1}{2}\|w\|^2$ used to balance the model complexity and training error term; set $\varepsilon$ as insensitive loss function; set $\xi$ as relaxation factor; $0 \leq v < 1$, then constant.

In this experiment, we use the dual equation of (3) to introduce the kernel function method with the correlation.

As for the dual issue;

$$\text{Min}\, W\left(\alpha^{(*)}\right) = \frac{1}{2}\sum_{i,j=1}^{l}\left(\alpha_i^* - \alpha_i\right)\left(\alpha_j^* - \alpha_j\right)K\left(x_i \cdot x_j\right) - \sum_{i=1}^{l}\left(\alpha_i^* - \alpha_i\right)y_i \tag{5}$$

The constraints are:

$$\sum_{i=1}^{l}\left(\alpha_i - \alpha_i^*\right) = 0, \alpha_i^{(*)} \in \left[0, \frac{C}{l}\right], i = 1, 2, \cdots l \tag{6}$$

$$\sum_{i=1}^{l}\left(\alpha_i + \alpha_i^*\right) \leq C \cdot v$$

The nonlinear mapping equation obtained by solving the convex quadratic programming can be expressed as the following Eq. (7):

$$f(x) = \sum_{i=1}^{l}\left(a_i^* - a_i\right)K(x_i, x) + b \tag{7}$$

Here, $\alpha_i^*$ and $\alpha_i$ are Lagrange key factor; $K(x_i, x_j) = \phi(x_i)\phi(x_j)$ is the core function.

## Propagation Potential of Metal Ions Based on $v$-SVR with PH Value Chaotic Sequence Prediction

### Model Predictive Analysis

The experimental results show that the removal rate of copper ions is obtained by the experiment. Through the phase space reconstruction, support vector machine training sample set $\{x(i), i = 1, 2, \cdots, N\}$ can be obtained $\tau$ by the change of PH value and $m$. For the chaotic sequence of $l$ changes in the delay of the embedded dimension $x(i)$, the point $x(n+1)$ to be predicted is that we need to establish the mapping $f : R^m \rightarrow R$ to satisfy the following equation:

$$x(n+1) = f(x(n), x(n-\tau), \cdots x(n-(m-1)\tau)) \tag{8}$$

The pH differential sequence is cut into two parts, the former is used to train the predictor, the parameters are estimated and analyzed, and the rest of the data is used to verify the validity of the model. The training sample data can then be constructed, marked as;

$$\Pi = \left\{((x(n-\tau), x(n-2\tau), \cdots, x(n-m\tau)), x(n)) \in R^m \times R \middle| m < n \leq \frac{l}{\tau}\right\} \tag{9}$$

The regression function of the support vector machine is obtained:

$$y(n) = \sum_{i=1}^{l-m\tau} (\alpha_i^* - \alpha_i) K(x(i) \cdot x(i_n)) + b, n = m\tau + 1, \cdots, l \tag{10}$$

And further predict the model, in line with the following formula:

$$y(l+1) = \sum_{i=1}^{l-m\tau} (\alpha_i^* - \alpha_i) K(x(i) \cdot x(i_{l+1})) + b \tag{11}$$

Here,

$$x(i_{l+1}) = \{x(l+1-m\tau), x(l+1-(m-1)\tau), \cdots x(l+1-\tau)\}$$

The p-step prediction model is;

$$y(l+p) = \sum_{i=1}^{l-m\tau} (\alpha_i^* - \alpha_i) K(x(i) \cdot x(i_{l+p})) + b \tag{12}$$

Here,
$x(i_{l+p}) = \{x(l+p-m\tau), x(l+p-(m-1)\tau), \cdots x(l+p-\tau)\}$

### Evaluation Indicators

In order to evaluate the prediction accuracy of the model, the root mean square relative error (RMSRE) is used to evaluate and evaluate the prediction effect of the model.

$$\text{RMSRE} = \sqrt{\frac{1}{N-l} \sum_{i=l}^{N} \left(\frac{x(i) - y(i_p)}{x(i)}\right)^2} \tag{13}$$

Here, $y(i_p)$—$x(i)$ 的 $p$ step previous value.

### Forecasting Steps

(1) The saturated embedding dimension, the delay variation $m$, $\tau$ and the Lyapunov exponent of the sequence are calculated according to the difference between the measured value of the copper ion adsorption removal rate and the PH value difference in the solution system, and the maximum predictable scale of the desired copper ion adsorption removal rate And its scope;
(2) Reconstruction of phase space, according to (9) type to carry out sample training;
(3) According to the combination of one method and grid search method, we choose the appropriate kernel function and the parameter C and v of kernel function.
(4) Using Yang Xinghua [2] and others proposed rolling prediction method to carry out model predictive analysis. It is assumed that the differential sequence of the PH value is predicted and analyzed $\{x(i), i = 1, 2, \cdots, N\}$, and the predicted number of steps is $k$ ($k$ is determined according to the actual situation). According to the $n$ sequence of the previous results, the primary task of rolling prediction is to use The k-order of the $n$-th order of the $n$-$m$ order changes the $k$-order of the value after the occurrence of the value of n; when the $k$-valued order is obtained, the $k$ new order is used to replace the preceding $k$-order One step prediction calculation analysis, and then get the next $k$ predictions.

## Analysis of the Results of the Discussion

In this paper, using the indoor test measured values, set the PH value of the difference between the variation interval $\Delta t = 0.05$, according to Grassberger-Procaccia algorithm, set $m = 20$. By using the Wolf algorithm, the maximum Lyapunov exponent of the sequence can be obtained by the analysis and analysis. It can be concluded that the maximum Lyapunov exponent $LE_1 = 0.024097 > 0$ when $\tau = \Delta t = 0.05$, which indicates that the sequence is the chaotic

**Table 1** The prediction results

| PH value | Lab value/% | | Model value/% | $C^2$/% | | PH value | Lab value/% | | Model value/% | $C^2$/% | |
|---|---|---|---|---|---|---|---|---|---|---|---|
| | Case 1 | Case 2 | | Case 1 | Case 2 | | Case 1 | Case 2 | | Case 1 | Case 2 |
| 4.55 | 99.79 | 99.91 | 99.85 | 0.06 | 0.06 | 5.05 | 99.97 | 99.99 | 99.96 | 0.03 | 0.01 |
| 4.60 | 99.81 | 99.94 | 99.87 | 0.07 | 0.06 | 5.10 | 99.97 | 99.91 | 99.94 | 0.03 | 0.03 |
| 4.65 | 99.85 | 99.85 | 99.83 | 0.02 | 0.02 | 5.15 | 99.96 | 99.98 | 99.98 | 0.00 | 0.02 |
| 4.70 | 99.83 | 99.94 | 99.91 | 0.03 | 0.08 | 5.20 | 99.94 | 99.89 | 99.90 | 0.01 | 0.04 |
| 4.75 | 99.90 | 99.93 | 99.91 | 0.02 | 0.01 | 5.25 | 99.99 | 99.94 | 99.95 | 0.01 | 0.04 |
| 4.80 | 99.89 | 99.87 | 99.90 | 0.03 | 0.01 | 5.30 | 99.96 | 99.99 | 99.98 | 0.01 | 0.02 |
| 4.85 | 99.91 | 99.92 | 99.92 | 0.00 | 0.01 | 5.35 | 99.91 | 99.97 | 99.98 | 0.01 | 0.07 |
| 4.90 | 99.78 | 99.89 | 99.81 | 0.08 | 0.03 | 5.40 | 99.85 | 99.98 | 99.92 | 0.06 | 0.07 |
| 4.95 | 99.98 | 99.99 | 99.98 | 0.01 | 0.00 | 5.45 | 99.99 | 99.92 | 99.98 | 0.06 | 0.01 |
| 5.00 | 99.99 | 99.99 | 99.99 | 0.00 | 0.00 | 5.50 | 99.98 | 99.99 | 99.91 | 0.08 | 0.07 |

RMSRE = 0.0149

sequence of PH value, and it can be seen that the longest forecast $T_m = 1/LE_1 = 40$. The practical significance is that when the predictive analysis is carried out using the measured data of the change sequence, the maximum prediction interval is 0–40 when the precision loss is limited. The concept of the maximum predictable PH value change scale is taken to help quantify the predictability of copper potential potential and provide a theoretical support for the prediction of metal ion potential potential changes more accurately.

The phase reconstruction is carried out by using the regression support vector machine to establish the mapping relationship $f: R^m \to R$ between the window $x(t_n) = \{x(t_n - \tau), x(t_n - 2\tau), \cdots x(t_n - m\tau)\}$ and the output of the dynamic PH value. In order to obtain the appropriate prediction model equation and $y(t_n) = (x(t_n))$ related parameters, we use the polynomial kernel function, the radial basis function and the sigmoid kernel function respectively to test the reliability of the analysis model. The optimal kernel function is the radial basis function, which is based on the method of left one method and grid search method. The optimal parameter is set to $C = 1024$, $\sigma^2 = 0.15$, and $v = 0.35$. This calculation is done on the basis of autonomous programming optimization free software libsvm 2.5.

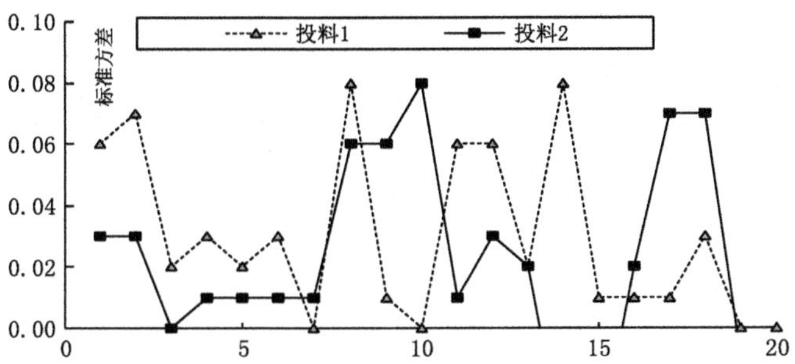


**Fig. 2** Comparison of predicted deformations with measured deformations

In this paper, the first 20 data of two kinds of different kinds of adsorbents were used as the training sample set, and the data of metal ion wandering potential were predicted and analyzed by rolling prediction method. The results are shown in Table 1, and the predicted values are compared with the measured standard variance. The results are shown in Fig. As can be seen from Table 1, the relative error of the model is relatively small, RMSRE = 0.0149. It can be seen from Fig. 2 that the prediction method using this method is very accurate. At the same time, it can be seen that the position where the relative error is more obvious is at the inflection point where the potential potential changes sharply, and the prediction is better in the steady state. There are many reasons for the dramatic increase in the potential potential of the copper ion potential, which may be due to experimental disturbances (e.g., laboratory ventilation, oscillator temperature control, recovery), solution temperature change, or due to the release of adsorbent absorbent Creep stage. In this case, the pure pH differential sequence is generally difficult to accurately determine. It can be seen that the prediction effect of the combination of the potentials of the $v$-SVM support vector machine based on the phase space reconstruction and the autonomous programming is reasonable, reliable and satisfying. It is pointed out that when the predicted value is different from the measured value of the measured test, or when the measured value of the measured test deviates from the prediction curve, the change of the adsorption rate of the metal ion is analyzed, and the measurement error and man-made In the case of interference disturbances, appropriate remedial or optimization measures should be taken.

## Conclusion

(1) This paper discusses the highly prediction method of copper ion adsorption and removal rate in modified steel slag adsorption system based on phase space reconstruction $\nu$-SVR. Through the phase space reconstruction, support an embedded space, can be restored in the context of the equivalent of the original system dynamics, take full advantage of the data contained in the objective sequence of the rules. Improved regression support vector machine $\nu$-SVR has good nonlinear function approximation ability, especially when the training sample is limited, it has good generalization ability. The results of the predicted analysis and the measured results are RMSRE = 0.0149, which indicates that the prediction method has high precision.

(2) Based on the results of $\nu$-SVR algorithm, it is found that the standard error dispersion of pickled-modified steel slag is larger than that of ordinary mechanical-modified steel slag system.

(3) The method is practical and easy to operate. It provides a more favorable way for the dynamic inversion of metal ion adsorption treatment and the optimal design of environmental engineering. At the same time, the stability of the adsorption system is analyzed displacement prediction system provides a practical and effective method.

## References

1. Zhongsheng Yi et al. Support Vector Machine Classification of Aquatic Toxicity Model of Organic Compounds[J]. Guangxi Science 2006, 13(1) p 31–34.
2. Xinghua Yang et al. AlignLoo Model Selection Method for Regularized Least Squares Classification[J]. Control and decision making. 2006.21(1) p 7–12.
3. Prediction of Chromatographic Retention of Phenylsulfonyl Naphthenate Compounds by Support Vector Machine[J]. Journal of Liaoning university. 2006, 31(1) p 52–55.
4. Wanzhao Cui et al. Prediction of chaotic time series by support vector machine[J]. Journal of Physics. 2004, 53(10) p 3303–3310.
5. Xunkai Wei et al. Analysis and Application of Time Series Forecasting Model Based on Support Vector Machine[J]. Systems Engineering and Electronics. 2005, 27(3) p 529–532.
6. Jun Liu et al. GA Optimization Support Vector Machine for Chaotic Time Series Prediction[J]. Journal of University of Science and Technology of China. 2005, 35(2) p 258–263.
7. Han Liu et al. Chaotic Time Series Nonlinear Prediction Based on Support Vector Machine[J]. System Engineering Theory and Practice. 2005, p 94–99.
8. Schölkopf B.,Smola A.,Williamson R.C. Bartlett P.L. New support vector algorithms[J]. Neural Computation, 2000, 12(5) p 1207–1245.
9. Naiyang Deng et al. A New Method in Data Mining - Support Vector Machine[M]. Beijing, Science Edition, 2004.
10. Support vector machine for aromatics compounds on aromatics receptor QSAR studies [J]. Journal of environmental science. 2005, 26(1) p 124–129.

# Author Index




D. Orlov et al. (eds.), *Magnesium Technology 2018*, The Minerals, Metals & Materials Series, https://doi.org/10.1007/978-3-319-72332-7

# Subject Index




D. Orlov et al. (eds.), *Magnesium Technology 2018*, The Minerals, Metals & Materials Series, https://doi.org/10.1007/978-3-319-72332-7

**N**



**O**



**P**



**Q**



**R**



**S**

Printed by Printforce, the Netherlands